AF663714

Heat Treatment of Metals

Volume 2

Contents of Volume 1

Heat Treatment of Metals

Volume 2

J. L. SMITH MS, PhD
G. M. RUSSEL M Tech
S. C. BHATIA BE

CBS Publishers & Distributors Pvt. Ltd.
New Delhi • Bangalore • Pune • Chennai • Cochin

ISBN : 978-81-239-1645-3

First Edition : 2008
Reprint : 2009, 2012

Published by Satish Kumar Jain and produced by V.K. Jain for
CBS Publishers & Distributors Pvt. Ltd.,
CBS Plaza, 4819/XI Prahlad Street, 24 Ansari Road, Daryaganj,
New Delhi - 110002, India. • Website: www.cbspd.com
e-mail: delhi@cbspd.com, cbspubs@airtelmail.in
Ph.: 23289259, 23266861, 23266867 • Fax: 011-23243014

Branches:

- ***Bengaluru:*** Seema House, 2975, 17th Cross, K.R. Road,
  Bansankari 2nd Stage, Bengaluru - 560070 Ph.: +91-80-26771678/79
  • E-mail: cbsbng@gmail.com, bangalore@cbspd.com
- ***Pune:*** Bhuruk Prestige, Sr. No. 52/12/2+1+3/2,
  Narhe, Haveli (Near Katraj-Dehu Road By-pass), Pune - 411051
  Ph.: +91-20-64704058/59, 32342277 • E-mail: pune@cbspd.com
- ***Kochi:*** 36/14, Kalluvilakam, Lissie Hospital Road,
  Kochi - 682018, Kerala • Ph.: +91-484-4059061-65
  Fax: +91-484-4059065 • E-mail: cochin@cbspd.com
- ***Chennai:*** 20, West Park Road, Shenoy Nagar, Chennai - 600030
  Ph.: +91-44-26260666, 26208620 E-mail: chennai@cbspd.com

Printed at :
India Binding House, Noida (UP)

Preface

Heat treatment is the controlled heating and cooling of metals to alter their physical and mechanical properties without changing the product shape. Heat treatment is sometimes done inadvertently due to manufacturing processes that either heat or cool the metal such as welding or forming.

Heat Treatment is often associated with increasing the strength of material, but it can also be used to alter certain manufacturability objectives such as improve machining, improve formability, restore ductility after a cold working operation. Thus it is a very enabling manufacturing process that cannot only help other manufacturing process, but can also improve product performance by increasing strength or other desirable characteristics.

Steels are particularly suitable for heat treatment, since they respond well to heat treatment and the commercial use of steels exceeds that of any other material. Steels are heat treated for one of the following reasons: (i) softening; (ii) hardening; and (iii) material modification.

This volume dealing with heat treating furnaces and equipments, thermomechanical treatments, heat treatment of cast iron, tool and stainless steel and heat treatment of non-ferrous metals and alloys and inspection and quality control is the second volume of heat treatment of metals.

This reference book is intended for practising engineers—metallurgical, mechanical, mining, material science and persons working in steel plants, rolling mills, automobile and mechanical industries.

This reference book is an effort to put together various new techniques of heat treatment of metals. Section I deals with heat treating furnaces and equipments. Chapter 1 focuses on heat treating furnaces which are in two broad categories, batch furnaces and continuous furnaces. Chapter 2 provides information on salt bath equipment which is well adapted to heat treatment of tool steels, as well as to treatment of nonferrous alloys. Chapter 3 concentrates on fluidised-bed equipment. Principles of fluidised-bed heat treating along with transfer aspects and surface treatments are discussed. Chapter 4 is an in-depth study on heat treating in vacuum furnaces and auxiliary equipment. Vacuum heat treating is a relatively new development in metallurgical processing and consists of thermally treating metals in heat enclosures that are evacuated to partial pressures compatible with the particular metals and processes. Vacuum is substituted for the more commonly used protective gas atmospheres during either part or all of the heat treatment. Chapters 5 and 6 elucidates on temperature control and atmosphere control. Various temperature and control systems used in heat treating such as temperature sensors and measurement instruments and set point programmers are discussed. The purpose of atmosphere control is to maintain consistent levels of constituents in a protective atmosphere. Chapter 7 is confine to furnace safety as heat treating furnaces require safety procedures common to all industrial installations.

Section II deals with thermomechanical treatment. Chapter 8 concentrates on low and high temperature thermomechanical heat treatment which is a kind of heat treatment including plastic deformation, which increases the density of defects and thus affects the formation of the structure the phase transformations occurring during thermal action. Chapter 9 provides information on thermomechanical treatment of martensite-hardenable steel.

Section III describes heat treatment of cast iron. Cast irons may be compared with steels in their reactions to hardening. However, because cast irons (except white iron) contain graphite and substantially higher percentages of silicon, they require higher austenitising temperatures. Chapter 10 is devoted to heat treating of cast irons: a review. Chapter 11 is an in-depth study of heat treatment of gray irons. The heat treatment most frequently applied to gray iron, with the possible exception of stress relieving is annealing. Annealing of gray iron consists of heating the iron to a temperature high enough to soften it or to minimise or eliminate massive eutectic carbides, thus improving its machinability. Chapter 12 and 13 elucidates the heat treatment of ductile irons and malleable irons. Various heat treatment methods such as normalising, hardening and tempering are discussed. Chapter 14 deals with heat treating of austenitic irons which are employed in a variety of applications, because the alloying elements in these iron develop metallurgical properties superior to those of conventional gray and ductile irons.

Section IV describes heat treatment of tool and stainless steel. It is fundamental to review the general principles controlling heating and cooling of tool steels for hardening or softening and to consider briefly the selection and operation of equipment necessary to carry out such procedures. Temperatures used for heat treatment of tool steels are perhaps of the widest range of any metallurgical product, ranging from sub-zero temperature used in special treatments to promote stabilisation of size in gauges upto temperatures of 2400°F (1315°C) used in heat treating high speed steels. Other factors that make tool steels of extreme interest from the viewpoint of heat treating operations involve the control of the surface hardness, hardness distribution or hardenability and distortion or movement that occurs during the quenching and tempering operations. Chapter 15 focuses on basic concepts of heat treatment of tool steels. Various heat treatment aspects of tool steels along with the advantages and disadvantages are discussed. Chapter 16 is confine to equipments and procedures for heat treatment of tool steel. The furnaces used in heating tool steels such as oil, gas and electric furnaces along with controlled atmosphere are discussed. Chapter 17 is an indepth study of specific classes of tool steels. Heat treating procedures vary significantly among classes of tool steel with the intended application. The preferred heat treating and hardening procedures, as well as mechanical properties are discussed in this chapter with respect to water and air-hardening tool steels, oil hardening and high carbon, high chromium cold work steels, low alloy and special purpose high-speed tool steels and shock resisting tool steels. Chapter 18 provides information on control of distortion in tool steels as tool steel parts includes all irreversible changes in size and shape that result from processing, from heat treatment and from temperature variations and loading in service. Nature and causes of distortion, size distortion, shape distortion in tool steels and special techniques for controlling shape distortion are discussed. Chapter 19 elucidates heat treating of stainless steels which serves to produce changes in physical condition, mechanical properties and residual stress level and to restore maximum corrosion resistance when that property has been adversely affected by previous fabrication or heating.

Section V deals with heat treatment of nonferrous metals and alloys. Chapter 20 is an indepth study of heat treating of aluminium alloys. The heat treating as applied to aluminium alloys is restricted to specific operations employed to increase strength and hardness of the precipitation hardenable wrought and cast alloys. Chapter 21 highlights heat treating of copper alloys which include homogenising, annealing, stress relieving, solution treating, precipitation hardening and quench hardening and tempering. Chapter 22 includes heat treating of lead and magnesium alloys. Lead normally is considered to be unresponsive to heat treatment. Yet, some means of strengthening lead and lead alloys may be required for some applications as magnesium alloys usually are heat treated either to improve mechanical properties or as a means of conditioning for specific fabricating operations. The type of heat treatment selected depends on alloy composition and form (cast or wrought) and on anticipated services conditions. Various heat treating procedures along with applications and test results are discussed. Chapter 23 elucidates heat treating of nickel and nickel alloys which may be subjected to five principal types of heat treatment, depending on chemical composition, fabrication requirements and intended service. These methods include—annealing, stress relieving, stress equalising, solution treating and age hardening. Chapter 24 is an indepth study of heat treating of heat-resisting alloys. Procedures and process control for heat treating, the principal types of heat-resisting alloys and refractory metals are discussed.

Section VI focuses on inspection and quality control in heat treatment. The capability of manufacturing process is determined by the output of the process. In quality control, the product is examined to determine whether or not the processing accomplished was what was specified in the design. Chapter 25 insists on measurement and inspection. Measurement, that is act of measuring or being measured is the fundamental activity of inspection. The intent of inspection is to ensure that what is being manufactured will conform to the specifications of the product. Various standards of measurement and inspection methods of measurement along with measuring instruments are discussed. Chapter 26 addresses process capability and quality control. Process capability studies use statistical and analytical tools as does quality control, except that the results are directed at the machines used in the processing rather than the output or products from the processes.

Section VII deals with special topics. Savings can be achieved in a variety of ways, including the introduction of more advanced technological approaches whereby expenditure of the necessary capital can yield corresponding energy saving. Considering this Chapter 27 focuses on energy-efficient operations in heat treatment. Various energy-saving strategies along with modification of heat treating practices along with equipment considerations are discussed. Chapter 28 is an indepth testing processes to evaluate metallic conditions. The evaluation processes can be established to determine specific mechanical properties, flaws in metals and inspect quality of fabrication and develop new alloys.

This reference book is throughout supplemented with diagrams and tables. The treatment of all topics is in a congent, lucid style aimed at enabling the reader to grasp the information quickly and easily. The glossary, references and index which are given in the end will prove useful for the readers.

Smith and Russel are consultants in material science and metallurgy and have been involved with heat treatment industry for over two decades. They have also published various research papers.

J. L. SMITH
G. M. RUSSEL
S. C. BHATIA

Contents

SECTION IV

SECTION V

SECTION VI

SECTION VII

Contents in Detail

SECTION II

SECTION III

SECTION IV

SECTION V

SECTION VI

SECTION VII

SECTION I

Heat Treating Furnaces and Equipments

Chapter 1

Heat Treating Furnaces

INTRODUCTION

Furnaces commonly used in heat treating are classified in two broad categories, batch furnaces and continuous furnaces. In batch furnaces, workpieces normally are loaded and unloaded manually. A continuous furnace has an automatic conveying system that provides a constant workload, in kilograms (pounds) per hour, to the unit.

TYPES OF FURNACES

Batch Furnaces

The basic batch furnace normally consists of an insulated chamber with an external reinforced steel shell, a heating system for the chamber and one or more access doors to the heated chamber. Standard batch furnaces such as box, car-bottom and pit types are most commonly used when a wide variety of heat-hold-cool temperature cycles are required.

With the addition of powered work-handling systems—integral quench tanks, slow-cool chambers and some automatic controls—the basic box-type batch furnace is upgraded to a semi-continuous batch furnace, which is a commonly used-piece of heat treating equipment. The design of a semi-continuous batch furnace is shown in Fig 1.1. The use of batch equipment for heat treating usually requires considerable labour for loading, handling and unloading of the work. High labour costs dictated by the process must always be considered in the selection of batch equipment.

Uses of batch furnaces

Batch furnaces are normally used:

1. To heat treat low volumes of parts, in terms of kilograms (pounds) per hour.
2. To handle special parts for which it would be difficult to adapt a conveying system for continuous handling (long drill rods processed in a pit furnace, for example).
3. To process large parts in small numbers, e.g., stress relief or annealing of large weldments or castings in a car-bottom-type furnace.
4. To carburise parts that require heavy case depths and long cycle times; for example, work done in integral-quench batch carburisers, such as gears, or rock bits or work done in pit-type carburisers such as drills rods or bearing races.

5. To process various parts requiring a wide range of heat treat cycles that can readily be changed, either manually or automatically.

Batch processing is especially appropriate when the work must be heated from room temperature to a maximum temperature at controlled rates, held at temperature and cooled at controlled rates. For example, car-type furnaces are used for critical stress-relief work or carbon baking in saggers.

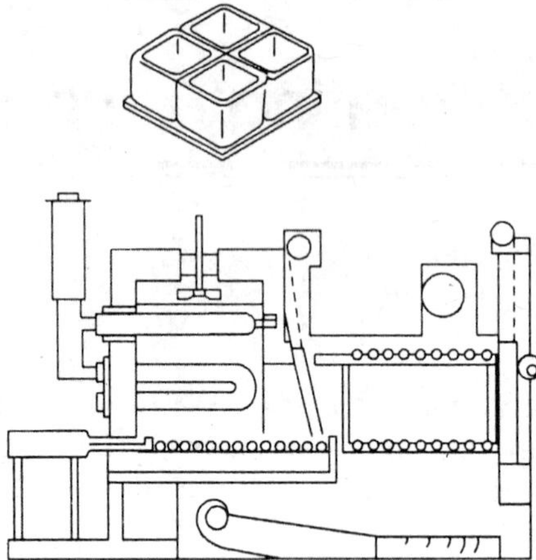

Fig. 1.1. Semi-continuous batch furnace.

Continuous Furnaces

Continuous furnaces consist of the same basic components as batch furnaces: an insulated chamber, heating system and access doors. Fig. 1.2 shows one configuration of a continuous furnace. Three common types of continuous furnaces are pusher, belt conveyor and roller hearth. With pusher furnaces, the total workload, in kilograms (pounds) per hour, consists of the work, trays and any fixtures required.

In belt-conveyor furnaces, both the conveyor and work must be heated unless belt is returned inside the furnace. Roller-hearth furnaces require that both tray and work be heated. Less common types of continuous furnaces are shaker hearth, rotary hearth and walking beam.

Continuous furnaces normally are employed when high part volumes are processed in consistent cycles. In addition to providing more uniform part quality, continuous furnaces greatly reduce the manual labour content inherent in batch systems. With a continuous flow of work through the furnace, the equipment can be integrated readily into the material flow required in a manufacturing process. Continuous furnace systems also are readily adaptable to automation.

Uses of continuous furnaces

Continuous furnaces are normally used:

1. To heat treat work to a consistent heat treat cycle and work quality. (Once a continuous furnace is properly adjusted for such variables as cycle time, zone temperatures and atmosphere integrity, the work produced will be of consistent quality. All work passing through the various furnace zones and chambers are subject to the same conditions).
2. To process large volumes of parts, in kilograms (pounds) per hour, although some continuous furnaces are designed to handle relatively low volumes. These furnaces are employed when continuous processing is required, regardless of the kilograms per hour. (A small sintering furnace

rated at or below 45 kg/hr. (100 lb/hr.) is one example. Continuous furnaces, however, are considered high-production units).

3. To maintain uniform in-plant material flow. Many times, a continuous furnace can be integrated with other elements in the manufacturing process. If loading and unloading of a product can be easily automated, an automatic system can be designed to receive the parts, load them into the continuous equipment, process the parts, unload them, and return them to the manufacturing process in a specific or random manner.

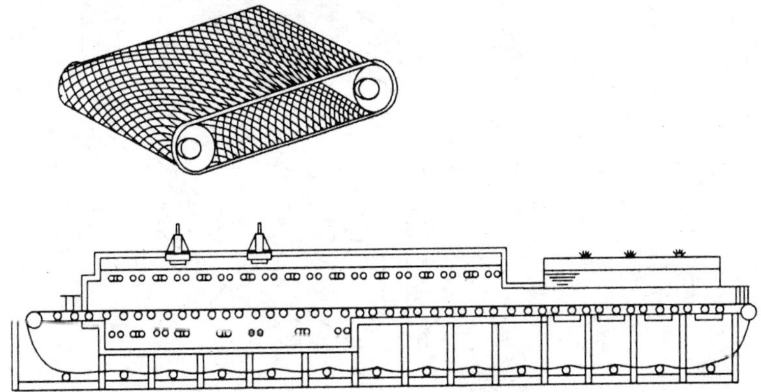

Fig.1.2. Continuous furnace.

Pusher Furnaces

A pusher furnace uses the 'tray-on-tray' concept to move work through the furnace, as shown schematically in Fig. 1.3. A pusher mechanism pushed a solid row of trays from the charge end until a tray is properly located and proven at the discharge end for removal. Fig. 1.4 shows a typical pusher mechanism. On a timed basis, the trays are successively moved through the furnace. Cycle time through the furnace is varied by changing push intervals.

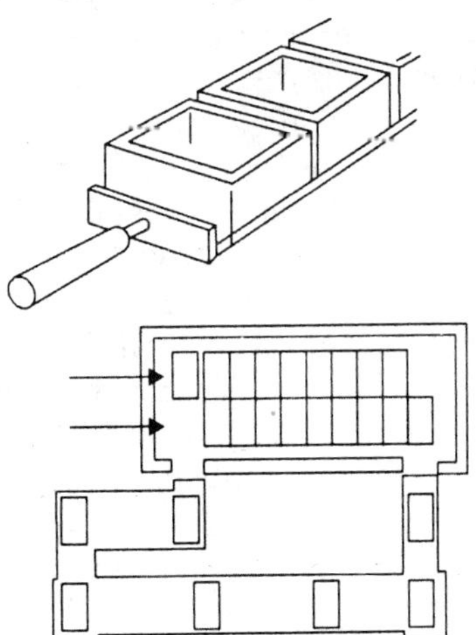

Fig.1.3. Pusher furnaces.

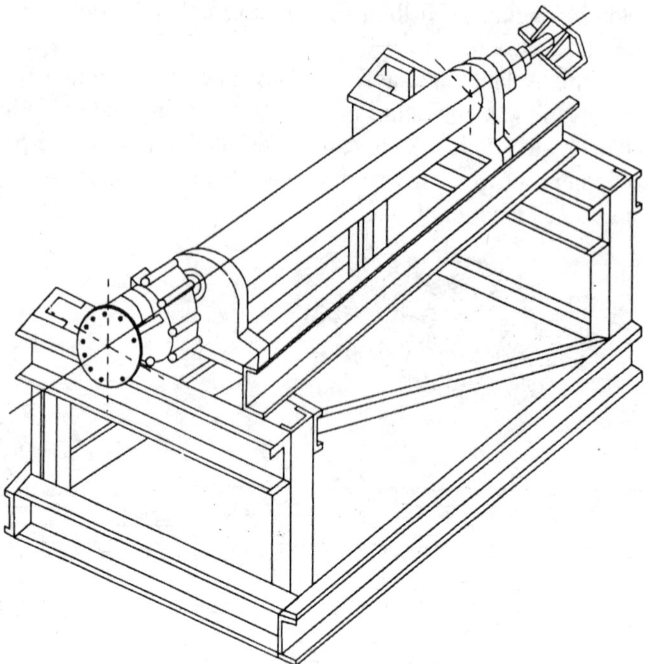

Fig.1.4. Main pusher.

Skid-rail furnaces

In a skid-rail pusher furnace, the work is placed on flat, normally reversible cast-alloy grid trays. The trays in turn are supported through the furnace on skid rails. The total gross load on the tray determines the number of rails necessary to minimise wear by maintaining the bearing pressure between the tray and the rail within acceptable limits. In certain applications, particularly when an endothermic or an enriched endothermic atmosphere is used, the skid rails are lubricated by the atmosphere and the coefficient of friction is reduced, decreasing wear and increasing tray and skid-rail life. Skid rails are used normally for light-to-moderate tray loadings. Skid rails cast from nickel-chrome heat-resisting alloy have been commonly used in pusher furnaces. The skid rails normally are supported by and anchored to a series of cast-alloy pier caps and pedestals. Because of the ever-increasing cost of nickel-chrome alloys, cast-alloy skid rails are being replaced where possible by less expensive silicon carbide refractory rails. Silicon carbide skid rails are moulded and pre-fired into various rectangular shapes and are then bricked into the lower piers or furnace floor. With the rails resting on edge, the cast-alloy grid tray is thus supported on two or more rail faces, each normally 64 mm (2½ inch) wide. Because silicon carbide in contact with a cast-alloy has a lower coefficient of friction than alloy on alloy, it makes excellent skid rails. However, in designing these rails, precautions should be taken to eliminate severe temperature gradients or thermal shock, both of which could cause the rails to fracture.

Roller rails

Roller rails are used to support and guide the trays as they are moved through the furnace. They are supported and anchored in a manner similar to that of cast-alloy skid rails, but the mechanical advantage of the wheel and axle reduces the pushing force required to move the load.

In certain instances, rails, wheels and axles are all made of a cast-alloy. The use of a dissimilar material for the axle, however, can reduce the coefficient of friction and also can eliminate the natural 'galling' effect of alloy on alloy. The cast-alloy roller rail tray usually has one or more runners and guides on the underside to keep the tray centered on the roller rails. Because the tray underside is not flat, it cannot be moved readily at 90° to its normal travel and a transfer carriage would be used. Also, the tray normally is not reversible. Because required pushing force is reduced, roller rails are used mostly for heavy tray loads and long pushes. To keep wear in the wheel or axle to an acceptable minimum, the total load per wheel (line pressure), in kilograms (pounds) per inch, must be kept within acceptable limits.

Buggy trays

Buggy trays make use of the mechanical advantage of the wheel and axle in a manner similar to that of the roller rail. Wheels and axles attached to the underside of the tray are guided through the furnace in refractory or alloy troughs. The tray cannot readily be moved at 90° to its normal travel and a carriage would be required for such movement. Although buggy trays can be used for extremely heavy loads, the maximum recommended load per wheel should not be exceeded, to keep wear within acceptable limits.

Walking-Beam Furnaces

The hearth through a walking-beam furnace consists basically of two sets of support rails. One such design employs two sets of rails, one stationary and the other movable. With this system, the moving rails lift the work from the stationary rails, move it forward and then lower it back onto the stationary rails. The moving rails then return to the starting position. Fig. 1.5 shows a typical walking-beam furnace design. In this system, the moving rail can move in an elliptical or rectangular pattern. The frequency of lift and length of stroke determine the total processing time. In another type of walking-beam furnace, both sets of rails move. One set of rails moves up and down and the second set moves forward and backward. This system is known as a true 'rectilinear' motion walking beam.

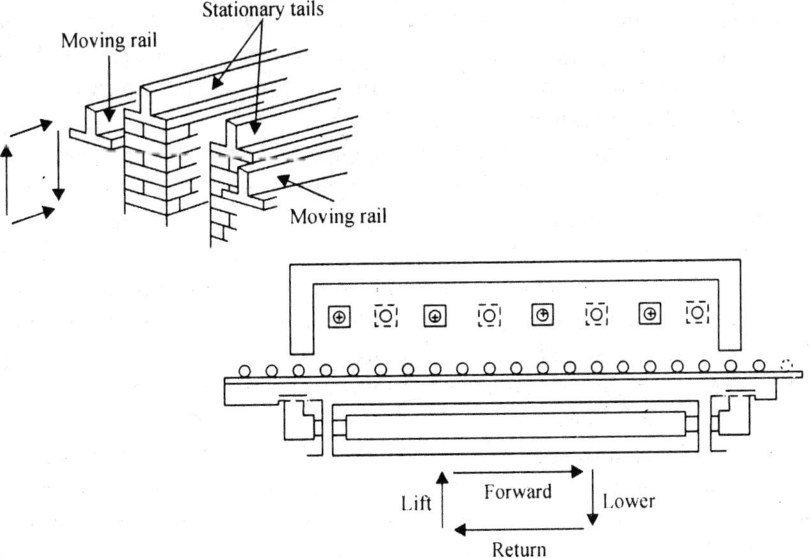

Fig.1.5. Walking-beam furnaces.

The sequence normally is as follows: the lifting beam moves up and the travelling beam moves in reverse, then the lifting beam moves down and the travelling beam moved forward. The work is thus sequenced through the furnace.

Uses of walking-beam furnace

Walking beams traditionally have been used in steel mills in re-heat furnace hearth systems for slabs and billets. Walking-beam systems can be built ruggedly to move extremely heavy loads. In heat treating operations, walking beams have been used successfully, with flat-top beams to carry such work as flat plates or trays or with pocketed-top beams to carry unstable parts such as rollers or shafts.

Advantages of walking-beam furnaces

In this type of furnace, only the work being processed has to be heated, because normally trays or fixtures are not needed, as in a pusher furnace. Because, the work is never skidded, friction is reduced for heavy loads.

The system can be loaded and unloaded automatically. A part can be picked from a specific spot and placed in a specific spot by using the walking-beam mechanism. Equipment is self-emptying on shutdown.

Disadvantages of walking-beam furnaces

The mechanisms are usually much more expensive than for pusher-type systems. On large high-temperature slab or billet re-heat furnaces, there is a dramatic increase in holding losses and fuel consumption due to the water-cooled insulated walking-beam rail system.

Car Furnaces

The car furnace, also called a 'bogie hearth', is normally considered an extremely large batch furnace. The bottom (or floor) of the furnace is constructed as an insulated movable car that is moved out of the furnace for loading and unloading, as shown in Fig. 1.6. When in position inside the furnace, the car is sealed to the furnace structure with sealing troughs or solid seals. Furnace cars can be self-propelled with the motor drive mounted directly on the car, or they can be moved in and out by a floor-mounted drive with a continuous chain or a rack-and-pinion drive. Most car furnaces are non-atmosphere type due mainly to the difficulty in sealing the car.

Heating systems normally are either direct-fired or electrically heated with resistance elements. With direct-fired systems, it has proven advantageous to design a pressure-control system controlling the flues. With the large difference in fuel burned during the heat-up portion of the cycle as compared to the soak portion of the cycle, it is extremely difficult to maintain a minimum acceptable furnace pressure with constant flues. Most car furnaces are heated from room temperature with the load already in the furnace. A typical cycle would be to heat from room temperature to a control temperature at a specific rate, hold at the control temperature for a specified time, and then slowly cool to discharge temperature at a specified rate. Many packaged temperature-control systems on the market today are capable of allowing a variable programmed cycle.

The use of ceramic fibre insulation in a car furnace allows greater control of furnace temperature when following a programmed cycle. Because ceramic fibre has minimal heat storage capacity compared to hard refractories, it will heat and cool at faster rates. Also, less total heat is required to bring the furnace to the desired temperature. Further, continuous cyclic heating and cooling has no effect on ceramic fibre lining.

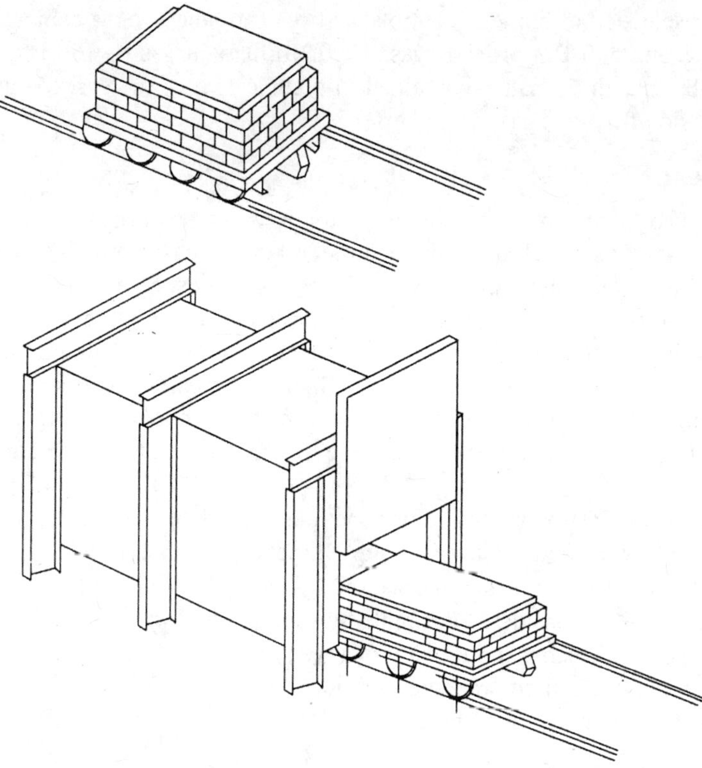

Fig.1.6. Car batch furnace.

Car furnaces are used from the lower stress-relieving ranges around 540°C (1000°F) to temperatures of over 1095°C (2000°F) for certain applications. Because many of the larger car furnaces are installed outdoors, increased allowances should be made for thermal holding losses caused by winds and other changes in ambient conditions.

Direct-Fired Furnace Equipment

With direct-fired furnace equipment, work being processed is directly exposed to the products of combustion, normally referred to as flue products. To minimise the scaling (oxide) effect on the work, the flue products can be controlled or varied by adjusting the fuel-air ratio of the combustion system. Although fuel-air adjustments can be made manually, more precise control can be achieved automatically by a wide variety of fuel-air ratio control systems on the market today. When direct-fired burner equipment is used in a heat treating furnace, the parts being processed often are in some primary or intermediate stage of manufacture. The oxide formed is not detrimental to the part because it will be removed later in the manufacturing process. For example, cold forged parts are annealed between successive drawing operations, and rough castings are annealed prior to machining.

Gas-fired equipment

Gaseous fuel used in heat treating furnaces can be natural gas, straight propane, a propane-air mix or a relatively low-Btu manufactured gas.

With the proper selection of burners, controls, orifices and pipe sizes, a combustion system can be designed to operate on 2500 Btu propane gas, 1000 Btu natural gas or 160 Btu producer gas. The number refers to the quantity of Btu's contained in a cubic foot of the gas. Manual adjustments are required for conversion from one gas to another.

Oil-fired equipment

Almost any grade of oil that can be satisfactorily atomised can be burned in direct-fired equipment. Lower viscosity oils such as diesel fuel and No. 2 fuel oil can be easily atomised with 28 gram (16 oz) cold (room temperature) air. These are probably the fuel oils most commonly used for heat treating. Even with easily atomised oils, caution should be employed in using them on flame-supervised in furnaces operating below 760°C (1400°F) with interrupted pilots. At low oil flows and excess air conditions, nuisance shutdowns can occur from the flame supervision devices. In certain instances, as dictated by the National Fire Prevention Association, 'constant pilots' may be used to eliminate the shutdowns. Insurance carriers must approve the use of constant pilots for the particular application, however.

Heavier grades of oil must be atomised by a method other than low-pressure 28 gram (16 oz) air. Normally, high pressure air and steam are used. Burners that can be fired by either gas or oil are available. In most instances, oil is used as the stand-by fuel, to be used in peak periods when natural gas supplies are curtailed. Oil is considered desirable by some in the forging industry because it creates a 'softer' scale on the billet, which is more easily removed in forging. High vanadium or sulphur content in fuel oil burned in the direct-fired process can reduce the useful life of various furnace components, especially the nickel-chrome heat-resisting alloys.

Advantages of fuel-fired furnaces

These furnaces are the easiest to adjust or alter the connected input. Normally, a simple orifice change is all that is required. Recuperator heat-saving devices can be added easily and controlled cooling can be initiated easily with proper design of combustion systems. These systems also have faster heat-up times because excess control factors can be added inexpensively.

Disadvantages of fuel-fired systems

The following disadvantages are common to fuel-fired furnaces:

1. Requires extensive ventilation systems.
2. Potential explosion or fire hazard.
3. Requires more manpower for start-up and shutdown.
4. Adjustment more difficult to maintain, resulting in excessive fuel use.

ELECTRICALLY HEATED FURNACE EQUIPMENT

Electrically heated furnaces are commonly found in all temperature ranges: from low-temperature tempering furnaces, through the heat treating range and up to forging temperatures. The basic consideration in selecting the type of heating element is to determine whether the elements are to be the 'open' type, which are exposed to the furnace environment, or the 'indirect' type, which are protected from the furnace environment by some means such as a radiant tube, muffle or retort. Factors affecting this decision are furnace atmosphere, need to protect the element from mechanical damage and space required for placement of the element.

Material Selection

Almost all furnace atmosphere other than air will in some way affect the over-all performance and subsequent life of each type of heating-element material. Manufacturers of heating-element materials provide charts that allow designers to predict the material performance with any of the given atmosphere. Each heating-element material can be exposed to the different furnace atmospheres with varying degrees of success.

The notable exception is with a carburising type atmosphere. The conventional nickel-chrome strip heating element does not perform well in a carburising atmosphere because the element itself carburises, affecting element performance. Generally, in a carburising atmosphere, heating elements are placed inside radiant tubes. However, some alternative elements are designed specifically to operate when exposed in a carburising atmosphere.

The selection of 'open' or 'indirect' elements is a choice also determined by the need to protect the element against mechanical damage from parts being heated, from accumulations of metallic scale or from broken refractories. In furnaces where bottom heat is mandatory and scale can be formed readily on the parts, or where parts may fall from a tray or conveyor, electric elements should be protected in radiant tubes below the hearth. Open elements may be used throughout the upper portion of the furnace.

In some furnace designs, the physical space available determines the design of the element.

A further consideration is whether to use an element material other than nickel-chrome strip or rod. Silicon carbide (Globor) elements or molybdenum disilicide rod elements have been used with success when directly exposed to various atmospheres, although the former normally is not recommended for use in a carburising atmosphere. Silicon carbide elements have been used on occasion inside radiant tubes for protection against carburising atmospheres.

Metallic Resistance Heating Elements

The following are general types of furnaces, with a description of the kinds of heating elements used in each.

Low-temperature furnaces with 'open' elements

The temperature range of this type of furnace varies from approximately 150 to 675°C (300 to 1250°F), and the furnace is normally a re-circulated-wind type. The simplest type of heating element is a commercially available duct heater, usually full-voltage, 440 V or 220 V heaters. These are quite useful when the designer can stay within the manufacturer's limitations. Maximum use temperature is normally limited to 400°C (750°F) .

The heater should be large enough to cover the entire re-circulated wind-duct cross-section, but designs are limited to the heater sizes available. The design watt density for this type of duct heater is normally 34000 W/m^2 (22 W/inch^2). Watt density is the expression commonly used for the connected power of each element in watts divided by its total surface area in square inches. Watts per square meter (square inch) is an important design consideration and the allowances vary greatly with temperature, type of element and furnace.

As an alternative to the commercial unit, a custom-built duct heater can be used. A steel or alloy frame can then be designed to completly fill the air-duct cross-section and such a unit can be removed easily through a sidewall bulkhead.

The nickel-chrome ribbon material is attached to ceramic insulators mounted in tyres. A common element material is 35Ni-18Cr-44Fe, and it is normally selected in the lighter gauge thicknesses and narrower widths. A typical cross-section for the ribbon material would be 13 by 0.8 mm (0.50 by 0.030 inch).

Variable with temperature and wind flow, the design watt density would be in the 23250 to 46500 W/m^2 (15 to 30 W/inch^2) range. This heater is also normally designed to operate at full-line voltage of 440 or 220 V.

High temperature furnace with 'open' elements

The temperature range of this type of furnace varies from approximately 675 to 955°C (1250 to 1750°F), and the furnace is normally a radiant-heating type.

Where large wall areas are available inside a furnace, a common method of mounting the nickel-chrome strip element is to attach it in a serpentine pattern from insulated alloy or ceramic anchors normally on the vertical walls only.

With this design especially at the higher temperatures, the structural strength of the element material and configuration must be considered. The element must support itself at operating temperatures without excessive droop or warping. Sufficient warping to cause the element to touch at various points could shorten the element's effective length, decrease the resistance and cause premature failure due to excessive currents and watt densities.

On larger furnaces with accessible wall areas, maintenance on this design element, although done from inside the 'cold' furnace, is relatively easy. On smaller furnaces, accessibility for replacement of wall elements becomes a problem.

Other types of modular or drawer-type elements are available, which makes element removal and maintenance much easier.

The element strip material used with open elements is generally one of the following types: 80 Ni-20Cr, 68Ni-20Cr or 35Ni-18Cr-44Fe. The nickel-chrome element is generally selected in the heavier gauge thicknesses and wider widths. A typical cross-section range would be from 1.3 to 2.3 mm (0.050 to 0.090 inch) thick and 19 to 38 mm (¾ to 1 ½ inch) wide. Variable with temperature and location of elements, the design watt density would be 12400 to 23250 W/m^2 (8 to 15 W/inch^2).

An alternative to the nickel-chrome strip element is the cast nickel-chrome heating element. This element had good structural strength, stability and resistance to atmospheric attack. The quality control necessary in the manufacture of this element has made it slightly less flexible and popular than nickel-chrome strip. The casting must have uniform density and cross-section to ensure a guaranteed resistance without danger of hot spots. Castings made with the 'lost-wax' process generally meet the quality requirements. Low voltages at the element and high currents tend to make the control hardware and wiring quite expensive. A common cast element material would be 35Ni-15Cr, and the watt-density range again would be 12400 to 15500 W/m^2 (8 to 10 W/inch^2).

Non-Metallic Resistance Heating Elements

In general, the non-metallic heating elements are used in furnaces operating above 1010°C (1850°F). Silicon carbide elements are generally used in temperature ranges of 1010°C (1850°F) and above. They tend to be very fragile, so care should be taken in the design to allow for proper support and freedom to expand and contract.

Silicon carbide elements undergo resistance increase with age; thus, to maintain constant power over the life of these elements, it is necessary to have a voltage adjustment available, usually with a step-transformer. The useful life of a silicon carbide element is usually established at the point when its resistance has increased four times. To maintain constant power would mean that a total voltage demand of twice the initial voltage would be required, because power is equal to E/R^2.

Silicon carbide elements are provided in various diameters and lengths, with published 'hot' resistances. Design watt densities vary with such factors as temperature and atmosphere, but with silicon carbide elements, conservatively designed watt densities result in better element life.

In a sintering furnace operating at 1150°C (2100°F) with an endothermic atmosphere, a design watt density of 31000 to 46500 W/m^2 (20 to 30 W/inch^2) would be considered appropriate. Molybdenum disilicide elements are commonly formed in U-shaped rod configurations and normally are mounted vertically. The published maximum temperature-use range in air is above 1650°C (3000°F), which covers all furnace temperatures up through the forging range.

Element location is an important consideration because these elements are designed to operate at very high watt densities and related high thermal heads. These elements undergo a high resistance change from cold to hot, because resistance increases with temperature. The control hardware and wiring must be properly designed to handle the high initial currents. A typical selection for a 955°C (1750°F) furnace with an endothermic atmosphere would be a 9 mm (0.35 inch) rod element, rated at 2,44,900 W/m^2 (158 W/inch^2) with an element temperature of 1430°C (2610°F).

Advantages of electrically heated furnaces

The following are advantages associated with electrically heated furnaces:

1. Systems are clean and free of the pollution normally found with fuel-fired systems.
2. Cooler plant environment without exhaust stacks and hoods.
3. Quieter because of the absence of blowers and combustion noise.
4. More uniform heat pattern from grid elements or side-to-side uniformity on electric radiant tubes.
5. No exhaust system required to affect building air pressures; no make-up air system required.
6. Does not require purge or flame safety systems.
7. Electrical power availability.

Disadvantages of electrically heated furnaces

The following are disadvantages of electrically heated furnaces:

1. Inflexible system makes changing connected heating capacities or varying individual element capacities in the same zone difficult.
2. High initial equipment costs.
3. With numerous electric furnaces in a plant and no form of peak-demand control, the user pays at a high demand rate for all power.
4. Higher operating costs.
5. Higher maintenance costs.
6. Cool down times are longer because no combustion air is available.

RADIANT-TUBE-HEATED FURNACE EQUIPMENT

With fuel-fired, radiant-tube-heated furnaces, the work chamber is protected from the products of combustion. With a radiant-tube furnace, the work chamber normally contains a manufactured controlled atmosphere as dictated by the process. There are, however, cases where the chamber remains filled with air, and the only purpose of the radiant tubes is to protect the work from the high dew point flue products. Electrically heated radiant tubes normally are used to protect the heating-element material from attack by the furnace atmosphere.

Gas-Fired Radiant Tubes

Gas-fired radiant tubes as shown in Fig. 1.7, are by far the most common type of fuel-fired indirect method of heating. This is mainly due to the wide availability of natural gas. The proper selection of furnace parts such as burner components, controls, orifices and piping, will allow the same radiant tube to be fired with a wide variety of gases: natural gas, propane-air mix, straight propane and certain low-Btu manufactured gases.

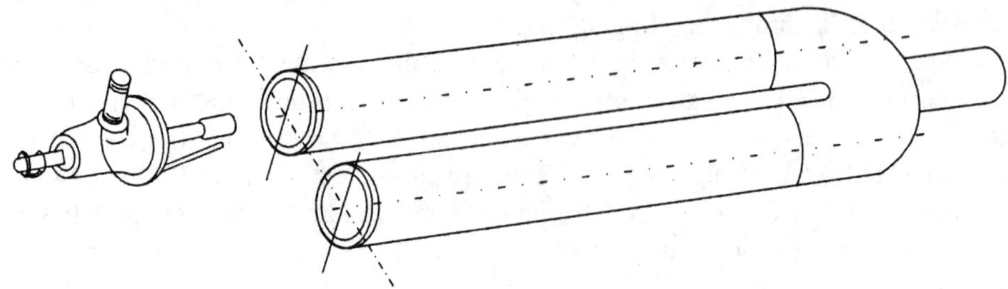

Fig. 1.7. Burner with radiant tube.

Radiant-tube burners are of two basic types, sealed-head and open-type burners. A particular advantage of the sealed-head burner is that it is readily recuperable, by using the products of combustion and an air-to-flue-gas heat exchanger to pre-heat the combustion air prior to its entering the burner. This advantage can result in considerable fuel savings. Radiant tubes are normally constructed from centrifugally cast tubing of various diameters, with wall thicknesses of 6 and 8 mm (1/4 and 5/16 inch). In many cases, tubes fabricated from 3 and 5 mm (1/8 and 3/16 inch) wrought alloy are used.

With the high metal temperatures attained in radiant tubes in most heat treating furnaces, the alloys commonly used are the higher-grade cast alloys HT, HK and NA22H or the wrought alloys 330, 601 and Incoloy 800.

Oil-Fired Radiant Tubes

Straigth oil-fired radiant tubes are somewhat uncommon and are used mainly where an adequate supply of gaseous fuel is not available. Radiant tubes equipped to burn both oil and gas are more common with the oil, usually No. 2 fuel oil, used as the stand-by fuel.

High vanadium and sulphur content in oil has a great effect on the useful life of nickel-chrome heat-resisting alloys. With the high temperatures attained inside the radiant tubes acting as a catalyst, attack on the alloy is accentuated.

When oil-fired radiant tubes are used, the construction and types of materials used are similar to those described for the gas-fired radiant tubes. Recuperation is also possible with certain sealed-head oil burners. The burner manufacturer should be consulted regarding possible damage to or problems with the atomising system from the pre-heated combustion air, normally supplied at 370 to 540°C (700 to 1000°F).

Electrically Heated Radiant Tubes

With this design, the radiant tube protects the resistance heating element from the furnace atmosphere. A common design, as shown in Fig. 1.8, uses nickel-chrome alloy rod inside the radiant tube. These rods are formed into hairpin shapes supported and contained by ceramic spacer discs.

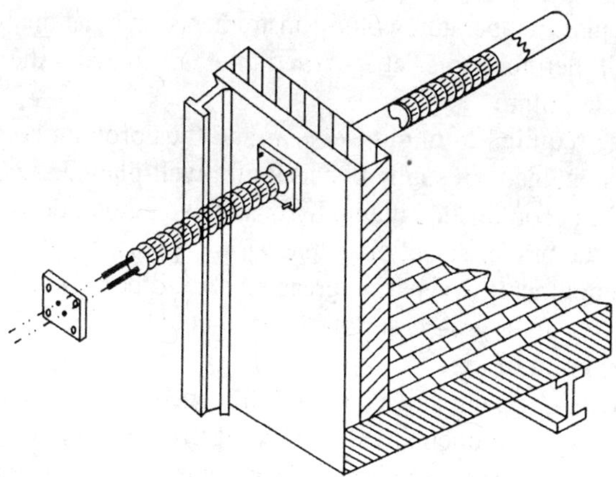

Fig. 1.8. Electric radiant tube.

With the heating element contained in a tube, it is very important to conservatively select the heating element watt density. The design watt density is a direct function of furnace temperature and varies from 18600 to 46500 W/m^2 (12 to 30 W/inch^2). Some designs use nickel-chrome strip material rather than rod in similar hairpin patterns.

If a gas-fired 'U' tube furnace is to be converted to electrically heated radiant tubes, it is desirable to replace the 'U' tube with two straight tubes for improved uniformity and reduced watt density. Thus, the conversion from natural gas to electricity can be accomplished without loss in production capacity.

Other types of electric radiant tubes are available, where the radiant tube itself becomes the resistive heating element. As with cast elements, however, the quality and condition of the radiant tube will determine whether it functions properly as a resistance element. Silicon carbide elements have also been used inside radiant tubes to protect them from carburising atmosphere.

GENERAL FURNACE MAINTENANCE

Maintenance on a furnace should be performed on a regular basis to prevent unscheduled shutdowns. In most plants, major maintenance is performed once each year, normally during the plant vacation period.

Because unscheduled maintenance is very disruptive to production, especially in plants without back-up heat treating capabilities, some manufacturers prefer a number of smaller furnaces rather than a single large furnace capable of handling all production.

Many components of furnaces must be considered as consumable items, although lifespan normally can be predicted from accurate maintenance records. Many furnace owners regularly inspect and change internal furnace components such as radiant tubes, retorts and electric heating elements. Trying to get a few more months of life from a certain internal component could result in an unscheduled shutdown and an extended loss of production can occur if a replacement part is not readily available. Most furnace equipment manufacturers provide a recommended list of spare parts that the furnace owner should maintain in stock to ensure reasonably uninterrupted production.

In addition to the consumable items that have to be regularly replaced, many furnace components must be adjusted and/or calibrated at regular intervals to maintain the efficiency and accuracy of the heat treating operation once each shift, daily, weekly, monthly or annually. Components that require regular monitoring for adjustment and calibration are mainly those that control the quality of the heat treating process, such as thermocouples, temperature-control instruments and gas analysers. For example, some furnace operators change all thermocouples at specified intervals to avoid the gradual deterioration that occurs prior to indiscriminate failure.

In addition, maintenance required to minimise wear and thus prolong component life must also be considered. This form of maintenance usually consists of a well-planned and documented lubrication schedule. Lubrication can be accomplished manually or with an automatic system.

Care should be taken in the proper selection of the various greases and oils, to ensure compatibility with the various furnace components. Excessive greasing should be avoided, as it will limit the life of the bearing seals. Excessively long grease supply lines should also be avoided because grease may harden before it reaches the point of use.

On manual systems, lubrication points should be carefully coded. Many special greases are incompatible with each other and lubricating components with the wrong grease can have disastrous results.

For major maintenance such as complete refractory replacement or complete re-building of mechanisms and systems, it is normally best to consult the original-equipment manufacturer. Because this type of maintenance usually occurs when the equipment is quite old, the manufacturer will normally make recommendations that will improve the equipment and upgrade the system to the present state-of-the-art and to comply with current recognised safety standards.

Chapter 2

Salt Bath Equipment

INTRODUCTION

Salt baths are used in a wide variety of commerical heat treating operations, including cyaniding, liquid carburising, liquid nitriding, austempering, martempering and tempering applications. Salt bath equipment is well adapted to heat treatment of tool steels, as well as to treatment of non-ferrous alloys. Advantages of using salt bath equipment include thermal control and rapid heating rates. Applications of the various furnace designs and auxiliary equipment to specific heat treating processes are described in other chapters.

EXTERNALLY HEATED FURNACES

Externally heated salt bath furnaces may be fired by gas or oil or heated by means of electrical resistance elements. Typical gas or oil-fired furnaces that are commonly used in liquid carburising applications are generally lower in initial cost than electrode or resistance heated furnaces and are simple to install and operate. To contain the molten carburising salt, fuel-fired furnaces employ a steel or alloy pot, which may be either round or rectangular. Heat is applied by two or more self-cooling burners that fire tangentially between the outer wall of the pot and the inner surface of the furnace lining. The hot gases are vented through a flue located near the top for atmosphere-type burners, or near the bottom for pressure-type burners and for atmosphere-type burners for which the flue is connected to a stack about 1 to 2 m (3.3 to 6.6 ft) high, to maintain negative pressure in the firing chamber.

The combustion chamber is lined with firebrick and additional insulation is required. A steel casing completely surrounds all sides of the furnace housing and provides adequate safety in the event of pot failure.

Electrical Resistance Furnaces

Electrical resistance furnaces for neutral heating or liquid carburising are less widely used than furnaces fired by gas or oil. These furnaces are heated by a series of resistance heaters surrounding the salt pot. For this reason, pot failure may result in the total destruction of the electrical heating elements. Operating temperatures below 900°C (1650°F) are used to prevent pot failure.

Salt Pots

Salt pots are usually supported from a flange; consequently, pot size is limited by the strength of the flange material. Round pots for gas or oil-fired furnaces normally range from 250 to 900 mm (9.8 to 35.4 inch) in diameter and from 200 to 750 mm (7.9 to 29.5 inch) in depth; they are about 10 mm (0.4 inch) thick. Larger sizes have been built for special applications and have operated successfully. Pots larger than about 350 mm (13.8 inch) in diameter and 450 mm (17.7 inch) deep are rerely used for electrical resistance furnaces. Although it is physically possible to support the bottom of a large pot on a refractory pier, excessive temperature gradients may result.

Pots may be press formed from a single piece of low-carbon steel or iron-nickel-chromium alloy; a composition of Fe-35Ni-15Cr is usually preferred for the latter. Less expensive welded pots may be fabricated from either of these materials.

In a well-designed furnace, the life of round alloy pots will vary with maximum operating temperature about as follows:

Temperature		*Service life,*
°C	*°F*	*months*
840	1550	9 to 12
870	1600	6 to 9
900	1650	3 to 6
920	1690	1 to 3

In one installation, the placement of an additional control thermocouple in the combustion chamber to prevent the temperature of the chamber from exceeding 1095°C (2000°F) served to extend the life of high-temperature (HT) alloy pots to 2 years (previous life has been 6 months). Pot temperature was maintained at 900°C (1650°F) during a work week of 120 hours (24 hours/day, 5 days/week).

Temperature

Temperature of the salt is measured and indicated by a thermocouple and suitable pyrometer. Operating within the range from 790 to 920°C (1455 to 1690°F), the externally fired furnace may vary as much as 10°C (18°F) above and below the set temperature when using on-off or high-low control systems. This is considered acceptable for many applications. Where closer control of temperature is required, a proportional control system, which will hold temperature variations to less than ±5°C (± 9°F) should be used.

Design and Operating Factors

In the design of fuel-fired furnaces, ample space must be provided for combustion so that the flame will not impinge on the pot. If flame impingement is unavoidable, the pot should be rotated slightly at least once a week. Rotating the pot and using a sleeve reduce local deterioration in the region of flame impingement and prolong service life of the pot. The combustion chamber atmosphere also has important effects on pot life. A system with a control range from high fire to low fire is preferable to an on-off system, because the latter allows air to enter the combustion chamber during the 'off' portion of the cycle, thereby increasing the rate of sealing of the outer surfaces of the pot.

Electrical resistance heated furnaces should be equipped with a second pyrometer controller whose thermocouple is in the heating chamber. This will prevent overheating of the resistance elements, particularly during melt-down, when the thermocouple that controls the temperature of the main bath is insulated by unmelted salt. Because heating elements and refractories are severely attacked by salt, all salt must be kept out of the combustion chamber. For this purpose, a mixture of high-temperature refractory cement, with long-fibre asbestos for strength, may be used to seal joints where the pot flange rests on the retaining ring at the top of the furnace.

Externally heated pots should be started on low fire—low heat input—regardless of the method of heating. Once the salt appears to melt around the top, heat can be gradually increased to high fire to complete melt-down. Excessive heating of the sidewalls or pot bottom during start-up may create pressures sufficient to expel salt violently from the pot. For added safety, the pot should be covered during melt-down, either with a cover or with an unfastened steel plate.

The waste heat of flue gases may be fed to an adjacent chamber and used to pre-heat work. Flue gases should always be visible to the operator. The presence of bluish white or white fumes at the vent are an indication that salts have entered the combustion chamber; prompt corrective action is required.

Advantages and disadvantages

Because of the ease with which they can be re-started, externally heated furnaces are well suited to intermittent operations. Another advantage of furnaces of this type is that a single furnace can be used for a variety of applications by simply changing the pot for one containing the proper salt composition.

Externally heated furnaces have several characteristics, however, that limit their usefulness in certain operations. They are less adaptable to close and uniform temperature control because they dissipate heat by convection, creating temperature gradients in the bath. Also, the temperature lag of the thermocouple and the recovery time of the furnace may result in over-shooting and undershooting of a desired temperature control point by 14°C (25°F). In addition to requiring an exhaust system for generated flue gases, externally heated furnaces may overheat at the bottom and sidewalls in re-starting creating a pressure build-up in the thermally expanding molten salt that may cause an explosion. Finally, externally heated furnaces are seldom practical for continuous high-volume production, because of the limitations of pots with respect to size and maximum operating temperature. High maintenance cost is also a factor.

IMMERSED-ELECTRODE FURNACES

Introduction of the immersed-electrode furnace greatly extended the useful range and capacity of molten carburising baths. The electrodes can be removed and replaced without bailing the furnace. This design is also suitable for neutral heating, as well as for cyanide and non-cyanide carburising processes. The molten salt is contained in a steel or ceramic pot surrounded by suitable insulating materials, which separate it from an exterior casing or framework of heavy-gage steel. The salt is heated by passing alternating current through it with immersed electrodes. As a result of the resistance built up to passage of current through salt, heat is generated within the salt itself. This heat is quickly dissipated by a downward stirring action created by the electrodes. The electrodes are attached by copper connectors to a transformer that converts the line voltage of the plant to a much lower secondary voltage (approximately 9 to 12 V) across the electrodes. Temperature is measured and automatically controlled by a system containing a thermocouple, pyrometer, relay and magnetic contactor.

The depth of salt pots for immersed-electrode furnaces is usually limited to 0.6 m (2 ft) for metal pots; ceramic pot depth is nearly unrestricted. Furnaces with pots up to 4.5 m (14.8 ft) in length, and with a power input of 360 kW, are presently in operation. They have a capacity to heat a workload of about 320 kg/hr. (705 lb/hr.). In contrast, smaller units with salt pots having a work space measuring 230 by

180 by 350 mm (9 inch by 7 inch) deep, and with 15 kW power input, can heat about 22 kg/hr. (50 lb/ hr.) to 920°C(1690°F).

Advantages and Disadvantages

Immersed-electrode furnaces do not require the use of iron-chromium-nickel alloy pots. Under normal operating conditions, the life of a carbon steel pot is usually 1 year or more. Carbon steel pots of welded construction, set into insulating brick but not cemented in place, give the following service life:

Maximum operating temperature °C	°F	*Service life, years*
840	1545	2 to 3
870	1600	1½ to 2
900	1650	1 to 1½
920	1690	1

These furnaces require minimum floor space and maintenance and can be used for all types of carburising salts. Electrodes made of alloy steel should have an average service life equivalent to that indicated for steel pots in the above table. Worn electrodes can be replaced while the furnace is in operation.

Depending on the positioning of electrodes, temperature control to within ±3°C (±5°F) is easily obtained with immersed-electrode furnaces. Heat is generated within the bath, and over-shooting is readily avoided. These furnaces lend themselves to mechanisation and are suitable for high-volume production the range of 815 to 1300°C (1500 to 2370°F).

Maximum pot size is not restricted. Pots may vary in length and width to suit requirements, and multiple pairs of electrodes can be installed to furnish the necessary heating capacity.

The immersed-electrode furnace is not recommended for intermittent operation. Depending on furnace size, re-heating the salt charge may require a day or more. Pots are not intended to be interchangeable. Removal of the pot usually involves replacement of the surrounding insulation.

STEEL POT FURNACES

Some metal treating processes are performed in salt compounds that cannot be contained in a ceramic liner. For these applications, furnace manufacturers make use of a welded steel pot with immersed electrodes. This type of furnace is suitable for special applications such as case hardening in straight cyanide baths, tempering, and marquenching.

The steel pot has a sloped back wall, which produces a 'bottom heating' effect resulting in better circulation and uniform temperature. This is accomplished by sloping the electrodes as shown in Figs. 2.1 and 2.2. As the current passes through the salt between the electrodes, the salt is heated, decreasing its density and causing it to rise toward the bath surface. Control of the rate of rise of the salt is effectively obtained by decreasing the distance from the electrodes to the steel pot. At the lower extremity of the electrode, the current enters the metal pot upon leaving the electrode to follow a shorter path to the other electrode. This arrangement ensures current flow through the salt along the entire electrode length. Due to the close proximity of the lower portion of the electrode to the pot, most of the heating is done in the lower part of the bath. This is the desired method of heating any liquid.

The metal pots are made of either plain steel or hot-dipped aluminised steel depending on the application. Thicknesses range from 12 to 38 mm (½ to 1½ inch). Reinforcing members for light plate, usually angular in shape, are welded around the top. Where depth of the pot so requires, additional members are used at the mid-section.

The pot is housed in an insulated 229 mm (9 inch) thick wall furnace with either a brick outside wall contained in a rigid welded steel frame or in a steel clad frame, depending on personal preference. In either type of construction, the frame is self supporting on a lattice formed by welding channels or beams to the underside of a steel base plate. The pot is supported on an insulated refractory pedestal.

Electrode Arrangement

Immersed electrodes are made of either mild steel or an alloy 'hot' leg welded to mild steel 'cold' leg. As previously mentioned, these are shaped to follow approximately the slope of the pot wall. The portion of the electrode that crosses over the top of the salt bath and is connected to the plant power source is referred to as the 'cold' leg. This is welded to the 'hot' leg, or the portion of the electorde that is immersed in the bath, with sufficient weld cross-section to provide the necessary current conductor capacity (Fig. 2.1). The shanks are drilled and tapped at the tinned terminal connection end for water cooling when necessary. If the latter is not required, the electrical connection is water cooled. Suitable clamping devices are used to facilitate electrode replacement.

Single-phase operation

Several electrode arrangements can be used, depending on the size of the bath. If only two electrodes are required, they are normally positioned on the sloped wall side and at least 127 mm (5 inch) apart. Three electrodes are usually placed so that the centre electrode, equal in size to two of the other electrodes, is used as a common conductor with equal current paths to each of the outer electrodes. More than three electrodes would be arranged in multiple groups.

Three-phase operation

Three electrodes used and spaced in a manner similar to the spacing described above. They are connected to three single-phase transformers that have 'Y' connected secondaries and 'delta' connected primaries. The current flows from the electrodes to the pot, which is the neutral point. Several variations of three-phase connections are used, depending on the type of furnace and load requirements.

All accessories such as starting units, transformers, sludging tools and secondary connectors are the same for steel pot immersed-electrode furnaces as for ceramic furnaces.

SUBMERGED-ELECTRODE FURNACES

In a typical electrically heated submerged-electrode furnace, the frame is made of heavy angle iron with a steel plate at the base beneath the brickwork. The outer brickwork consists of hollow ceramic tile or common building brick. The salt pot is made of burned alumina firebrick. Castable insulating refractory fills the space between the sidewalls and the ceramic pot.

When salt is melted in the pot, it penetrates the refractory until it becomes cool enough to freeze. The resulting shell of solidified salt retains the liquid salt in the furnace. If the refractory develops a crack, bath temperature must be lowered to permit salt to solidify in the crack.

Water-cooled electrodes are in contact with liquid salt in the pot and are sealed in the refractory walls by frozen salt. Current travels between the electrodes, which are flush with the side-walls. The path of current travel extends a few inches above the top of the electrodes.

Start-up and Shutdown

The submerged-electrode furnace can be started by adding molten salt from another furnace or by using a gas-fired torch or electric starting coil to melt a pool of salt that will wet both electrodes and provide molten salt for the current path. After the current path has been established in the molten salt between the electrodes, salt may be added to bring the bath up to working level. Additional salt will be required to maintain this level because a small amount, ≃5 per cent, will seep into the brickwork and freeze.

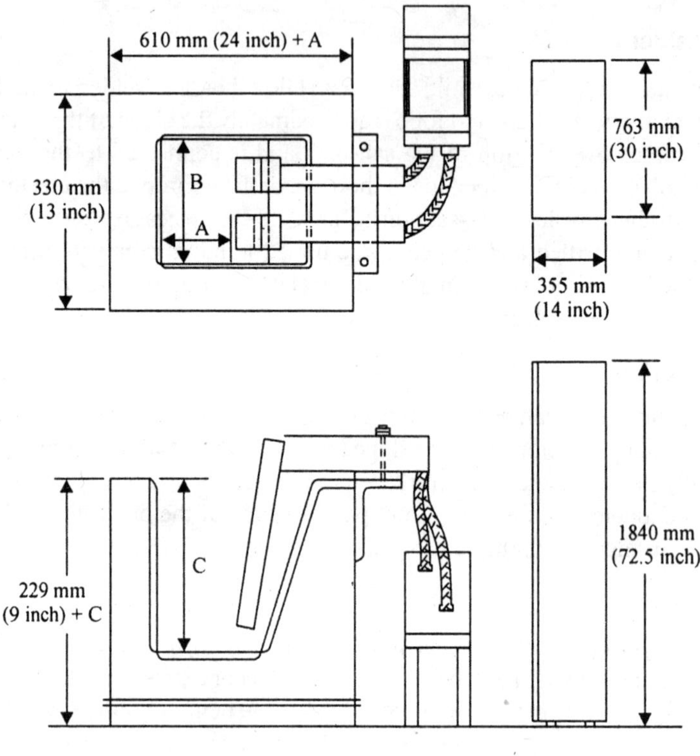

Typical standard sizes

Temperature range		*Working dimensions*						*Input*	*Heating capacity*	
		(A) Length		*(B) Width*		*(C) Depth*				
°C	°F	*mm*	*inch*	*mm*	*inch*	*mm*	*inch*	*kW*	*kg/hr.*	*lb/hr.*
540 to 150	1000 to 300	457	18	457	18	610	24	25	45	100
540 to 150	1000 to 300	457	18	686	27	610	24	25	68	150
540 to 150	1000 to 300	610	24	914	36	762	30	50	159	350

Fig. 2.1 Metal pot, immersed-electrode salt bath furnace for ferrous tempering and isothermal annealing.

If the furnace must be shut down, the molten salt should be bailed from the furnace before it freezes. However, if the salt is allowed to remain in the furnace, a resistance heated starting coil should be submerged in the bottom of the furnace while the salt is still molten. This coil remains in the frozen salt and it is connected to the transformer leads to start up the furnace.

Advantages and disadvantages

In common with immersed-electrode furnaces, submerged-electrode furnaces require minimum floor space and maintenance and are highly adaptable to mechanisation.

Because the submerged-electrode furnace employs water to cool the electrodes and transformer, it may be operated at 50 per cent overload without over-heating the transformer, whereas the immersed-electrode furnace, being air cooled, should not be operated at an overload above 10 per cent.

Because a ceramic pot is used, unexpected pot failure is rare with submerged-electrode furnaces, and the furnaces can be rebuilt on a planned schedule during annual shutdowns. In common with other electrical equipment, submerged-electrode furnaces are at a disadvantage where electric power rates are high, but this can be overcome to some extent by working the furnace in non-peak periods when lower power rates are applicable.

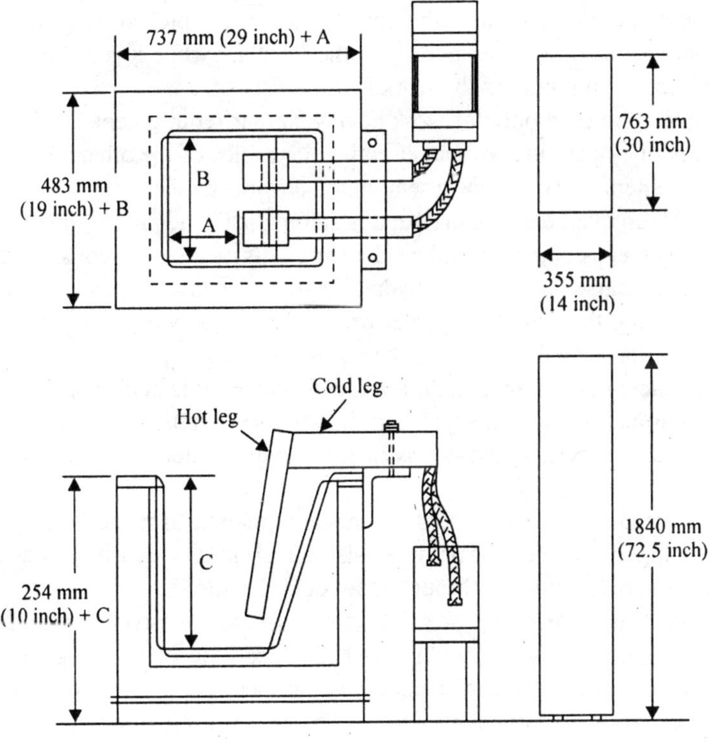

Typical standard sizes

Temperature range °C	°F	Working dimensions (A) Length mm	inch	(B) Width mm	inch	(C) Depth mm	inch	Input kW	Heating capacity kg/hr.	lb/hr.
955 to 650	1750 to 1200	305	12	305	12	455	18	25	34	75
955 to 650	1750 to 1200	305	12	455	18	610	24	40	68	150
955 to 650	1750 to 1200	455	18	610	24	610	24	75	159	350

Fig. 2.2. Metal pot, immersed-electrode salt bath furnace for liquid carburising, cyaniding and carbonate baths.

Because of the erosive effects on ceramic pots of water-soluble salts with high sodium carbonate or high sodium cyanide contents, submerged-electrode furnaces can be used only with low-cyanide, low-carbonate salts. Baths with high cyanide or carbonate salt require a modified basic brick. The furnace with modified brick and submerged alloy electrodes provides many years of service in non-cyanide and cyanide operations. To increase furnace life, the furnace has a modified basic brick lining for use with the basic carburising salts. The alloy electrodes are replaced with continuing graphite electrodes. The electrodes are renewed as they become consumed without disconnecting them or even shutting off the power.

AUTOMATIC AND SEMI-AUTOMATIC LINES

A fully automatic 'jack-rabbit' mechanism used for salt bath carburising and hardening. The mechanism has synchronised, continuous chain conveyors that carry the work through the various operations. Work suspended from horizontal bars is moved through baths at the proper speed by a main conveyor. Transfer conveyors carry the work from bath to bath. Completed work is picked up by a third conveyor and is dried by warm air in the enclosed upper portion of the structure while it is being returned to the loading point. One operator loads and unloads the work.

This mechanism can be used only for work having similar requirements. It does not permit varying the time cycle of any one operation without affecting the cycles of the other synchronised operations.

Where part requirements are different, a semi-automatic mechanism may be used. Work is transferred between operations by an overhead monorail and is automatically advanced through the carburising and tempering furnaces by means of a push-pull mechanism. This mechanism consists of two parallel beams with reciprocating push bars. Driven either hydraulically or electrically, the bars carry the dogs, which are spaced to advance the fixtures at the center of the furnace only a part of the stroke, while the end fixtures are advanced through the entire stroke. By closed spacing of the work at the center of the furnace and wider spacing at the ends, high productive capacity is achieved with ease of loading and unloading. A semi-automatic line of this type permits variations in the time cycling of any one operation without affecting other operations, and is less likely to require modification if the work requirements change.

The use of automated hoists makes possible the combination of austempering, martempering ,and tempering or carburising in one line. One or more hoists travel simultaneously back and forth automatically advancing the fixture carriers of work through the required stations.

The hoist movement is controlled by a solid state programmable control with functions that would normally require hundreds of relays, counters, switches and extensive wiring. Once programmed, the controller performs the desired commands and functions specified by the user. Time cycles, sequences, drills, and skips are easily entered or changed to meet metallurgical requirements. For instance, parts can be programmed to be carburised, air cooled, washed, rinsed, and returned for unloading. A push-button command then returns the program to standard processing.

Parts suitable for fully automatic or semi-automatic installations are those that can be fixtured by wiring, racking or placing in baskets and that do not present problems in either buoyancy or drainage.

ISOTHERMAL QUENCH FURNACES

Isothermal quench furnace systems were designed to eliminate the occurrence of chloride carry-over from the austenitising bath to the quench bath, through salt separation and uniform vertical lamellar flow agitation. The three most common approaches to alleviating the salt concentration are chemical, temperature and gravity separation.

Chemical Precipitation

Chemical agents have been used to attempt to lower the solubility of the salts that precipitate into the quench. When the salts settle to the bottom of the quench tank, they are removed as sludge. This method offers little success in that the precipitate that forms is fine-textured and buoyant and thus tends to remain in suspension rather than precipitating out.

Temperature Precipitation

The elimination of carry-over salts has also been attempted by continuously pumping salt through a small auxiliary chamber whose temperature is maintained at a lower level than the main chamber. As the salt is processed through the auxiliary chamber, chlorides are continuously precipitated out.

Although this method appears practical, a fundamental error exists in its application. The salt is cooled by air blown through a space between the pot and outer shell of the precipitation chamber. Air is blown through this space to maintain the temperature levels of the main chamber and precipitation chambers. The moving air cools the pot walls below the salt precipitation point so that the salt freezes and cakes to the sides. Salt build-up continues until the bath is unusable. Consequently, depending on the level of salt concentration, the bath would have to be shut down, possibly after only few weeks of operation, to remove the remaining molten salt and chip away the caked salt.

Gravity Separation

This system of carry-over salt removal also uses a two-chamber design. The caking problem is eliminated by heavily insulating the pot walls at all points and using an internal air-water heat exchanger. Because the pot walls and salt are at the same temperature, there is no caking action. The chloride salts settle into an easily removable shallow pan at the bottom of the precipitating chamber, or if they are fine textured and buoyant, the salts float to the top of the tanks and are easily skimmed off.

The main advantages of two-chamber gravity equipment separation include:

1. Easily removable variable speed propeller-type agitator with suitable baffling to provide vertical lamellar flow within the quench area, therefore, ensuring maximum quench power and minimum distortion.
2. Separate chloride precipitation chamber with adjustable weirs to maintain a low chloride level and subsequently high quenching power.
3. Easily removable internal heat exchanger to maintain quench temperature and precipitate chlorides.
4. Easily removable settling pan to ensure maximum efficiency in removal of chlorides.
5. Heavily insulated pot and precipitation chamber to eliminate salt caking on walls.

Furnace Heating

Generally, either gas or electricity may be used for heating isothermal quenching furnaces. When gas heating is desired, immersion tubes are recommended because the tubes are usually made of mild steel and provide long service life. Further, if the pot should develop a leak, the insulation and other shell will contain the salt. If a furnace with an externally heated pot were to develop a leak, the nitrate-nitrite salt would drip on the hot refractory and may cause a fire hazard. One or more immersion tubes normally are used, depending on bath size. Generally, they will have nozzle-mix sealed-in burners and will be available to FM or FIA specification.

Electric heating may be by one of the following methods depending on the maximum operating temperature:

1. Sheathed resistance strip heaters are mounted externally to the side walls near the bottom. Maximum operating temperature is 425°C (800°C). They are easily removable through an insulated plug-type door. Protection against overshooting is achieved by locating a sensing device in close proximity to the heaters. The sensors operate directly on line voltage.
2. Sheathed resistance immersion heaters have a maximum operating temperature of 425°C (800°F). They can operate without a transformer, but are susceptible to premature burnout due to sludge accumulation or operator tampering and abuse.
3. Immersed-electrode heaters operate in the same manner as electrode pot furnaces for carburising and tempering.

Furnace Construction

The pot is fabricated from firebox-quality steel plate double welded inside and out and properly supported to maintain its shape. Steel plate offers adequate resistance to chemical attack by the standard alkaline nitrate-nitrite salts at normal austempering and martempering temperatures. The pot is insulated with 102 to 152 mm (4 to 6 inch) of slab-type mineral insulation to prevent the chloride saturated nitrate salt from freezing to the sidewalls or bottom. The insulation is externally contained by a continuously welded outer steel shell. The shell is reinforced to ensure retention of the original shape and dimensions throughout its designed operating temperature range.

Chapter 3

Fluidised-Bed Equipment

INTRODUCTION

Fluidised-bed techniques are not new. An American patent describes the roasting of minerals under fluidised-bed conditions. Other established applications include potter's clay and miner's hydraulic slurries. Systems of fluidised solid particles, such as quicksand, occur in nature.

Early attempts to use fluidised beds in heat treatment of metals were limited in the temperatures that could be employed. Electrically heated furnaces capable of maintaining fluidised beds at temperatures up to 500°C (930°F) could be produced commercially, but difficulties were encountered when attempts were made to attain higher temperatures. A principal problem was the high rate at which refractory distributors, which distribute the hot fluidising gases, were consumed.

In early gas-fired fluidised-bed furnace design, gas entered the base of the container after being mixed with air to make it ignitable at the point of entry. With newer designs, the mixtures are introduced separately and so cannot be ignited accidentally. This design eliminates the danger of explosion at the point of entry. The surface of the bed is heated first and the heating of surface particles causes progressive ignition downward through the container until the entire contents of the bed achieves uniform heat treating temperature. Newer furnace designs extend fluidised-bed technology into the higher temperature ranges required for most common heat treatments.

PRINCIPLES OF FLUIDISED-BED HEAT TREATING

In fluidisation, a bed of dry, finely divided particles, typically aluminium oxide in the heat treating context, is made to behave like a liquid by a moving gas fed upward through the bed. A gas-fluidised bed is considered a dense-phase fluidised bed when it exhibits a clearly defined upper limit or surface. At a sufficiently high fluid-flow rate, however, the terminal velocity of the solids is exceeded, the bed goes into motion and the upper surface of the bed disappears. This state constitutes a disperse, dilute or lean-phase fluidised bed with pneumatic transport of solids. The general types of fluidised beds are shown in Fig. 3.1 . The majority of beds used for heat treatment are of the aggregative or bubbling type.

Although the properties of solid and fluid alone determine the quality of fluidisation (that is, whether smooth or bubbling fluidisation occurs), many factors influence the rate of solid mixing, the size of bubbles and the extent of heterogeneity in the bed. These factors include bed geometry; gas-flow rate; type of gas distributor; and internal-vessel features such as screens, baffles and heat exchangers.

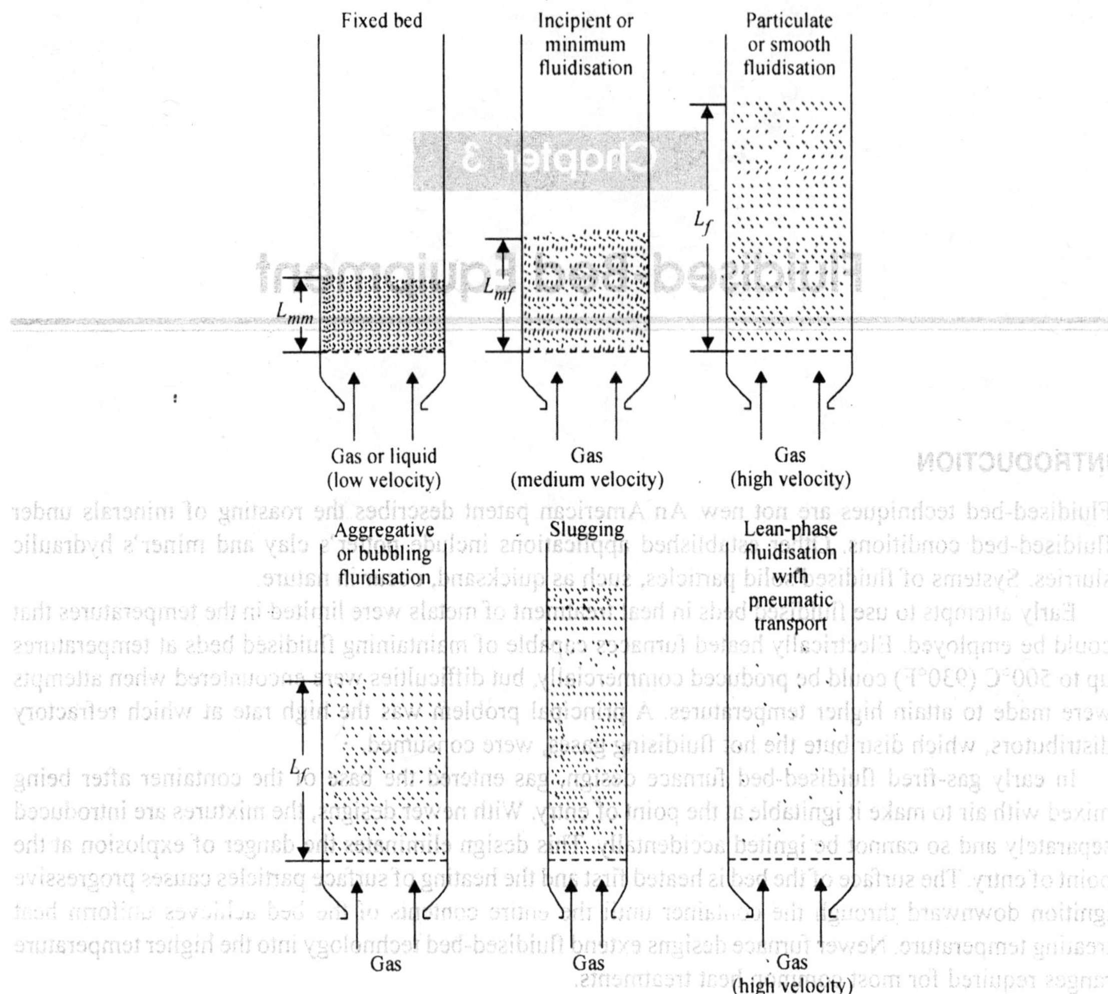

Fig. 3.1. Various types of contacting in fluidised beds.

Determination of Fluidisation Velocity

In determining the quality of fluidisation, a diagram of pressure drop (Δp) versus velocity (μo) is useful as a rough indication when visual observation is not possible. A well-fluidised bed will behave as shown in the diagram in Fig. 3.2, which has two distinct zones. In the first, at relatively low flow rates in a packed bed, the pressure drop is approximately proportional to gas velocity, usually reaching a maximum value (Δp_{max}) slightly higher than the static pressure of the bed. With a further increase in gas velocity, the packed bed suddenly 'unlocks' and becomes fluid-like. When gas velocity increases beyond minimum fluidisation (μ_{mf}), the bed expands and gas bubbles rise, resulting in a heterogeneous bed. This is the second zone, in which despite a rise in gas flow, the pressure drop remains practically unchanged. The dense gas-solid phase is well aerated and can deform easily without appreciable resistance. In its hydrodynamic behaviour, the dense phase can be likened to a liquid.

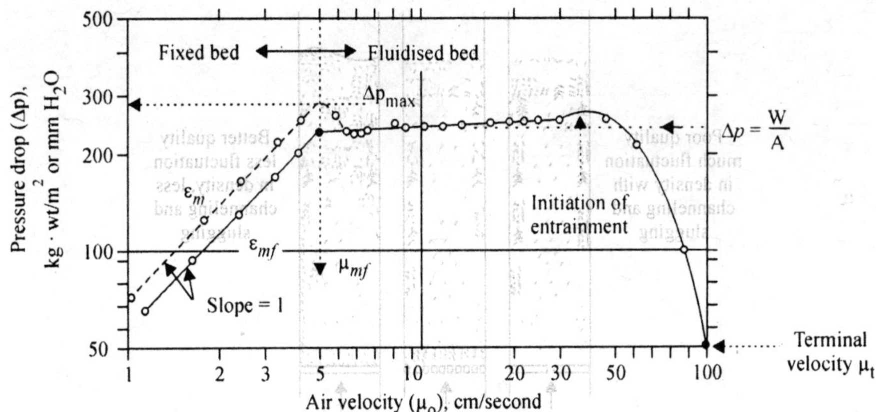

Fig. 3.2. Pressure drop versus gas velocity for a bed of uniform-size particles.

If a gas is introduced into the bottom of a tank containing a liquid of low viscosity, the pressure required for injection is roughly the static pressure of the liquid and is independent of the flow rate of the gas. The constancy in pressure drop in both the bubbling liquid and the bubbling fluidised bed may be taken intuitively to be analogous.

The diagrams in Fig. 3.3 show poorly fluidised beds. The large pressure fluctuations in Fig. 3.3a suggest a slugging bed. In Fig. 3.3b, an absence of the characteristic sharp change in slope at minimum fluidisation and the abnormally low pressure drop suggest incomplete contacting, with particles only partly fluidised.

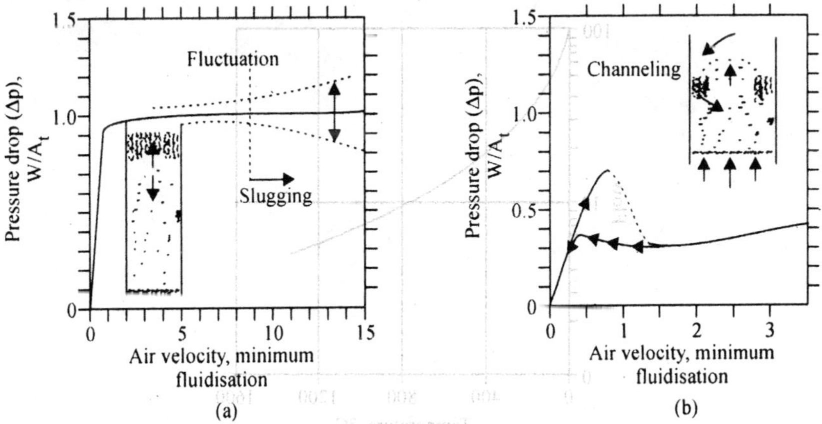

Fig. 3.3. Pressure-drop diagrams for poorly fluidised beds.

One of the most important factors influencing the quality of fluidisation is the type of distributor. Fig. 3.4 illustrates this schematically.

Temperature Effect on Minimum Fluidisation Velocity

One of the most important parameters of a fluidised bed is the minimum fluidisation velocity. In simplified terms, minimum fluidisation velocity (μ_{mf}) approximates to a function of the square of the particle diameter (d) and a linear function of particle mass (p) as follows:

$$\mu_{mf} = d^2p$$

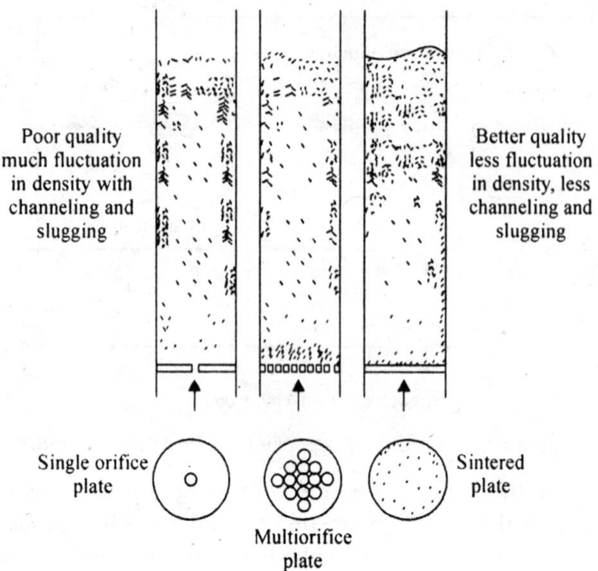

Fig. 3.4. Quality of fluidisation as influenced by type of gas distributor.

In the design of heat treating furnaces, the effect of temperature must be considered. Fig. 3.5 shows how the flow of gas required for fluidisation decreases rapidly with increases in temperature.

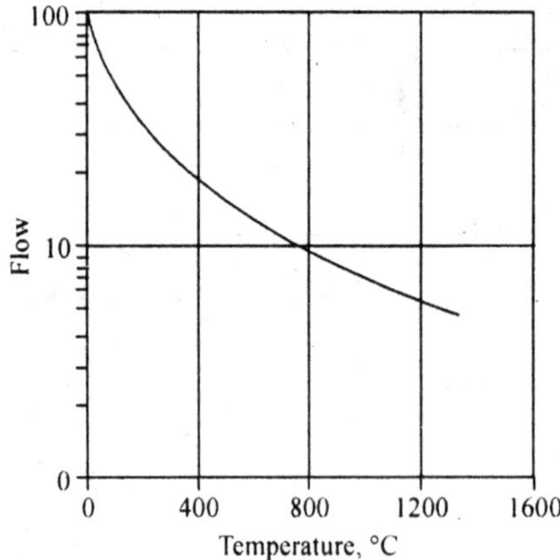

Fig. 3.5. Effect of temperature on the flow corresponding to minimum fluidisation for particles 0.1 mm (0.004 inch) in diameter having an apparent density of 2.

Defluidisation

One of the common misconceptions about fluidised beds is that, because of their principle of operation, they are not well suited for large solid parts with horizontal surfaces that remain stationary in the bed. With parts of this type, a cap of nonfluidised particles collects on the horizontal surfaces, forming a thermal screen. The higher the temperature of operation, however, the greater the energy and agitation of the bed and the smaller the likelihood that the bed will collapse.

Moreover, various methods can be used to overcome this apparent disadvantage and these are designed into most fluidised beds. These methods are:

1. Movement of the part being treated.
2. Introduction of additional agitation in the zone of fluidisation around the part, either by localised injection of fluidising gas or by careful design of the outline of the basket that holds the parts.
3. Increased fluidising velocity.

Selective Heat Treatment

Bed collapse can be turned to advantage for special heat treatments where one area must be hard and tough while the remainder must be soft and more ductile, as in the case of the engineered parts of the shape described above. In this case, after uniform heating, the product is submerged in a fluidised quenching bed, with the part to be hardened facing downwards. The top horizontal surface becomes covered with a cap of particles that form a thermal screen which retards the vigorous cooling caused by the fluidised bed.

HEAT TRANSFER IN FLUIDISED BEDS

An important characteristic of fluidised beds is high-efficiency heat transfer. The turbulent motion and rapid circulation of the particles in the fluid furnace provide a heat-transfer efficiency comparable with that of conventional salt or lead-bath equipment.

The heat-transfer coefficient of a fluidised bed is typically between 120 and 1200 W/m^2·°C (21 and 210 Btu/ft^2·hr.·°F). The turbulent motion and rapid circulation rate of the particles and the extremely high solid-gas interfacial area account for this feature. The following factors are important in heat transfer.

Particle Diameter

Of all the parameters that affect the heat-transfer coefficient in fluidised beds, particle diameter exerts the greatest influence. Particle diameter generally should be as small as possible; however, below a certain size, entrainment and carry-out effects may cause problems.

Bed Material

The governing physical property of any bed material is its density. There appears to be an optimum density for bed materials —about 1280 to 1600 kg/m^3 (80 to 100 lb/ft^3). High-density materials tend to produce lower heat-transfer coefficients and also require more power for fluidisation. Carry-out problems occur with low-density materials. Other properties, such as thermal conductivity and specific heat, are less important.

Fluidisation Velocity of Gas

It is essential to use the optimum flow rate that provides maximum heat-transfer rates for a particular particle density and diameter. Generally, this flow rate is considered to be between two and three times the minimum fluidisation velocity. Too high a velocity leads to particle entrainment, high consumption of fluidising gas and poor heat transfer; too low a velocity leads to poor heat transfer and lack of uniformity in processing.

Heating Rates

Relative heating rates of a 16-mm (0.6-inch) steel bar in salt, in lead, in a fluidised bed and in a conventional furnace are illustrated in Fig. 3.6a; relative cooling rates for air, oil, water and a fluidised bed are shown in Fig. 3.6b. Fig. 3.7 presents heating and recovery rates for a fluidised bed. Results of both hardening and isothermal quenching of type D3 tool steel with salt baths and with fluidised beds are given in Table 3.1. The difference between the two installations in total time for final heating and holding resulted from a difference in preheating conditions.

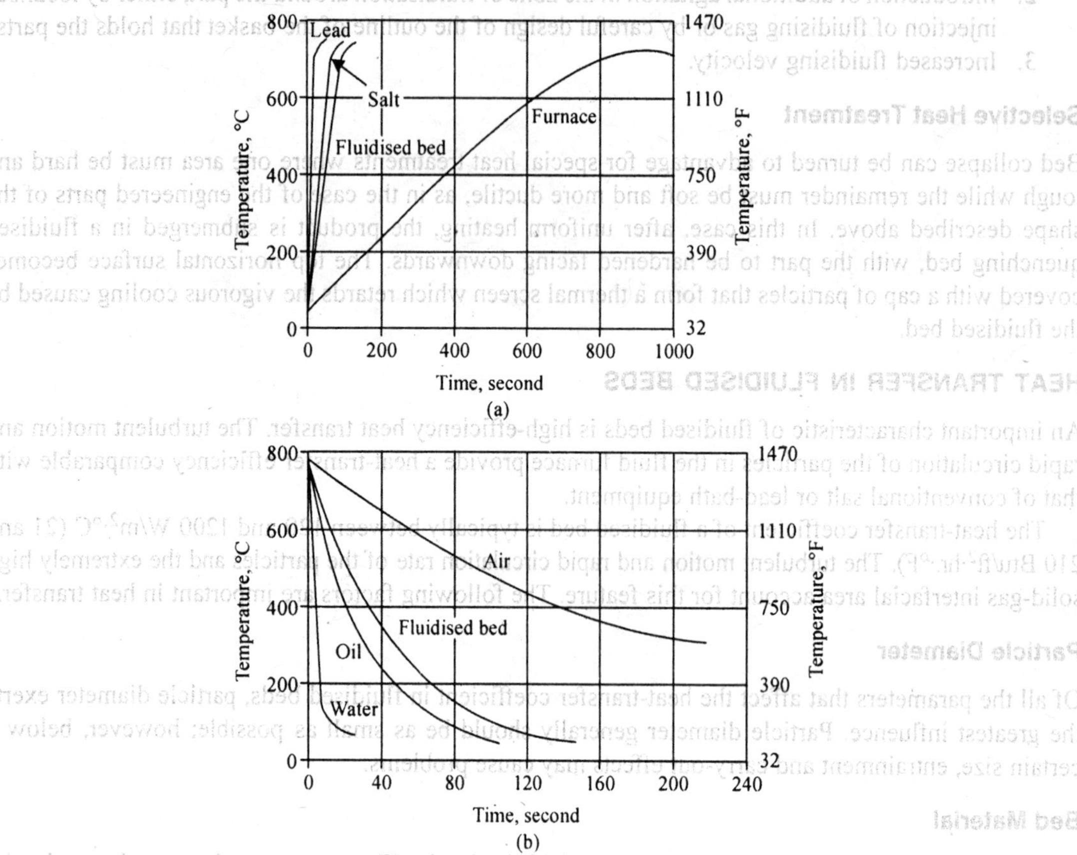

Fig. 3.6. Relative heat-transfer rates.

Table 3.1. Comparison of the effects of hardening and isothermal quenching of type D3 tool steel in salt baths and in fluidised beds.

Heating or cooling medium	*Diameter of test pieces*		*Pre-heating temperature*		*Total time for final heating and holding at 960 °C (1760 °F), minute*	*Hardness, HRC*	
	mm	*inch*	*°C*	*°F*		*At surface*	*At center*
Salt bath	80	3.2	500	930	44	65.5	65
Fluidised bed (a)	80	3.2	490	915	51	65	65
Salt bath	40	1.6	540	1000	36	64.5	64
Fluidised bed	40	1.6	500	930	41	64.5	64

(a) Small parts of the same steel but with a diameter of 8 mm (0.3 inch) were treated at the same time; hardness of these parts was 66 HRC.

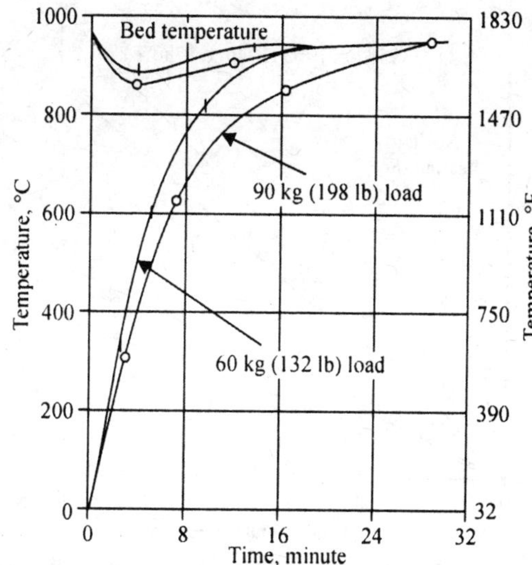

Fig. 3.7. Recovery rates for 25 mm (1 inch) diam steel parts in a 0.3·m^3 (10 ft^3) fluidised bed.

CONTROL OF ATMOSPHERES

A full range of atmospheres can be used within the work zones of fluidised beds. The volume of gas used is clearly dictated by particle size, temperature of operation and optimum fluidisation velocity. However, it can be shown that, with careful design and use of low-cost carrier gases such as nitrogen, even low-temperature surface treatments can be both effective and economical. In addition, one of the major advantages of a fluidised bed is that expensive gas need not be consumed while there is no work in the bed. Atmosphere conditioning is rapid: within about 30 to 60 seconds after an inert gas is introduced into the bed, the purity of the atmosphere is equivalent to that of the gas supply. In internal-combustion fluidised beds, two types of atmospheres can be obtained, as discussed below.

Reducing or Oxidising Atmosphere

Adjustment of the gas/air mixture to the bed so that it is either gas-rich or oxidising causes some decarburisation or oxidation reactions in the materials being processed (the gas-rich mixture produces somewhat less severe reactions). However, these are time-dependent reactions and because of the rapid heating rates of parts being processed and the subsequent short immersion times needed to obtain the correct structure and through hardness, little surface effect other than discolouration and slight scaling is exhibited in section sizes up to 25 mm (1 inch). For larger sizes, the user must be aware of surface reactions that can occur, particularly as the processing temperature increases. Fig. 3.8 illustrates the relative decarburisation bands for steels held in a fluidised bed.

Neutral Hardening and Carburising

In electrically heated or special gas-heated beds, atmospheres for neutral hardening of tool steels or carburising of low-carbon steels can be used for bed flotation. This will allow oxygen free heating of tool steels. However, caution must be exercised during transport of workpieces to the quench tank, to prevent decarburisation or oxidation. For carburising, the atmosphere can be collected, rejuvenated and recycled for efficient usage of the carbon-carrying vehicle.

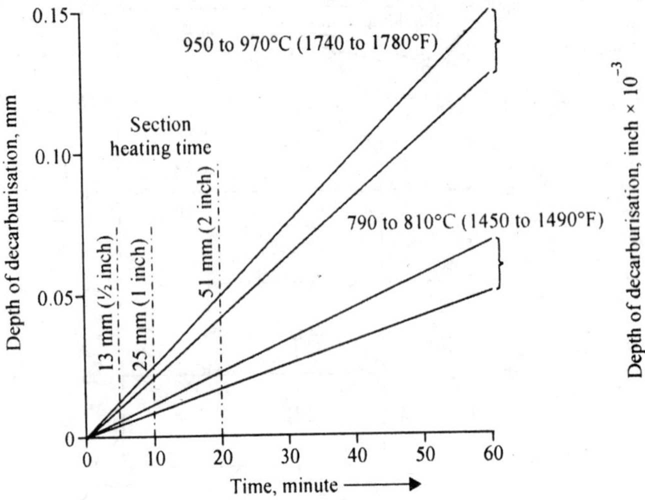

Steels used: type O1 and type D3 tool steels and 0.75 per cent C plain carbon steel.

Fig. 3.8. Representative decarburisation bands for steel held in a fluidised bed.

SURFACE TREATMENTS

Fluidised beds, using atmospheres composed of ammonia, endothermic gas and oxygen, or similar combinations, are capable of performing low-temperature nitrocarburising treatments equivalent to conventional salt-bath processes or other atmosphere processes. High-speed tools oxynitrided in a fluidised bed are comparable to similar tools treated by the more conventional gaseous process. Carburising and carbonitriding in a fluidised bed can yield results similar to those achieved in conventional atmosphere furnaces. Mixtures of propane and air produced the results shown in Fig. 3.9 which compares the case depths obtained on SAE 8620 steel bearing rings carburised in a fluidised bed and by the conventional atmosphere process. An effective case depth of 1 mm (0.04 inch) was achieved in 1.5 hours, using the fluidised-bed technique.

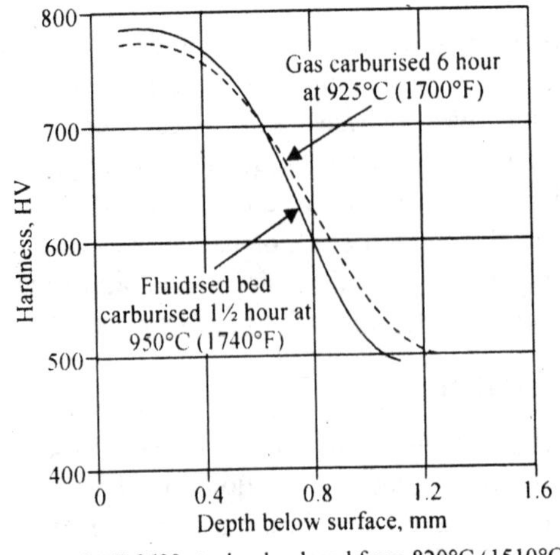

SAE 8620 steel, rehardened from 820°C (1510°C)

Fig. 3.9. Comparison of hardness profiles obtained by fluidised-bed and conventional gas carburising.

Fundamental work on this process is still being performed, but sufficient knowledge exists to compare the mechanisms of conventional gas carburising and the fluidised-bed process.

Conventional Gas Carburising

Carburising occurs through catalytic decomposition of CO according to

$$CO + H_2 \rightarrow C_{Fe} + H_2O$$

Propane enrichment aids this reaction according to

$$C_3H_8 + 3CO_2 \rightarrow 6CO + 4H_2$$

and

$$C_3H_8 + 3H_2O \rightarrow 3CO + 7H_2$$

Fluidised-Bed Carburising

The relatively large volumes of propane consumed during fluidised-bed carburising, together with high gas velocities, favour carburisation by the thermal decomposition of propane to precipitate carbon in accordance with

$$C_3H_8 \rightarrow C\downarrow + 2CH_4$$

The amount of carbon precipitated is proportional to the number of carbon atoms in the hydrocarbon fuel gas; that is, propane forms more carbon than methane. In addition, the purity of propane is important, especially with respect to unsaturated hydrocarbon content, which increases its carbon-forming capability.

The precipitated carbon reacts instantaneously with the oxidising products of combustion:

$$C_3H_8 + 5O_2 \rightleftharpoons 3CO_2 + 4H_2O$$

to form carbon monoxide and hydrogen:

$$C + H_2O \rightarrow CO + H_2$$

and

$$C + CO_2 \rightarrow 2CO$$

Carburisation then proceeds by the catalytic decomposition of CO by H_2 as in conventional carburising. It is possible that carburisation is further complemented by thermal dissociation of the methane formed during carbon precipitation:

$$CH_4 \rightarrow C_{Fe} + 2H_2$$

The carbon potential of the atmosphere varies with the air-to-gas ratio. For each type of hydrocarbon gas, a relationship can be established among air-to-gas ratio, temperature and carbon potential. Control of the reaction and carbon potential of the atmosphere by conventional gas analysis is possible and fluidised-bed furnaces are equipped with sample ports and probes so that suitable measurements can be taken.

TYPES OF FURNACES FOR HEAT TREATING WITH FLUIDISED BEDS

The type of fluidised bed most widely used for heat treatment is the densephase type, although units have been constructed based on the dispersed-phase bed, with particle circulation for heat treatment of long, thin metal parts such as shafts and plates. In a typical dense-phase fluidised bed, the parts to be treated are submerged in a bed of fine solid particles held in suspension, without any particle entrainment, by an upward flow of gas.

Liberation of adequate quantities of heat within fluidised beds is a prime consideration in adapting them for metal processing. Because transfer of heat from the bed to the workpiece is usually much more efficient than transfer of heat from the heat source to the fluidising medium, the greatest difficulty is encountered in transferring suitable quantities of heat to the fluidising medium. In addition, the major part of the heat loss from any practical fluidised system is the heat content of the spent fluidising gas. In instances in which thermal efficiency is unduly influenced by this factor, recirculation of the fluidising gas or installation of a recuperative system may be justified and each has been used in practical applications. Heat input to a fluidised bed can be achieved by several different methods which are discussed below.

Internal-Resistance-Heated Fluidised Beds

In this type of unit, the gas and particles are heated by suitably sheathed internal-resistance-heated elements (Fig. 3.10). For high-temperature operation between 500 and 1000°C (930 and 1830°F), silicon carbide elements can be used, but they must be sheathed to prevent reactions between the elements and the bed material. At lower temperatures, a mineral-insulated heater with an integral metal sheath can be used, particularly where a heater with greater structural integrity is required. With a system of this type, it is essential to ensure that there are no areas of poor fluidisation close to the element. Such areas cause localised overheating and buildup of fused material on the heater, resulting in failure. This problem normally can be avoided and such heaters provide a simple method of heating fluidised beds up to approximately 500°C (930°F). However, heat-up rates (from cold to operating temperature) and heat-recovery rates are relatively slow in this type of bed, particularly for higher temperatures.

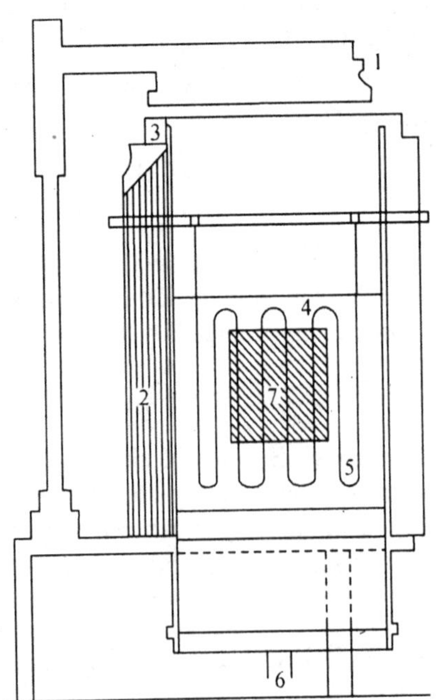

(1) Pivoting cover in two parts. (2) Insulating lagging.
(3) Refractory material. (4) Fluidised bed.
(5) Heating elements. (6) Intake for fluidising gas.
(7) Parts to be treated.

Fig. 3.10 Fluidised-bed furnace with internal heating by electrical-resistance elements.

External-Resistance-Heated Fluidised Beds

A fluidised bed contained in a heat-resisting pot can be heated by external resistance elements, as shown in Fig. 3.11. Waste-heat recovery can be used to increase thermal efficiency and the fluidising gas can be maintained at any desired composition. Although this appears to be a good method of applying heat to the fluidised bed, there is a severe limitation on the rate of heat input that can be achieved through the wall of the pot. Heat-up rates from cold to operating temperatures of 700 to 800°C (1290 to 1470°F) can be as long as 5 to 6 hours.

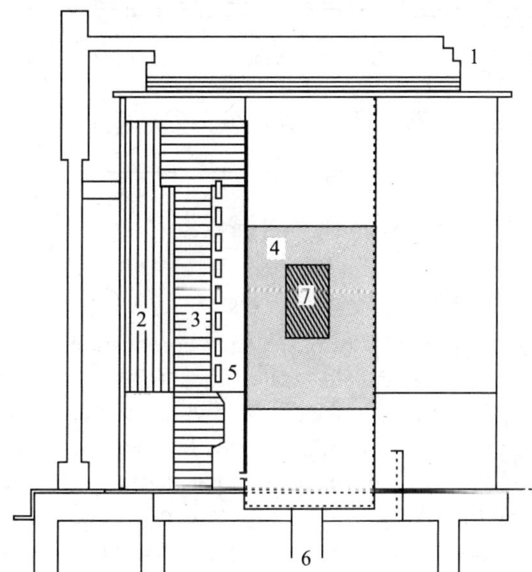

(1) Pivoting cover. (2) Insulating lagging. (3) Refractory material. (4) Fluidised bed. (5) Resistance elements. (6) Intake for fluidising gas (air or nitrogen). (7) Parts to be treated.

Fig. 3.11. Fluidised-bed furnace with external heating by electrical-resistance elements.

Direct-Resistance-Heated Fluidised Beds

Direct resistance heated fluidised beds employ an electrically conducting material such as carbon powder or silicon carbide as the bed material. Power is applied directly to the bed by means of electrodes and heat is generated within the bed by direct passage of an electric current. However, the current distribution is influenced by a metallic workpiece situated in a region of current flow. This heating method has been shown to operate satisfactorily at temperature up to 1300°C (2370°F).

External-Combustion-Heated Fluidised Beds

A fluidised bed contained in a heat-resisting pot can be heated by external gas firing (Fig. 3.12). In this arrangement, a fuel/air mixture is introduced through a standard commercial burner. The burner can be controlled very accurately down to low temperatures for low-temperature tempering. The products of combustion are passed through a perforated metallic plate to achieve flotation of the bed. By introduction of tempering air after mixing and burning of the bed, the burner can be operated stoichiometrically for optimum efficiency and prevention of aldehyde formation.

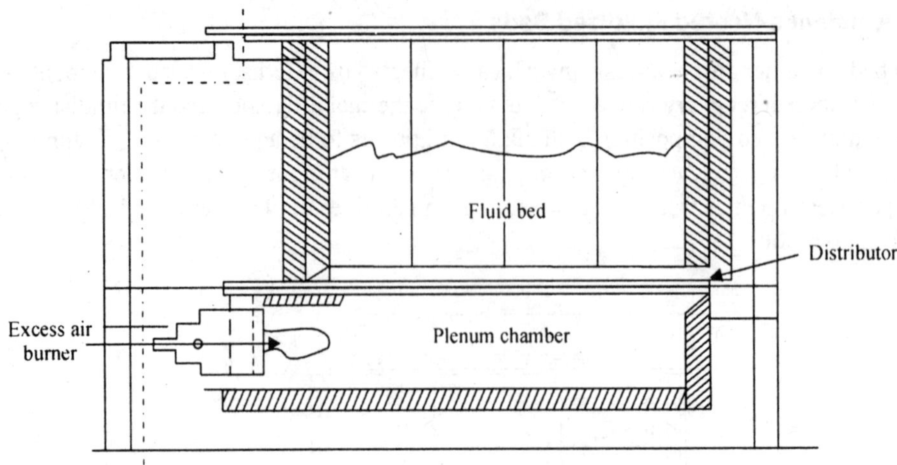

Fig. 3.12 Externally fired furnace.

Submerged-Combustion Fluidised Beds

The technique of submerged combustion consists of passing the combustion products directly through the mass to be heated. This method provides an excellent rate of heat transfer and is now well established for a wide range of liquid heating applications, from heating of swimming pools to concentration of acid solutions. Application of this method to the heating of a fluidised bed requires that the burner be used in such a way as to provide strong agitation of the suspended particles, thereby achieving the desired properties of outstandingly good heat transfer and uniformity of bed temperature.

Equipment developed for this purpose consists essentially of a burner, two concentric tubes and a particle separator. A suitable gas mixture is fed through the burner into the central tube, where it is ignited. The flame develops in the tube and the combustion products escape at its lower end, where they impart heat to the suspended particles before moving up through the annular space between the two tubes. As they rise, a quantity of particles is entrained. These are separated from the gas stream by the deflector plate and fall back into the bed by virtue of gravity. Fig. 3.13 illustrates a system that incorporates submerged combustion together with a controlled atmosphere for low-temperature treatment of metals.

Internal-Combustion Gas-fired Fluidised Beds

A major development in the heating of fluidised beds occurred when an air/gas mixture was used for fluidisation and ignited in the bed, generating heat by internal combustion. Prior to this breakthrough, many technical difficulties prevented use of this mode of fluidised-bed heating. A typical furnace design incorporating this technique is shown in Fig. 3.14.

The advantage of this system is that the bed is fluidised by burning gases and thus the heat is generated within the bed. In gas-fired fluidised beds, the supporting gas or fluidising medium is a near-stoichiometric mixture of gas and air. This combustible mixture is ignited above the bed and quickly imparts its heat to the particles, which in turn heat the incoming gas further down the bed. After a period, combustion takes place spontaneously within the bed, being complete within the first 25 mm (1 inch) of the diffuser once the spontaneous combustion temperature for the gas being used is reached. This temperature commonly varies between 600 and 800°C (1110 and 1470°F). If the vessel is well insulated, the bed temperature can rise to a theoretical combustion temperature and heat-up rates from cold to 800°C (1470°F) are typically between 1 and 1½ hr. However, the following problems are inherent to the basic technique and must be considered:

1. The bed is fluidised by burning gases. To obtain good temperature control and optimum fluidising conditions, however, it is desirable that the fuel-input rate and fluidising velocity be independently variable.
2. Combustion is somewhat unstable below the spontaneous combustion temperature.
3. Very high temperatures can occur in the immediate vicinity of the distributor/diffuser tile. When the bed is incorrectly fluidised, so that this heat cannot be removed from the top of the distributor, theoretical flame temperatures are achieved with consequent deterioration of the distributor. The thermal stresses of expansion and contraction on the distributor tile at these high temperatures tend, even with the best fixing techniques available, to cause failure of joints, which enhances the problem.

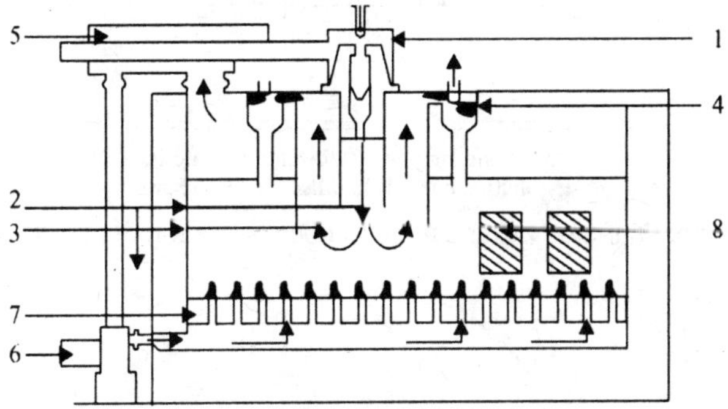

(1) Burner. (2) Combustion tube. (3) Tube through which combustion gases and particles rise. (4) Particle separators. (5) Heat exchanger. (6) Gas recycle compressor for fluidisation. (7) Distributor plate. (8) Parts to be treated.

Fig. 3.13. Controlled-atmosphere fluidised-bed furnace heated by submerged combustion.

Two-Stage Internal-Combustion Gas-Fired Fluidised Beds

The basic problem of separating control of heat input from control of fluidising velocity has been overcome in two alternative designs, shown in Fig. 3.15. In both designs, the initial heat-up from cold to operating temperature is carried out by two-stage internal combustion. A noncombustible mixture of gas and air is introduced beneath the distributor tile. Secondary air is added to make up a stoichiometric or slightly gas-rich mixture immediately above the tile by means of jet holes drilled into heat resisting tubes. This is done to reduce the possibility of explosion and to avoid high flame temperatures at the surface of the tile. The technique has an adverse effect on good fluidisation, but this is unimportant during initial heatup, in which the prime objective is to raise the bed to operating temperature as quickly as possible.

Once the bed has reached operating temperature, the remaining problem is that of isolating heat-up control from control of the fluidising velocity. This is achieved in two ways:

Three-chamber design

In this design (Fig. 3.15a) ,the heat-control outer chambers are separated from the treatment zone by a muffle. The fluidising velocity and atmosphere are independently controlled in the inner chamber, while the outer two zones are still supplying heat by internal combustion. To achieve adequate heat input, fluidisation levels in these outer chambers are above the optimum for heat transfer and surface reactions, but this fact is relatively unimportant.

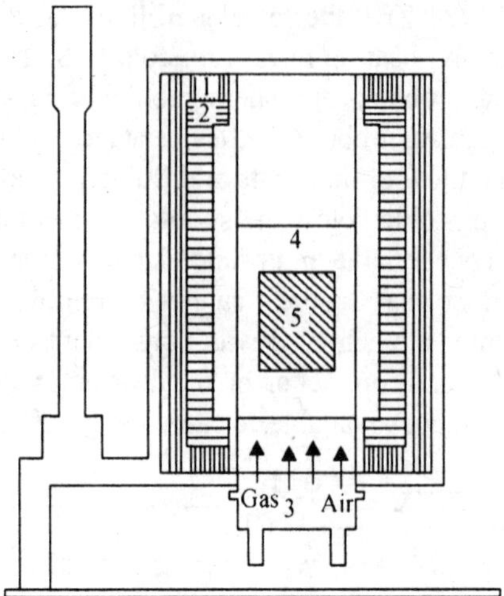

(1) Insulating lagging. (2) Refractory material.
(3) Air and gas distribution box. (4) Fluidised bed. (5) Parts to be treated.

Fig. 3.14. Gas-fired fluidised-bed furnace with internal combustion.

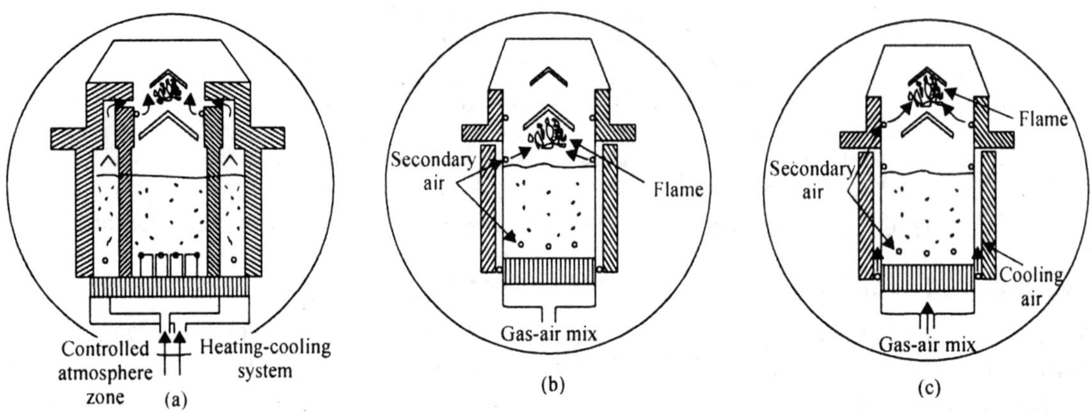

(a) Three-chamber design. (b) Back-radiation design in heating mode. (c) Back-radiation design in cooling mode.

Fig. 3.15 Two-stage, gas-fired, internal-combustion fluidised beds.

Back-radiation design

When fuelrich gases are permitted to burn by the injection of secondary air immediately above the control chamber of the fluidised bed, a back-radiation effect causes a rise in bed temperature. This design (shown operating in the heating/controlling and cooling modes in Figs. 3.15b and 3.15c makes use of this effect and at the same time, utilises heat that is normally dissipated when gases are burned outside the furnace. It is therefore more economical in fuel usage. In principle, the gas-rich mixture is supplied to the central chamber and extra air is added to produce stoichiometric conditions during initial heating of the bed. When cold work is loaded for treatment, the extra air is injected above the bed to produce a radiating flame and recover bed temperature. Should bed temperature exceed set temperature, the extra air is switched to the outside of the furnace wall to provide cooling and finally is mixed with the rich gas/air to produce combustion at the top of the specially constructed hood.

APPLICATIONS OF FLUIDISED-BED FURNACES

The potential applications of fluidised-bed technology to heat treating are many. Fig. 3.16 specifies those applications in which fluidised beds can compete with conventional furnaces.

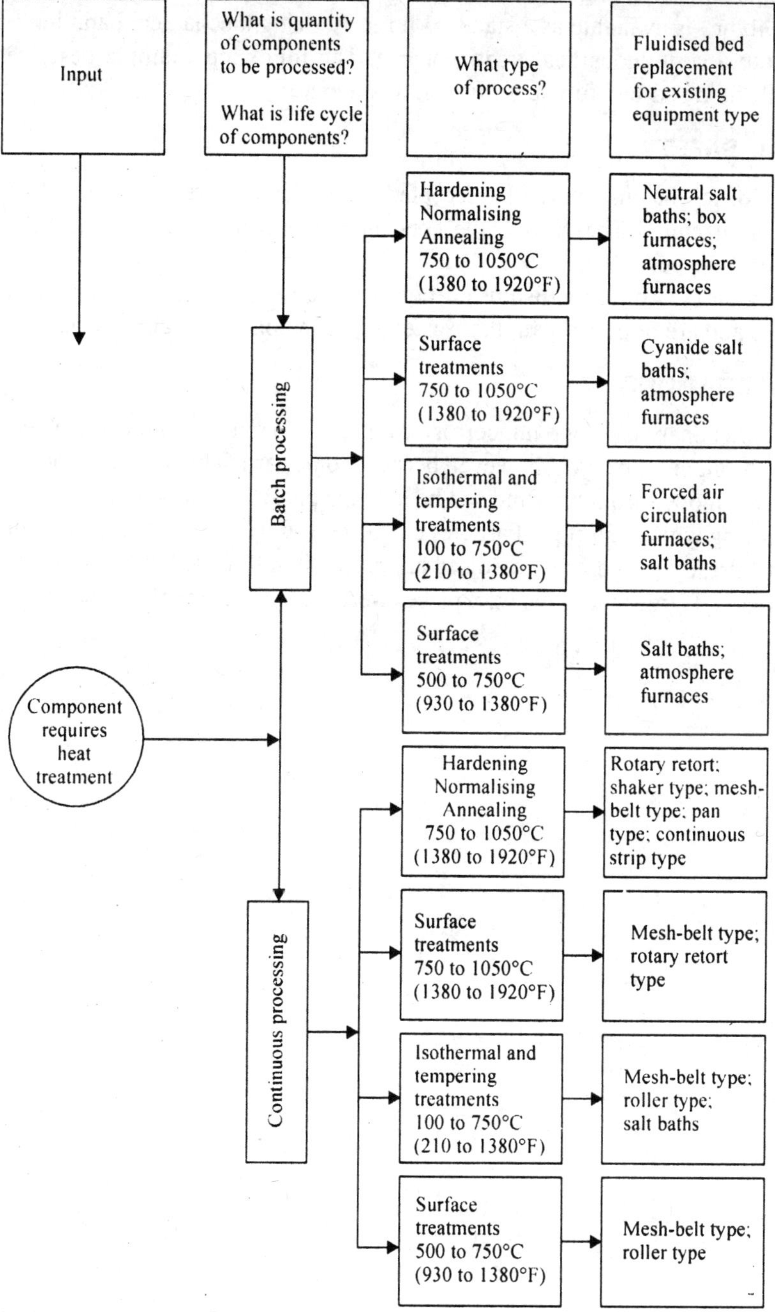

Fig. 3.16 Fluidised-bed application: Decision model.

Applications of fluidised-bed furnaces to heat treatment of metals include continuous units for all types of wire and strip processing (patenting, austenitising, annealing, tempering, quenching, etc.); continuous rotary types for fasteners, bearings and other small parts; and all configurations of batch-type units for general heat treating applications. A typical batch-type unit with an output of approximately 150 kg/hr. (330 lb/hr.) is available as a standard furnace. Using mechanical handling equipment, it can be automated into a continuous heat treatment line. The following example describes how one firm decided to install fluidised-bed furnaces for heat treatment.

OPERATIONAL SAFETY

As for all forms of gas heating, normally accepted safety devices are incorporated on the majority of beds presently manufactured. The 'flexible tile' concept ensures that any failure of joints does not influence the performance of the bed.

Parts carrying surface oil or moisture do not create an explosion risk because the contaminants simply vapourise and are removed with the waste gas, as in conventional furnaces.

CLEANING OPERATIONS

Fluidised solids are nonabrasive and noncorrosive and they do not wet immersed objects. There is some drag-out loss of the aluminum oxide, however, because some particles accumulate on flat surfaces occurs as workloads are removed from the fluidised bed. These particles can be removed in part by agitation, bouncing or blowing with an air pipe. Particles can be reused by being dried, sieved and returned to the bed. When parts already scaled or preoxidised are placed in a fluidised bed, particles tend to adhere to the scale to a greater degree than if the workpieces were clean. These particles can be removed by water spraying.

Chapter 4

Heat Treating in Vacuum Furnaces and Auxiliary Equipment

INTRODUCTION

Vacuum heat treating is a relatively new development in metallurgical processing. Vacuum heat processing consists of thermally treating metals in heated enclosures that are evacuated to partial pressures compatible with the particular metals and processes. Vacuum is substituted for the more commonly used protective gas atmospheres during either part or all of the heat treatment. Furnace equipment used in vacuum heat treatment differs widely in size, shape, construction and method of loading.

Although originally developed for the processing of electron tube materials and refractory metals for aerospace applications, vacuum furnaces are now employed in brazing, sintering, heat treating and diffusion bonding of metals. Vacuum furnaces also are used for annealing, carburising, heating and quenching, tempering and stress relieving. Furnaces for vacuum heat treating are equipped for workloads ranging from several pounds up to 100 tonnes and heated working chambers range in size from 0.03 m^3 (1 ft^3) up to hundreds of cubic feet. Although most vacuum furnaces are batch-type installations, continuous vacuum furnaces with multiple zones for purging, preheating, high-temperature processing and cooling by gas or liquid quenching also are used. Vacuum heat treating furnaces also:

1. Prevent surface reactions, such as oxidation or decarburisation on workpieces, thus retaining a clean surface intact.
2. Remove surface contaminants such as oxide films and residual traces of lubricants resulting from fabricating operations.
3. Add a substance to the surface layers of the work (carburisation, for example).
4. Remove dissolved contaminating substances from metals, using the degassing effect of a vacuum (removal of hydrogen and oxygen from titanium, for example).
5. Join metals by brazing or diffusion bonding.

VACUUM MEASUREMENTS

A theoretical or ideal vacuum is an empty space that does not contain either vapours, particles, gases or other matter and consequently has no atmospheric pressure. Because this condition does not exist, even in outerspace, an ideal vacuum cannot be achieved. Therefore, a manufactured vacuum is expressed in relative terms of pressure compared with the standard atmospheric pressure surrounding the earth. Standard atmosphere pressure at sea level, 45°C latitude and 0°C has the following values: 760 mm Hg, 760000 μ or μm Hg, 29.921 in. Hg or 14.696 psi.

The most common units of pressure used in measuring vacuum or partial pressure are as follows: millimetres of mercury (μm Hg); microns (μ Hg) or micrometres of mercury (μm Hg); and torr (1 torr equals 1 mm Hg). Microns (μ) are established and well-known units; however, micrometres (μm) are exactly equal and consistent with metric practice. These pressure or vacuum values refer to the height of a mercury column sustained by the differential between atmospheric pressure and an attained level of vacuum or partial pressure. Fig. 4.1 provides a comparison of vacuum ranges in relation to atmosphere. Note the normal operating range for vacuum heat treating.

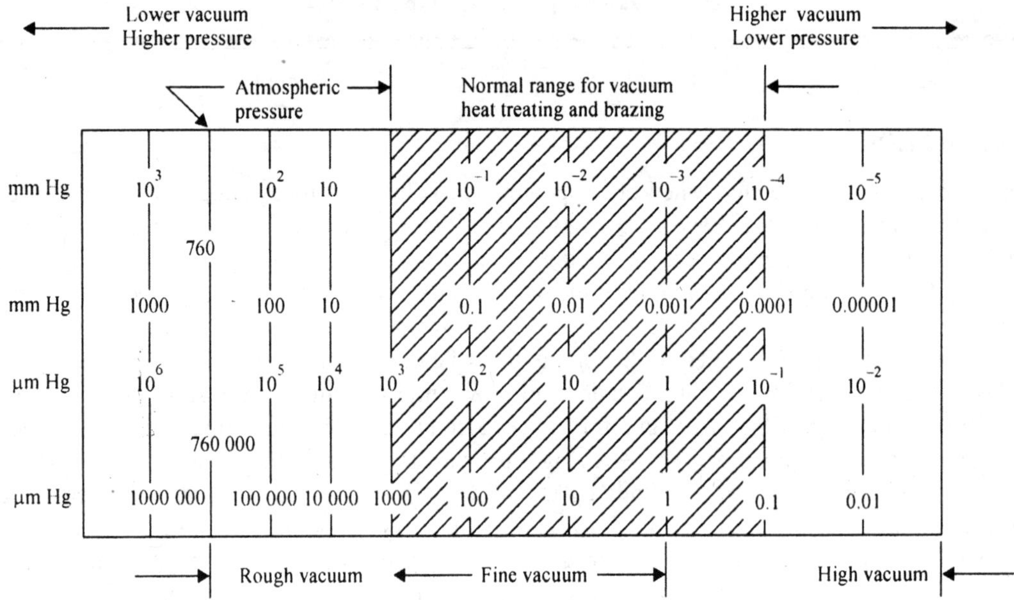

Fig. 4.1. Range of pressures and their relation to atmosphere.

COMPARISON OF VACUUM AND ATMOSPHERE FURNACE PROCESSING

Unlike conventional atmosphere heat treatment, vacuum heat treatment does not require control of carbon potential of prepared atmospheres and related furnace conditioning requirements. In atmosphere heat treating, the water-vapour content, or dew point, of a protective gas atmosphere is often the most critically controlled variable in addition to the temperature and time of processing. Time and temperature must be controlled in vacuum furnaces as well; however, unlike atmosphere heat treating furnaces, a vacuum furnace contains only very small amounts of residual gas or atmosphere. In a leak proof vacuum furnace, analysis of the residual atmosphere at a vacuum of about 10^{-3} torr indicates that less than 0.1 per cent of the original air remains. The residual gases consist mainly of water vapour, with the remainder comprised of primarily organic vapours from the seals, vacuum greases and vacuum oils. The oxygen content at 10^{-3} torr is less than 1 ppm. If all of the residual gas in the vacuum furnace were converted to water vapour, the water vapour content would be approximately 1.5 ppm or equal to that of a gas with a dew point of about –80°C (– 110°F). At a vacuum level of 10^{-4} torr, the equivalent dew point of gas is estimated to be approximately –90°C (–130°F) or less.

These low dew point equivalents compare favourably with the driest inert gases available from highly efficient gas-dehydration equipment. With suitable pumping systems, the concentration of oxygen and water vapour can be reduced to lower levels than that achieved in inert or reducing-gas atmospheres.

After a vacuum heat treating furnace has been evacuated, gaseous reactions such as those encountered with atmosphere heat treatment are virtually eliminated. Moreover, the vacuum extracts many gases, surface contaminants and processing lubricants that would be difficult and costly to remove by any other method. Gases drawn from the metal surface into the vacuum surrounding the charge are trapped by the vacuum pumps and exhausted from the system. This advantage of a vacuum system is of greater significance when parts with complex shapes, blind holes or deep recesses are being heat treated. A complete purging of such parts in a protective atmosphere requires an extended purging period. Even long-time purging, however, may not ensure complete removal of entrapped air and other contaminants.

When more thorough purging is required, the furance can be evacuated with a simple vacuum system and the enclosure or retort can be backfilled with the desired protective atmosphere. This method markedly reduces the amount of protective atmosphere and time required to produce satisfactory results.

VOLATILISATION AND DISSOCIATION

In vacuum technology, the reactions are related principally to the physical chemistry of the materials and the laws of physics, whereas atmosphere processes are controlled largely by laws governing physical chemistry. Therefore, vapour pressures of elements and the required vacuum or backfill pressure levels become important considerations in the use of a vacuum furnace.

Vapour pressure, which is the gas pressure exerted when a substance is in equilibrium with its own vapour, increases rapidly with temperature, because the amplitude of molecular vibration increases with temperature. Some molecules in the outer surface of the solid material have higher energies than others and they escape as free molecules or vapour. If a solid substance is contained in an enclosure devoid of any other material, molecules will continue to escape from the solid surface until their rate of escape is exactly balanced by the rate of condensation or recapture of the gaseous molecules. The equilibrium pressure developed is the vapour pressure of the substance at that temperature. Vapour pressure of a metal is dependent on temperature only.

It is normally desirable to use a vacuum-temperature combination that accelerates desorption of gases without producing vapourisation of more volatile alloy constituents. Alloys with high concentrations of volatile elements, such as brass, are not heat treated in vacuum furnaces.

If brass were heated in a vacuum at a temperature of 540°C (1000°F) and a vacuum level of the order of 0.1 μm Hg, the zinc component will vapourise (volatilise) and the brass will be soon converted to copper sponge.

Metals such as aluminium, lead, zinc and magnesium have relatively high vapour pressures; if heated above a temperature at which the vapour pressure of the element exceeds the pressure in the furnace, they will evaporate or sublime rapidly. Thus, high-vacuum heat treatment is not applicable to some metals and alloys. To handle certain metals and alloys properly, the pressure must be limited to the 'fine' vacuum range (Fig. 4.1), or a backfill partial pressure must be employed.

Alloys with lower concentrations of volatile elements can be processed in vacuum by using a backfill pressure that exceeds the sublimation pressure of the element at the temperature involved. A backfill pressure of a few hundred μm Hg at temperatures about 980°C (1800°F) precludes the vapourisation of elements such as chromium, copper and manganese from steels processed at these higher temperatures.

For example, if pure manganese were heated to approximately 790°C (1455°F) at a pressure of 10^{-4} torr, it would vapourise (Fig. 4.2). If the material was held at a higher temperature or lower pressure for an adequate period of time, the metals would become depleted and eventually disappear and the vapours would condense on the colder areas of the furnace and/or pumping system. Backfill or higher partial pressures greatly slow the rate of evapouration or volatilisation. The vapour pressures of carbon and selected pure metals, as related to temperature, are shown in Fig. 4.2 and Table 4.1.

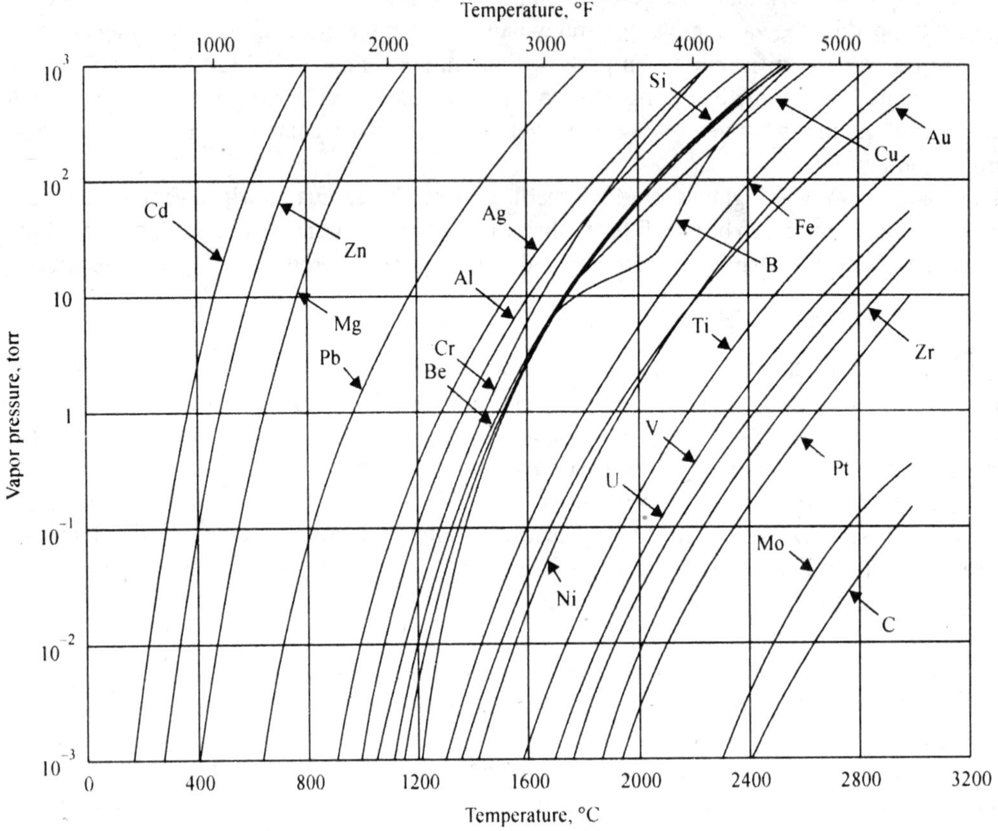

Fig. 4.2. Vapour pressure versus temperature for carbon and various pure metals.

Table 4.1. Vapour pressures of various elements.

	Vapour pressure at:									
	10^{-4} mm Hg 0.1 μm		*10^{-3} mm Hg 1.0 μm*		*10^{-2} mm Hg 10 μm*		*10^{-1} mm Hg 100 μm*		*760 mm Hg 760,000 μm*	
Element	°C	°F	°C	°F	°C	°F	°C	°F	°C	°F
Aluminum	808	1486	890	1632	997	1825	1124	2053	2085	3733
Antimony	525	977	595	1103	678	1251	780	1434	1441	2624
Arsenic	–	–	220	428	–	–	310	590	610	1130
Barium	544	1011	626	1157	717	1321	830	1524	1404	2557

(Contd...)

Element	Vapour pressure at: 10^{-4} mm Hg 0.1 μm °C	°F	10^{-3} mm Hg 1.0 μm °C	°F	10^{-2} mm Hg 10 μm °C	°F	10^{-1} mm Hg 100 μm °C	°F	760 mm Hg 7,60,000 μm °C	°F
Beryllium	1030	1884	1131	2066	1247	2275	1396	2543	–	–
Bismuth	537	997	609	1128	699	1290	721	1328	1421	2588
Boron	1141	2084	1240	2262	1356	2471	1490	2712	–	–
Cadmium	180	356	220	428	264	507	321	610	766	1409
Calcium	463	865	528	982	605	1121	701	1292	1488	2709
Carbon	2290	4150	2473	4480	2683	4858	2928	5299	4831	8721
Cerium	1092	1996	1191	2174	1306	2381	1440	2622	–	–
Caesium	74	165	110	230	153	307	207	405	691	1274
Chromium	993	1818	1091	1994	1206	2201	1343	2448	2484	4500
Cobalt	1363	2484	1495	2721	1650	3000	1834	3331	–	–
Copper	1036	1895	1142	2086	1274	2323	1433	2610	2764	5003
Gallium	860	1578	966	1769	1094	1999	1249	2278	–	–
Germanium	997	1825	1113	2034	1252	2284	1422	2590	–	–
Gold	1191	2174	1317	2401	1466	2669	1647	2995	2999	5425
Indium	747	1375	841	1544	953	1746	1089	1990	–	–
Iridium	2156	3909	2342	4244	2558	4633	2813	5092	–	–
Iron	1196	2183	1311	2390	1448	2637	1604	2916	2737	4955
Lanthanum	1126	2057	1243	2268	1382	2518	1550	2820	–	–
Lead	548	1018	620	1148	718	1324	821	1508	1745	3171
Lithium	378	711	439	822	514	957	608	1125	1373	2502
Magnesium	331	628	380	716	443	829	515	959	1108	2025
Manganese	792	1456	878	1612	981	1796	1021	1868	2153	3904
Molybdenum	2097	3803	2297	4163	2535	4591	3011	5448	5573	10056
Nickel	1258	2295	1372	2500	1511	2750	1680	3054	2734	4950
Niobium	2357	4271	2541	4602	–	–	–	–	–	–
Osmium	2266	4107	2453	4444	2669	4833	2922	5288	–	–
Palladium	1272	2320	1406	2161	1567	2851	1760	3198	–	–
Platinum	1745	3171	1905	3459	2092	3794	2295	4159	4411	7965
Potassium	123	253	161	322	207	405	265	509	643	1189
Rhodium	1816	3299	1973	3580	2151	3900	2359	4274	–	–
Rubidium	88	190	123	253	165	329	217	423	679	1254
Ruthenium	3704	3736	2232	4046	2433	4408	2668	4831	–	–
Scandium	1162	2122	1283	2340	1424	2593	1596	2903	–	–
Silicon	1117	2041	1224	2233	1344	2449	1486	2705	2289	4149
Silver	848	1558	921	1688	1048	1917	1161	2120	2214	4014
Sodium	195	383	238	460	291	556	356	673	893	1638

(Contd...)

	Vapour pressure at:									
	10^{-4} mm Hg 0.1 μm		10^{-3} mm Hg 1.0 μm		10^{-2} mm Hg 10 μm		10^{-1} mm Hg 100 μm		760 mm Hg 7,60,000 μm	
Element	°C	°F	°C	°F	°C	°F	°C	°F	°C	°F
Strontium	413	775	475	887	549	1020	639	1182	1385	2523
Tantalum	2601	4710	2822	5108	–	–	–	–	–	–
Thallium	461	862	500	932	607	1123	661	1220	1458	2655
Thorium	1833	3328	2000	3630	2198	3985	2433	4408	–	–
Tin	923	1692	1011	1850	1190	2172	1271	2318	2272	4118
Titanium	1250	2280	1385	2523	1547	2815	1725	3168	–	–
Tungsten	2769	5013	3019	5461	3312	5988	–	–	5932	10701
Uranium	1586	2885	1731	3146	1899	3448	2099	3808	–	–
Vanadium	1587	2887	1726	3137	1889	3430	2081	3774	–	–
Yttrium	1363	2484	1495	2721	2967	3000	1834	3331	–	–
Zinc	248	478	290	554	343	649	405	761	908	1665
Ziroconium	1661	3020	1818	3301	2003	3634	2214	4014	–	–

The vapour pressure of metals is fixed with probable values at a given temperature and the temperature at which the solid is in equilibrium with its own vapour descends as the pressure to which it is exposed descends. For example: iron must be heated to 2737°C (4955°F) at atmosphere before its vapour pressure is greater than atmosphere (760 mm Hg); this point is reached at 1311°C (2390°F) at a pressure of 1 mm (10^{-3} μm Hg).

Alloy Vapour Pressures

Vapour pressures of pure metals are constant, well-established values; the vapour pressure of a given alloy is variable depending upon conditions. The vapour pressure of an alloy is governed in part by a law analogous to Dalton's law of partial pressures: the total vapour pressure of an alloy, under ideal conditions, is equal to the sum of the partial vapour pressures of its constituents. However, the partial pressure of each element in the alloy is lower than its normal vapour pressure and is proportional to its concentration.

In processing at temperatures where the vapour pressures of more volatile minor constituents are still in the micron range, alloys behave in accordance with Dalton's law. For example, if pure manganese is heated to 790°C (1455°F), its vapour pressure reaches 10^{-4} mm Hg; it will be impossible to evacuate to lower pressures without evaporating all of the manganese. However, when manganese is alloyed with other elements, as a solid solution in iron, for example, its effective vapour pressure is lowered. The total vapour pressure for the alloy is the sum of vapour pressures of each individual element multiplied by its concentration in the alloy. The vapour pressure of manganese in a 1 per cent Mn alloy at 790°C (1455°F) is about 10^{-6} mm Hg.

Many metals form compounds by reaction with oxygen, hydrogen and nitrogen. These reactions are usually exothermic and the possibility for dissociation of the resulting compound increases with higher temperatures. Some oxides, such as water, vapourise at temperatures so low that dissociation occurs only in the vapour phase. For an oxide, nitride or hydride that remains a solid over a wide range of temperatures, a dissociation pressure exists at any temperature that represents an equilibrium between the compound, the gas and the metal. All metallic compounds decompose into constituent elements when heated to sufficiently high temperatures. However, many of the metal oxides are quite stable, requiring low pressures at high temperatures to effect dissociation. It is impractical to dissociate many of these compounds due to the combination of vacuum level and temperature required. When a metal oxide dissociates, the metals remains and the oxygen is evacuated. For example, chromium oxide will dissociate in a 10^{-5} torr vacuum at 1300°C (2370°F). Dissociation of a metal oxide usually depends more on temperature than on pressure. Most oxides can be dissociated under normal operating vacuum levels at approximately their reduction temperature in a highly reducing hydrogen atmosphere.

The nitrides and hydrides often have higher dissociation pressures, making many of them unstable when heated in a vacuum. For this reason, vacuum heat treating can be used both to dissociate these compounds and to remove the evolved gas without disturbing the base metal.

When oxidised surfaces brighten during vacuum heat treating, the mechanism involved is not believed to be simply thermal dissociation of the oxide. Bright surfaces do not discolour, or become brighter, when they are exposed to a vacuum atmosphere that is theoretically oxidising. A metal surface can be maintained almost free from visible oxidation at a partial pressure several decades higher than suggested by theoretical calculations. The following theories have been proposed to explain this apparent anomaly:

1. The solution and diffusion rate for oxygen exceeds its surface absorption rate.
2. Oxide nucleation occurs at discrete sites rather than as a continuous film.
3. The effective concentration of oxygen is reduced by carbon and hydrogen in the solid metal and by the vacuum atmosphere.

FURNACE EQUIPMENT

Although conventional atmosphere furnaces can be adapted for vacuum heat treating by adding a vacuum-tight retort connected to a suitable pumping system, furnace equipment developed especially for vacuum heat treating is generally used.

Vacuum furnaces can be grouped into three basic designs: top-loading, or pit furnaces; bottom-loading, or bell furnaces; and horizontal-loading, or box furnaces. Furnace designs can be varied to fit a wide variety of processing requirements by changing the chamber length, adding internal doors, circulating fans or recirculating gas systems and internal quenching systems.

Every vacuum furnace, regardless of its end use and basic hot or cold wall design, requires: (i) heating elements controlled to proper processing temperatures and cooling rates; (ii) suitable vacuum enclosures with access opening; (iii) vacuum pumping system; and (iv) instrumentation to monitor and display critical processing data. Production furnaces may be single-chamber units, batch-type units or multichamber, semicontinuous units.

HOT WALL VACUUM FURNACES

Vacuum furnaces are classified according to the location of the heating and insulating components. Hot wall furnaces were the first type to be designed. Because of the demand of the heat treating industry for higher temperatures, lower pressures, rapid heating and cooling capabilities and higher production rates, hot wall vacuum furnaces have become essentially obsolete and have largely been replaced by cold wall vacuum furnaces.

The entire vacuum vessel is heated by external heating elements in the hot wall construction. The heat is contained by insulation materials similar to the materials used in electrically heated atmosphere heat treating furnaces. Hot wall furnaces have limited utilisation because of slow heating and cooling capabilities. They are also limited in temperature because strength of materials is reduced at elevated temperature. However, hot wall equipment is readily adaptable to low-temperature operations not exceeding 980°C (1800°F), with moderate sized chambers.

The double-pump modification of the hot wall furnace permits the construction of larger vessels and the use of higher operating temperatures. This system incorporates a second vacuum vessel outside the vacuum retort to maintain a roughing vacuum during the heating cycle. This removes the stress of the atmospheric pressure on the heated retort or vacuum vessel.

Bell-Type Furnace

A bell-type hot wall furnace is shown in Fig. 4.3. The workload is placed on an elevated refractory metals hearth that rests on an insulated base clad with an alloy plate material. A water-cooled circumferential flange and vacuum gasket are located on the vacuum-tight base cover adjacent to the heated zone but in an unheated area. A retort made with a heavy walled heat-resisting alloy covers the work load. A flange at the bottom of the retort fits on top of the base gasket to provide a vacuum-tight enclosure. The bell-shaped furnace equipped with internal electrical heating elements is lowered into position over the retort by a vertical hoist. The vacuum pumping system is connected through the insulated base.

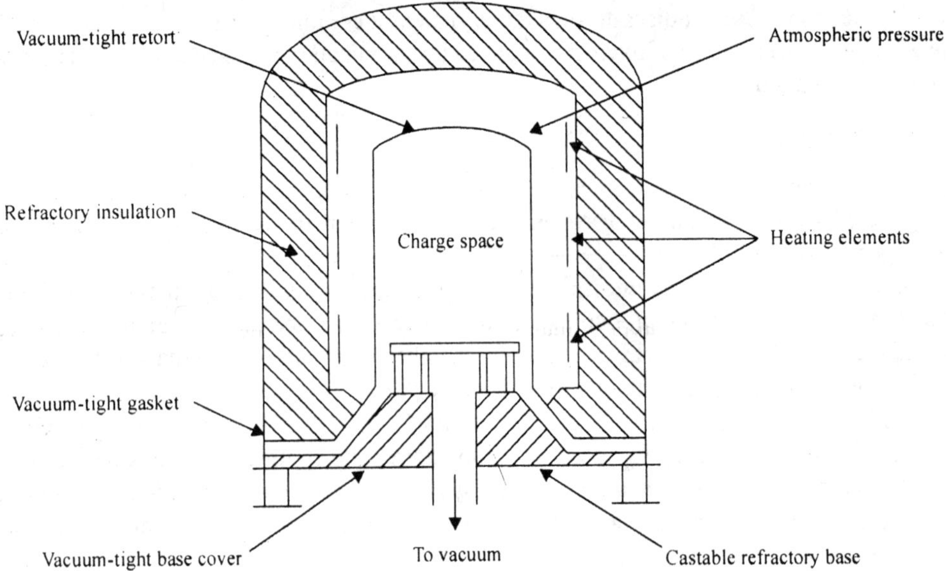

Fig. 4.3. Bell-type hot wall vacuum furnace.

Because this furnace cannot be heated or cooled rapidly, even when the bell furnace is removed, production rates and the number of thermal cycles within a given time period are limited. Moreover, because the hot retort must support the entire pressure of the external atmosphere, its wall must be quite heavy. Practical operating temperatures for a furnace of this type generally are limited to approximately 925°C (1700°F).

Pit-Type Furnace

Fig. 4.4 shows a schematic drawing of a pit-type hot wall furnace. The workload is placed in a top-loading muffle or retort made from a heat-resisting alloy. The upper end of the retort is provided with a water-cooled flange and vacuum gasket that interlock with a flange on the upper part of the furnace above the heated zone.

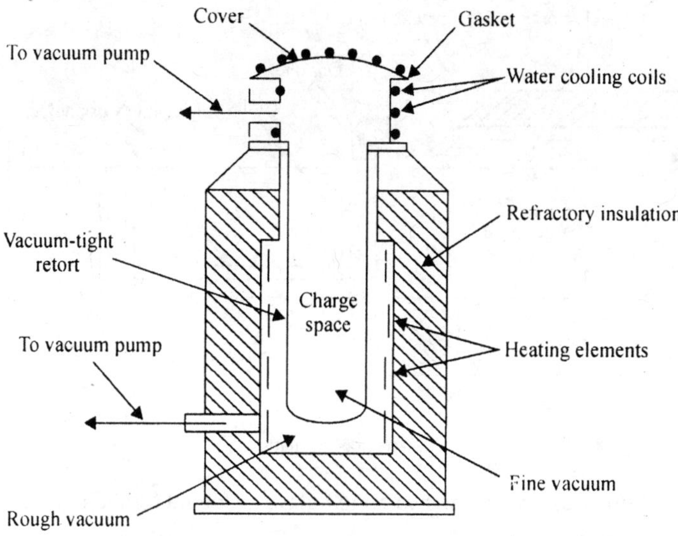

Fig. 4.4. Pit-type hot wall vacuum furnace.

The muffle is lowered into the furnace by an overhead hoist, providing vacuum connections for the furnace and retort. With this construction, the space between the muffle and heating furnace can be evacuated by a roughing pump so that the pressure at the exterior surface of the muffle is essentially zero. This evacuation permits the use of a muffle with a much thinner wall and raises the maximum operating temperature of the system to approximately 1175°C (2150°F). When the heating cycle is completed, inert gas is bled into the retort and air is bled simultaneously into the heating furnace so that the pressures remain balanced on both sides of the retort.

The retort then can be removed from the furnace to a cooling stand and another retort can be inserted in the hot furnace. This construction increases heating and cooling flexibility, which in turn increases cycle frequency and production.

Horizontal and Vertical Two-zone Hot-Wall Vacuum Furnaces

Horizontal and vertical two-zone hot-wall vacuum furnaces are shown in Fig. 4.5 and 4.6. In both types, the heat-resisting alloy muffle is extended much further beyond the heating section of the furnace. This extended section has a water-cooled jacket to provide accelerated cooling.

In the horizontal furnace, the charge is carried on an alloy hearth that can be moved in and out of the heated zone by a push rod extending through a seal in the outer fixed end of the muffle. This hearth has vertical heat baffles or multiple radiation shields at each end to confine the heat to the heated portion of the muffle. An equivalent means of transferring the workload is necessary in the vertical furnace as well, although this is not shown in Fig. 4.6.

By using this technique, the charge can be cooled much faster, because it is only necessary to remove the heat from the hearth and workload and not from the hot end of the muffle. This increased cooling rate permits hardening of air-hardening steels and is adaptable to certain solution treatments not possible with other hot wall furnaces.

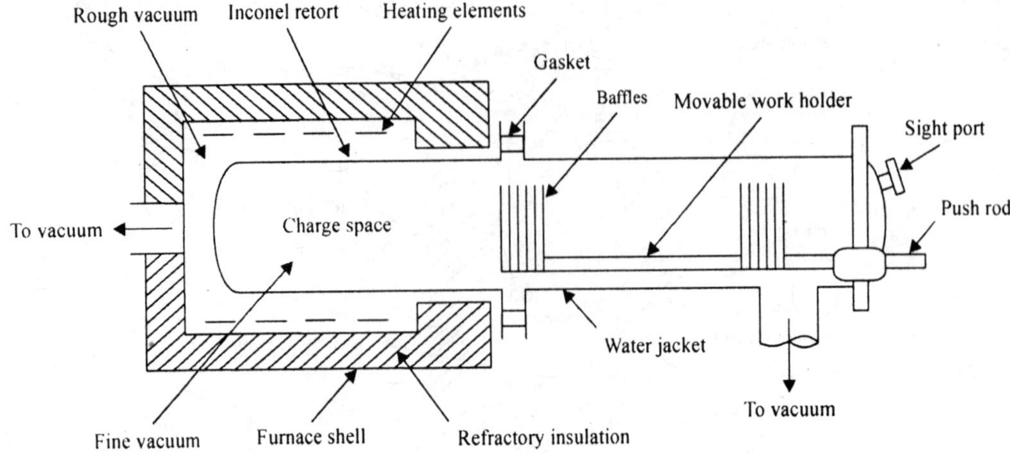

Fig. 4.5. Horizontal, two-zone hot wall vacuum furnace.

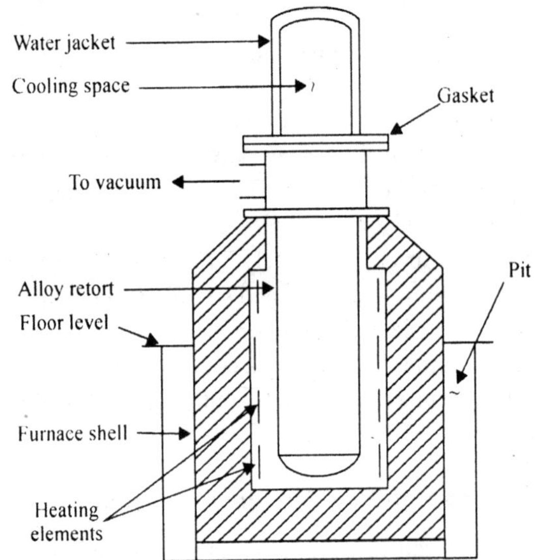

Fig. 4.6. Vertical, two-zone hot wall vacuum furnace.

COLD WALL FURNACES

Cold wall furnace units consist of a water-cooled vacuum vessel maintained near ambient temperature during high-temperature operations. Consequently, because the operating temperature does not affect the strength of the vessel material, large units can be constructed for use at high operating temperatures.

In the cold wall design, the water-cooled vacuum vessel contains and supports the internal insulation, the electrical heating elements and the hearth upon which the workload rests. The vacuum acts as: (i) a substitute for the normal heat treating atmosphere to protect the workload; (ii) an insulating medium in the furnace, because the thermal conductivity of a vacuum is essentially zero; and (iii) an effective protective coating around the heating elements, the heat shields and the supporting hearth.

The use of a vacuum as the insulating medium has permitted the use of multiple radiation shields of very low mass or special lightweight ceramic insulations that facilitate rapid heating and cooling. Rapid rates of heating and cooling are important because each treatment cycle is started at ambient temperature and must be cooled to ambient temperature at completion. As the protective medium, vacuum has permitted use of materials such as graphite, tungsten, molybdenum or tantalum, for heating elements and hot furnace structures. Such materials normally cannot be used in other furnace constructions without more elaborate, expensive and sometimes hazardous protective atmosphere environment.

Batch Vacuum Furnaces

The workload remains stationary inside the furnace during heating in batch operation; however, in semicontinuous vacuum furnaces with multiple chambers, the workload is moved within the vacuum usually after completion of a processing step or segment. By using high melting-point materials in furnace structures, extremely rapid rates of heating and high temperatures can be attained. Heat is transferred to the workload almost entirely by radiation.

Radiation cooling in the hot furnace is extremely slow, however. To reduce furnace time and shorten quenching rates, pressure in the vacuum chamber usually is increased to either just below atmospheric pressure or up to five times the atmospheric pressure by introducing a pure inert gas such as nitrogen or argon. This gas is rapidly recirculated within the furnace, through cooling coils or through an external heat exchanger and back into the furnace, with high-powered large-capacity gas pumps.

Advantages

Advantages of batch cold wall vacuum furnaces compared to atmosphere furnaces include:

1. Bright, oxide-free treatment of most metals and alloys.
2. Outgassing and purging of entrapped volumes of gas.
3. Retained surface finish.
4. Uniform heating.
5. Removal of surface volatiles.
6. No heat added to local environment.
7. No chemical effect on furnace or work during treatment.

Disadvantages

Disadvantages of batch cold wall vacuum furnaces compared to atmosphere furnaces include:

1. Vapourisation of alloying elements with high vapour pressure.
2. Inert gas required to effect quench cycle.

Other advantages of batch cold wall vacuum furnaces include:

1. Instant pushbutton start from cold.
2. Low pollution.
3. Wide operating temperature range.
4. Easy maintenance.
5. Safe operation.
6. Fully automatic processing.
7. Wide range of programmable heating and cooling rates.
8. Minimal distortion of treated work.
9. High heating rates and temperatures resulting from use of high-melting point materials.

10. Complete shutdown when not in use; no need to maintain heat to maintain low dew point.
11. Blanketing gas usually not required during heating.

Other disadvantages of batch cold wall vacuum furnaces include:

1. Furnace must withstand atmospheric pressure and be leak-tight.
2. High capital cost.

Types of Cold Wall Vacuum Furnaces

Bottom-loading furnaces

As shown in Fig. 4.7, the furnace is stationary and elevated well above floor level. The bottom descends to floor level for ease of loading. The work is loaded on trays that are placed on the hearth by a fork lift when the bottom is in the lowered position. Such furnaces are built to handle large heavy loads and are cooled rapidly by a high-velocity internal or external circulating gas system.

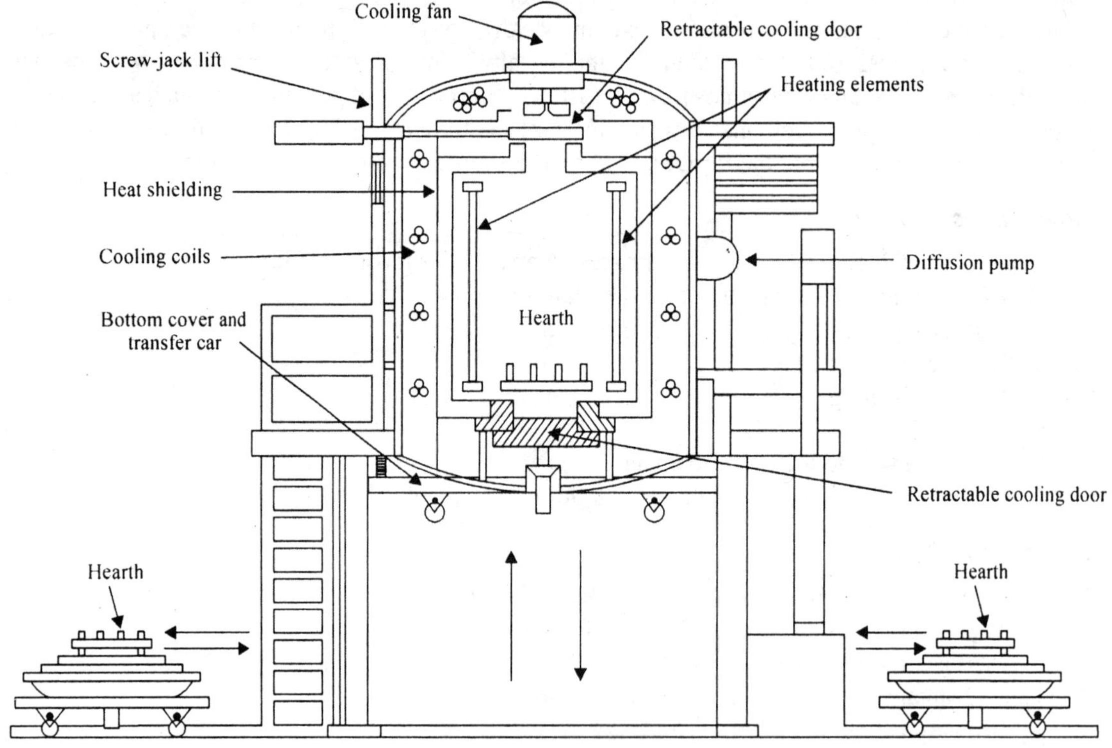

Fig. 4.7. Bottom-loading cold wall vacuum furnace.

Top-loading furnaces

Top-loading furnaces, as shown in Fig. 4.8, are not as widely used as the bottom-loading furnace. However, they are useful in processing long and relatively thin workpieces such as slender shafts. Workpieces are suspended vertically from hangers attached to the removable top of the furnace. Adequate head room and a vertical hoist are required. These furnaces are cooled in the same way as the bottom-loading units.

Horizontal-loading furnaces

A box-type horizontal furnace consists of a gastight cylindrical shell with circular convex end plates. In some designs, both of the end plates are hinged to permit easy access to the furnace interior. Furnaces also are constructed with a stationary rear plate or with a second hinged access door at the front that is smaller than the main front plate. The cylindrical shell and the end plates are water cooled by copper coils soldered to the exterior surface or through the use of double-wall construction. Depending on the intended use, the gastight shell may be made of stainless or carbon steel. The movable end plates are sealed by O-rings at the end faces of the cylindrical section. The pressure of the outside atmosphere on the convex ends supplies the pressure for vacuum sealing. Usually, auxiliary clamps are provided to supply sealing pressure and to prevent the door from becoming unsealed when positive pressure inside the furnace is used during inert gas quenching.

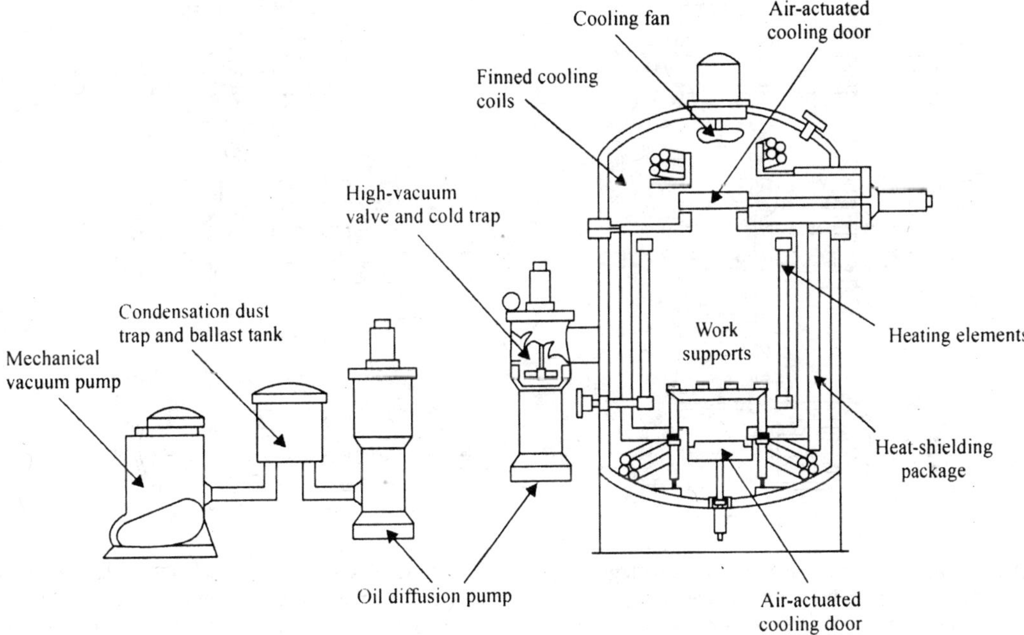

Fig. 4.8. Top-loading cold wall vacuum furnace.

Many horizontal-loading furnaces are equipped with a special lifting and transfer truck that is stationed in front of the furnace. Frequently, this truck rolls on rails so that it is permanently aligned with the heating chamber. A hydraulic fork lift raises the work basket and the truck moves forward to transfer the basket into the furnace where it is lowered onto the hearth or work pedestal. Such a mechanism avoids possible damage to the interior of the furnace, which could occur if transfer is attempted without controlled movement. Horizontal-loading furnaces may have multiple chambers, depending on the heat treating operation to be performed (Fig. 4.9). Special systems have been designed to transfer workloads inside multiple-chamber furnaces the roller-hearth and pusher-type furnace designs can be adapted for vacuum furnaces.

The hearth is supported on wheels that roll on rails which are installed below and outside the heated zone and are protected by movable heat baffles. The longitudinal motion can be supplied by a sealed push rod extending through. The furnace wall to an air or hydraulic cylinder or by an internal chain driven conveyor in the cool area.

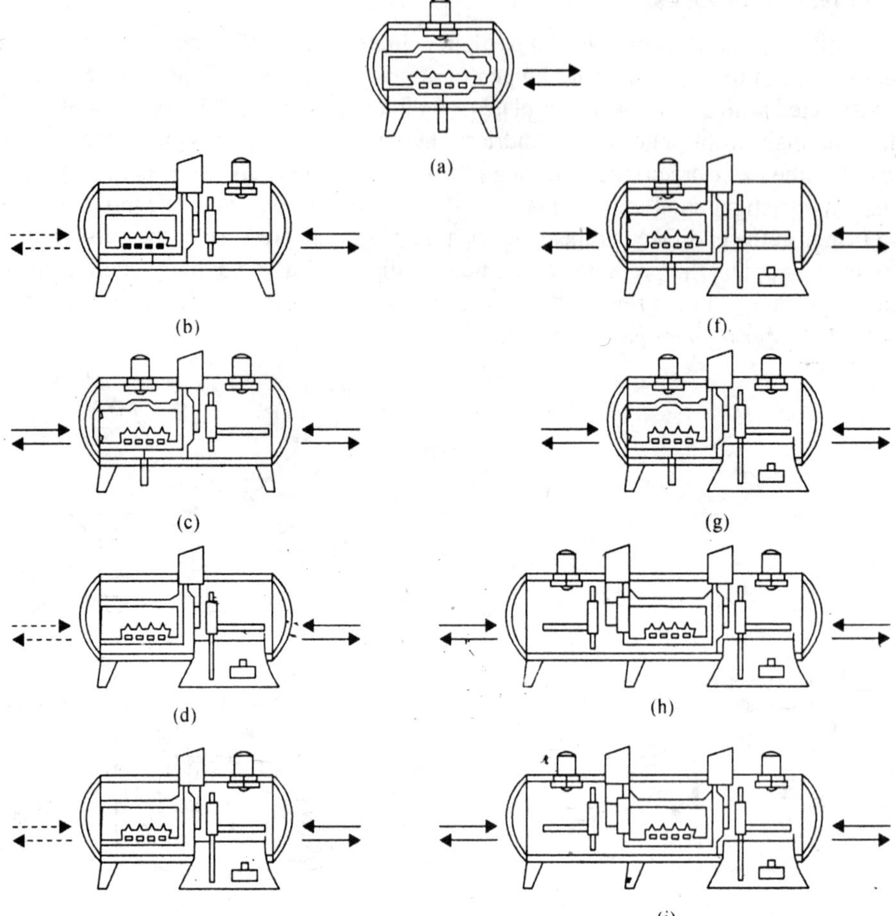

(a) VFC: Single-chamber vacuum furnace with gas/fan quenching. (b) RVF: Two-chamber vacuum furnace; one chamber for heating with integral chamber for gas/fan quenching only and mechanism for internal in and out transverse movement of workload to and from heat chamber. Unit can be loaded and unloaded only from cooling or quenching chamber. (c) RVFCF: Same as (b) but gas/fan cooling is also included in heat chamber and unit can also be loaded or unloaded from either chamber. (d) RVOQ: Two-chamber vacuum furnace; one chamber for heating with integral second chamber for oil quenching only and mechanism for internal movement of workload to and from heat zone. Unit can be loaded/unloaded from oil quenching zone only. (e) RVFOQ: Same as (d) but gas/fan cooling is also included only in quenching or cooling chamber. (f) RVFCOQ: Same as (d) but gas/fan cooling is also included only in the heating chamber and unit can be loaded or unloaded from either chamber. (g) RVFCFOQ: Same as (d) but gas/fan cooling is included in both chambers and unit can be loaded or unloaded from either chamber. (h) FRVOQ: Three-chamber vacuum furnace; middle chamber is for heating only. One end chamber contains gas/fan quenching only and internal mechanism for work movement to and from heat zone. Other end chamber contains oil quenching only and also an internal mechanisms for work movement to and from heat zone. Unit can be loaded or unloaded from either end chamber. (i) FRVF: Three-chamber vacuum furnace with middle chamber for heating only. End chambers are both alike in containing gas/fan quenching only and internal mechanism for workload movement to and from heat zone. Unit can be unloaded or loaded from either end.

Fig. 4.9. Horizontal Vacuum furnace configurations.

Another method of work transfer within a horizontally loaded furnace uses an internal rack-and pinion drive. An overhead chain-driven conveyor can also be used. The work trays may also be transferred longitudinally to the second chamber hearth by a lifting mechanism installed in the unheated chamber that is only exposed to the furnace heat during the short transfer time. Rack-and-pinion drives and pneumatic cylinders often are used to execute vertical elevator movements in and out of liquid quench tanks, as well as to open and close internal heat shields and vertical doors.

A horizontally loaded vacuum furnace equipped with radiation shields is shown, in a vertical section, in Fig. 4.10. The workload is exposed directly to radiation from the electrical heating elements. The multiple radiation shields are made of thin sheets of heat resisting material, such as molybdenum, in parallel layers between the heating elements and the chamber shell. An alternative construction, where thermal insulating material is used instead of radiation shields is shown in Fig. 4.11. This insulation may be a fibre fill, graphite felt or special low-density fibre ceramic material.

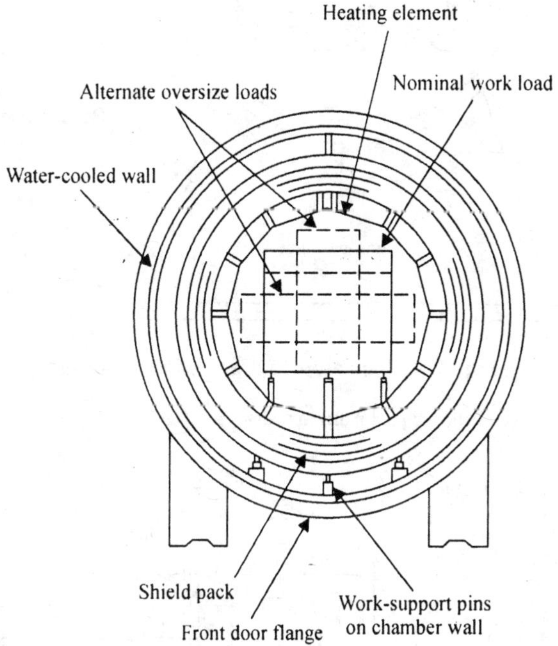

Fig. 4.10. Radiation-shield cold wall vacuum furnace.

Cross-sectional views of a three chamber oil-quench furnace are shown in Fig. 4.12. The front chamber is equipped with internal cooling coils and a circulating fan for accelerated gas cooling. The centre chamber is the heating chamber, which can be sealed at both ends during the heating cycle by internal moving heat shields and doors equipped with O-rings. The third chamber contains the oil quench and the vertical transport system required to immerse the work in the circulated quenching oil.

Semicontinuous furnaces are constructed of modular units of three or more vacuum chambers. Each unit has a work carrier transfer system, an internal assembly of heating elements and shield package, pumping system and temperature-controlled power system. Isolation locks or doors at each end of the vacuum heating or brazing environment separate these modular units from the entry and exit vacuum vestibules. These vestibules, in turn, have doors for access to and from atmospheric pressure and other assembly operations. Usually included with the furnace are fans for fast cooling at the exit end of the furnace and an overhead conveyor system to transfer work carriers from the exit to the furnace entrance.

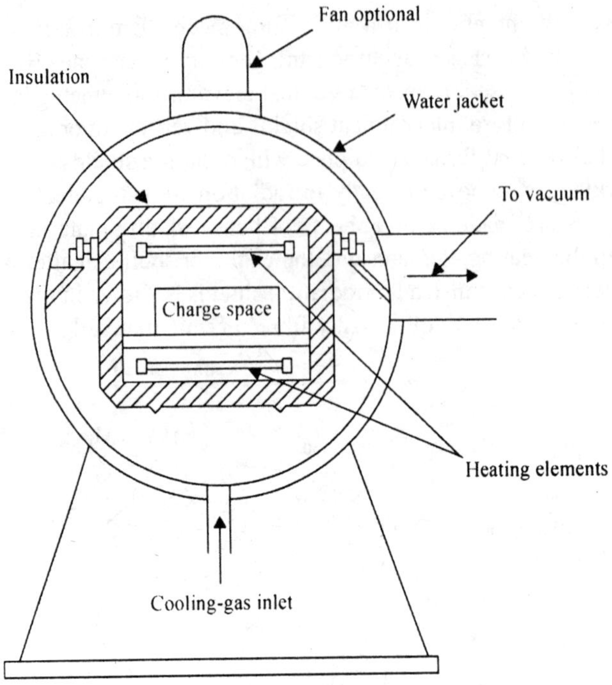

Fig. 4.11. Insulated cold wall vacuum furnace.

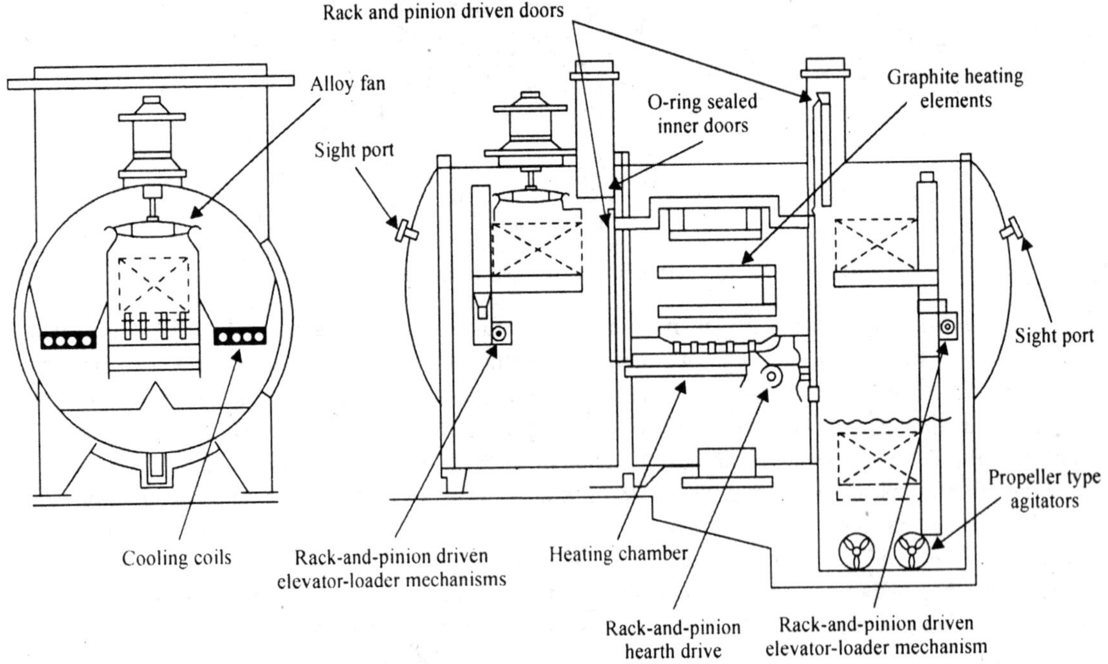

Fig. 4.12. Three-chamber vacuum oil quench furnace.

Work carrier loading and unloading stations are incorporated in the external overhead conveyor system. A high volume production semicontinuous vacuum furnace is shown in Fig. 4.13. These furnaces are used for fluxless brasing of aluminum heat exchangers at production rates of 100 to 250 parts per hour. For high-volume production and easy flow of workloads, semicontinuous vacuum furnaces are equipped with electrical controls that can also be computer programmed for automatic operation. The internal and external work carrier transfer systems, door operations, pumping systems, heating systems, gas backfill and external cooling, loading and unloading stations are all electrically interlocked and controlled for completely automatic operation. Automatically controlled mechanical loading and unloading of the work carriers can also be incorporated into the complete system.

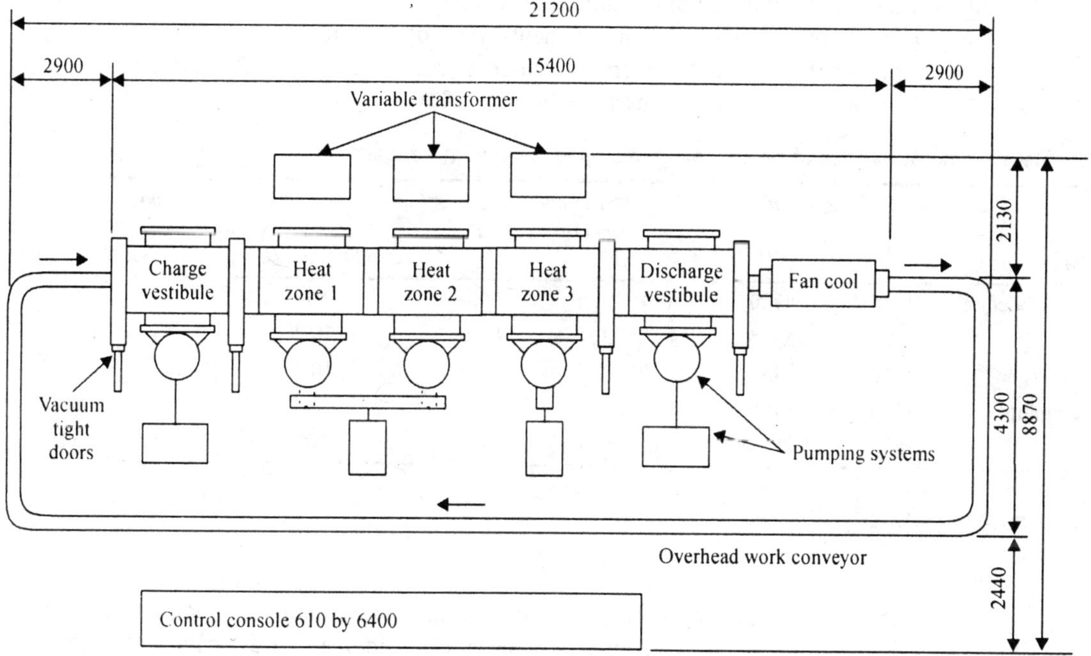

Fig. 4.13. Layout of an automatic vacuum furnace for fluxless aluminum brazing.

HEATING ELEMENTS

Resistance heating and induction heating are the two most common methods of heating within the cold wall furnace. When vacuum furnaces are heated inductively, a graphite cylinder is used as a susceptor; the graphite is heated by induction and radiates the heat to the work inside the cylinder. When heating is provided by the more common resistance elements, the heat transfer is also completed by radiation; therefore, the active heating surface should be large enough to effect a rapid transfer of heat.

Essentially all vacuum furnaces employ three-phase 60-Hz power supplies. Three types of power supplies and controls are used:

1. Controllable variable autotrans formers.
2. Saturable core reactors.
3. Silicon-controlled rectifiers (SCR).

Low voltage should be employed in the vacuum chamber because a high electrical potential can produce short circuiting of the elements by ionising the residual gases within the chamber.

Resistance heating elements operating in a vacuum do not require oxidation-resistant properties equal to those required in oxidising atmospheres. To improve operating efficiency, resistance heating elements are heated to higher temperatures than are the elements used in conventional furnaces because the transfer of radiant energy is proportional to the fourth power of the absolute temperature. Higher temperatures require heating elements with low vapour pressures to ensure long life. Materials meeting these requirements are:

1. Refractory metals, such as tungsten, molybdenum and tantalum.
2. Pure solid graphite in the form of bar, rod or tube.
3. Pure graphite cloth woven from fine filaments of pyrolysed graphite.
4. Chromium-nickel elements for operating temperatures less than 980°C (1800°F).

Properties of these materials are compared with iron in Table 4.2.

Table 4.2. Characteristics of heating elements used in vacuum furnaces.

	Melting point		*Upper operating temperature limit*		*Vapour pressure torr, at:*	
Material	*°C*	*°F*	*°C*	*°F*	*1600°C (2910°F)*	*1800°C (2910°F)°F*
Molybdenum	2625	4760	1700	3100	10^{-8}	10^{-6}
Tantalum	2996	5425	2500	4530	10^{-11}	10^{-9}
Tungsten	3410	6170	2800	5070	10^{-13}	10^{-11}
Graphite	3700	6700	2500	4530	10^{-13}	10^{-10}
Iron	1535	2800	–	–	10^{-1}	1

Refractory Metals

Tungsten

Tungsten is capable of withstanding higher operating temperatures than the other refractory metals (see Table 4.2). As a heating element material, it is used as a thin sheet or as sections of woven wire screen. Wire screen is less subject to damage from thermal stresses that occur during heating or cooling.

Molybdenum

Molybdenum, in the form of solid rod, strip or thin sheet material, is the most widely used metallic element. Material in sheet form is normally preferred because the electrical power density (watts per square inch of radiating surface) is low compared to cylindrical rod, resulting in lower operating temperatures and thus longer service life for the elements. Also, thermal expansion and contraction and the resulting stresses are handled more easily from a design standpoint. However, thin sheets are subject to mechanical damage. Hangers and supports for metallic heating elements must have good insulating properties and must be chemically stable at the temperatures and pressures encountered in service. Heating elements must not be restrained by the support system from free movement during the thermal cycle.

Molybdenum undergoes a large change in electrical resistance between room temperature and the normal operating temperatures; consequently, the design of the power supply controls the current during the early stages of heating to avoid damaging the elements, as well as furnace heating rates.

Solid Graphite Heaters

All metals lose some strength when heated, whereas crystalline carbon in the form of graphite increases in strength as the temperature increases. Pure graphite in the form of flat bar and rod is less expensive than other high-temperature metallic resistors. Graphite also has a much lower heat expansion coefficient and is more resistant to thermal shock than most metallic materials.

As shown in Table 4.2, graphite also has a high melting point and a low vapour pressure; thus, it is an excellent choice for a vacuum furnace heating element material. In processes where possible minute concentrations of carbonaceous material in the vacuum atmosphere will not have an adverse effect on workpieces, graphite resistors are commonly chosen. Moreover, the presence of incandescent carbon may provide additional cleansing or gettering action with respect to oxygen or water molecules in the residual vacuum atmosphere. This subtle action is not provided by any of the metallic resistors.

Replacement of graphite resistors may be less expensive than for metallic resistors because, elements often can be replaced without disturbing the surrounding insulation and because electrical connections are mechanical and require no welding.

Graphite Cloth Heaters

A third type of material used for vacuum heating elements is a cloth composed of fine graphite fibres. This material is made from rayon cloth pyrolysed at high temperature to convert the carbon in the rayon to crystalline graphite. The cloth is strong and very flexible. It can be cut with ordinary scissors to the desired size and shape. Because the cloth is flexible, the supporting system can be simplified considerably. The ends usually are clamped in graphite electrodes. This graphite cloth also is available as a hollow cylinder with solid graphite ends of high electrical conductivity to form a self-supporting electrode with a large radiating surface area.

HEAT INSULATION

Part of the insulation in a cold wall vacuum furnace is provided by the vacuum itself. Where space is essentially void, there can be virtually no heating by convection or conduction. Radiant energy emitted by the resistors in all directions generally is confined to the desired heating zone by one of three designs of radiation shielding.

Metallic Shielding

Multiple concentric layers of thin heat-resisting sheet metal reflect the energy back into the heating chamber of the furnace. Approximately 6 mm (¼ inch) spacing is maintained between these sheets. Thin wire coil springs are sometimes used as spacers between adjacent surfaces. Inner shields are sometimes fabricated from molybdenum, because of its superior heat-resisting qualities; the next two layers may be stainless steel, with the outer layer composed of nickel steel or plain carbon steel, depending on the maximum operating temperature of the furnace. As the number of the shield layers increases, the efficiency of the shielding increases; however, the insulating effect of each added shield decreases. Consequently, no more than five or six shields or layers are used. Approximately 85 per cent efficiency can be achieved in this manner, even with a material of 0.5 emissivity. The effect on efficiency of the number of insulating layers and the emissivity of the sheeting material is illustrated in Fig. 4.14. Efficiency declines as the originally bright clean metal surfaces become affected by the deposition of oxide or sooty films. This decline must be considered during the design of the furnace system.

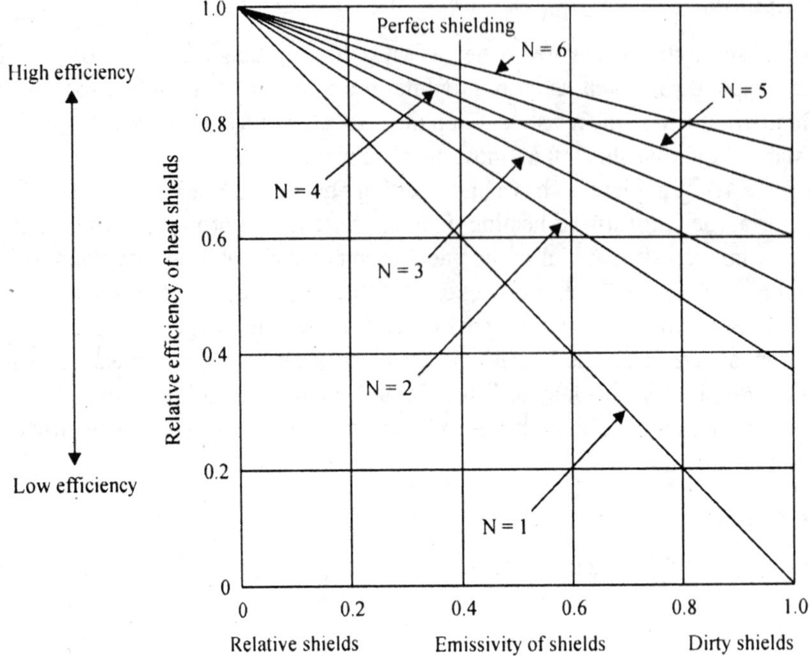

Fig. 4.14. Relation of heat shield efficiency and emissivity of sheet metal for various numbers, N, of sheets.

Although slightly faster evacuation rates and higher ultimate vacuum levels are usually obtainable with designs incorporating radiation shields, these advantages are realised only if the furnace is subsequently used for a singular process application. These advantages readily revert to disadvantages when radiation-shielded units are used for multiprocess purposes.

Sandwich Construction

Sandwich construction resembles the wall of a conventional low-temperature furnace and has proven to be an efficient design. One or more layers of graphite felt or high-temperature refractory fibre fill are placed between the inner and outer sheet metal walls. For instance, the inner wall may be molybdenum and the outer wall may be stainless steel. A vacuum between the fibres reduces thermal conductivity of the wool packing to a fraction of what it would be in an oxidising atmosphere. Thermally insulated cold wall vacuum furnaces are better suited. to conventional vacuum heat treating involving multiple process and temperature requirements. Comparable evacuation rates and ultimate vacuum levels are realised with much less frequent need for vacuum outgassing or hydrogen reduction cycles compared to the radiation shield design.

Multilayer Graphite

This design utilises multiple layers of high-purity graphite felt fastened to an outer cage of refractory metal by molybdenum clips. This material has a density of only about 80 kg/m^3 (5 lb/ft^3) and it functions as a thermal insulator in much the same way as refractory fibre fill. Because it is a felted material with inherent cohesion, it does not require support by sheet metal walls as does the fibre fill. Its emissivity is near unity—about 0.98. Graphite felt is relatively inexpensive, compared to molybdenum and is easy to replace. This method, like sandwich construction, relies upon an insulating material with low conductivity in a wall heated from one side.

Insulation Maintenance

All types of insulation can be accidentally contaminated. In many cases, foreign materials such as lubricants used in deep drawing operations, volatile metals or even cotton gloves may be inadvertently carried into vacuum furnaces with the workload. Contamination can result in extensive damage to radiation shields and considerable downtime if mechanical cleaning of the shield is necessary. In an insulated furnace that does not depend on reflectivity, a thorough outgassing will normally return the furnace to operating conditions within several hours.

The efficiency of multiple radiation shields decreases rapidly as the shields become dirty and nonreflective and also as emissivity increases. One method of cleaning molybdenum shields consists of heating the furnace to above the operating temperature after backfilling with dry hydrogen to a pressure of 1 torr. A second method consists of carefully cleaning them with clean compressed air.

An effective method of protecting the innermost radiation shield is to overlay the molybdenum with a layer of graphite foil. This material is essentially pure graphite that has been rolled into a thin sheet by a special process with dry hydrogen. It is held in place by mechanical fasteners. Any contamination from splattering of brazing filler metal or evapouration of volatile materials will be deposited on this inert material instead of the molybdenum shield. Graphite foil can be applied to almost any other metal lining surface as well.

GAS QUENCHING

The inability to obtain the high cooling rates necessary for many metallurgical operations severely limited early hot wall vacuum furnaces. Development of a retort that could be transferred from the hot furnace to a special cooling stand provided a somewhat inconvenient solution.

With the development of the cold wall vacuum furnace, the cooling problem was overcome to some extent by radiation shielding of multiple thin metal sheets, thermal insulation of very low mass such as refractory fibre or special foam refractory brick or graphite fibre felt. Because of the low heat content of these insulating media, most of the heat to be dissipated is contained in the workload itself. Cooling of the work remains quite slow because of the absence of convective heat transfer in a vacuum. Backfilling the chamber with an inert gas is necessary to promote conductive transfer of heat from the work to the water-cooled shell. Fans made from stainless steel or other heat-resisting alloys can be installed above the insulated work area to promote the convective circulation of the backfill atmosphere. Baffles or bungs of insulating material usually must be moved from above and below the workload to improve the access of the cooling atmosphere to the hot work.

Finned copper coils cooled by circulating water can be installed between the insulating medium and the shell of the furnace to aid in the heat transfer, rather than relying entirely upon the cold furnace shell as the transfer surface.

Circulation by an internal fan is often used because it provides the fastest gas cooling, but it is not likely to provide uniform cooling throughout the load. An inert gas circulating system with an internal manifold is shown in Fig. 4.15. To increase the rate of heat removal, the mass-flow coefficient must be increased. Mass flow is the product of the mass of gas moved times its velocity. To increase this coefficient, the velocity of the gas circulating through the load can be increased or the gas pressure in the system can be increased so that a greater number of gas molecules are present per unit volume.

Internal pressures used during inert gas quenching range from 100 to 650 torr on furnace that do not have positive pressure clamps on the doors. For even more rapid gas cooling with the proper furnace construction, positive pressures up to five times atmospheric pressure have been used.

Because of technical advances in fan motors, door-sealing methods and related equipment, it has become common to cool loads with gas fans at pressures up to 1275 torr in conventional units and to several times atmosphere pressure in furnaces with turbine recirculation systems. Because the cooling rate is proportional to absolute pressure, it is increased considerably by positive pressure cooling.

By backfilling to pressures above atmospheric pressure, for example, 70 kPa (10 psi) gauge, rather than slightly below atmospheric pressure, the cooling time can be shortened by as much as 30 per cent (Fig. 4.15). This development has greatly enhanced the use of vacuum equipment for air-hardening tool steels with decreased over-all cycle time.

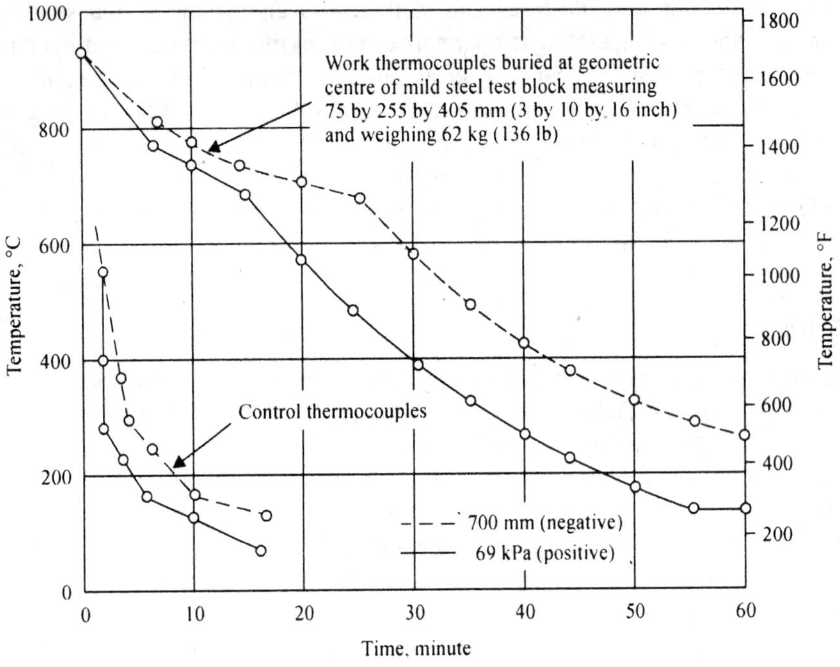

Fig. 4.15. Comparison of cooling rates and backfill pressures.

The proportional increase of quenching speed with increasing gas pressure up to two times atmospheric pressure also applies to 5 atm, if the time delay in reaching this pressure and starting the fast gas circulation is minimal. Fast backfilling is promoted by a compact design and a special turbine system that builds up the forced circulation very rapidly. Figure 4.16 illustrates the effect of pressure on quenching speeds from a hardening temperature to 550°C (1010°F). An increase in pressure from 1 to 2 atm will decrease cooling time by 60 second or 50 per cent. An increase from four to five times atmospheric pressure would give a cooling time decrease of 6 second or 20 per cent.

Liquid Quenching

When even faster cooling rates are needed, furnaces are available with liquid quench capabilities. This usually requires an arrangement where the quenching is done in a chamber isolated from the heating chamber. In one such design, a horizontal loading furnace has a rear heating chamber, a forward cooling chamber, with a vacuum sealing door between them. When the heating cycle is completed, the work is transferred to the front chamber. After the sealing door is closed, the workload is lowered into the quench tank by an elevator.

The quenching liquid is agitated vigorously by propellers and the heat absorbed by the quenching medium is removed by a heat exchanger similar to that used in many atmosphere heat treating furnaces. The chamber that houses the quench tank also can be equipped with a circulating fan and cooling coils to provide forced convection gas cooling or liquid quenching. Quenching is usually completed at slightly below atmospheric pressure within an inert gas backfill.

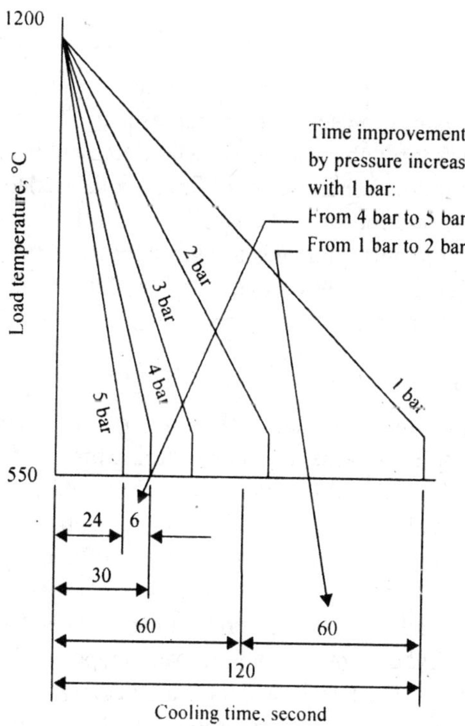

Fig. 4.16. Relationship between gas pressure and quenching rate.

Considerable care must be used to ensure that trays and fixtures that carry the workload into the quenchant are thoroughly cleaned before reuse. Organic residues on trays and workload will continually outgas and destroy the quality of the vacuum if not removed.

GASES USED FOR BACKFILLING

Gases used for backfilling vacuum furnaces are argon, nitrogen, helium and in rare cases, hydrogen. Argon and nitrogen can be obtained as compressed gases or as condensed liquids stored in cryogenic containers. Helium and hydrogen are available only as compressed gases. If appreciable amounts of argon or nitrogen are to be used, it is much less expensive to purchase the gas in liquid form. Very high purity gases are available and some typical analyses of these gases are listed in Table 4.3.

Table 4.3. Typical analysis of backfill gases.

Gas	Purity, %	Impurity, ppm O_2	N_2	CO_2	CO	H_2	Hydro- Carbons	Carbon- aceous gas	Dew point °C	°F	Thermal conductivity at 0°C (32°F) (a), K × 10^3	Relative cooling rate (b)
Argon	99.9995	2	2	1	–	–	1	–	–79	--110	0.04	0.74
Nitrogen	99.9993	3	–	–	–	–	–	1	–79	–110	0.06	1.0
Helium	99.998	1	10	–	–	1	1	–	–62	–80	0.34	1.03
Hydrogen	99.9	10	1500	1	2	–	25	–	–59	–75	0.42	1.4

(a) Cal·cm/°C·cm^2
(b) Relative to nitrogen as 1.0

When a gas is purchased in a liquid state, it is stored under pressure in a large tank with a safety pressure relief valve that allows gas to escape to the atmosphere if the internal pressure exceeds a set value. Liquid gas leaves the tank through a vapouriser where ambient heat supplies the necessary heat of vapourisation. Frequently, this vapouriser is a series of finned coils used to increase heat transfer. Because of the excellent insulation provided for the tank, usually very little gas is lost by venting. Sometimes a reservoir or storage tank of appreciable volume is used to store the vapourised gas at a nominal pressure to avoid sudden surges of pressure in the cryogenic system and to ensure that sufficient volume is available when backfilling is in progress.

WORKLOAD SUPPORT

The workload in most vacuum furnaces is placed on a tray or in a basket to facilitate loading. Because these fixtures are heated and cooled at rather rapid rates during processing, the material and design must allow for the cycles of thermal stresses. Molybdenum is often used for support fixtures. If moderately high temperatures are used, as with many tool and die steels, austenitic stainless steel trays may be used. However, work baskets of Inconel alloys are often used with sufficiently long service life in processing high speed tool steels.

The work-supporting tray usually rests on a graphite or metallic hearth. Frequently, this hearth consists of three or four horizontal molybdenum or graphite bars supported by heat-resisting piers from the furnace shell below. Some hearths used ceramic bars and others use ceramic alumina rolls on which the work tray can be rolled. The end bearings for such rolls should be located outside the intense heat zone of the furnace.

Graphite hearths will react with stainless steel and Inconel alloys to form a eutectic melting at approximately 1125°C (2060°F). To prevent a graphite hearth from reacting with a work basket or a workload placed directly upon the hearth, a thin sheet of a ceramic material capable of resisting high temperatures can be placed between the hearth and the workpiece. Some graphite hearth blocks have a longitudinal groove at the top. An alumina ceramic rod or tube is placed in the groove and supports the work or the work basket, thus separating the graphite and the metallic workpiece.

Molybdenum will react with nickel to form a eutectic melting at approximately 1315°C (2400°F); therefore, it is necessary to separate nickel-bearing alloys from a molybdenum hearth at this temperature. Slabs of honeycomb alumina ceramic are available for this purpose. Nickel and titanium form a eutectic melting at approximately 955°C (1750°F). Therefore, alloys of these metals must be kept separated if temperatures this high are contemplated.

VACUUM CHAMBERS

The prime requisites in the design, fabrication and operation of a vacuum chamber are prevention of leakage from the outside atmosphere and the assurance that the planned material processing will be completed without damage to the chamber or product. Leakage is a serious consideration even though the pressure within the vessel may be held at the required vacuum, because the continual entrance of air into the evacuated chamber can be harmful to the product or internal components. If a high-capacity pump is used, the pump may well be able to maintain a low pressure inside the furnace regardless of a substantial air leak constantly bringing oxygen into the furnace, which can react with the surface of the workload. Many operators of vacuum equipment routinely conduct leak tests before energising the heating elements. This is particularly useful when the workload consists of very expensive material. For example, a furnace is evacuated to 10^{-4} torr for at least 1 hour. The vacuum valve connecting the furnace to the pump is closed and the rise in internal furnace pressure in a specified time interval is checked. It is difficult to set specific limits on leakage tolerance, because this tolerance depends upon the type of material being processed and the duration of the heating cycle. Common industrial practice with vacuum furnaces is to establish permissible leakage rates in terms of microns or micrometres per hour. A rate of rise between 10 and 25 μm Hg per hour generally is an acceptable specification for most industrial work, but some more stringent applications can require less.

Generally, equipment manufacturers specify rate-of-rise measurements of 10 μm Hg per hour or less for an empty, clean, cold and thoroughly out gassed system. This measurement is usually obtained by checking the furnace after a vacuum cooling from a heat-up to a temperature sufficient to dry out or bake out the internal components. This procedure eliminates almost completely the outgassing effects of workload and furnace components on rate-of-rise measurements.

Once a vacuum furnace has been properly installed and put into operation, the heat treater usually needs only to be concerned with preventing and correcting the leakage of air at seals around the doors, sight ports, pumping ports, electrical and water feed lines, mechanical rotating and sliding seals for introducing mechanical force and motion into the chamber and any other point on the vessel that is subject to opening and closing.

Unless a vacuum furnace has been built for service at vacuum levels of 10^{-6} torr or less, typical elastomer seals used in most applications will be worn out if subjected to overheating, to repeated openings and closings or to rotating or sliding motions. When seals are replaced, all flanges must be properly aligned to ensure complete seating of the seal and sharp edges or burrs must be removed to prevent physical damage. A good grade of vacuum grease lubricant should be used sparingly to facilitate placement of the seals and to improve the sealing.

The best and most expensive vacuum chambers are constructed of an oxidations-resistant material such as a 300-series stainless steel. This material must be used for vacuum service in the pressure range of 10^{-6} torr or less or when the vacuum chambers must operate as a hot wall or retort unit. A less expensive carbon steel, adequately coated, is effective in cold wall chambers for most heat treating applications. However, unprotected carbon steel should not be used in construction of the vacuum chamber. When carbon steel is exposed to a vacuum, it can be so thoroughly degassed that oxidation can occur very rapidly when re-exposed to atmospheric conditions. Oxidation that occurs in carbon steel chambers is permanent and progressive because the vacuum chamber is completely water jacketed and does not achieve temperatures necessary for surface cleaning.

The efficiency of the pumping system can be affected by the material used to build the vacuum chamber. For example, a stainless steel chamber with a total volume of 0.3 m^3 (10 ft^3) will require

10 minute for evacuation to a pressure of 10^{-5} torr by a 152-mm (6 inch) diffusion pump. A chamber with the same volume constructed of unprotected carbon steel can require evacuation times of up to 30 minute to achieve the same pressure. If the carbon steel surfaces are oxidised, evacuation times of up to 2 hours can be required.

PUMPING SYSTEMS

Vacuum vessels are evacuated by various types of pumping systems that depend, to a great extent, on the pressure range needed for processing. An adequate vacuum pumping system must attain the specified pressure and must have sufficient capacity to handle the processing gas load, not only at the ultimate pressure but at all intermediate pressures during the pumpdown cycle. Pumping systems are usually divided into two subsystem: the roughing pump and the high-vacuum pump. For certain requirements, a single pumping system is sufficient for the entire range and cycle. Pumps usually are classified as mechanical pumps or diffusion pumps. The choice of pump or combination of pumps depend largely on the pressure and gas volume or pumping rates required for a specific process and size of vacuum vessel. The vacuum system shown in Fig. 4.17 consists of a mechanical forepump that can be connected directly to the vacuum vessel by closing the high-vacuum valve, opening the roughing valve and closing the foreline valve. This procedure isolates the diffusion pump from the rest of the system. The diffusion pump interior can then be pumped free of air by its mechanical holding pump.

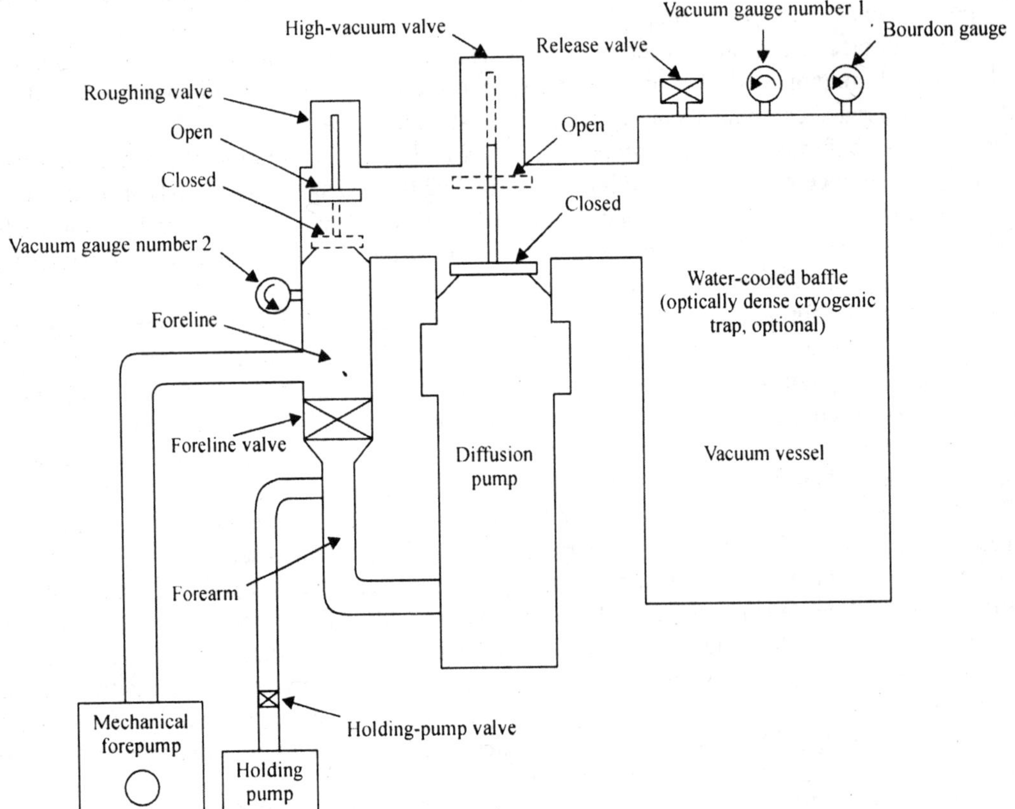

Fig. 4.17. Typical vacuum system containing mechanical and vapour pumps with interconnected valving and piping.

When the pressure as shown on vacuum gauge No. 2 has been reduced to a level where the diffusion pump can operate efficiently, the roughing valve is closed and the foreline valve and the high-vacuum valve are both opened. The residual gas in the vacuum vessel can then expand continuously into the region of the diffusion pump, through the foreline to the forepump and then to the atmosphere. The water-cooled baffle or other optically dense trap atop the diffusion pump prevents diffusion pump oil from diffusing backward into the vacuum system. Vapours from the system are condensed in the liquid nitrogen cryogenic trap, if one is used, when a very high vacuum (very low pressure) of 10^{-6} torr or below is desired. Vacuum gauge No. 1 measures the pressure in the vacuum vessel. It should be a more sensitive gauge than No. 2. The release valve controls backfilling of the system to atmospheric pressure up to the high-vacuum valve.

Mechanical Pumps

Mechanical pumps operate on the fluid-flow principle and are primarily positive displacement pumps with suitable seals to permit operation at low pressures. Piston pumps or rotary blowers in various pumping speed ratings are available. Vacuum system levels down to 25 μmHg can be obtained with oil-sealed rotary mechanical pumps. Depending on the application, they may be called roughing or forepumps. They can discharge directly into the air against normal atmospheric pressure. A portion of air in the closed system expands into and is trapped within a chamber of the pump. This volume of air is then compressed by the movement of vanes or a piston in the interior of the pump and expelled through a port equipped with a check valve This process is repeated and with each cycle, a portion of the remaining air volume in the closed system is expelled. When the back leakage through the pump or leaks in the wall of the container equal the rate of air removal by the pump, the closed-vessel internal pressure remains constant. A nonmechanical limitation of the ultimate pressure of a mechanical pump is the vapour pressure of the oil itself. Most commercial pump oils are controlled grades of petroleum type SAE 20, which have a vapour pressure of about 1 μm Hg.

When air containing moisture is compressed in the interior of the pump, water may condense and contaminate the oil, consequently affecting the ultimate pressure attainable. One method used to prevent condensate is to introduce enough pure dry air at the start of the cycle to ensure that the resulting moisture level of the trapped air is below the level at which compressive condensation can occur. Called gas ballasting, this method alleviates the moisture condensation problem.

Diffusion Pumps

When the pressure in the vacuum chamber becomes so low and the molecules so few that the path typically traversed by a gas molecule exceeds the dimensions of the chamber, the remaining gas molecule, collide more often with the walls than with each other. At higher pressures, the constant and frequent collisions of adjacent gas molecules and the resulting elastic rebounding effectively scatter and expand the gas to quickly fill any new volume created. At lower pressures, this effect nearly disappears and the remaining gas is difficult to pump with fluid-flow positive-displacement mechanical pumps.

At these pressures, it is necessary to allow molecules to diffuse randomly into the throat of the pump, then to impart a preferred direction of motion to the molecules by momentum transfer. For pumping at a vacuum system level below 10^{-3} torr, a vapour diffusion pump generally is used. Pumping action is directed by a high-velocity stream of heavy molecules in the form of a pump fluid, usually oil. The heavy molecules strike the gas molecules and push them in the desired downward direction toward the outlet of the pump. To be effective, the inlet pressure of the diffusion pump should be below 1 μm Hg so that the vapour stream is operating in nearly empty space except for the occasional diffusing gas molecule.

A schematic diagram of a three-nozzle vapour-diffusion pump is shown in Fig. 4.18. Vapour from a liquid held in a closed boiler heated at the bottom is forced upward inside the boiler. The vapour passes quickly through a narrow circumferential opening in the nozzles at a downward angle. Molecules of gas that stray from the vacuum chamber above the pump toward the vapour jet streaming from the nozzles encounter the downward-directed stream of heavy molecules. The over-all effect is to compress the gas molecules and force them downward to a point where they can be removed by the mechanical forepump.

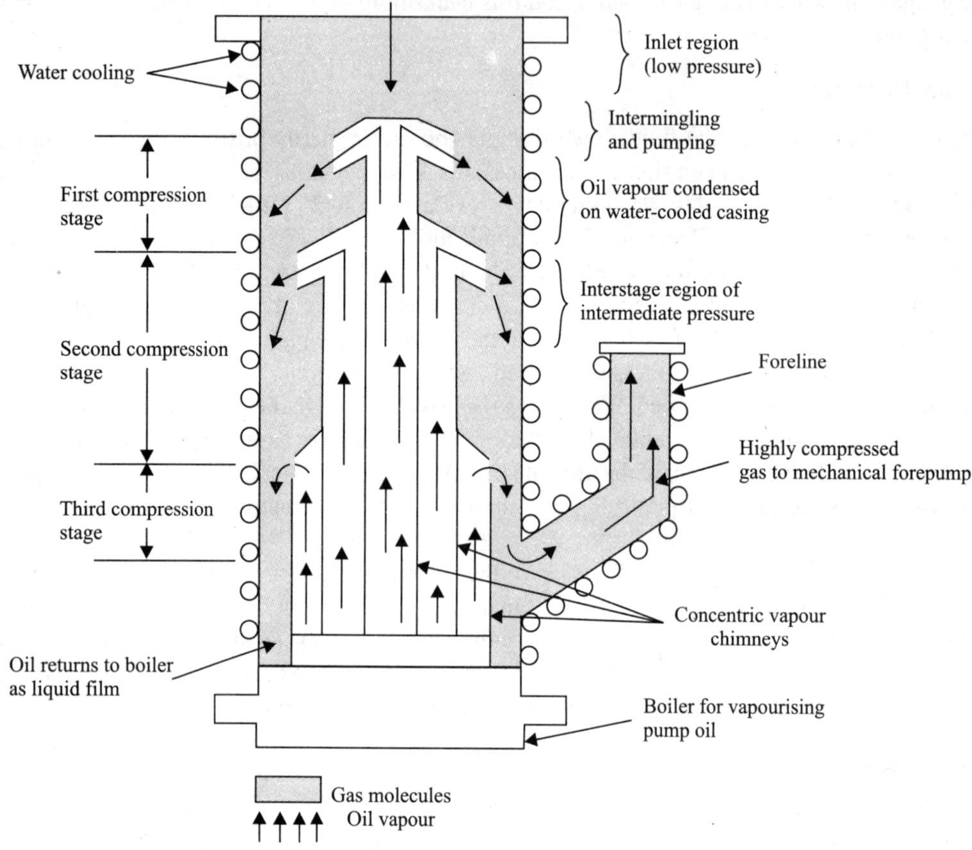

Fig. 4.18. Oil vapour diffusion pump.

Pump vapour is condensed on the cooled inner walls of the pump and returns as a liquid to the boiler. Maximum velocity is imparted to the gas molecules by using a liquid composed of heavy molecules in the boiler. Efficiency is improved by using several nozzles in line, one above the other. Special highly stable liquids with very low vapour pressure are required for such diffusion pumps.

Backstreaming is the movement of molecules of the pump fluid above the inlet flange of the diffusion pump and in the direction of the vacuum chamber. The rate of backstreaming increases rapidly as the inlet pressure exceeds 1 μm Hg and depends on the size and type of diffusion pump. Back-streaming also can be caused by exceeding the fore pressure limit, which is about 500 μm Hg, of most diffusion pumps.

Backstreaming can be reduced by interposing opaque baffles between the diffusion pump throat and the vacuum chamber. Most backstreaming originates at the top jet of the pump and can be reduced by placing a water-cooled cap above the jet. The oil vapour condenses on the cap and drips back into the pump. By using a cold trap placed above the throat of the diffusion pump, it is possible to reduce oil backstreaming further. The cold trap consists of an arrangement of baffles and water-cooled or refrigerant-cooled walls that provide an opaque path through the trap. A refrigerant such as liquid nitrogen may be used. Molecules of oil condense on the surface of the cooled baffle and remain trapped. Such traps also attract condensable vapours that may be present in the vacuum chamber. Water vapour, the most common contaminant present in high-vacuum systems, may be removed successfully in this way. The ability of the trap to serve as a pump for condensable vapours increases as the temperature of the refrigerant decreases.

Other Pumping Systems

Many types of vacuum pumps may be used depending on the pressure to be maintained for a given vacuum process. The rotary mechanical pump and the oil vapour-diffusion pump are commonly used for most vacuum metallurgical processes. Other types of pumps include steam ejector pumps, oil booster pumps and cryogenic pumps. These pumps are used in vacuum processing for such functions as degassing and drying in conjunction with a rotary pump.

Steam ejector pumps have an operating range from 1 to 10 µm Hg. Although steam ejector pumps eliminate the need for a mechanical pump, they require a large volume of steam to operate. Because their principal advantage is the capability of removing large volumes of gas vapour at low vacuum levels, they are suitable for laboratory work and for very large vacuum vessels, especially if noxious vapour are to be pumped. Oil booster pumps have an operating range from 1000 to 1 µm Hg and require a backing pump. These pumps tend to introduce excessive back-streaming of oil vapour into the vacuum vessel and are used mainly when high pumping capacities are desired above 10 µm Hg. They can be used in conjunction with an oil diffusion pump to achieve lower pressures while maintaining high pumping capacities.

Cryogenic pumps condense gas molecules on refrigerated surfaces at –195°C (–320°F) or lower. These pumps are regenerated periodically by heating the condensing surfaces to vent the accumulation of condensed gases. Cryogenic pumps are often used instead of diffusion pumps to pump high volumes of water vapour and other condensable gases in the 10^{-3} to 10^{-8} mm Hg range.

Valves

Gas flow of the vacuum pumping system is controlled by three specific types of valves. High-vacuum valves isolate the oil diffusion pumps from the vessel and butterfly and gate valves are used as roughing, foreline and holding-line valves. All of these valves have apertures that provide a minimum gas flow impedance and special seals for vacuum service.

CONTROL INSTRUMENTS

Except for vacuum gauging instruments, the instruments employed in vacuum equipment processing are similar to those used in other heat treating operations. Once vacuum control has been attained, heating cycles can be initiated either manually or automatically by the temperature recording instrument.

Temperature Control

Temperature control systems consist of a primary sensing device, the control instrument and the final control element. The optical pyrometer is seldom used, because it is not adaptable for automatic control.

The total radiation pyrometer is adaptable to automatic control, but its accuracy may be seriously impaired by intervening media such as gas, smoke or a discoloured sight-glass window. Sublimation of materials within the furnace may cause deposition of metallic vapours on the sight-glass window, thereby reducing the radiation reaching the detector.

Thermocouples are typically used as temperature-sensing devices, although their performance varies from a vacuum to an oxidising atmosphere. Unless the hot junction of the thermocouple is attached to the part being measured, heat transfer to the thermocouple is based almost completely on radiation. In air or other furnace atmosphere, a thermocouple receives heat by conduction and convection. For this reason, the response time of a thermocouple in a vacuum is much slower than in air. A change in air gap of 0.3 mm (0.001 inch) between the hot junction of the thermocouple and the part being measured can change the response time of the thermocouple significantly.

Control thermocouples made from nickel-nickel molybdenum are satisfactory for many heat treating applications up to 1290°C (2350°F). Noble metal thermocouples such as platinum-platinum rhodium are used to 1650°C (3000°F) and tungsten-rhenium thermocouples are used to 2205°C (4000°F). The thermocouples are sometimes used with unsheathed (bare wire) hot junctions, which reduces lag time tremendously. In many applications, particularly with platinum, thermocouples are sheathed in an adequate ceramic or metal protection tube. Below 1095°C (2000°F), sheathed base-metal thermocouples can be used to measure different locations in the load.

Because rapid heating rates can be achieved in most vacuum furnaces, it is important to determine temperatures at various locations in the workload. For instance, a large die may overheat in certain locations within the furnace. Information on uniformity of heating also is required when processing smaller parts loaded in baskets.

To bring thermocouple lead wires through the exterior shell of the furnace, special feed-through fixtures are provided. These fixtures are electrically insulated and vacuum tight and they constitute a permanent portion of the furnace. The ends of the lead wires inside the furnace have ceramic-insulated quick-disconnecting terminals to which workpiece thermocouples can be connected. Chromel-alumel thermocouples often are used to monitor workpieces if the temperature to be measured is not above 1175°C (2150°F) and the time at temperatures above 980°C (1800°F) is not excessively long.

Pressure Control

Instruments used to measure and record the pressure inside a vacuum-processing chamber fall into two classifications: those that measure the pressure hydrostatically and those that sense some physical characteristic of the gas which bears a definite relationship to the pressure. A Bourdon gauge accurately and continuously indicates the pressure from 1 atm to approximately 0.01 atm and can be used effectively to monitor the roughing cycle and the performance of the roughing pump.

Pressures of a few torr down to 10^{-4} torr can be measured periodically with a McLeod gauge, which samples the gas and compresses it to a small calibrated volume. It then registers the ratio of its initial and final volume and this ratio is an indication of the pressure at which the gas sample was taken. Special expensive diaphragm gauges are available for measuring pressures continuously from atmospheric down to the range of 10^{-4} torr.

The second classification of gauges includes those that sense the thermal and electrical conductivity of the gas. These physical characteristics bear a direct relationship to pressure.

The thermal conductivity gauges are widely used for most vacuum metallurgical processing because they are relatively inexpensive and can monitor vacuum levels continuously between 1 and 1000 μm Hg.

The thermal conductivity or convective heat transfer, of a gas is essentially constant as pressure is reduced until a pressure of about 1 torr (1000 μm Hg) is reached. From that point, the conductivity declines until, at a pressure somewhat less than 1 μm Hg, there is almost no heat transfer by the gas molecules to the surface of the gauge walls and the thermal conductivity becomes virtually zero. This trend for a typical common gas is shown in Fig. 4.19.

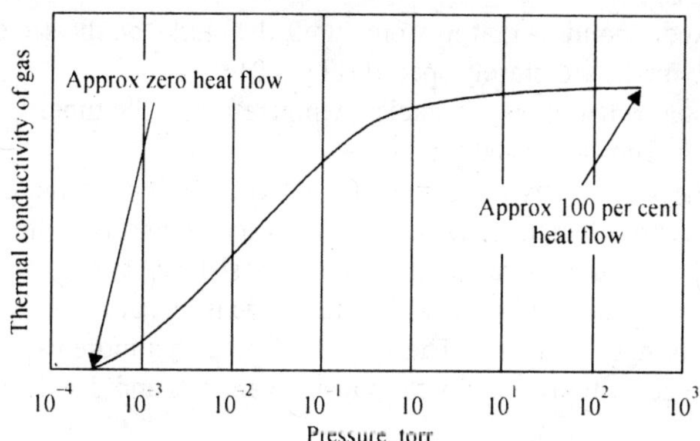

Fig. 4.19. Relation of thermal conductivity of gas and pressure.

Thermal conductivity varies considerably from 1 to 1000 μm Hg (10^{-3} to 1 torr), but varies little above that pressure; this behaviour constitutes the operational principle of the thermal conductivity gauge. Two gauges utilise this principle, the thermocouple and the Pirani. The radiation or thermocouple gauge is based on the principle that, as the amount of gas in a vessel decreases, the temperature of a constantly heated wire increases because less heat is radiated to the surrounding environment.

The wire temperature is measured by a fine wire thermocouple attached to the midpoint of the heated wire. The maximum temperature of the wire is about 115°C (240°F) and this temperature is reached at pressures of 1 μm Hg or less. A typical thermocouple gage is shown in Fig. 4.20.

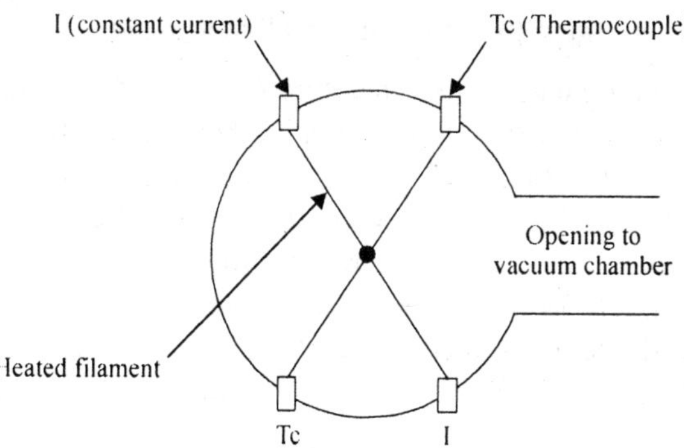

Fig. 4.20. Thermocouple type thermal conductivity vacuum gage.

Advantages of this gauge are:

1. It is comparatively inexpensive for both circuit and tubes.
2. It samples continuously.

3. Because of the relatively low wire temperature, it can be exposed to air for years without damage or danger of burnout.
4. The signal can be used to activate relays or other remote controls.

Disadvantages of this gauge are:

1. Because different gases have different thermal conductivity values, the gage requires calibration for each type of gas.
2. The scale is markedly nonlinear; at low pressures, the marks on the scale are spread widely; in higher ranges, the marks are closely spaced (Fig. 4.21).
3. Output depends somewhat upon the ambient temperature of the tube.
4. The useful range is from 1 to 1000 μm Hg.

The Pirani gauge (Fig. 4.21) uses the same thermal conductivity principle as the radiation or thermocouple gauge. The wire filament is exposed to the vacuum and is heated by a constant current source. As the temperature of the wire varies, its electrical resistance varies. This change in resistance upsets the balance of a Wheatstone bridge, of which the gauge is one arm. The vacuum pressure is thus measured in terms of the bridge unbalance. The system is somewhat more expensive and complicated than the thermocouple gauge, with essentially the same advantages and disadvantages.

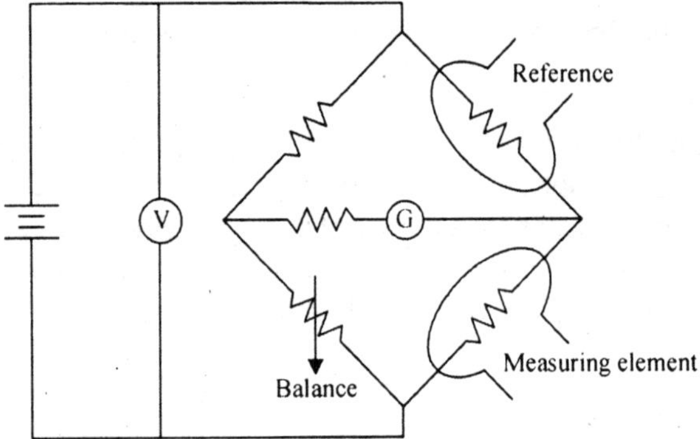

Fig. 4.21. Schematic of a Pirani gauge.

The thermocouple and Pirani gauges do not measure pressures of less than 1 μm Hg; therefore, a different type of gauge is necessary for systems operating below this level. For vacuums from 1 μm Hg (10^{-3} torr) to 1 pm Hg (10^{-6} μm Hg or 10^{-9} torr), the hot filament ionisation gauge can be used. The sensing element of the gauge resembles a triode vacuum tube. Fig. 4.22 shows the three elements of a hot filament ionisation gauge. A heated filament emits electrons that are attracted to the loosely wound and positively charged grid. The electrons strike residual molecules of gas in the vacuum. These collisions may remove an electron from the gas molecule, resulting in a positively charged gas ion, which is attracted to the collector. The positive ion current flow from the collector to ground is a measure of the gas pressure or vacuum.

The glass ionisation gauge, which originated in the radio tube industry, usually is calibrated with dry nitrogen. Its response varies slightly with other gases, which must be considered when great accuracy is required. Another type of ion gauge is the cold cathode, which depends on the measurement of an ion current produced by a high-voltage discharge. The cathode in the sensing element releases electrons that spiral through a magnetic field toward the anode.

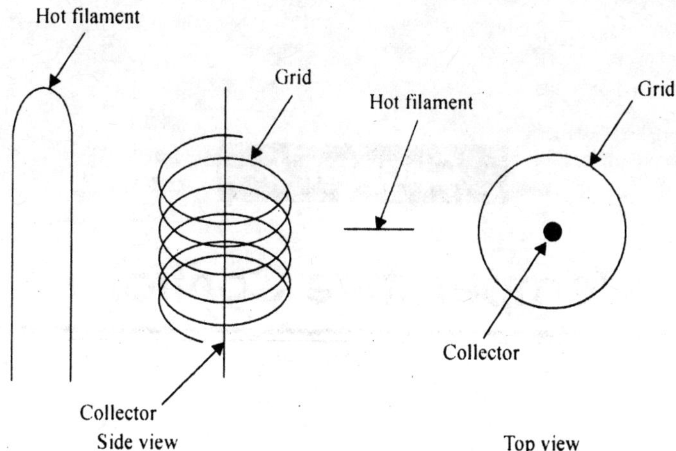

Fig. 4.22. Schematic arrangement of hot-filament ionisation vacuum gauge.

This spiralling motion lengthens the distance that electrons travel between cathode and anode, thus increasing the probability of collision with gas molecules and formation of positive ions. Formation of ions varies linearly with pressure and the ion current indirectly indicates the pressure. These rugged gauges are well suited for production applications, but are more expensive than hot filament gauges. They cannot be degassed as easily, however and they are more readily contaminated and less accurate than hot filament gauges. Output is linear below 10^{-3} torr and the usable range for measuring vacuum is between 10^{-2} to 10^{-6} torr.

Chapter 5

Temperature Control

INTRODUCTION

Temperature instrumentation and control systems used in heat treating include temperature sensors, controllers, final control elements, measurement instruments and set-point programmers. A basic control loop includes a temperature sensor, controller and final control element. The heat process temperature is detected by a temperature sensor that generates a signal proportional to the process temperature. This actual temperature is compared to the desired temperature determined by the controller set point. Based on this comparison, the controller develops an output signal that adjusts the final control element regulating heat flow to the heat treating process.

BASIC CONTROL LOOP AND AUXILIARY DEVICES

The basic control loop is illustrated schematically in Fig. 5.1. Auxiliary devices used with this basic control loop include measurement instruments and set-point programmers (Fig. 5.2) . The measurement instrument monitors the same temperature sensor as that used by the controller. The set-point programmer automatically varies the controller set point to provide a temperature cycle or temperature programme in accordance with an established plan. A variation of the basic control loop frequently used in the metals industry is the recorder-controller configuration illustrated in Fig. 5.3. The recorder measurement of the temperature sensor signal is compared to the recorder set point. Based on this comparison signal, the integral controller regulates the process. The basic control loop may incorporate a temperature transmitter to amplify the temperature sensor signals. This amplification is desirable, when the control instrument is remotely located from the temperature sensor. This amplification also helps to avoid interaction between two or more instruments using a common sensor. For example, the temperature signal may be transmitted to a remote central control room where it is connected to several controllers, a recorder, a data acquisition system or a digital computer.

The various measurement and control loop configurations must function properly to achieve optimum operation of the system. Although the necessity of this co-ordination is obvious, one or more elements of the system are commonly neglected. For example, if the temperature sensor, which may be only 1 per cent of the instrumentation cost, is out of calibration or not properly located, the performance of the complete system is degraded. Poor input into even the most sophisticated temperature measurement and control systems will produce poor output. Temperature is one of the metric system's seven basic units. According to *Maxwell*, 'the temperature of a body is its thermal state considered with reference to its ability to communicate heat to other bodies". Methods or modes of communicating heat are convection, conduction and radiation. An understanding of these modes of heat transfer is essential to measurement and control of temperature.

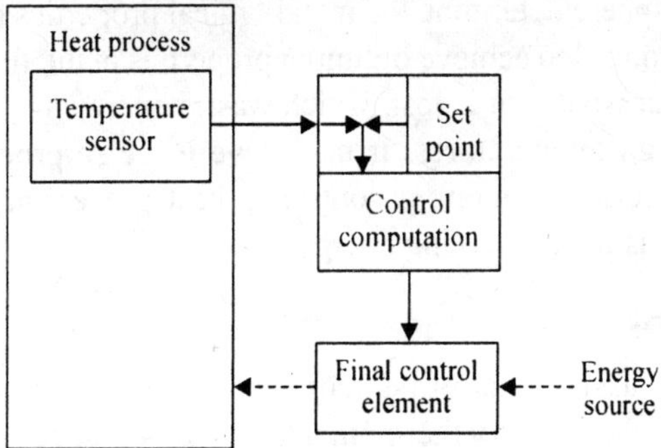

Fig. 5.1. Basic control loop.

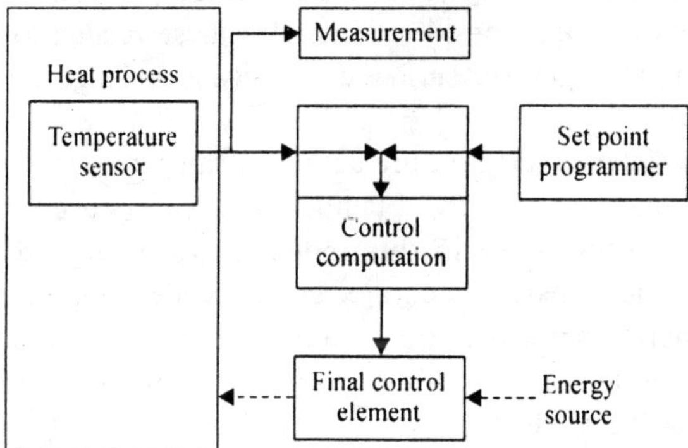

Fig. 5.2. Basic control loop with auxiliary devices.

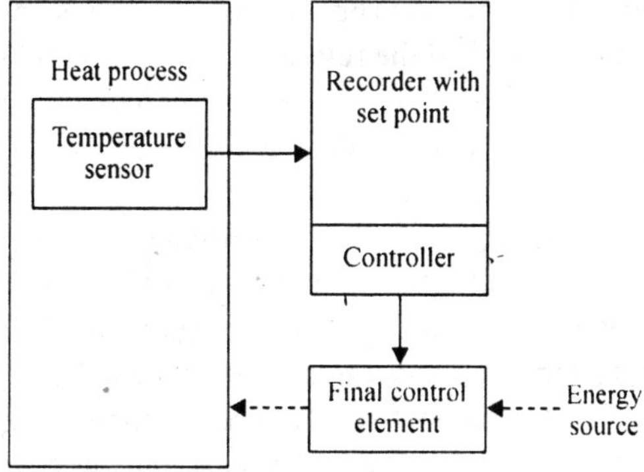

Fig. 5.3. Basic control loop with recorder-controller.

The majority of heat processing applications in the metals industry—including virtually all heat treating— are in the temperature range of –100 to 1370°C (–150 to 2500°F).

Important temperature-related properties of metals include such factors as thermal conductivity, thermal expansion, electrical resistivity, ductility, hardness and oxidation resistance. In heat treating applications, temperature, time and atmosphere determine the metallurgical properties obtained. Consequently, these parameters require precise control to achieve optimum properties in the final product. If the temperature is too high, for example, excessive heat is lost, which wastes energy. If the product is not permanently damaged by excessively high temperatures, it may have to be re-processed, thus doubling energy consumption. Therefore, precise temperature control in heat processing is mandatory for producing acceptable products as well as for conserving energy.

TEMPERATURE SENSORS

Temperature sensors are classified as either contact or non-contact and as either electrical or non- electrical. For example, the mercury thermometer is a non-contact, non-electrical type. Because of high-temperature requirements and the trend toward electrical instrumentation, most heat treating applications require electrical sensors. When selecting a contact sensor, certain variables should be considered, such as cost, temperature range, useful life, accuracy, size and response speed. Selection of non-contact sensors involves similar considerations, plus those related to radiation factors. With an optical pyrometer, for example, target size, surface emissivity, focal length and sighting path interference could be examined.

Thermocouples and resistance temperature detectors (also known as resistance thermometers) are the most important contact-type electrical temperature sensors used in the metals industry. Well over 90 per cent of the sensors used in this industry are estimated to be thermocouples. A thermocouple is rugged, inexpensive, accurate, covers wide temperature ranges and is fast in response. A resistance thermometer is more accurate and stable than a thermocouple. However, the resistance thermometer is more expensive, slower in response and limited to lower temperatures, typically 540°C (1000°F).

THERMOCOUPLES

Thermocouples consist of two dissimilar wires that are metallurgically homogeneous. They are joined at one end, called the measuring or hot junction. The other end, which is connected to the copper wire of the measuring instrument circuitry, is called the reference or cold junction. The electrical signal output in millivolts is proportional to the difference in temperature between the measuring junction (hot) and the reference junction (cold). The different types of thermocouples, classified by their metallurgical compositions, have differing output signal calibrations.

The thermocouple wire combination, with or without insulators, is called a thermocouple element. A complete thermocouple assembly includes the element, and a protection tube or well, to protect the element from contamination and provide mechanical strength (Fig. 5.4) . In addition, the assembly provides mounting fittings and a terminal junction in a head assembly.

Extension lead wires, which match the thermoelectric characteristics of the thermocouple elements, are used to connect these terminal junctions to the instrumentation. Fig. 5.5 shows a simple thermocouple with lead wires and reference junction. The lower sketch of this figure shows details of a complete thermocouple assembly in a protection tube.

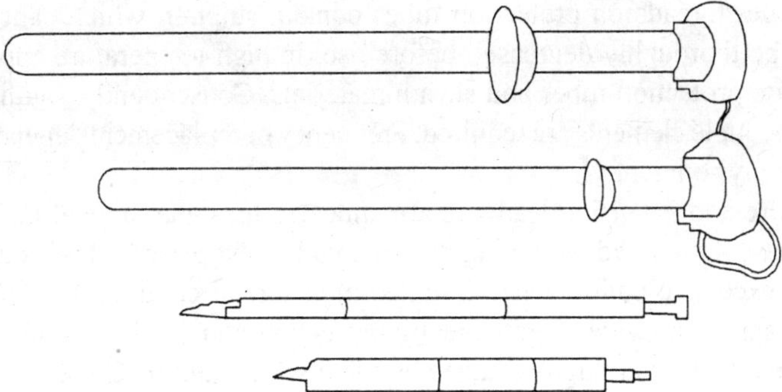

Fig. 5.4. Thermocouple elements (two lower photos); complete thermocouples (upper photos).

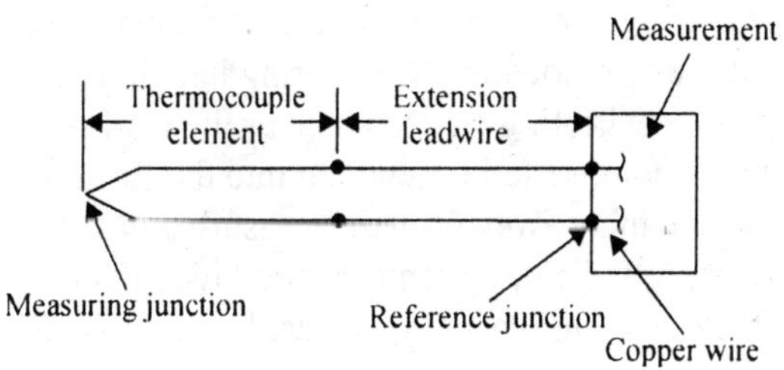

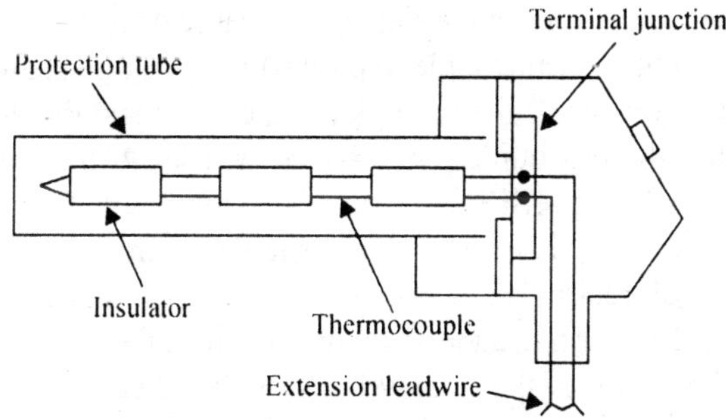

Fig. 5.5. Simple thermocouple (upper view); and cut-away of a thermocouple assembly (lower view).

The operating life of a thermocouple depends on its operating temperature, time at the operating temperature, ambient atmosphere and number of experienced high-to-low temperature cycles. An erroneous low output signal from a defective thermocouple indicates a measured temperature lower than the actual process temperature. Thus, if the temperature control is based on this faulty measurement, the process temperature will be too high. This makes wastes energy and could damage both the work in process and the furnace equipment.

Heat treating operations that rely on thermocouples require numerous precautionary measures. For example, thermocouple error due to contamination normally is characterised by a low output signal. Contaminated wires do not meet the requirement for metallurgical homogeneity. Contamination comes from touching the wire with bare hands during assembly, dirty insulators, oils and dirt inside the protection tube or the atmosphere used in the process. Consequently, cleanliness is important in assembling thermocouples. Insulators that become contaminated in normal service should not be re-used. Even the cutting oils used to cut threads on protection tubes contain sulphur, which can cause contamination. These tubes should be thoroughly degreased before use. In high-temperature applications, the furnace atmospheres penetrate protection tubes and sheath materials. Consequently, routine replacement of the contaminated thermocouple elements are required. Frequency of replacement depends largely on operating temperature and the environment.

Thermocouple wire and extension leadwire are annealed for metallurgical homogeneity. Therefore, the wire should not be cold worked by flexing it, as in making sharp bends. The extension leadwire should not be strained with excessive pulling when installed in electrical conduit. A separate metal conduit is often necessary, because it provides mechanically protection and shields the extension leadwire from picking up electrical signals from high-voltage electrical wiring and machinery.

Thermocouple Location

Location of thermocouple measuring junctions must allow exposure to the correct heat process temperature. Normally, the tip of the thermocouple (measuring junction) is placed near the work in the furnace. Thus, it is exposed to the same heating environment as the workpieces. A common error occurs when the thermocouple assembly is not inserted far enough into the furnace. Under this condition, heat is conducted by the thermocouple assembly away from the measuring junction, toward the colder furnace wall. Consequently, the thermocouple indicates a temperature lower than the actual work temperature. One way to determine correct location is to place an adjustable flange on the thermocouple protection tube. The correct insertion depth then can be determined by adjusting the flange to achieve the maximum temperature and the minimum insertion depth, while the furnace is in operation. However, thermocouples never should be located near burners or heating elements, because radiant heat may cause erroneous readings.

The speed of response of the thermocouple to process temperature changes should be as fast or faster than that of the work in the furnace. If the workpieces are small and the thermocouple is contained in a heavy protection tube, the thermocouple temperature will lag behind the actual work temperature. Consequently, workpieces will be over-heated when they are initially brought up to control temperature.

Furnace temperature uniformity is important and should be checked before installing the permanent control thermocouple. If an air recirculating system is used, an average temperature in the duct work can be measured by installing three or more thermocouples across the duct width connected in parallel. A single thermocouple in the centre of the duct typically will indicate a higher temperature because this location is normally the hottest.

Types of Thermocouples

Thermocouples most commonly used in heat processing applications are listed in Table 5.1. Table 5.2 gives temperature limits for thermocouples in oxidising atmospheres and Table 5.3 contains limits of error as a percentage of the temperature readings for these thermocouple types.

Selection of a specific kind of thermocouple is based primarily on the required temperature range and cost. However, the choice also is affected by its metallurgical properties. For example, type J is superior to type K in a reducing atmosphere and type K is superior to type J in an oxidising atmosphere. Type K is very susceptible to sulphur contamination, such as sulphur dioxide in combustion gases. Type K should be used in a large protection tube or one with a ventilated design. If the ambient atmosphere has a low oxygen concentration, the resulting green-rot corrosion results in a low output singnal. Green-rot is caused by preferential oxidation of chromium at low oxygen concentration. A new thermocouple has been designed that attempts to overcome some of the limitations of the type K device.

Table 5.1. Comparison of thermocouple types.

Type	Usable temperature range °C	°F	Advantages	Restrictions
J (Iron-constantan) iron;	–185 to 870	–300 to 1600	Comparatively inexpensive; suitable for continuous service to 870 °C (1600°F) in neutral or reducing atmospheres	Maximum upper limit in oxidising atmosphere is 760°C (1400°F), due to the oxidation of the protection tubes shoud be used above 480°C (900°F); protection tubes should always be used in a contaminating medium
K (nickel, chromium-nickel, aluminium)	–20 to 1370	0 to 2500	Suitable for oxidising atmospheres; in higher temperature ranges, provides a more mechanincally and thermally rugged unit than platinum, rhodium platinum, and longer life than iron-constantan	Especially vulnerable to reducing atmospheres, requiring substantial protection when used
T (copper-constantan)	–185 to 370	300 to 700	Resists atmosphere corrosion; applicable in reducing or oxidising atmospheres below 315°C (600°F); its stability makes it useful at sub-zero temperatures; has high conformity to published calibration data	Copper oxidises above 315°C (600°F)
E (nickel, chromium-constantan)	–185 to 870	–300 to 1600	Has high thermoelectric power; both elements are highly corrosion-resistant, permitting use in oxidising atmospheres; does not corrode at sub-zero temperatures	Stability is unsatisfactory in reducing atmospheres
S (platinum, 10% rhodium-platinum) R (platinum, 13% rhodium-platinum)	–20 to 1480	0 to 2700	Usable in oxidising atmospheres; provides a higher usable range than type k; frequently more practical than non-contact pyrometers; has high conformity to published calibration data	Easily contaminated in other than oxidising atmospheres
B (platinum, 30% rhodium-platinum, 6% rhodium)	870 to 1650	1600 to 3000	Better stability than types S or R; increased mechanical strength; usable to higher temperatures than types S or R; reference -junction compensation is not required if junction temperature does not exceed 65°C (150°F)	Available in standard grade only; high temperature limit requires the use of alumina insulators and protection tubes; easily contaminated in other than oxidising atmospheres

Table 5.2. Temperature limits for thermocouples in oxidising atmospheres.

Upper temperature limits are a function of wire diameter; because temperature tends to have deleterious effects on thermocouples, the larger the amount of material in thermocouple cross-section, the longer it can be expected to last.

Type	Thermocoupl Material	Condition	AWG (American Wire Gauge) 8 °C	8 °F	14 °C	14 °F	16 °C	16 °F	20 °C	20 °F	24 °C	24 °F	30 °C	30 °F
J	Iron-constantan	Bare	650	1200	480	900	480	900	425	800	345	650	315	600
		Protected	760	1400	595	1100	595	1100	480	900	370	700	370	700
K	Nickel, chromium	Bare	1095	2000	925	1700	925	1700	870	1600	760	1400	705	1300
	nickle, aluminium	Protected	1260	2300	1095	2000	1095	2000	980	1800	870	1600	815	1500
T	Copper-constantan	Bare	315	600	315	600	260	500	205	400	205	400	205	400
		Protected	370	700	370	700	315	600	260	500	205	400	205	400
E	Nickel, chromium-constantan	Bare	760	1400	595	1100	595	1100	480	900	370	700	370	700
		Protected	870	1600	650	1200	650	1200	540	1000	425	800	425	800
S and R	Platinum, rhodium-platinum	Protected	–	–	–	–	–	–	1540	2800	1480	2700	1315	2400
B	Platinum, 30% rhodium platinum, 6% rhodium	Protected	–	–	–	–	–	–	–	–	1705	3100	–	–

The two major thermocouple design classifications are: (i) the compacted ceramic; and (ii) the protection tube, or well-type, with integral element insulators.

The two major thermocouple design classifications are: (i) the compacted ceramic; and (ii) the protection tube, or well-type, with integral element insulators.

Compacted Ceramic Thermocouples

Compacted ceramic thermocouples use a ceramic powder to insulate the thermocouple wires. These wires are compacted inside a metal sheath providing protection from the environment. This compacted ceramic thermocouple is small in outside diameter, provides fast response and can be bent easily to suit installation requirements. It can be used at high pressures and high temperatures. Two types of compacted ceramic thermocouples, grounded and ungrounded, are shown in Fig. 5.6 . The metal sheath has a plug of the same metal welded in place as the tip. In the grounded type, the end of the thermocouple is permitted to touch this plug, providing more rapid response. The ungrounded type is used when required by the measurement and/or control circuit electronics.

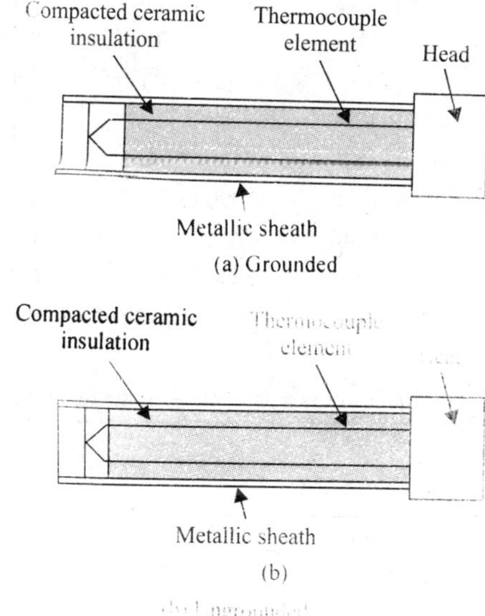

Fig. 5.6. Grounded and ungrounded types of compacted ceramic thermocouples.

Protection Tubes and Wells

Thermocouple elements with insulators and compact ceramic elements are used inside protection tubes and wells. The thermocouple elements with insulators are less expensive than the compacted ceramic elements and can be manufactured by users. Protective tubes and wells are used to provide mechanical support and protect the thermocouple from the process environment. Protection tubes are made from metallic or ceramic materials and are used in air or other gaseous atmospheres (Table 5.4). These atmospheres are near atmospheric pressure, such as those used in heat treating furnaces.

Wells are metallic and made of either drilled metal bar stock (one-piece construction) or from welded pipes. Most often, wells are used to protect the thermocouple in liquid or gaseous environments with pressures greater than 330 kPa (50 psi) for the drilled design. Typically, they are used in quench tanks, hydraulic systems or steam lines.

Table 5.3. Limits of error for thermocouples.

ANSI type	Type of thermocouple	Temperature range °C	°F	Limits of error(a) Standard	Special
J	Iron and constantan	–190 to –75	–310 to –100	–	± 2%
		–75 to + 315	–100 to + 600	± 2°C (± 4°F)	± 1°C (± 2°F)
		315 to 425	600 to 800	± 2°C (± 4°F)	± 0.33%
		425 to 760	800 to 1400	± 0.75%	± 0.33%
K	Nickel, chromium	0 to 275	32 to 530	± 2°C (± 4°F)	± 1°C (± 2°F)
	and nickel, aluminium	275 to 1260	30 to 2300	± 0.75%	± 0.38%
T	Copper and constantan	–185 to –60	–300 to –75	–	± 1%
		–100 to –60	–150 to –75	± 2%	± 1%
		–60 to + 95	–75 to + 200	± 1°C (± 1.5°C)	± 0.5% °C (±0.75°F
		95 to 370	200 to 700	± 0.75%	± 0.38%
E	Nickel, chromium	0 to 315	32 to 600	± 2°C(± 3°F)	± 1°C (± 2°F)
	and constantan	315 to 870	600 to 1600	± 0.5%	± 0.38%
S	Platinum, 10%	–15 to +540	0 to 1000	± 1.5°C (± 2.5°F)	± 1°C (± 1.5°F)
	rhodium and platinum	540 to 1480	1000 to 2700	± 0.25%	± 0.15%
R	Platinum, 13% rhodium	–15 to +540	0 to 1000	± 1.5°C (± 2.5°F)	–
	and platinum	540 to 1480	1000 to 2700	± 0.25%	–
B	Platinum, 30% rhodium	870 to 1705	1600 to 3100	± 0.5%	–
	and platinum, 6% rhodium				

(a) When expressed as a percentage, the limit of error is a percentage of the temperature reading, not of the range.

Table 5.4. Protection tube and thermowell materials.

Material (a)	Recommended maximum temperature °C	°F	Description
Metal and metal-ceramic materials			
Carbon steel	540	1000	Satisfactory in any except corrosive atmospheres
Yoloy (Ni-Cu alloy steel)	705	1300	Resistant to corrosion in both oxidising and reducing environments; ideally suited for use in condensate return lines, salt-water and brine solutions, condenser water lines, vent and waste piping, or corrosive water lines
Cast iron	705	1300	Generally more useful than Yoloy in the chemical industry; resistant to concentrated sulphuric acid and caustics solution; can be used to 870°C (1600°F) in reducing atmospheres
304 stainless steel (18Cr-8Ni)	870	1600	Resistant to oxidation and corrosion; generally used in wet-process applications such as steam lines, oil refineries and chemical solutions; resists nitric acids well, halogen acids poorly, and the sulphuric acids moderately
316 stainless steel (18Cr-8Ni-2Mo)	870	1600	Superior to 304 stainless steel in corrosion resistance; resists pitting in phosphoric and acetic acids

(Contd...)

Material (a)	*Recommended maximum temperature* °C	°F	*Description*
446 stainless steel (28Cr, Fe)	1095	2000	Excellent corrosion resistance at high temperatures; used extensively in general-purpose alloy tubes; highly resistant to sulphur attack
Nickel	1095	2000	Resistant to attack by many chemicals at high temperatures; principally used for hot caustic and molten-salt baths; should not be used where sulphur is present
Inconel 600 (80Ni-15Cr)	1150	2100	For general high-temperature use; has greater mechanical strength than 446 stainless steel; should not be used in sulphur atmospheres
R-Monel (67Ni-30Cu)	480	900	Used where high strength and resistance to corrosion are required, such as sea water, dilute sulphuric acid and strong caustic solutions
F-11 (1¼Cr-½Mo)	595	1100	Generally used in power plants, on water and steam line applications; available in thermo-wells only
F-22 (2¼Cr-1Mo)	595	1100	Same as F-11, except has better oxidation resistance
LT-1 (77Cr-23Al_3O_4)	1370	2500	For high-temperature applications; intermediate between metal and ceramic tubes, has good thermal and mechanical shock resistance
Ceramic materials L and N fyrestan	1540	2800	A highly refractory porcelain, normally used for primary protection and often for secondary protection. L and N Fyrestan tubes can be substituted in many applications for nickel or chromium-alloy tubes, and although not as strong as metal alloys, are fully as gas-tight. Having a softening point of 1650°C (3000°F), these tubes can be mounted vertically at a temperature of 1540°C (2800°F). When mounted horizontally, the recommended maximum temperatures are 1510, 1455, 1370 and 1260°C (2750, 2650) 2500 and 2300°F) for unsupported lengths of 75, 150, 300 and 450 mm (3, 6, 12 and 18 inch), respectively. The tubes are impervious to air to 1650°C (3000°F) and to dry hydrogen to 1400°C (2550°F). Their low rate of thermal expansion gives good thermal shock resistance
High-purity alumina	1870	3400	A highly refractory 99 + % pure alumina, normally used for primary or secondary protection. Alumina tubes have excellent deformation resistance and superior resistance to reducing atmospheres and chemical reactions at temperatures as high as 1870°C (3400°F). (If the tube is mounted horizontally, it must be supported at temperatures above 1650°C (3000°F). Its excellent resistance to oxidising and reducing atmospheres, as well as its exceptional stability at elevated temperatures, makes this material the ideal choice for protection of precious-metal thermocouples—especially type B. These tubes are highly resistant to corrosive alkaline vapours and aluminium

(Contd...)

Material (a)	Recommended maximum temperature °C	°F	Description
			chloride vapours and are stable to acids alkalies molten metals, most glass fluxes, molten salts and slags. The tubes are also impermeable to most gases under the conditions found in industrial furnaces
Silicon carbide refractory	1650	3000	Highly resistant to the cutting action of flames and gases and the corrosive effects of SO_2, tubes of silicon carbide refractory can be used for primary as well as secondary protection for temperatures upto 1650°C (3000°F). Ceramic-bonded silicon carbide features high thermal conductivity, high thermal shock resistance, high strength and low permeability at differential pressures less than 50 to 75 mm (2 to 3 inch) H_2O

(a) Materials are, in general, arranged in the order of their increasing resistance to oxidation, increasing mechanical strength at elevated temperatures and increasing limit of useful temperatures.

If the insulation between the thermocouple wires in the element or extension leadwire breaks down from heat, contamination or mechanical damage, a low output signal results. This signal indicates an erroneously low temperature. This type of error results in process operation at an excessive temperature. Valuable energy is lost, and the work can be damaged or destroyed.

Thermocouple Applications

A typical thermocouple construction includes the thermocouple wire element with insulators mounted in a protection tube. The thermocouple head provides terminals for a junction between the thermocouple element and lead wire used for connection to instrumentation.

A typical application is a car-bottom annealing furnace used up to 1095°C (2000°F) with an air atmosphere. An 8-gauge type K thermocouple in a ceramic protection tube is selected because it is useful up to 1260°F (2300°F) in an oxidising atmosphere and is less expensive than the noble metal types R, S and B. A 16-gauge type K leadwire with asbestos insulation is selected because it matches the type K element up to 205°C (400°F), suitable for a heat treating area and is strong enough for pulling through a conduit. Type K should be used for all the furnace measurements to maintain uniformity; that is, recording and control of furnace temperatures, plus the high limit safety control and work temperature.

The furnace temperature thermocouple element should be 8 gage for maximum life and long enough so that its tip or measuring junction can be inserted well into the furnace cavity while the head is far enough back from the furnace wall to prevent its rising above 205°C (400°F) . A ceramic protection tube will meet the process temperature requirements and is strong enough not to sag if mounted horizontally. An adjustable flange permits changing the depth of insertion in the furnace to the optimum depth. For example, when the furnace is at operating temperature, the thermocouple may be slowly withdrawn until its temperature indicator starts to drop, showing it is too close to the cold furnace wall. The adjustable flange may then be locked in place on the protection tube.

Location of thermocouples depends on furnace design but, in any event, must be inserted where representative temperatures exist. These are a few typical considerations in selecting thermocouple specifications.

RESISTANCE TEMPERATURE DETECTORS

Resistance temperature detectors are contact-type sensors. Their electrical resistance is proportional to temperature. Typical detector materials are platinum, copper and nickel. They are more stable and accurate than thermocouples, but even the platinum detectors have an upper temperature limit of approximately 540°C (1000°F), which reduces their usage in the metals industry. Because their signal is relatively large, resistance temperature detectors can be used with instruments having short temperature ranges of 5 to 10°C (10 to 20°F).

Resistance temperature detectors are normally larger in size and slower in response than thermocouples. However, the new thin-film deposited detectors minimise this disadvantage, which characterises the conventional wirewound detectors.

Resistance temperature detectors are used typically in quenching systems, low-temperature annealing furnaces and tempering furnaces. Because resistance temperature detectors are very accurate and stable, temperature errors are usually due to incorrect installation or damaged detectors.

NON-CONTACT SENSORS

Radiation sensors are non-contact type temperature sensors used with radiation pyrometers. One type of radiation pyrometer is the optical pyrometer. Temperature measurements made with these devices are based on different properties of the radiant energy emitted from a hot target source. These measurements, which compare the apparent brightness of a standard electrical filament with the target, or the unknown, are commonly called the disappearing filament type. The human eye makes the comparison. A manual adjustment of the filament current is made until the apparent brightness of the filament matches that of the target; that is, it disappears. The measuring circuitry of the parameter converts the electrical filament current value to an indicated temperature value. This measurement, made in the visible spectrum at 0.655 μm (microns), is a narrow band measurement. Optical pyrometers generally are used to calibrate radiation sensors.

Radiation sensor detectors

Radiation sensor detectors respond to radiant energy. They are classified as total radiation (wide-band) or narrow-band types, depending on the width of the radiation wavelength band to which they respond. Total radiation sensors use thermal detectors and narrow-band sensors use photoelectric detectors. Radiant energy from metals is characterised at lower temperatures as red hot (with longer wavelengths). At higher temperatures, it is characterised as white hot (with shorter wavelengths). Radiation wavelengths in microns are shown at the top of Table 5.5. The bottom of the table lists detectors and the wavelengths for which they are used.

The total amount of energy transmitted from the hot body to the cold body detector is a function of the difference between the fourth powers of their absolute temperatures. The hot body may radiate or reflect the energy it receives, depending on its characteristics. If the hot body radiates 100 per cent of the received energy, it is called a black body, with an emissivity of 1. If it radiates only 50 per cent of the energy, the hot body has an emissivity of 0.5. Consequently, the emissivity of the hot body affects the measurements of its temperature when using radiation sensors.

One of the factors that affects the energy per unit of time emitted from a material is its surface characteristics. For example, a heated brick wall will emit 98 per cent of its energy related to its temperature (emissivity of 0.98), whereas polished aluminium, at the same temperature, emits only 8 per cent of its energy (emissivity of 0.08). At 150°C (300°F), clean gray steel emits 60 per cent of its energy, but steel that has been buffed to a bright finish emits only 11 per cent. Thus, the emissivity factor must be taken into consideration when using radiation sensors. In addition, energy in the infrared region can be absorbed by gases or dust particles between the hot body and the sensor, including dirt on the sensor window. Thus, smoke, water vapour, carbon dioxide and other gases will affect the measurement. When a radiation measurement is made, these factors must be either eliminated or minimised.

Table 5.5. Radiation wavelengths for ultraviolet, visible and infrared light.

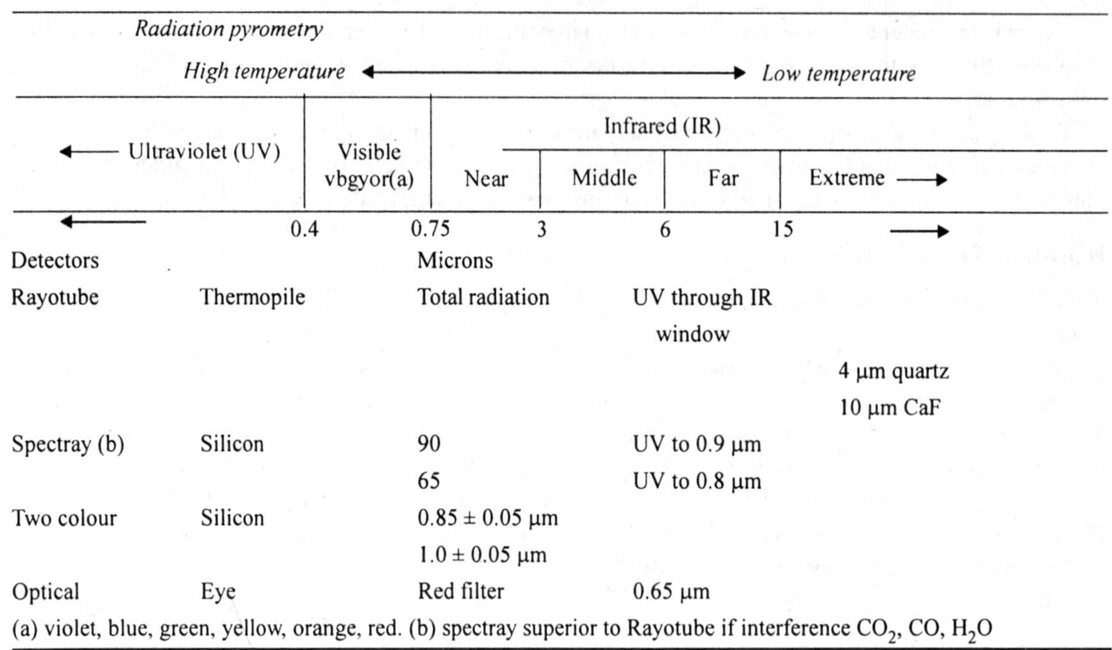

Radiation pyrometry

High temperature ←——→ *Low temperature*

← Ultraviolet (UV)	Visible vbgyor(a)	Infrared (IR)			
		Near	Middle	Far	Extreme →
← 0.4	0.75	3	6	15	→

Detectors		Microns		
Rayotube	Thermopile	Total radiation	UV through IR window	
				4 µm quartz
				10 µm CaF
Spectray (b)	Silicon	90	UV to 0.9 µm	
		65	UV to 0.8 µm	
Two colour	Silicon	0.85 ± 0.05 µm		
		1.0 ± 0.05 µm		
Optical	Eye	Red filter	0.65 µm	

(a) violet, blue, green, yellow, orange, red. (b) spectray superior to Rayotube if interference CO_2, CO, H_2O

Radiation devices are available in various ranges such as 425 to 980°C, 760 to 1650°C and 1010 to 2200°C (800 to 1800°F, 1400 to 3000°F and 1850 to 4000°F, respectively). They also are made with varying response speeds to meet the different application demands. For example, detectors having a high response speed (0.01 seconds) are useful when the target object is moving or when the temperature of the target object varies rapidly. Slow radiation sensor response speeds are useful for filtering out rapid or erratic changes in temperature, such as those given by swirling flames.

Total Radiation Sensors

Total radiation sensors are sensitive to broad bands of radiation wavelengths of approximately 0.3 to 10.5 µm (microns). In these wavelengths, the upper limit is determined by the window used in the radiation sensor, in this instance, calcium fluoride. If a quartz window is used, the transmission of longer wavelengths is limited, and the span is 0.3 to 4 µm. The radiant energy is focused on the centre of a thermopile detector, which consists of one or more thermocouples in series. The measuring junctions of the thermocouples are at the centre of the thermopile. The thermopile converts the radiant energy received from the hot body target to an electrical signal. This signal is proportional to temperature as a fourth power function, a non linear output signal. The sensor signal output can be increased by using thermopile detectors with more thermocouples. However, this increased mass reduces the detector response to temperature changes. Thermopile designs, using thin-film deposited thermopiles, have minimised this limitation and increased the signal output. Narrow-band radiation sensors of the photovoltaic type, using silicon diode detectors, respond to radiant energy in the range of 0.3 to 1.0 μm. They are characterised by a large signal output, fast response speed (0.001seconds) and a very non-linear output signal (12th power) because of their narrow-band response.

A variety of sensor designs are used for specific applications to compensate for the characteristics of specific detectors. Consequently, the following comparison between total-radiation (wide-band) and narrow-band sensors is a generalisation. Wide-band sensors respond to both high and low temperatures, while narrow-band sensors responds only to high temperatures. Wide-band sensors are slower in response, which is sometimes a disadvantage. This slower response is advantageous when signal averaging is desired. Signal averaging facilitates control or reading a temperature indication on a recorder or digital display. Wide-band sensors have low signal outputs, which limits the use of optical filters needed to make them spectrally selective. This low signal output characteristic also limits the minimum target size that can be used. The signal output characteristics, 12th power, or narrow-band, versus 4th power or wide-band sensors, are an important consideration. If the target emissivity is reduced, or if energy is lost due to interference in the optical path, the error will be less for the narrow-band sensor.

For example, referring to Fig. 5.7, describing the black body curves (emissivity 1.0) for T^{12} and T^4 sensors, both produce 18 mV at 1650°C (3000 °F) . At a 1370°C (2500°F) target temperature, with emissivity changing from 1.0 to 0.5, 50 per cent less radiant energy and an apparent temperature loss results. Under black body conditions at 1370°C (2500°F) , the T^{12} sensor produces 2.90 mV and the T^4 sensor produces 8.75 mV. Loss of millivolts, due to the emittance change, produces apparent temperatures of 1290°C (2350°F) for the T^{12} sensor and 1140°C (2080°F) for the T^4 sensor. As shown in Fig. 5.7, the ratio of these changes is 150/421 or approximately one-third. Therefore, the error for the narrow-band sensor is one-third that of the wide-band sensor. Erroneous temperature measurements especially low ones, result in high-temperature operation, which wastes energy. Low -temperature errors in measuring by radiation methods can result from conditions of the target source and from optical path interference.

Target source measurement errors occur primarily due to emissivity factors or the size of the target. For example, a given metal with an unoxidised (polished) surface will have a lower emissivity than an oxidised surface, resulting in a lower indicated temperature than the actual temperature. Loose scale has a high emissivity but is lower in temperature than the metal to which it is attached, because it is a poor conductor. If the target size does not fill the radiation sensor's optical field, a lower than actual temperature is indicated. For example, the target alignment of hot rod stock in a mill may vary as it passes the sensor.

Optical Path Interference

Optical path interference can be produced in several ways. Radiant energy is absorbed by water vapour and carbon dioxide gas, steam from water quenching operations, or products of combustion from the heating process. These processes are frequently located near high-temperature measurement systems. For these conditions, the narrow-band sensor is not affected because the absorption bands for water vapour and carbon dioxide are at wavelengths longer than 1.0 μm. Other gases may cause interference, depending on their spectral absorption bands compared with the sensor's spectral response.

Radiant energy also is lost because of dust and smoke in the optical path. One way to minimise such interference is to use an open-end target tube purged with dry air. Another method is the use of a closed-end target tube with its tip at the temperature of the work. Radiation sensors are used typically to control annealing furnaces as well as brazing and forge furnaces. They also are used for blast furnace stoves, salt pots, checker bricks, rolling and strip mills and induction-heating processes. One or more of the following conditions justify the use of non-contact sensors instead of contact sensors.

1. Temperatures are too high for contact sensors.
2. Work is moving too fast for detection by contact sensors.
3. Required response rate is too fast for contact sensors.

Selection of the best radiation detector is normally a compromise based on specific application requirements. However, when emissivity varies, the narrow band is generally the best choice because the amount of error is decreased (Fig. 5.7). For instance, hot slabs of steel, which pass through the sighting field of a radiation sensor, provide a target with variable emissivity. Under these conditions, a narrow-band sensor will minimise the error. In other instances, as when temperatures are lower or emissivity is more nearly constant, a wide-band sensor is a better choice. Temperature measurement and control is a highly specialised field and technical assistance usually can be obtained from the instrument manufactures.

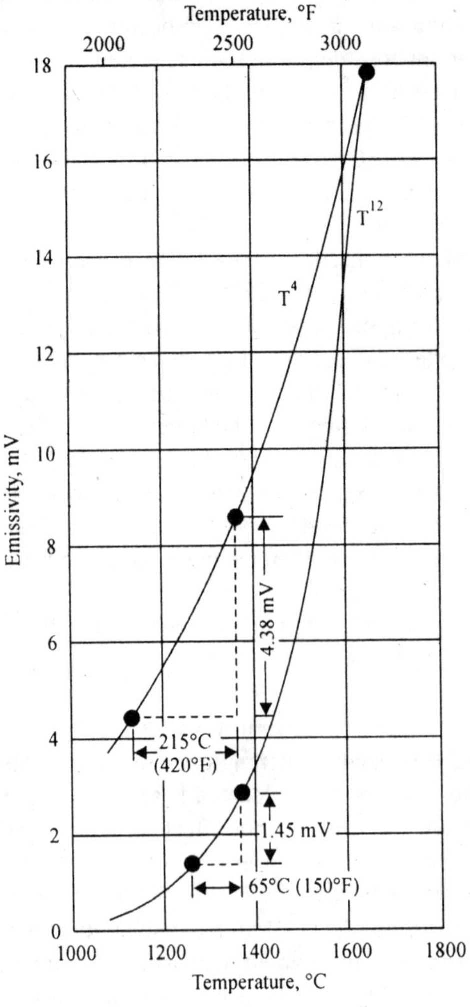

Fig. 5.7. Sensor comparison of T^{12} versus T^4 outputs.

MEASUREMENT AND CONTROL INSTRUMENTS

Measurement instruments measure the output signal of the temperature sensor and convert it to a temperature indication or recording in engineering units. Transmitters are used in some measurement systems to amplify and condition the temperature signal. The accuracy of the measurement depends greatly on the accuracy of the temperature sensor and the connecting leadwire. The accuracy of the measurement instrument is defined in its specification under referenced conditions for its power supply, ambient conditions (temperature and humidity), electrical noise rejection and maximum source impedance. The accuracy of the transmitter has similar qualifications.

Measurement instruments are classified by their displays, analog or digital and whether they are recording or non-recording types. Analog displays include meters and motor-driven pointers. Analog strip chart or round chart recorders include analog temperature indication. Digital displays are available with or without digital printers. In general, digital equipment, is more accurate than analog equipment, but specifications must be checked in each case. The analog versus digital choice in displays and recording usually depends on user preference and specific applications.

Temperature measurement instruments incorporate reference junction compensation in their circuitry for thermocouple measurements and emissivity compensation for non-contact radiation sensors. Reference junction compensation automatically adjusts the measurement depending on the temperature at the junction between the thermocouple wire and the copper wire of the instrument's measuring circuit. Radiation-type temperature measurements require an emissivity compensator. This compensator makes a calibration adjustment by comparing measurement of the same target with an optical parameter or calibrated thermocouple. The emissivity compensator is adjusted to make the radiation pyrometer measurement indication agree with the reference calibration instrument.

In small installations, the resolution and clarity of digital displays is often preferable. In larger installation, where many analog displays can be scanned easily, analog records often are preferred. Sometimes, analog records are required because trends are easier to observe. Digital displays are convenient when doing furnace surveys, such as checking temperature uniformity. Analog records provide compact historical temperature data showing precisely how the process has operated over a long time cycle. These data are useful in monitoring the performance of the control instrumentation and the total process. They are also useful for making a process analysis before and after energy-saving procedures have been implemented.

Controllers

A temperature controller should provide just enough energy to satisfy process requirements, even though operating conditions vary. Variations include changes in process load, fuel characteristics and ambient temperature. Thus, controller requirements are more stringent when process requirements are demanding and especially when operating conditions vary significantly.

In operation, the controller set point that represents the desired temperature is compared with the process or actual temperature. The controller's stability and its sensitivity to the difference between desired and actual temperatures are critical. Based on this comparison, the controller regulates the energy flow to the process.

The two basic types of control are the two-position or on-off type and the proportioning, or modulating type.These two basic types exist in many variations.

On-off Types

These type are inexpensive, simple to operate and easy to maintain. On-off types, however, use energy inefficiently and increase the cost of heater and brickwork maintenance, particularly when controlling high-temperature processes where heat transfer lags are long. The process temperature cycles above and below the controller set point as the controller turns the heat input on and off. During start-up, the heat remains on until the temperature reaches the controller set point and then turns off. Consequently, the thermal inertia of the process, equipment and work load, causes an overshoot beyond the set point and energy is wasted.

After the temperature has stabilised in a characteristic on-off cycle, energy is still wasted because of the effects of heat transfer. Assuming that the temperature cycles are equally distributed above and below the controller set point, the average actual temperature will equal the set-point temperature. However, heat transfer losses or wasted energy, which result from losses above the set point, exceed the savings that result from temperatures below the set point. Thus, a net loss is produced, even though the average actual temperature is equal to the set point. These losses increase as operating temperature increases.

The magnitude of the temperature cycle, that is, the magnitude of the deviation from the set point is directly related to the heat transfer characteristics and measuring lags of a particular process. Therefore, if the temperature measurement responds rapidly to changes in heat input, departure from the set point will be small. Conversely, when response is slow, substantial temperature deviations will result.

The effect of process demand on two-position (on-off) control is illustrated in Fig. 5.8 . The curve in Fig. 5.8a shows temperature cycles evenly above and below the set point. In other words, the heater on and off times are equal. For a higher process demand, such as more heat required to maintain a given temperature, the response is shown by the curve in Fig. 5.8b. In this case, the on time is greater. The temperature cycles are at a lower average value in the curve in Fig. 5.8c. The average value is higher than the set point. The curve in Fig. 5.8d shows the effects reversed for a lower process demand, which again brings the average to very near the set point. The relationship between heat supply, the controlled variable and the set point at about one-half load conditions also is shown in Fig. 5.8e. Cyclic process temperatures cause cyclic process pressures in enclosed gas atmosphere furnaces and ovens. This condition can cause breathing, or inspiration of ambient air, resulting in additional energy losses.

In summary, on-off controllers have the advantage of low cost and simple installation. However, they should be considered only for relatively non-critical processes.

Proportioning or Modulating Controllers

When process temperatures are critical, or when large quantities of energy are consumed, proportioning control should be considered. Compared to on-off controllers, proportioning or modulating controllers are more expensive and sophisticated. They are less expensive to operate, however, because they use energy more efficiently. The process temperature is consistently maintained at the controller set-point temperature, because the controller adjusts the heat input to match the heat demand of the process. Any changes in the process demand temperature will be matched by corresponding changes of the heat input to offset the temperature changes and maintain the desired temperature. Proportioning controllers can be designed to minimise or eliminate set-point overshoot, which is caused by the thermal inertia of the furnace and work load when the process is initially brought from ambient to operating temperature. Normally the process heat head, if not controlled, causes the temperature to overshoot.

The process-control response characteristics vary significantly for various processes. They even can vary for the same process when different operating conditions exist. For example, a furnace might be idled at 120°C (250°F) and operated at 1095°C (2000°F). The proportioning controller, in response to the error signal comparison (set-point temperature versus actual temperature), performs a computation to solve the control equation. This computation includes proportional, integral and derivative terms; thus, the controller is referred to as a PID or three-mode controller. The proportional term is proportional band or gain, and its units are per cent proportional band or gain. The integral term is re-set and its units are repeated per minute. The derivative term is rate and its units are minutes. These turning adjustments (proportional band, re-set and rate) match the controller response to the process dynamics. Proportional band relateed to the size of the error signal. Re-set relates to the error signal size and the length of time it has existed. Rate relates to the error signal speed of change. Consequently, the controller, when properly tuned, in response to these error-signal characteristics, continuously adjusts the energy input to the process, to maintain the desired controller set-point temperature. A properly tuned proportional controller will reduce energy losses significantly by closely maintaining the desired temperature.

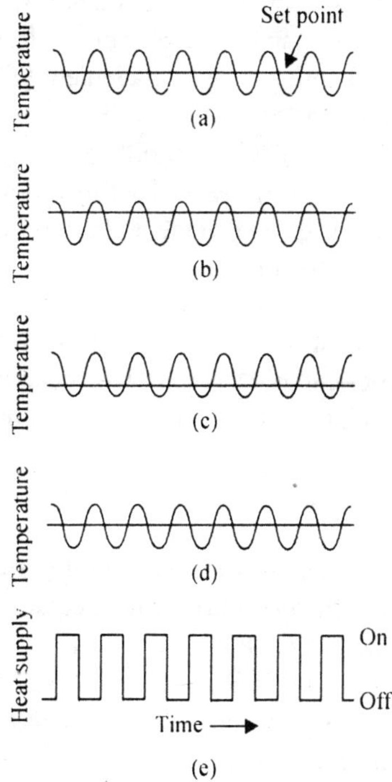

Fig. 5.8. Effect of process demand on two-position (on-off) control.

In a typical proportional control unit, proportional controllers, which are classified according to their outputs, include current-adjusting type (CAT), position-adjusting type (PAT), and duration-adjusting type (DAT). The CAT may be used with an electric power regulator, such as a silicon-controlled rectifier (SCR), to control electrically heated processes. It also can be used with an electropneumatic converter to regulate an air-operated valve. The PAT may be used with an electric motor drive unit to regulate a lever-operated valve. The PAT may be used with an electrical contactor, electric motor drive unit valve combination or solenoid-operated valve.

Typical applications include the use of CAT controller to regulate an SCR. The SCR, in turn, regulates electrical power to heater elements in a furnace. A PAT controller typically is used with electric-motor drive to operate a valve on a gas-fired furnace. The DAT controller is used with an electric contactor to regulate electric power to heater elements. Electric contactors are less expensive than SCR's and are adequate for slow-response, high-temperature furnaces. DAT control also is used to operate gas valves on radiant tube-fired furnaces to achieve uniform heating throughout the tube.

The process temperature, even with PID control, tends to overshoot the control set-point during start-up. This action is because of the system's thermal inertia. If the temperature sensor is of heavy construction, its slow response speed will compound this condition. A feature called Approach Action may be specified in the controller to minimise this problem, if the controller is properly tuned. This function also can be achieved with a feature called re-set limiting.

Final Elements

The final element of the control loop responds to the controller output by regulating the energy flow. It may consist of one or more components, such as valves, dampers, power regulators, heater elements, converters or actuators. Some desirable characteristics include fast response speed, linear response, high sensitivity, adequate power and reliability. The final element should have a linear response over its full range; that is, a change in the final element should produce a corresponding change in temperature. The sizing of the valves, power regulators and heaters should match the process requirements to optimise control performance. The final element may provide feedback devices, such as slidewires and limit switches, for use in the control system. The final element also may be designed to drive to a limit or to lock in its last position, in the event of a power failure.

The following list includes some final elements used with electric controllers:

1. An electric motor-valve combination used to regulate flow of steam or hot water.
2. An electric motor-valve-burner combination used to regulate fuel combustion.
3. Pneumatic valves used to regulate fuel. An electropneumatic converter may be used to convert the output signal from the controller.
4. Electrical resistance-type heating elements may be used with contactors, saturable core reactors, silicon-controlled rectifiers and triacs.

Design of the final element in any energy-efficient control system is extremely critical, because it regulates the energy flow. For example, butterfly valves are frequently selected instead of proportioning valves, because the butterfly valves cost less. However, their flow characteristics are non-linear, which makes control over the entire range more difficult. In butterfly valves, a large flow change occurs near the closed valve position and a small change occurs near the open position. A non-linear toggle linkage between the electric motor drive unit and valve can reduce this non-linearity and provide improved control.

The curves in Fig. 5.9 show the effect of toggle action on flow when using a non-linear valve, compared with flow without toggle linkage. The non-linear toggle linkage between the electric motor drive unit and the valve can reduce this non-linearity, thus providing better control. The non-linear linkage provides small increments of travel in the lower portion of its range. Greater increments of travel are provided in the upper portion of its range, for equal increments of drive mechanism travel.

Maximum toggle action is obtained if the drive mechanism crank and link rod are within 5° of dead centre when the controlled device crank arm and link rod approach a 90° angle. This situation is illustrated in Fig. 5.10 in which a valve serves as the controlling device. The control slidewire limits the drive mechanism crank movement to only 85°. Reduced toggle effect is obtained when the drive mechanism crank arm link rod angle is increased approximately 5 to 30°.

Set-point Programmers

A temperature cycle or temperature programme is required in many heat treating applications. As the temperature changes, the metallurgical structure of the work being processed passes through phase changes or structure changes. Because these changes occur slowly, the process temperature should be changed slowly to avoid internal stresses, which cause distortion or cracking of the metal parts. Automatic control of the temperature versus time cycle is achieved by using a set-point programmer. This device provides the set-point signal to a three-mode proportioning controller. This automatic control of the complete temperature cycle ensures that the process uses a minimum amount of energy.

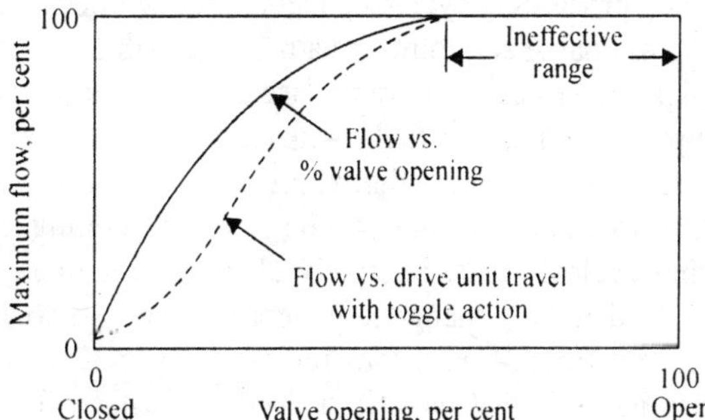

Fig. 5.9. Curves showing effect of toggle action on flow when using a non-linear valve.

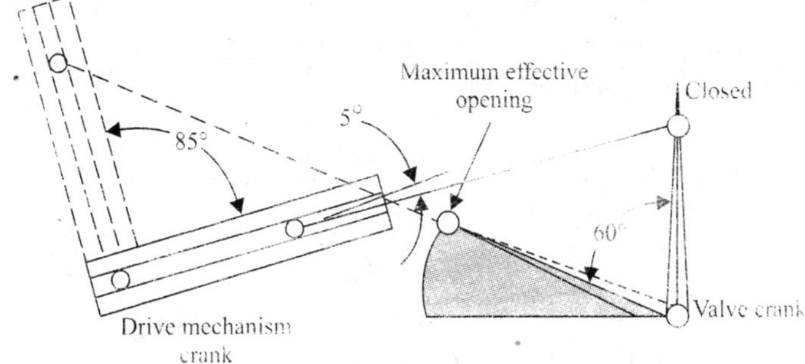

Fig. 5.10. Schematic view of linkage arrangement for maximum toggle action.

Most temperature programmes specify linear heating and cooling rates with holding periods, or constant temperature plateaus. The set-points may be motorised, cam actuated, curve follower or digital types. The cam actuated and curve follower types also may be used for non-linear programmes. Equipment selection is dictated by the complexity of the desired temperature programme and the need for specific features, including degree of accuracy and economics.

To provide an automated temperature programme, the three-mode controller set-point is provided by the programmer set-point. The controller computes the proper control action by comparing the programmer set point (desired temperature) to the process (actual temperature).

The accuracy of the set-point programmer is very important when evaluating its energy saving potential. However, flexibility, convenience, simplicity and cost are also significant factors in the selection of a programmer.

The digital set-point programmer provides exceptional flexibility, accuracy and resolution based on microprocessor technology. Microprocessor systems are compact, reliable, inexpensive computer systems, custom designed for specific applications. Programme control in heat treating applications provides the temperature programme. The logic control regulates the sequence of events programme, such as operating doors, conveyors, fans and atmosphere gas additions, in the desired sequence and for the desired time periods. Digital set-point programmers provide more expanded set-point control capabilities and their logic control eliminates the inflexible and costly hard-wired cabinet of relays used in the past.

A powerful digital programmer providing two set-point outputs and logic control of powerful programmer, working with such signals as contact closures from the heat treating equipment, can completely control the complex processing cycles in accordance with established programmes.

The combination of a programmable set-point and a programmable logic control provides a flexible method to completely control processes that require precise time-temperature control. Additional energy efficiency can be gained by programming the fuel-to-air ratio for fuel-fired furnaces. The temperature and fuel-to-air ratio can be programmed from a digital programmer with two set-point outputs.

Typical heat treating cycles include annealing, normalising, hardening, carburising, tempering and nitriding. Heating and cooling cycles can be relatively elaborate for these processes. A heat treating cycle for heating, holding and quenching manganese steel castings is shown graphically in Fig. 5.11. This process was conducted in a three, car-bottom furnace. Fig. 5.12 is a graphic presentation of a programmed cycle for hardening tool steel parts in a vacuum furnace. The temperature and vacuum conditions are shown in the upper portion of the Fig.5.12. Set points 1 and 2 are shown as regions just below the centre of the figure. Segments of the logic sequence numbered 1 to 8 are shown as bars which indicate the time for each. The events of the sequence are identified at the lower right of Fig. 5.12.

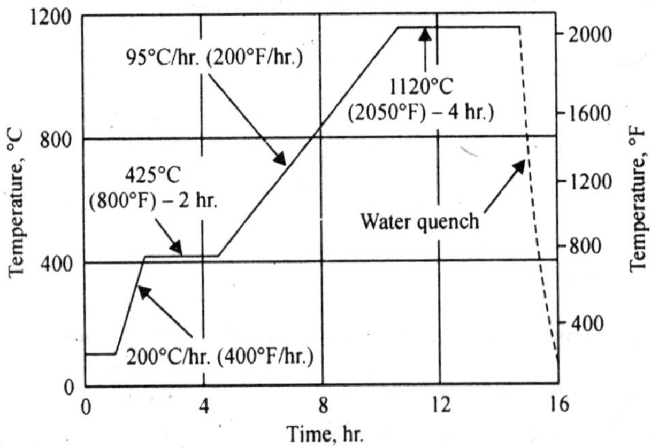

Fig. 5.11. Heat treating cycle for manganese steel castings.

CHARACTERISTICS OF THE PROCESS, PRODUCT AND PRODUCTION

A control system regulates the energy input to the process to achieve desired heating and cooling rates and maintains soak temperatures by matching heat input to heat losses. If these heat losses were always constant, a pre-set heat supply could keep temperature in balance indefinitely.

In practice, this never happens. Furnaces must be started up, idled and operated at varying production rates and temperatures. To handle such changes efficiently, control systems must constantly re-adjust heat input so that the work temperature is maintained despite changing operating conditions.

Two principal characteristics of the instrumentation and the process equipment influence the selection of a control system—lag and thermal inertia. In a heat treating furnace, refractory, muffles or retorts through which the heat must flow re-present heat barriers. The greater their number, the greater the heat transfer resistance or lag, with the result that there is a longer delay between the heat input change and a change in product temperature. After this mass is heated—refractory, muffles, retorts—it also represents thermal inertia.

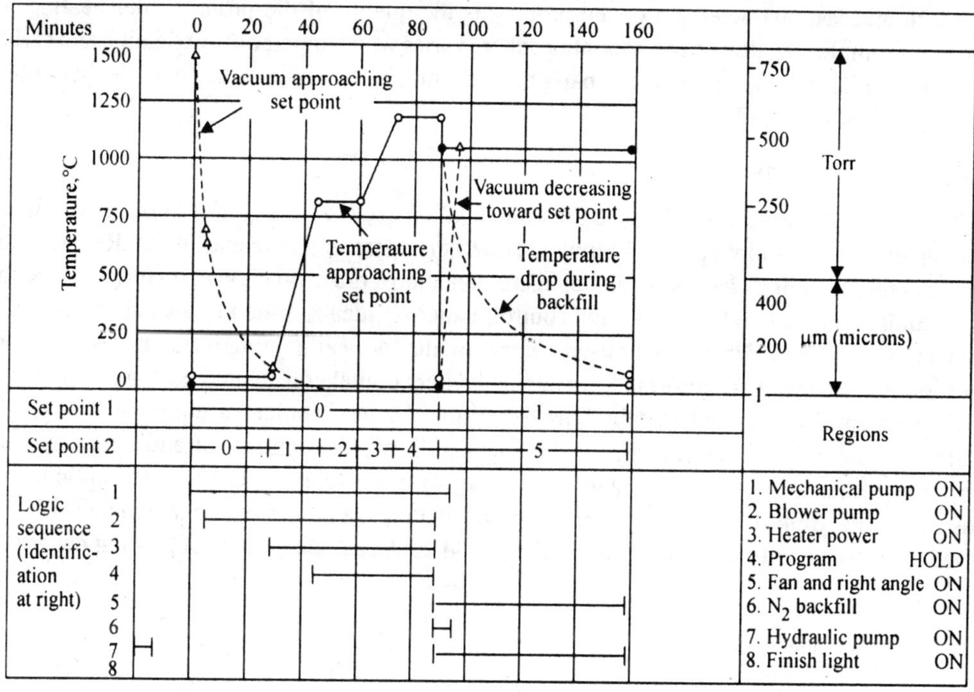

Fig. 5.12. Cycle for hardening tool steel parts in a vacuum furnace.

Thermal inertia of a process is its capacity to continue a temperature change in the same direction after the heat supply is altered, before it begins to move in the opposite direction. The time it takes to turn a process around determines the extent of its thermal inertia. A furnace with a large, heavy muffle has a high heat storage capacity. Heat heads will build up rapidly during heat input periods and the momentum from the latent heat continues to push product temperature up after the heat is turned off. The converse is also true. Product temperature will continue to drop after the heat is turned on. In short, the furnace is sluggish and the temperature swings are wide and slow.

In addition to the resistance to heat transfer and thermal inertia of the product and process equipment, there are five characteristics of the product to be considered in selection of a temperature control system. The first characteristic of the product is its state. This is important in deciding the degree of acceptable temperature tolerances. For example, a solid product such as steel or ceramics may have a temperature tolerance of ±3, 6, 8°C (±5, 10, 15°F) or even more. The second characteristic is the material of the product whether it is ferrous or non-ferrous. Most non-ferrous and high-alloy steel products have temperature tolerances of ±3 or 6°C (±5 or 10°F). On the other hand, low carbon steel and other ferrous metals may have a temperature tolerance of ±8 to 14°C (±15 to 25°F).

The third characteristic—assuming the furnace is properly designed for the job—is the mass and shape of the product. This determines the responsibility the control system must assume for uniform heat penetration. The mass and shape of a large roll of steel influences the selection of not only the furnace design but also the further selection of the control system required to maintain uniform heat from top to bottom and side to side. The fourth characteristic of the product is its critical temperature. The nearer the product is heated to its critical temperature, the more accurate and responsive must be the control system.

The fifth and last characteristic of the product is the quality of the product after heating. In heat treating, for example, the effect of furnace atmosphere varies with temperature. Oxidation and carburising occur faster at higher temperatures. For parts of uniform quality, temperature must be held uniformly throughout the furnace load.

Instrumentation Trends

Digital instrumentation is replacing analog instrumentation because its superior performance, flexibility and simplicity provide a more powerful tool, useful to the operator and management. Reduced costs for microprocessor technology has made digital instrumentation a practical choice. A completely automated process requires interactive loop and logic control, which is ideally suited to the capabilities of digital instrumentation. A trend toward centralised control rooms for heat treating installations with computer supervision and management information systems favours digital control and data acquisition systems.

Individual applications and installations determine the best choice of equipment. A single-loop proportioning controller provides digital implementation of PID control. In addition, this particular controller can change its tuning constants (P, I and D) in response to changes for optimum control because it 'learns' from its experiences. For large installations, a multiple loop and logic control system designed to interface with computers (Fig. 5.13) could control a complete heat treating department.

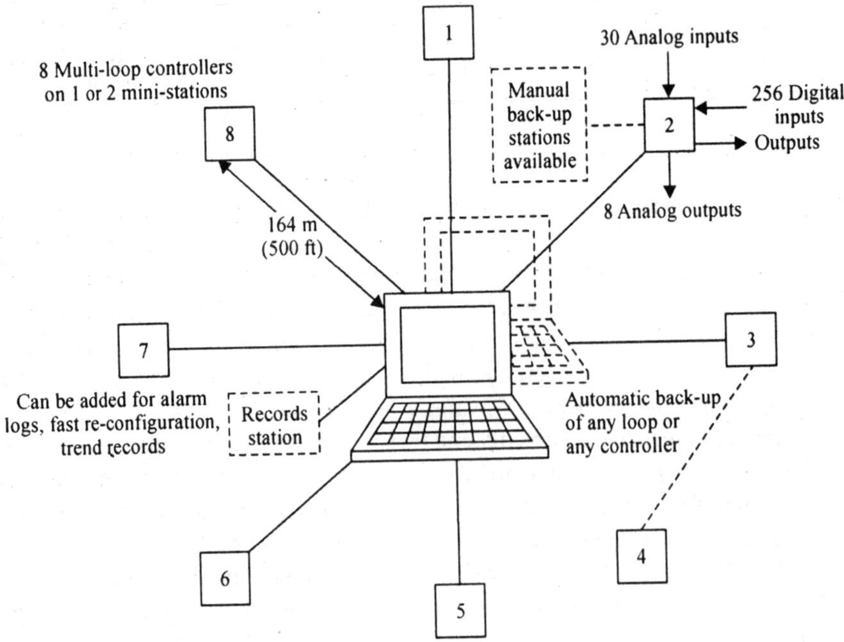

Fig. 5.13. Multiple-loop logic control system.

Individual multiple loop and logic controllers controlling individual furnaces or groups of furnaces are connected to a central video display and control station. This configuration of instrumentation is called a distributed control system. An operator can view all the loops, logic and alarms on the video display and make changes from the control station.

Distributed control systems are consistent with the concept of the completely automated plant, monitored and supervised with a central computer system. This type of installation integrates all plant operations, as required, to achieve optimum productivity in response to demands.

Chapter 6

Atmosphere Control

INTRODUCTION

The purpose of atmosphere control is to maintain consistent levels of constituents in a protective atmosphere and/or to determine the changes in those levels that are required in order to produce a desired result under a given set of conditions. Controls are required for various heat treating operations in many different atmospheres. All methods of atmosphere control can effectively be divided into two groups: those involving control of the atmosphere-generating system, and those involving control of the atmosphere once it is inside the furnace. Both are important in maintaining a controlled condition throughout the heating process. Most operations performed in the heat processing industry can be done under endothermic or exothermic protective atmospheres. Effective control of the generators that produce such atmospheres may be accomplished by means of certain types of controllers—namely, combustibles controllers, infrared CO_2 controllers and oxygen probes. The required accuracy of control is directly proportional to the economics of the individual atmosphere system.

Most common among combustibles controllers is one that operates on the principle of catalytic combustion. A sample drawn from the gas generator or from the furnace is mixed with air (to supply oxygen) and then is passed over a detector. Any combustibles in the gas burn catalytically, raising the temperature of the detector and increasing the electrical resistance. The detectors in the catalytic chamber make up half of a balanced electrical circuit. A corresponding imbalance in the bridge resistance develops, and the resultant electrical output voltage proportional to the concentration of total combustibles in the sample is read out on the panel meter.

An infrared-type analyser can be utilised for specific analysis of simple or complex mixtures of vapours and liquids and can be used to detect any components that absorb infrared energy. An infrared-type analysing system could be of a design utilising two separate or twin infrared radiation cells. One is referred to as a reference source, and gives a known signal in response to the reference gas. The other cell is used as a sample cell, and would vary from the known reference cell depending upon the elements in the gas contained inside that cell. When the gases in both cells are the same, they are in balance. If the gas absorbing infrared radiation in the sample cell is increased, more infrared radiation is absorbed, and consequently the beams become unequal. Movement of a membrane within the detector varies a condenser's electrical capacity, in turn resulting in an electrical signal proportional to the difference between the two beams. The signal is amplified and fed into an indicating meter.

These same principles are utilised for both furnace control and atmosphere generator control. Infrared analysers may be used to control systems other than carbon monoxide and carbon dioxide in furnace applications. For example, another form of atmosphere control would be dew-point control (water vapour). Because many heat treating systems are based on carbon dioxide analysis, infrared control is very common. Carbon dioxide control and/or dew point control can be used for determining carbon potential in carburising atmospheres. Infrared analysers can also be used to monitor ammonia in nitriding atmospheres and carbon monoxide in other applications. Some applications involving heat processes require the use of analysers to determine the oxygen content of a given processed gas. The most common practice for oxygen analysis in heat treating furnaces is to determine the lack of oxygen. Depending on the level of oxygen desired or permitted in the processed gas, specific ranges of concentration may have to be determined.

The magnetic oxygen analyser is a common analysing system used with heat treating systems. Oxygen has an affinity for magnetic fields; most other gases do not. By adding a sample gas to a magnetic field and a detector, the change in resistance can be measured in a cell. The output from this cell is fed to a meter or a recorder and quite often is used as a permissive circuit in an atmosphere-control system.

As mentioned previously, use of a dew-point measuring system is another method of controlling the carbon potential of a furnace atmosphere. It can also be utilised to determine the water-vapour content in any given atmosphere. One type of dew-point recorder consists of an element with two parallel electrical conductors helically wound on an insulated tube and connected to electrical power. The insulation of the tube is saturated with a hydroscopic salt and absorption of the moisture by the salt permits the electrical current to flow between the conductors, causing heating to an equilibrium point where moisture absorption ceases. That point where equilibrium exists is fed to an indicator and corresponds to the dew point of the major sample. The above-mentioned control systems can be modified and incorporated together to form an automatic control system for any given set of conditions, utilising normal instrumentation and final control devices. The system is ultimately a product of the needs of the user and justifiable economics. As metallurgical reproducibility of any given process becomes more important, the need for atmosphere control to produce chemical stability becomes a necessity. A second imperative not to be overlooked in the application of atmosphere-control devices is the safety aspects of operating atmosphere generators and industrial furnaces.

ANALYSIS AND CONTROL OF ENDOTHERMIC AND EXOTHERMIC ATMOSPHERES

For success in metallurgical atmosphere control, gas analysis instrumentation must be applied where the atmosphere starts—at the gas generator. This is true for endothermic as well as exothermic atmosphere applications. Analysers and associated control systems applied to a furnace should never be expected to correct deficiencies in the basic atmosphere produced by the gas generator. Instrumentation dedicated to the furnace is designed to provide information and control relative to operation of the furnace and to the product being processed. Likewise, instrumentation dedicated to the generator is designed to provide information and control relative to operation of the generator and to the atmosphere produced by it.

Because the heat treated product emerges from the furnace, there is a tendency to emphasise furnace atmosphere control and overlook the generator. In many cases, automatic atmosphere control of the gas generator will resolve heat treating atmosphere problems, simplify control of the furnace atmosphere, and in some cases reduce furnace monitoring to occasional spot checks.

Endothermic gas is used in heat treating furnaces as a protective atmosphere for hardening, stress relieving, carbon restoration and carburising. The most recognised guides to endothermic generator operation are CO_2 and dew point. The ultimate guide is CO_2 and dew point. Any deviation from the relationship between the two indicates a generator problem, such as a leak or a carbon-coated catalyst. Fig. 6.1 shows proper carbon dioxide/dew point relationship.

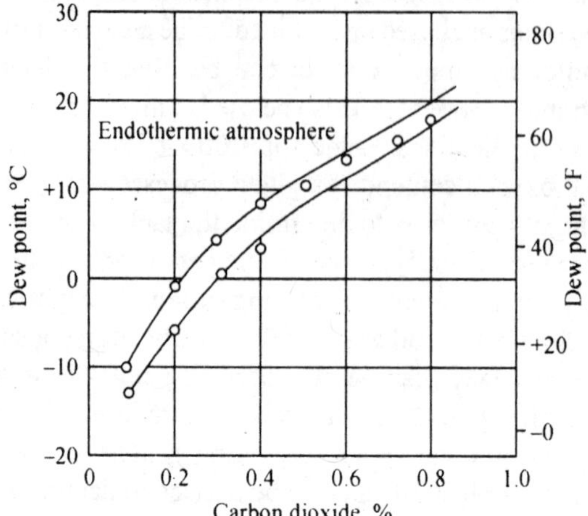

The generator in each plant was operated at a different temperature, all within the range from 1005 to 1095°C (1840 to 2000°F).

Fig. 6.1. Variation in the relation between dew point and carbon dioxide in generation of an endothermic atmosphere, as obtained from four plants.

The nominal operating ranges for endothermic-atmosphere generators are –7 to +16°C (20 to 60 °F) dew point and 0.2 to 0.7 per cent CO_2. Generally, maintenance of an endothermic generator can be reduced by operating the generator near the lower end of the CO_2 range or near the higher end of the dew point range within the limits required by the furnace process. As a rule of thumb, the dew point will rise approximately 6°C (10°F) between a sample taken at the generator and a sample taken at the furnace before enriching gases are admitted to produce the desired carbon potential within the furnace chamber.

Typical Endothermic Generator Problems

The following list describes typical endothermic generator problems that can be identified, corrected, or avoided through application of proper analysis and control:

1. Carbon on catalyst.
2. Worn or clogged carburetor or mixer.
3. Poor temperature control.
4. Combustion products leaking into retort.
5. Barometric pressure change.
6. Humidity change.
7. Hydrocarbons from quench oil or engine exhaust drawn into generator with air.
8. Change in natural gas composition.
9. Sticking regulators.
10. Dirty flowscope orifices.

When it is determined that carbon is present on the catalyst, a 'burn-out' procedure is followed to restore catalyst activity. Burn-out is accomplished by turning off the natural gas and allowing air to pass through the catalyst. Oxygen combines with the excess carbon to form CO_2. Using the CO_2 analyser to follow the decreases in CO_2 during burn-out allows prompt determination of burn-out endpoint. Operators often allow burn-out to continue for excessive periods of time. Following the CO_2 decrease shows that burn-out usually can be completed in less than one hour. The necessity for burning out a generator is greatly reduced when the generator is controlled by CO_2 or dew point analysis, because large upsets commonly induced by manual adjustments are eliminated. Reports have been received indicating that endothermic generators have been controlled at 0.20 per cent CO_2 (dew point of – 4 °C or + 25 °F) for one year without burn-out. This is exceptional and somewhat contingent on generator design.

Exothermic-type atmospheres are used where inert atmospheres are required. Applications are common in the metal treating industry for bright annealing of copper, annealing of motor laminations, processing of aluminium, and annealing of coils of wire and steel sheet. Exothermic atmospheres are also used as an inert blanket in production of glass, in food processing and in rubber curing. Various chemical processes and storage facilities employ inert atmospheres produced by exothermic-gas generators.

Typical analyses of various exothermic atmospheres are listed below, along with recommended analysers. Total combustibles analysers are applied most frequently to monitor and/or control exothermic atmospheres. Very rich (20 per cent or higher total combustibles) and very lean (less than 1 per cent total combustibles) exothermic atmospheres are most accurately monitored and/or controlled by an infrared analyser sensitised to an appropriate range of carbon monoxide.

Rich exothermic atmosphere

1. 10 to 25 per cent combustibles (approximately equal parts CO and H_2), 5 per cent CO_2, 3 per cent H_2O, rem N_2.
2. Recommended analysers to control and/or monitor rich exothermic atmosphere:
   (a) Infrared analyser calibrated for 0 to 10 or 20 per cent carbon monoxide.
   (b) Total combustibles analyser (catalytic type) calibrated for 0 to 15 per cent, 20 or 25 per cent combustibles.

Lean exothermic atmosphere

1. 1 to 10 per cent combustibles (approximately equal parts CO and H_2), 12 per cent CO_2, 3 per cent H_2O rem N_2.
2. Recommended analysers to control and/or monitor lean exothermic atmosphere:
   (a) Total combustibles analyser (catalytic type) calibrated for 0 to 1, 2, 5 or 10 per cent combustibles.
   (b) Infrared analyser calibrated for 1, 2, 5 or 10 per cent carbon monoxide. Oxygen analyser for 0 to 1.0 per cent or 2 per cent oxygen to monitor lean atmospheres to ensure minimum oxygen.

Nitrogen (lean exothermic) atmosphere

1. 0.2 per cent combustibles, 0.1 per cent oxygen, 12 per cent CO_2, H_2O rem N_2.
2. Recommended analysers to control and/or monitor nitrogen atmospheres:
   (a) Recommended analysers are similar to lean exothermic analysers but with lower ranges specified for greater accuracy.
   (b) Infrared analysers calibrated for 0 to 0.1 per cent or 0 to 0.5 per cent carbon monoxide are generally preferred over total combustibles analysers when best accuracy is required.

Exothermic atmospheres with CO_2 and H_2O removed

1. Exothermic atmospheres may be passed through MEA (monoethyl amine) or a molecular sieve scrubber to remove CO_2 (<1000 ppm) and H_2O to produce N_2 or N_2 + combustibles atmosphere;

in addition to applying the above analysers, an infrared analyser calibrated for 0 to 0.1 per cent carbon dioxide provides an excellent means of monitoring CO_2 scrubber efficiency as well as prompt indication of carbon dioxide breakthrough.

Figs. 6.2 and 6.3 summarise application of analysers to exothermic and endothermic-gas generators, indicate analyser sample take-off locations and show proper locations for installation of control valves for automatic atmosphere control. The following adjustment procedure should be followed for proper set-up of a control valve and motor when they are first installed on a gas generator for automatic control.

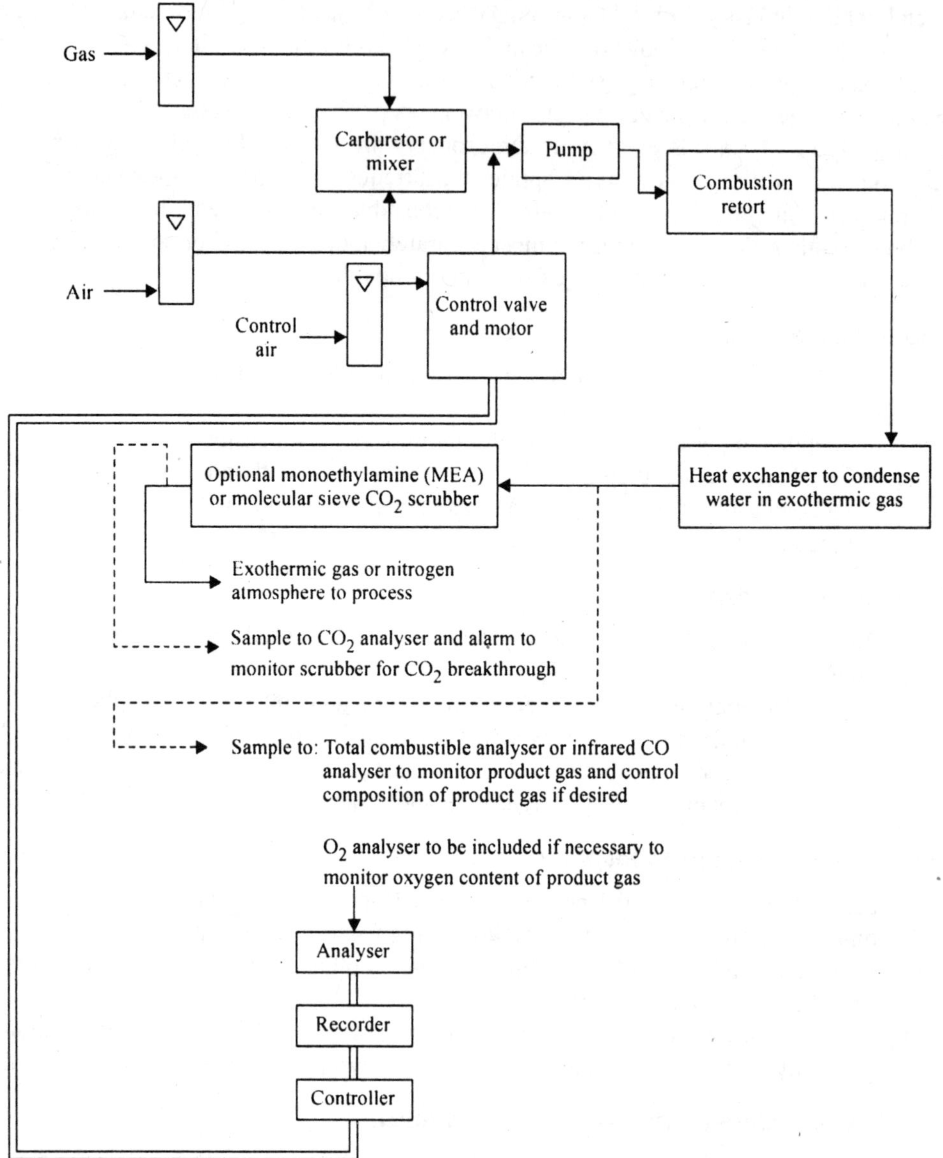

Fig 6.2. Typical analyser and controller applied to exothermic-gas or nitrogen generator.

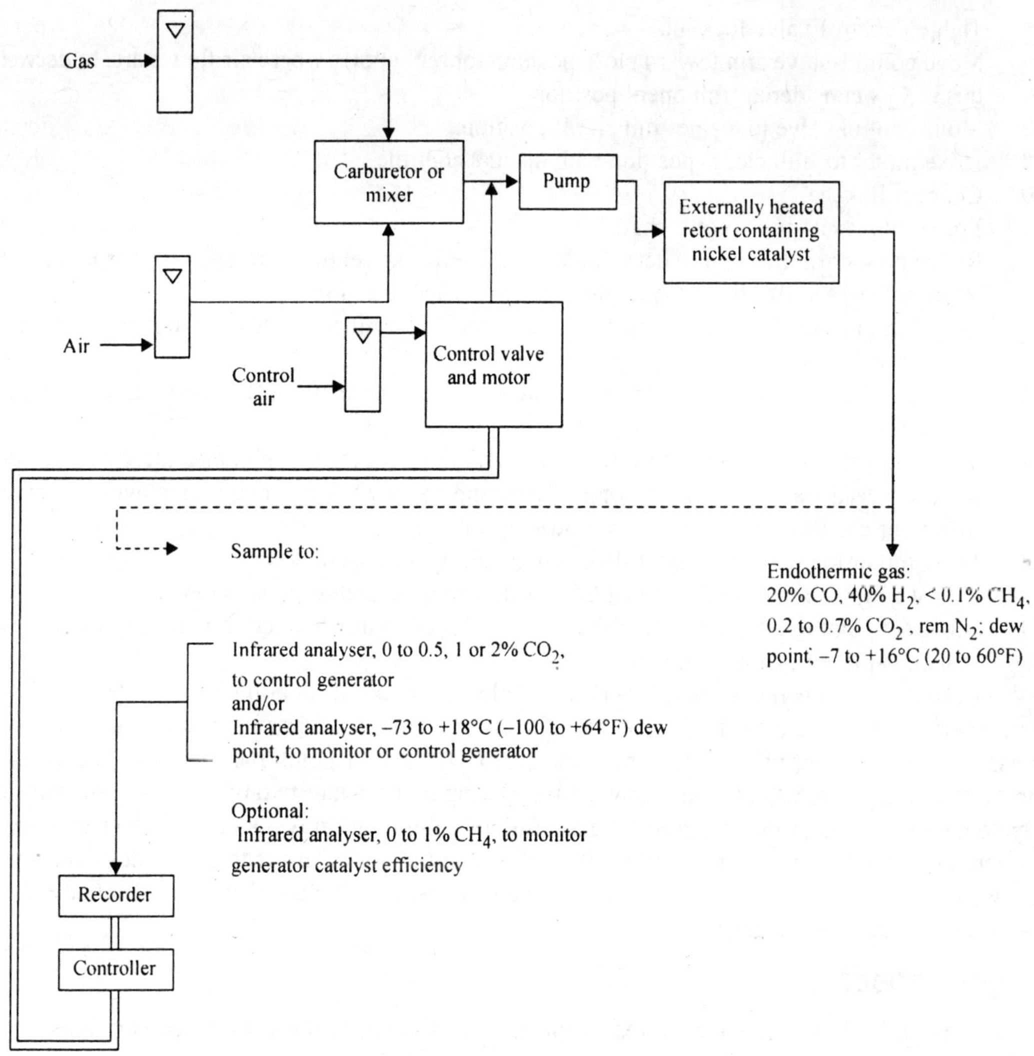

Fig 6.3. Typical analyser and controller applied to endothermic-gas generator.

Control Valve and Motor Adjustment Procedure

The effectiveness of an automatic atmosphere analysis control system is directly related to the proper set-up of the control valve and motor. Proper-control valve port size and driven-motor speed are important to system success. The following procedure is provided to avoid problems that are easily overlooked during initial set-up of control valves and drive motors.

1. Disconnect control-valve arm from motor linkage.
2. Loosen control-valve locknut.
3. Move control-valve arm to open position.
4. Adjust control-valve port size to pass amount of air equal to 10 per cent of air normally flowing into generator.

5. Tighten control-valve locknut.
6. Move control-valve arm toward close position, observe point where air flow starts to decrease, this is now considered 'full open' position.
7. Move control valve to locate 'full close' position.
8. Drive motor to 'full close' position with manual control.
9. Connect linkage.
10. Loosen linkage arm at motor shaft.
11. Rotate motor arm by hand to determine that full valve travel (close to open) corresponds to full motor travel. Adjust linkage as required to obtain above action.
12. Drive motor from 'full close' position toward open. Valve must start to open immediately without 'dead travel'.
13. Continue to drive motor to 'full open' position. Valve must arrive at 'full open' as motor reaches end of travel.
14. Drive motor from 'full open' toward close. Valve must start to move immediately wihtout 'dead travel'. 'Dead travel' is when motor moves and valve does not move or moves in opposite direction; condition is corrected by adjusting linkage.
15. Drive motor to mid-position, air flow is one-half of 'full open' setting.
16. Adjust generator carburetor so that CO_2 or dew point is at desired set point.
17. Move set point to a higher setting and observe direction in which controller drives valve when system is placed in automatic control mode.

Valve should drive open when analysis point is below set point. If set point is moved below analysis point, valve should drive closed. If valve drives in wrong direction, wires to motor must be reversed. Increasing air to endothermic generator must increase CO_2 and dew point. Increasing air to exothermic generator must decrease CO or total combustibles. During first week or two of operation, control valve on endothermic generator may keep seeking a position at full motor travel and it will be necessary for the operator to mannually re-set to mid-positon, adjust carburetor or mixer to obtain desired set point and switch back to automatic. This is a normal sequence caused by gradual cleaning and conditioning of catalyst in the endothermic generator.

OXYGEN PROBES

The oxygen probe is perhaps the newest technique used today in the heat processing industry for monitoring and controlling different atmospheres such as endothermic gas, exothermic gas, steam with small traces of hydrogen, direct-firing (fuel-rich) atmospheres, and hydrogen-bearing gases with low to medium water contents. Oxygen probes have been in use long enough to provide significant operating experience in heat treatment process control. Sufficient output data have been accumulated and correlated to provide practical usages that offer significant advantages over other forms of monitoring and control such as infrared and dew point analysers.

The oxygen probe is based in theory on a hot ceramic electrochemical cell. The probe will respond to oxygen, hydrogen, carbon monoxide, water and carbon dioxide and thus can determine the oxidisation potential of a gas. The output of the oxygen probe is a direct measurement of the oxidation potential of the atmosphere at the process temperature of the furnace. Therefore, when the probe temperature is close to the furnace temperature, the response of the probe is a direct indication of whether the atmosphere will oxidise or reduce steel, provided that the composition of the atmosphere with regard to the proportions of carbon gases and hydrogen is known. Under such conditions, the probe will give a reliable indication of the oxidation/reduction situation for all furnace temperatures. Thus, with simple mechanical methods, additive gases or liquids can be introduced into the furnace in order to control the oxidation/reduction potential of the atmosphere.

An oxygen probe is a closed-end tube usually constructed of lime-stabilised zirconia or yttriastabilised material for temperatures up to 1600°C (2900°F). When such a probe is subjected to elevated temperatures, the non-porous sheath material acts as a solid electrolyte which permits the passage of oxygen ions when the inner and outer surface are subjected to atmospheres of different oxygen partial pressures—for example, a reference gas (such as air, because its O_2 content is constant at 20.9 per cent by volume at sea level) and the process furnace atmosphere, respectively. The electromotive force (emf) thus generated and measured via the electrodes attached to the sheath, is related directly to and provides an accurate quantification of, atmosphere characteristics in terms of its oxidising/reducing or in some endothermic-atmosphere applications, carburising/decarburising tendencies at a known temperature. The electrodes mentioned above are in physical contact with the zirconia on both the inside and outside of the tube, and usually are constructed of platinum because of its superior chemical resistance at elevated temperatures. When the probe is subjected to elevated temperatures, there is a difference between the oxygen partial pressures on the two sides of the probe, and electricity will flow through a circuit connecting the two sides. This flow of electricity is from the higher pressure to the lower. If the oxygen pressure on one side is known, the oxygen pressure on the other side can be determined:

$$E = K \times T \log \frac{[O_2]\text{known}}{[O_2]\text{unknown}}$$

where T is absolute temperature, $[O_2]$ is oxygen partial pressure, K is a constant and E is the electromotive force generated. This measurement can be made using a simple panel meter without electronics.

Oxygen probes have been extensively used for control of carburising furnaces. Accuracy of carbon-potential control through use of oxygen probes is estimated to be approximately ± 0.05 per cent carbon in actual practice because of the limitations of temperature control and temperature variations in typical heat treating furnaces. Because of the fast response rate, on/off control systems utilising solenoid valves for regulating propane, natural gas, or liquid enrichment are adequate for batch furnaces. *In situ* or *ex situ* probes control carbon potential by adjusting the set point on the control instrument to the desired carbon level wanted in the furnace. The oxygen probe supplies to the control station an electrical signal related to the carbon potential . High and low deviation contacts are adjusted through the control instrument and either contact is made by the electrical signal from the probe. The low deviation contact controls the solenoid valve that supplies enriching gas or liquid to the furnace. The high-deviation contact can be made to control the solenoid valve to add air or an oxidiser to the furnace. Carbon-concentration reproducibility of ±0.02 per cent is frequently achieved using systems of this type, provided that the cycle, temperature and furnace conditions remain constant.

Continuous carburising furnaces and straight hardening furnaces usually use proportional control in conjunction with the oxygen-probe control. Experience in control of continuous furnaces has shown that proportional control is normally required in order to cope with disturbances involving movement of work, opening of doors and the like. Compensation for these disturbances are rapid because of the rapid response rate of the oxygen probe. In the proportional-control system, the output voltage is supplied to a two-mode controller, and the controller positions a motorised or pneumatic control valve that regulates the flow of enriching gas to the furnace. Such carbon control is advantageous when short treatment cycles of 20 minutes or less are used, because compensations of 2 to 3 minutes or less are required for disturbances in such applications.

Oxygen probes often are used to monitor a furnace atmosphere while control is accomplished manually by adjusting flowmeters on enriching lines to the furnace. This is often used when the cost of control systems cannot be justified. The cost of replacing the consumable probe and the cost of simple on/off control systems must be weighed against human error, process accuracy and reproducibility.

Endothermic generators also can be controlled by use of the oxygen probe. *Ex situ* probes are generally used because of the difficulty of placing an *in situ* probe in the retort of the generator. In the case of the *ex situ* probe, a sample is taken from the output line of the generator. The electrical signal from the probe is wired to the control instrument that activates the solenoid valve located on the air-bypass line to the mixer. Thus, the air/gas ratio is automatically adjusted to give the desired properties of the endothermic gas. Stainless steel , nickel, brass, titanium, aluminium, Incoloy and other metals and alloys are frequently annealed using exothermic gas. The reducing power of the gas necessary to protect these metals varies considerably; in general, copper alloys require a lean exothermic gas and stainless steels require a rich gas.

Usual practice for control of such a process is to adjust the gas composition by varying the air/gas ratio prior to combustion in the gas generator. However, in practical experiments and in theoretical calculations, it is shown that it should be possible to control the air/fuel ratio closely by measurement of oxygen potential in either the gas generator or the annealing furnace.

By means of the oxygen probe, for any atmosphere, it should be possible to provide very early warnings of the ingress of air into the furnace from leaking seals or of ingress of water from water-cooled bearings or fans. It is also anticipated that this instrument may replace the conventional oxygen analyser as a device for providing very early warnings of possible hazardous conditions in the furnace.

Chapter 7

Furnace Safety

INTRODUCTION

Heat treating furnaces require safety procedures common to all industrial installations, but in addition, they have requirements specific to the use of high-temperature energy sources and potentially explosive gases and liquids used as aids to chemical processing. Because these heat treating processes require careful control for optimum technical results as well as for safety, proper training of operating personnel is a primary consideration. Proper equipment design is also critical.

The information presented here is not intended to be interpreted as a safety standard but is offered only as a set of guidelines.

All equipment should be installed and operated with awareness of the potentials for fire and explosion and the hazards to operators and equipment. Equipment designs should ensure reliable, safe operation over the expected maximum life of the equipment.

FUEL-FIRED FURNACES

Fuel-fired furnaces for heat treating have several major control requirements that depend on: (i) whether the process must be direct or indirect fired; (ii) whether heat treating is to be done under a particular pressure or vacuum, or in a controlled atmosphere; and (iii) whether the product uses some special type of pre-coat or laminant. In all situations, there are fundamental control variables and instrumentation is available to identify and control change and drift, thus achieving the desired results.

The main control elements are the three requirements for proper combustion: a source of heat, an oxidising agent and time.

Fuel-fired furnaces can be automated to the extent that manual intervention is not required for normal operation. Many processes and operations require manual control, however; thus, furnace controls range from almost completely manual devices to highly sophisticated computer-controlled devices.

The major control variables and types of instrumentation used in each instance are described below, in sequence, from start-up through final cycle control.

Electrical Power for Fuel-Fired Furnaces

The safe use of electrical energy employed in heat treating control processes requires adherence to National Electrical Codes and to local requirements of states and communities. Good practice dictates that a circuit breaker be positioned within view of the operator. Control cabinets must be designed to ensure that operators cannot inadvertently become a path to ground. Wiring type should be based on the environment of use, and wiring for all motor and control circuits should be contained in appropriate conduits. Numbers of wires within specific conduits should be governed by the fact that elevated temperatures may be encountered. All enclosures for electrical apparatus should be designed to protect the contents from the environment. The furnace itself should be grounded for proper control.

The source of electrical power to the furnace installation should be equipped with fuses. Each motor also should be equipped individually with fuses and protected with thermal-overload elements based on operating temperatures. Motor selection should be based on such conditions of use as temperature, weather, dust, dirt, atmosphere and humidity. Manufacturers should be consulted on motor design and selection. Control-panel power should be fused and provided with externally operated switches to allow safe entry by authorised personnel.

Electronic-signal wiring from the flame-safety circuits should have its own conduit, free from the 'noise' and induction present in normal power and control circuits. Thermocouple wiring also should be contained in its own conduit to avoid creating induced errors from such random sources as power lines, signal wiring, motors and ballasts.

Control Circuits for Fuel-Fired Furnaces

Combustion-air blower control

Combustion-air blowers must be interlocked with the combustion-limit circuits to shut down the process in the event of failure. The flow of combustion air must always be proven before and during a processing cycle with two independent sources of information. The motor should be protected from short circuits with fuses and from overheating or amperage draw with thermal breakers (heaters). The motor starter should be wired so that it will disconnect when any phase is interrupted or when the motor malfunctions. It should not be assumed that the blower is providing combustion air just because the blower motor is operating; combustion-air flow must be proven. At one time, an end switch, or rotary switch, on the motor was a common indirect method of gaining this information. A better method is to use a pressure switch in the air line for direct sensing. A sail or flag switch, although not quite as good because of the mechanical movements required, also can be used to sense air flow directly.

Gas-pressure control

Fuel must arrive at the burner in the correct quantity and at the correct time for safe combustion. Fuel pressure thus must be proven within an allowable range. Gas-pressure switches for both high and low gas limits are installed in the main gas lines. Visual pressure gauges also are helpful to operators in setting burners and in verifying that the fuel is being supplied within the range desired and that pressure limit switches are not malfunctioning. Mercury-wetted relay pressure switches are recommended for their ease of set-up and maintenance and for their reliability.

Pressure regulators

Pressure of gaseous fuel is most commonly regulated by pressure-regulating diaphragms. Good, safe design normally requires one regulator for pilot fuel and one regulator for main-burner fuel. The pilot gas, if taken from the main fuel line, should be drawn from a point between the gas supply and the regulator for the main fuel. Thus, the pilot and main burner can be set up optimally, safely and independently. The regulators should be vented to a safe location outside the plant to ensure safety if a regulator diaphragm is damaged in service. Good practice and manufacturers' recommendations show that diaphragm life can be substantial if regulators are shielded from thermal radiation and are used below their maximum design limits. Positive lock-up regulators are recommended to prevent downstream pressure build-up during shutdown periods.

Valves

Blocking valves normally are closed valves that are energised only by the combustion-control circuits. The pilot-gas blocking valve is placed downstream of the pressure regulator and a hand-operated gas cock. A pipe union should be inserted just ahead of the electrically operated blocking valve to allow safe removal if repairs are needed. The blocking valve is opened to the pilot assembly only after the furnace is purged.

Purging of Fuel-Fired Furnaces

The furnace must be purged of any possible combustible materials. This is best accomplished by opening the furnace doors, which should be equipped with a limit switch to ensure that they are opened adequately. Once the doors are open, the combustion blower or exhaust fans can be timed to allow for a minimum of four changes of air in the combustion chamber. This purge cycle is required for safe start-up and is standard practice for all well-managed operations.

Pilot Control

Pilot assemblies can be of either the atmospheric or the blast type. The atmospheric type is similar to an atmospheric burner, in which the air is inspirated from the atmosphere by the gas stream.

In the blast type, air and gas are brought to a mixer under pressure. The gas is then reduced to atmospheric pressure and pulled into the mixer by the pressurised air stream. This is the most positive means of pilot-gas control.

Ignition

For ignition trials, a high-voltage transformer is used in conjunction with a spark plug designed for the pilot or burner assembly. The control circuit causes the pilot valve to open and a spark to be produced. The spark continues for a short period (normally 15 seconds) and establishes a flame, which can be detected. If the flame is not established, because the flame or signal is inadequate, the cycle returns to the purging stage.

The voltages normally employed are approximately 5000 to 6000 V and the high-voltage transformer is normally mounted on the furnace and grounded to it. The spark in turn is grounded to the pilot assembly, then to the furnace; hence, a well-grounded furnace is an important safety requirement.

Flame Detection

Thermocouple junction

A thermocouple junction placed in intimate contact with the pilot flame is perhaps the most common means of flame detection, but thermocouples are useful only on very small pilot assemblies or burners, and they are not useful after a burner becomes hot. The flame may no longer be present, but a hot burner block or refractory may retain heat and slow the rate of thermocouple cooling. Thus, thermocouple junctions are not recommended except for quench-tank heaters of the constant-pilot, open-grid burner design or for small atmospheric burners.

Flame electrodes

Flame electrodes which are small anodes of heat-resisting alloy placed in intimate contact with the normal pilot flame, work on the principle that flame causes ionisation within the burner atmosphere and thus allows a circuit to be formed to ground. The flow of a minute amount of current, at low voltage, is sufficient to sense and communicate the presence of a flame.

Flame electrodes are common on all industrial heat treating furnaces where the flame is kept on-ratio or slightly oxidising. The flame electrode tends to become carbon coated in a reducing flame, a condition that can cause nuisance shutdowns.

Ultraviolet scanners

Ultraviolet scanners are the third common device for sensing flame. They are normally dependable if the lens viewing the flame is kept clean. The UV scanners must not be used in any application where ultraviolet light is present from a source other than the burner in question. The UV scanner is a useful and practical device for any clean-flame, clean-furnace operation, if it is located and aimed properly. A flow of clean, filtered cooling air across the scanner face aids in keeping it clean and cool, extending scanner life appreciably.

Depending on the burner used, the application, and property-insurance requirements, it may be necessary to monitor both the main burner and the pilot flame independently.

Common and serious errors in flame detection are made by operators who circumvent flame-safety equipment rather than correcting the usually minor problems that cause nuisance shutdowns. Flame-safety equipment that uses totally enclosed relays is recommended in favour of types with accessible relays that may be kept open, for example, with a piece of paper. This point, however, trivial it may seem, has been profoundly recognised by those firms who have lost operators, furnaces and product as a result of poorly designed flame-safety equipment that can be circumvented easily. Any employee found tampering with this equipment should receive disciplinary action, and all employees should be trained in use of flame-safety equipment.

Burner Operation

The main fuel supply for fuel-fired heat treating furnaces normally is natural gas, propane-air, propane, butane or one of the fuel oils. Although this discussion centers on natural gas, the same principles apply to the other gases and oils.

The main gas valve may be of the manual-reset type, requiring an operator, or may be fully automatic. The manual type usually is preferred when the furnace is run intermittently or when operators must perform some other function, such as opening doors. When the operator opens the valve, he is in effect making a conscious decision that conditions are ready for the main burner heat. The valve may be made automatic when the furnace is designed and interlocked to preclude an unsafe condition.

For furnaces with capacities greater than 422 MJ/hr. (4,00,000 Btu/hr.), it is recommended that a second blocking valve be inserted into the main gas line and that a normally open vent valve be installed between the main blocking valves. The vent valve should be vented out of doors away from any openings such as windows, air intakes and exhaust louvers in the building walls or roof. This 'double block and bleed' arrangement prevents faulty valves or dirty valve seats from causing leakage of fuel into the furnace between operating cycles. It is not totally foolproof: the open vent may malfunction and allow valuable fuel to leak into the atmosphere during operations, or the internal mechanisms may malfunction because their design is unsuited to operating factors such as environmental conditions. Good preventive maintenance is required. One survey showed that 1 out of every 10 blocking valves was faulty after operating for no more than 10 years.

Burner control

The gas-air ratio ordinarily is controlled to about 10 parts air to 1 part natural gas for good combustion efficiency. There are several devices involved in control of this ratio. Typically, the amount of blower air is varied by a butterfly valve to satisfy the demands of a temperature-control device. A pulse or static pressure line is connected from the combustion air line, downstream of the butterfly valve, to a proportionator valve located in the gas line. The gas is then regulated by the ratio-control valve in proportion to the air flow, and the air-to-gas ratio remains constant throughout the firing range. The devices used to regulate the ratio fall into two broad categories: the diaphragm type, or proportionator, which uses the pulse line to keep air and gas at a specified ratio; and the mechanical-linkage type. Both are effective and common, but the diaphragm type is the more positive, because there are no linkages that can slip and require adjustment. Also, if air lines should become dirty, resulting in a lessening of air pressure, the gas pressure will follow, maintaining the correct ratio.

Ratio control alone is not sufficient to ensure safe start-up of the main burners. It is recommended that the burners be set to a low firing rate when the main burner is started. This may be done either automatically or manually. Once the main burners have been started, the furnace doors may be shut and the furnace brought up to temperature.

Temperature Control

Temperature-control devices fall into two categories: primary controls and process limiting devices. Safe operation, especially when furnace practices require long cycles and little operator attention, dictates that limits be placed on the process to ensure adequate alarm and perhaps to shutdown the operation to prevent destruction of the product, the furnace or the plant itself. Whether an analog device, a strip chart recorder, digital readout or a printout is used is a matter of operator preference and depends on the nature of the product.

The typical temperature sensor is either a thermocouple or a resistive temperature device (RTD). The thermocouple is most common. Several types of thermocouple junctions are available, with the choice depending on such factors as temperature range and furnace atmosphere. They are comparatively inexpensive and can be easily protected from atmospheres with protective 'wells', which are immersion tubes that project into the furnace zone to be controlled. RTDs, although more accurate than thermocouples by factors ranging from 10 to 1 up to 50 to 1, are expensive and less rugged. For most purposes, thermocouples are satisfactory. Some firms are using heat-flow sensing to remotely ascertain interior temperatures and to provide an element of redundancy for protection of furnaces and their contents. Good temperature-sensing devices will detect failure of a thermocouple or RTD, cause the process firing rate to be reduced to its minimum rate, and perhaps provide an alarm.

Furnace temperature can be regulated by one of two very common procedures. Simple high and low firing rates are used when temperature can be allowed to vary within a fairly large range. More common, in heat treating, is the use of proportional control, wherein the temperature is held nearly constant through the use of a bridge circuit. This circuit balances the signal between the controller and the butterfly valve and holds the latter at the proper opening to maintain the desired temperature. The latter scheme, although more costly, is required for close control.

Waste-Heat Recovery

Recuperative devices used for conserving energy present special problems in safety and control instrumentation. Typically, these recuperators use the products of combustion for pre-heating of the combustion air. Shell and tube heat exchangers are normally used in this type of arrangement. Because pre-heated air becomes less dense, the air temperature must be sensed, and control of the gas-to-air ratio must be based on this temperature-density function. Experience with these devices has shown that such factors as poor design, poor gasketing, leaks in heat-exchange surfaces, overheating of burner blocks and failure to allow for expansion and contraction have caused numerous operational problems that constitute safety hazards. Tracking of air-gas ratios can be affected by leakage of mechanical seals, and products of combustion can enter the combustion air. Thus, it is recommended that oxygen analysers be used periodically to check the combustion air immediately ahead of the burner block. This analysis will reduce the likelihood of erroneous and perhaps hazardous conditions in the furnace, will give clues to potential design changes needed and will give warning of part deterioration. Further, it is recommended that heat treating operators monitor the room atmosphere for carbon monoxide on a periodic basis. Although exhaust may be provided for products of combustion and sufficient air exchanges may be occurring to satisfy state regulations, there may be a temperature inversion that can cause leaking of products of combustion inside the building, resulting in a potentially dangerous situation.

Supervisory Gas-Cock System

A supervisory gas-cock system is used to ensure a safe lightoff procedure on a manually ignited, multiburner furnace that does not have flame-safety equipment and a programmed sequence of piloting the main burners.

The system consists of specially designed gas valves that have inlet and outlet passages for a checking pressure medium such as air or gas. Air or gas—usually air from a combustion blower—can pass only through the valve when the valve is fully closed. When the individual burner valves and the main gas line valve are closed, the air flow enters a pressure switch that closes and completes an electrical or pneumatic circuit. This allows the main gas valve, usually of the manual-reset type, to be opened. The burners are then individually manually ignited.

Supervisory gas-cock systems are used on radiant-tube furnaces and other furnaces where flame-safety systems are difficult to apply. Fewer of these systems are being used on new furnaces, because most burners now are adapted to flame safety and automatic ignition.

ELECTRIC FURNACES

Electric-furnace installations are made up of various electrical and mechanical components, many of which are water cooled and equipped with protective devices.

Furnace manufacturers generally issue instructions concerning safe practices, and these instructions should never be ignored. Potential hazards can be avoided by ensuring that operating personnel are trained thoroughly and that installations conform to safety practices and local codes.

Original equipment usually contains devices for preventing overloads and short circuits. In addition, ground detectors and surge detectors protect motor-generator units from faulty coil or transformer installations at heating stations and from breakdown of insulation in the generator windings.

Protective devices commonly used with induction-heating radio-frequency generators are as follows:

1. Door interlocks.
2. Grounding devices to ground high-voltage circuit when furnace doors are open.

3. Warning lights.
4. Warning signs.
5. Circuit breaker for entire unit.
6. Overload relays.
7. Water-flow switches.
8. Water-temperature switches.
9. Time-delay relays (tube warmup).
10. Grid overload relays.
11. Control-circuit overload relays.
12. Arc gaps on blocking and tank capacitors.
13. Surge protection.
14. Electronic crowbar.

Operators should become familiar with these safety devices and should inspect them periodically to ensure that they are in good working condition.

Electrical Power

Although motor generators account for the largest total power output of installed induction-heating equipment, vacuum-tube oscillators probably occur in the greatest numbers of units. Many small vacuum-tube oscillators are required to account for as many kilowatts as one 1250 kW, 3 kilocycle motor-generator set. Although many vacuum-tube oscillators for induction heating are made in small sizes, 25 kW and 50 kW outputs are also common ratings. Some have been constructed for special applications with ratings as high as 500 kW. Many small composite (custom-built, or home-made) vacuum tube units are in use also.

Power Interlocks

Most systems produced by reputable manufacturers are designed to be completely 'fail-safe'. These systems also have interlocks that automatically shutdown the power supply if a fault develops during operation. The system cannot be re-started until the fault is corrected. Interlocking systems also are used to increase production of induction heating machines.

Induction-heating machines are expensive; therefore, steps are taken to keep them busy as much of the time as possible. If a hardening process, for example, requires 5 seconds heating, followed by 5 seconds of quenching (before the part is moved from the inductor) and if another 5 seconds are required on the average for the operator to load and unload the part, then the generator itself is only in use one-third of the time. Production can be increased appreciably by having the same basic equipment supply two or more individual work coils (Fig. 7.1). This can be done by arranging the control circuit so that, if one of the work coils is demanding heat, none of the others can be started. If an operator pushes the start button at one station while another is heating, a relay withholds the actual start of heating until the first station has completed its high-frequency power demand cycle.

It is also customary in the case of interlocked multistation operation of motor-generator equipment to pre-set the alternator field current for the various stations. Each station has its own field-adjusting autotransformer or potentiometer, which is automatically switched into the circuit when that station has the power. This is feasible because, with only one station on at a time, it is not necessary to use the same voltage at each.

Fixtures

As introduction of automated systems increases in induction heating, the need for safety controls increases beyond the greater need for such devices as power interlocks.

For example, with highly automated induction-heating machines, a part completely foreign to the part to be treated may enter the automatic feed hopper or bin. If the fixture tries to feed this part into the coil, mechanical jamming and damage may result. If the part does get pushed into the coil itself, assorted problems may result, especially if the unwanted part is larger than the inside diameter of the coil.

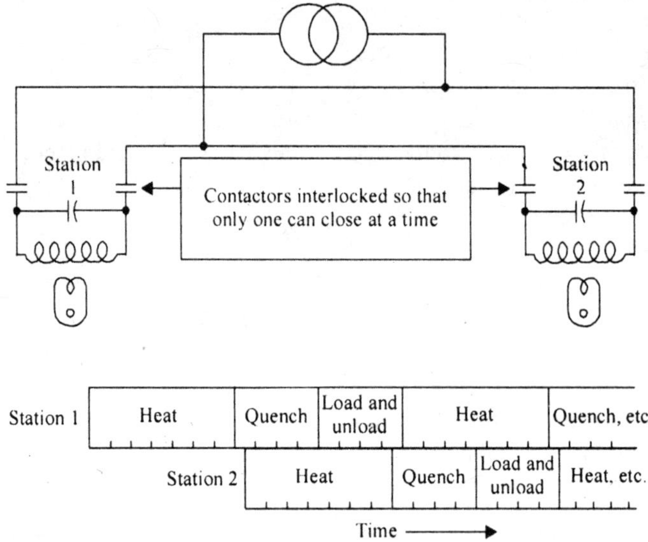

Assume that cycle is the same for both stations: heating, 7 seconds; quenching, 4 seconds; loading and unloading, 4 seconds.

Fig. 7.1. Interlocked multistation operation of two work coils on an induction-heating machine.

The greater the degree of automation, the greater the necessity for safety devices that further complicate the machine. Some automatic induction-heating machines incorporate templates through which the workpiece must pass before being fed to the more delicate parts of the mechanism. If the part touches the template, an electric circuit stops the machine until the error is manually corrected.

Spurious Radiation

Industrial heating equipment using radio-frequency energy is, in many instances, governed by rules of the Federal Communications Commission (FCC). FCC rules apply to induction-heating equipment operating at 10 kHz or above. Any operation whatsoever in frequency bands reserved for international distress equipment is strictly prohibited. One of these bands is from 490 to 510 kHz.

The best way for a user of induction-heating equipment to become familiar with the FCC rules is to acquire a copy of Part 18 of the rules and study them carefully.

Vacuum tube oscillators and some motor-generator equipment fall into the FCC-designated categories. The rules state that operation must either be within certain narrow frequency bands (in which any amount of energy may be radiated) or be restricted in field strengths. All such equipment must be certified by a competent engineer, such certification being based upon actual measurements of field strength made around the equipment. In some circumstances, prototype models of industrial high-frequency heating equipment may be tested at the manufacturer's plant and a certificate issued to cover other equipment of the same design. Even though a specific piece of equipment may have been properly certified, and even though its spurious radiation may fall below the prescribed limits, FCC rules state that, if it interferes with communications equipment, further corrective action must be taken. The mere existence of a certificate, therefore, does not necessarily absolve the user of further responsibility.

In terms of output, induction-heating machines sometimes rival the largest communications transmitters. The frequencies, and harmonics thereof, used by many induction-heating oscillators fall within the range utilised by their more delicate counterparts in radio and television services. If only a small portion of the power output of high-frequency heating machines were to be broadcast as unwanted (spurious) radiation, the results would be catastrophic.

Historically, dielectric heating machines have caused more interference than have induction-heating machines; they operate at higher frequencies and are more difficult to shield. However, induction-heating machines, especially vacuum tube oscillators, have also caused trouble. It is necessary to observe certain precautions in their design and operation to ensure that they do not create interference. Reputable manufacturers of induction-heating equipment take precautions to protect users of their equipment from this type of trouble. They house their equipment in heavy steel cabinets, and provide instructions which, if followed by the user, will ensure conformity with FCC rules. However, it must be emphasised that the final legal responsibility for a piece of equipment lies with the user.

Maintenance

Electrical heat treating equipment is expensive and stand-by equipment generally is not maintained. Thus, preventive maintenance is critical, and ready availability of replacement parts is highly desirable.

Dust, dirt, moisture and high ambient temperatures are the primary causes of electrical equipment failures; these conditions are commonly present in industrial locations where induction-heating units are installed. In any maintenance programme, warnings should be highly visible and clearly stated. The following is a typical warning.

'If the interlocks are disabled and the main circuit breaker is on with the door open, potentially lethal voltages are exposed. There is always 460 V AC present behind the control circuit breaker and on the line side of the main circuit breaker: care should be exercised at all times when the door is open. Power should be removed by opening the feed breaker or disconnect switch external to supply before working within cabinet. Solid-state circuit breaker board and isolator board are connected directly to 300 V AC. Turn off all breakers and allow one minute for capacitors to discharge before working on these boards'.

SPECIAL HEAT TREATING PROCESSES

Certain special heat treating processes using such systems as lasers, electron-beam heating, plasma carburising and ion nitriding have their own unique safety requirements in addition to standard safeguards associated with high-temperature processing.

Safety of personnel is paramount, but safety of equipment often ensures personnel safety. Thus, proper care and use of equipment cannot be over stressed and frequently becomes almost synonymous with safety.

In this section, ion nitriding and plasma carburising are used as examples of special processes and the safety precautions related to them. For all heat treating systems, however, special safety problems can be solved through sound training programmes for operators and through effective and regular maintenance.

Ion Nitriding

In this system, also known as 'glow-discharge' nitriding and as 'ionitriding', parts are connected to the cathode for processing, and the retort is the anode. After the retort is evacuated of atmospheric gases, nitrogen and hydrogen are bled slowly into it. The glow discharge is produced when the parts are heated by electric current to approximately 500°C (930°F), although specified temperatures can be as high as 565°C (1050°F).

The retort becomes heated by radiation from the parts; additional heat is not required. The glow discharge ionises the nitrogen and the electrical potential accelerates the movement of the ionised nitrogen toward the parts.

Although ion nitriding is faster and produces a more ductile and fatigue-resistant case with less white layer, the extra handling and precautions it requires is an important factor in over-all cost.

Plasma Carburising

The normal range of electrical power used for plasma-arc processing is 25 to 50 kW. Most systems are of 'fail safe' design and are interlocked to shutdown the power supply automatically if a fault develops during operation.

One of the most serious hazards associated with plasma-arc operation is radiation caused by electromagnetic high-electron excitation. Such radiation ranges from radio frequencies to the far ultraviolet, and it includes infrared and visible radiant-energy light rays.

The radiation produced by the plasma is capable of producing severe eye and skin burns. The plasma should never be observed with the naked eye.

Furnace Protection

The primary safety feature of a surface ion nitriding furnace is the arc-control system. Successful ion processing requires application of 300 to 1000 V DC to the workpiece. In these voltage ranges, the potential for an arc to form between the cathode and the anode is quite high during the initial part of the cycle. These arcs can be quite small; under certain circumstances, however, major arcs can occur that may be powerful enough to rupture vessel walls.

One method of controlling arcing and protecting equipment is to place a large resistance in series between the furnace vessel and the power supply. This prevents overloading of the power supply. An electrical device used to construct a type of 'arc-shutdown' circuit is a Saturable Core Reactor (SCR), which operates with a low-resistance load at steady voltage.

When an arc begins to form there is an initial rapid increase in current prior to formation of the arc. Upon arc formation, the voltage drops drastically (dv/dt); the SCR senses the voltage change (dv/dt increase) and increases the resistance, thus protecting the power supply. Additionally, the SCR is normally connected to another electrical device, such as a bridge rectifier. Once the change in voltage signifying arc formation is detected, the rectifier damps the current supply. This damping is normally sufficient to shutdown the arc by allowing re-distribution of the energy on the portion of the part that was arcing.

This type of circuit reacts after the arc has formed, however, and the potential for ruined workpieces or holes in anodes or vessel walls remains. This is a particular problem if the unit requires operators to manually shutdown the power supply after observing dead shorts, formed by misloaded parts that create short circuits.

Arc Suppression

Some equipment used in ion nitriding and plasma carburising does not rely on an arc-shutdown circuit as the primary safety factor. Rather, true arc-suppression circuits are employed. Such a circuit senses the change in current just prior to the formation of an arc and shuts the power off completely, preventing the arc from forming. The power is then proportionally ramped back on, allowing re-distribution of energy.

In one system, a counter in the controlling microprocessor tallies the number of times the power is turned off and on.

If the potential for arcing is too large, as determined by the logic pre-programmed into the microprocessor as part of the executive command package, the system shuts down, prints a fault message on both the cathode-ray tube and data logger, sounds an alarm and siren, and lights a warning light. All this occurs automatically without an arc forming and protects both the part and the equipment. The microprocessor detects a dead short by sensing and reporting the rapid frequency of shutdown/start-up cycles and prints a different 'short' warning message, also with alarms, siren and lights. This sequencing is also totally automatic and requires no operator interface. The equipment has a back-up system that operates on the dv/dt principle. Additionally, all leads through the vessel wall that could carry current to the power supply are triple-protected. As an example, the thermocouple has primary protection through an outer ceramic insulator. This is followed by a second powder ceramic insulator. The third protection is a high-voltage isolating amplifier between the thermocouple and the processor.

Thermocouples

The problems associated with passing a thermocouple, or any other lead, through a vessel wall in ion-processing equipment are as follows:

1. The lead can become metallised, creating a pathway for catastrophic arcing. This problem becomes increasingly severe as a function of vessel use.
2. The material of construction used for seals through the vessel wall is quite critical. The conditions of ion processing can affect the sealing materials, allowing increased leakage as a function of vessel use.
3. Temperature uniformity throughout the entire workload in the vessel is quite critical. Proper design of thermocouple insulators that are calibrated to give true temperature readings, as well as proper design of fixturing, is necessary to achieve uniform temperatures.

Fixture Design

Design of the fixture is critical to the successful application of ion processing. Poorly designed fixtures can allow overheated or underheated parts. If equipment without an arc-suppression circuit is used, parts can be ruined because of poor fixture design. The fixture often allows simple masking and through proper design, can minimise or eliminate the hollow-cathode effect. This effect is signified by either: (i) failure of the glow to uniformly penetrate the interior surface of a hole or cylinder; and (ii) overheating and possible melting of the part, caused by overlapping glows.

ATMOSPHERE FURNACES

Atmosphere furnaces must be considered in any discussion of furnace safety because of the potential explosion hazard produced by introduction of special flammable atmospheres.

Although many of these furnaces are supplied with 'inert' gas purging and stand-by emergency purging, training of operators in manual 'burn-out' procedures is extremely important in the event of failure of automatic controls. These emergency instructions may vary with equipment design, and thus the importance of consulting and understanding emergency procedures, as outlined by the furnace manufacturer's instructions, should not be minimised.

Protective Controls

Protective devices should be installed and interlocked and should include the following:

1. A safety shut-off valve on the atmosphere supply line to the furnace.
2. An atmosphere gas-supply monitoring device that permits the operator to visually determine the adequacy of atmosphere gas flow at all times.
3. A sufficient number of temperature-monitoring devices to determine temperature in all zones of the furnace; these should be interlocked to prevent opening of the atmosphere-gas-supply safety shut-off valve until all zones are at or above 760°C (1400°F).
4. An automatic safety shut-off valve for flame curtain burner supply gas; this should be interlocked to prevent opening of the valve when furnace temperature is below 760°C (1400°F).
5. Audible and/or visual alarms to alert the furnace operator of abnormal conditions
6. Manual door-opening facilities to permit operator control in the event of power failure.

Operator Training

The most essential safety consideration is the selection of alert and competent operators. Their knowledge and training are vital to continued safe operation. New operators should be instructed thoroughly and required to demonstrate an adequate understanding of the equipment and its operations.

Regular operators should receive scheduled retraining to maintain a high level of proficiency and effectiveness, and all operators should have ready access to operating instructions at all times. An outline of these instructions should be posted near the furnace.

Operating instructions generally are provided by the equipment manufacturer, and these instructions include schematic piping and wiring diagrams. All such instructions should include procedures for light-up, shut-down, emergencies and maintenance.

Operator training should include instructions in:

1. Combustion of air-gas mixtures
2. Explosion hazards
3. Sources of ignition and ignition temperature
4. Atmosphere gas analysis
5. Handling of flammable atmosphere gases
6. Handling of toxic atmosphere gases
7. Functions of control and safety devices
8. Purpose and basic principles of atmosphere-gas generators.

This listing is intended only to serve as a guideline.

PROCESS COOLING

Heat treating of metals includes controlled cooling or quenching of the heated metal; metals are cooled from the specific treatment temperature in a variety of media which include air, oils, salts, water and synthetic fluids.

As a general rule, furnace equipment does not include instrumentation for control of the safety aspects of the quench media. Normal practice in layout of plant and facilities will provide for isolation of air-cooling areas, for the usual pedestrian protection at pits and for the necessary building protection should an uncontrolled conflagration occur due to the quenching operation. However, certain equipment can be specified to ensure safe and controllable operation in specific cooling processes.

The greatest concern exists for fires associated with oil quenching. All of the ingredients for a dangerous fire—fuel, oxygen and a source of ignition—exist at the surface of an oil tank.

The most common type of fire occurs when movement of a hot workpiece is obstructed as it enters the quench oil. The result is sustained ignition and vapourisation that continues as the liquid is locally heated above its flash point. Prompt immersion removes the source of local vapourisation, and local flashing is extinguished by normal agitation of the oil.

A second type of fire occurs when the main body of oil is heated above the flash point because of malfunction of heating or cooling equipment, or when the quench load is greater than that for which the system was designed. When an ignition source is supplied, the resultant fire soon reaches full intensity and is very difficult to extinguish.

A third and less likely type of fire occurs because of material-handling accidents that involve spills on or near heated furnaces or cooling equipment.

Equipment is available for detection of fires and release of control media. Automatic water-spray systems are usually recommended in buildings of fire-resistant or non-combustible construction, and areas adjacent to quench-oil tanks can be protected with automatic sprinklers. Quench oil and water should not be mixed, however. High-value oils can be protected with automatic carbon dioxide or dry chemical systems. In general, these automatic systems are specified for centralised large-capacity quench-oil systems.

Special equipment is available for totally enclosed systems that operate in special atmosphere furnaces. Because water and quench oil do not mix, however, this incompatibility can be a source of trouble in these units. Water in oil rapidly turns to steam when locally heated beyond 100°C (212°F). This steam can cause violent boilover, increased pressure within the enclosed system and forcible ejection of burning oil from small openings. Commercial safety equipment capable of detecting even small quantities of water in oil can be arranged to alert the furnace operator and to interrupt the quench process.

Quench-tank heating systems should be equipped with all of the safety devices normally used in conjunction with the particular heating method chosen. Overtemperature safety systems are essential. In addition, system interlocks between heating media actuators and both agitators and pumps will prevent local overheating of the bath.

MECHANICAL EQUIPMENT

Material handling cannot be separated from other safety considerations associated with heat treating operations. Many mechanical operations must be performed before, after and sometimes during the actual temperature-induced transformation that usually occurs while the work is in the heat treating furnace. Doors must be opened and closed, conveyors started, rolls advanced and mechanical handling equipment activated. All of these mechanical operations constitute hazards of varying degrees of severity that must be evaluated. Some are serious enough to warrant the introduction of safety equipment to preclude serious injury of personnel and damage of expensive furnace equipment.

Furnace doors are often interlocked with other components of a facility through use of limit switches that prevent inappropriate opening or closing. For instance, in an atmosphere furnace, inner doors cannot be opened until proper vestibule ambient conditions have been restored after parts are removed. Interlocks on doors may include complex designs to prevent inadvertent opening during a power failure, on a hydrogen atmosphere furnace. There are simpler designs that prevent closing of a door on a simple normalising furnace before the extractor is removed. Some door interlocks are connected to pressure or temperature-sensing devices. Such devices are used with vacuum furnaces in which inadvertent exposure of the molybdenum-graphite heating elements to air at high temperature would be disastrous. Moreover, without door interlocks, roller hearth or walking-beam furnaces could advance workpieces into an unopened discharge-zone door.

Moving parts of furnaces represent potential hazards that can be neutralised rather easily with simple time-delay relays connected to audible alarms, which in turn are connected to the start buttons of electric motors. For instance, the conveyors on some large continuous furnaces cannot be advanced until an alarm has been sounded and a timer has allowed sufficient time for workmen to stand clear.

SECTION II

Thermomechanical Treatment

Chapter 8

Low and High Temperature Thermomechanical Heat Treatment

INTRODUCTION

Plastic deformation can change the pattern of distribution and increase the density of imperfections, such as dislocations, vacancies, stacking faults, low- and high-angle boundaries in a crystal structure. Since crystal lattice defects produce a strong effect on the formation of structure in alloys during phase transformations, plastic deformation before or during phase transformations can be used to form the optimum structure in a heat-treated alloy.

Thermomechanical treatment is a kind of heat treatment including plastic deformation which increases the density of defects and thus affects the formation of the structure in the phase transformations occurring during the thermal action.

Consequently, not every combination of straining, heating and cooling can be called thermomechanical treatment. For example, if plastic deformation is carried out after all operations of heat treatment, we are dealing with common heat treatment followed with plastic working rather than with thermo-mechanical treatment. Such plastic deformation, for instance, cold rolling after ageing, may produce strain-hardening and increase strength properties but has no effect on the structure formed through phase transformations, since these transformations have taken place before the plastic deformation was applied.

If plastic deformation was effected prior to heat treatment and produced no decisive effect on the final structure formed in an alloy during phase transformations, it should be regarded as a simple combination of plastic deformation and heat treatment, rather than thermomechanical treatment. For example, cold rolling followed by heating for quenching, during which re-crystallisation takes place, are not components of thermomechanical treatment, since the re-crystallised structure has a low density of crystal imperfections.

The processes of plastic deformation and heat treatment in thermo-mechanical treatment may be either combined in a single technological operation (i.e., carried out simultaneously) or made with a time interval of up to a few days one after the other. It is only important that phase transformations take place under the conditions of increased density of the lattice defects formed through plastic deformation.

At present, various methods of thermomechanical treatment are being used or tested in the industry; all of them include plastic deformation which produces a decisive effect on the formation of the structure in alloys on ageing or during polymorphic, mainly martensitic, transformations.

As has been noted, thermomechanical treatment (thermoplastic working) is a combination of two methods of strengthening—plastic deformation and phase transformations.

As applied to steels, thermomechanical treatment (TMT) consists in strain hardening of austenite followed by the transformation of austenite into another structure.

Processes of Thermomechanical Treatment

The most common processes of thermomechanical treatment are as follows (Fig. 8.1).

High-temperature thermomechanical treatment (HTMT)

In high-temperature thermomechanical treatment (HTMT), austenite is deformed at temperatures above the re-crystallisation point (t_r in Fig. 8.1), usually above the critical points and then cooled quickly.

Low-temperature thermomechanical treatment (LTMT)

In low-temperature thermomechanical treatment (LTMT) austenite is deformed at a temperture below t_r, i.e., in the unstable state of undercooling below the critical points and then cooled to cause phase transformations.

Preliminary thermomechanical treatment (PTMT)

In preliminary thermomechanical treatment (PTMT), the steel is strain hardened prior to being heat treated and then heated up quickly to the austenitic state and cooled, but so as to retain to some or other extent the strain-hardening effect.

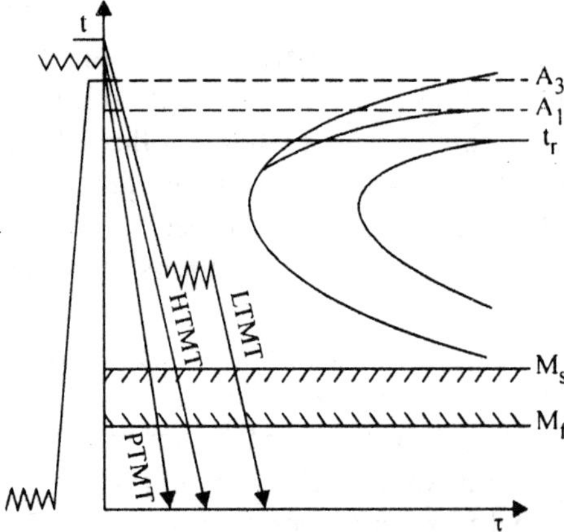

Fig. 8.1. Classification diagram of thermomechanical treatment.

There are also many other varieties of thermomechanical treatment differing in the conditions of heating or cooling, type of deformation and other minor details.

In all types of thermomechanical treatment the changes that are caused in austenite by plastic deformation are transferred (either partially or completely) to martensite.

Thus, a steel is strengthened due to the martensitic reaction and also due to the structural defects produced in deformed austenite and inherited by martensite.

A typical procedure of thermomechanical treatment is low-temperature thermomechanical treatment. Austenite is undercooled to 500-600°C and deformed to cause the strain-hardening effect. A heavier deformation, and therefore, a greater strain-hardening effect, produce a higher strength of martensite (Fig. 8.2). Very heavy deformations (80-90 per cent) are employed to obtain the maximum strength. Since deformation is made at a low temperature (substantially lower than the common temperatures for hot working), the steel retains a high ductility (more strictly, a high resistance to deformation), so that the necessity of a heavy deforming is associated with certain technological difficulties.

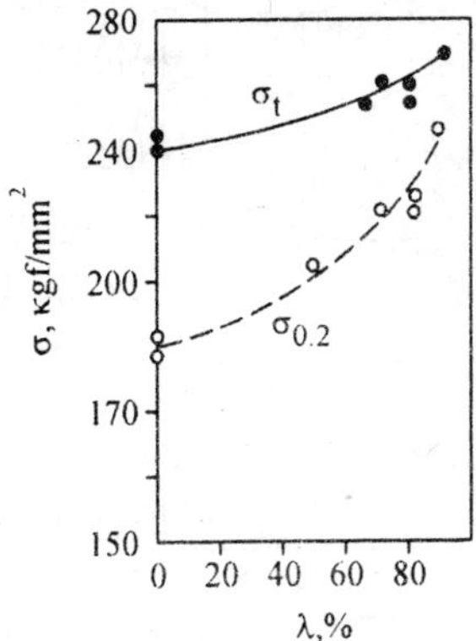

Fig. 8.2. Effect of deformation (λ) in low-temperature thermomechanical treatment on strength of steel (0.3% C, 2.2% Cr, 1.5% Ni, 0.3% V, and 0.3% Mo).

THERMOMECHANICAL TREATMENT OF AGEING ALLOYS

Fig. 8.3 shows the principal schemes of thermomechanical treatment (TMT) applicable with ageing alloys. The toothed lines denote plastic deformation.

Low-Temperature ThermoMechanical Treatment (LTMT)

Low-temperature thermomechanical treatment (LTMT) of ageing alloys is the oldest and most widely used method of thermo-mechanical treatment. It is mainly used to enhance the strength properties of alloys. With LTMT, an alloy is first quenched by a common technique and then subjected to cold deformation prior to ageing (Fig. 8.3).

As compared with ageing without preliminary deformation, LTHT yields higher values of ultimate strength and yield strength but lowers ductility indices.

Fig. 8.4 shows how the degree of cold deformation can affect the hardness of a quenched nickel alloy (curve 1) and of the same alloy after deformation and ageing (curve 2).

The hardening effect produced by LTMT may be explained by two causes. Firstly, the cold deformation produces strain-hardening, so that the subsequent precipitation hardening begins from a higher initial hardness level of an alloy. Secondly and most important, cold deformation enhances the effect of precipitation hardening. For instance, the hardening effect produced in Nimonic-90 alloy through ageing at 450 °C without cold strain-hardening is very low—only 15 kgf/mm^2. With increasing degree of cold deformation, the hardening effect on ageing increases continuously (see the deverging curves 1 and 2 in Fig. 8.4. With 90 per cent reduction, the gain in hardness produced by ageing has amounted to 175 kgf/mm^2. Consequently, cold strain-hardening in the case considered has enhanced the hardening effect on ageing by one order of magnitude. Such a strong hardening effect on LTMT, compared with hardening by a common technique (quenching followed with ageing) is a relatively rare phenomenon and may be explained by the fact that an ageing temperature of 450°C is too low for the Nimonic alloy, so that the precipitation from the supersaturated solid solution in the absence of cold strain-hardening can proceed only very feebly. If ageing upon quenching is done at a temperature that is optimal for the maximum hardening effect (around 700 °C), the effect produced by cold strain-hardening will be appreciably weaker.

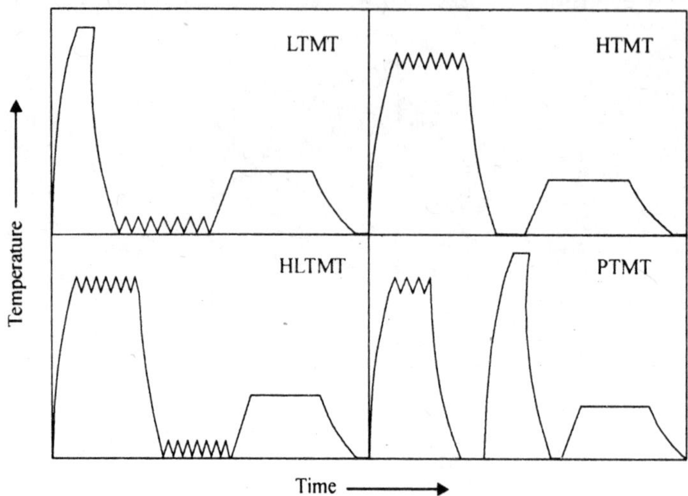

Fig. 8.3. Schemes of thermomechanical treatment of ageing alloys.

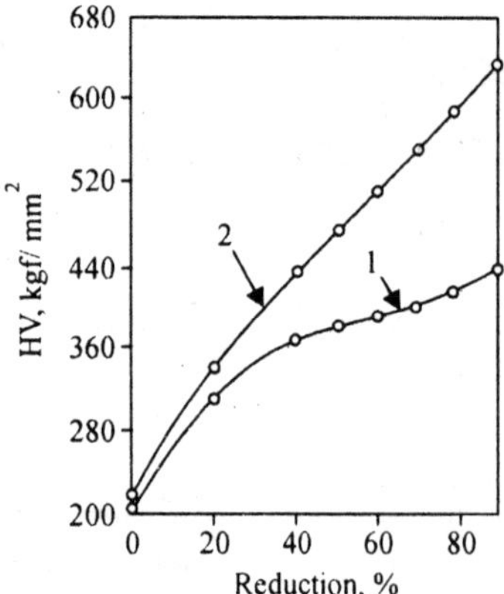

1. Cold-drawn.
2. Deformed and aged at 450°C for 16 hr.

Fig. 8.4. Effect of reduction in drawing after quenching from 1000°C on hardness of cold-drawn and aged 4 mm wire made of Niminic-90 alloy.

In the first approximation it may be stated that cold strain-hardening by increasing the density of imperfections in crystals of a supersaturated solution makes the latter less stable thermodynamically and accelerates ageing. Experimental evidence and a more detailed analysis show, however, that the effect of strain-hardening on ageing may be quite complicated. The nature of this effect depends on the conditions of quenching, straining and ageing, the nature of alloy and for the same alloy, on the type of precipitates in ageing.

Let us first consider the effect of cold deformation on zone ageing. It might seem that deformation should accelerate zone ageing, since it increases the density of dislocations and the concentration of vacancies. It should be noted, however, that, firstly, zones nucleate homogeneously, but not at dislocations, and secondly, dislocations are effective sinks for vacancies. A very strong plastic deformation can increase vacancy concentration (the ratio of the number of vacancies to that of atoms) only by 10^{-6}, whereas quenching can produce an appreciably greater supersaturation of the lattice with vacancies: the concentration of quenching vacancies may reach 10^{-4}.

The 'sweeping off' of a large number of quenching vacancies by gliding dislocations during plastic deformation of a quenched alloy can fully outweigh the relatively small supply of vacancies which are generated by jogs of gliding dislocations. As a result, plastic deformation of a aquenched alloy slows down zone ageing, since excess vacancies play an important part in the diffusion transfer of atoms of the dissolved element to the growing zones.

In alloys having a low concentration of a dissolved element, cold strain-hardening can slow down the formation of zones on ageing, since atoms of that element are attracted to dislocations and thus the matrix is depleted of that element.

A slower growth of zones in a strain-hardened alloy manifests itself in decelerated hardening and a lower gain in strength on ageing. Such a result is obtained, for instance, on cold deformation of a freshly quenched duralumin. That is why zone ageing is not commonly used as the final operation in LTMT.

If the metal is cooled at a relatively moderate rate on quenching, for instance, in air, and the lattice is not supersaturated strongly with quenching vacancies, the strain-hardening may cause no decelerating effect on zone ageing. Moreover, strain-hardening can then even accelerate zone ageing, since it increases the concentration of mobile vacancies owing to the fact that vacancy clusters that have had time to form during the slower quench cooling are being destroyed by dislocations. Besides, accelerated zone ageing under such conditions may also result from the generation of vacancies by jogs of gliding dislocations. In such cases LTMT with zone ageing can be employed for additional hardening of alloys.

Of incomparably greater interest is LTMT with phase ageing. Dislocations facilitate the nucleation of phase precipitates. Because of this, phase ageing proceeds quicker in a strain-hardended alloy having an increased dislocation density, and therefore, the density of precipitates and correspondingly, the hardening effect turns to be higher than on ageing of the alloy without strain-hardening. The higher the degree cf cold deformation, the greater will be the dislocation density, and correspondingly, the higher will be the density of precipitates and the hardening effect on phase ageing (Fig. 8.4).

The effectiveness of LTMT is decided by the strengthening phase that precipitates on ageing. For instance; the additional hardening effect produced by straining prior to artificial ageing is greater in Al-Cu-Mg alloys (where the strengthening phase is S′) than in Al-Cu alloys (where the strengthening phase is θ′).

Usually, heating for ageing after cold deformation is not accompanied by re-crystallisation with only recovery and polygonisation processes developing, which diminish somewhat the hardening effect provided by LTMT. It should be born in mind that these processes and those of precipitation from the solution can mutually influence each other: precipitates retard the polygonisation, while the latter if it has had time to proceed, varies the density and pattern of distribution of precipitates.

The effect of cold strain-hardening may be not only in accelerated precipitation from the solution and increased density of precipitates. It can also cause the formation of a phase that cannot precipitate under the same ageing conditions without strain-hardening. It even may be capable of altering the sequence in which the phases appear on ageing.

For instance, during ageing of Al + 4 per cent Cu alloy at 150 °C without strain-hardening, θ′-phase can be detected radiographically in 15 days, while θ′-phase ($CuAl_2$) does not appear at all. In ageing at the same temperature after cold rolling with 90 per cent reduction, θ-phase appears first (in 5 minutes) and soon after it, θ′ phase (in 30 minutes). The appearance of the metastable phase after the precipitaion of the stable one contradicts the common sequence of formation of phases on precipitaion from solid solution and may be explained as follows:

Cold deformation with a high reduction distorts the lattice of the solid solution in the whole bulk of crystals to such an extent that semi-coherent θ′-phase precipitates cannot nucleate. On the other hand, nucleation of θ-phase precipitates is facilitated in portions which have been distorted especially strongly. After a certain while, the processes of recovery and polygonisation which take place at the temperature of ageing make the lattice in some portions of crystals so perfect that nucleation of semi-coherent θ′-phase precipitates becomes possible there (provided that precipitation with the formation of θ-phase still has had no time to proceed in those portions). The formation of incoherent precipitates of a stable phase instead of semi-choerent precipitates of the metastable phase during precipitation from a strongly strain-hardened solid solution can diminish the hardening effect on ageing. For this reason, the strength of aged alloy after a heavy deformation may turn to be lower than that of a less deformed alloy, notwithstanding the higher strength level of the former before ageing.

If the solid solution has underwent partial precipitation before cold deformation, this may have an effect on the kinetics of final ageing and on the properties of the alloy. As shown in Fig. 8.5, with cold deformation of freshly quenched Al + 4 per cent Cu alloy, the maximum of hardness in ageing at 160 °C is obtained in 20–30 hours (curve 1), but if the same deformation is applied upon natural ageing, the maximum hardening in ageing at 160 °C is attained in 8–10 hours. This accelerating effect on artificial ageing may be explained by the fact that dislocations are stopped by G.P. zones during plastic deformation of a naturally aged alloy, so that on subsequent artificial ageing there are copper-enriched portions near the dislocations, which facilitate the nucleation of θ′-phase. Consequently, it is advantageous to deform alloy after natural ageing in order to accelerate the subsequent artificial ageing.

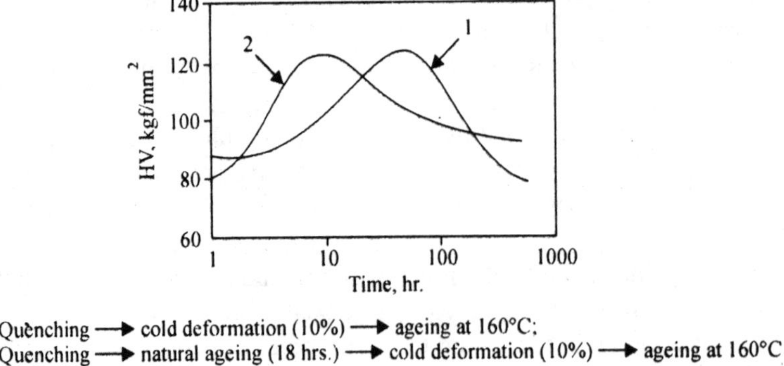

Fig.8.5. Dependence of hardness of Al + 4% Cu alloy on time of ageing at 160 °C after LTMT by various methods.

Pre-ageing (prior to plastic deformation) cannot only accelerate the final ageing, but also enhance the final hardening effect. One of the probable causes may be that plastic deformation of a preliminarily aged alloy forms a more favourable structure characterised by a high density of uniformly distributed dislocations.

In view of what has been said above, a time interval must be allowed between quenching and cold deformation in LTMT of naturally ageing alloys. For instance, in LTMT of sheets and tubes made of A31 alloy, this interval must be at least 2 hours in order to obtain the maximum hardening effect.

Pre-ageing can be either natural or artificial. It may be recommended to try on various alloys more complicated methods of LTMT of the type of quenching → ageing → cold deformation → ageing.

An increased strength can also be obtained by using common cold deformation after ageing (without LTMT). As compared with this common method, LTMT ensures a higher ductility for the same strength, as also lower residual stresses and a structure possessing a higher thermal stability.

At present, LTMT is widely used in the manufacture of semi-products and products from ageing copper, aluminium and austenitic alloys.

Cold rolling of beryllium bronze before ageing increases the yield limit of the metal additionally by 20 per cent. Electrode copper alloys of the type of chrome-zirconium bronzes can partially retain the hardening effect produced by LTMT at elevated temperatures at which contact-welding electrodes operate. The degree of cold deformation of ageing copper alloys in LTMT may be from 10 to 80 per cent.

With LTMT of austenitic spring alloys, with deformation of 50 per cent, their yield limit $\sigma_{0.002}$ turns to be 20–30 per cent higher than in the case of quenching and ageing.

For some types of aluminium alloys, LTMT with deformations is very promising. For instance, LTMT of sheets and tubes of alloy by the scheme: quenching → natural ageing for at least 2 hrs. → 60–90 per cent cold deformation → ageing at 200°C, can increase the ultimate strength by 25–30 per cent compared with that obtained on common heat treatment (quenching + ageing).

LTMT can be resorted to not only for increasing the ultimate strenght. Aluminium alloys of the type AMr6 are recommended to be cold-deformed by 30-40 per cent, followed with ageing at 220°C for few hours. In this process, polygonisation takes place in the first 20 minutes of heating after the deformation, after which β-phase (Al_3Mg_2) precipitates uniformly at sub-boundaries. This sturcture possesses an increased resistance to stress corrosion.

Semiproducts (shaped sections, tubes, sheets, etc.,) made of ageing aluminium alloys are always straightened by stretching or planishing after quenching. The degree of deformation in straightening is not high, usually not more than 3 per cent. However, even a low cold deformation, 1-3 per cent, can strongly enhance the hardening effect in subsequent artificial ageing. For instance, the yield limit of duralumin aged at 190°C can be increased from 40 to 45.5 kgf/cm^2 by stretching by 1.5 per cent before ageing.

The process by the scheme: quenching → stretching → ageing is not usually called thermo-mechanical treatment, though it actually is a typical low-temperature thermomechanical treatment.

Owing to simple techniques and high effectiveness of the process, the number of various ageing alloys and the assortment of articles subjected to LTMT will widen in the future. However, when resorting to LTMT, one should always consider the possible adverse effects, such as a reduced ductility which is characteristic of most alloys, a lower creep resistance in some aluminium alloys, anisotropy of properties etc.

Low-temperature thermomechanical treatment has found no wide application which is not surprising if we recall that it increases the strength at the expense of ductility.

HTMT is not a typical procedure of thermomechanical treatment, at least because the strain-hardening in austenite is not retained in pure form up to the martensitic transformation. As follows from its definition, HTMT is effected above the point of re-crystallisation, so that re-crystallisation processes are likely to occur if the steel is not cooled immediately below t_r (which is difficult to do in practice).

Practically, and this is not so bad, there is a pause, i.e., a time interval from the end of deformation to the beginning of quenching during which re-crystallisation of austenite takes place. The best results are obtained when the pause is sufficient to complete the first stage of re-crystallisation, i.e., to remove the strain-hardening effect and form fine grains of re-crystallised austenite. A longer pause (holding time) can result in grain coarsening and impairment of steel properties. The time of the pause evidently depends on steel composition, temperature, deformation and other factors. Since this type of HTMT produces no strain-hardening effect in the deformed metal, there is no strengthening (increasing of σ_t, $\sigma_{0.2}$ and hardness) in the final steel, but the refinement of grain lowers the ductile-to-brittle transition temperature and increases the percentage of ductile componenet in the fracture in high-strength steels, which results in better indices of ductility (ψ) and toughness (a_n) in conventional tests (Fig. 8.6).

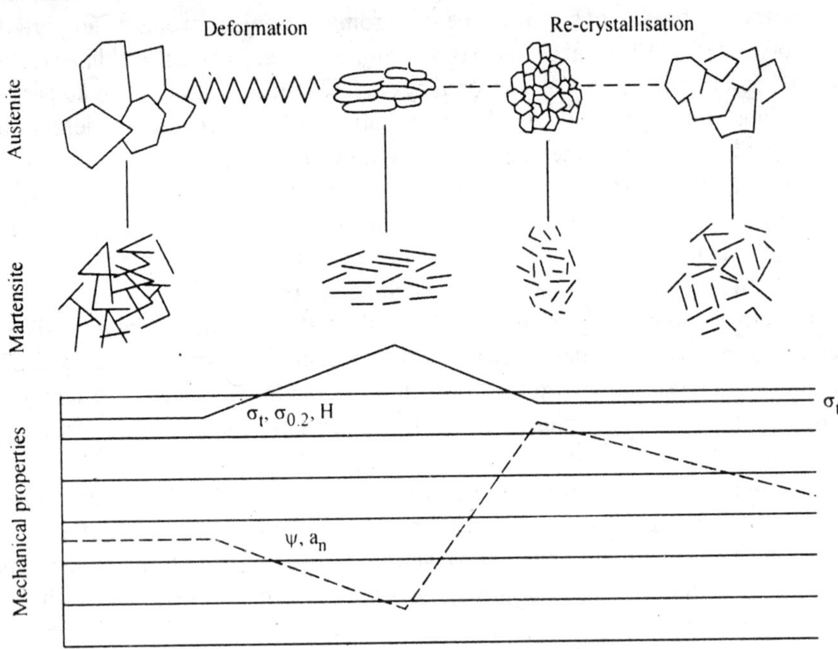

Fig. 8.6. Effect of high-temperature thermomechanical treatment on mechanical properties of steel.

This effect of HTMT can be obtained at a relatively slight deformation (from 20 to 40 per cent).

High-temperature Thermomechanical Treatment (HTMT)

High-temperature thermomechanical treatment includes hot deformation, quenching from deformation temperature, and ageing .

Hot deformation increases the density of dislocations and produces hot strain-hardening effect, which can be partially or fully removed owing to the development of dynamic polygonisation and dynamic re-crystallisation.

The stress-strain curve of the process has an ascending section of creep stress, which corresponds to the stage of hot strain-hardening, and a descending stress section, which is due to the development first of polygonisation and then of re-crystallisation.

With dynamic polygonisation (as also with the common static polygonisation which is observed in heating after cold deformation) the formation and migration of low-angle boundaries are controlled by dislocation climb. In contrast to static polygonisation, however, dislocations are continously 'pumped' into the body of sub-grains under the action of the stresses applied in the course of hot straining. A similar difference exists between static and dynamic re-crystallisation. In the course of hot deformation, the processes of hardening (increasing the dislocation density) and softening (decreasing the dislocation density in polygonisation and re-crystallisation) alternate continuously.

Depending on the nature of alloy, temperature, rate, degree and pattern of straining, the metal at the end of hot deformation may be either strain-hardened or have a polygonised, re-crystallised or mixed-type (partially polygonised and partially re-crystallised) structure.

A fully re-crystallised structure with the minimum dislocation density corresponds to the most stable state. If the structure has been not re-crystallised, or has been re-crystallised only partially at the end of hot deformation, there forms a stimulus to static re-crystallisation. Upon straining with a high degree of deformation, a very quick re-crystallisation without any induction period may be observed, since the nuclei of re-crystallised grains have fromed already in the course of hot deformation. This process is called metadynamic re-crystallisation.

The aim of HTMT is to form a supersaturated solid solution with non-recrystallised structure upon hot deformation and quenching, i.e., a structure having an increased density of imperfections (sub-grain boundaries and free dislocations). An alloy of such a structure obtains better mechanical properties on subsequent ageing. In most cases the polygonised matrix of the quenched alloy is the best choice. When using HTMT, at least three requirements must be fulfilled: (i) a non-recrystallised structure must form at the end of hot deformation; (ii) probable recrystallisation upon hot deformation must be prevented; and (iii) the degree of supersaturation of the solid solution must be sufficient for subsequent ageing.

If the first two requirements have not been fulfilled and the quenched alloy is fully re-crystallised, the actual process is quenching from the temperature of deformation heating, rather than HTMT. Such a process, which combines hot deformation with heating for quenching, is profitable, but connot improve the properties of the metal compared with a common heat treatment in which the metal is specially heated for quenching. The formation of polygonised sturcture at the end of hot deformation is favoured by increased temperature, reduced rate and degree of deformation, and by the deformation pattern approaching uniform compression (as in pressing).

The use of HTMT is limited by the factors as follows:

An alloy may have a very narrow range of temperatures of heating for quenching, so that it is practically impossible to maintain the temperature of hot plastic working in the same narrow range.

The optimum temperature range of hot deformation may be substantially lower than the range of temperatures of heating for queching. For instance, the rate of deformation in pressing of aluminium alloys can be the lower (without the appearance of surface cracks), the higher is the temperature. If it is required to increase substantially the temperature of hot deformation (to the temperature level of heating for quenching), the rate of deformation must be inevitably reduced, which lowers the productivity.

The precipitation from the undercooled solution in an alloy of non-recrystallised structure can proceed more quickly than in one fully re-crystallised. If such alloy has a low hardenability, the latter turns to be even lower under the conditions of HTMT, so that thorough quenching will be impossible.

With an alloy having a high susceptibility to re-crystallisation, it may be difficult to form unre-crystallised sturcture to the end of deformation and retain it in the period between the end of deformation and the moment when the temperature will drop down to t^s_r in cooling.

HTMT is used to improve the strength properties of alloys compared with the properties they have in re-crystallised state. The strength increases on HTMT owing to the total increase in the density of imperfections already in as-quenched state and to a more uniform precipitation from the supersaturated solution in the body of grains in ageing (the strengthening phase precipitates at sub-grain boudaries and at individual dislocations inside sub-grains).

A more uniform precipitation from the solution and very fine grains with curved boundaries, which are typical of many alloys, ensure a high level of ductility upon HTMT. In contrast to LTMT, which noticeably lowers the ductility, the additional hardening effect produced by HTMT is obtained with the ductility being virtually unchanged. In aluminium alloys, the ductility and impact toughness can even increase upon HTMT. The hardening effect produced by HTMT can be retained at higher temperatures than that obtained through LTMT. An increased heat resistance of alloys on HTMT may be attirbuted to the toothed shape of grain boundaries, which inhibit intergranular fracture, and to the apearance of sub-grain boundaries locked by precipitated particles.

HTMT has found some applications in the industry for increasing the heat resistance of parts made of precipitation-hardening austenitic steels or nickel and titanium alloys and intended for operation at relatively low temperatures. At high operating temperatures, HTMT may give poorer results than common heat treatment. For instance, Nimonic alloy has σ_{100} = 80 kgf/mm^2 at 550°C upon quenching and ageing and σ_{100}=92 kgf/mm^2 upon HTMT. At 750°C, an inverse picture is obtained: σ_{100} = 30 kgf/mm^2 on quenching and ageing and 27 kgf/mm^2 on HTMT.

HTMT of titanium alloy grade BT8 by the scheme: deformation at 970 °C → water quenching → ageing at 570 °C for 2 hours, increases the ultimate strength at room temperature by 20-40 kgf/mm^2 , with the ductility and impact toughness remaining at the same level as upon quenching with ageing. At 450°C, the additional hardening effect produed by HTMT still does not appear.

Among aluminium alloys, only A1-Mg-Si and Al-Zn-Mg alloys are often subjected to HTMT. Aluminium may be characterised by a high stacking fault energy, so that polygonisation can occur in it quite easily. Pressing usually forms a very stable polygonised structure in aluminium alloys, so that special measures to form and retain this structure in HTMT are not needed.

The aluminium alloys are characterised by a wide range of temperatures of heating for quenching (from 350°C to 500°C for Al-Zn-Mg alloys) and possess a very high hardenability. For these reasons, they can be subjected to HTMT without any difficulties. With pressing of thin-walled shaped sections (up to 12 mm thick), air cooling provides through-hardening of the metal. With a larger thickness of shaped sections made of 1915 alloy, they are cooled by water directly in the press.

With ageing alloys, HTMT is used industrially substantially less frequently than LTMT, mainly because of the processing limitations mentioned above and the lower hardening effect.

A process has been recommended for industrial application, which is a combination of HTMT and LTMT and has been called high-low-temperature thermomechanical treatment (HLTMT). It comprises quenching from the deformation temperature, followed with cold deformation and ageing (Fig. 8.3). The process gives a higher ultimate strength and a lower ductility than HTMT.

Preliminary Thermomechanical Treatment (PTMT)

In preliminary thermomechanical treatment (PTMT), the processes are such that a semi-product, which has an unre-crystalised structure upon hot deformation, retains this structure during heating for quenching. PTMT differs from HTMT in that the operations of hot deformation and heating for quenching are performed separately (Fig. 8.3).

PTMT is widely employed in the manufacture of semi-products from aluminium alloys. It has been long known that pressed semi-products of alloys of the type of duralumin, Avial, etc., have an appreciably higher strength than rolled or forged products. The phenomenon has been called extrusion effect. The difference in strength values is due to that pressed semi-products have an unrecrystallised structure upon quenching, while the structure of rolled and forged products is recrystallised. As has been found later, hot-rolled sheets and stampings of some alloys may also be in unre-crystallised state upon quenching and thus have an increased strength.

It has been proposed to replace the term 'extrusion effect' and some other synonymous terms by the term 'structural strengthening', which is understood as an increase in the strength of heat-treated semi-products due to the retention of unre-crystallised structure upon quenching. The properties of an alloy in the state of sturctural strengthening are usually compared with those it has in re-crystallised state. The gain in the ultimate strength and yield strength in ageing aluminium alloys due to the retention of unre-crystallised structure may be from 10 to 40 per cent. This gain may appear already in a freshly quenched state owing to an increased density of imperfections (dislocations locked in sub-boundaries and individual dislocations). Ageing of an unre-crystallised alloy produces a higher hardening effect by increasing the difference between the properties of re-crystallised and unre-crystallised semi-products. The ductility is, however, higher in alloys in re-crystallised state. For that reason, semiproducts are produced either with a re-crystallised structure and increased ductility or with polygonised structure and increased strength.

Both in PTMT and HTMT, the formation of an unre-crystallised structure is fovoured by higher temperature and lower rate of deformation, a lower degree of deformation and by the pattern of deformation approaching uniform compression. As has been noted earlier, polygonisation can develop intensively in aluminium alloys and produce a stable network of sub-boundaries, because of which an unre-crystallised structure can be easily formed in them, especially by pressing.

Additions of manganese, zirconium and other antire-crystallisers, which form disperse aluminides (Al_6Mn, Al_3Zr, etc.), increase the temperature of the beginning of re-crystallisation and promote the formation of uncrystallised structure on hot deformation. What is most important is that these additions which are contained in many aluminium alloys favour the retention of uncrystallised structure in subsequent heating for quenching.

Structural strengthening of aluminium alloys (or PHT) has found a wide application in large tonnage mass production of semi-products and is an example of effective industrial use of thermomechanical treatment.

Chapter 9

Thermomechanical Treatment of Martensite-Hardenable Steel

INTRODUCTION

When mechanical work is applied to steel in the metastable austenitic condition, a substantial increase in tensile strength and yield strength takes place. The technique of thermal-mechanical processing extends the strength level of high hardenability steels, such as AISI 4340, from former limits of the order of 3,00,000 lb/inch^2 tensile strength resulting from conventional quench and temper methods to tensile strengths in excess of 4,00,000 lb/inch^2 with satisfactory ductility. An improved technique, known as the Ausform Process, consists of mechanical working in the bainite temperature range of the isothermal transformation diagram, followed immediately by oil quenching to prevent the formation of non-martensitic transformation products.

The resultant microstructure consists of fine martensitic plates, the size and dispersion of which is governed by prior austenitic grain size and the magnitude of plastic deformation. Test results on an alloy steel containing 0.63 per cent carbon indicate a tensile strength of 4,64,000 lb/inch^2, a yield strength of about 3,20,000 lb/inch^2 and an elongation of 8 per cent in 2 inch after tempering at 212°F (100°C). Tempering at 600°F (315°C) caused an increase in yield strength to 4,00,000 lb/inch^2 with reduction in elongation to about 4 per cent in 2 inch.

LOW-TEMPERATURE THERMO-MECHANICAL TREATMENT (LTHT)

With low-temperature thermo-mechanical treatment (LTHT), undercooled austenite is deformed in the temperature range where it has an elevated stability, but always below the point of the beginning of re-crystallisation and then transforms into martensite (Fig. 9.1). After that, a low-temperature tempering is made. The effect of strong strengthening owing to plastic deformation of undercooled austenite followed with quenching from the temperature of straining was discovered by *Lips* and *Zuilen* of the USA in 1954. This process, called ausforming, allowed the ultimate strength of structural steels to be increased up to 280–330 kgf/mm^2 at δ = 5–7 per cent.

The values of ductility and impact toughness were not lower, and in some cases even higher, than those obtained on a common heat treatment which ensured σ_t = 180–220 kgf/mm^2. It is quite understandable that ausforming, which produced superstrong steels, attracted an enormous interest.

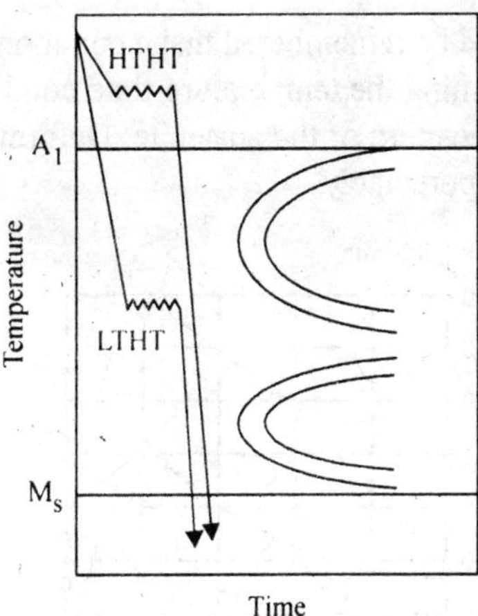

Fig. 9.1 Schemes of high-temperature (HTHT) and low-temperature (LTHT) thermo-mechanical treatment of martensite-hardened alloy steel.

The cause why steels are strengthened on LTHT lies in that the martensite inherits the dislocation structure of the deformed austenite.

Cold deformation cannot produce a high strengthening in martensitic steel, since the martensite, which contains much carbon, is brittle and cannot be deformed with high reductions. On the contrary, austenite can be heavily deformed at temperatures below the point of beginning of re-crystallisation. Such a deformation can strongly increase the total density of dislocations in austenite, with the formation of dense dislocation tangles and a cellular structure.

With martensitic transformation, neighbours of any atom in the austenite remain such in the martensite. Therefore, dislocations do not disappear in the $\gamma \rightarrow \alpha$-transformation by the martensitic mechanism, but are 'passed over' from the initial phase to the new one, i.e., the martensite inherits the sub-structure of deformed austenite. A very high density of dislocations in martensite, which are locked by carbon atoms and carbon precipitates, is responsible for the record figures of ultimate strength on LTHT of steels.

The large refinement of martensite crystals explains why the steel in the high-strength state can have an acceptable level of ductility indices.

LTHT is practically applicable only to alloy steels possessing an appreciable stability of undercooled austenite.

To obtain record figures of ultimate strength (up to 330 kgf/mm^2), alloy steels can be subjected to LTHT by the scheme: austenitisation with heating above Ac_3 $\rightarrow$ undercooling of austenite down to 600–400°C $\rightarrow$ plastic working with deformation up to 90 per cent $\rightarrow$ martensite hardening $\rightarrow$ tempering at 100–200°C.

The gain in strength properties on LTHT depends on the degree and temperature of straining, the temperature of tempering, the content of carbon, and other factors. The strengthening effect on LTHT increases continuously with increasing degree of deformation.

Fig. 9.2 shows how the temperature of rolling in LTHT can affect the properties of Cr-Ni-Mo steel. The rise in the strength properties of steel on lowering the temperature of deformation is caused by increased strain hardening of the austenite. The loss in strength properties on rolling at 400°C is due to bainitic transformation taking place during deformation. Since the non-martensitic products of the transformation which are formed in LTHT can lower the strength, the process must be carried out so as to prevent their appearance. It should be remembered that a common C-diagram cannot provide us with precise quantitative data for establishing the temperature-time conditions of deforming in LTHT, since deformation accelerates the decomposition of the austenite. Deformation of undercooled austenite can shorten the induction period quite appreciably.

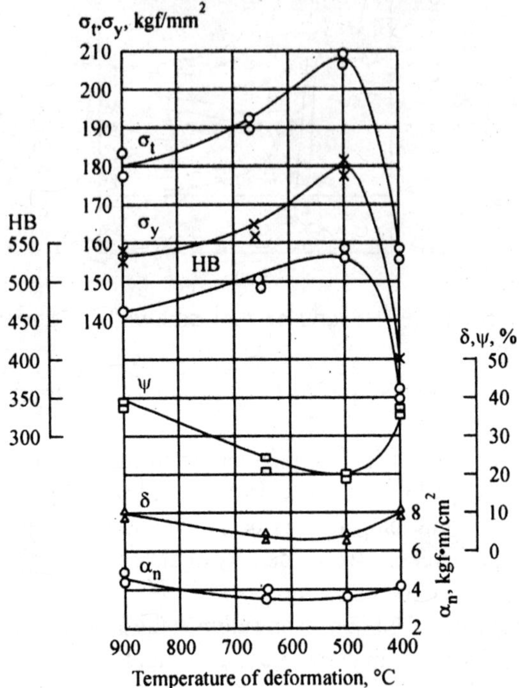

Fig. 9.2. Effect of rolling temperature on mechanical properties of steel 30XHMA upon LTHT. Conditions of LTHT: austenitisation at 1150°C → chilling → rolling with 50 per cent reduction → oil quenching → tempering at 200°C or 4 hrs.

The optimum mechanical properties upon LTHT of structural steels are obtained after low-temperature tempering (at 100–200°C). With increasing temperature of tempering, the strengthening effect produced by LTHT is gradually lost.

The highest effect from LTHT is obtained in steels with 0.4–0.5 per cent C. With a higher carbon content, the relative elongation and ultimate strength can diminish owing to embrittlement.

Industrial application of LTHT of steels encounters some difficulties linked with the necessity to use powerful equipment for plastic working, since the steel must be subjected to heavy deformations (not less than 50 per cent) at temperatures at which its strain resistance is still quite high.

Another essential drawback of LTHT is that a highly strengthened steel has a low resistance to brittle fracture. With an increase in the density of dislocations in martensite, which is needed to produce a high strengthening effect, the resistance of the metal of crack propagation (which is the most important characteristic of structural materials) does not change or even lowers somewhat upon LTHT.

In view of the need to use powerful special equipment for deforming and also because the treated steel has a too low brittle fracture resistance, which is insufficient for modern structures, low-temperature thermo-mechanical treatment of steels is unlikely to find wide application in the industry.

HIGH-TEMPERATURE THERMO-MECHANICAL TREATMENT (HTHT)

With HTHT, austenite is deformed in the temperature range where it is thermodynamically stable, then quenched to martensite (Fig. 9.1) and finally subjected to low-temperature tempering.

The main aim of common heat treatment from deformation (rolling or forging) heating is to avoid special heating for hardening and owing to this, obtain a certain effect. The main purpose of HTHT is, however, to improve the mechanical properties of steel.

In HTHT, martensite inherits the sub-structure of the unrecrystallised austenite that has been formed through hot deformation.

The structural changes on hot deformation have already been discussed. As the degree of deformation is increased at a constant temperature, the density of dislocations in austenite increases, so that only strengthening (hot strain hardening), is initially produced in the metal, after which the processes of softening (dynamic polygonisation and dynamic re-crystallisation) develop. Austenite may be characterised by a low stacking fault energy, and therefore, extended dislocations cannot easily leave their slip planes. Thus, austenite can be strongly strain-hardened and then can re-crystallise quickly.

The best complex of mechanical properties is obtained in steel when martensite forms from austenite having a well developed polygonised structure. It has been revealed by diffraction electron microscopy that martensitic crystals fully inherit the dislocation sub-boundaries from hot deformed austenite. Dislocation tangles can also be inherited by martensite.

As a result of fragmentation of martensitic crystals by sub-grain boundaries, the ultimate strength and yield strength of a steel increase on HTHT. The sub-structure of the martensite obtained from hot-deformed austenite is such that HTHT can increase the strength without changing the ductility indices, such as relative reduction (Fig. 9.3). As compared with common hardening and tempering, however, HTHT of steel can give higher indices of ductility with the same ultimate strength.

HTHT of machine-building carbon, low-alloyed and medium alloyed steels gives the following complex of mechanical properties σ_t = 220–260 kgf/mm^2 ; $\sigma_{0.2}$ = 190–220 kgf/mm^2 ; δ = 7–8 per cent and ψ = 25–40 per cent.

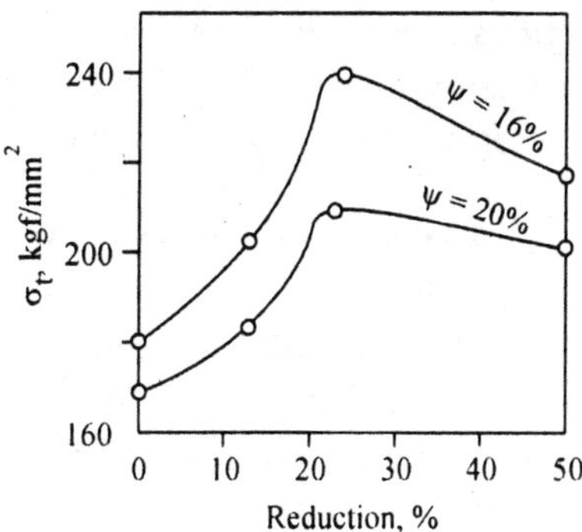

Fig. 9.3. Effect of deformation ratio in HTHT on the ultimate strength of steel at constant relative reduction.

A steel upon HTHT has lower strength properties and higher ductility properties than upon LTHT.

HTHT of steel can be used to diminish drastically or, in some cases, even remove almost fully both reversible and irreversible temper embrittlement (Fig. 9.4). It causes a change from intergranular fracture, which is typical of the state of temper embrittlement, to intragranular fracture.

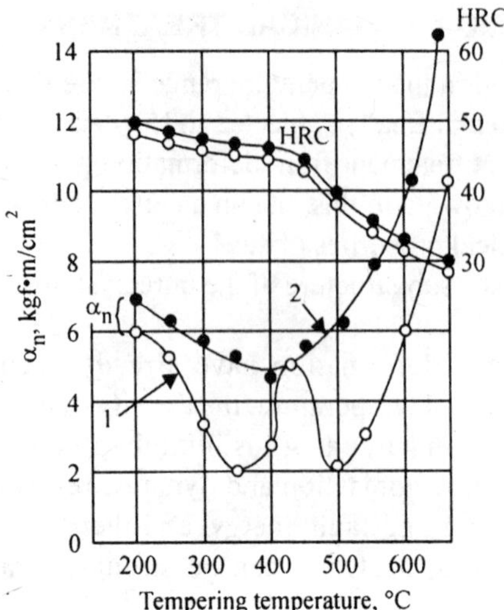

Fig. 9.4. Dependence of impact toughness on temperature of tempering after common hardening (1) and after HTHT (2) of steel 40XH4.

The ultimate strength can easily be raised up to 220 kgf/mm^2 without resorting to HTHT, simply by increasing the content of carbon in steel and carrying out common hardening and low-temperature tempering. But the steel upon the common heat treatment will be characterised by a very high susceptibility to brittle fracture and a high sensitivity to sharp notches at the surface of articles. Fatigue cracks in such a steel can propagate quickly owing to stress localisation at the ends of cracks, thus causing sudden brittle fracture at stresses well below the yield limit. A steel having high values of ultimate strength and yield limit, but a low resistance to crack propagation cannot be used in heavily stressed structures.

The most essential merit of HTHT is its remarkable capability for increasing simultaneously both the strength indices and the resistance to crack propagation (fracture toughness).

In martensitic crystals fragmented by sub-boundaries, dislocation tangles that form on straining are smaller, i.e., the peaks of locals stresses are lower. Besides, with a developed sub-granular structure, the stresses localised at the ends of cracks can relax easily. The lower level of peak stressed and their greater capacity of relaxing are responsible for a high resistance to crack propagation in steels subjected to HTHT.

The conditions for HTHT are so chosen that the austenite has a well developed polygonised structure at the beginning of martensitic transformation.

The degree of hot deformation in HTHT must not be excessively high, otherwise re-crystallisation might develop and lower the strengthening effect Fig. 9.3. With increasing rate of deformation, hot strain-hardening is enhanced, which, combined with deformation heating, may cause re-crystallisation.

The use of multiple deformation with the total reduction equal to a heavy deformation in one pass, can facilitate the process of deforming and promote dynamic polygonisation.

The temperature, degree and rate of deformation and the number of deformation stages should be selected at the optimum level for each grade of steel and each kind and size of product, in order to obtain a developed polygonised structure in the austenite.

Carbon and low-alloy steels should be quenched immediately after deforming, so as to prevent static re-crystallisation and retain the polygonised structure up to the beginning of martensitic transformation.

In medium-alloyed and especially in high-alloyed steels, the climb of dislocations, and therefore, polygonisation can proceed only slowly. Such steels upon deforming may be in a state of hot strain-hardening and have an unpolygonised structure, which, if inherited by martensite, will not ensure the required level of mechanical properties. For these reasons, a strictly specified holding interval between deformation and hardening stages is required for such steels, so as to allow sufficient time for the development of static polygonisation.

Of large interest is the phenomenon discovered by *Bernstein*, consisting in that the strengthening effect produced by HTHT is inherited upon repeated heat treatment. It has turned out that the strengthening produced by HTHT is retained if the steel is re-hardened with a short holding at the temperature of heating for hardening or if the steel strengthened by HTHT is first subjected to high-temperature tempering and then re-hardened. For instance, the ultimate strength of steel Grade 37XH3A after HTHT (25 per cent deformation at 950°C $\rightarrow$ hardening $\rightarrow$ tempering at 100°C) is 250 kgf/mm^2. If the steel is tempered at 500°C for 30 minutes, hardened from 900°C for 2 minutes and tempered at 100°C, its ultimate strength will again be around 250 kgf/mm^2. Thus, the sub-structure formed on hot deforming of the austenite, has been inherited once on the $\gamma \rightarrow \alpha$-transformation during HTHT, retained to a certain extent during the high-temperature tempering, and again inherited twice during re-hardening on the $\alpha \rightarrow \gamma$-and $\gamma \rightarrow \alpha$-transformations. The mechanism of inheritance of substructure on $\alpha \rightarrow \gamma$-transformations still remains unclear. A short holding time of the second hardening operation prevents the development of re-crystallisation in the austenite, which otherwise might destroy the polygonised structure, and therefore, the strengthening effect provided by the proceeding HTHT.

The phenomenon of inherited strengthening in repeated short time heating can widen the range of application of HTHT. For instances, steel can be hardened at the steelmaking works directly from the rolling temperature and then subjected to high-temperature tempering to improve its machinability. The final products are hardened with a short holding time and subjected to a low temperature tempering and thus acquire the increased strength that has been imparted to the metal on HTHT.

Notwithstanding the lower strengthening effect produced, HTHT has undeniable advantages over LTHT. It simultaneously increases both the ultimate strength and fracture toughness, can be performed more easily (no special powerful equipment is required for deformation), can be applied to carbon and low-alloy steels, as well as to alloy steels having an increased stability of undercooled austenite.

The above discussion of the pattern of changes of the structure and properties of steels on HTHT shows that this kind of thermomechanical treatment is one of the most promising ways for improving the structural strength of steels.

PRELIMINARY THERMO-MECHANICAL TREATMENT (PHT)

Preliminary thermo-mechanical treatment is carried out by the scheme: cold plastic deformation $\rightarrow$ pre-re-crystallisational heating $\rightarrow$ hardening with accelerated heating and short holding time $\rightarrow$ tempering.

The initial structure before deformation is ferrite-carbide mixture. Cold deformation increases the density of dislocations which rearrange in pre-recrystallisation heating to form a polygonised structure of ferrite. In subsequent hardening with accelerated heating and short holding time, the sub-structure is inherited on the $\alpha \rightarrow \gamma$-and-$\gamma \rightarrow \alpha$-transformation and martensitic crystals are fragmented. Therefore, the strengthening mechanism in PHT is the same as the mechanism that gives an increased strength on repeated hardening after HTHT.

The sub-structure formed by hot deforming of the austenite in HTHT is more stable than that obtained on pre-recrystallisational heating after cold deformation. For that reason, the polygonised structure and therefore, the strengthening effect on repeated hardening after HTHT are inherited more fully than in the process of preliminary thermo-mechanical treatment. PHT is attractive by the ease with which it can be performed. The time interval between cold deformation and heating may be of any length. Deformation requires no special equipment. PHT with short salt-bath or induction heating before hardening is an effective method for strengthening cold-rolled sheets and thin walled tubes.

DEFECTS DUE TO HEAT TREATMENT OF STEELS

The main types of rejects in annealing and normalising are due to faulty regulation of heating temperature, including overheating, burning, underheating, etc.

Due to the application of perfected pyrometry and automatic equipment, the above types of rejects have been practically eliminated in up-to-date heat treating departments.

The chief cause of quenching defects is high residual (internal) stresses occurring in hardened articles. These stresses may cause distortion, warping and even cracking.

The types of defects encountered in the heat treatment of steel, their causes and measures for their prevention or correction are listed in Table 9.1.

Table 9.1. Principal defects in steel due to faulty heat treatment and measures for their elimination.

Defects and its characteristics	*Causes*	*Measures for avoiding or correcting the defect*
Overheating—coarse-grained microstructure and fracture, Widmanstatten structure in the annealed steel, coarse crystallic martensite in the hardened steel. Reduced ductility and, especially, impact strength	Heating for long periods at temperatures considerably exceeding the normal values	Elimination: 1. For slight overheating-apply normal annealing and normalising. 2. For considerable overheating-apply double annealing or normalising (first annealing or normalising at a temperature of Ac_3 + 100°C to 150°C, second at normal temperature)
Burning—grain boundaries have: 1. Regions enriched in carbon—1st stage of burning 2. Non-oxidised cavities and blowholes—2nd stage 3. Iron oxide inclusions—3rd stage Stone-like fracture Low ductility	Heating for a long period at a high temperature in an oxidising atmosphere or at temperatures near the melting point	1. 1st stage—use homogenising followed by double annealing 2. 2nd stage—use forging followed by annealing. 3. 3rd stage—defect cannot be remedied.
Oxidation: Thick layer of scale on the surface of the steel article	Oxidising atmosphere in the heating furnace	Prevention of the defect: 1. Heat in furnaces with reducing, neutral or protective atmospheres 2. Heat in boxes with used carburising agent or cast iron chips 3. Heat in molten salt baths
Decarburisation—loss of carbon in the surface layers of the article Lower hardness after quenching, lower fatigue limit	Ditto	Ditto Correction of the defect—machine off the decarburised layer if the machining allowance is sufficiently large

(Contd...)

Defects and its characteristics	*Causes*	*Measures for avoiding or correcting the defect*
Excessive hardness of hot-worked annealed steel	Excessive cooling rate for ordinary annealing or insufficient holding time for isothermal annealing	Repeat annealing with cooling at specified rate
Black fracture–free graphite (annealing carbon) inclusions in the steel	Excessive heating time and slow cooling after annealing	Heat the steel to a high temperature and forge it thoroughly
Quenching cracks. Usually appear at the grain boundaries and are zigzag in form. They may be external and internal. Cracks may be longitudinal, arc-like, etc	1. Internal stresses due to volume changes at low temperatures, caused by the martensitic transformation 2. The martensitic transformation is not simultaneous throughout the article being hardened	This defect cannot be corrected. It may be prevented by: 1. Avoiding sharp projections, sharp corners, and sudden transitions from thick to thin sections in designing the article 2. The article should be free of stresses before being hardened; stresses should be relieved by previous annealing 3. Heat to the minimum suitable temperature for hardening 4. Cool slowly in the martensitic range (apply interrupted quenching, oil quenching, martempering, etc.) 5. Apply austempering 6. Temper immediately after quenching
Deformation and volume (dimension) changes after hardening	The volume of the steel increases due to the martensitic transformation. The higher the hardenability of the steel, the more severe the deformation in hardening will be	1. Use alloy steels that are only slightly deformed by quenching 2. Cool slowly in the martensitic range 3. Apply surface hardening (when feasible)
Warping–asymmetrical deformation	1. Volume changes in cooling (of the article in quenching) (or heating) 2. Nonuniform heating (or cooling) of the article 3. Internal stresses in the article before heating. 4. The article is lowered into the quenching bath in an inclined position	Ditto and also; 1. Anneal, normalise or temper at a high temperature before hardening 2. Heat uniformly for hardening 3. Quench as uniformly as possible in hardening 4. Hold the article in the proper position when lowering it into the quenching bath 5. Use special quenching jigs. The defect may be corrected by straightening or grinding (if the grinding allowance is sufficiently large)
Insufficient hardness after quenching	1. Hardening temperature is too low 2. Holding time is insufficient at the hardening temperature 3. Cooling rate is too low	The defect may be corrected by normalising or annealing followed by hardening with the proper specified procedure

(Contd...)

Defects and its characteristics	*Causes*	*Measures for avoiding or correcting the defect*
Soft spots—zones on the surface of hardened articles with lower hardness	1. Vapour accumulates (forming bubbles) on the surface of the quenched articles reducing the cooling rate at the given places 2. Localised decarburisation 3. In homogeneity of initial structure in solidification of the steel	The defect may be prevented by: 1. Using a more effective quenching medium 2. Obtaining a more homogeneous structure, employing annealing or normalisation before hardening 3. Protecting against decarburisation in heating The defect may be corrected by normalising followed by quenching at a higher cooling rate
Excessive hardness after tempering.	Low temperature or insufficient holding time in tempering	Corrected by a second tempering with proper temperature and holding time
Insufficient hardness after tempering	Tempering temperature was too high	Anneal, re-harden and temper at normal temperature
Erosion—reduction in size of the article or distortion of its form due to loss of metal from its surface	Chemical reaction of chlorous salts and oxidation of metals heated in molten salt baths	Preventive measures: 1. Deoxidise salt bath with borax or ferrosilicon 2. Properly position the article in reference to the electrodes of electrical salt baths
Corrosion—pitting or groove-type destruction of the surface of an article	In heating in molten salt baths: 1. High content of sulphuric salts (over 0.7–0.8 per cent) 2. Bath has become rich in oxygen or iron oxides. In heating in flame furnaces–non-uniform scale formation	Preventive measures: 1. Carefully control salt composition 2. Deoxidise the bath; 3. Eliminate the oxidising atmosphere when heating in flame furnaces

Notes:

1. Factors conducive to cracking are: sharp corners, large volume changes at low temperatures and low cohesive strength. All factors that favour hardenability are also conducive to cracking.

2. Deformation in hardening may be substantially reduced by employing special heat treating procedures which enable tempered martensite and an increased amount of retained austenite to be obtained after hardening. This is achieved by increasing the hardening temperature to a certain extent and by slow cooling in the temperature range of martensitic transformation.

Overheated, Burnt and Underannealed Structures

When the steel is heated well above the upper critical temperature large austenite crystals form and slow cooling give rise to the Widmanstätten type of structure, with its characteristic lack of both ductility and resistance to shock. This is known as an overheated structure and it can be refined by re-heating the steel to just above the upper critical point. Surface decarburisation usually occurs during overheating.

During the Second World War, aircraft engine makers were troubled with overheating (above 1250°C) in drop-stampings made from alloy steels. In the hardened and tempered condition the fractured surface shows dull facets. The minimum overheating temperature depends on the 'purity' of the steel and is substantially lower in general for electric steel than for open-hearth steel. The overheated structure in these alloy steels occurs when they are cooled at an intermediate rate from the high temperature—at faster or slower rates the overheated structure may be eliminated. This together with the fact that the overheating temperature is very materially raised in the presence of high contents of MnS and inclusions suggests that this overheating is concerned in some way with a diffusion and precipitation process, involving MnS. Such overheating can occur in an atmosphere free from oxygen, thus emphasising the difference between overheating and burning.

As the steel approaches the solidus temperature, incipient fusion and oxidation take place at the grain boundaries. Such a steel is said to be burnt and it is characterised by the presence of brittle iron oxide films which render the steel unfit for service, except as scrap for re-melting.

Quench Cracks

The volume changes which occur when austenite is cooled are:

1. Expansion when gamma iron transforms to ferrite.
2. Contraction when cementite is precipitated.
3. Normal thermal contraction.

When a steel is quenched these volume changes occur very rapidly and unevenly throughout the specimen. The outside cools most quickly, and is mainly martensitic, in which contraction (2) has not occurred. The centre may be troostitic and contraction (2) started. Stresses are set up which may cause the metal either to distort or to crack if the ductility is insufficient for plastic flow to occur. Such cracks may occur some time after quenching or in the early stages of tempering.

Quench cracks are liable to occur:

1. Due to presence of non-metallic inclusion, cementite masses, etc.
2. When austenite is coarse grained due to high quenching temperature.
3. Owing to uneven quenching.
4. In pieces of irregular section and when sharp re-entrant angle are present in the design.

Relation of design to heat treatment

The relation of design to heat-treatment is very important. Articles of irregular section need special care. When a steel has been chosen which needs a water-quench, then the designer must use generous fillets in the corners and a uniform section should be aimed at. This can sometimes be obtained by boring out metal from bulky parts without materially affecting the design; examples are given in Fig. 9.5. A hole drilled from the side to meet a central hole may cause cracking and it should be drilled right through and temporarily stopped up with asbestos wool during heat-treatment. A crack would also form at the junction of the solid gear with the shaft and there would be a serious danger of cracks at the roots of the teeth owing to the great change in size of section. This design could be improved by machining the metal away under the rim to make a cross-section of uniform mass.

Low melting point compounds

Low melting-point compounds, such as iron sulphide in steel, nickel sulphide in nickel, are a common cause of cracked welds, especially in welding free-cutting steels. The addition of sufficient manganese largely overcomes the troubles in steel by forming a high melting-point sulphide.

Ranges of low ductility

Ranges of low ductility occur in some materials within certain temperature ranges as shown in Fig. 9.6. If welds are stressed at about 700°C cracks may develop.

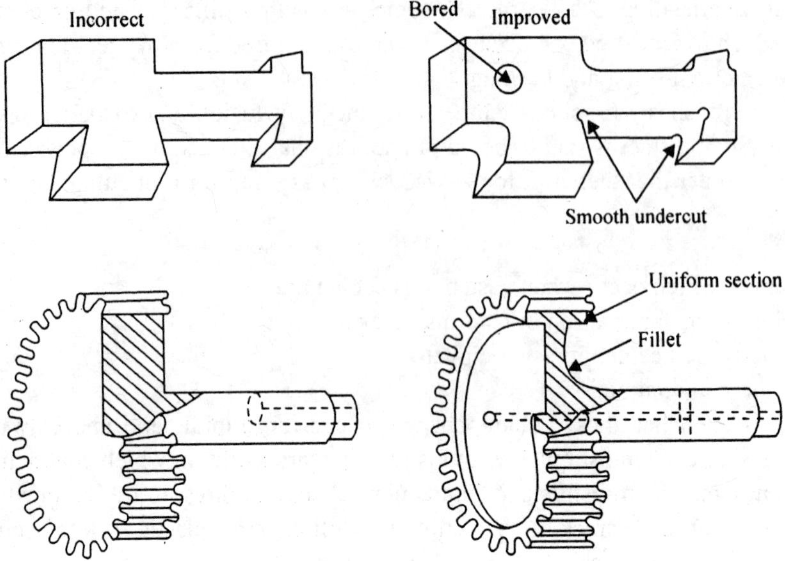

Fig. 9.5. The relation of design to heat-treatment.

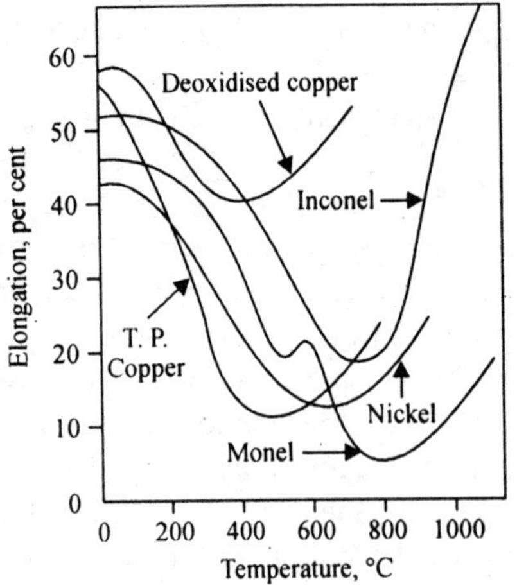

Fig. 9.6. Ranges of low ductility.

Effect of welding on the plate

Effect of welding on the plate is also important, since the operation causes temperature gradients in the plates, the limits of which are the melting-point of the metal and room temperature. In the case of steel a portion of the plate will be within the critical ranges.

If such a specimen is cooled slowly the ferrite which was taken into solution by the austenite is deposited at the boundaries of the small austenite crystals and is followed by the formation of pearlite. The areas of the latter are very much smaller than those originally in the specimen and each is surrounded by fine ferrite networks, which can be distinguished from the original skeleton of ferrite. This structure is illustrated diagrammatically in Fig. 9.7. This treatment is also known as under-annealing.

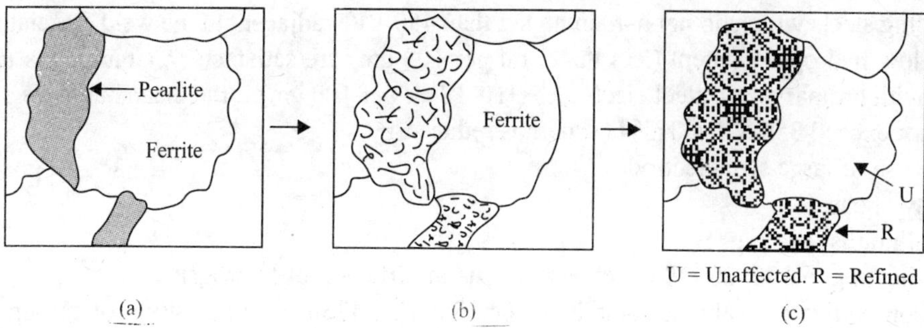

Fig. 9.7. Diagrammatic illustration of formation of structure (a) original; (b) between A_1 and A_3; (c) after cooling.

When the plates which have to be welded are in a cold-worked condition due either to slight cold-rolling or to deformation during fabrication, re-crystallisation effects arise. At inclusions lamellar tearing can occur.

Welding High Tensile Steels

Increasing use has been made of welding in the fabrication of low alloy steels with yield points up to 430 N/mm^2 and also of oil-or-air hardening steels, e.g., armour. While carbon and alloys render the welds more liable to hot cracks, the most important defect is likely to be cracks in the hardened zone of the plate, i.e., hard metal cracks which form at or near room temperature.

The factors which influence this cracking are:

Constraint

Constraint, such as in rigid structures which sets up multi-axial stresses in the weld. These stresses increase with thicker plates and are influenced by the design and the welding technique used.

Formation of a brittle martensite zone

In formation of a brittle martensite zone adjacent to the weld, its hardness is greater with increase of carbon contents in particular, of alloy content and rate of cooling. To minimise cracking the end of the martensite reaction should be above 290°C.

Hydrogen

Hydrogen dissolved in the weld bead and its diffusion into the parent metal. Atomic hydrogen is soluble in austenite at all temperature but practically insoluble in cold ferrite or martensite. When the heat affected zone transforms, therefore, atomic hydrogen is rejected and tends to collect in inclusions, pores and sub-microscopic defects to form molecular hydrogen which cannot diffuse away and therefore builds up a high-pressure additive to the contraction stresses. Alternatively the hydrogen may reduce the cohesive strength of the steel, thus enabling a brittle fracture to occur with little strain. Hair-line cracking in alloy steel ingots is a related phenomenon.

One method of overcoming the trouble due to hydrogen has been the use of 18/8/3 Cr—Ni—Mo austenitic arc-welding electrodes, having a flux covering with the minimum of water containing substances. The austenite acts as reservoir for the hydrogen which is moderately low (12 ml per 100 gm). The strength of the weld is about 700 N/mm^2.

For welding steels which do not harden higher than 450 VPN adjacent to the weld, special mild steel welds with low hydrogen content (less that 3 ml per 100 gm) are satisfactory. Cracking is unlikely to occur even with ordinary mild steel electrodes (H = 18 ml per 100 gm) if the mean hardness of the hard zone does not exceed 350 VPN. This hardness is reduced by:

1. The use of large size electrodes.
2. Heavy beads.
3. Continuous welding.
4. Pre-heating to 100–200°C which also assists in diffusion of hydrogen.

The carbon equivalent value is some item specified (BS 4360) to fix the limit of carbon and alloy content in regard to welding.

$$CE = C + \frac{Mn}{6} + \frac{Cr + Mo + V}{5} + \frac{Ni + Cu}{15}$$

CE < 0.14 no special precautions with rutile electrodes.

0.41–0.45 use low hydrogen electrodes or pre-heat.

>0.45 use low hydrogen electrodes + pre-heat.

For critical steels it is helpful to pre-coat the surface of the joint with austenitic weld metal preferably of the 25–Cr 20–Ni type.

The stray arcing of an electrode on high tensile steels is liable to produce a localised hard zone from which can initiate a fatigue crack.

The deposition of mild steel over 18/8 austenitic welds is liable to produce cracks in the mild steel which is hardened by the pick-up of alloying elements.

The WI Controlled Thermal Severity (CTS) test is shown in Fig. 9.8. A square plate of the required material is fillet welded along two opposite sides to the base plate. The remaining two sides are then welded up with the electrodes to be tested, the second weld being laid down immediately after the first. The assembly is allowed to stand for a period of time, after which the welds are sectioned and examined for cracks.

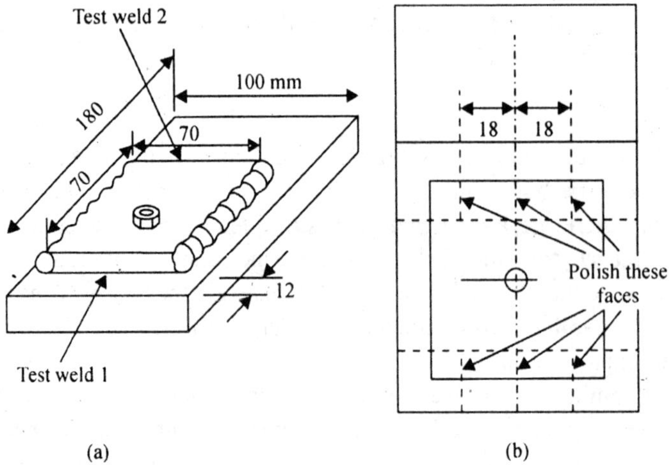

Fig. 9.8. CTS weldability test.

The severity of the test may be varied by altering the thickness of the plates, the hydrogen level in the test welds and the composition of the weld metal. The cooling rate is designated by means of a thermal severity number (TSN). TSN 1 is the thermal severity corresponding to heat flow along a single steel plate 6 mm thick. TSN 2 is obtained in butt weld between two 6 mm plates, while in a 6 mm tee joint, where there are three heat flow paths, the thermal severity number is 3. The TSN number is also increased in proportion to the plate thickness, so that a CTS testpiece in 12 mm plate would have TSN 6.

Lamellar tearing sometimes occurs in structural steels parallel to the surface of a plate beneath heavy fillet welds. The cracking usually follows planes of non-metallic inclusions within the plate giving a step-like fracture appearance in transverse sections.

SECTION III

Heat Treatment of Cast Iron

Chapter 10

Heat Treating of Cast Irons: A Review

INTRODUCTION

Cast irons may be compared with steels in their reactions to hardening. However, because cast irons (except white iron) contain graphite and substantially higher percentages of silicon, they require higher austenitising temperatures. The graphitising effect of silicon is so powerful that an unalloyed gray iron may become completely graphitised below the A_1 temperature during heating for austenitising. Thus, some high-silicon irons require longer intervals at the austenitising temperature for re-absorption of the desired carbon content in austenite. This interval is extended because silicon retards the absorption of carbon in austenite. The lower-silicon irons respond best to heat treatment. The response of the cast iron matrix to hardening depends on the carbon and alloy content of austenite at the austenitising temperature. The higher the austenitising temperature above A_1, the more carbon is dissolved in austenite and the higher the as-quenched hardness. This increase in hardness and dissolved carbon above A_1 and the decrease with graphitisation during heating below A_1, are shown in Table 10.1.

Table 10.1. Effects of quenching temperature on combined carbon content and as-quenched hardness of gray irons. (Specimens, 30 mm (1.2 inch) diam bars 50 mm (2 inch) long, were quenched in oil from temperatures shown after holding for 1 hr.)

	Unalloyed iron(a)		*Cr-Ni-Mo iron(b)*	
Condition	*Combined carbon, %*	*Hardness, HB*	*Combined carbon, %*	*Hardness, HB*
As cast	0.69	217	0.70	255
Quenched from:				
650°C (1200°F)	0.54	207	0.65	250
675°C (1250°F)	0.38	187	0.63	241
705°C (1300°F)	0.09	170	0.59	229
730°C (1350°F)	0.09	143	0.47	217
760°C (1400°F)	Nil	137	0.45	197
790°C (1450°F)	0.05	143	0.42	207
815°C (1500°F)	0.47	269	0.60	444
845°C (1550°F)	0.59	444	0.69	514
870°C (1600°F)	0.67	477	0.76	601

(a) Total carbon content, 3.19%; and (b) total carbon content, 3.10%.

HARDNESS MEASUREMENTS

Conventional hardness measurements on cast irons always indicate lower values than the true hardness of the metal matrix. This discrepancy, which is more pronounced in gray iron than in ductile and malleable irons, occurs because conventional hardness readings are composite values that reflect the hardnesses of both the matrix material and soft graphite. Provided that the test impression is made on a smooth, flat surface of adequate size, a Brinell hardness tester employing a 10 mm ball penetrator and a 3000 kg load will determine the average hardness of malleable irons with reasonable accuracy. The hardness value, by virtue of the size of the ball penetrator, is representative of the gross structure under test.

In determining the hardness of small castings, it is often impossible to use a Brinell tester and a Rockwell tester must be employed. Conversions have been developed to compensate for the discrepancies inherent in hardness measurements made on malleable irons with the smaller Rockwell penetrator. Figs 10.1a and 10.1b show conversions from Brinell (HB) to Rockwell B and G (HRB and HRG) scales for malleable and pearlitic malleable irons, respectively. Fig. 10.1b also shows HRC equivalents for HB values, for pearlitic malleable. These conversions are generally accepted by producers of malleable iron.

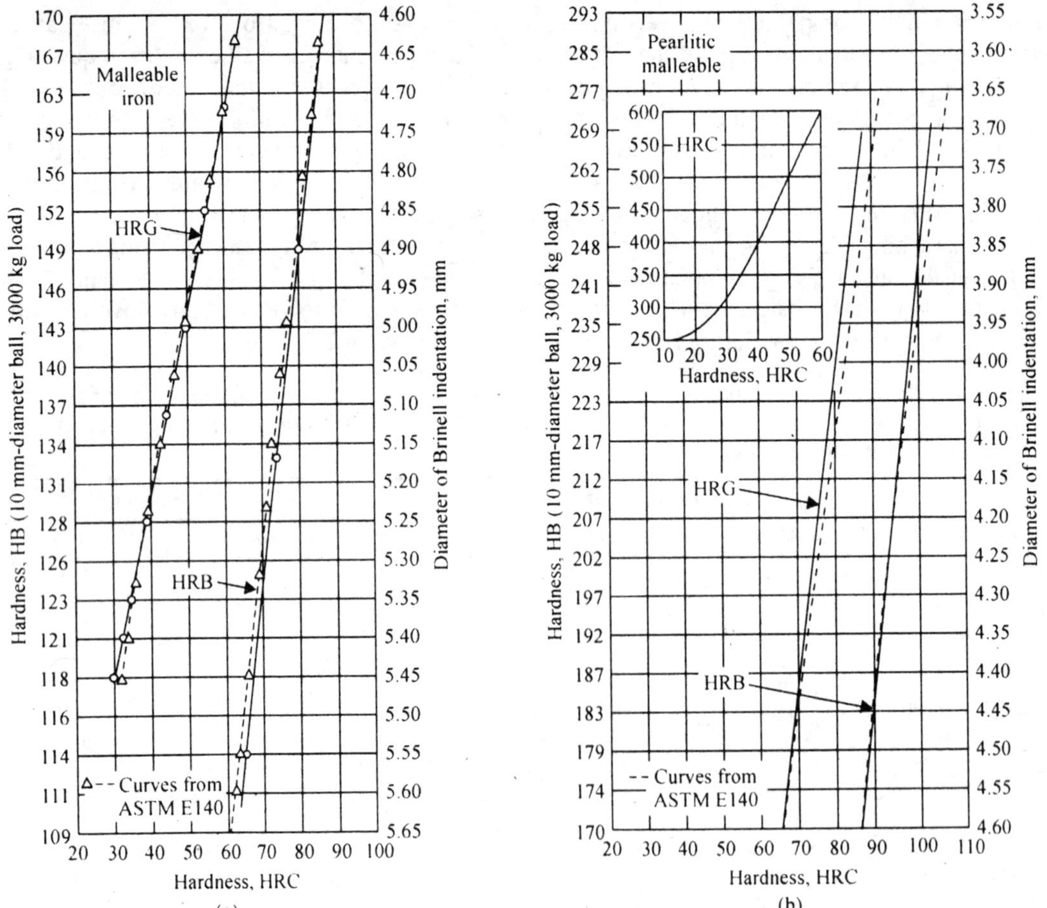

Fig. 10.1. Hardness conversions: (a) conversion from HB to HRB and HRG scales for malleable iron; and (b) conversion from HB to HRB, HRC and HRG scales for pearlitic malleable iron.

Matrix hardness is determined more accurately by Brinell than by Rockwell hardness tests, although both require correction to obtain true matrix hardness. The microhardness test is the most accurate method for determining matrix hardness.

Comparative hardness readings obtained on ten quenched and tempered ductile irons are given in Table 10.2. Observed HRC readings ranged from 3.8 to 8.3 points lower than those converted from microhardness readings.

Table 10.2. Comparative hardness values for quenched and tempered ductile irons.

Iron	*HB(a)*	*HRC converted from HB(b)*	*Observed HRC(c)*	*Microhardness HV(d)*	*HRC converted from HV(b)*	*HRC converted from HV minus observed HRC*
1	415	44.5	44.4	527	50.9	6.5
2	444	47.2	45.0	521	50.6	5.6
3	444	47.2	45.7	530	51.1	5.4
4	444	47.2	47.6	593	54.9	7.3
5	461	48.8	46.7	595	55.0	8.3
6	461	48.8	48.3	560	53.0	4.7
7	461	48.8	49.1	581	54.2	5.1
8	477	50.3	49.6	572	53.7	4.1
9	477	50.3	50.1	618	56.2	6.1
10	555	55.6	53.4	637	57.2	3.8

(a) Average of three readings for each iron; (b) values based on SAE-ASM-ASTM hardness conversions for steel; (c) average of five readings of each iron; and (d) average of a minimum of five readings for each iron, 100 kg load.

Fig. 10.2a shows the relation between observed HRC readings and those converted from microhardness values for five gray irons of different carbon equivalents. Hardness measurements were taken at two laboratories after quenching and after tempering of each iron. The data in Fig. 10.2a show why the observed values obtained by conventional hardness testing may be misleading and help to explain the good wear resistance of gray irons with apparently low hardness. Note that there is a correlation with carbon equivalent for all five irons tested and that the discrepancy between observed and converted hardness values diminished at the lower hardness level.

Another comparison between observed and converted HRC values for gray and ductile irons is shown in Fig. 10.2b. These irons were quenched in water from 900°C (1650°F) and tempered at 425°C (800°F) for 2 hours.

HEAT TREATING EQUIPMENT

Heat treating equipment for hardening cast iron should be selected with an understanding of the differences between the hardening characteristics of cast iron and those of steel. For example, cast iron (especially gray iron) is more likely than steel to crack on heating if it is not heated uniformly. In the austenitising range, metallurgical reactions within cast irons are more complex than those that occur in steels; thus, heat treating of cast iron requires more accurate temperature control. In properly balanced alloys, the hardenability of cast iron increases with alloy content, particularly manganese, vanadium and molybdenum contents. Chromium content increases hardenability, but it also complicates casting production.

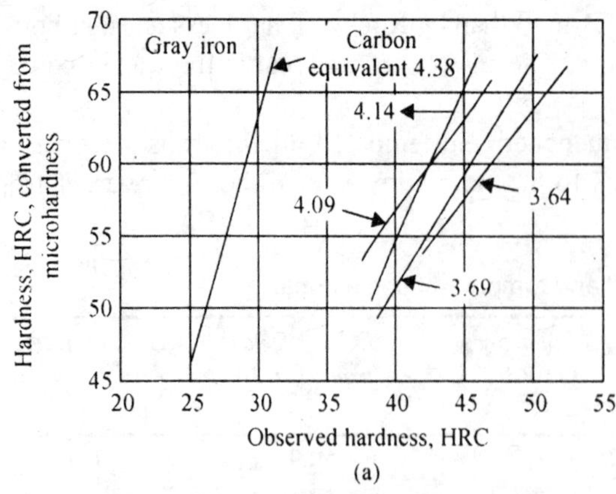

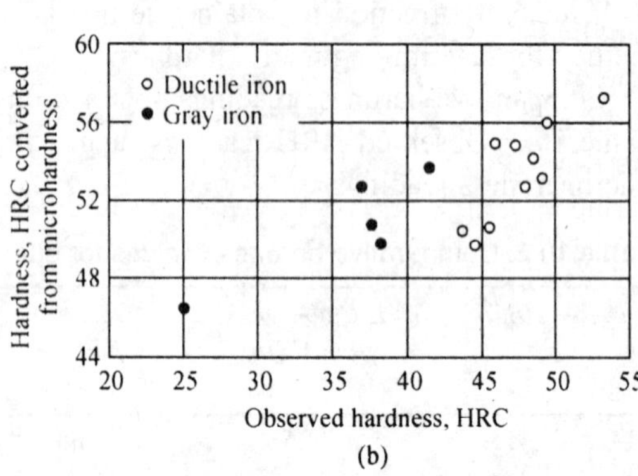

(a) Relation, as influenced by carbon equivalent, for gray iron containing type 3 graphite. (b) Relation for gray and ductile irons quenched in water from 900°C (1650°F) and tempered 2 hours at 425°C (800°F).

Fig. 10.2. Relations between observed and converted hardness values for gray and ductile irons.

For simplicity of processing and to avoid the costs associated with the use of alloy iron, quenching facilities and procedures should, wherever possible, be arranged so as to take maximum advantage of the hardening characteristics of low-alloy or unalloyed irons. Finally, cast irons increase in volume during hardening to a much greater extent than do steels and this added growth contributes to heat treating problems. Table 10.3 summarises the principal advantages and disadvantages of three types of heating equipment that may be considered for use in heat treating cast irons.

Table 10.3. Advantages and disadvantages of equipments used for heat treating of cast iron.

Advantages	*Disadvantages*
Direct-fired equipment	
Lowest initial, operating and maintenance costs	Forms oxides, making cleaning difficult
Comparatively easy to operate and mechanise	Cannot be used for machined castings
Controlled-atmosphere equipment	
Lower operating cost than salt bath	Entails higher operating costs than direct-fired equipment
Atmospheres can be adjusted more readily than salt composition, to suit work	Requires specially trained personnel to operate and maintain the atmosphere-generating equipment
No contamination of quench medium	Equipment vulnerable to atmosphere leaks
Machined castings can be heat treated	Fixturing needed to prevent distortion
Comparatively easy to mechanise	
Prevents decarburisation and scaling	
Salt bath equipment	
High heating rate	High operating costs
Fixturing for preventing distortion less essential than for other methods	Salt composition less readily changed than atmosphere, to suit work
Machined castings can be heat treated	Contamination of quench medium
Salt greatly aids in removal of dirt and sand	Salt must be removed after treatment
Prevents decarburisation and scaling	Not easily mechanised

Advantages and disadvantages must be weighed also in selecting methods and equipment for work handling. Belt furnaces provide more uniform heating and quenching than is obtained in processes employing baskets or trays. Baskets and trays, however, permit easier fixturing for control of distortion, particularly for small parts. A major disadvantage of trays and baskets is the high cost of the alloy materials of which they must be made because of the high temperatures to which they are subjected during austenitising for hardening. They also require more energy because they are removed from the furnace for re-loading and permitted to cool down.

Heating rate

In general, it is recommended that all sections of a casting be heated uniformly, to minimise the thermal stresses that can cause cracking. The danger of cracking is greatest at temperatures up to 425°C (800°F) and diminishes thereafter. The rate of heating that can be tolerated for gray iron depends on the complexity of the casting and the amount of residual stress in the as-cast condition.

Continuous furnaces with zoned temperatures and pre-heating furnaces, are used to handle complex gray iron castings. Batch furnaces, in which the work is immediately exposed to high temperature, have been used for simpler castings whose shapes and section sizes ensure uniform heating. Salt baths have been used satisfactorily for castings of uniform wall thickness, such as cylinder sleeves, but the rate of heating in molten salt (about five or more times the rate achieved in air) is too fast for gray iron castings for complex shape.

Temperature Control

To avoid thermal gradients, both furnace temperature and heating rate must be uniform. Large castings, in particular, require a high degree of temperature uniformity to avoid cracking; such uniformity is best provided by furnaces with indirect firing and good circulation of atmosphere. Direct firing on a casting has been known to cause cracking and may make that area of the casting softer.

All cast irons except white iron contain graphite, some of which will dissolve in the matrix at elevated temperature. In heating for hardening, iron carbide breaks down into ferrite plus additional graphite at temperatures in the range of about 650 to 790°C (1200 to 1450°F). From 790 to about 845°C (1450 to about 1550°F), the iron achieves a state of equilibrium if held at any one temperature and is made up of ferrite, austenite with a fixed content of dissolved carbon and graphite. At temperatures above about 845°C (1550°F), all ferrite is transformed to austenite with a dissolved carbon content that increases with increasing temperature.

Although, with reasonable temperature uniformity, steel will produce the same structures over a wide temperature range, cast iron may undergo structural alterations with every temperature increment. Therefore, furnaces for hardening of cast iron should have adequate atmosphere circulation and good proportioning control to ensure uniform furnace temperatures and a narrow control range. Limiting temperature controls also are desirable.

Heating Media

Air, controlled atmosphere and molten salt are in general use for hardening of cast iron. Because cast iron is subject to decarburisation and scaling at heat treating temperatures, controlled atmospheres or molten salt should be used if either of these conditions is objectionable. Salt-bath equipment is less costly to install than controlled-atmosphere equipment, but is generally more costly to operate. In one high-production martempering installation, the decrease in operating costs that was realised by switching from high-temperature salt to controlled atmosphere offset the increase in equipment costs in less than one year. Growth during hardening is directly influenced by the heating medium used. In general, permanent growth is attributed to the breakdown of iron carbide into iron and graphite, with resultant volume expansion and to internal oxidation of the iron. Even when heating and cooling rates are fairly high and the iron is held for only short times at the temperatures at which these effects occur, the heating medium may still exert some influence on the amount of growth that takes place. The growth in different heating media of specimens from three different 120 mm (4 ¾ inch) diameter gray iron cylinder castings of the same composition is given in Table 10.4. Because these castings were ferritic prior to hardening, dimensional results were not affected by the breakdown of iron carbide during heat treatment.

Table 10.4. Effect of heating medium on growth of gray iron cylinders.
(Cylinders were 120 mm (4 ¾ inch) in diameter. Composition of iron: 3.38 total C, 2.10 Si, 0.65 Mn, 0.40 Cr, 0.55 Cu + Ni).

Test(a)	*Atmosphere or heating medium*	*Heating time minute*	*Temperature °C*	*°F*	*Average growth, mm/mm (inch/inch)*
1	Charcoal-base(b)	80	845	1550	0.00048
2	Lean endothermic(c)	60	870	1600	0.0019
3	Lean endothermic(c)	60	870	1600	0.0021
4	Air	60	870	1600	0.0026
5	Air	60	870	1600	0.0026
6	Molten salt(d)	10	845	1550	0.0027

(a) Test specimens were taken from three castings: specimens for tests 1 and 6 were taken from the first casting, 2 and 4 from the second and 3 and 5 from the third. Specimens were fully annealed before hardening; final hardness range of oil-quenched specimens, 45 to 50 HRC; (b) approximate composition: 64 N_2, 35 CO, 1 H_2; dew point, –29°C (–20°F); (c) approximate composition: 45 N_2, 20 CO, 35 H_2, 0.4 CO_2, 0.3 CH_4; dew point, 10°C (50°F); and (d) composition: 48 to 52 NaCl, 48 to 52 KCl.

From the data in Table 10.4 it is apparent that the oxidising characteristics of gaseous media have the greatest influence on growth of cast iron heated in these media. It is significant that the growth that occurred in the non-oxidising charcoal-base atmosphere was similar to that of steel during hardening.

The high growth in molten salt may be attributed to thermal upsetting caused by rapid heating. Characteristically, thermal upsetting reduces the length of a casting and increases its girth.

Quenching Facilities

Selection of a quenching medium is closely related to the complexity of the shape to be quenched, which largely determines the probability of quench cracking. Because many castings are of complex design, hot oil and molten-salt quenching are used much more frequently than water quenching. When castings with heavy sections are being quenched, agitation of the quenching bath may be necessary.

When poor quenching facilities are used, increased alloy additions may be necessary to achieve uniform hardening. However, alloy additions may complicate production of castings in the foundry and may increase casting costs.

Chapter 11

Heat Treatment of Gray Irons

INTRODUCTION

The heat treatment most frequently applied to gray iron, with the possible exception of stress relieving, is annealing. Annealing of gray iron consists of heating the iron to a temperature high enough to soften it and/or to minimise or eliminate massive eutectic carbides, thus improving its machinability. This heat treatment reduces mechanical properties substantially, however. It will reduce the grade level approximately to the next lower grade: for example, the properties of a class 40 gray iron will be diminished to those of a class 30 gray iron. Fig. 11.1 shows the effect of full annealing on tensile strength of class 30 gray iron arbitration bars. The degree of reduction of properties depends on the annealing temperature, the time at temperature and the alloy composition of the iron.

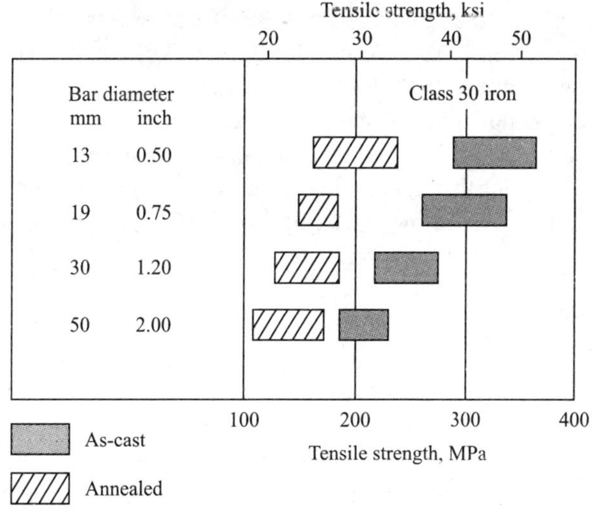

Specimens were arbitration bars from 31 heats. Bars were annealed at 925°C (1700°F) for 2 hour plus 1 hour per 25 mm (1 inch) of section over 25 mm, and cooled at a maximum rate of 160°C/hr. (285°F/hr.) from 925 to 565°C (1700 to 1050°F). Cooling continued from 565°C at a maximum rate of 130°C/hr. (230°F/hr.) to 200°C (390°F); bars were then air cooled to room temperature.

Fig. 11.1. Effect of annealing on tensile strength of class 30 Annealing

ANNEALING OF GRAY IRON

Gray iron commonly is subjected to one of three annealing treatments, each of which involves heating to a different range of temperature. These treatments are ferritising annealing, medium (or 'full') annealing and graphitising annealing.

Ferritising Annealing

For an unalloyed or low-alloy cast iron of normal composition, when the only result desired is the conversion of pearlitic carbide to ferrite and graphite for improved machinability, it generally is unnecessary to heat the casting above the transformation range. Up to approximately 595°C (1100°F), the effect of short times at temperature on the structure of gray iron is insignificant. As the temperature increases above 595°C (1100°F), the rate at which iron carbide decomposes to ferrite plus graphite increases markedly, reaching a maximum at the lower transformation temperature (about 760°C, or 1400°F, for unalloyed or low-alloy iron). The structure of unalloyed gray iron in the As-cast condition and after being held for 1 hr. at 760°C (1400°F) per inch of section and heating to a higher temperature for the same period of time may be detrimental to the annealing process if it causes partial or complete transformation to austenite.

For most gray irons, a ferritising annealing temperature between 705 and 760°C (1300 and 1400°F) is recommended. The furnace temperature profile must be such that castings are sure to reach the set temperatures. Precise temperatures within this range depend on the exact composition of the iron. When machining properties are of primary importance, it in advisable to anneal several samples at various temperatures between 705 and 760°C in order to determine the temperature that yields the lowest final hardness. The casting must be held at temperature long enough to allow the graphitising process to proceed to completion. At temperatures below 705°C (1300°F), an excessively long holding time usually is required. At temperatures between 705 and 760°C (1300 and 1400°F), holding time varies with chemical composition, and may be as short as 10 minutes for unalloyed irons. If an unusually low rate of cooling is employed, the time at temperature may be still further reduced.

Although the rate of cooling *per se* is not of great importance to the annealing process, slow cooling is recommended if the stress relief that automatically occurs during annealing is to be retained as the casting cools to room temperature. A cooling rate as high as 110°C/hr. (200°F/hr.) down to 290°C (550°F) is satisfactory for all except the most complex castings.

Medium ('Full') Annealing

Medium (full) annealing is usually performed at temperatures between 790 and 900°C (1450 and 1650°F). This treatment is used when a ferritising anneal would be ineffective because of the high alloy content of a particular iron. It is recommended, however, that the efficacy of temperatures at or below 760°C (1400°F) be tested before a higher annealing temperature is adopted as part of a standard procedure.

Holding times comparable to those used in ferritising annealing are usually employed. When the high temperatures of medium annealing are used, however, the casting must be cooled slowly through the transformation range, from about 790 to 675°C (1450 to 1250°F).

Graphitising Annealing

If the microstructure of gray iron contains massive carbide particles, higher annealing temperatures are necessary. Graphitising annealing may have the purpose simply of converting massive carbide to pearlite and graphite, although in some applications it may be desired to carry out a ferritising annealing treatment to provide maximum machinability.

Production of free carbide that must later be removed by annealing is, except with pipe and permanent mould castings, almost always an accident resulting from inadequate inoculation or the presence of excess carbide formers, which inhibit normal graphitisation: thus, the annealing process is not considered to be part of the normal production cycle.

To break down massive carbide with reasonable speed, temperatures of at least 870°C (1600°F) are required. With each additional 55°C (100°F) increment in holding temperature, the rate of carbide decomposition doubles: consequently, it is general practice to employ holding temperature of 900 to 955°C (1650 to 1750°F). However, at 925°C (1700°F) and above, the phosphide eutectic present in irons containing 0.10 per cent P or more may melt.

The holding time at temperature may vary from a few minutes to several hours. The chill (white iron) in some high-silicon, high-carbon irons can be eliminated in as little as 15 minute at 940°C (1720°F). In all applications, unless a controlled-atmosphere furnace is used, the time at temperature should be as short as possible, because at these high temperatures gray iron is susceptible to scaling if moisture is present in the furnace atmosphere.

The cooling rate chosen depends on the final use of the iron. If the principal object of the treatment is to break down carbides, and it is desired to retain maximum strength and wear resistance, the casting should be air cooled from the annealing temperature to about 540°C (1000°F), to promote formation of a pearlitic structure. If maximum machinability is the object, the casting should be furnace cooled to 540°C (1100°F) and special care should be exerted to ensure slow cooling from 540°C (1100°F) to about 290°C (550°F) at not more than 110°C C/h (200°F/hr.) is recommended, to minimise residual stresses.

Effect of Alloy Content on Time at Temperature

Certain elements, such as carbon and silicon, will accelerate decomposition of pearlite and massive carbide at annealing temperatures: therefore, when these elements are present in sufficient percentages, the time at annealing temperature may be reduced. In an investigation of the decomposition of pearlite at various temperatures in irons containing 1.93 and 2.68 per cent Si, it was determined that the pearlite always broke down more rapidly in the higher-silicon iron, and that this iron could be effectively annealed over a greater temperature range. For example, at an annealing temperature of 750°C (1380°F), complete breakdown of pearlite occurred in the higher-silicon iron in 10 minute, whereas 45 minute was required for the lower-silicon iron. This shows the pronounced effect of silicon as an aid to diffusion of carbon to the flakes present in the iron.

On the other hand, both the carbide-stabilising elements (vanadium, chromium and manganese) and the pearlite-stabilising elements (phosphorus, nickel and copper) retard the rate of pearlite decomposition. The percentage increases in the time required to decompose pearlite that are effected by 0.10 per cent additions of five of these elements are as follows:

Manganese	60 per cent
Nickel	30 per cent
Copper	30 per cent
Chromium	200 per cent
Phosphorus	30 per cent

NORMALISING

Gray iron is normalised by being heated to a temperature above the transformation range, held at this temperature for a period of about 1hour per inch of maximum section thickness, and cooled in still air to room temperature. Normalising may be used to enhance mechanical properties, such as hardness and tensile strength, or to restore As-cast properties that have been modified by another heating process, such as graphitising or the pre-heating and post-heating associated with repair welding.

The temperature range for normalising gray iron is approximately 885 to 925°C (1625 to 1700°F). Heating temperature has a marked effect on microstructure and on mechanical properties such as hardness and tensile strength. This is demonstrated in Table 11.1, which presents data on specimens 30 mm in diameter by 178 mm (1.2 inch, in diameter by 7 inch) of unalloyed and alloy gray irons that were heated for 1hour at temperatures ranging from 540 to 980°C (1000 to 1800°F) in increments of 55°C (100°F). Specimens of both irons were cooled in still air from each of these temperatures. As indicated by the separation of the table between 760 and 815°C (1400 and 1500°F), air cooling from 760°C or below softens the As-cast structure. Temperatures of 815°C and higher are above the transformation temperature; hence, the matrix of the As-cast iron is converted to austenite on heating and is transformed during air cooling to pearlite that varies in fineness depending on the maximum temperature (the normalising temperature) and the alloy content. For the alloy iron, the higher the normalising temperature, the harder and stronger was the normalised structure. For the unalloyed iron, all normalising temperatures produced the same hardness and strength and all structures produced were softer than the As-cast material. Thus, normalising is a function of alloy content, which affects pearlite spacing.

Table 11.1. Effect of air cooling from various temperatures on typical properties of gray irons.

Condition (a)		Unalloyed iron (b) Hardness, HB	Tensile strength MPa	ksi	Alloyed iron (c) Hardness, HB	Tensile strength MPa	ksi	Combined carbon, %
As-cast		207	265	38.1	212	265	38.7	0.84
Air cooled from:								
°C	°F							
540	1000	202	210	30.4	212	275	39.7	0.82
595	1100	190	255	36.8	210	275	40.0	0.86
650	1200	138	195	28.2	202	265	38.6	0.80
705	1300	125	180	26.4	187	265	38.4	0.81
760	1400	131	190	27.4	170	235	34.2	0.60
815	1500	152	205	29.7	212	295	42.6	0.76
870	1600	152	205	29.7	217	305	44.2	0.81
925	1700	152	205	29.6	223	290	42.4	0.82
980	1800	152	210	30.1	255	340	49.0	0.80

(a) Specimens 30 mm in diameter by 180 mm (1.2 inch in diameter by 7 inch) were held at temperature for 1 hr. before being cooled in still air to room temperature; (b) As-cast composition: 3.15 total C, 0.54 combined C, 2.59 Si, 0.09 P, 0.135 S, 0.88 Mn, 0.01 Cr, 0.10 Ni; and (c) As-cast composition: 3.33 total C, 0.84 combined C, 2.27 Si, 0.76 P, 0.122 S, 0.72 Mn, 0.44 Cr, 0.36 Ni, 0.28 Mo.

Some control of hardness can be exercised during normalising by allowing castings to cool in the furnace to a temperature below the normalising temperature. Fig. 11.2 shows the results obtained with gray iron rings that were heated to 955°C (1750°F), then furnace cooled to different temperatures before being removed from the furnace and cooled in air. These data also indicate that annealing can be accomplished by cooling castings in the furnace to 650°C (1200°F) and then air cooling. However, if stress-free castings are desired, they should be cooled in the furnace to below 455°C (850°F) before being removed.

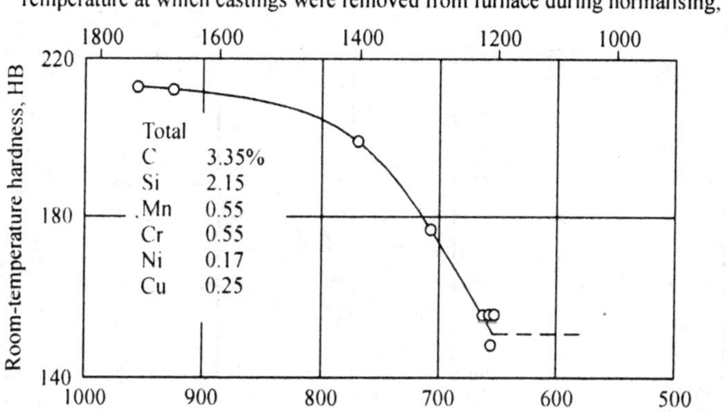

Effect of temperature at start of air cooling on hardness of normalised gray iron rings 120 mm (4¾ inch) in OD, 38 mm (1½ inch) high and 13 mm (½ inch) thick

Fig. 11.2. Room-temperature hardness of gray iron after normalising.

Additional effects of normalising, as a function of alloy content and carbon equivalent, are shown in Table 11.2. Bars 1, 3, 4, 6 and 7 are essentially free of alloying elements, except for residual amounts. Bars 1 and 3, characterised by high As-cast strength and low carbon equivalent, virtually regained their As-cast strength as a result of normalising for 1½ hour at 900°C (1650°F), air cooling, and stress relieving at 540°C (1000°F). The same treatment lowered the strength of bars 4, 6 and 7, all of which had higher carbon equivalents. Bar 2 showed an increase in strength because of the high stabilising effect of the molybdenum and nickel contents. Bar 5, despite a high carbon equivalent, greatly exceeded as-cast strength because of its chromium, molybdenum and nickel contents.

The effect of alloy content on hardness after normalising is shown in Fig. 11.3 for two alloy irons with different carbon equivalents and nickel and chromium contents. Again it is evident that alloy content has a stabilising effect in the graphitising annealing range and serves to increase hardness when the austenitising temperature ranges from about 790 to 980°C (1450 to 1800°F).

Thus it can he concluded that normalising serves to restore As-cast properties to gray iron—or, if carbon equivalent is sufficiently low, even causes these properties to be exceeded—and that the alloying elements chromium, molybdenum and nickel enhance the strengthening effect of normalising.

HARDENING AND TEMPERING

Gary irons are hardened and tempered to improve their mechanical properties, particularly strength and wear resistance. After being hardened and tempered, these irons usually exhibit wear resistance approximately five times as great as that of pearlitic gray irons.

Table 11.2. Influence of alloy content and carbon equivalent on typical properties of gray irons before and after normailising.

Specimens 30 mm (1.2 inch) in diameter were normalised at 900°C (1650°F), then stress relieved at 540°C (1000°F).

	Composition, %										As-cast			Normalised		
										Carbon equivalent,	Tensile strength		Hardness,	Tensile strength		Hardness,
Bar	*C*	*Si*	*P*	*S*	*Mn*	*Cr*	*Ni*	*Mo*	*Cu*	*%*	*MPa*	*ksi*	*HB*	*MPa*	*ksi*	*HB*
1	2.71	2.00	0.13	0.031	0.46	0.076	0.061	0.059	–	3.37	405	58.9	241	380	55.0	241
2	3.25	2.03	0.02	0.031	0.67	0.085	0.80	0.30	0.22	3.93	380	54.8	241	425	62.0	255
3	2.66	1.90	0.03	0.018	0.63	0.063	0.092	0.042	–	3.27	400	57.7	255	385	55.7	241
4	3.15	2.20	0.38	0.018	0.44	0.074	0.071	0.071	0.39	3.88	295	93.0	229	235	34.3	179
5	3.45	2.16	0.09	0.077	0.84	0.39	1.21	0.50	0.10	4.17	250	36.2	248	405	58.5	311
6	3.31	2.10	0.39	0.070	0.41	0.069	0.08	0.055	0.44	4.01	275	39.9	212	200	29.2	163
7	3.42	2.44	0.42	0.058	0.56	0.063	0.058	0.057	0.108	4.23	215	31.1	187	180	26.0	143

Table 11.3. Effect of austenitising temperature on hardness of various oil-quenched gray irons.

Composition, %									Hardness, HB, of	Hardness, HB, after oil quenching from			
TC(a)	*CC(b)*	*Si*	*P*	*S*	*Mn*	*Cr*	*Ni*	*Mo*	As-cast *iron*	*790°C (1450°F)*	*815°C (1500°F)*	*845°C (1550°F)*	*870°C (1600°F)*
3.19	0.69	1.70	0.216	0.097	0.76	0.03	–	0.013	217	159	269	444	477
3.10	0.70	2.05	–	–	0.80	0.27	0.37	0.45	255	207	444	514	601
3.20	0.58	1.76	0.187	0.054	0.64	0.005	Trace	0.48	223	311	477	486	529
3.22	0.53	2.02	0.114	0.067	0.66	0.02	1.21	0.52	241	355	469	486	460
3.21	0.60	2.24	0.114	0.071	0.67	0.50	0.06	0.52	235	208	487	520	512
3.36	0.61	1.96	0.158	0.070	0.74	0.35	0.52	0.47	235	370	477	480	465

(a) Total carbon; and (b) combined carbon.

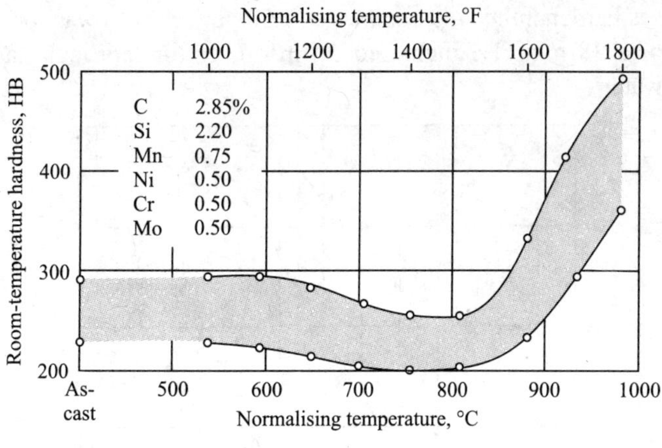

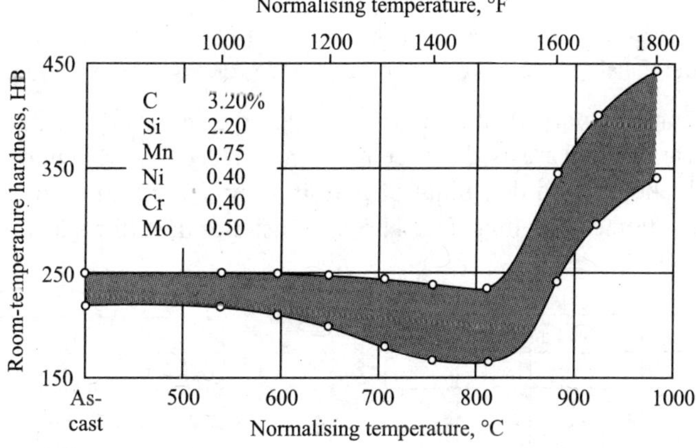

Production and experimental data

Fig. 11.3. Room-temperature hardness of two different alloy gray irons held at normalising temperature 1 hour for each 25 mm (1 inch) of thickness and air cooled on wire-mesh screen.

Furnace or salt-bath hardening can be applied to a wider variety of gray irons than either flame or induction hardening. In flame and induction hardening, a relatively large content of combined carbon is required because of the extremely short period available for solution of carbon in austenite. In furnace or salt-bath hardening, however, the casting can be held at a temperature above the transformation range for as long as necessary and even an iron initially containing no combined carbon can be hardened.

Unalloyed gray iron of low combined carbon content must be austenitised for a relatively long time. Figure 11.4 shows the relation of hardness to holding time for small quenched specimens. With increased time, more carbon is dissolved in austenite, and hardness after quenching is increased.

Because of its higher silicon content, an unalloyed gray iron with a combined carbon content of 0.60 per cent exhibits higher hardenability than a carbon steel with the same carbon content. However, because of the effect of silicon in reducing the solubility of carbon in austenite, unalloyed irons with higher silicon contents necessarily require higher austenitising temperatures to attain maximum hardenability.

Manganese increases hardenability. Approximately 1.50 per cent Mn was found to be sufficient for thorough hardening of a 38 mm (1½ inch) section in oil, or for thorough hardening of a 64 mm (2 ½ inch) section in water.

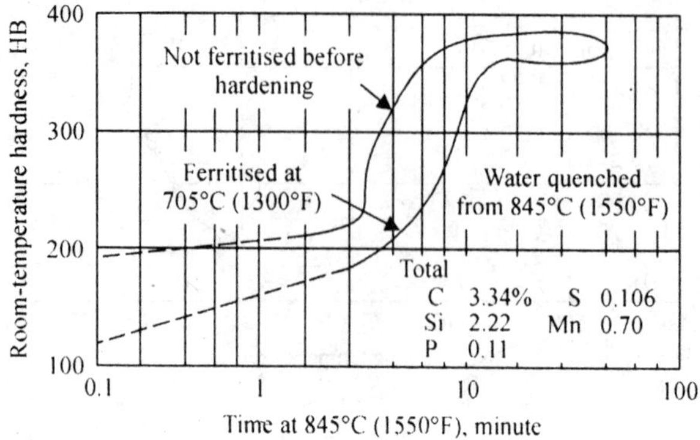

Specimens were 32 mm (1¼ inch) in diameter by 19 mm (¾ inch) thick

Fig. 11.4. Effect of austenitising time on room-temperature hardness of quenched gray iron specimens.

Manganese, vanadium and molybdenum are the recognised elements for increasing the hardenability of gray iron, chromium and nickel are used to accentuate the effect of molybdenum. Although chromium, by itself, does not influence the hardenability of gray iron, its contribution to carbide stabilisation is important, particularly in flame hardening. Fig. 11.5 demonstrates the effects of various combinations of elements on hardenability.

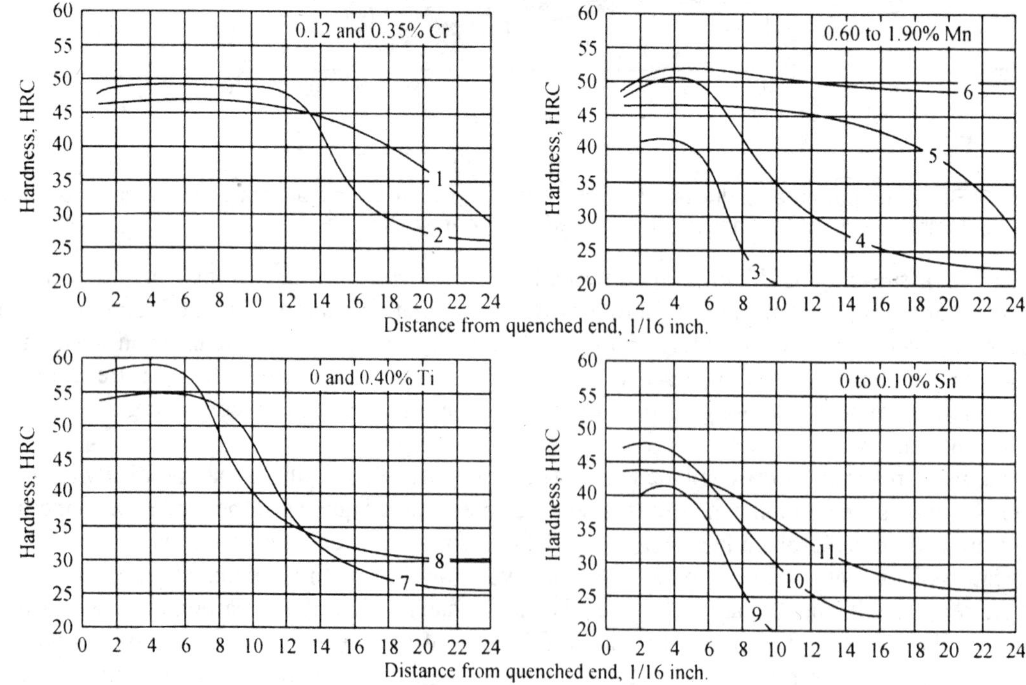

Fig. 11.5. Effects of alloying elements on hardenability of gray iron.

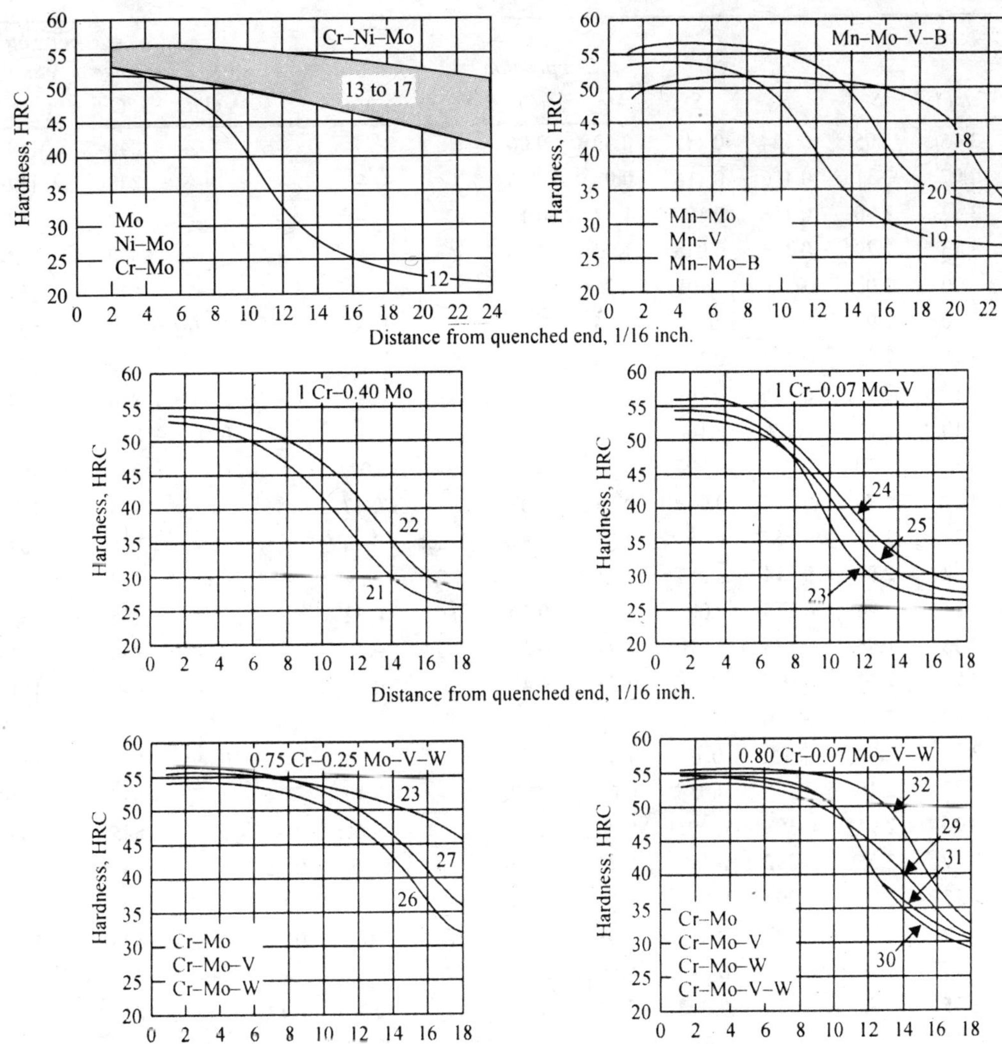

Iron 1 through 20 were induction-furnace melted and poured into standard dry sand molds 30 mm (1.2 inch) in diameter. Alloy constituents for irons 21 through 32 were added to 205-kg (450-lb) ladles, from which the metal was pured into automotive-camshaft production molds. Standard end-quench hardenability specimens were machined from these castings, austentised for 1 hour at the temperatures indicated and then water quenched.

Fig. 11.5. (Continued). Effects of alloying elements on hardenability of gray iron.

					Composition, %						*Austenitising temperature*	
Iron	*Total C*	*Si*	*P*	*S*	*Mn*	*Cr*	*Ni*	*Mo*	*V*	*Other*	*°C*	*°F*
1	3.30	1.40	0.116	0.10	1.47	0.12	–	–	–	–	840	1540
2	3.30	1.90	0.116	0.10	1.43	0.35	–	–	–	–	840	1540

(Contd...)

Iron	Total C	Si	P	S	Mn	Cr	Ni	Mo	V	Other	Austenitising temperature °C	°F
3	3.15	2.05	0.124	0.112	0.60	0.06	–	–	–	–	840	1540
4	2.97	2.31	0.116	0.116	0.92	0.06	–	–	–	–	840	1540
5	3.42	1.90	0.116	0.100	1.47	0.12	–	–	–	–	840	1540
6	3.13	2.29	0.116	0.018	1.90	0.08	–	–	–	–	840	1540
7	3.00	2.00	0.15	0.10	1.25	–	–	–	–	–	840	1540
8	3.00	2.00	0.15	0.10	1.25	–	–	–	–	0.40 Ti	840	1540
9	3.15	2.05	0.124	0.112	0.60	–	–	–	–	–	840	1540
10	3.10	2.25	0.120	0.160	0.65	–	–	–	–	0.05 Sn	840	1540
11	3.10	2.25	0.120	0.160	0.65	–	–	–	–	0.10 Sn	840	1540
12	3.19	1.70	0.216	0.097	0.76	0.03	–	0.013	–	–	855	1575
13	3.22	1.73	0.212	0.089	0.75	0.03	–	0.47	–	–	855	1575
14	3.20	1.76	0.187	0.054	0.64	0.005	Trace	0.48	–	–	855	1575
15	3.22	2.02	0.114	0.067	0.66	0.02	1.21	0.52	–	–	8.55	1575
16	3.21	2.24	0.114	0.071	0.67	0.50	0.06	0.52	–	–	855	1575
17	3.36	1.96	0.158	0.070	0.74	0.35	0.52	0.47	–	–	855	1575
18	3.21	2.01	0.15	0.10	1.53	0.40	–	0.13	–	–	840	1540
19	3.20	2.00	0.15	0.10	1.25	0.40	–	–	0.05	–	840	1540
20	3.10	2.09	0.15	0.10	1.46	0.44	–	0.14	–	0.095B	840	1540
21	3.22	2.10	0.108	0.088	0.68	0.97	–	0.40	–	–	845	1550
22	3.20	2.15	0.108	0.093	0.70	1.00	–	0.41	–	–	845	1550
23	3.19	2.55	0.092	0.090	0.71	0.96	–	0.054	0.16	–	845	1550
24	3.17	2.20	0.094	0.092	0.66	0.95	–	0.069	0.081	–	845	1550
25	3.19	2.20	0.092	0.092	0.68	0.93	–	0.075	0.27	–	845	1550
26	3.17	1.90	0.080	0.094	0.65	0.73	–	0.19	–	–	845	1550
27	3.25	1.85	0.074	0.092	0.65	0.77	–	0.30	0.13	–	845	1550
28	3.21	1.90	0.069	0.100	0.70	0.75	–	0.28	–	0.40 W	845	1550
29	3.20	2.20	0.096	0.090	0.68	0.94	–	0.047	0.13	0.75 W	845	1550
30	3.12	1.80	0.074	0.090	0.69	0.75	–	0.064	–	–	845	1550
31	3.18	1.80	0.073	0.090	0.68	0.77	–	0.091	0.12	–	845	1550
32	3.14	1.70	0.079	0.090	0.69	0.77	–	0.071	–	0.37 W	845	1550

Austenitising

In hardening gray iron, the casting is heated to a temperature high enough to promote formation of austenite, held at that temperature until the desired a mount of carbon has been dissolved and then quenched at a suitable rate. Heating for austenitising may be accomplished in a salt bath or in an electrically heated, gas-fired or oil-fired furnace.

The temperature to which the casting must be heated is determined by the transformation range of the particular gray iron of which it is made. The transformation range can extend more than 55°C (100°F) above the A_1 (transformation-start) temperature. A formula for determining the approximate A_1 transformation temperature of unalloyed gray iron is:

°C: 730 + 28.0 (per cent Si) – 25.0 (per cent Mn)

°F: 1345 + 50.4 (per cent Si) – 45.0 (per cent Mn)

Chromium raises the transformation range of gray iron. In high-nickel, high silicon irons, for example, each per cent of chromium raises the transformation range by about 40°C (72°F). Nickel, on the other hand, lowers the critical range. In a gray iron containing from 4 to 5 per cent Ni, the upper limit of the transformation range is about 710°C (1310°F).

Provided that recommended limits are not exceeded, the higher the casting is heated above the transformation range, the greater will be the amount of carbon dissolved in the austenite (Fig. 11.6) and the higher will be the hardness of the casting after quenching Table 11.3. In practice, temperatures as much as 95°C (175°F) higher than the calculated A_1 transformation temperature are used to ensure full austenitising. However, excessively high temperatures should be avoided, because quenching from such high temperatures increases the danger of distortion and cracking and promotes retention of austenite, particularly in alloyed irons.

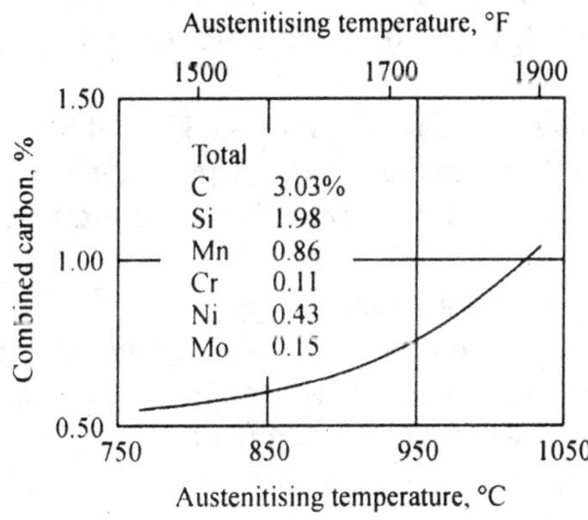

Fig. 11.6. Increase in combined carbon with increase in austenitising temperature for gray iron.

Castings should be heated through the lower temperature range slowly, in order that cracking may be avoided. Above a range of 595 to 650°C (1100 to 1200°F)—that is, above the stress-relieving range heating may be as rapid as desired. In fact, time may be saved by heating the casting slowly to about 650°C in one furnace and then transferring it to a second furnace and bringing it rapidly up to the austenitising temperature. As little as 20 minutes per inch of section may be sufficient time at temperature.

Quenching

Oil is the quenching medium most frequently used for gray iron. Water generally is not a satisfactory quenching medium for furnace heated gray iron; It extracts heat so rapidly that distortion and cracking are likely in all except small parts of simple design. Recently developed water soluble polymer quenches can provide the convenience of water quenching along with lower cooling rates, which can minimise thermal shock.

The least severe quenching medium is air. Unalloyed or low-alloy gray iron castings usually cannot be air quenched, because the cooling rate is not high enough to form martensite. However, for irons of high alloy content, forced-air quenching is frequently the most desirable cooling method.

A casting of non-uniform section should be quenched in such a way that the heavier section enters the quenching bath first. During quenching, agitation is desirable because it ensures even temperature distribution in the bath and improves quenching efficiency. Because As-quenched castings at room temperature are extremely sensitive to cracking, they should be removed from the quench bath as soon as their temperature falls to about 150°C (300°F) and tempered immediately.

Tempering

After quenching, castings usually are tempered at temperatures well below the transformation range for about 1 hour per inch of thickest section. As the quenched iron is tempered, its hardness decreases, whereas it usually gains in strength and toughness (Figs. 11.7 and 11.8).

Applications

Examples of the quenching and tempering temperatures employed, and results obtained, for four different production parts made of gray iron are as follows:

Example 1. Unalloyed gray iron valve guides were heated in an atmosphere-controlled furnace and held for 1 hr. at 885°C (1625°F). The guides were then quenched in oil maintained at about 60°C (140°F). An as-quenched hardness of 45 to 50 HRC was obtained. After tempering at 480°C (900°F), hardness was 30 to 34 HRC.

Example 2. Valve guides made of gray iron containing 3.40 C, 2,40 Si, 0.21 Cr and 0.50 Cu were heated to 870°C (1600°F) and held at this temperature for 1 hr. After being oil quenched from 870°C, the guides were tempered for 1 hr. at 495°C (925°F). Hardness distributions after quenching and after tempering, for 25 guides, are shown in Fig.11.9.

Example 3. Automotive valve lifter castings made of gray iron containing 3.10 to 3.30 C, 2.10 to 2.40 Si, 1.00 to 1.25 Cr, 0.40 to 0.70 Ni and 0.50 to 0.70 Mo were heated in a gas-fired radiant-tube conveyor furnace. A controlled atmosphere of endothermic generator gas was used to prevent decarburisation of the machined surfaces. The castings were held for 45 minute at 855°C (1570°F), quenched in oil at 55°C (130°F), then tempered for 3 hour at 150°C (300°F) to a hardness of 55 to 61 HRC. The hardness readings were made 1.6 mm ($\frac{1}{16}$ inch) off centre of the tappet face. A distribution of hardness values representing castings from 25 heats is given in Fig. 11.9.

Example 4. Ring-shape castings 64 mm (2.5 inch) in outside diameter by 19 mm (0.75 inch) high by 6.4 to 13 mm (0.25 to 0.5 inch) in wall thickness, made of a hardenable gray iron, were heat treated in groups and sampled for response to heat treatment. These castings were heat treated by holding for ½ hour at 850°C (1560°F) followed by oil quenching. Hardness data for 20 consecutive lots representing 40 heats are shown in Fig. 11.9. Each casting was tested at and opposite the gate.

Strength

Tempering increases the tensile strength of hardened gray iron. Although the tempering temperature that will yield maximum strength depends on chemical composition and increases with higher alloy content, the high-strength tempering range for both unalloyed and alloyed gray irons is approximately 370 to 510°C (700 to 950°F).

Other properties

Tempering temperatures near 370°C (700°F) increase the impact resistance of hardened low silicon unalloyed gray iron (Fig. 11.7). If alloying elements are present, higher tempering temperatures may be necessary.

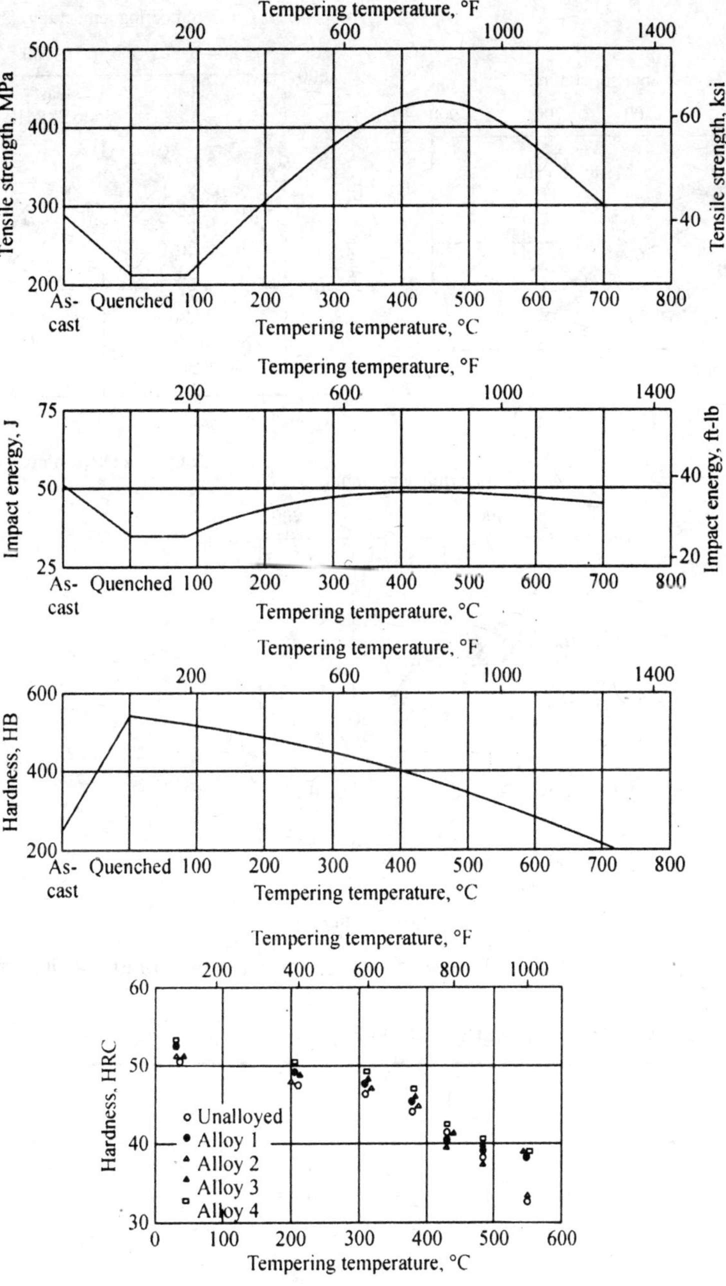

(a) (c) Changes in mechanical properties of hardened low-silicon unalloyed gray iron.
(b) Hardness of gray iron specimens quenched in oil from 870°C (1600°F) and tempered. Each point on chart represents an average of five hardness readings. For compositions of irons, see table 11.4

Fig. 11.7. Effect of tempering temperature on gray iron.

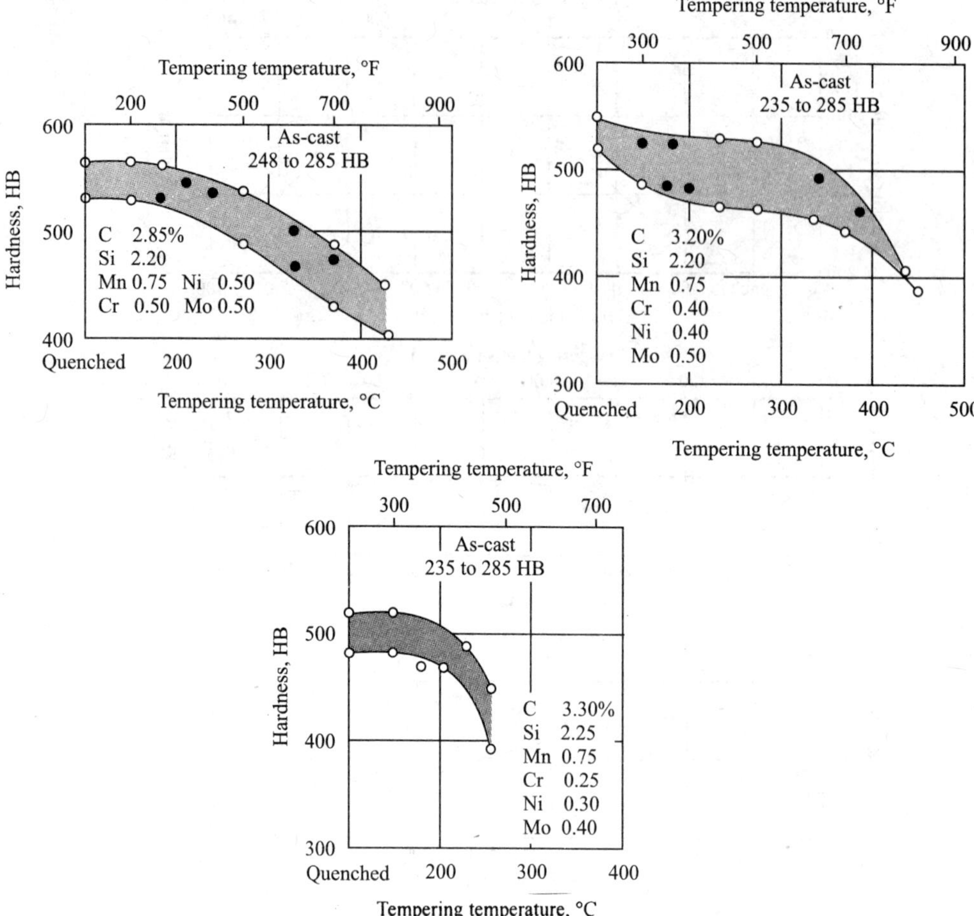

Fig. 11.8. Influence of alloy content on hardness of quenched and tempered gray iron test castings.

Table 11.4. Composition of gray irons in Fig. 11.7d.

	Composition, %					
Iron	*TC(a)*	*CC(b)*	*Si*	*Cr*	*Ni*	*Mo*
Unalloyed	3.20	0.80	2.43	0.13	0.05	0.17
1	3.23	0.70	2.55	0.58	0.06	0.12
2	3.21	0.83	1.90	0.08	0.78	0.27
3	3.29	0.79	2.58	0.24	0.10	0.55
4	3.02	0.75	2.38	0.40	0.07	0.43

(a) Total carbon; and (b) combined carbon.

Although considerable improvements in tensile strength may be brought about by hardening and tempering, proportional improvements in fatigue properties are rarely obtained.

The modulus of elasticity of a low-alloy (1.32 per cent Ni, 0.44 per cent Cr) gray iron was found to be moderately improved by quench hardening and tempering:

Condition	*Modulus of elasticity*	
	GPa	*ksi*
As cast	122	17,700
Oil quenched	112	16,200
Tempered at 300°C (570°F)	131	19,000

The modulus of rupture increased markedly with the use of higher tempering temperatures. A similar increase was noted for the transverse breaking load and deflection. In an alloy iron containing 4.5 per cent Ni and 1.5 per cent Cr, transverse breaking load was increased substantially, even with tempering temperatures as low as 205°C (400°F).

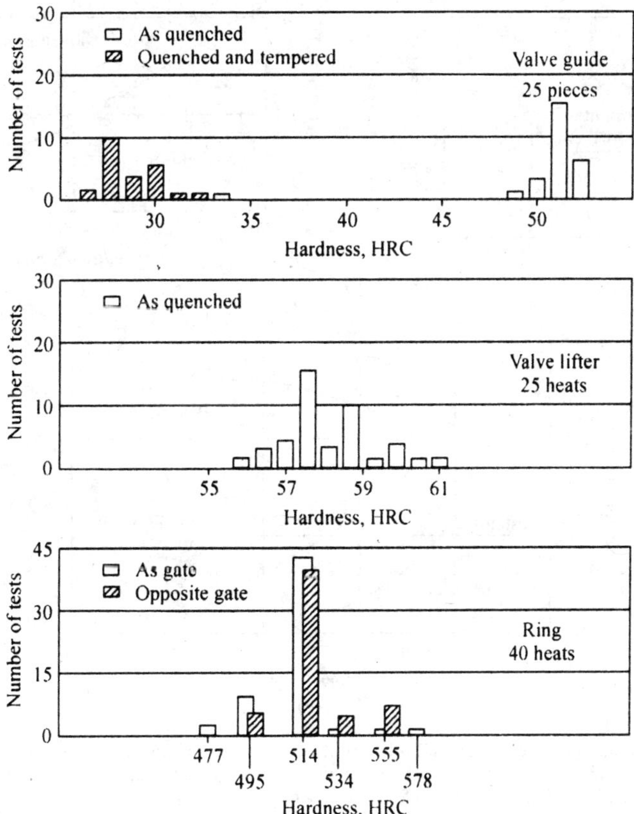

Fig. 11.9. Distribution of hardness for quenched and tempered production parts made of gray iron.

AUSTEMPERING

In austempering, the microstructural end product of the gray iron matrix formed below the pearlite range but above the martensite range is an acicular or bainitic ferrite plus varying amounts of retained austenite depending on the transformation temperature. As shown in Fig. 11.10a, the iron is quenched from a temperature above the transformation range in a hot quenching bath and is maintained in the bath at constant temperature until transformation has ceased.

In all hot quenching processes, the temperatures to which castings must be heated for austenitising and the required holding times at temperature prior to quenching in the hot bath correspond to the temperatures and time used in conventional hardening—that is, temperatures between 760 and 900°C (1400 and 1650°F) and holding times of 20 minutes per inch of thickness. A characteristic curve for hardness as related to isothermal transformation temperature is shown in Fig. 11.10b.

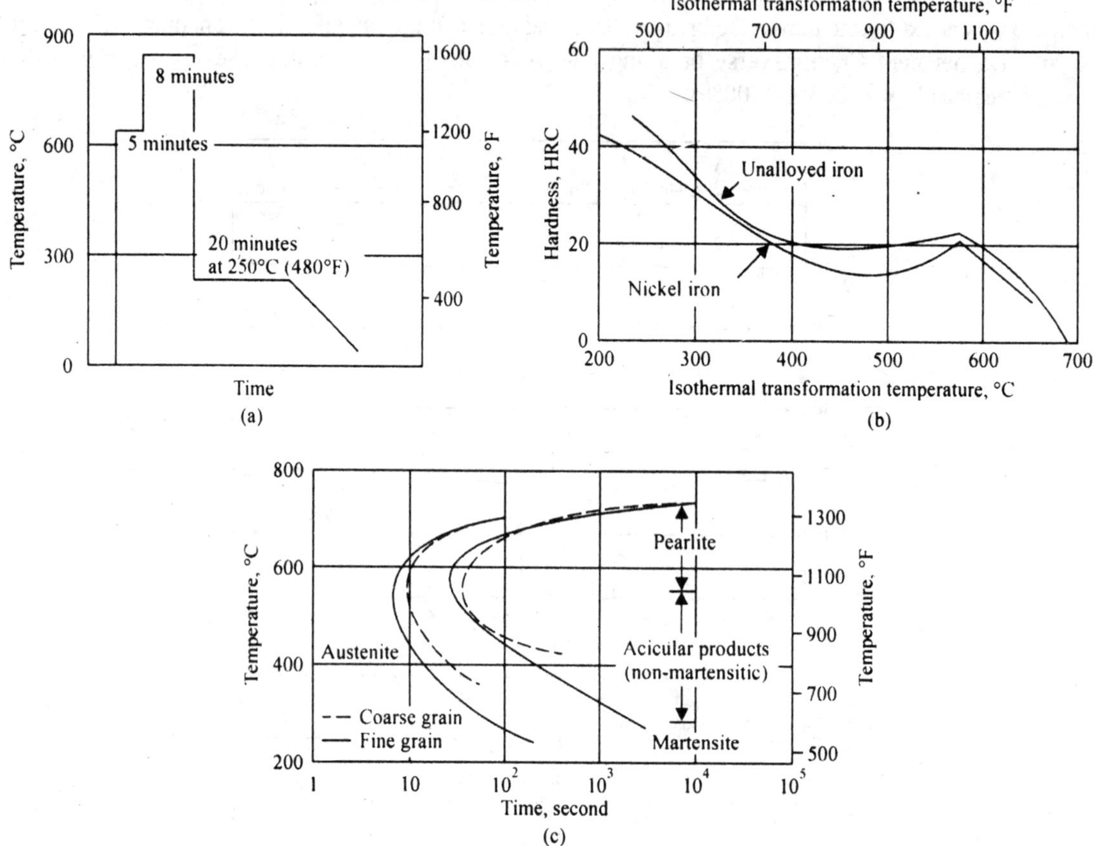

(a) Schematic representation of one austempering cycle, (b) Effect of isothermal transformation temperature on hardness of austempered gray irons. Holding times were sufficient to complete transformation, (c) Time-temperature transformation curves for gray iron containing 3.75 C, 2.9 Si, 0.4 P, 0.065 S and 0.55 Mn.

Fig. 11.10. Austempering of gray iron.

Gray iron is usually quenched in salt, oil or lead baths at 230 to 425°C (450 to 800°F) for austempering. When high hardness and wear resistance are the ultimate aim of this treatment, the temperature of the quench bath is usually held between 230 and 290°C (450 and 550°F).

The holding time required for maximum transformation is determined by the temperature of the quenching bath and the composition of the iron. The influence of austenite grain size in gray iron on holding time is usually small, as shown in Fig 11.10c. The effect of iron composition on the holding time may be considerable. Alloy additions, such as nickel, chromium and molybdenum, increase the time required for transformation.

The section thickness and shape of the casting often limit the use of austempering, because cooling must be fast enough to prevent any transformation of austenite until the casting reaches the temperature of the bath. Gray iron cylinder liners containing a minimum of 1.5 per cent total Cr, Ni, Mo and Cu have been austempered at about 260°C (500°F) at rates of 35 to 75 liners per hour, depending on size.

MARTEMPERING

Martempering is used to produce martensite without developing the high stresses that usually accompany its formation. It is similar to conventional hardening, except that distortion is minimised. Nevertheless, the characteristic brittleness of the martensite remains in a gray iron casting after martempering and martempered castings are nearly always tempered. As shown in Fig. 11.11 the casting is quenched from above the transformation range in a salt, oil or lead bath, held in the bath at a temperature slightly above the range at which martensite forms (205 to 260°C or 400 to 500°F for unalloyed irons) only until the casting has reached the bath temperature and then cooled to room temperature.

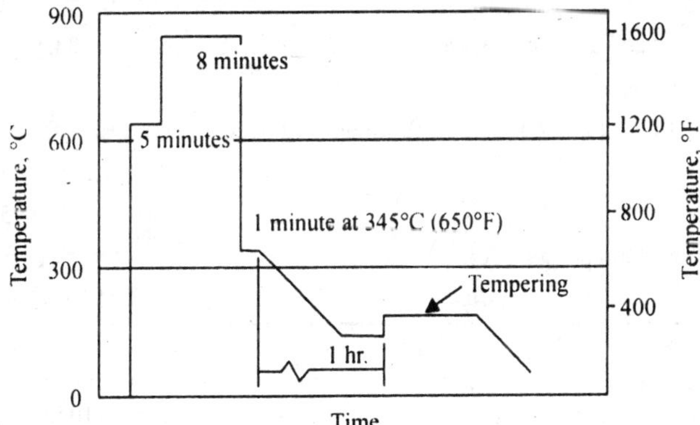

Fig. 11.11. Schematic representation of one martempering and tempering cycle for gray iron.

If a wholly martensitic structure is desired, the casting must be held in the hot quench bath only long enough to permit it to reach the temperature of the bath. Thus, the size and shape of the casting dictate the duration of martempering.

The following example illustrates the use of martempering.

Example 5

Cylinder liners (sleeves) for diesel and heavy-duty gasoline engines, made of gray iron containing 3.25 to 3.50 C, 2.00 to 2.25 Si, 0.25 max P, 0.12 max S, 0.55 to 0.80 Mn and 0.30 to 0.40 Cr, were fully annealed at 870°C (1600°F) in a neutral atmosphere for a total of 1h (Table 11.5 and Fig. 11.12). They were quenched for 1 minute in molten nitrate-nitrite salt at 245°C (475°F) and air cooled for 30 minute After being washed to remove adhering salt, the casting were tempered for 2 hour at 205°C (400°F).

Austenitising for martempering was accomplished in a pusher type continuous furnace. Sixteen sleeves were loaded on each rack, which rests on a furnace tray. On emerging from the furnace, the rack was pushed from the tray onto a quench elevator that immediately descended into the salt bath: thus, the furnace tray did not come in contact with the molten salt. The sleeves and rack were automatically raised from the salt and remained above the bath for drainage until the next rack was pushed into position, at which point the drained rack was pushed onto a slowly moving conveyor for the 30 minute cooling cycle. After cooling, the rack and sleeves were moved onto the washer conveyor, and after being washed, they were deposited on the tempering furnace conveyor. Equipment details are summarised in Table 11.5.

Table 11.5. Equipment requirements for martempering cast iron cylinder liners.

Cylinder liners were austenitised in a radiant-tube gas-fired single-row pusher furnace containing 27 U-type radiant tubes and accommodating 13 trays per cycle. Each tray measured 610 by 710 mm (24 by 28 inch). Austenitising temperature was 870°C (1600°F). An endothermic atmosphere of 34 m^3/hr. (1200 ft^3/hr.) was controlled to 0.2 to 0.4 per cent CO_2 by Orsat analysis.

Production requirements	
Weight of each piece	3.8 kg (8.3 lb)
Number of pieces per push	16
Number of pushes per hour	11
Weight of empty rack	32 kg (70 lb)
Total weight of material heated and quenched per hour	1010 kg (2231 lb)
Martempering furnace requirements	
Salt pot, 1.83 by 1.98 by 1.83 m (72 by 78 by 72 inch). Immersion heated, 6.6 m^3 (234 ft^3) (a)	
Type of salt	Nitrate-nitrite (b)
Operating temperature of salt	245°C (475°F)
Agitation, by 230 mm (9 inch) diam propeller(c)	One 7½-hp variable speed motor
Cooling system	Three ½-hp fans(d)

(a) Pot, containing about 11.8 tonne (13 tonnes) of molten salt, was heated by 24 immersion rods, each rated at 5 kW; (b) melting point of salt, 145°C (290°F). No chloride-separation chamber was required, because parts were transferred to salt from a controlled-atmosphere hardening furnace; (c) propeller was in a 255 mm (10 inch)-diam pipe connected to the quench funnel; and (d) fans force outdoor air through ducts in furnace walls.

The As-cast sleeves had a graphitic structure predominantly of type A and a graphite flake size of 4 to 6. During the hardening cycle, the matrix dissolved about 0.70 per cent C, and a martensitic structure resulted. The hardness range after hardening was 45 to 50 HRC; microhardness of the matrix was more than 60 HRC.

Martempering replaced conventional oil quenching, during which each cylinder liner had been placed on a mandrel in order to preserve shape. Despite the use of the mandrel, out-of-roundness after hardening (Fig. 11.12) and size variations were about twice as great for oil-quenched liners as for those that were martempered. Martempering also eliminated the small percentage of cracked sleeves previously encountered and reduced the amount of grinding stock required for finishing.

If final dimensional accuracy is important for martempered parts, allowance for growth must be made prior to heat treatment. The uniformity of growth and the allowance required depend on the condition of the iron prior to hardening. Annealing the cylinder sleeves discussed in Example 5 at 870°C (1600°F) and furnace cooling prior to martempering resulted in growth of 0.03 to 0.38 mm/mm (0.0010 to 0.0015 inch/inch).

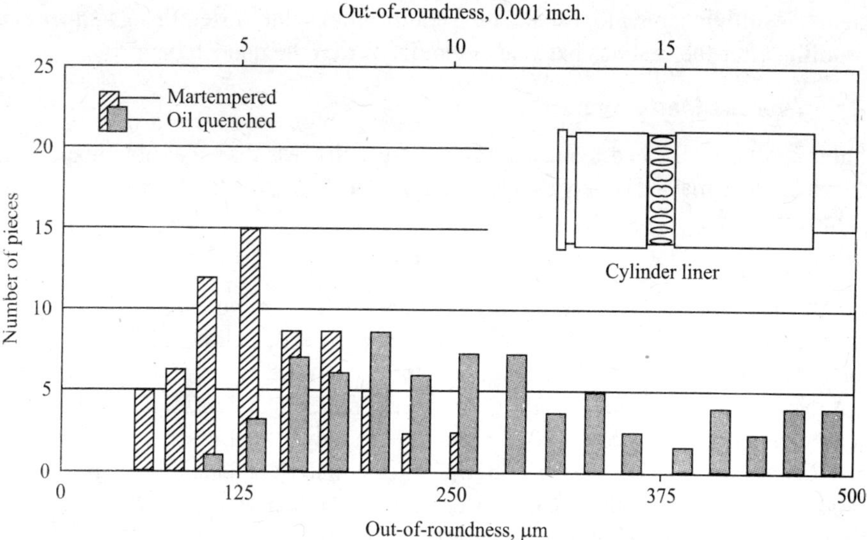

Before being measured, liners were furnace tempered for 2 hour at 205°C (400°F)

Fig. 11.12. Distortion in gray iron cylinder liners after martempering and after conventional oil quenching.

This was only half as much growth as was encountered when as cast or normalised pearlitic structures were martempered. Time at temperature during the annealing cycle was not critical in the range of 2 to 8 hour. The effect of annealing temperature on growth of 305 mm (12 inch) long cast specimens during subsequent martempering is shown in Fig. 11.13.

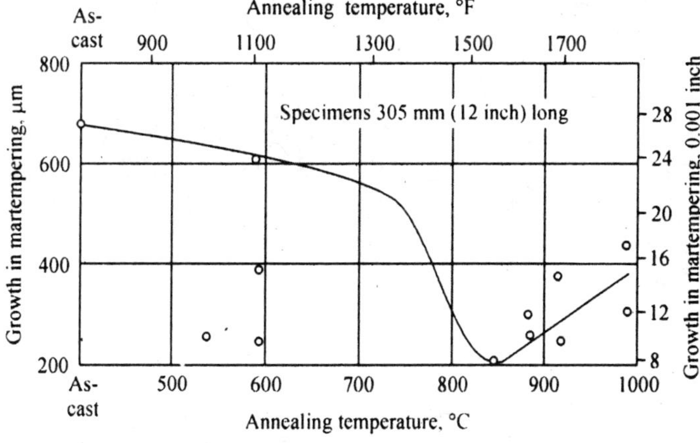

Heat treatment consisted of holding at the annealing temperature for 2 hour and cooling in the furnace, re-heating to 870°C (1600°F)in a neutral endothermic atmosphere, quenching for 1 minute in molten salt at 245°C (475°F), air cooling for 30 minute and tempering at 205°C (1400°F) for 2 hour As-casts specimens were not annealed before hardening.

Fig. 11.13. Effect of annealing temperature on growth of gray iron during martempering.

Some distortion, as differentiated from growth, occurs during martempering. Conditions that may promote excessive distortion are; residual stresses from casting, machining or rapid cooling during prior heat treatments; insufficient time for establishing equilibrium at the austenitising temperature; and drafts during air cooling after the castings have been removed from the quench bath.

Austempering versus Martempering

The maximum hardness obtained by austempering is usually less than that obtained by martempering, although this difference may be largely cancelled during the tempering treatment that usually is necessary following martempering.

Example 6

A high-strength, low-silicon processed iron martempered for 1 minute at 260°C (500°F) exhibited a hardness of 555 HB, whereas the same iron austempered for 30 minute at 275°C (530°F) showed a hardness of only 444 HB. After being tempered for 30 minute at 390°C (730°F), the martempered material had about the same hardness as the austempered material, but its impact resistance (Charpy test, 15-mm-square specimen, knife edges 70 mm apart) was only 14.2 J or 10.5 ft.lb, whereas that of the austempered material was 23.7 J or 17.5 ft.lb (the As-cast material had an impact resistance of 19.7 J or 14.5 ft.lb, and a hardness of 255 HB). Both austempering and martempering can result in less distortion and growth than conventional oil quenching and tempering.

Example 7

When quenched in oil at 95°C (200°F) and tempered at 290°C (550°F), diesel-engine cylinder liners exhibited an average distortion of 0.26 mm (0.0101 inch) and cracked occasionally. When austempered, the same liners had an average distortion of 0.07 mm (0.0026 inch) and when martempered they displayed an average distortion of 0.06 mm (0.0025 inch). The average maximum growth for these liners was as follows:

Oil quenched and tempered	0.71 mm (0.0279 inch)
Austempered	0.25 mm (0.0099 inch)
Martempered	0.23 mm (0.0089 inch)

The austempering cycle consisted of pre-heating for 5 minute at 650°C (1200°F), austenitising for 8 min. at 870°C (1600°F), quenching to 250°C (480°F) and holding for 20 minute and cooling in air. The martempering cycle comprised pre-heating for 5 minute at 650°C (1200°F), austenitising for 8 minute at 855°C (1575°F), quenching for 1 minute in agitated salt at 345°C (650°F), cooling in air for a minimum of 1 hour and tempering at 175°C (350°F).

FLAME HARDENING

Flame hardening is the method of surface hardening most commonly applied to gray iron. After flame hardening, a gray iron casting consists of: (i) a hard, wear-resistant outer layer of martensite; and (ii) a core of softer gray iron, which during treatment did not reach the A_1, transformation temperature (in fact, the unhardened metal immediately below the hardened case, which has been to some extent heated by the flame, may even be partially annealed during flame hardening if it is unalloyed).

Recommended Composition

Both unalloyed and alloyed gray irons can be successfully flame hardened. However, some compositions yield much better results than others. One of the most important aspects of composition is the combined carbon content, which should be in the range of 0.50 to 0.70 per cent, although irons with as little as 0.40 per cent combined carbon can be flame hardened. In general flame hardening is not recommended for irons that contain more than 0.80 per cent combined carbon, because such irons (mottled or white irons) may crack in surface hardening.

The stability of the micro constituents from which the carbon precipitates is a factor in determining the hardness of flame-hardened iron.

Example 8

One user has observed that the stability of microconstituents containing combined carbon is indicated by the hardness of test specimens annealed at 845°C (1550°F). A test of four pearlitic irons from different sources, containing 3.30 max. C, 1.40 to 1.60 Si, 0.80 to 1.10 Mn, and 0.15 to 0.20 Cr, gave the following results:

	Hardness, HB		
Iron	*As-cast*	*Annealed*	*Hardness(a), HRC*
A	220	200	47–50
B	190	180	47–50
C	200	140	35–38
D	200	120	20–25

(a) After flame hardening As-cast iron

To obtain better stabilisation of these microconstituents, 0.15 to 0.30 per cent Mo was added, and the chromium and manganese contents were increased.

For maximum hardness, it is advisable to employ an iron containing as small an amount of total carbon as is consistent with the production of sound casting free from any danger of cracking. The coarse graphite flakes typical of high-carbon irons should be avoided because they may be burned out during flame heating thereby producing a porous and unattractive surface.

Because silicon promotes formation of graphite and of a low combined carbon content, a relatively low silicon content also is advisable. Several users report that silicon content should not exceed 2 per cent in any iron submitted to a flame-hardening operation, although alloy irons containing 2.4 per cent Si can be hardened successfully. They also recommend that manganese content be held in the range of 0.80 to 1.00 per cent, to enhance depth of hardening. Gray iron to be flame hardened should be as free as possible from porosity and from foreign matter such as sand or slag, because porosity and even small inclusions of foreign matter can produce a rough surface or result in cracking after hardening. Rough casting surfaces should be sand or shot blasted prior to heat treatment, because skin or scale on the surface acts as a heat insulator and reduces the effectiveness of flame hardening.

Effects of Alloying Elements

In general, alloy gray irons can be flame hardened with greater ease than can unalloyed iron-spartly because alloy irons can be heat treated successfully over a wider temperature range, thus minimising the necessity for precise pyrometric control. Also, the hardenability, or depth of martensitic case, is increased by addition of most alloying elements.

Final hardness also may be increased by alloying additions. The maximum hardness obtainable by flame hardening an unalloyed gray iron containing approximately 3.0 per cent total carbon, 1.7 per cent Si and 0.60 to 0.80 per cent Mn ranges from 400 to 500 HB. This is because the Brinell hardness value for gray iron is an average of the hardness of the matrix and that of the relatively soft graphite flakes. Actually, the matrix hardness on which wear resistance depends approximates 600 HB. With the addition of 2.5 per cent Ni and 0.5 per cent Cr, an average surface hardness of 550 HB can be obtained. The same result has been achieved using 1.0 to 1.5 per cent Ni and 0.25 per cent Mo. Small additions of chromium are particularly valuable in preventing softening and ensuring retention of a high content of combined carbon during austenitising for hardening. Automotive camshafts containing 1 per cent Cr, 0.50 per cent Mo and 0.8 per cent Mn are easily flame hardened to 52 HRC to required depth. These parts are not tempered or stress relieved.

Stress Relieving

Whenever practicable or economically feasible, flame-hardened castings should be stress relieved at 150 to 205°C (300 to 400°F) in a furnace, in hot oil, or by passing a flame over the hardened surface. Such a treatment will minimise distortion or cracking and will increase the toughness of the hardened layer.

Stress relieving at 150°C (300°F) for 7 hours was found to remove 25 to 40 per cent of the residual stresses in a flame-hardened casting, while reducing the hardness of the surface by only 2 to 5 points on the HRA scale. Although stress relieving is desirable, it can often be safely omitted.

Hardness

The surface of flame-hardened gray iron typically has a somewhat lower hardness than the metal immediately below the surface (Fig. 11.14). This decrease in hardness may be caused by retention of relatively soft austenite at the surface. Surface hardness often can be raised by heating in the range from 195 to 250°C (380 to 480°F).

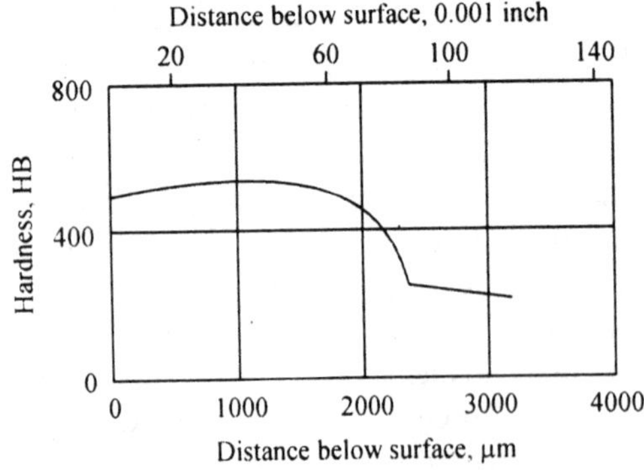

Fig. 11.14. Typical hardness gradient produced in gray iron by flame hardening.

The depth and microstructure of the hardened layer depend on two other factors in addition to temperature: (i) the amount of carbon and alloying elements in solution when the flame-hardened surface is quenched; and (ii) the efficiency of quenching. If softness is due to the presence of austenite in the microstructure, subjecting the part to –40°C (–40°F) for 1 hour will transform the austenite and increase the hardness.

Fatigue Strength

Fatigue strength is usually increased by surface hardening, because the treatment induces compressive stresses at the surface. For example, flame or induction heating followed by water quenching induces high compressive stresses (over 205 MPa, or 30 ksi) in the fillet areas of crankshafts; however, if self-quenching instead of water quenching is used, undesirable tensile stresses may result at the surface. This degree of improvement in fatigue strength cannot be produced by thorough hardening.

Quenching

The various methods of flame hardening influence selection of the quenching medium. In the progressive method, only non-flammable media such as water, soluble-oil mixtures, and solutions of polyvinyl alcohol in water can be employed. Conventional oil cannot be used because of the fire hazard. In spot hardening or spinning methods, in which the flame head is withdrawn from the part before quenching, parts are quenched conventionally by immersion in hot oil.

When quenching is done with water, the water should be at about 32°C (90°F) for best results. Lower quenching rates, such as those obtained with 5 to 15 per cent soluble-oil mixtures, compressed air, or compressed air and water at low pressure, are employed to prevent cracking. Air quenching is especially suited to highly alloyed cast irons because of their susceptibility to cracking.

INDUCTION HARDENING

Gray iron castings can be surface hardened by the induction method when the number of castings to be processed is large enough to warrant the relatively high equipment cost and the need for special induction coils. For details of this process and its application to gray iron, see 'Induction Hardening and Tempering'.

STRESS RELIEVING

Gray iron in the As-cast condition contains residual stresses (unless the iron is cooled in the mould, in which case, much of the solidification stress is removed), because cooling (and therefore contraction) proceeds at different rates throughout various sections of a casting. The resultant residual stresses may reduce strength, cause distortion, and in some extreme cases even result in failure or cracking. The magnitudes of these stresses depend on the shape and section size of the casting, on the casting technique employed, on the composition and properties of the material cast, and on whether the casting has been stress relieved.

Temperature

Temperature of stress relieving usually is well below the range for transformation of pearlite to austenite. The effects of stress relieving at 650°C (1200°F) for 6 hours on the tensile strength and hardness of gray irons of classes 25, 35 and 50 are plotted in Fig. 11.15. As indicated, the properties of the class 25 iron were affected considerably more than were those of the class 35 and class 50 irons.

Fig. 11.16 shows the effect of stress-relieving temperature on hardness of unalloyed and alloyed gray irons. For maximum relief of stress with minimum decomposition of carbide in unalloyed irons, a temperature range of 540 to 565°C (1000 to 1050°F) is desirable. Figure 11.17 indicates that from 75 to 85 per cent of the residual stress can be removed by holding for 1h in this range. Other investigations (for example., Fig. 11.18 indicate that the curves in Fig. 11.17 are applicable to gray irons over a wide range of composition.

When almost complete stress relief (over 85 per cent) is required in unalloyed iron, a minimum temperature of 595°C (1100°F) can be employed; however, some sacrifice in strength, hardness and wear resistance is likely. Fortunately, the soft irons of higher carbon equivalent normally exhibit lower level of residual stress and comparatively low creep resistance, which facilitates stress relief. In fact, irons of high carbon equivalent can be satisfactorily stress relieved at the lower end of the suggested temperature range—that is, at about 510°C (950°F). Low-alloy gray irons usually require higher stress-relieving temperature—on the order of 595 to 650°C (1100 to 1200°F), depending on alloy content.

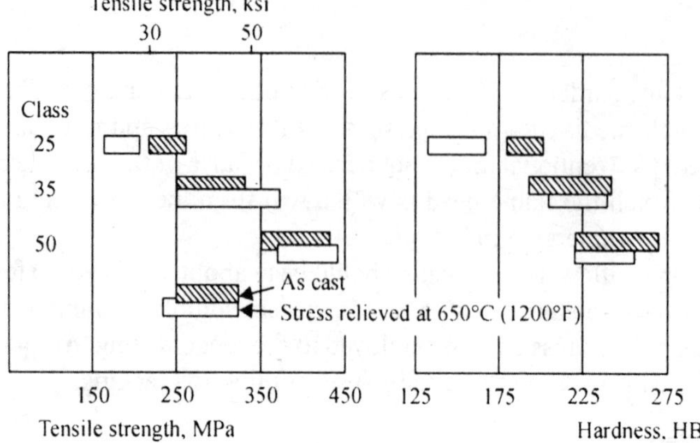

Gray iron bars, 30 mm (1.2 inch) in diameter, were stress relieved for 6 hours at 650°C (1200°F) in a car-bottom furnace approximately 13 by 4 by 3m (42 by 13 by 9 ft); total furnace time was 43¾ hours.

Fig. 11.15. Effects of stress relieving on tensile strength and hardness of gray iron.

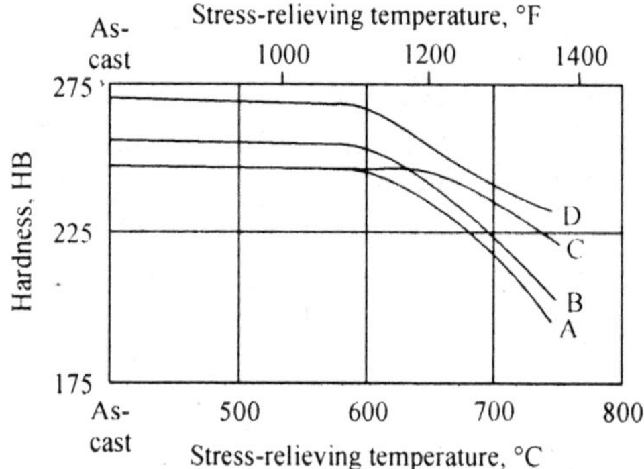

Iron	Composition, %					
	TC(a)	*CC(b)*	*Si*	*Cr*	*Ni*	*Mo*
A	3.20	0.80	2.43	0.13	0.05	0.17
B	3.29	0.79	2.58	0.24	0.10	0.55
C	3.23	0.70	2.55	0.58	0.06	0.12
D	3.02	0.75	2.38	0.40	0.07	0.43

Bar specimens 30 mm (1.2 inch) in diameter were held for 1 hr. at indicated temperature and then air cooled.

Fig. 11.16. Effect of stress-relieving temperature on hardness of gray irons.

Quantitative data concerning the effects of alloying elements on optimum stress-relieving temperature are meager. However, it has been reported that in one instance the addition of as little as 0.14 per cent Cr to a 3.20 C, 2.01 Si iron permitted exposure of the iron to a temperature of about 650°C (1200°F) for 1 hour without sacrifice in room temperature tensile strength. Figure 11.19 shows the effect of temperature and time on relief of stresses for seven low-alloy irons, and the tabulation below the graphs indicates that these irons, depending on shakeout time, can be stress relieved for 8 hours at 620°C (1150°F) with no adverse effect on hardness.

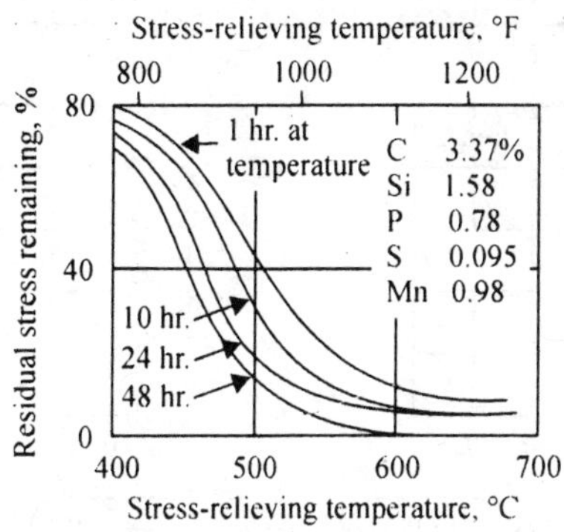

Fig. 11.17. Effect of stress-relieving temperature on residual stress in gray iron.

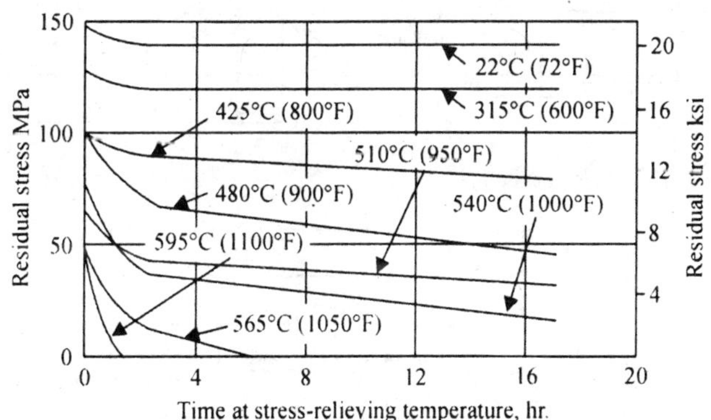

Composition of iron: 2.72 C, 1.97 Si, 0.141 P, 0.080 S, 0.51 Mn

Fig. 11.18. Effect of stress-relieving temperature and time on residual stress in gray iron.

The following stress-relieving temperatures, based on normal shakeout time in the foundry, can be recommended.

Unalloyed or alloyed without Cr	510 to 565°C (950 to 1050°F)
0.15 to 0.30 per cent Cr	595 to 620°C (1100 to 1150°F)
Over 0.30 per cent Cr	620 to 650°C (1150 to 1200°F)

If the service requirements of a casting demand a particularly low residual stress, temperatures about 28°C (50°F) above those listed may be used. When these higher temperatures are used, it is advisable, if hardness and strength are critical, to check the hardness of the stress relieved casting to determine whether an unacceptable decrease in hardness or strength has taken place.

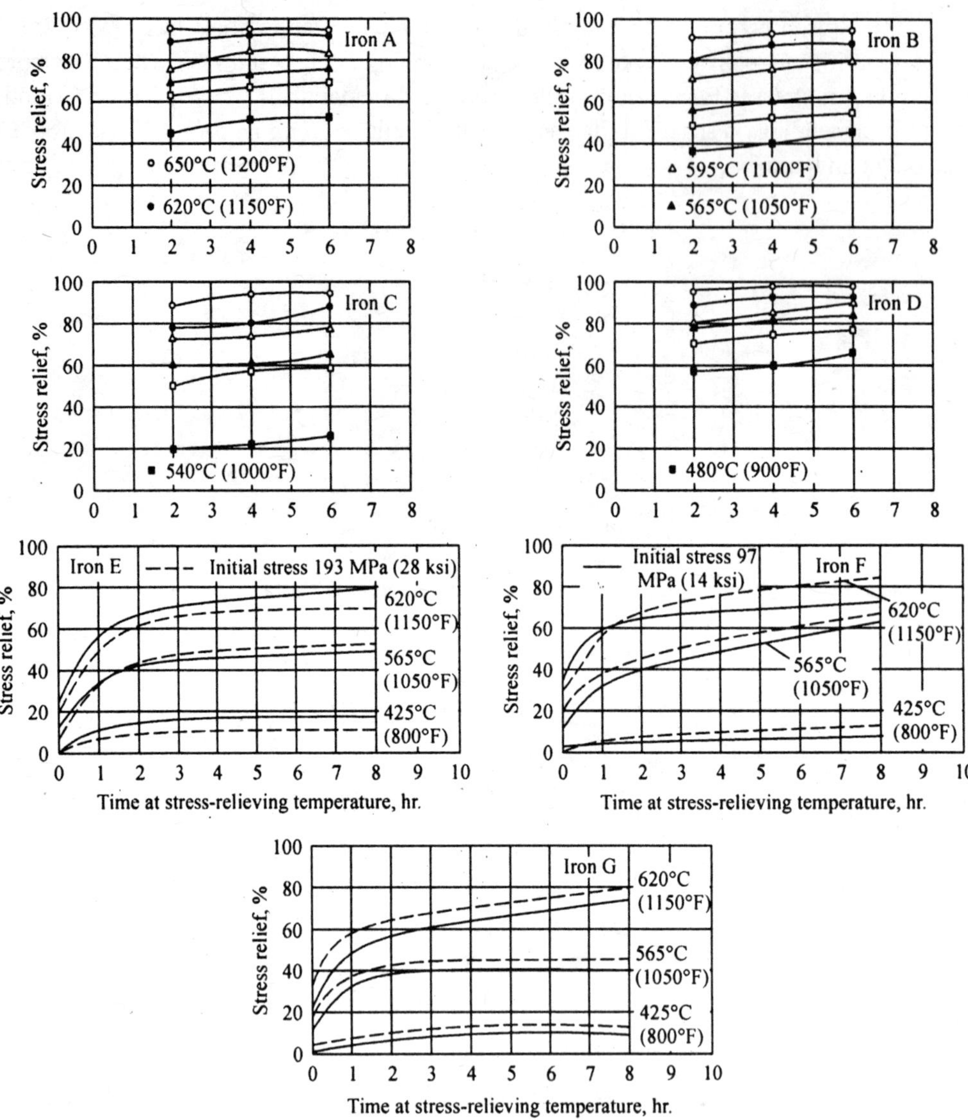

Fig. 11.19. Effect of stress-relieving time and temperature on degree of stress-relief obtained in low-alloy gray irons.

Iron	*C*	*Si*	*P*	*S*	*Mn*	*Ni*	*Cr*	*Mo*	*Cu*	*V*	*Hardness, HRB* *Before stress relieving*	*After stress relieving for 8 hrs. 620°C (1150°F)*
	Composition, %											
A	2.93	2.14	0.110	0.57	0.47	0.35	0.10	–	–	–	98	94
B	3.43	2.12	0.104	0.70	0.81	0.34	0.18	0.23	–	–	98	94

(Contd...)

Iron	C	Si	P	S	Mn	Ni	Cr	Mo	Cu	V	Hardness, HRB Before stress relieving	After stress relieving for 8 hrs. 620°C (1150°F)
	Composition, %											
C	3.24	2.55	0.107	0.62	0.87	0.51	0.20	0.22	–	–	95	95
D	3.91	1.43	0.54	0.25	0.32	1.56	0.06	–	–	–	82	80
E	3.18	2.13	0.73	0.125	0.70	1.03	0.33	0.65	–	–	98	98
F	3.12	1.76	0.075	0.097	0.78	1.02	0.41	0.58	–	–	94	95
G	2.78	1.77	0.065	0.135	0.55	0.36	0.10	0.33	0.46	0.04	96	96

Compositions are given in table above, which also shows negligible effect of maximum stress-relieving conditions on hardness.

Rate of Heating

The rate at which gray iron castings are heated for stress relief depends on the shape and size of the part but, except for the most complex shapes, is not especially critical. When a batch-type furnace is employed, it is of the utmost importance that furnace temperature does not exceed 95°C (200°F) at the time of loading. After the furnace is loaded, the heating rate may be fairly high. For example, it is common practice to heat to 620°C (1150°F) in about 3 hours hold at temperature for 1 hour and cool to 315°C (600°F) in about 4 hours, before removing castings from the furnace and permitting them to cool in air. These conditions apply also to continuous furnaces in which the various temperature zones can be controlled to avoid introducing additional thermal stress in the castings. It is imperative that flame impingement on castings, which may result in variations in hardness, be avoided.

Rate of Cooling

If a casting is allowed to cool rapidly from the stress-relieving temperature to room temperature, new stresses may be developed and the object of maximum stress elimination will not be fully achieved. For this reason, slow cooling from the stress-relieving temperature, at least in the upper temperature range, is an essential part of stress relieving.

It is generally recommended that castings be furnace cooled to 315°C (600°F) or lower before being allowed to cool in air; for castings of intricate design, it may even be advisable to continue furnace cooling until a temperature of about 95°C (200°F) has been reached. Most commercial furnaces cool slowly enough to meet all requirements.

Applications

The following examples illustrate the use of stress relieving to eliminate distortion and cracking.

Example 9. A flatness tolerance of 0.1 mm (0.004 inch) could not be maintained after machining As-cast clutch plates. These parts were made of gray iron containing 3.40 C (0.70 to 0.90 combined carbon), 1.30 to 1.80 Si, 0.25 to 0.40 Cr, 0.30 to 0.40 Mo and nickel and copper as required for a minimum tensile strength of 275 MPa (40 ksi) and a hardness of 207 to 255 HB. After these clutch plates were stress relieved at 620°C (1150°F) for 2 hours, the tolerance could be held.

Example 10. Necessitated by a rush order, a change in procedure from a slow cool in the mould to a fast shakeout resulted in high residual stress in cast transmission cases. These parts were made of gray iron containing 3.10 to 3.40 C, 2.15 to 2.35 Si and 0.20 to 0.40 Cr. Some of the first castings produced after the change in procedure developed cracks up to 1.3 mm (0.050 inch) wide during machining; others, although they survived machining without cracking, developed cracks during a final hot washing operation. This problem was eliminated for the remaining pieces from this order by stress relieving them at 620°C (1150°F) for 2 hours. A return to the original cooling and shakeout practice on subsequent orders eliminated cracking and the need for stress relieving.

Example 11. Hot tears and out-of-roundness occurred in As-cast rings, and additional distortion in the flat plane of these rings occurred during machining. A change in the method of gating these castings equalised temperature differentials and controlled shrinkage distribution, thus eliminating the hot tears and the As-cast distortion. The distortion occurring during machining was eliminated by stress relieving the rings at 370°C (700°F) for 2 hours.

Example 12. Table 11.6 shows the results of an evaluation of different cooling methods and stress-relieving treatments on residual stresses in 225 kg (500 lb) diesel-engine cylinder blocks. These blocks were made of gray iron containing 3.25 C, 2.20 Si, and 0.30 Cr, and having a tensile strength of 240 MPa (35 ksi).

Table 11.6. Effect of shakeout practice and stress relieving on residual stress in 225 kg (500 lb) diesel-engine cylinder blocks.

	Close-in		
Practice	*mm*	*inch*	*Stress relief, %*
1. Shakeout after 6 hours; cores in place while cooling to 27°C (80°F); total time, 16 hrs.	4.1	0.160	Basis for evaluation
2. Same as above, except cores also shaken out at 6 hr.	1.9	0.076	52
3. Cooled in mould 16 hours before shakeout	1.6	0.064	60
4. Practice 1, followed by stress relieving at 540°C (1000°C) for 2 hours furnace cooling to 370°C(700°F)	1.5 to 1.9	0.060 to 0.076	52 to 62
5. Practice 1, followed by stress relieving at 620°C (1150°F) for 2 hours furnace cooling to 370°C (700°F)	0.3 to 0.4	0.012 to 0.015	91

Two vertical lines were scribed on one end of each block after shakeout. The distance between these lines was measured before and after vertical saw cuts (parallel to the scribed lines) were made through the end of each block to the first cylinder bore. The amount of close-in between the two vertical lines after sawing was a relative measure of the residual stress.

The results (Table 11.6) indicate that the complete shakeout (including cores), allowing inner sections to cool faster, materially reduced stresses, as did the procedure of leaving the casting in the mould to slow down the cooling of external sections. This latter procedure is effective if the mass of metal is sufficient to keep the external sections at a high temperature throughout—620°C (1150°F) or higher—while equilibrium between outer and inner sections is taking place. As indicated, stress relieving at 540°C (1000°F) was substantially less effective than stress relieving at 620°C (1150°F).

Other examples of castings that require stress relief are pump volutes, scrolls and casings. These parts require two stress-relieving treatments: first, in the As-cast condition, to minimise distortion during machining, and second, after rough machining, to minimise distortion during final machining.

Chapter 12

Heat Treatment of Ductile Irons

INTRODUCTION

When maximum ductility and good machinability are desired and high strength is not required, ductile iron castings are generally given a full ferritise anneal. The microstructure is thus converted to ferrite and spheroidal graphite. Manganese and phosphorus, and alloying elements such as chromium, nickel and molybdenum, should be as low as possible if superior machinability is desired, because these elements retard the annealing process.

Two different annealing cycles may be used satisfactorily. Selection of one or the other will depend on the type of heat treating equipment to be used.

Hold at 900 to 955°C (1650 to 1750°F) for 1 hour plus 1 hour or more per inch of section thickness. For thin-section casting containing 2.20 to 2.70 per cent Si, holding at 955°C (1750°F) for 1 to 3 hours is sufficient. In heavy-section castings where chill has formed on corners, holding at 955°C (1750°F) for 3 to 8 hours may be required. Cool to 690°C (1275°F) in any convenient manner (but uniformly, if residual stress is to be avoided), and hold at 690°C (1275°F) for 5 hours plus 1 hour per inch of casting section.

Hold at 900 to 955°C (1650 to 1750°F) as above, but furnace cool to 650°C (1200°F) so that the cooling rate between 790 and 650°C(1450 and 1200°F) does not exceed 20°C/hour (35°F/hour).

A shorter, sub-critical annealing cycle can be used when carbides can be tolerated and maximum impact properties are not required. A reduction of as much as 50 per cent in impact strength and an increase of 30°C (50°F) in transition temperature will result if prior austenitising heat treatment is omitted. In this treatment, castings are heated to 705°C (1300°F) and held for 5 hours plus 1 hour per inch of maximum thickness. Castings should be furnace cooled to at least 595°C (1100°F). The influence of sub-critical annealing at 705°C (1300°F) for various periods of time on the hardness of four ductile irons is shown in Fig. 12.1.

The effect of time at temperature on decomposition of primary carbide for 6 mm (¼ inch) plate castings with a nominal composition of 3.15 per cent TC, 2.66 per cent Si, 0.40 per cent Mn, 0.15 per cent P, 0.10 per cent Cu and 0.04 per cent Mg is given in Fig. 12.2. These carbides usually break down readily in a manner similar to castings that have been post-inoculated during manufacture.

Certain carbide-forming elements, mainly chromium, form primary carbides that are very difficult, if not impossible, to decompose. For example, the presence of 0.26 per cent chromium results in primary carbides that cannot be broken down in 2 to 20 hours heat treatments at 925°C (1700°F). The resulting matrix after pearlite breakdown is ferritic-carbidic with only 5 per cent elongation. Other carbide stabilisers are molybdenum contents over 0.3 per cent, copper over 1.0 per cent, 0.002 per cent or more boron and vanadium and tungsten.

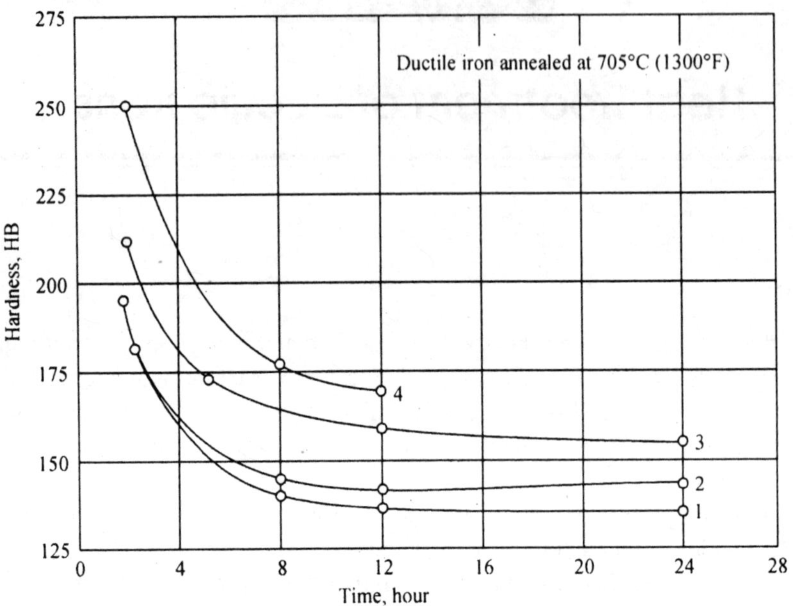

Fig. 12.1. Influence of time at sub-critical annealing temperature on hardness.

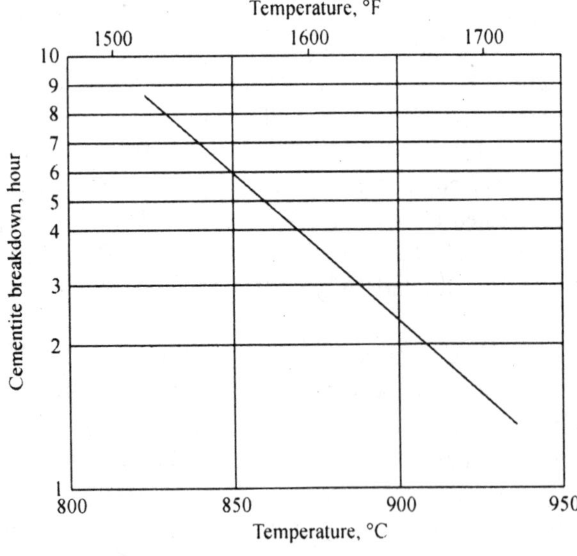

Fig. 12.2. Effect of time at temperature on decomposition of primary carbide.

NORMALISING OF DUCTILE IRON

Normalising can result in a considerable improvement in tensile properties and may be used in the production of ductile iron of types 100–70–03 and 120–90–02. The microstructure obtained by normalising will depend on the composition of the castings and the cooling rate. The cooling rate depends on the mass of the casting, but it also may be influenced by the temperature and movement of the surrounding air during cooling. Normalising generally produces a homogeneous structure of fine pearlite, if the metal is not too high in silicon content and has at least a moderate manganese content (0.3 to 0.5 per cent or higher). Heavier castings should contain alloying elements such as nickel, molybdenum and additional manganese for development of a fully pearlitic structure after normalising. Lighter castings made of alloyed iron may be martensitic after normalising.

The normalising temperature is usually between 870 and 940°C (1600 and 1725°F). The standard time at temperature of 1 hour per inch of section thickness or 1 hour minimum is satisfactory, but shorter times often can be used successfully. The following temperatures and minimum holding times are recommended for the normalising of unalloyed ductile iron:

Section			*Time*
mm	*inch*	*Temperature*	*hr.*
Under 13	Under ½	870°C min (1600°F minutes)	1
13 to 25	½ to 1	940°C (1725°F)	1
Over 25	Over 1	940°C (1725°F)	2

These temperatures and times will vary with composition especially with silicon and chromium contents. As shown in Fig. 12.3a, the temperature of ductile iron when it is removed from the furnace for air cooling during normalising strongly affects hardness. Because of increasing amounts of ferrite in the resultant structures, hardness decreases markedly as the normalising temperature drops below 845°C (1550°F). Alloying elements have an important function in ductile iron castings that are to be normalised. Moderate additions increase hardenability sufficiently so that a microstructure of fine pearlite may be obtained in normalised castings with sections up to 150 mm (6 inch) thick, provided the cooling rate has been adequate. Nickel or copper are preferred as alloying elements because they strengthen the iron in all section thicknesses and do not form carbides. Molybdenum, in combination with nickel, is effective in increasing hardness and strength. The influence of various nickel contents and several combinations of alloying elements on the hardness obtained by normalising various sections is shown in Fig. 12.3b.

Normalising is commonly followed by tempering to attain the desired hardness and relieve residual stresses that develop upon air cooling when various parts of a casting with different section sizes cool at different rates. The effect of tempering on hardness and tensile properties depends on the composition of the iron and on the hardness level that was obtained in normalising. In general, pearlitic structures, such as result from normalising, soften less then the harder martensitic structures that are obtained by liquid quenching. Fig. 12.3 (c) shows the effect of tempering 25 mm (1 inch) cubes for 1 hour at 510 and 620°C (950 and 1150°F) on the hardness of a series of unalloyed, nickel, and nickel-molybdenum ductile irons after normalising. Tempering after normalising is also used to obtain high toughness and impact resistance along with high tensile properties. This consists of re-heating to temperatures of 425 to 650°C (800 to 1200°F) and holding at the desired temperature for 1 hour per inch of cross section. These temperatures are varied within the above range to meet specification limits. Fig. 12.4 shows the effect of tempering temperature on hardness of normalised ductile iron.

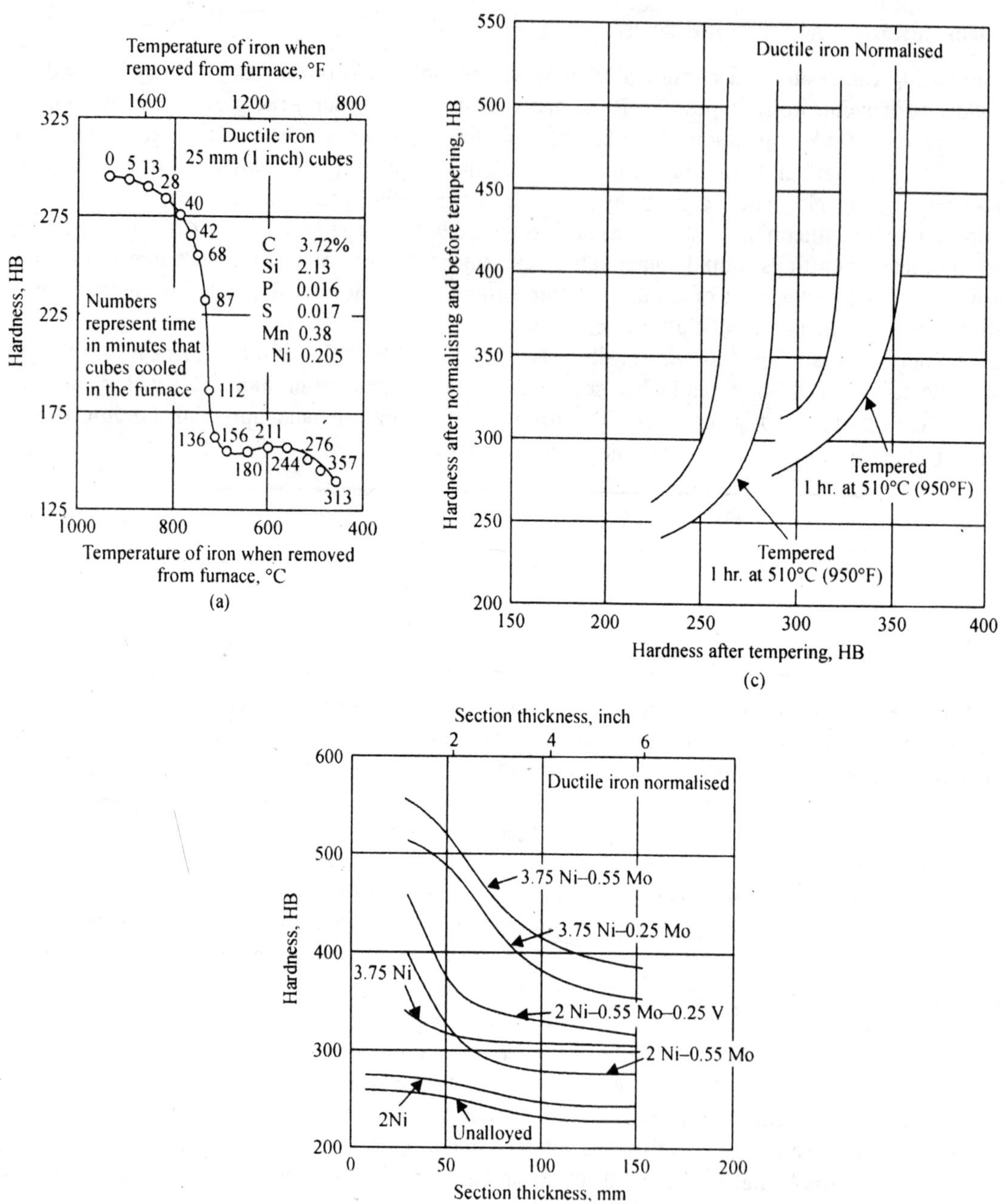

Fig. 12.3. Effect of variables on hardness after normalising and after normalising and tempering of ductile iron. (a) Effect of temperature from which ductile iron was air cooled during normalising. Specimens (25 mm or 1 inch cubes) were held at 925°C (1700°F) for 3 hours and then cooled in the furnace to various temperatures in the time indicated before being cooled in air; (b) effect of alloy content and section thickness; and (c) effect of tempering after normalising, for unalloyed, nickel and nickel-molybdenum ductile irons.

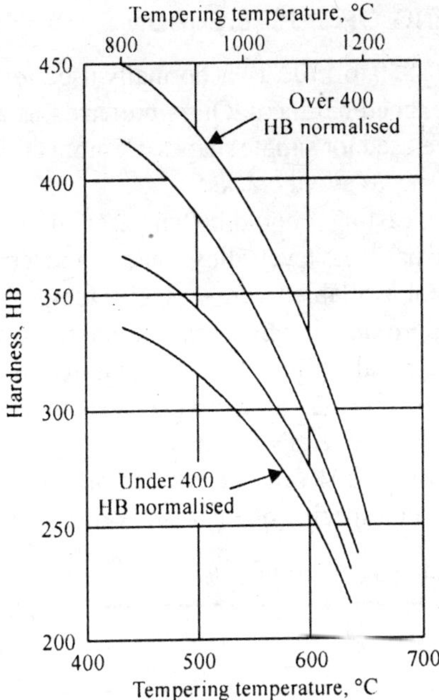

Fig. 12.4. Hardness of normalised ductile iron tempered at various temperatures.

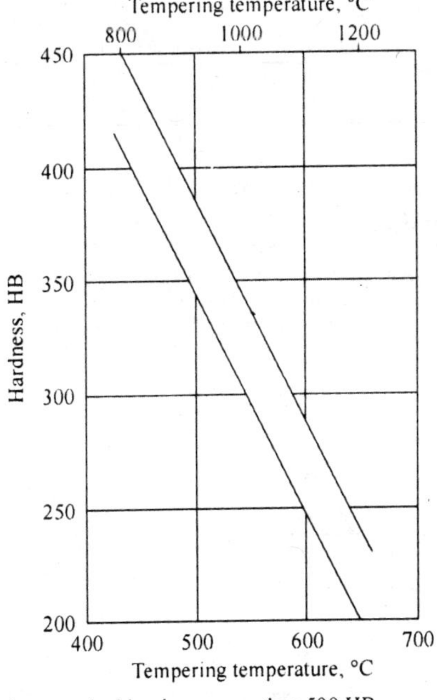

Fig. 12.5. Hardness of oil-quenched ductile iron tempered for 1 hour.

HARDENING AND TEMPERING OF DUCTILE IRON

A temperature of 845 to 925°C (1550 to 1700°F) is normally used for austenitising commercial castings and produces the highest as-quenched hardness. Oil is preferred as a quenching medium, to minimise stresses, but water or brine may be used for simple shapes. Complicated castings may have to be quenched in oil at 80 to 100°C (180 to 210°F) to avoid cracks.

To relieve quenching stresses, castings should be tampered immediately after quenching. Tempered hardness depends on as-quenched hardness level, alloy content, and tempering time as well as temperature. Precise data on ductile irons is not available to draw precise tempering curves such as those for steels. Fig. 12.6 can be used as a first approximation by the heat treater. More accurate control of hardness can be attained by close control of material composition and heat treating cycle.

Severity of quench is shown in Table 12.1 in Grossman's H values. The higher the H value, the faster the quench. Geometry of the part will generally dictate the quenching medium because of stress-induced cracking during quenching. Grossman's H values are based on the diameter of round bar steel that will be half martensitic at the centre after a specific quench. The values are used as a measure of hardenability.

Table 12.1. Comparison of quenching mediums to water (a).

	Grossman H values			
Type of circulation	*Air*	*Oil*	*Water*	*Brine*
No circulation of liquid or agitation of piece	0.02	0.25–0.30	0.9–1.0	2.0
Mild circulation		0.30–0.35	1.0–1.1	2.0–2.2
Moderate circulation		0.35–0.40	1.2–1.3	–
Good circulation		0.40–0.50	1.4–1.5	–
Strong circulation		0.50–0.80	1.6–2.0	–
Violent circulation		0.80–1.10	4.0	5.0

(a) At 15°C (65°F).

The influence of the austenitising temperature on the hardness of water quenching 13 mm (½ inch) cubes of ductile iron is shown in Fig. 12.6.

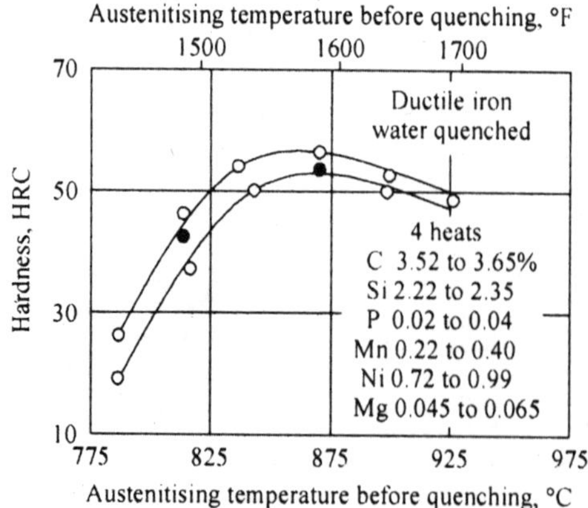

Each value represents the average of three hardness reading. Specimens (13 mm or ½ inch cubes) were heated in air for 1 hour and water quenched.

Fig. 12.6. Influence of austenitising temperature on hardness of ductile iron.

These data show that the highest range of hardness (55 to 57 HRC) was obtained with austenitising temperatures between 845 and 870°C (1550 and 1600°F). Specimens quenched from 925°C (1700°F) contained enough retained austenite to lower the hardness to 47 HRC.

Time at austenitising temperature also is important for obtaining full hardness during quenching. This was indicated by further tests with 13 mm (½ inch) cubes of ductile iron; both as-cast and annealed cubes were heated in salt at 870°C (1600°F) and quenched for a total of 2 minutes contained 30 to 35 per cent ferrite in the microstructure and developed a hardness of 32 to 45 HRC. After 4 minutes of heating, 12 to 15 per cent ferrite was retained, and the hardness varied form 44 to 51 HRC. When the cubes were heated for a total of 10 minutes or more, the microstructure did not contain ferrite, and full hardness. (53 to 57 HRC) was developed.

Within their normal ranges, the total carbon and silicon contents have a minor influence on hardenability. The combined carbon content in the matrix depends on the austenitising temperature; manganese has a moderate influence. The effect of varying amounts of carbon, silicon and manganese on the hardenability of ductile irons is shown in Fig. 12.7

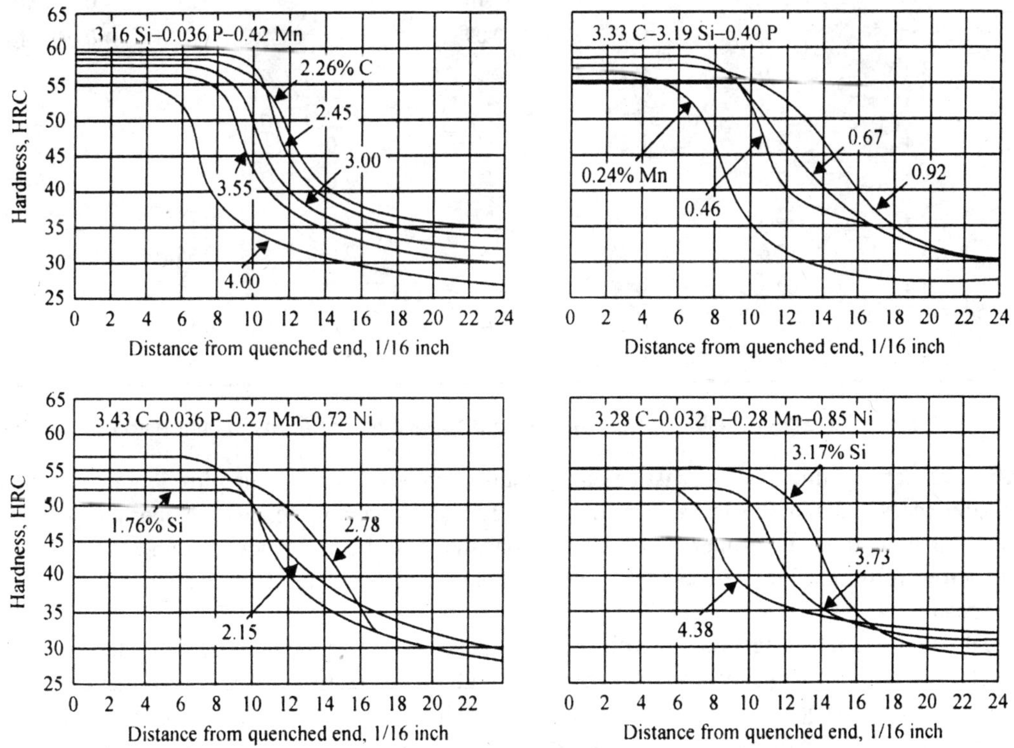

Fig. 12.7. Effect of varying amounts of carbon, silicon and manganese on hardenability of ductile irons.

A wide variety of hardenabilities can be obtained by the addition of alloying elements, either singly or in combination. The effects of nickel and molybdenum on Jominy end-quench curves for ductile iron are shown in Fig. 12.8.

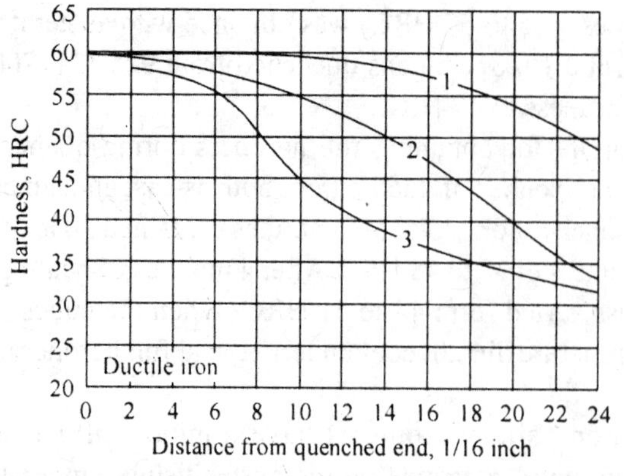

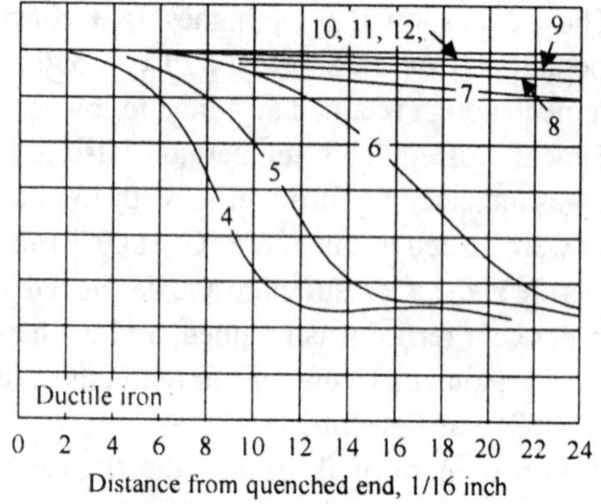

Iron	*TC*	*Si*	*Mn*	*Cr*	*Ni*	*Mo*	*Other*
1	3.52	2.74	0.37	0.10	–	0.41	0.061 Mg
2	3.47	2.69	0.37	0.10	–	0.23	0.050 Mg
3	3.54	2.55	0.37	0.09	–	0.022	0.049 Mg
4	3.4	2.6	0.26	–	–	–	–
5	3.4	2.6	0.26	–	1.0	–	–
6	3.4	2.6	0.26	–	2.0	–	–
7	3.4	2.6	0.26	–	4.0	–	–
8	3.4	2.6	0.26	–	2.0	0.25	–
9	3.4	2.6	0.26	–	2.0	0.55	–
10	3.4	2.6	0.26	–	3.75	0.25	–
11	3.4	2.6	0.26	–	3.75	0.55	–
12	3.4	2.6	0.26	–	2.0	0.55	0.25 V

Fig. 12.8. Effect of alloying elements on end-quench hardenability of ductile iron.

After quenching, ductile iron castings are usually tempered for 1 hour plus 1 hour per inch of section thickness to obtain the desired hardness. Fig. 12.9 shows the influence of tempering temperature on the hardness and tensile properties of ductile irons that have been quenched from 870°C (1600°F). Higher-alloy irons require the use of somewhat higher tempering temperatures in order to obtain similar results.

The distribution of hardness readings obtained on quenched and tempered ductile iron levers is shown in Fig. 12.10a. Fig. 12.10b shows the hardness distribution for ductile iron gears as quenched and as tempered. Austempering and martempering practices are based on isothermal transformation diagrams. The diagram for the unalloyed ductile iron in Fig. 12.11 provides information about the austenite transformation process. Transformation of austenite to bainite in unalloyed ductile iron is completed in about 2 hours at austempering temperature.

The effects of alloying elements on the isothermal transformation of ductile iron can be expected to be similar to their effects in eutectoid steel. However, when carbide-forming alloys are used in large quantities, a part of any alloying element may be present as an insoluble carbide and is not effective in increasing the hardenability.

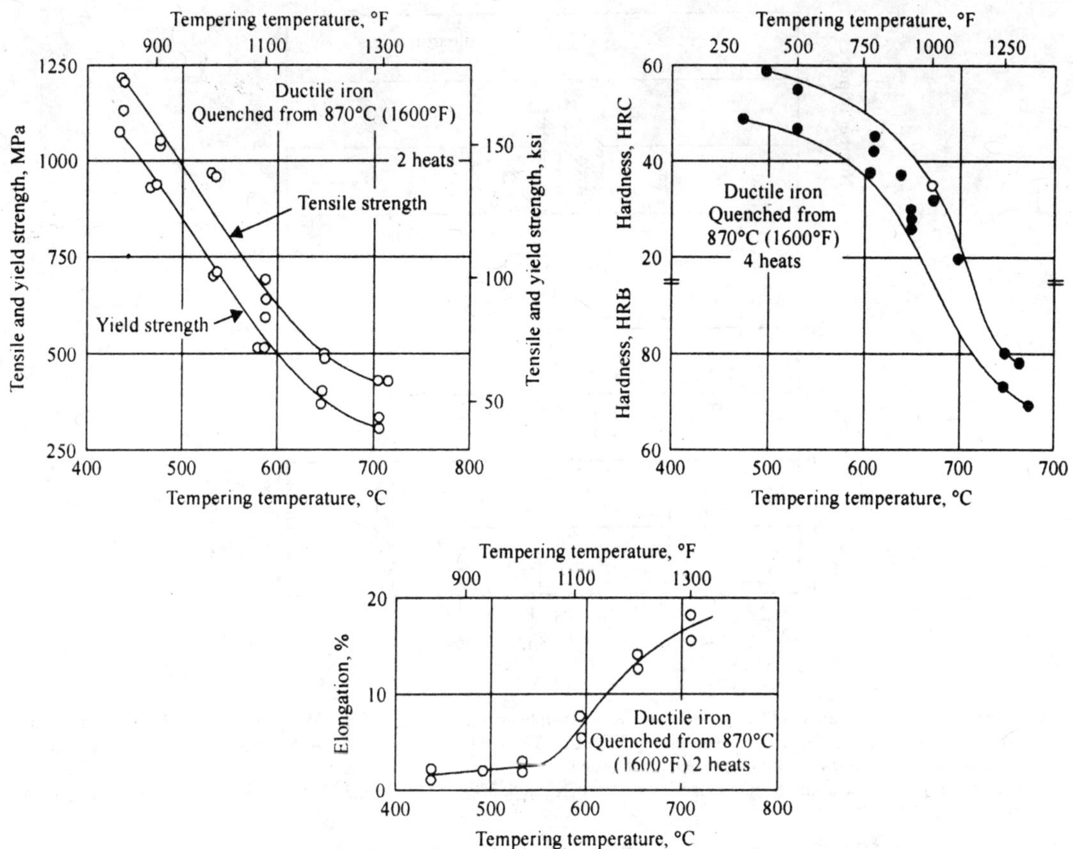

Data represent irons from four heats with composition ranges of: 3.52 to 3.68 per cent C, 2.28 to 2.35 per cent Si, 0.02 to 0.04 per cent P, 0.22 to 0.41 per cent Mn, 0.69 to 0.99 per cent Ni and 0.045 to 0.065 per cent Mg. Data for tensile strength, tensile yield strength and elongation are for irons (from two of these heats) that contained 0.91 and 0.99% Ni.

Fig. 12.9. Influence of tempering temperature on mechanical properties of ductile iron quenched from 870°C (1600°F), tempered 2 hour

Surface Hardening of Ductile Iron

Ductile iron responds readily to surface hardening by flame or induction processes. Because of the short heating cycle in these processes, the pearlitic types of ductile iron, 80-60-03 and 100-70-03, are preferred. Irons without free ferrite in their microstructure respond almost instantly to flame or induction heating and require very little holding time at the austenitising temperature in order to be fully hardened.

With a moderate amount of free ferrite, the response may be satisfactory, but an entirely ferritic matrix, typical of the grades with high ductility, requires several minutes at 870°C (1600°F) to be fully hardened by subsequent cooling. A matrix microstructure of fine pearlite, readily obtained by normalising, has quick response to surface hardening and provides excellent core support for the hardened case.

Castings that are predominantly pearlitic in the as-cast condition (grade 80-55-06), should be given a relatively high temperature stress relief for a short time, for example, tempering at 595 to 650°C (1100 to 1200°F) for 1 hour per inch of section thickness. This treatment will remove virtually all of the internal or residual stresses from the casting before it undergoes a localised hardening treatment

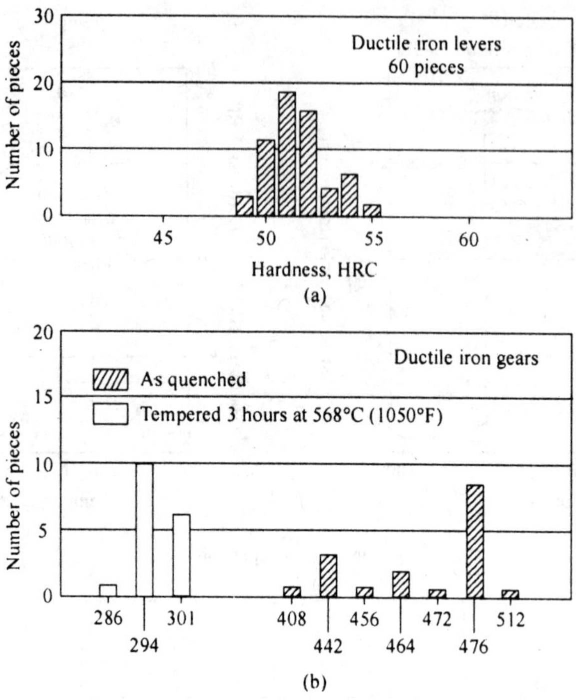

(a) levers austenitised is salt at 870°C (1600°F) for ½ hr., then tempered in air at 175°C (350°F) for 2 hours Levers were made from 2 heats of electric-furnace iron containing 3.5% C, 2.3% Si 0.03% Mn, 0.75% Ni, and 0.06% Mg. Data represent average of three hardness readings on each piece tested.

(b) Gears quenched in oil from 885°C (1625°F) then tempered at 565°C (1050°F) for 3 hours Hardness of the as-cast gears ranged from 170 to 228 HB.

Fig. 12.10. Hardness distribution for quenched and tempered ductile iron parts.

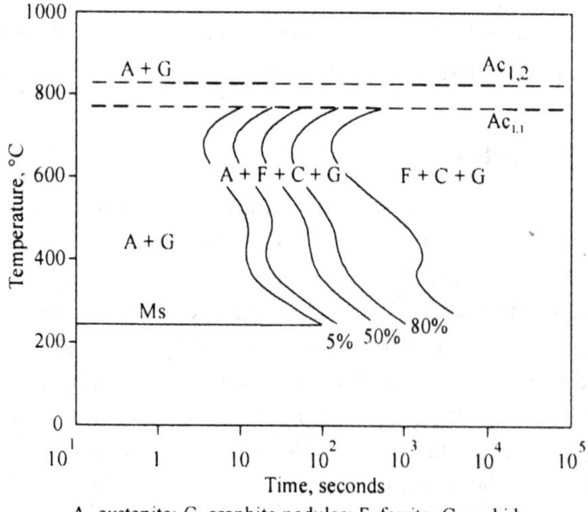

A, austenite; G, graphite nodules; F, ferrite; C, carbides

Fig. 12.11. Isothermal transformation diagram for an unalloyed ductile iron containing 3.3 per cent C, 2.52 per cent Si.

With proper technique and control of temperature between 845 and 870°C (1550 and 1600°F), the following ranges of surface hardness of ductile iron can be expected in commercial production:

1. Ductile iron, fully annealed (ferritic), water quenched behind the flame, 35 to 45 HRC.
2. Ductile iron, predominantly ferritic (partly pearlitic), stress relieved prior to heating and self quenched, 40 to 45 HRC.
3. Ductile iron, predominantly ferritic (partly pearlitic), stress relieved prior to heating, water quenched, 50 to 55 HRC.
4. Ductile iron, mostly pearlitic, stress relieved before heating, water quenched, 58 to 62 HRC.

Heating time and temperature, amount of dissolved carbon, section size, and rate of quench help to determine final hardness values.

Flame or induction hardened ductile iron castings have been used for heavy-duty applications such as rolls for cold working titanium, ring gears for papermill drives, crankshafts, and large sprockets for chain drives.

Stress Relieving of Ductile Irons

When not otherwise heat treated, complex engineering castings of ductile iron may be stress relieved at 510 to 675°C (950 to 1250°F). Temperatures at the lower end of this range are satisfactory for many applications; temperatures at the higher end will eliminate virtually all residual stress (Fig. 12.12), but will also effect some reduction in hardness and tensile strength. Recommended ranges of stress-relieving temperature for various types of ductile iron are as follows:

1. Unalloyed 510 to 565°C (950 to 1050°F)
2. Low-alloy 565 to 595°C (1050 to 1100°F)
3. High-alloy 595 to 650°C (1100 to 1200°F)
4. Austenitic 620 to 675°C (1150 to 1250°F)

The required time at temperature will depend on the temperature used, the complexity of the casting, and the completeness of stress relief that is desired (Fig. 12.12), but 1 hour plus 1 hour per inch of section thickness is good practice.

Cooling should be uniform, to avoid re-introducing stresses. Castings should be furnace cooled to 290°C (550°F), after which they can be air cooled (in most instances: however, austenitic iron can be uniformly air cooled from the stress-relieving temperature).

Effect of Heat Treatment on Fatigue Strength

The long life fatigue properties of as-cast ductile irons are improved by heat treatment, but not in the same proportion as the monotonic properties. This is demonstrated by the endurance ratios (fatigue strength/ultimate tensile strength) shown in Table 12.2. In heat treating to improve properties, the proper composition and temperature must be selected to ensure the optimum improvement. For example, one cause of the low endurance ratios in quenched and tempered ductile irons is the precipitation of secondary graphite throughout the matrix on tempering. The amount of secondary graphite can be controlled by composition (primarily carbon and silicon) and tempering temperature (increases with temperature).

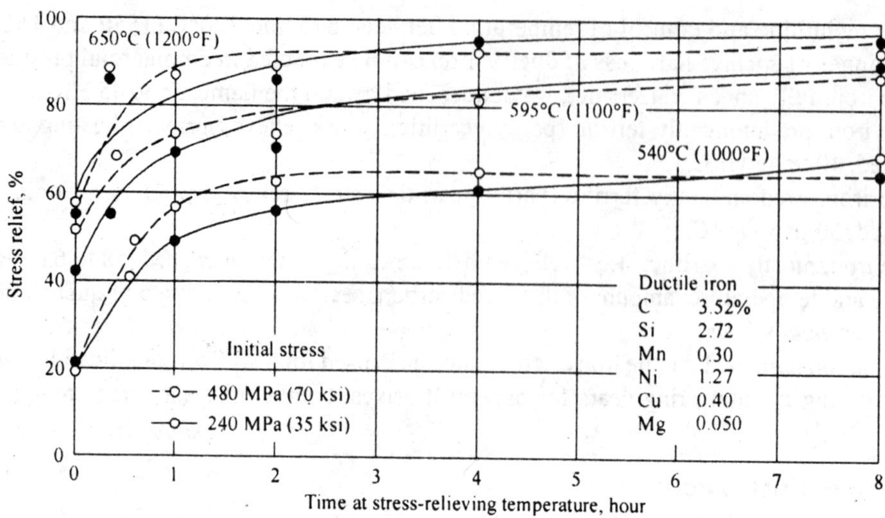

Initial hardness was 102 to 103 HRB. Hardness after holding at 540, 595 and 650°C (1000, 1100 and 1200°F) for 8 hours was 102 to 104, 101 to 103, and 90 to 93 HRB, respectively.

Fig. 12.12. Stress relief obtained in ductile iron held at three temperature ½ to 8 hours

Table 12.2. Fatigue results on heat treated ductile irons.

Type	*Hardness, HB*	*Ultimate tensile strength*		*Elongation, %*	*Fatigue strength*		*Endurance ratio*
		MPa	*ksi*		*MPa*	*ksi*	
As-cast	255	664	96	3.0	278	40	0.42
Quenched and tempered (a)	295	927	134	4.0	340	49	0.37
Quenched and tempered (b)	350	1029	149	1.5	340	49	0.33
As-cast	267	659	96	2.0	301	44	0.46
Normalised (c)	325	1041	151	4.0	340	49	0.33

(a) Oil quenched from 900°C (1650°F); tempered 2 hours at 600°C (1110°F); (b) oil quenched from 900°C (1650°F); tempered 2 hours at 550°C (1020°F); and (c) normalised from 900°C (1650°F).

Chapter 13

Heat Treating of Malleable Irons

INTRODUCTION

Ferritic and pearlitic malleable irons are both produced by annealing white iron of controlled composition. Thus, annealing is an essential part of the manufacturing process for these irons.

The annealing treatment involves three important steps. The first causes nucleation of graphite and is initiated during heating to a high holding temperature and occurs very early during the holding period.

The second step consists of holding at 900 to 970°C (1650 to 1780°F); this step is called first-stage graphitisation (FSG). During FSG, massive carbides are eliminated from the iron structure. At this point, the iron is rapidly cooled to 725 to 740°C (1340 to 1360°F) prior to entering second-stage graphitisation.

The third step in the annealing treatment consists of slow cooling through the allotropic transformation range of the iron; this step is called second-stage graphitisation (SSG). During SSG, a completely ferritic matrix free of pearlite and carbides is obtained when the cooling rate is 2 to 17°C/hr. (3 to 30°F/hr.). This cooling rate, which depends on the silicon content of the iron and the temper carbon nodule count, may be increased to 85°C/min. (150°F/min.) to form a pearlitic matrix. Oil quenching from the first-stage graphitisation temperature following completion of that step will produce a martensitic matrix.

HARDENING AND TEMPERING OF PEARLITIC MALLEABLE IRON

A typical procedure for producing a hardened pearlitic malleable iron consists of (i) air quenching castings after first-stage annealing, which results in retention of about 0.75 per cent combined carbon in the matrix; (ii) re-heating and holding for 1 hour at 845 to 870°C (1550 to 1600°F) to re-austenitise the matrix; and then (iii) quenching in heated (80 to 105°C; 180 to 220°F) and agitated oil, thus developing a matrix consisting of martensite and bainite with a hardness of 555 to 627 HB. Fig. 13.1 shows the effects of austenitising, temperature, quenching medium and manganese content on the hardness of ferritic and pearlitic malleable iron both before and after heat treating. If direct oil quenching is used, caution must be exercised to prevent cracking due to high combined carbon. Increasing the austenitising temperature increases the amount of dissolved carbon, which is measured as combined carbon in the matrix after quenching to room temperature. (Fig. 13.2). Higher austenitising temperatures (900 to 930°C; 1650 to 1700°F) result in a more homogeneous austenite, which is desirable for a more uniform martensite but which can result in a greater tendency toward distortion or cracking. Temperature and time of tempering to attain a specified hardness may be selected from curves such as those shown in Fig. 13.3.

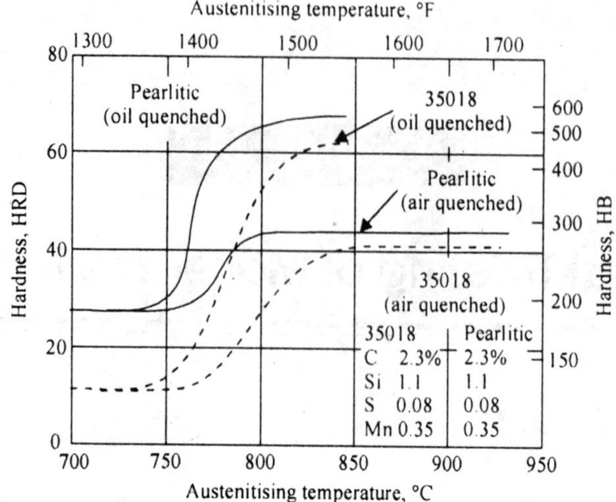

Fig. 13.1. Effects of austenitising temperature, quenching medium and manganese content on hardness of as-quenched malleable iron.

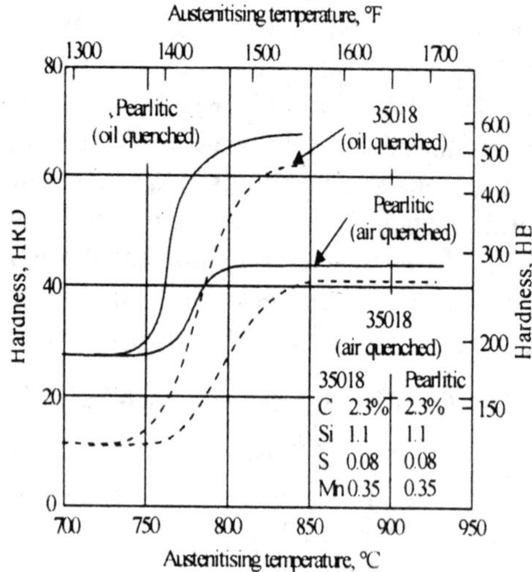

Fig. 13.2. Effect of austenitising temperature on combined carbon content of pearlitic malleable iron.

Hardened and tempered pearlitic malleable iron can be produced also from fully annealed ferritic malleable iron, the matrix of which is essentially carbon-free; graphite can be dissolved in austenite by holding at 845 to 870°C (1550 to 1600°F) for a time sufficiently long for production of an austenite matrix of uniform carbon content. In general, the combined carbon content of the matrix produced by this procedure is slightly lower than that of a pearlitic malleable made by air quenching directly from the first-stage graphitisation temperature (arrested annealing) and therefore the final tempering temperatures required for the development of specific hardnesses are lower (Fig. 13.4).

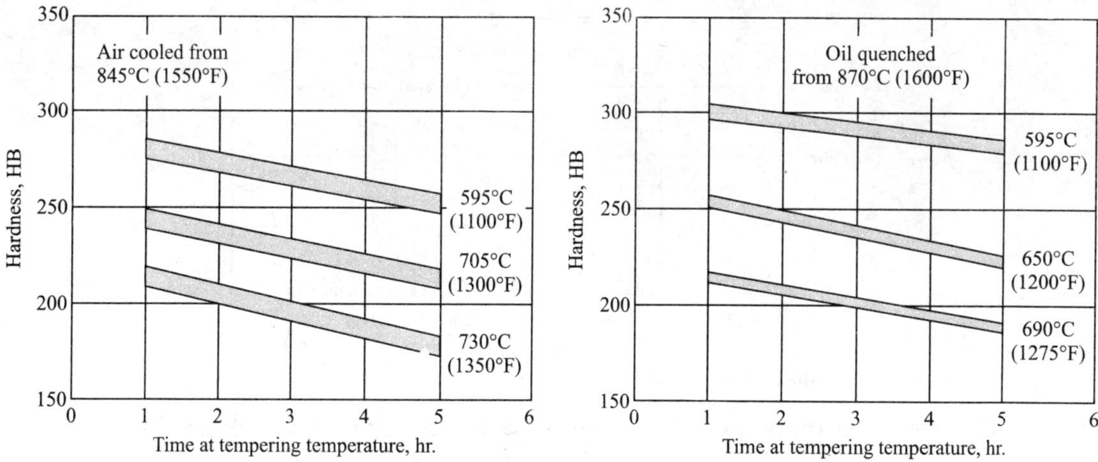

Fig. 13.3. Influence of time and tempering temperature on room-temperature hardness of pearlitic malleable iron.

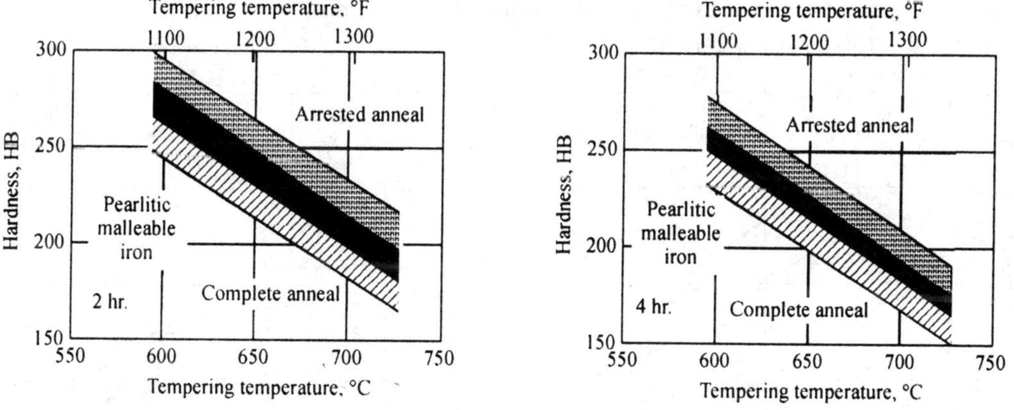

Fig. 13.4. Room-temperature hardness of tempered pearlitic malleable iron produced by arrested annealing and by complete-ferritise or ferritic annealing and re-hardening.

Tempering

Tempering treatments consist of cycles of no less than 2 hours at temperature to ensure uniformity of product. Tempering times must also be adjusted for section thickness and quenched microstructures. Fine pearlite and bainite require longer tempering times than that for martensite. In general, final hardness is controlled with process controls approximately the same as those encountered in the heat treatment of medium-carbon and higher-carbon steels. This is particularly true when the specification requires final hardnesses in the range from 241 to 321 HB.

Fig. 13.5 shows the control of hardness attained by one foundry producing quenched and tempered sleeve yokes of grade 80002 pearlitic malleable iron. These yokes were heated at 870°C (1600°F) for 30 minutes, then quenched in oil at 65°C (150°F) and tempered at 650°C (1200°F) for 2 hours.

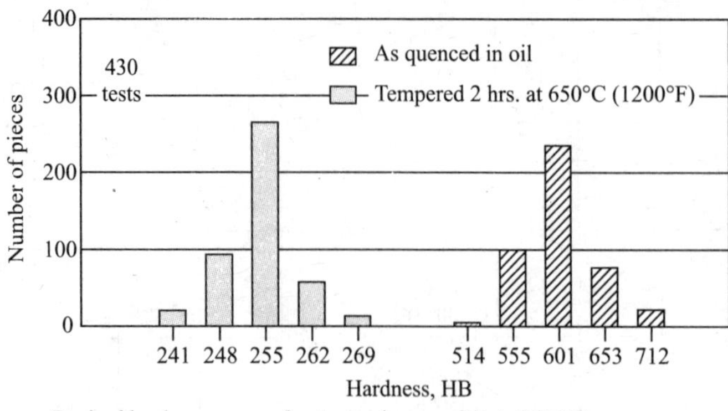

Fig. 13.5. Room-temperature hardness distribution for quenched and tempered pearlitic malleable iron sleeve yokes.

The effects of tempering on the hardness of alloyed and unalloyed malleable irons are shown in Fig. 13.6; these data illustrate the beneficial effects of alloying on as-quenched hardness and stability at elevated temperature. During all tempering treatments, carbide has a tendency to decompose, with consequent deposition of graphite on existing temper carbon nodules. This tendency is least at the lower tempering temperatures or in suitably alloyed pearlitic malleable irons.

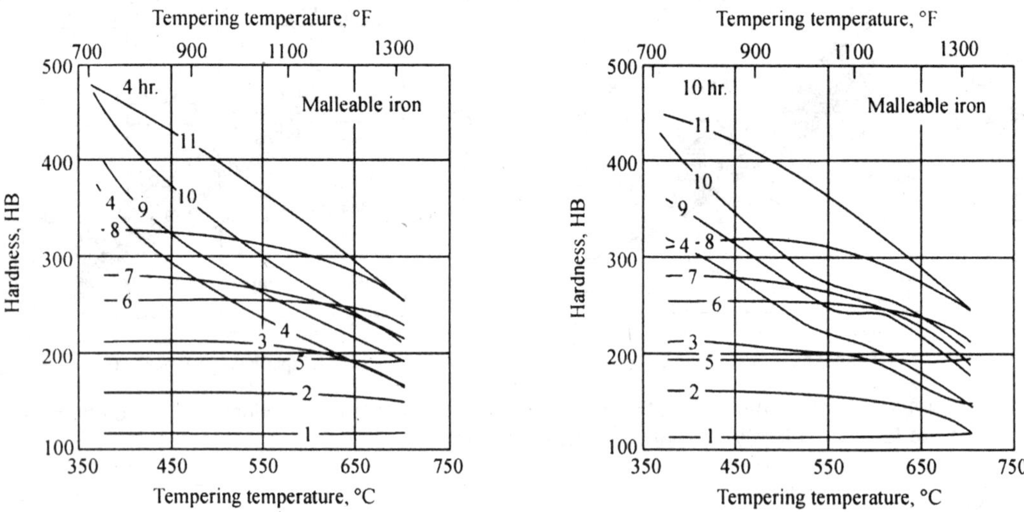

Fig. 13.6. Effect of tempering temperature and time on hardness of ferritic and pearlitic malleable irons in the as-received and the re-heated-and-quenched conditions.

Iron	*Material*	Composition, %					*Alloying and prior heat treatment*	*Hardness, HB*
		TC	*Si*	*S*	*Mn*	*Mo*		
1	Standard (ferritic) grade 32510	2.40	1.80	0.072	0.30	–	Unalloyed; fully malleablised	116

(Contd...)

Iron	Material	Composition, % TC	Si	S	Mn	Mo	Alloying and prior heat treatment	Hardness, HB
2	Pearlitic malleable iron, grade 45007	2.40	1.80	0.072	0.30	–	Unalloyed; air quenched from 925°C (1700°F), tempered 8 hr. at 695°C (1280°F)	156
3	Pearlitic malleable iron, grade 60003	2.40	1.80	0.072	0.30	–	Unalloyed; oil quenched from 870°C (1600°F), tempered 3 hrs. at 650°C (1200°F)	212
4	Oil-quenched malleable iron	2.40	1.80	0.072	0.30	–	Unalloyed; oil quenched from 870°C (1600°F), not tempered	444
5	Pearlitic malleable iron, grade 45010	2.40	1.80	0.076	0.90	–	Alloyed (Mn); air quenched from 940°C (1720°F), tempered 34 hr. at 715°C (1320°F)	192
6	Pearlitic malleable iron, grade 80002	2.40	1.80	0.072	0.90	0.45	Alloyed (Mn and Mo); air quenched from 940°C (1720°F), tempered 12 hr. at 620°C (1150°F)	262
7	Air-quenched alloyed malleable iron	2.40	1.80	0.079	0.90	–	Alloyed (Mn); air quenched from 925°C (1700°F), not tempered	285
8	Air-quenched alloyed malleable iron	2.40	1.80	0.076	1.10	–	Alloyed (Mn); air quenched from 925°C (1700°F), not tempered	321
9	Oil-quenched alloyed malleable iron	2.40	1.80	0.079	0.90	–	Alloyed (Mn); oil quenched from 830°C (1525°F), not tempered	514
10	Oil-quenched alloyed malleable iron	2.40	1.80	0.076	1.10	–	Alloyed (Mn); oil quenched from 830°C (1525°F), not tempered	578
11	Air-quenched alloyed malleable iron	2.40	1.80	0.072	0.90	0.45	Alloyed (Mn and Mo); air quenched from 940°C (1720°F), not tempered	514

Martempering and Tempering

Martempering and tempering develops mechanical properties similar to those resulting from conventional oil quenching and tempering—typical tensile strength, 862 MPa (125 ksi); yield strength, 758 MPa (110 ksi); and hardness, 300 HB.

Pearlitic malleable iron castings that are susceptible to cracking when quenched in warm oil (40 to 95°C; 100 to 200°F) from the austenitising temperature may be safely quenched in salt or oil at about 205°C (400°F). Elevator camshafts varying in length from 0.3 to 0.45 m (12 to 18 inch) and various sizes of wear-chain components are examples of martempered pearlitic malleable iron.

BAINITIC HEAT TREATMENT OF PEARLITIC MALLEABLE IRON

Both upper and lower bainite can be formed in pearlitic malleable iron with a marked increase in tensile strength and hardness but with decrease in ductility. A pearlitic malleable iron (2.6C-1.4Si-0.5Mn-0.11S), annealed at 930°C (1700°F) for 16 hours, air quenched and tempered at 680°C (1250°F) for 4 hours, developed an ultimate tensile strength of 650 MPa (94.2 ksi), a yield strength of 460 MPa (66.5 ksi) and a 3.4 per cent elongation at 217 HB.

This same iron austenitised at 900°C (1650°F) in molten salt for 1 hour, quenched in molten salt at 295°C (560°F) for 3 hours and air cooled, gave an ultimate strength of 995 MPa (144.2 ksi), a yield strength of 920 MPa (133.4 ksi) and a 1 per cent elongation at 388 HB.

SURFACE HARDENING OF PEARLITIC MALLEABLE IRON

Fully pearlitic malleable iron may be surface hardened by either induction heating and quenching or flame heating and quenching. Laser and electron beam techniques also have been used for hardening selected areas on the surface of pearlitic and ferritic malleable iron castings that are free from decarburisation. Generally, hardness in the range from 55 to 60 HRC is attainable, with the depth of penetration being controlled by the rate of heating and by the temperature developed at the surface of the part being hardened. In induction hardening, this is accomplished by close regulation of power output, operating frequency, heating time and alloy content of the iron.

The maximum hardness obtainable in the matrix of a properly hardened part is 67 HRC; however, as has been mentioned previously, conventional hardness measurements show less than the true matrix hardness because of the temper carbon nodules that are averaged into the hardness. Generally, a casting with a matrix microhardness of 67 HRC will have about 62 HRC average hardness, as measured with the standard Rockwell tester.

Rocker arms and clutch hubs are examples of automotive production parts that are surface hardened by induction. Flame hardening is not employed for this application because it would result in distortion that would interfere with their operation. The two examples that follow describe successful application of induction and flame hardening to other production parts.

Grade 45010 pearlitic malleable iron was used for tools for crimping electrical connectors. The jaws of these tools as originally designed were inserts made of hardened tool steel. Shell mould casting these jaws and heat treating them to pearlitic malleable iron made it possible for the jaws to be integrally cast to the desired intricate contour and to be hardened by induction heating and water quenching to provide required wear resistance. Hardening was thus restricted to the jaws and the strength and toughness inherent in grade 45010 were maintained in the arms of the handles.

Flame hardening has been used effectively on pearlitic malleable iron pinion spacers that support the cups of roller bearings. The ends of the pinion spacers were flame hardened to file hard (58 HRC or more) to a depth of about 2.4 mm ($\frac{3}{32}$ inch). This treatment eliminated service failures.

Chapter 14

Heat Treating of Austenitic Irons

INTRODUCTION

Austenitic cast irons are employed in a variety of applications because the alloying elements in these irons develop metallurgical properties superior to those of conventional gray and ductile irons. Nickel, chromium and copper promote corrosion resistance and nickel and chromium improve wear resistance.

Corrosion resistance is improved because the austenitic phase is stabilised at room temperature by the alloying elements and is more corrosion resistant than pearlite. Austenite is stable at high temperatures and normally transforms into pearlite and carbide phases during cooling.

FLAKE-GRAPHITE CORROSION-RESISTANT AUSTENITIC CAST IRONS

These cast irons exhibit properties that include: (i) resistance to corrosion by alkalis, acids, salts, oils and foods; (ii) high-temperature oxidation resistance; (iii) high electrical resistivity; (iv) non-magnetic characteristics; (v) abrasive wear resistance; (vi) uniform thermal expansion; and (vii) moderate strength and toughness.

Many of the irons combine several of these properties, which result from the high-temperature austenite phase; this phase is stabilised at room temperature by the high alloy content and contains a uniform dispersion of carbides. The compositions of flake-graphite corrosion-resistant austenitic cast irons are given in Table 14.1; typical mechanical properties of these irons are presented in Table 14.2.

Table 14.1. Compositions of flake-graphite corrosion-resistant austenitic cast irons.

	Composition %					
Type	*TC(a)*	*Si*	*Mn*	*Ni*	*Cu*	*Cr*
1(b)	3.00 max	1.00–2.80	0.50–1.50	13.50 –17.50	5.50-7.50	1.50-2.50
1b	3.00 max	1.00–2.80	0.50–1.50	13.50–17.50	5.50–7.50	2.50–3.50
2(c)	3.00 max	1.00–2.80	0.50–1.50	18.00–22.00	0.50 max	1.50–2.50
2b	3.00 max	1.00–2.80	0.50–1.50	18.00–22.00	0.50 max	3.00–6.00(d)
3	2.60 max	1.00–2.00	0.50–1.50	28.00–32.00	0.50 max	2.50–3.50
4	2.60 max	5.00–6.00	0.50–1.50	29.00–32.00	0.50 max	4.50–5.50

(Contd...)

			Composition %			
Type	*TC(a)*	*Si*	*Mn*	*Ni*	*Cu*	*Cr*
5	2.40 max	1.00–2.00	0.50–1.50	34.00–36.00	0.50 max	0.10 max(e)
6(f)	3.00 max	1.50–2.50	0.50–1.50	18.00–22.00	3.50–5.50	1.00–2.00

(a) Total carbon; (b) Type 1 is recommended for applications in which the presence of copper offers corrosion-resistance advantages; (c) Type 2 is recommended for applications in which copper contamination cannot be tolerated, such as handling of foods or caustics; (d) Where some machining is required, 3.0 to 4.0 Cr is recommended; (e) Where increased hardness, strength and heat resistance are desired and where increased expansivity can be tolerated, Cr may be increased to 2.5 to 3.0 per cent; and (f) Type 6 also contains 1.0 per cent Mo.

Table 14.2. Typical mechanical properties of flake-graphite corrosion resistant austenitic cast irons.

Type	*Tensile strength(a)*		*Hardness,*
	MPa	*ksi*	*HB(b)*
1	170	25	131–183
1b	205	30	149–212
2	170	25	118–174
2b	205	30	171-248
3	170	25	118-159
4	170	25	149-212
5	140	20	99–124
6	170	25	124–174

(a) Minimum; and (b) 3000-kg load.

These alloys are susceptible to work hardening during machining and require careful cooling from the casting operation and/or subsequent heat treating operations to minimise the initial stresses and the rate of work hardening during metal-removal operations. Castings that have not been heat treated may cause 'chattering' during machining.

Stress Relieving

For most applications, it is recommended that austenitic cast irons be stress relieved at 620 to 675°C (1150 to 1250°F), for 1 hour per inch of section, to remove residual stresses resulting from casting or machining, or both. Stress relieving should follow rough maching, particularly for castings that must conform to close dimensional tolerances, that have been extensively welded, or that are to be exposed to high stresses in service.

Holding of castings at 480°C (900°F) for 1 hour per inch of thickness will remove about 60 per cent of the stress; stress relieving at 675°C (1250°F) will remove almost 95 per cent. It is usually acceptable to cool castings in air at a rate of 1 to 2 hours per inch of section thickness, although furnace cooling produces maximum stress relief. Stress relieving does not affect tensile strength, hardness or ductility.

Spheroidise Annealing

Castings with hardnesses above 190 HB may be softened by heating to 980 to 1040°C (1800 to 1900°F) for ½ to 5 hours, except those alloys containing 4 per cent or more chromium. Excessive carbides cause this high hardness and may occur in rapidly cooled castings and thin sections. Annealing dissolves or spheroidises carbides. Although it lowers hardness, spheroidise annealing does not adversely affect strength.

High-Temperature Stabilisation

Except for castings of alloy type 1, which are not recommended for service above 430°C (800°F), castings used for either static or cyclic service at 480°C (900°F) or above should be given a stabilisation heat treatment. This treatment consists of holding at 760°C (1400°F) for 4 hours minimum or at 870°C (1600°F) for 2 hours minimum, furnace cooling to 540°C (1000°F) and then cooling in air. This treatment stabilises the microstructure and minimises growth and distortion at high temperatures as carbon diffuses from the austenite and from dissolving carbides and re-precipitates as larger-volume graphite flakes. Thus, it is usually advisable to stabilise castings prior to final machining.

Dimensional Stabilisation

This treatment normally is limited to castings that require true dimensional stability, such as those used in precision machinery or scientific instruments. The treatment is not applicable to castings of type 1 alloys. Other alloys may be dimensionally stabilised by the following treatments:

1. Heat to 870°C (1600°F) and hold for 2 hours minimum plus 1 hour per inch of section.
2. Furnace cool, at a maximum rate of 50°C/hr., (100°F/hr.), to 540°C (1000°F).
3. Hold at 540°C (1000°F) for 1 hour per inch of section and then cool uniformly in air.
4. After rough machining, re-heat to 455 to 480°C (850 to 900°F) and hold for 1 hour per inch of section and cool uniformly in air.
5. Finish machine and re-heat to 260 to 315°C (500 to 600°F) and cool uniformly in air.

Solution treating

Although this treatment is seldom used, high-temperature quenching is capable of producing higher-than-normal strength levels and slightly higher hardnesses by dissolving some carbon in austenite at elevated temperatures and by preventing precipitation of the carbon by rapid cooling. This treatment consists of heating to 925 to 1010°C (1700 to 1850°F) and quenching in oil or water. Because no metallurgical phase change occurs, the possibility of cracking is lessened.

NODULAR-GRAPHITE CORROSION-RESISTANT AUSTENITIC CAST IRONS

These irons, the compositions of which are given in Table 14.3, are similar to those listed in Table 14.1 but are treated with magnesium to produce spheroidal (nodular) graphite, which results in higher values of strength and ductility. To ensure satisfactory graphite nodularity, sulphur content is held to 0.02 per cent or less and residual magnesium content should exceed 0.03 per cent. The procedures and temperatures of the heat treatments for these alloys are identical to those described for flake-graphite corrosion-resistant austenitic cast irons. The mechanical properties of several of these nodular austenitic irons are listed in Table 14.3.

Table 14.3. Compositions of nodular-graphite corrosion-resistant austenitic cast irons.

			Composition %			
Type	*TC(a)*	*Si*	*Mn*	*P*	*Ni*	*Cr*
D-2	3.00 max	1.50 to 3.00	0.70 to 1.25	0.08 max	18.0 to 22.0	1.75 to 2.75
D-2b	3.00 max	1.50 to 3.00	0.70 to 1.25	0.08 max	18.0 to 22.0	2.75 to 4.00

(Contd...)

	Composition %					
Type	*TC(a)*	*Si*	*Mn*	*P*	*Ni*	*Cr*
D-2c	2.90 max	1.00 to 3.00	1.80 to 2.40	0.08 max	21.0 to 24.0	0.50 max
D-3	2.60 max	1.00 to 2.80	1.00 max	0.08 max	28.0 to 32.0	2.50 to 3.50
D-3a	2.60 max	1.00 to 2.80	1.00 max	0.08 max	28.0 to 32.0	1.00 to 1.50
D-4	2.60 max	5.00 to 6.00	1.00 max	0.08 max	28.0 to 32.0	4.50 to 5.50
D-5	2.60 max	1.00 to 2.80	1.00 max	0.08 max	34.0 to 36.0	0.10 max
D-5b	2.40 max	1.00 to 2.80	1.00 max	0.08 max	34.0 to 36.0	2.00 to 3.00

(a) Total carbon

Table 14.4. Typical mechanical properties of nodular-graphite corrosion-resistant austenitic cast irons.

	Minimum tensile strength		Minimum yield strength		Minimum	Hardness,
Type	*MPa*	*ksi*	*MPa*	*ksi*	*elongation (a), %*	*HB*
D-2	400	58	205	30	8	139 to 202
D-2b	400	58	205	30	7	148 to 211
D-3	380	55	205	30	6	139 to 202
D-4	415	60	–	–	–	202 to 273
D-5	380	55	205	30	20	131 to 185

(a) In 50 mm or 2 inch.

ABRASION-RESISTANT NICKEL-CHROMIUM WHITE IRONS

These special white irons posses outstanding abrasion resistance and are widely used in the mining, power, cement, paint, steel and foundry industries. As shown in Table 14.5, these irons contain less nickel, silicon and manganese than the austenitic alloys. This composition results in an entirely carbidic structure.

Table 14.5. Compositions of abrasion-resistant nickel-chromium white irons.

	Composition range, %	
Element	*Type I (regular)*	*Type II (high-strength)*
Total carbon	3.00–3.60	2.90 max
Silicon	0.40–0.70(a)	0.40–0.70(a)
Manganese	0.40–0.70	0.40–0.70
Sulphur	0.15 max	0.15 max
Phosphorus	0.40 max	0.40 max
Nickel	4.00–4.75	4.00–4.75
Chromium	1.40–3.50	1.40–3.50

(a) Silicon content may exceed 0.70% in thin-wall castings or in composite castings poured over gray iron cores.

Because of their carbidic structures, abrasion-resistant Ni-Cr white irons must be stress relieved at 200 to 230°C (400 to 450°F) before being placed in service. This treatment relieves the transformation stresses present in castings after cooling to room temperature and increases strength and impact resistance with no loss in abrasion resistance.

HIGH-CHROMIUM, LOW CARBON WHITE IRONS

The chemical composition ranges for high-chromium, low-carbon white irons are given in Table 14.6. These abrasion-resistant irons are used in the manufacture of ceramics and of equipment for dredging, oil-well drilling, coke and ore handling and shot blasting.

Table 14.6. Composition ranges for high-chromium, low carbon white irons.

Element	*Composition range, %*
Carbon	2.25 to 2.85
Silicon	0.25 to 1.00
Manganese	0.50 to 1.25
Chromium	24.0 to 30.0
Vanadium(a)	0.25 to 2.00
Molybdenum(a)	0 to 6.50

(a) Optional.

Annealing

Although castings of these irons usually are finished primarily by grinding, they can be annealed to a hardness sufficiently low to permit simple machining. The annealing treatment consists of heating slowly to 760 to 790°C (1400 to 1450°F), holding at temperature for 12 to 24 hours and then furnace cooling in still air. Hardnesses as low as 350 to 450 HB can be obtained by this treatment.

Austenitising

For maximum wear resistance under conditions of extreme rubbing or scouring, an austenitising heat treatment is effective. The room-temperature austenite produced by this treatment readily work hardens by rubbing or mild impingement.

This transformation is somewhat analogous to the behaviour of austenitic high-manganese steel. Austenitising consists of heating to 1095 to 1105°C (2000 to 2020°F) and cooling in air.

SECTION IV

Heat Treatment of Tool and Stainless Steel

Chapter 15

Basic Concepts of Heat Treatment of Tool Steels

INTRODUCTION

Because tool steels are of such diversified composition, heat treatment of tool steels is a vast, complex and interesting subject in which principles are more easily remembered than are the many facts regarding treatment of each of the separate steels. It is fundamental to review the general principles controlling heating and cooling of tool steels for hardening or softening and to consider briefly the selection and operation of equipment necessary to carry out such procedures.

Temperatures used for heat treatment of tool steels are perhaps of the widest range of any metallurgical product, ranging from sub-zero temperatures used in special treatments to promote stabilisation of size in gages, upto temperatures of 2400°F (1315°C) used in heat treating high speed steels. Likewise, throughout the many types of tool steels, there will be variations in method of quenching that involve all of the commercially possible rates of cooling. Thus, shallow-hardening carbon tool steels are quenched in brine, the most rapid quenching medium, while many tool steels are of such high alloy content that cooling in still air provides full hardness. Temperature of the quenching bath may also vary over a wide range. Water and brine quenching baths are usually maintained at as low as temperature is possible; in many instances, hot quenching treatments are used for high-alloy steels in which the bath temperature may reach 1000 to 1100°F. The range of tempering temperatures for the many tool steel compositions is also large. Carbon tool steels and low alloy steels are often tempered as low as 250 to 300°F, while high speed steels and hot working die steels may be tempered as high as 1100 or 1200°F (650°C).

Carbon is the most important single element in controlling the properties of all tool steels despite the fact that many steels owe their special properties to the presence of certain alloying elements. A change of carbon content of several points in most tool steels will exert a greater influence over properties after heat treatment than a similar or even larger change in the percentage of any other element present. The carbon content of tool steels must, therefore, be controlled very closely. Many steels are made with very high carbon contents to promote increased wear resistance; if this carbon content is lowered, the purpose is defeated. These facts, coupled with the recognition that in the finished tool the surface of the steel does the actual work, dictate that close attention be paid to the control of the surface carbon content of the steel during all heat treating processes. Other factors that make tool steels of extreme interest from the viewpoint of heat treating operations involve the control of the surface hardness, hardness distribution or hardenability and distortion or movement that occurs during the quenching and tempering operations.

PRINCIPLES OF HEAT TREATING

The basis for heat treatment of all steels lies in the fact that, when they are heated, a structural change takes place. The mixture of ferrite and carbide that exists in steel at low temperatures transforms to austenite on passing the critical temperature during heating. Austenite dissolves much larger amounts of carbon than ferrite does and on cooling from the heat treating temperature, the size and distribution of ferrite-carbide aggregates can be controlled through the cooling procedure. Through control of the distribution of ferrite and carbide in steel, we are able to obtain the widest variation of mechanical properties possible in any structural or engineering material.

The phenomenon of the formation of austenite on heating with the attendant solution of a large portion of the carbide phase and the phenomenon of the subsequent formation of a controlled ferrite-cementite aggregate on cooling are represented basically by the diagrams listed below:

1. Iron-carbon diagram shown in Fig. 15.2.
2. 'S-curve' shown in Fig. 15.3.

The terms 'S-curve', 'C-curve', 'TTT (time-temperature-transformation) curve' and 'isothermal transformation diagram' are synonymous and all are used to describe the type of diagram shown in Fig. 15.3.

Both of these diagrams are useful in predicting structures that are obtainable with various combinations of heating and cooling cycles. In addition, by the use of the S-curve, the effect of variations of the rate of cooling can be predicted. ·

Transformations on Heating

When a sample of carbon steel containing approximately 0.8 per cent carbonC is heated, a decalescence occurs on passing through the critical temperature, as shown in Fig 15.1. This decalescence is associated with the absorption of heat occurring when ferrite changes to austenite. By heating a series of steels of varying carbon contents to elevated temperatures, a series of such transformation temperatures is obtained that, combined with the results of dilatometric and metallographic investigations, yields the iron-carbon diagram (Fig.15.2).The iron carbon diagram indicates the stable phases or structures at any combination of temperature and composition for iron-carbon alloys.

Thus, referring to Fig. 15.2 at a temperature of 1600°F (870°C), it is noted that 0.80 per cent carbon steel is composed entirely of austenite, while at the same temperature a 1.30 per cent carbon steel is composed of austenite plus a certain quantity of cementite or carbide. Further, note that at room temperature or any temperature upto 1335°F (725°C), any commercial steel containing above 0.05 per cent carbon is composed of ferrite and cementite (or carbide). The amount of carbide phase present increases as the carbon content increases. For purposes of describing the heat treatment of steels, it is not necessary to extend the diagrams beyond commercial heat treating temperatures and thus the upper part of the diagram is of no value for our purposes. It does, however, indicate the relationship between the temperature and composition at which melting, either partial or complete, occurs.

With respect to carbon tool steels, which usually contain above 0.6 per cent carbon (although a majority contain from 0.90 to 1.1 per cent carbon), austenite begins to form at a temperature of approximately 1335°F (725°C). However, as the carbon content increases over the indicated range, higher heat treating temperatures must be used if all of the carbide is to be dissolved and if the austenite is to be homogeneous. Commercial practice, however, rarely makes use of such high heat treating temperatures for hypereutectoid steels, except in certain normalising operations.

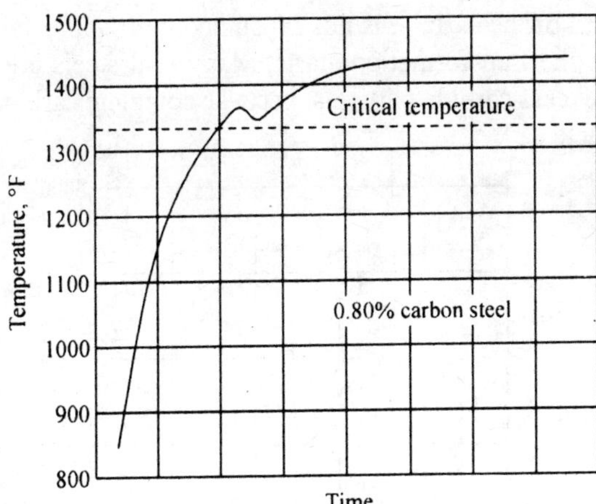

Fig. 15.1. Typical simple heating curve for 0.80 per cent carbon steel, showing decalescence caused by marked absorption of heat on passing through critical temperature. Extent of decalescence depends on rate of heating and size of specimen.

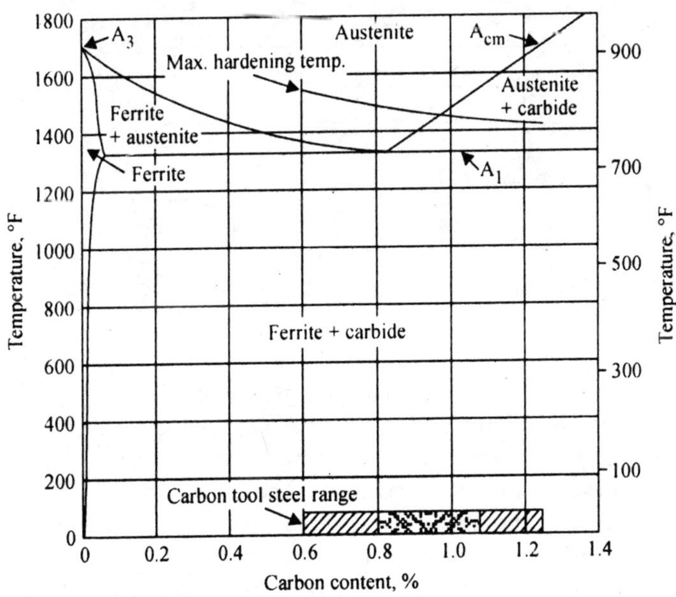

Fig. 15.2. A section of the iron-iron carbide diagram showing the approximate range of carbon contents found in carbon tool steels and the maximum hardening temperature commonly used.

At first the iron-carbon diagram might be considered inadequate when alloy steels are studied. It is true that a great volume of tool steels have appreciable alloy contents, some reaching high proportions. However, just as carbon steels are the basis for discussing all other steels, so is the iron-carbon diagram the basis for our knowledge of transformations occurring not only in carbon steels but in other steels as well. Virtually all of the tool steels with which we will be concerned in this chapter are hypereutectoid in nature. When considering alloy steels, this term indicates that when the ferrite and carbide react on heating to form austenite, all of the carbide is not consumed until appreciably higher temperatures are reached than that at which the transformation starts. Many tool steels are characterised by a certain quantity of undissolved or excess carbide particles, because complete carbide-solution temperatures are not used in commercial practice.

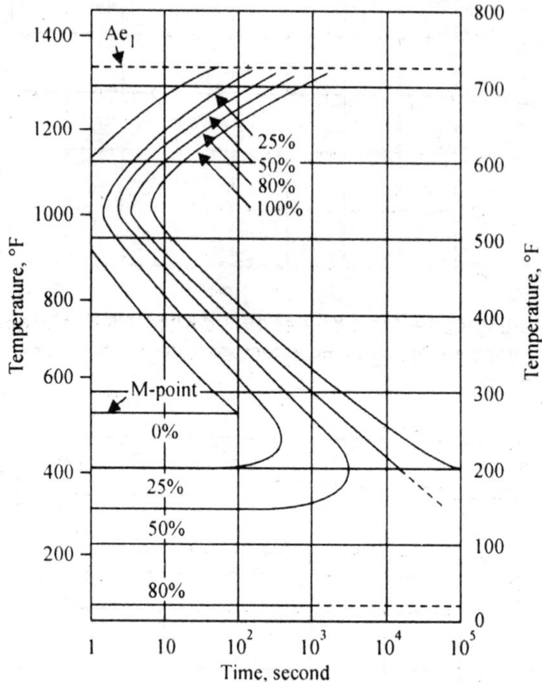

Fig. 15.3. The 'S-curve' for a plain carbon steel (0.80 per cent carbon), showing the time required for transformation at each temperature.

Consideration of the squence of events, when a hypereutectoid carbon steel is heated from room temperature to any temperature within the austenite range, exemplifies the interpretation of heating and cooling phenomena by means of the iron-carbon diagram. On heating, no changes should occur until the temperature of 1335°F (725°C) is reached. At this temperature, if the heating rate is very slow, some of the ferrite reacts with some of the carbide and forms small pools of austenite. As the temperature is further raised at a hypothetically chosen slow rate, these austenite pools increase in size at the expense of the ferrite and caıbide and new austenite pools form in other parts or critical temperature, all of the ferrite should be consumed and the structure existing at this temperature consists of austenite grains with undissolved carbide embedded throughout. Each of these grains of austenite started from a small nucleus that grew until it impinged upon other growing grains. At the time at which the ferrite has just been consumed, the impingement of the austenite grains, of course, determines their size and this size is what is termed the initial austenite grain size. Methods of measuring this size and later grain sizes after any growth has occurred have been discussed previously. As long as the excess carbide occurs in large quantities in conjunction with these austenite grains, growth of the grains will not take place to any appreciable extent, although a slight increase will be noted.

As the temperature is further raised beyond the A_{cm} line, the diagram indicates that pure austenite should be the only existing phase in the steel. Thus, after the austenite has initially formed and the specimen is heated to the A_{cm} temperature, the carbide is under a continuous process of solution. This solution, however, is by no means complete at the exact moment of heating above the A_{cm} line and in some instances, it is necessary to hold for appreciable times at temperatures above 1700°F (925°C) in order to achiever even a partial degree of homogenisation of the austenite with respect to these carbide particles. When the carbide has been almost completely consumed by solution in the austenite, its restricting influence on grain size is no longer present and the grains will grow rapidly. Both the solution of the carbide and the growth of these grains appear to be related to the deoxidation practice and forging practice used in manufacturing steel.

If at any temperature in the austenite or in the austenite and carbide regions of the diagram the heating is stopped, the transformations do not necessarily stop. Thus, if a hypereutectoid steel was heated to 1400°F (760°C), a certain amount of austenite and carbide exists at the time this temperature was reached. However, in most instances, less austenite and more carbide are present than would be predicted from the equilibrium diagram. During holding at 1400°F (760°C), the steel would tend to approach equilibrium and the constituents would adjust their ratio to conform with the indications of the iron-carbide diagram. Likewise, on heating a 1.00 per cent carbon steel to 1700°F (925°C), carbide will very likely exist when the temperature is reached. As the steel is held at this temperature for longer times, the equilibrium structure will be approached and 100 per cent austenite should result.

In the preceding paragraph, the term equilibrium diagram was introduced. This implies that the diagram holds only when the equilibrium state has been achieved. In most metallurgical processes involving the heating of solid metals, equilibrium is never achieved instantaneously and perhaps in only a few instances is it even approached. However, by use of a long holding time at temperature or its equivalent, use of extremely slow heating and cooling rates, an approach to equilibrium is obtained that is sufficient for all practical purposes. Let us see how this concept of equilibrium or better, how the lack of equilibrium during simple heating and cooling operations affects our considerations of the heating and cooling process.

In most commercial heating processes, heating is not done at an extremely slow rate and, therefore, when a temperature of 1335°F (725°C) is reached and even exceeded to some extent, no austenite will form although the diagram indicates that it should. At some higher temperature, however, the characteristic decalescence associated with the absorption of heat and the formation of austenite is noted. On heating in an ordinary semi-muffle furnace, this transformation is apparent at a temperature between 1340 and 1350°F (725 and 730°C). If, however, the heating is conducted in a salt or lead bath so that the heating rate is extremely rapid, the transformation does not occur until a temperature between 1360 and 1370°F (735 to 745°C) is reached. Thus, as the heating rate increases, the temperature at which the transformation to austenite occurs is raised. The minimum temperature at which the transformation can start on heating is 1335°F, the equilibrium critical temperature.

Transformations on Cooling

Similar logic can be applied to problems of cooling steel from temperatures in the austenite or austenite-plus-carbide range. The equilibrium diagram indicates that, on cooling a steel containing 0.8 to 0.9 per cent carbon, the austenite should decompose to form ferrite and carbide aggregates at a temperature of 1335°F (725°C). In a hypereutectoid carbon steel, however, some of the excess carbide that has been dissolved at the heat treating temperature can precipitate above 1335°F (725°C) if the cooling is slow. If the austenite has been made homogeneous by heating to a temperature above the A_{cm} and the carbide has been dissolved to a large extent, the precipitation of cementite on cooling will have a great tendency to take place at the austenite grain boundaries. If, however, the carbide has not been completely dissolved and large quantities remain in the form of rounded particles throughout the matrix, carbide precipitation will take place on these pre-existing points and the network of cementite surrounding the grain boundary will not form.

At a temperature of 1335°F (725°C), the transformation from austenite to ferrite and carbide does not take place as indicated by the equilibrium diagram because again there is a 'lag' similar to that discussed in considering the formation of austenite. If the steel were cooled at the lowest possible rate, the transformation from austenite to ferrite and carbide would probably take place between 1310 and 1325°F (710 and 720°C). However, under any normal commercial conditions of cooling, the transformation probably takes place well below 1300°F (705°C). Again as the rate of cooling increases, the temperature at which the transformation takes place decreases, that is, moves farther away from the temperature indicated by the equilibrium diagram. The amount of lag on cooling, however, is much greater than it is on heating.

Rationalisation of this phenomenon is most easily made by a study of Fig. 4.3 and the 'S-curve', which is constructed or determined as follows:

Samples of the steel in question (for our purposes a plain carbon eutectoid steel) are heated to the austenite range, held for the desired length of time, rapidly cooled to below the critical (usually by quenching in a hot bath), and held at some constant temperature. The extent of the decomposition of austenite is determined as a function of the holding time at the hot bath temperature, as shown in Fig. 15.4. These curves are typical of reactions that involve diffusion and apply also to the formation of austenite previously discussed. The reaction takes place over a period of time—indeed, the austenite does not begin to decompose at an appreciable rate until after an 'incubation period'.

If the steel is cooled to and held at another temperature below the critical, a different rate of austenite decomposition (compare curve 2, Fig. 15.4 and curve 1, Fig 15.4) will result. By conducting such experiments over the whole temperature range from the critical temperature to 500°F, the time required to cause 0.5 per cent reaction and 99.5 per cent reaction ('beginning' and 'end' of the reaction) can be determined for each temperature and assembled into the 'S-curve'. Therefore, the left-hand curve of Fig 15.3 indicates the time at which austenite begins to decompose at any temperature and the right-hand curve indicates the time at which the decomposition is practically complete. Since such a wide range of time is encountered, it is convenient to use a logarithmic time scale.

The features of this curve that are important are as follows:

1. Exceptionally long times are needed to transform the austenite at temperatures just below the critical temperature. Thus, on cooling, unless the cooling rate is exceptionally slow, the steel will not be held at these temperatures for a sufficient length of time to cause the transformation to occur. The transformation will then occur at a temperature lower than expected.
2. The time required to cause the transformation to occur decreases rapidly as the temperature is lowered to 1000°F (540°C).
3. There is a maximum in the rate of transformation occurring near 1000°F (540°C). The time to complete the transformation at this temperature is small for plain carbon steels and generally is about one second or less.
4. Below this temperature, the rate decreases again and at 500°F, it often requires 10 to 15 minutes to start the decomposition of the austenite.

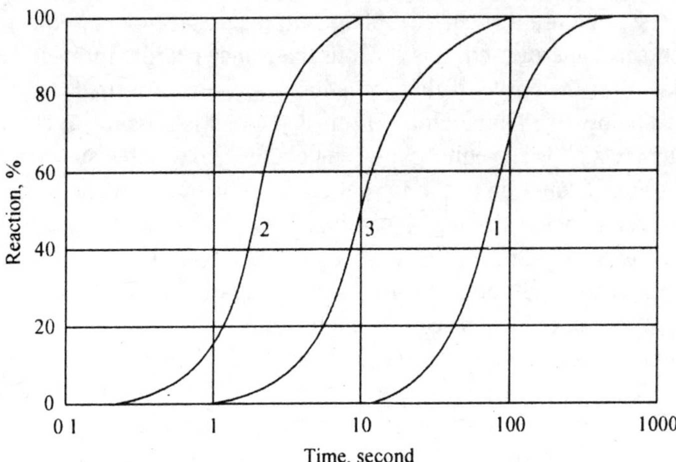

Fig. 15.4. Typical isothermal reaction curves obtained at (1) 1300°F, (2) 1100°F and (3) 800°F for a eutectoid steel. Such data are used to construct the S-curve.

The structure that is produced at each of these decomposition temperatures is also of special importance and interest. If the steel consists of homogeneous austenite before cooling (austenite in which no excess carbide is present), at all temperatures from 1335 to 1000°F (725 to 540°C) lamellar pearlite is the decomposition product. At temperature near the critical , the distance between the carbide plates in the pearlite, called the pearlite spacing, is relatively large and therefore, the product is relatively soft, having a hardness of about Rockwell carbon 10 to 15. As the temperature of the transformation is lowered, however, the pearlite spacing decreases and the product becomes correspondingly harder, rising to a value to Rockwell carbon 40 to 45 at the temperature of the fastest rate of reaction. This portion of the curve in which the fastest rate of reaction is indicated is commonly called the 'knee' of the curve.

At temperatures below the 'knee', lamellar pearlite does not form—instead an acicular product forms that is evidently a fine mixture of ferrite and carbide. This product, bainite, has a hardness that increases as the temperature of transformation is lowered from Rockwell carbon 45 to Rockwell carbon 60.

It is important to mention also that in the temperature range from the critical temperature to the 'knee' of the curve, spheroidite is often found in the completely transformed structure. It has been definitely established that spheroidite can form directly from austenite only if the austenite is heterogeneous, that is, if undissolved carbide particles are present upon which further precipitation can take place. This is of special importance in the annealing tool steels, most of which are hypereutectoid, at temperatures between 1335°F (725°C) and the A_{cm} line. In this instance, the tendency for the formation of rounded carbide particles is so great that carbide plates associated with lamellar pearlite structures cannot grow. This spheroidite will form directly from austenite only, however, if the temperature of reaction is within 50° of the critical, or in annealing operations, if the cooling rate is relatively slow. As the temperature of transformation becomes closer to the 'knee' of the curve, even though the initial austenite was markedly heterogeneous, pearlite will form. Since it takes an appreciable time for pearlite to form at elevated temperatures, the carbide plates in the pearlite spheroidise at a relatively rapid rate and it is not uncommon to find upto 20 per cent spheroidite in samples that have been cooled at a slow rate or that have been held long times at temperatures between 1200 and 1300°F (650 and 705°C) for transformation. The hardness of spheroidite is always less than that of pearlite and thus, when any spheroidite forms, softer structures are obtained. The softest structure that can be obtained in hypereutectoid carbon steels is one in which the carbide is very coarsely spheroidised. It is obvious that, since pearlite forms in such short time at the temperature of the 'knee' of the curve, if the formation of bainite is desired, the steel must cool at an extremely rapid rate to below this temperature. If this is done, it is possible to suppress completely the formation of pearlite and retain the austenite to low temperatures, in which instance it will either transform to bainite or to martensite. Discussion of the formation of the latter structures is necessary to complete the picture of the transformations occurring on cooling. It will be assumed that the steel has been heated into the austenite range and has been subsequently cooled at a rate that has prevented the formation of pearlite. In order to cause the formation of bainite, the cooling must be stopped or appreciably retarded, since the rate of bainite formation decreases as the temperature decreases.

If, on the other hand, cooling is continued, austenite will begin to decompose to martensite at approximately 400°F (205°C) for a 1.00 per cent carbon steel. This martensite, the principal constituent of fully hardened steel, consists of numerous microscopic plates, each of which forms from the parent austenite by a process of sudden shear in less than one millionth of a second. Martensite has a body-centered tetragonal lattice (in contrast to the body-centered cubic lattice of ferrite) and has the same chemical composition as the parent asutenite.

Unlike the pearlite or bainite transformation, martensite plates will form only as the temperature decreases; a certain amount of new martensite appears at each temperature range throughout the cooling process. If the cooling is stopped after a certain amount of martensite has formed, transformation will stop for all practical purposes. Further cooling is then necessary to initiate the transformation once more. This sequence of events is indicated by the horizontal lines on the S-curve, Fig. 15.3 and may be studied in greater detail by the martensite transformation curve 1, Fig. 15.5. The temperature at which austenite begins to decompose to martensite on cooling is called the M_s point. On further cooling below M_s, martensite formation continues through room temperature to sub-zero temperatures and eventually stops short of consuming all the remaining austenite at a second temperature designated the M_f point. The resulting as-quenched hardness is a function of the carbon content of the martensite (inherited from the austenite) and the quantity of relatively soft austenite retained at the completion of cooling. Although micro-hardness tests indicate that retained austenite is nearly as soft as annealed steel, the relatively large indentor size used in the Rockwell hardness test does not reveal an appreciable hardness loss until the austenite exceeds about 20 per cent of the microstructure.

Austenite retained after the cooling process is completed has been a subject of considerable discussion in recent years. Retained austenite is unstable and may transform to fresh martensite under the influence of cold work or deformation (as in Hadfield's manganese steel), during tempering operations or simply during ageing at room temperature, causing unexpected dimensional change, brittleness or cracking. In other instances, large percentages of retained austenite may prevent the attainment of desired hardness because the austenite transforms to a product softer than martensite during subsequent tempering operations.The amount of retained austenite depends largely on the M_s temperature and the quenching bath temperature; it is possible to obtain a functional relationship between these variables that enables prediction of the per cent of retained austenite; the higher the M_s, the lower the amount of retained austenite for a given quenching bath temperature. This M_s temperature is determined by the composition of the austenite prior to the quench and thus, is a function of both the steel composition and the austenitising temperature. Increasing carbon content sharply lowers the M_s; the common alloying elements have a similar but less drastic influence. Higher austenitising temperatures lower the M_s by dissolving more of the excess carbides in the austenite prior to the quench. The M_f temperature, on the other hand, appears to be fairly constant at –250°F (–155°C) for carbon and low-alloy steel and –150°F (–100°C) for high speed and high-carbon high-chromium steels, regardless of the position of the M_s point. Actually, the formation of a few additional martensite plates has been observed after cooling to 8°F above absolute zero (425°F = 4 K), but the temperatures above M_f may be considered the stopping points for all practical purposes.

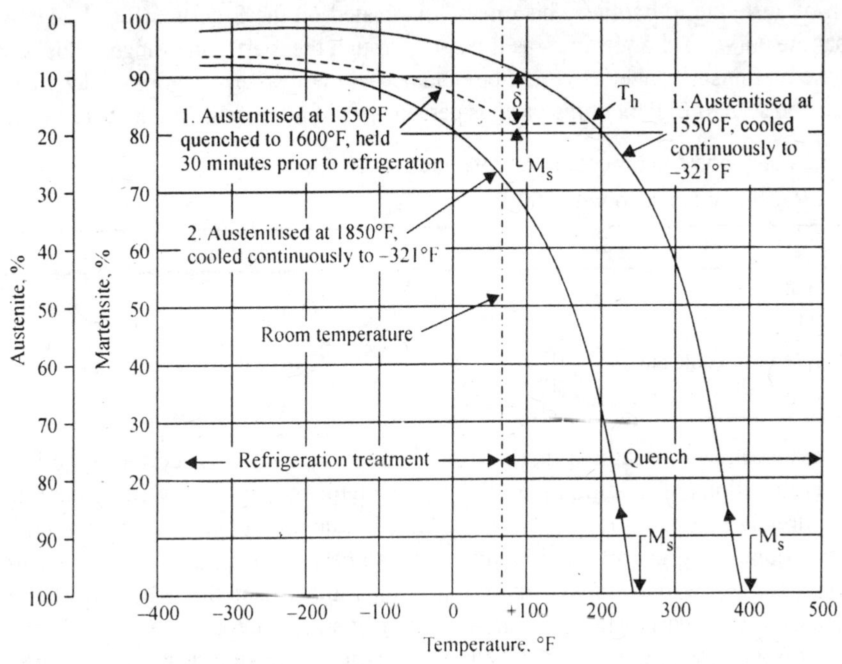

Fig. 15.5. Martensite transformation curves for a low-alloy tool steel.

In addition to the influence of M_s (composition and austenitising temperature) on the quantity of retained austenite, a phenomenon known as stabilisation may also appear either during the cooling process or during re-heating for tempering. Stabilisation can occur on interrupting the quench within the austenite-to-martensite transformation range or on slow cooling through this range and results in greater quantities of retained austenite that would have been obtained by more rapid continuous cooling. For example, in curve 1 of Fig. 15.5, a 30 minutes hold in hot oil at 160°F (70°C) (T_h) stabilised the austenite against additional martensite formation on further cooling until the temperature had reached 80°F (27°C) (M_s') and subsequent transformation followed curve 3. Over this temperature interval, the 8 per cent martensite (delta) that would have formed on continuous cooling was retained as austenite and the steel reached room temperature with 15 per cent rather than 7 per cent retained austenite. Even a drastic refrigeration treatment would not remove much of the additional austenite resulting from the hot quench treatment. As measured by the temperature interval (M_s'– T_h), this stabilisation effect is strongest at room temperature (that is, holding prior to refrigeration), becomes progressively weaker at higher hot quench temperatures and disappears completely at holding temperatures above M_s.

The degree of stabilisation is also influenced by time, rising sharply during the first few minutes at the holding temperature but showing little increase with holding times over 30 minutes. Because of this time dependence, some stabilisation may be observed during slow continuous cooling through the martensite range. This effect is illustrated for a low-alloy steel in Table 15.1, in which various quenchants provided various cooling rates through the martensite range. The oil quench resulted in slight stabilisation and a noticeable increase in retained austenite, while the still slower air cool nearly doubled the percentage of retained austenite. The hold in the 450°F (230°C) salt quench produced no stabilisation in itself since this temperature was above M_s for steel. Although slow cooling or holding in the temperature range below M_s tends to stabilise austenite against further martensite formation, it conversely tends to accelerate the transformation of austenite to bainite. This effect is indicated on the S-curve, (Fig. 15.3), by a 'bending back' of the bainite curve to shorter times as it passes through the martensite range. Thus, slower cooling rates through the martensite range, as would be obtained in the heavier sections of deep-hardening dies, may result in mixed structures containing various percentages of martensite, retained austenite and bainite.

Table. 15.1. Effect of quenching procedure on retained austenite.
(1.0C, 1.5Cr, 0.2V steel, austenitised at 1550°F[1(a)])

Treatment	*Retained austenite, %*
Water quenched to 68°F	5.8
Oil quenched to 68°F	7.0
Salt quenched to 450°F, air cooled to 68°F	10.6
Salt quenched to 450°F, water quenched to 68°F	6.1

There are several mixed structures that can be obtained by cooling at certain specified rates or by using certain combinations of holding times and holding temperatures followed by cooling at different rates. One has already been mentioned, that is, a mixed structure containing pearlite and spheroidite. Mixed structures containing pearlite and bainite can also be formed in steel. These are invariably the result of holding for isothermal transformation at temperatures near the 'knee' of the curve. Mixed structures of pearlite and martensite are quite common in steel. These are produced in the following manner. If the cooling rate is not sufficiently rapid to completely suppress the formation of pearlite, nodules of pearlite will form at the austenite grain boundaries and start to grow. It must be stipulated, however, that the cooling is not sufficiently slow to allow this transformation to go to completion. After a certain amount of pearlite has formed, therefore, the reaction stops; steel continues to cool and martensite forms from the remaining austenite. These structures are, therefore, found commonly in the transition zone between the case and the core of shallow or medium-hardening steels. Finally, mixed structures of bainite and martensite can be obtained by a process to be described later.

In the preceding discussion of transformations on cooling, the S-curve has been employed to discuss the different decomposition products of austenite. While this is completely adequate for descriptive purposes, it is well to recall that the S-curve is determined by an isothermal transformation technique and cannot strictly apply to transformations on continuous cooling, such as occur in the actual quenching process. A more satisfactory method of describing the transformation that occur on cooling is afforded by the 'CCT (continuous cooling transformation) diagram'. This diagram is more difficult to determine than the more familiar S-curve and, hence, is available for only a limited number of steels.

RE-HEATING TO BELOW THE CRITICAL TEMPERATURE (TEMPERING)

If any of the structures described above are re-heated, a radical change takes place. This change, regardless of the initial structure, is always in one direction, that is, toward complete spheroidisation of the steel. Of course, if the steel is re-heated rapidly to above the critical temperature, austenite will form before this spheroidisation can occur to an appreciable extent. Since the reactions occurring on re-heating all lead to an increased spheroidisation of the steel, they all lead to softening. If the carbide is almost completely precipitated in the initial structure (in spheroidite, pearlite or bainite structures), the spheroidisation that occurs will consist solely of an agglomeration process that depends only on the diffusion of carbon through the ferrite. The rate of this spheroidisation process, therefore, increases rapidly as the temperature increases and the process depends on time.

The re-heating of martensite structures in which the carbide is not completely precipitated is more complex. It is generally agreed that martensite is a body-centered tetragonal lattice that contains 'in solution' considerable excess carbon and perhaps minute carbide particles which, although precipitated, are probably no larger than several molecules in size. Practically all tool steels in the quenched condition contain in addition to this martensite appreciable quantities (1 to 30 per cent) of retained austenite. This retained austenite is more prevalent in alloy steels in which the temperature at which martensite begins to form on cooling (M_s point) is depressed over that in plain carbon steels. There are serveral changes that must, therefore, occur in order to promote the formation of a structure consisting of large carbide particles in a body-centered cubic matrix.

The sequence of changes that occurs on tempering and ultimately results in a return to the annealed condition can be divided into several overlapping stages. These structural changes are schematically indicated at the bottom of Fig. 15.6, which also presents the hardness changes found on re-heating quenched and refrigerated 0.40, 0.80 and 1.40 per cent carbon alloys. On heating at 150°F (66°C) for 1 hour, the body-centered tetragonal structure of primary martensite begins to release the carbon held in supersaturated solution, permitting the lattice to decrease in tetragonality and resulting in an over-all contraction in volume. In the temperature range upto about 400°F (205°C), this carbon tends to precipitate within the martensite grains in the form of a transitional iron carbide of hexagonal structure, called epsilon carbide.

Of primary interest in this transformation is the increase in hardness, or at least retarded softening, achieved during its early stages (Fig. 15.6). Since, the matrix of 0.25 to 0.30 per cent carbon martensite should be softer than in an as-quenched 0.40 per cent carbon alloy, the fine initial precipitate of epsilon carbide must contribute a marked hardening effect to maintain the hardness shown in the 250 to 300°F (120 to 150°C) tempering range. Although part of the hardness increase in the 1.40 per cent carbon alloy might be attributed to a transformation of retained austenite, microhardness tests of individual martensite needles in a quenched and tempered 1.70 per cent carbon alloy (Fig. 15.7) indicate a pronounced hardening effect in martensite alone after tempering up to 300°F (150°C). This reaction in which high-carbon primary martensite is gradually replaced by precipitated epsilon carbide and low-carbon martensite has been designated the first stage of tempering. Fig. 15.7 also indicates a slight increase in hardness of retained austenite during tempering below 400°F. Stark and Lement, however, have shown that this slight hardness increase is from the formation of 'surface martensite' and not from any precipitation within the retained austenite.

Further heating above 400°F (205°C) initiates the precipitation of cementite (Fe_3C), which gradually replaces the epsilon carbide in the low-carbon martensite matrix. At 450 to 550°F (230 to 290°C), this cementite forms thin elongated films at the martensite boundaries, which are believed responsible for the minimum in notch toughness of carbon and low-alloy steels tempered in this range. From 550 to 1000°F (290 to 540°C), the cementite gradually coarsens and coalesces, depleting the matrix in carbon until it is entirely body-centered cubic ferrite. This precipitation and growth of cementite, accompanied by a steady contraction and more rapid softening, is called the third stage of tempering. In medium to high-alloy steels, re-heating at temperatures above 1000°F (540°C) introduces still a fourth stage of tempering in which the cementite dissolves in the highly alloyed ferrite matrix and complex alloy carbides are simultaneously precipitated. This precipitation of alloy carbides, together with the conditioning and transformation of highly alloyed retained austenite (described below), is believed reponsible for secondary hardening in high-alloy tool steels.

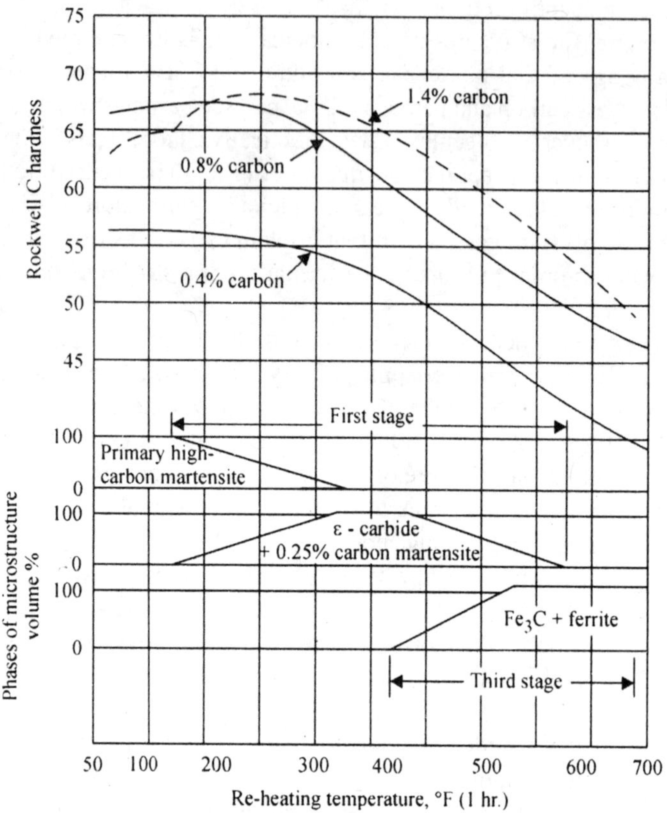

Fig. 15.6. Hardness and approximate structural changes on re-heating quenched and refrigerated iron-carbon alloys.

The second stage of tempering involves the transformation of retained austenite to either bainite or martensite, depending on the composition and treatment of the steel. This stage may occur simultaneously with any of the aforementioned stages of martensite decomposition and always results in a marked expansion that opposes the contractions due to martensite decompostion. To avoid confusion, it is convenient to designate the products of retained austenite transformation on re-heating as 'secondary bainite' and 'secondary martensite' to distinguish them from similar phases formed during the initial cooling from the austenitising temperature. In carbon and low-alloy steels, the austenite begins to transform to secondary bainite on re-heating 350 to 400°F (175 to 250°C) and is usually completely transformed after 1 hour at 600°F (315°C) or longer times at lower temperatures. Figure 15.7 indicates that this secondary bainite assumes an identical hardness with that of martensite tempered at the same temperature, thus providing a desirable structural uniformity. If, to retain maximum hardness, the re-heating temperature is kept too low to convert all of the austenite to secondary bainite, it will be found that the remaining austenite is resistant to removal by any subsequent refrigeration treatment. This austenite stabilisation on re-heating, similar to that discussed previously with regard to quenching, may be studied in Fig. 15.8. If a steel is quenched to room temperature and refrigerated immediately, its M_s' for further martensite formation is, of course, at room temperature and the martensite transformation curve is similar to curve 1, (Fig. 15.5). However, if the refrigeration treatment is delayed for 1 hr. at room temperature, stabilisation depresses M_s' to – 45°F, or – 43°C (Fig. 15.8), thereby decreasing the percentage of retained austenite transformed at sub-zero temperatures as in the curve 3, (Fig. 15.5). Re-heating for 1 hour at successively higher temperatures prior to refrigeration depresses M_s' even more until it falls below the M_f point, where essentially no austenite can be removed by the refrigeration treatment.

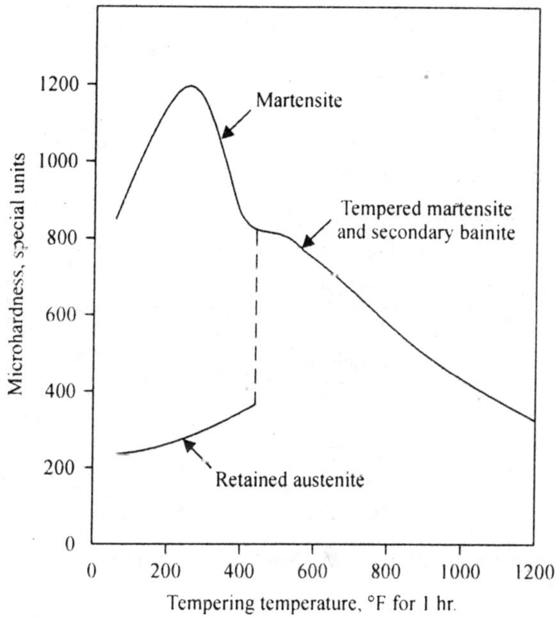

Fig. 15.7. Microhardness of individual phases in 1.70 per cent carbon iron-carbon alloy quenched from 2055°F (1125°C) and tempered for 1 hour at various temperatures.

When retained austenite high in carbon or alloy content exists, as in high-carbon high-chromium or high speed steels, it may not undergo decomposition to secondary bainite in the 400 to 600°F (205 to 315°C) temperature range. In this instance, the austenite remains stabilised until it reaches temperatures above 850°F (450°C), when a phenomenon known as conditioning occurs. Conditioning of highly alloyed retained austenite appears to be associated with the precipitation of complex (alloy) carbide particles, or at least a re-distribution of carbon and alloy atoms, that shifts the effective composition of the austenite to lower alloy concentrations and thereby permits its transformation. By holding for exceptionally long times at 950 to 1150°F (510 to 620°C), the conditioned austenite will transform directly to ferrite and carbide; of more commercial importance is the fact that it will readily transform to secondary martensite on cooling from the re-heating temperature. As shown in Fig. 15.8, M_s' rises rapidly above room temperature with re-heating temperatures above 950°F (510°C). For example, a tempering treatment of 1 hour at 1050°F (565°C) increases the M_s' of the retained austenite from 70°F (21°C) (after the original quench) to 470°F (245°C), thereby permitting a major part of the austenite to transform to secondary martensite during the subsequent cooling. Longer tempering times or a second tempering treatment will further raise the M_s' to a slight extent. The metallurgical basis for multiple tempering of high-alloy steels is apparent from this discussion. Until the retained austenite has been completely consumed, each cool from the tempering temperature involves the formation of new martensite that should be tempered again for maximum properties.

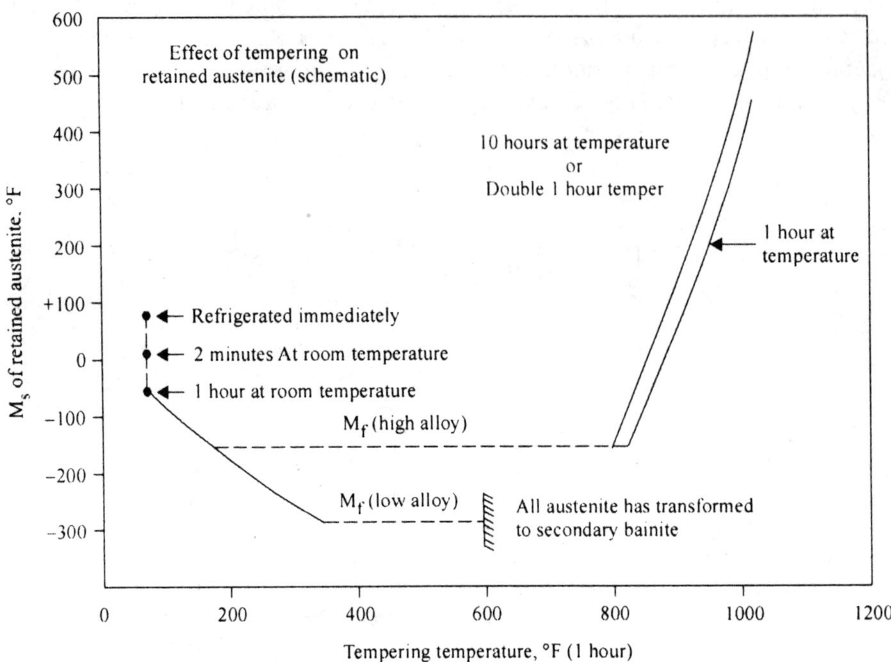

Fig. 15.8. Influence of tempering temperature on M_s' point for transformation of retained austenite to martensite on subsequent cooling.

The above-mentioned martensite decomposition and austenite transformation reactions that occur quite rapidly on heating have also been observed during ageing at room temperature for extended periods of time. The slight contractions and expansions accompanying these reactions are of primary concern in the manufacture of precision gages, but they may explain instances of unexpected cracking in quenched steels held for some time prior to tempering.

Thus far, nothing has been said concerning stresses that are produced within steel when these transformations occur. When austenite decomposes, a volume increase always occurs since each of its decomposition products is of greater specific volume. This expansion, especially when not occurring simultaneously thoughout the piece, leads to the formation of high internal stresses. On a microscopic scale, these stresses can be exceedingly high and nonuniform; in some instances, they may lead to the formation of minute cracks, especially when the product of the decomposition is martensite. Figure 15.9 illustrates in a simplified manner how stresses, causing distortion and danger of cracking, are created in a die during liquid quenching. The die, initially uniform in temperature, is rapidly cooled on the surface by contact with the quench, while the interior remains much hotter. As the temperature gets progressively lower, into the martensite formation range below 600°F, the cool surface begins to transform and expand placing the hot plastic interior under tension. After the surface layers complete transformation to a hard immobile condition, the interior reaches the martensite range and begins to expand. This can result in extremely high tensile stresses on the outer case, tending to produce distortion and possible cracking. Another source of high stress is the unequal cooling of heavy and thin sections during a liquid quench.

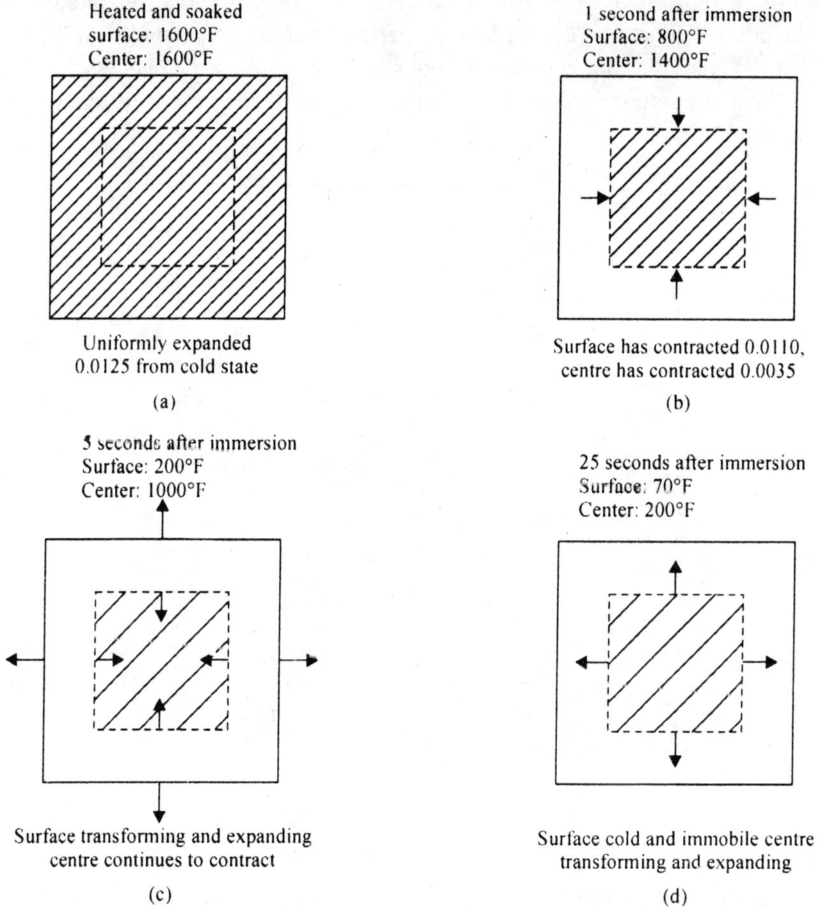

Fig. 15.9. Schematic illustration of the state of stress existing within a block of steel during quenching in water.

The tempering operation, in addition to causing structural changes, will tend to reduce the magnitude of these stresses.The effect of tempering temperature on the residual stress at the surface of a 2 by 2 by ½ inch block of type 410 (O1) after oil quenching is shown in Fig 15.10. Because of the formation of 'secondary' martensite (and the accompanying stresses) in those steels that are subject to conditioning, it is often desirable to temper more than once. In reality, since the stresses are known to be responsible for a large portion of the lack of toughness and ductility exhibited by quenched structures, tempering may be viewed as an operation in which their removal is a prime objective and in which the loss of hardness is a necessary although not always a desirable consequence.

Normalising

Normalising is a heat treating operation performed to modify the structure of the steel and make it more amenable to heat treatment. Normalising is usually conducted far enough above the critical range to carry all the carbides into solution; thus, temperature of normalising are generally above the A_1 or A_{cm}

line for the particular steel in question. After heating to this temperature and holding for a time proportional to the size of the piece, it is cooled in air. By means of this operation, excess carbides may be broken up and a more uniform distribution of carbides obtained. A finer grain size can be obtained and the structure of the steel can be controlled so that the hardening operation can be performed with the least difficulty. Normalising is usually done on forged pieces of tool steel but should not be necessary on steel that has been annealed by the manufacturer unless it has been subsequently forged. Steels that will air harden are not usually normalised.

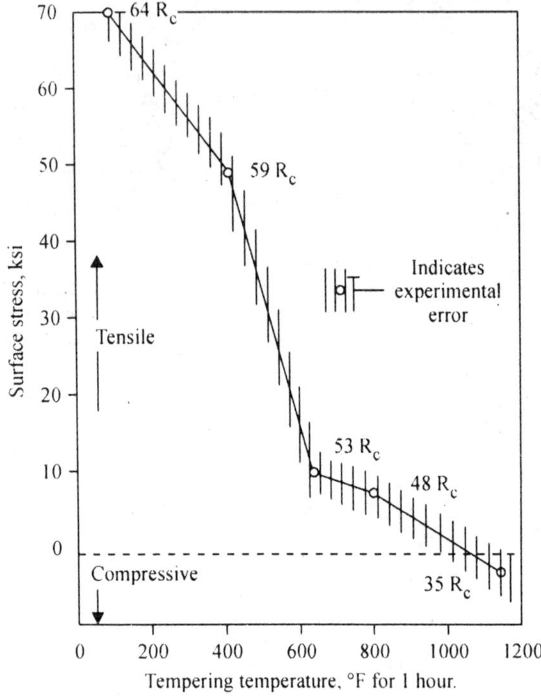

Fig. 15.10. Effect of tempering temperature on the residual surface stress in oil-quenched type 410 (O1).

Annealing

There are two types of annealing:

1. Full annealing.
2. Stress-relief annealing.

Full annealing is designed to establish the softest condition in the steel so that it can be machined after forging or so that it can be re-machined after hardening. For most tool steels, the annealing temperature is just slightly above the critical temperature and no attempt is made to dissolve all of the carbide. In fact, this might be detrimental; to completely soften the steel, it is necessary to produce a spheroidised structure. From the preceding discussion, it is known that a nonhomogeneous austenite is more conducive to the formation of such a structure than a homogeneous austenite. The time of holding is usually long and the heating rate is fairly slow; thus, it is customary to pack steel in an inert material and seal it from the effect of the atmospheric gases in the furnace. Packing materials used include cast iron chips, lime, mica, spent pitch coke, or sand to which a small percentage of carbon has been added. This packing material also plays an important part in preventing rapid changes of temperature, which are to be especially avoided in the re-heating of forged parts of air-hardening steels or the re-heating of previously hardened specimens. The presence of scale, as well as contact of the steel with the container, should be avoided because these will lead to decarburisation. The soaking time is generally from 1 to 4 hours and is extended as the alloy content of the metal is increased.

It is necessary, after the material has been held at the annealing temperature for the required time, to cool very slowly, preferably in the furnace or in mica or other heat-insulating material. Here again, the practice of packing the material in a relatively large volume of neutral material is an advantage, because it will prevent rapid cooling to some extent. In general, the slower the cool, particularly through the critical range, the softer the steel. The maximum rate of cooling of 50°F/hr. to a temperature of 1000°F (540°C) is suggested for most tool steels.

Since, the purpose of annealing is to obtain a completely spheroidised structure and since the preceding method usually requires a long time cycle, interest in short-time methods has been great. *Körber* and *Köster* described the rapid production of spheroidite in steels that were heated for short times above the critical temperature and then cooled slowly. *Payson*, *Hodapp* and *Leeder* extended this to include the use of an isothermal bath held just slightly under the critical temperature in place of the slow cooling. A typical cycle for a 1 per cent plain carbon tool steel to be annealed by this isothermal method would include the following steps:

1. Heat to 1350 to 1400°F (730 to 760°C) and hold until the structure consists of austenite plus non-lamellar (spheroidised) carbide.
2. Cool to 1250 to 1300°F (675 to 705°C) and hold for 1 to 5 hours until the desired softening has been achieved.

This method, however, requires that the control of temperature be very close and as such, is not well adapted to batch methods. If the temperature in the first step becomes too high, sufficient undissolved carbide may not remain to cause the formation of spheroidite on cooling. If the temperature in step 2 becomes too low, pearlite will form regardless of the amount of carbide undissolved. In recent years, the roller-hearth furnace with multiple temperature zones and controlled atmosphere has been successfully adapted to the isothermal annealing of many types of tool steel.

The oscillating method also has been used successfully to produce annealed structures in relatively short times. Here the temperature is raised and lowered through the critical temperature a number of times, followed by slow cooling to room temperature.

The second type of annealing, stress-relief annealing, otherwise known as sub-critical annealing, strain-relief annealing, or process annealing, is useful in removing residual stresses induced by cold working operations, including heavy machine work. Cold hubbing operations, bending, forming, coining, or even deep stamping marks, as well as cold drawing and heavy machining, will cause internal stresses in the material. It is often necessary and frequently desirable to eliminate these stresses before the tool is hardened. If a large amount of metal is removed by machining, the resulting stresses are likely to cause the tool to go out of shape (warp) during hardening and where great dimensional accuracy is required, the stress-relief anneal may be applied between the rough and finish machining operations.

Stress-relief annealing is usually carried out just below the critical temperature, that is, from 1100 to 1325°F(595 to 720°C) and is followed by furnace cooling. Lime, ash, or air cooling is also satisfactory, but there is always danger that with more rapid cooling additional stresses will be produced because of the non-uniform temperature gradients throughout the piece. This stress-relief anneal is often applied to cold drawn high speed steel rounds before machining operations are attempted.

Hardening

The hardening operation consists in heating the steel to a temperature in the austenite region or in the austenite-plus-carbide region and quenching at a rate that prevents the formation of pearlite, spheroidite or bainite and permits the formation of martensite. The quenching temperature, holding time at this temperature and quenching medium will depend mainly on the composition of the steel and on the size of the section being treated. The detailed procedure with respect to these variables will be considered later when each of the separate tool steels is described.

There are three precautions to be observed in heating tool steels for hardening:

1. Sufficient solution of carbides must occur to obtain full hardening. However, excessive carbide solution should be avoided or large quantities of austenite will be retained.
2. The grain size should be kept at a minimum in the majority of instances to ensure sufficient toughness.
3. The surface chemistry of the steel should be controlled within narrow limits.

Carbide solution

The amount of carbide that can be dissolved in austenite is proportional to the austenitising temperature, increasing progressively as the temperature is raised. Depending on its alloy content and the corresponding alloy content of the carbides, each grade of tool steel has a temperature range in which an optimum amount of this carbide is dissolved. If too little carbide is dissolved, both as-quenched and tempered hardness will be lowered; however, toughness may be improved. If too much carbide is dissolved, excessive grain growth may occur with a corresponding decrease of toughness. Thus, it can be appreciated that the aforementioned precautions are interdependent to a certain extent.

Grain size

Grain size of steel increases as the temperature of the steel increases above the critical temperature. The growth of the austenite grains is very gradual at temperatures near the critical temperature; in most tool steels, when sufficient carbide has been dissolved, a sudden growth is noted at elevated temperatures. Figure 15.11 illustrates typical grain-size-temperature curves for several heats of plain carbon and carbon-vanadium tool steel. It is apparent that the rate of growth varies widely even among steels of similar chemical composition. This is caused by a number of factors of which the melting and deoxidation practice are perhaps the most important. The rate of growth of the austenite grains is also controlled by the forging and rolling practice and by the initial structure prior to heating. Figure 15.12 illustrates the vast difference in the grain growth characteristics of several common tool steels. The presence of the various alloying elements in these steels is effective in altering the grain size and its variation with temperature, through alterations in the initial structure, in the critical temperature, in the rate of solution of the excess carbide and in the melting temperature (when very high hardening temperatures are considered) of the steel. The austenite grain size also varies with time at the hardening temperature. Figure 15.13 shows the relationship between the fracture grain size and the logarithm of the holding time for several plain carbon tool steels at three different temperatures. This relation holds for eutectoid steels, it holds equally well for steels that are slightly hypereutectoid up to a certain time period. Included in Fig. 15.13 are the results obtained with one hypereutectoid steel containing 0.19 per cent V. It is likely that some such relation exists for most tool steels when not heated to exceptionally high temperatures. The logarithmic nature of this effect indicates that, from a practical point of view, the first few minutes of the heating cycle cause the largest change in the grain size. Longer times are relatively less effective and it is possible that, in certain steels, no change in the grain size occurs after a finite period. Thus, longer soaking times at lower temperatures have been recommended for hardening high speed steel, since the increased solution of the carbide obtained is not associated with as great a grain size change as would be encountered by maintaining the same cycle by increasing the temperature.

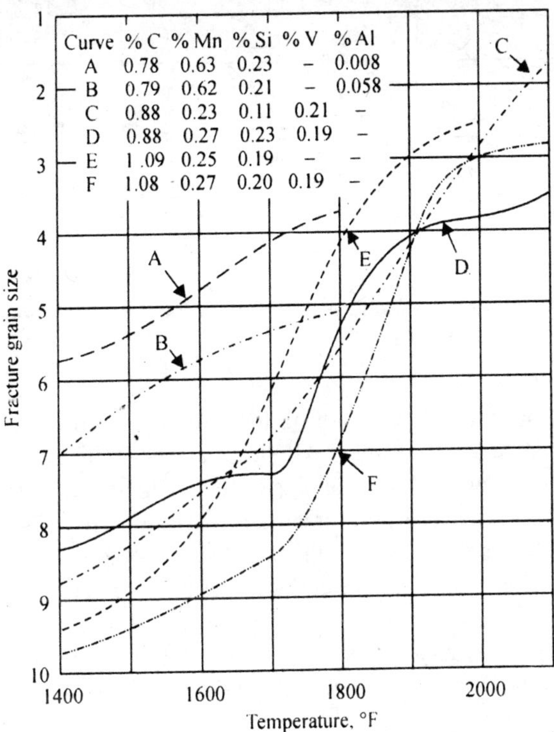

Fig. 15.11. Effect of temperature on the fracture grain size of several plain carbon and carbon-vanadium steels. Steels A and B are open-hearth steels. Steels C, D, E and F are basic electric furnace steels.

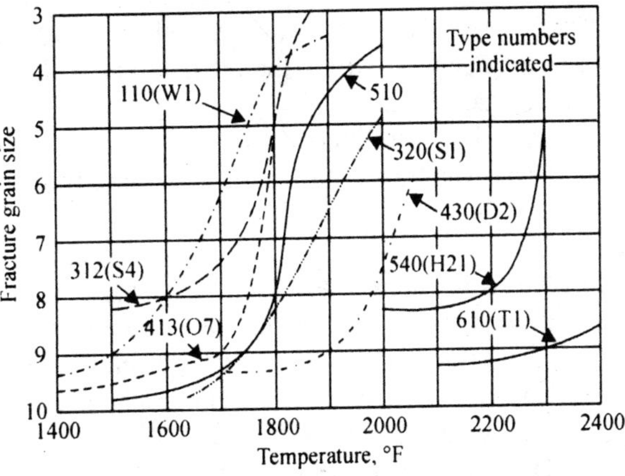

Fig. 15.12. Effect of temperature on the fracture grain size of several common tool steels.

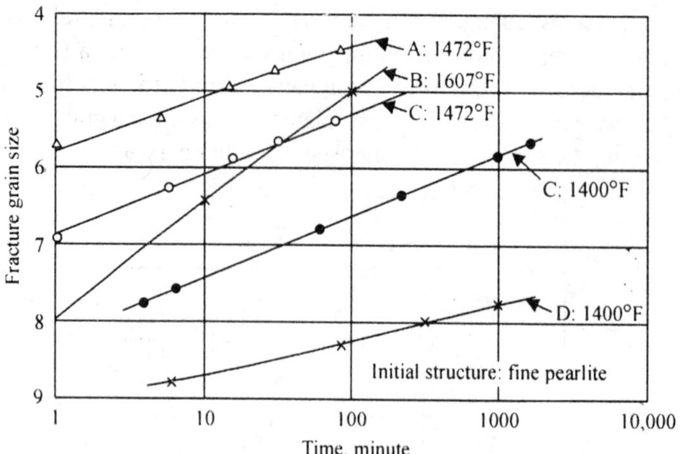

Fig. 15.13. Relationship between the fracture grain size and the austenitising time for carbon steels.

Surface chemistry

The third factor requiring special attention during the hardening process concerns the necessity for controlling the surface chemistry of the steel. All steels are characterised to a large degree by their chemical composition. Other things being equal, this chemical composition is in most instances directly responsible for the behaviour of the steel and for certain of its properties and distinguishes it from other grades. In most tool steels, where the surface of the tool does the major share of the work, it is important that the composition of the steel at and near the surface does not materially change. If carbon is lost from the surface (decarburisation), the surface zone may not harden, or on cooling, the transformation in the surface zone may occur at a materially higher temperature than in the center of the steel. The resulting stresses, often greater than the normal strsses developed during hardening, may lead to cracking. The control of the surface chemistry implies not only the prevention of the loss of carbon purposely added to the steel but also indicates that serious attention should be paid to the pick up of elements from the atomosphere of the furnace or from surrounding materials. Scaling of steel is a major example of such a pick up; in this instance, the added element is oxygen, which combines with the iron on the steel surface to form various iron oxides. The presence of such scale may prevent the attainment of the desired depth of hardening by altering the rate of cooling during the quench and it also may result in a change in dimensions that would render the steel useless.

The means commonly employed for controlling the surface chemistry include liquid baths or packing materials to prevent contact of the air with heated steel and the use of controlled atmospheres or vacuum.

Time element in hardening

To judge the length of time at hardening temperature, it is necessary that some means be available for estimating the exact moment at which the steel reaches the furnace temperature. One of the simplest means of doing this is to observe, through a small hole in the front of the furnace, the relation of the specimen to the surrounding refractories. When the specimen has reached the furnace temperature throughout, the colour will merge with that of the furnace refractories and it will be impossible to distinguish one from the other. The steel is said to have reached the 'pass out' and, as shown in Fig. 15.14 will have reached the furnace temperature for all practical purposes.

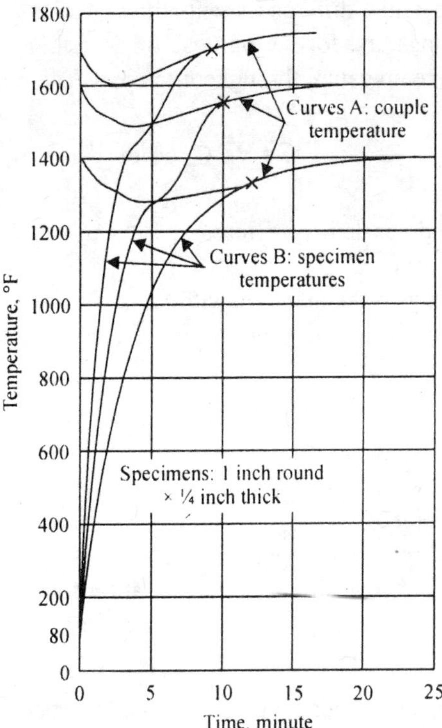

Fig. 15.14. Comparison of specimen temperature and furnace thermocouple temperature when heating small disks of 1per cent carbon steel in a semimuffle furnace. Furnace held at 1400, 1600 and 1700°F prior to insertion of samples. Points X on curves indicate the 'pass out' time and show that the specimen and thermocouple are at the same temperature for all practical purposes. The furnace may or may not have returned to the original temperature at this point.

At the 'pass out', the steel will not necessarily be at the desired hardening temperature. Soaking time at temperature is not considered to begin until both steel and furnace are at the hardening temperature. Use of thermocouples in the furnace chamber is one of the most common methods of judging temperatures in commercial heating furnaces. However, the temperature recorded by the thermocouple is not necessarily the exact temperature of the steel. It is almost impossible to determine the time at which the steel reaches the furnace temperature by examining the time-temperature chart obtained with a thermocouple unless the thermocouple is in actual contact with the sample. In pack hardening of steel, it is customary to place a small wire rod in one end of the box, inserted so that the end of the rod is approximately in the center of the box and close to the surface of the steel to be heated. By withdrawing this rod at frequent intervals and noting its colour against the background of the furnace refractories, it is possible to get a reasonably accurate estimation of the rate of heating of the contents of the box.

It would be useful to be able to predict the time required to heat a given piece of steel to the hardening temperature. Under certain conditions, Newton's law of heating affords a means of doing this. The rate of heating of good heat conductors, such as metals, is frequently limited by the transfer of heat from the surroundings (furnace and furnace atmosphere) to the surface and not by internal resistance to heat flow in the piece being heated. Temperature differences within the piece are small in comparison with those between surface and surroundings and for the purpose of describing the rate of heating, it is often justifiable to assume a uniform temperature throughout the part. The rate of heating absorption can be represented as:

$$Q = V\rho\, C_p\, dT/dt$$

where $V\rho$ = (volume times density) = mass

C_p = specific heat.

dT/dt = rate of change of temperature with time.

The rate of heat transfer from the surroundings to the surface (by convection and radiation) can be represented by:

$$Q = hA(T_f - T)$$

where, h = heat transfer coefficient

A = surface area

T_f = furnace temperature

T = temperature of the piece being heated.

Equating these two expressions gives the following relation for the rate of change of temperature with time:

$$dT/dt = hA/V\rho\, C_p(T_f - T)$$

Assuming constancy of the heat transfer coefficient, h and the specific heat, C_p, the equation can be integrated to yield:

$$t = \frac{2.3V\rho\, C_p}{hA} \log \frac{(T_f - T_o)}{T_f - T}$$

where T_o is the initial temperature of the piece when placed in the furnace. The last equation is an expression of Newton's law of heating. Note that the time required to heat a given value of $T_f - T_o/T_f - T$ is directly proportional to the volume to surface ratio (V/A). This ratio depends on the size and shape of the part. For a long cylinder, it is directly proportional to the diameter. Experimental confirmation of this relation between volume to surface ratio and time required for heating can be seen in Fig. 15.15, where heating curves for various size rounds of a plain carbon steel are shown. Note that by doubling the diameter, the time to heat to 1450°F is also doubled. Note also that for the 3 inch round, the difference in temperature between furnace and steel is always much greater than that between the surface and center of the bar. This is evidence of the good heat conductivity of steel and justifies the application of Newton's law. The above considerations suggest a practical method for estimating the time required to heat a piece of steel in a furnace. Once the time required to heat a given size and shape part to a particular temperature in a particular furnace is known, the time required to heat any other shape and size can be predicted. For example:

A bar 1 inch round by 6 inch long of high speed steel requires 10 minutes to heat from the pre-heat temperature of 1600°F to the hardening temperature of 2350°F in a controlled-atmosphere furnace. How long will it take to heat a 6 inch cube of the same material from the pre-heat temperature to the hardening temperature in the same furnace?

For the round bar, V/A = 0.23

For the cube, V/A = 1.00

By simple proportion,

$$\frac{t\,(\text{bar})}{t\,(\text{cube})} = \frac{V/A(\text{bar})}{V/A(\text{cube})}$$

or

$$t\,(\text{cube}) = (10)\,(1.00/0.23)$$
$$t\,(\text{cube}) = 43 \text{ minutes}$$

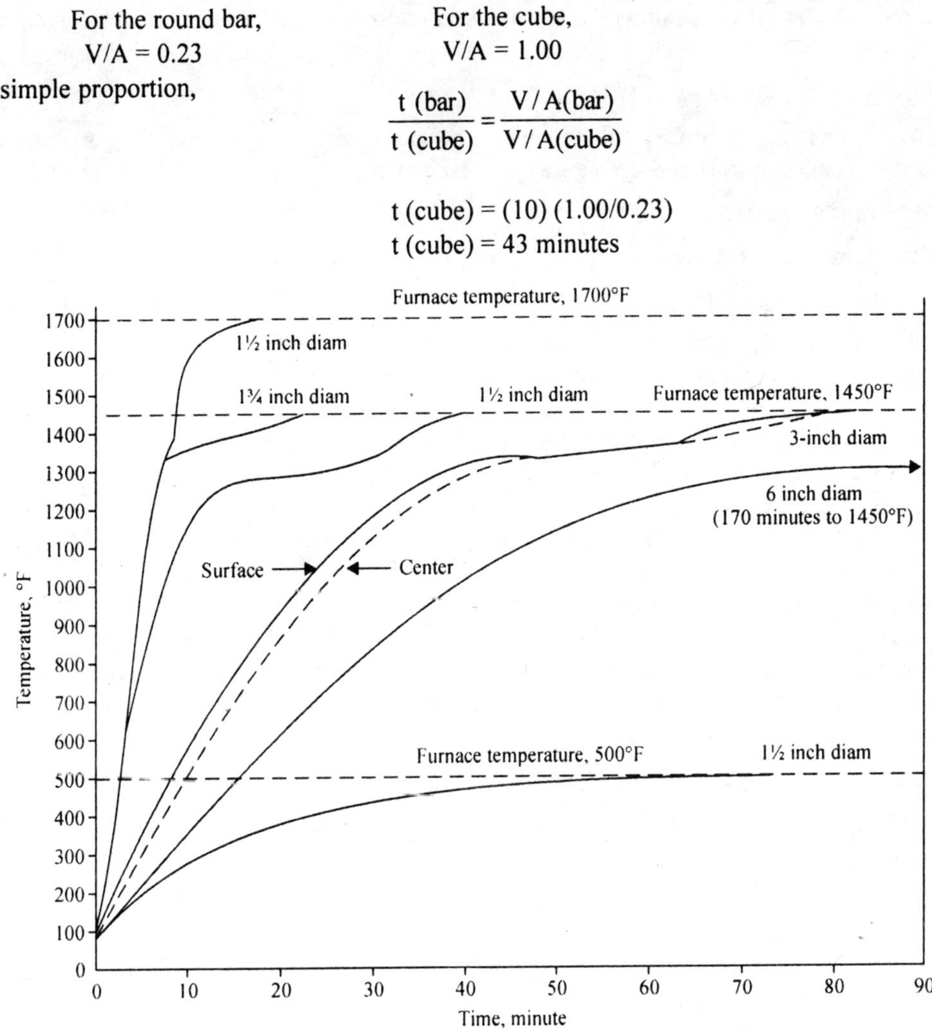

Fig. 15.15. Effect of several variables on the time required for steel to reach furnace temperature. Data apply to a specific furnace only. Furnace at indicated temperature prior to charging of samples.

Quenching the steel

After the steel is soaked for the desired length of time at the hardening temperature, it must be quenched. The transformation that subsequently occurs will lead to the formation of high stresses; thus, precautions must be taken to prevent them from reaching a magnitude that would exceed the strength of the steel and cause it to crack. Numerous factors other than the quench itself may contribute to quench cracking. Among the most important are:

Design of the tool

Sharp corners unrelieved by fillets, heavy sections adjoining light sections and lack of symmetry all exert a stress-concentrating effect that may result in quench cracking. In the event that the design cannot be altered to reduce stress concentrations, it is necessary to select a deeper-hardening steel that requires a less severe quench.

Machining marks or stamped identification marks

Sharp notches, whether introduced deliberately in the tool design or unintentionally as in machining, behave as stress concentrators and can cause quench cracking.

Overheating during hardening

The resultant large grain size frequently causes cracking.

Changes in surface chemistry

Either carburisation or decarburisation of the surface of the tool can produce cracking.

When the above factors have been considered and precautionary steps taken to ensure against quench cracking, it is necessary to quench in accordance with the two rules: (a) do not cool too rapidly during the final stages of the cooling process while martensite is forming; (b) cool the steel to a temperature that will ensure that sufficient martensite forms prior to heating the steel for tempering, but do not cool so low as to allow the formation of cracks. The temperature that must be attained on cooling before heating for tempering is thus a compromise in many instances and depends on the percentage of martensite that exists at each temperature, which in turn is influenced by the chemical composition of the steel and the heat treating temperature used. Certain steels are more brittle than others even when similar amounts of martensite are present and this must be considered. In most instances, the tempering operation should be started as soon as possible after quenching. It is not uncommon to find cracks in quenched articles that have been allowed to remain at room temperature for several hours, although such cracks were not present immediately after quenching.

Dimensional changes and internal stresses

Certain fundamental properties of steel govern the tendency of parts to change their size and shape when hardened. These include (Fig 15.16):

1. Thermal expansion of steel when heated.
2. Thermal contraction when cooled.
3. Slight contraction of steel as it passes through the critical temperature, changing to austenite on heating.
4. Larger expansion when austenite changes to martensite on cooling.

An additional factor of prime importance involves the decrease in yield strength of steel as the temperature is raised. The interaction of these properties in producing dimensional changes and internal stresses when quenching steels of different hardenability, different size and shape in relation to the surfaces that contact the coolant and steels with different transformation temperatures, at different rates of cooling, can be made clear by consideration of several simple examples.

One of the most common instances of dimensional change in steel articles is the warpage that occurs during the heating operation and that is often mistaken for a dimensional change occurring in the quench. Figure 15.16 shows the expansion of steel as it is heated and cooled and exemplifies the four factors listed above. It will be noted that on heating a relatively large expansion takes place and that at the transformation temperature, a slight contraction occurs. Since, this contraction occurs when the steel is hot and very plastic, it is not likely to lead to cracking. However, the continuous expansion on heating will cause heavy objects or long slim objects to warp if not properly supported, if the rate of heating is too great or if non-uniform heating occurs. It is necessary to emphasise the importance of proper support for any article to be heated to a high temperature, because at this temperature the lack of strength will, many times, allow the steel to bend under its own weight. Warpage and dimensional changes can also arise on heating from machining stresses and from unbalanced design. Slow rates of heating offer less danger of either occurring than fast rates of heating because at any one time the temperature gradients throughout the steels are less. When any piece of steel is cooled by quenching in water, there is a thermal gradient established from the outside to the center of the article (Fig. 15.9). The outer portions in contact with the liquid cool rapidly and contract (curve 2 of Fig. 15.16). Since the interior regions have not yet cooled, they tend to maintain their original dimensions, but since the two portions cannot separate, there will be established internal stresses in the steel—compressive in the interior regions and tensile in the outer portions. These stresses will cause the center to contract plastically (upset) rather than cause the expansion of the exterior regions, because the former is hotter and possesses a lower yield strength. This contraction will often lead to a bulging or a barreling of the surface of the article. On further cooling, the interior will not contract normally with temperature but, because of the plastic deformation that has taken place, will have a final length shorter than that of the surface zones. Again, the two parts may not separate and an equilibrium or balance will be reached in which the surface will be under compression (wanting to expand) and the center under tension (wanting to contract). It is important to note that, unless non-uniform plastic deformation has taken place, no internal stresses will exist at room temperature. During the hardening of tool steels, however, such non-uniform deformations will always arise and thus, stress will always be produced. The logic used in deducing the above course of events on cooling can be applied to a limited number of other instances. The results obtained in severel of these will be briefly stated at this time. In shallow-hardening steels, the expansion of the case during the transformation to martensite (curve 4 of Fig. 15.16) is operative in addition to the principles outlined above. This expansion of the exterior alters the final distribution only by increasing the value of the compressive stresses on the surface and the tensile stresses in the center. On the other hand, if the steel is deep hardening, the last transformation occurring during the quench will be the austenite-martensite change at the center. This center expansion will result in a partial alleviation of the stress in the surface and may, under certain conditions, reverse the stress system, placing the surface in tension and the center under compression.

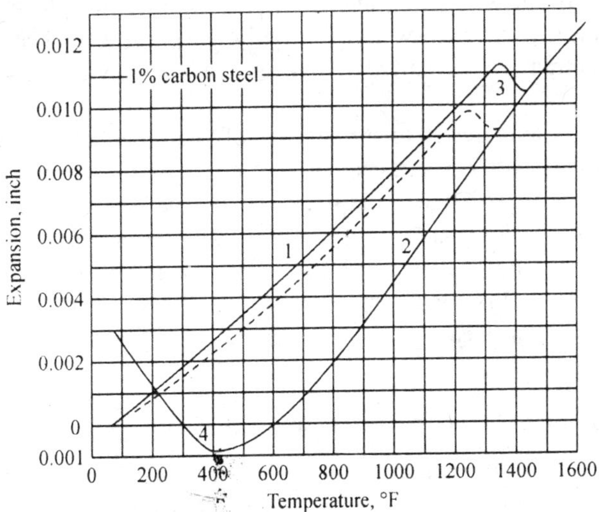

Fig.15.16. Dilation of a 1 per cent C steel on heating and quenching (full lines) and on slow cooling (dashed line).

The state of stress existing in shallow-hardening steels is similar to that produced in the quenching of rings, drawing dies or header dies in fixtures designed to harden the interior surfaces only. The compressive stresses in the inner surfaces often are of material aid in resisting imposed tensile stresses in operation. They are not to be desired, although almost always present in applications such as header dies in which longitudinal compressive stressses are often applied during operation. The likelihood of fracture under compression is less than under tension and the above situation is therefore not particularly detrimental.

If a ring of steel is quenched by complete submersion, the outer surfaces will cool most rapidly and cause upsetting of the center. Further cooling will cause normal contraction of the center and result generally in tensile stresses around the hole.

Tempering

The tempering operation follows the hardening operation and it is almost a necessity to apply it in the heat treatment of tool steels. The purpose of the tempering operation are to relieve the stresses that are set up during the transformation of austenite to martensite and during the non-uniform cooling of the steel to room temperature, to toughen the steel and sometimes to reduce the hardness. The latter is not always desired but must be tolerated if the stresses are to be relieved. The tempering operation consists in heating the quenched and hardened steel to a temperature below the critical temperature and holding for a period of time sufficient to bring about desired changes in stresses and structure. These changes will continue to occur with time, but again the process follows a logarithmic pattern (that is, the first hour is more important than the second hour and so on) and commercial tempering times at temperature will vary from ½ to 2½ hours. It is general practice to control the progress of the softening and tempering process by variations in the tempering temperature and not in the tempering time.

It should be emphasised that in the as-quenched condition a tool is likely to be quite brittle; therefore, it may crack if subjected to the thermal shock of placing it cold into a hot furnace. A better procedure would be to bring the tool up to the tempering temperature along with the furnace. Since, tempering is usually carried out at temperature where no visible radiation is given off by the steel, estimation of the time at which a piece reaches the tempering temperature is more difficult than in hardening. If the furnace is brought up slowly, the furnace thermocouple or an auxiliary couple located near the piece will give a fairly accurate indication as the 'levelling off' point is approached. The selection of a relatively long holding time will permit greater latitude in the estimation without causing undue variations in properties.

Tempering by colour is often referred to as a practice of 'old time' heat treaters. In this instance, the tempering operation is stopped when a certain colour is seen on the surface of the steel. The colours that appear are the result of a slight oxidation of the steel surface. The higher the temperature, the thicker is the oxide film and a series of colours are seen as the tempering increases. The use of such a method of controlling the tempering operation is restricted, as it is not conducive to uniformity. Only the surface temperature is obtained; the colours are different for different steels and since the time of holding or the rate of heating will materially influence the amount of oxidation that occurs at each temperature, the relationships may be altered.

Austempering and Other Hot Quenching Treatments

Austempering is a process that attempts to combine the principal features of both the hardening and tempering operations and eliminate some of the less desirable features of both. Small sections of steel are quenched in a molten bath maintained at an elevated temperature between the 'knee' of the S-curve and the temperature of the martensite point, generally between 500 and 700°F. Since quenching in hot baths results in a relatively slow rate of cooling compared to brine quenching, to prevent the formation of pearlite at the 'knee' of the S-curve, the process is necessarily restricted to thin sections of shallow hardening steels or moderate sections of deep-hardening steels. At this point, the austenite that is retained on quenching tends to transform to bainite, producing a structure with a hardness approximately equal to that obtained by quenching to room temperature and tempering to the same temperature but possessing a better toughness at hardness levels in excess of Rockwell C 55. Minute quenching cracks that may form during quenching to room temperature and reduce the strength of the steel are not encountered in the austempering process. At low hardness levels, however, the partially spheroidised structures obtained with the quench-temper treatment posses slightly better toughness than do the acicular structures of bainite associated with the austempering process.

One disadvantage of the austempering process, other than the restriction to this sections, involves the long time at the hot quenching temperature sometiems necessary to cause the complete transformation to bainite. The harder the material desired (the lower the bath temperature), the longer the time necessary. To overcome this, the transformation may be stopped at an intermediate point and the specimen may be quenched, transforming the remaining austenite to martensite. The whole may then be tempered to the desired temperature, producing a compromise in mechanical properties between those of quenched-and-tempered and austempered structures. Since, a large portion of the internal stresses in quenched steels may be attributed to rapid cooling through the range of temperatures in which martensite forms, they can be minimised by quenching the steel in a hot bath held just above or near the M_s point and then slowly cooling in air to room temperature (martempering). Other variations of this procedure include quenching in water or oil for a brief interval and air cooling, or quenching briefly in water followed by oil.

For high-alloy steels that must be quenched from high temperatures, i.e., above 1800°F (980°C), the severity of oil quenching may lead to dimensional changes or cracking even before transformations occur. Hot quenching to an intermediate temperature at which the austenite does not react in short times, followed by air cooling, is therefore beneficial. Thus, high speed steel is often quenched to 1000 to 1200°F (540 to 650°C) in salt and then cooled to room temperature in air.

Surface Treatments

In addition to the above processes employed in the conventional heat treatment of tool steel, many surface treatments are available that permit superficial hardening or the production of special wear-resistant surfaces for certain applications. These treatments include carburising, cyaniding, nitriding, Chapmanising, oxidising, hard chromium plating and flame or induction hardening. Specific examples of the application of these treatments for tool steels are included in the following sections.

Carburising

Carburising, or case hardening, is a process by which the carbon content in the surface layers is increased by heating the steel in contact with carbonaceous gases, solids or liquids. The purpose of carburising is either to permit the steel to be hardened or to provide a more wear-resistant surface to the steel article. Most tool steels contain enough carbon so that they are readily hardenable and have many hard carbides for wear resistance. Even those tool steels that have relatively low carbon contents, such as silicon tool steels or several of the hot working tool steels, achieve relatively high hardness and do not respond well to a carburising treatment. As a general rule then, the process of carburising has secondary importance in the heat treatment of most tool steels.

There are instances, however, in which carburising is applied to tools that must withstand exceptional wear. High speed steel and tungsten chisel steel have been carburised commercially to acheive greater wear resistance that is possible with the normal structure plus a shock-resistant core. In addition to these applications, carburising is a necessary part of the heat treatment of the low-carbon mould steels (die irons). These are very low-carbon steels or irons into which the desired die impression is sunk by hubbing. The working surface is then carburised and hardened.

In order to carburise steel, carbon must be available at the surface for absorption and diffusion inward. This is commonly accomplished by one of three means:

1. Contact with a carburising gas.
2. Contact with a carburising solid.
3. Contact with a carburising liquid.

In the first two instances, the carbon is supplied to the surface by a reducing gas such as carbon monoxide and in the third, it is supplied by a strongly reducing liquid-barium cyanide. By reacting with the steel at the surface, these compounds deposit carbon in the austenite or, in high-carbon steels, form Fe_3C or a complex carbide at the surface. As a result of this action, a larger concentration of carbon exists at the surface than in the interior of the steel (a carbon concentration gradient exists) and once this situation has been created, carbon will diffuse from the surface to the interior. There is a possibility that carbon monoxide deposits carbon in the nascent state on the steel surface, but since iron cannot be carburised with soot, the direct solution of carbon in austenite or the direct formation of carbide at high temperatures is a more plausible explanation for the carburising process.

The rate of carburising (the speed at which a certain steel absorbs carbon), usually measured by the depth of that carburised layer or by the increase in depth per unit time, increases rapidly as the temperature is raised. It is possible to control this depth by variation of both the temperature and the time of carburising.

In gas carburising, the gases commonly used in practice are:

1. Natural gas.
2. Natural gas partially burned to give about 8 per cent CO, 9 per cent H_2 and 4.5 per cent CO_2 mixed with natural gas.
3. Any hydrocarbon gas.

The pure gases tend to break down and deposit soot on the steel surface, interfering with the carburising process; thus, mixtures are generally recommended. For carburising low-carbon steels, temperatures of 1600 to 1750°F (870 to 955°C) and times of 8 to 28 hours are generally used.

Carburising with solid carburising material is called pack carburising. The steel is packed in suitable containers and completely surrounded by the carburising medium. The container and contents are then placed in a furnace held at the carburising temperature. Virtually every kind of solid carbonaceous material has been used. Usually mixtures of two or more substances, known as carburising compounds, have been found most suitable. These are generally compounds of two or more of the following ingredients: charcoal, charred bone, charred leather, bituminous coal, one or more chemicals to act as energisers and coke to act as a filler and decrease the shrinkage of the packing during the carburising process, The function of the energiser, which may consist of barium, calcium or sodium carbonate, barium oxide, calcium saccharate or sodium cyanide, is to form carbon dioxide, which reacts with carbon in the packing to increase the concentration of carbon monoxide, or to liberate nascent gases that will reduce the carbon dioxide content. The pack carburising method is generally used in the carburising of most tools steels. If, because of a very active carburising agent, a high carburising temperature, or a short carburising time, the carbon concentration gradient is very steep across the surface layers of the steel, brittleness of the case is likely to occur. This can be materially reduced by lowering the carbon concentration gradient.

Alternatively, if contact with the carburising medium is stopped but the steel is held for additional time at an elevated temperature, the high carbon contents existing at the immediate surface will be reduced by diffusion toward the center of the piece and the brittleness may be overcome.

The third method of carburising involves placing the steel in a liquid carburising medium. Although this process is not used often on tool steels, it may prove advantageous when selective carburising of certain portions of a tool is required. For example, the tips of punches may be selectively carburised without employing any 'stop off'. Liquid carburising baths differ from cyaniding baths in that certain alkaline earth metals (barium, calcium, strontium) are present that act as catalysers for the carburising process enabling the production of deep cases of high-carbon low-nitrogen content. Typical bath compositions are shown in Table 15.2.

The sodium cyanide in the bath does not produce the active cementation agent directly but reacts with the other components of the bath according to the following reactions:

$$2\ NaCN + BaCl_2 = 2NaCl + Ba(CN)_2$$

$$Ba(CN)_2 = BaCN_2 + C \text{ (in steel)}$$

$$2NaCN + O_2 = 2NaCNO$$

$$4NaCNO = Na_2CO_3 + 2NaCN + CO + 2N$$

Table 15.2. Compositions of liquid carburising baths.

	Composition, %	
Salt	*Light case*	*Heavy case*
Sodium cyanide	17 to 23	7 ½ to 12
Barium chloride	15 to 40	45 to 55
Other alkaline earths	0 to 3 ½	2 to 10
Potassium chloride	–	5 ½ to 20
Sodium chloride	20 to 30	0 to 15
Sodium carbonate	30 max.	30 max.
Sodium cyanate	1 max.	0.30 max.

The first two reactions combine to produce direct carburisation of the steel, while the last two reactions result in the formation of CO and N; the CO is also carburising under the reducing conditions of the bath. To keep absorption of nitrogen by the steel to a minimum, the temperature of the bath should be above 1650°F.

Precautions

The following precautions should be observed in handling a cyanide salt bath:

1. No volatile or reactive materials should be introduced into cyanide baths. Even the smallest amount of moisture will cause violent spattering and may result in severe burns if molten salt comes in contact with the skin.
2. Any oxidising agent such as nitrates and nitrite mixtures used for low temperature tempering or colouring baths will react explosively with cyanide baths. In certain instances, oxidising agents such as manganese dioxide (MnO_2) or anhydrous boric acid (B_2O_3) are introduced to accelerate the ageing of the bath and formation of cyanate. These are not generally recommended and should be handled with extreme caution.

3. At all times, if possible, melting of the bath should proceed from the top down. In externally heated furnaces in which this is not done, a cast iron wedge should be suspended in the center of the bath before it freezes. It should touch the bottom of the pot and the base should extend at least 4 inch above the bath surface. On removal, after complete solidification (but not before), this leaves an opening through which the gases are released and into which the salt will expand on re-heating.
4. Adequate ventilation or a hood should be provided to carry off the white fumes consisting mainly of sodium carbonate.

Nitriding

Nitrided steels have many advantages over untreated steels in that they show a high surface hardness, added corrosion resistance and a lack of distortion during treating. The hardness of most nitrided steels is difficult to measure because of the shallow cases which result. By use of the microcharacter or a Knoop indentor, it has been possible to estimate that the hardness of nitrided cases is approximately Rockwell C 70 to 74. Nitrided cases are somewhat brittle, but where wear resistance is desired, this is not an important matter. By careful preparation of the steel before the nitriding process, especially by prevention of any traces of decarburisation and by maintaining a relatively shallow nitrided case, it is possible to keep this brittleness to a minimum.

Nitriding has proved extremely useful on certain tool steels. In the field of high speed steels, nitriding is used on cutting tools to improve resistance to wear and seizing.

There are to two types of nitriding processes in commercial use: (i) the ammonia process; and (ii) liquid nitriding.

In contact with hot steel, ammonia decomposes to form nascent nitrogen and hydrogen. The nitrogen will be absorbed by the steel at the surface and will diffuse inward. Evidence of the presence of nitrogen in the layers adjacent to the surface is obtained by the appearance of characteristic needlelike nitrided particles in the microstructure. The ammonia process is generally carried out at around 980°F (525°C) for 10 to 90 hours. Longer nitriding times are necessary if deep cases are desired.

The liquid nitriding operation takes advantage of the activity of cyanide baths, the use of which has been previously discussed in connection with the carburising of steel. When the latter operations are conducted at temperatures above 1650°F, both carbon and nitrogen are absorbed by the steel. When these baths are used at low temperatures, absorption of nitrogen becomes predominant over absorption of carbon. Baths used for nitriding of steels are generally mixtures of sodium and potassium cyanide (60:40 or 70:30). These baths are not the same as those used for liquid carburising for they usually contain no alkaline earth metals and are generally operated at temperatures below the melting point of the latter. The sodium-potassium cyanide mixtures should not be used for carburising because their rate of decomposition is very high at the high temperatures. The melting point of the 60:40 mixture of sodium cyanide and potassium cyanide is approximately 925°F (495°C), but as with other baths of this nature, the melting point falls to below 800°F (425°C) as the cyanate content becomes appreciable through oxidation. When the bath is to be operated at 1050°F (565°C), the more economical 70:30 mixture may be used. A mixture of 55 per cent sodium cyanide, 25 per cent sodium chloride and 20 per cent sodium carbonate is suitable for nitriding but has a shorter life.

The chemistry of these baths is the same as that described previously for carburising baths. However, barium cyanide is not formed and the predominant bath reaction is the formation of sodium cyanate. Ageing of the baths is necessary to build the cyanate content to about 5 per cent. This ageing requires from 10 to 20 hours at 1050°F (565°C), shorter times being required in nickel-chromium pots than in pressed steel pots. Contamination of the bath with nickel, which may plate out on the work when alloy pots are used or when alloy thermocouple protection tubes are used, should be noted. When the carbonate content of these baths reaches a certain point, it will freeze out and may be removed. Acceleration of this freezing process may be made periodically by lowering the bath temperature to about 900°F (480°C).

The operation involves immersion of a sample in the molten bath maintained at a temperature between 850 and 1100°F (455 and 595°C) for approximately 2 hours. An increase in the nitriding temperature increases the depth of the case but decreases the effective hardness. The depth of the case can be improved and the hardness of the surface maintained by use of the duplex process, which involves heating for 2 hours at a high temperature—for example, 1025°F (550°C)—and then heating for 2 hours at a lower temperature, say 860°F (460°C). Such treatment gives an adequate case depth with maximum surface hardness. The use of liquid nitriding in this temperature range permits the nitriding of fully heat treated secondary-hardening tool steels with no apparent loss in core hardness because the operation is conducted at or below the usual tempering temperature and the times are no longer than those required in commercial tempering operations on these steels.

The hardness of a nitrided case shows good resistance to tempering. Figure 15.17 shows the hardness of this case as a function of the tempering temperature compared with the hardness of a plain carbon steel under similar conditions.

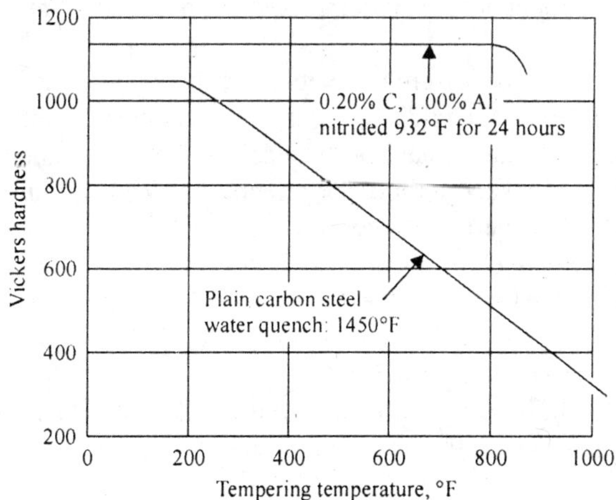

Fig. 15.17. Effect of tempering temperature on the room-temperature hardness of a nitrided 0.20 per cent carbon steel and a 1 per cent carbon tool steel. Treatments are indicated in the figure.

The steels that can be successfully nitrided are many. However, carbon steels are not entirely satisfactory, because the nitrided case fails to achieve high hardness. Commercially, this led to the development of nitriding steels that contain appreciable quantities of aluminium, vanadium, molybdenum or chromium, all of which materially increase the case depth, rate of penetration and hardness obtained. Since, many tool steels contain these elements, especially chromium and vanadium, they can be easily nitrided. However, structural nitriding steels and nitrided low-alloy tool steels cannot be successfully used in applications involving large pressures because the core is too soft. High-alloy steels containing these elements are satisfactory because they will resist core softening at the nitriding temperature.

Oxidation

Oxidation is applied occasionally to high speed steel cutting tools. Although not nearly as effective as chromium plating or nitriding, the oxide coating does effect a decrease in the coefficient of friction between the tool and the work.

Oxidation can be achieved by heating the tools to about 1050°F in a steam atmosphere or by use of a salt bath process. The latter process depends on the formation of a thin layer of black oxide of iron (Fe_3O_4) by immersing a previously chemically cleaned surface into an aqueous solution of sodium hydroxide and sodium nitrite (or other suitable oxidising materials) at a closely controlled temperature of about 285°F. The times of immersion or oxidation are from 5 to 20 minutes. Tests have indicated that the somewhat more expensive two-bath method of oxidising involving the use of one bath at 280 to 290°F and a slightly more concentrated bath at a temperature of 300 to 320°F is more satisfactory for obtaining a thorough penetration and complete coverage of the steel with the particular oxide being formed. In oxidising tools in salt baths, extremely close temperature control is essential and the baths must be kept boiling gently at all times. The concentration of the bath determines its boiling point and by closely watching the boiling temperature, the need for further additions of salt to the solution can be indicated. After immersion, the work should be agitated to break up water globules and equalise the temperature. In order to prevent undue chilling of the oxidising bath, it is advisable with large charges to pre-heat the material to about 160 to 180°F in another salt before immersion. The work should then be permitted to remain in the bath for about 5 to 30 minutes, depending on the intensity of the oxidation and the penetration desired.

Quite often after removing the work from the oxidising salt bath and rinsing in running cold or hot water, or after removing from the steam oxidising furnace, an oil dip is used that supposedly thoroughly permeates the oxide coating. Some claims have been made that this oil permeation immediately after the oxidising process is responsible for superior tool performance. At least it is known to prevent rusting on parts on which a complete oxide covering has not been achieved.

The black oxide coating will not chip, peel, crack, or dust off even on bending or cutting. Since, the temperatures of the salt method are low and since the steam method is conducted at the normal tempering temperature, there is no change in the hardness of high speed steel so treated. The thickness of the black oxide coating is generally less than 0.0002 inch.

Where, during operation, failure occurs by galling of the tool or die, the non-welding characteristics of an oxide-coated surface may materially increase tool life.

Hard chromium plating

Industrial chromium plating, or 'hard' chromium plating, can be applied to many tool steels to give longer life where failure of the tool usually occurs by wear and to give easier operation on certain types of taps, reamers, milling cutters and drawing dies. Chromium plating can also be used in building up worn parts of shafts, arbours, cutters, or dies. The plate that is applied by an electrolytic process is generally thicker than that used for purely decorative purposes and is applied directly to the base metal. Plate thicknesses for industrial purposes will range from 0.0001 to 0.040 inch. The hardness of the chromium plating varies widely, depending on the plating conditions and particularly on the plate thickness. An average deposit is said to have a hardness of approximately 950 to 1050 Vickers. The plate is quite brittle and should not be used on tools that involve shock. This brittleness appears to increase as the thickness of the plating increases. The toughness of the chromium plating can be increased without material loss of wear resistance by heating to temperature under 500°F (260°C), which is said to drive off the occluded hydrogen absorbed in the plating process. The chromium plate is really a solid solution of chromium and hydrogen and undoubtedly, the hardness of the plating can be attributed somewhat to the presence of hydrogen. After heating to 550°F (290°C), the hardness begins to drop rapidly and after holding at 1900°F (1040°C), the plating is fully annealed and has a Vickers hardness of only 150, exhibiting little wear resistance. The plate is then soft and ductile like cast chromium.

Two of the main advantages of chromium plating are its very low coefficient of friction and its ability to take a high finish that prevents binding and scoring of rubbing surfaces. The chromium plating resists all acids except hydrochloric acid, although it is slowly attacked by concentrated sulphuric acid and will retain a bright surface up to approximately 900°F (480°C). One precaution that is necessary in the use of tools or dies that have been chromium plated is the possibility of flaking if the plated object is subjected to wide temperature variations. The coefficient of expansion of the chromium plate is less than that of steel and most other metals. Rapid temperature changes will cause a variation in the expansion between the plate and the base and may cause the plate to flake and remove itself from the base metal.

Flame hardening

Flame hardening is a surface hardening method that, unlike those previously described, involves no change in chemical composition. In this method, the part to be hardened is rapidly heated to above the critical range of the steel with an oxyacetylene torch and is immediately cooled by spraying with water or by directing a jet of cold air against the heated surface. For some alloy steels that have air-hardening properties, the parts may be permitted to cool in the air after heating in the flame. There are many instances where parts are too large or where there is danger of excessive distortion if furnace hardening were to be employed. Here, the flame-hardening operation has been successful. The temperature of the flame is approximately 6300°F and it is usually maintained only ¼ inch from the work. Careful control of the machanics of the process is, therefore, necessary in order to prevent local overheating and melting. Because of the reducing nature of the flame, no decarburisation is possible by this method. Flame hardening is not customarily applied to high-alloy or high-carbon steels, because very large objects are seldom made from such steels and because in such steels danger of cracking exists if the heating rate is exceptionally rapid. Band and hack saw blades are, however, frequently flame hardened.

Induction hardening

A final method of surface treatment of metals that, like flame hardening, does not involve a change in chemical composition at the surface is induction hardening. By this method, the object or part to be hardened is inserted within a induction coil. A high-frequency current passed through this coil induces a current within the metal object and causes the surface layers to become heated rapidly to above the critical temperature. When the steel has reached a certain temperature, it is generally quenched with water sprays. The heating rate in induction hardening is rapid and the maximum temperature must be controlled by the time during which the primary current is maintained. Accurate timing is thus required and hardening is completed with a time expenditure of only a few seconds for each part. The depth of penetration of the hardened case depends on the frequency of the primary current. With a high frequency, the heat-affected zone of the metal is restricted to the immediate surface layers, while with a lower frequency, a deeper penetration of the heat is obtained. By this method, the control of the case depth is not as much a matter of hardenability as it is of the mechanics of the heating operation. Obviously, a deep-hardening steel will not harden throughout the cross-section if only the surface layers have been heated to above the critical temperature. The hardened cases in induction hardening are generally very shallow.

Chapter 16

Equipments and Procedure for Heat Treatment of Tool Steel

INTRODUCTION

Metallurgical requirements for the heating of tool steels whether for purposes of forging, annealing or hardening are extremely rigid and the cost of the tools or dies being processed is so large that usually economical or other considerations are overshadowed. These metallurgical requirements involve the control of the surface condition and chemistry, accurate control of the temperatures often up to 2400°F (1315°C) and the control of the heating and cooling rates. Special attention must be paid to these requirements in the design, construction and operation of these furnaces, especially those used for hardening, where the metallurgical factors involved become all-important.

A variety of types of furnaces are in use. These vary in size and design with the type of heating medium selected, the size of the articles to be heated and the quantity of material to be treated in a unit time. The maximum operating temperature and the type of atmosphere selected, both of which largely depend on the size, shape and chemical analysis of the tool or tool steel to be treated, will also be considered in determining the best type of furnace. It is not intended here to recommend any furnace type but simply to point out the limitations and ranges of usefulness of the many heating units now in use in the industry.

TYPE OF FURNACES

In general, the heating of tool steels is a batch-type operation. Continuous operations are used in annealing but seldom for hardening, although they may be found in the hardening of saw blades and other objects. The furnaces used in heating of tool steels can be divided into three types, depending on the heating medium:

1. Oil-fired furnaces.
2. Gas-fired furnaces.
3. Electric furnaces.

Oil and gas-fired furnaces can be divided into two types, depending on the type of firing. The furnaces may be direct fired, in which instance combustion takes place within and near the top of the heating chamber, or they may be indirect fired, in which instance a combustion zone is maintained separate from the part of the furnace in which the steel is placed by means of suitable baffles. The direct-fired type of furnace increases the difficulty of obtaining uniform heating and of controlling closely the temperature of the chamber itself. It is more commonly found in forging furnaces and furnaces for re-heating than in hardening installations. The indirect-fired type is more suitable with respect to these factors. These furnaces may be underfired with the combustion chamber in the bottom; they may be overfired with the combustion at the top, or they may be end or sidefired with combustion on the level with the heating chamber. In all of these instances, the hot gases circulate around the articles to be heated. Still another class is the radiant-tube type furnace, which may be gas, oil, or coal (pulverised) fired. In it, the fuel and the products of combustion are enclosed in alloy steel or refractory tubes that radiate heat to the work. The furnace is, therefore, a complete muffle type.

Gas-fired furnaces are relatively easy to regulate and possess the advantage of having a clean fuel. One of the main disadvantages of gas furnaces is that the fuel may vary considerably in composition from time to time in certain localities. Most gas-fired furnaces are equipped with multiple burners, which contribute to uniformity of temperature and facilitate control of the atmosphere surrounding the steel.

Several types of burners are available for the combustion of gases. The blast burners of the nozzle-mixing type, the proportioning type and high-pressure type generally use a separate air and gas control with automatic mixing and generally are all air inspirated. Furnaces can also be equipped with luminous or radiant-type burners that give a highly luminous flame.

Frequently, oil-fired furnaces are used in larger-size installations in which the hearth is over 2 by 3ft in size. Oil as a fuel allows a rapid heat input and a reasonable atmosphere control when manually operated burners are used. In some instances, oil burners do not throttle down as well as gas burners on electrically controlled furnaces and it may be more difficult to obtain close temperature control. Dirt should not be tolerated in the fuel oil to be used in order to eliminate clogging of the burners.

Electrically heated furnaces offer excellent temperature control since they can easily be made automatic. The main disadvantage to the installation of electric furnaces has been the fact that the only atmosphere available to surround the steel was air. This, of course, would scale and decarburise the steel in a short time at the elevated temperatures used unless packing of each part was performed. With the use of synthetic gaseous atmosphere, which are fed into the muffle of the electric furnace, however, electrically heated hardening furnaces are suitable for many types of work. The heat input in an electric furnace tends to be slow and this, coupled with the fact that the application of controlled atmospheres to exceedingly large furnaces is difficult, makes the use of electrically heated furnaces for forging and open annealing less popular than other types.

Furnaces for tempering tool steels generally operate on a recirculation principle in which either: (i) gas or oil is burned in a separate chamber—and the resulting products of combustion are circulated through the furnace by a fan or (ii) spiral electrical windings are suspended around the periphery of the furnace and a fan circulates heat throughout the charge. Since these furnaces generally operate at temperatures below 1100°F (595°C) and since decarburisation is practically absent at these temperatures and scaling is only slight, the problem of controlled atmosphere is negligible. Lead, salt and oil baths are also frequently used for the tempering of tool steels.

CONTROLLED ATMOSPHERE

The ideal atmosphere for the heating of any steel is one that neither oxidises (scales) the surface nor causes decarburisation or carburisation. The most straightforward methods for accomplishing these ends are either to exclude all atmospheres (heat treat in a vacuum) or surround the steel with an inert atmosphere (for example, argon or helium). Much effort has been devoted to the development of atmosphere derived from the combustion (or partial combustion) of ordinary fuels, such as natural gas.

Before proceeding to the discussion of atmosphere control, it is instructive to examine the atmospheres that surround a steel heated in a furnace when no deliberate control is attempted. In a full muffle furnace, whether heated electrically or by combustion of some fuel, the atmosphere is usually air containing a certain amount of water vapour. Both scaling and decarburisation occur to an extent determined mainly by the temperature of the furnace, the length of time at the furnace temperature and the chemical composition of the steel. On the other hand, in a fuel-fired semi-muffle or open furnace, the products of combustion come in direct contact with the surface of the steel and thus, the atmosphere consists of a mixture of CO_2, CO, H_2, H_2O, O_2, N_2 and certain hydrocarbons (in natural gas, principally CH_4). In addition, SO_2 may also be present if the fuel contains sulphur. The relative proportion of each gas is controlled by the air/gas ratio Fig. 16.1 shows how these proportions vary with air/gas ratio for the burning of natural gas. It will be noted that the water content indicated is constant. This water content varies materially, however, with the hydrogen content of the fuel and with the moisture content in the fuel and air supply.

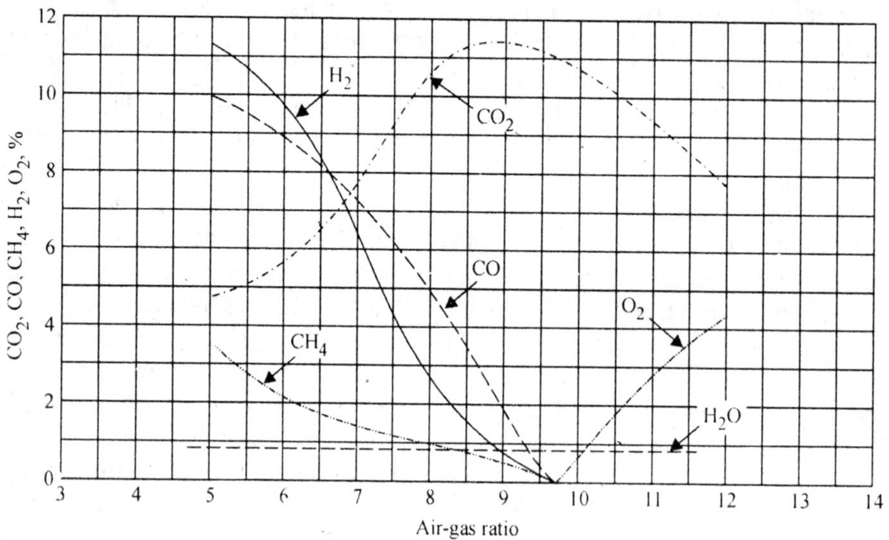

Fig. 16.1. Effect of variation in the air-gas ratio on the products of combustion of natural gas, after cooling to 40 °F.

The gas mixture produced by combustion is not generally in equilibrium with the steel and numerous reactions can occur. These reactions have been described by *Koebel* as follows:

$$Fe + H_2O \rightleftarrows FeO + H_2 \quad \text{... 16.1}$$

$$Fe + CO_2 \rightleftarrows FeO + CO \quad \text{... 16.2}$$

$$Fe_3C + CO_2 \rightleftarrows 2CO + 3Fe \quad \text{... 16.3}$$

$$3Fe + CH_4 \rightleftarrows 2H_2 + Fe_3C \quad \text{... 16.4}$$

$$Fe_3C + H_2O \rightleftarrows CO + H_2 + 3Fe \quad \text{... 16.5}$$

$$CO + H_2O \rightleftarrows CO_2 + H_2 \quad \text{... 16.6}$$

$$C + CO_2 \rightleftarrows 2CO \quad \text{... 16.7}$$

$$CH_4 \rightleftarrows C + 2H_2 \quad \text{... 16.8}$$

Reactions (16.1) and (16.2) produce scaling; reactions (16.3) and (16.5) cause decarburisation; reaction (16.4) results in carburisation; reactions (16.6), (16.7) and (16.8) involve chemical changes that occur in the gas phase. These reactions are reversible and by suitably controlling the percentages of the different gases, it is possible to favour either the forward or the reverse reaction. For example, by the law of mass action, an increase in the proportion of hydrogen will force reaction (16.1) to the left, thus tending to prohibit the formation of scale. Similarly, increasing the proportion of CO tends to prohibit scale formation according to reaction (16.2). To prevent scaling, therefore, high H_2 and CO contents should be combined with low H_2O and CO_2 contents.

Controlled Atmosphere in Open and Semi-muffle Furnaces

Manipulation of the air/gas ratio so as to increase CO and H_2 in accordance with Fig. 16.1 is the simplest method of minimising scaling. An atmosphere with 2 per cent CO is said to reduce the scaling 50 per cent and 3 per cent CO is reported to reduce it 75 per cent at any but the very high temperature. *Jominy* has stated that an atmosphere containing 15 per cent CO practically eliminates scaling, but such an atmosphere would be extremely wasteful of gas. Actually, there are two reasons why operating the furnace with less than the theoretical amount of air ('reducing' atmosphere) is not usually desirable:

1. Since the gas must also furnish heat for the furnace, the air/gas ratio cannot be too far from that giving no appreciable quantities of CO or O_2, because such a practice would only result in a considerable waste. As excess gas is increased, a point will finally be reached at which soot will start to form. Fig. 16.2 shows the loss in combustion efficiency with deviations from the theoretical air.
2. Even though scaling is inhibited, decarburisation is not and therefore a soft 'skin' is generally produced. In addition, a very thin scale is likely to be tighter and more difficult to remove on pickling or blasting.

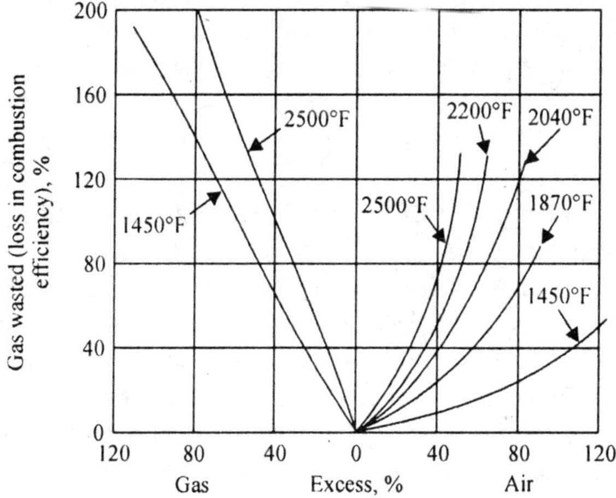

Fig. 16.2. Loss in combustion efficiency as a function of the amount of excess air or excess gas entering the burners and as a function of the operating temperature.

Surprisingly, an oxidising atmosphere (more than theoretical amount of air) can be employed for hardening certain types of tool steel. The reason for this is that oxidation (scaling) proceeds so fast that iron is removed as rapidly as carbon and no decarburisation layer exists. In fact, with plain carbon tool steels, *Stout* and *Aho* have indicated that an actual build-up of carbon may occur at the surface. This has also been shown by *Schlegel* in the heat treatment of high speed steel in oxidising atmospheres. At a slower rate of oxidation, carbon may be removed faster than iron and faster than diffusion of carbon from the inside can take place to replenish the depletion of carbon at the surface. Tool steels that are adapted to hardening in an oxidising atmosphere scale freely; the chromium steels such as types 420 (A2) and 430 (D2) cannot be hardened in this manner without excessive decarburisation. Steels that can be hardened satisfactorily in an oxidising atmosphere generally have low chromium (1 per cent or less) and do not require a high hardening temperature (no greater than 1600°F).

Although it is not generally recommended, tungsten high speed steel can be hardened in a semi-muffle or open furnace. Because high speed steel is treated at such high temperatures, an absence of oxygen is desired to minimise the scaling tendency and hence the hardening furnace is generally operated with a slight excess of gas. Pre-heating furnaces for high speed steel operating at 1400 to 1600°F use an oxidising atmosphere to form a thin adherent scale that appears to partially protect the steel from decarburisation.

When heating most tool steels for forging in semi-muffle furnaces, it is better to scale lightly than to carburise or decarburise. For hardening, the preferred controlled atmosphere would carburise to a depth of not more than 0.001 inch, which would normally be removed in finishing the tool.

Controlled Atmosphere in Muffle Furnaces (Electric or Gas Fired)

Since it has been found impractical to operate direct-fired fuel furnaces with sufficient CO and H_2 content to offset the scaling action of water vapour and CO_2, major developments in atmospheric controls have been confined to electric and gas-fired muffle furnaces. The aim is to maintain an atmosphere in the muffle that is approximately neutral so that no reaction between the steel and the atmosphere occurs. Such atmospheres include mixed gases obtained from the partial combustion of hydrocarbon gases or charcoal, dissociated ammonia and hydrogen from the electrolytic dissociation of water.

In electric furnaces equipped with a controlled atmosphere, the atmosphere is generated separately and plays no part in the heating of the furnace. In gas-fired furnaces, the same system may be used or, in certain installations, the waste gases from the combustion chamber may be treated to adjust their composition and then recirculated to provide the atmosphere.

The following discussion will briefly describe some of the more common protective atmosphere used in heat treating tool steels and their characteristics.

As described previously, the products of combustion of natural gas and air, although adjustable (by varying the air/gas ratio) to reduce scaling, tend to cause decarburisation. Referring to reactions (16.3) and (16.5) (the decarburisation reactions), it can be seen that removal of CO_2 and H_2O from the atmosphere will tend to reverse these reactions and may even cause carburisation. In fact, such an atmosphere (called a purified exothermic atmosphere) can be used for annealing or hardening tool steels free of scaling or decarburisation. Typical compositions of such atmospheres are shown in Table 16.1. The carburising potential can be adjusted by varying the dew point (per cent water vapour). A convenient method of adjusting dew point involves by-passing some of the wet CO_2-free gas around the drying mechanism.

It should be emphasised that any attempts to remove CO_2 or H_2O from the atmosphere are of little value if both constituents are not thoroughly removed. The water-gas reaction (16.6) is strongly catalysed by steel at high temperature and even if the moisture has been entirely removed from the gas before introduction into the heating chamber, the presence of CO_2 and hydrogen will cause more H_2O to form. Thus, for effective control of the furnace atmosphere, it is essential to know the dew point of the gases in contact with the steel; measurements made at the atmosphere generator are not reliable. The water-gas reaction, if allowed to proceed, will cause roughening of the surface of the metal, although no appreciable scale will form. If the reaction is completed before the gas is introduced into the furnace (by passing the gas over heated cast iron chips, for example), roughening can be prevented.

Table 16.1. Typical controlled atmospheres.

Type	*Air/natural gas ratio*	*Composition, %*						*% H_2O or dew point, °F*
		CO_2	O_2	CO	H_2	CH_4	N_2	
Purified exothermic (H_2O and CO_2 removed								
Rich	6:1	0.0	0.0	10.5	15.5	1.0	73.0	–40
Fairly lean	9.25:1	0.0	0.0	3.0	3.0	0.0	94.0	–40
Endothermic								
Dry	2.4:1	0.0	0.0	20.0	38.0	0.5	41.5	–10
Wet	3.5:1	2.0	0.0	17.0	28.0	0.0	53.0	Saturated
Charcoal gas (Charcoal burned above 1800°F)	–	1 to 3 or 0.10 to 0.50 if removed	–	25 to 35	1 to 3	0.50 to 1	66	0.60% after drying
Dissociated ammonia	–	–	–	–	75	–	25	–40
Diluted	–	–	–	–	5 to 10	–	90 to 95	–40

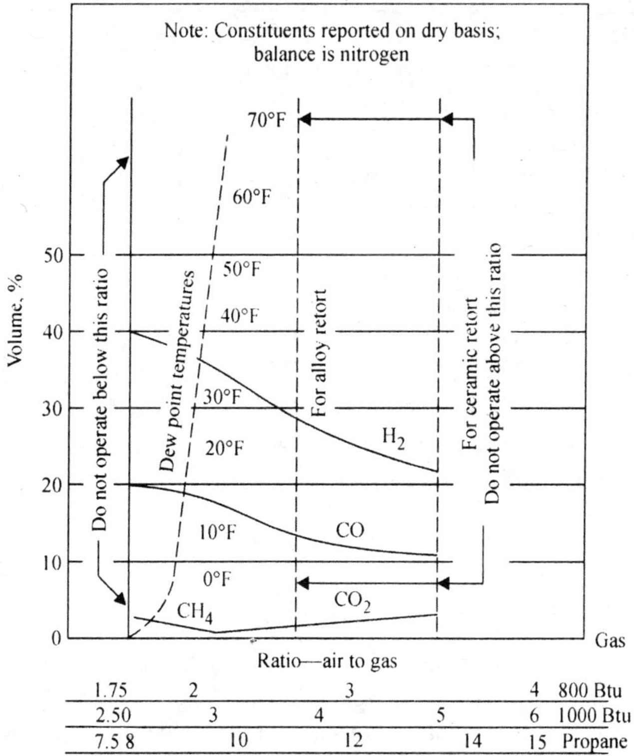

Fig. 16.3. Approximate analysis of endothermic generator gas when cracking various air-gas mixtures, the gases being either natural gas of 800 Btu per cu ft, 1000 Btu gas, or commercial propane.

The next type of atmosphere to be considered is the endothermic atmosphere, which, in many respects, is the most satisfactory. Since air/natural gas ratios below about 6:1 will not support combustion, the low endothermic ratios are reacted by externally supplied heat. Typical endothermic gas compositions are shown in Table 16.1 and the variation in composition with air/gas ratio is shown in Fig. 16.3.

By simply adjusting the air/gas ratio, an atmosphere containing extremely low CO_2 and H_2O can be produced; thus, no auxiliary equipment is necessary for removing these constituents, as is the case with an exothermic atmosphere. Endothermic atmospheres can be regulated (by controlling the dew point) to provide a constant carbon potential and thus are adaptable to annealing and hardening a wide variety of steels. Fig. 16.4 shows the dew point that must be maintained in an endothermic atmosphere so as to be in equilibrium with plain carbon steel of a given carbon content. For a given temperature, the lower the dew point, the higher the carburising tendency of the atmosphere. Curves in Fig. 16.5 show the dew point that should be maintained at a given temperature to prevent carburisation and decarburisation of common alloy steels. Table 16.2 shows the equilibrium dew point and the permissible range for hardening some common tool steels in an endothermic atmosphere.

Table 16.2. Dew point temperature in furnace using endothermic atmosphere for hardening some common types of tool steel.

Type	*Description*	*Hardening temp. °F*	*Dew point, °F* Min.	Equilib.	Max.
522 (H12)	Cr-W-Mo hot work die steel	1850	40	46.5	54
420 (A2)	5 Cr-1 Mo-1 C air-hardening steel	1750	40	45	50
310 (S2)	Si-Mn shock-resisting chisel steel	1550	40	52.5	60
410 (O1)	Manganese non-deforming oil-hardening die steel	1450	45	50	55
432 (D3)	High-carbon high-chromium oil-hardening steel	1750	20	–	70
321 (S1)	Low-tungsten chromium chisel steel	1750	40	42	45
430 (D2)	High-carbon high-chromium air-hardening steel	1850	20	–	70
340 (F2)	Tungsten fast-finishing steel	1550	23	27	33.5
120 (W2)	C-V shallow-hardening die steel	1500	45	50	55
413 (O7)	High-carbon low-tungsten oil-hardening steel	1550	25	31.5	36
232 (L6)	Cr-Ni-Mo tool steel	1500	50	60	65
610 (T1)	18-4-1 high speed steel	2350	0	8.5	10
630 (M1)	1½-9-4-1 W-Cr-Mo high speed steel	2250	5	12	15

Endothermic atmosphere are combustible and thus require cautious handling to avoid explosion. When these atmosphere are employed in furnaces operating below 1300 °F, there will be a tendency for soot to deposit on the work. The reaction responsible for this is the decomposition of CO to C and CO_2 which proceeds most rapidly between 900 and 1300 °F. Iron oxide catalyses this reaction and therefore, pieces should be free of scale before being placed in the furnace.

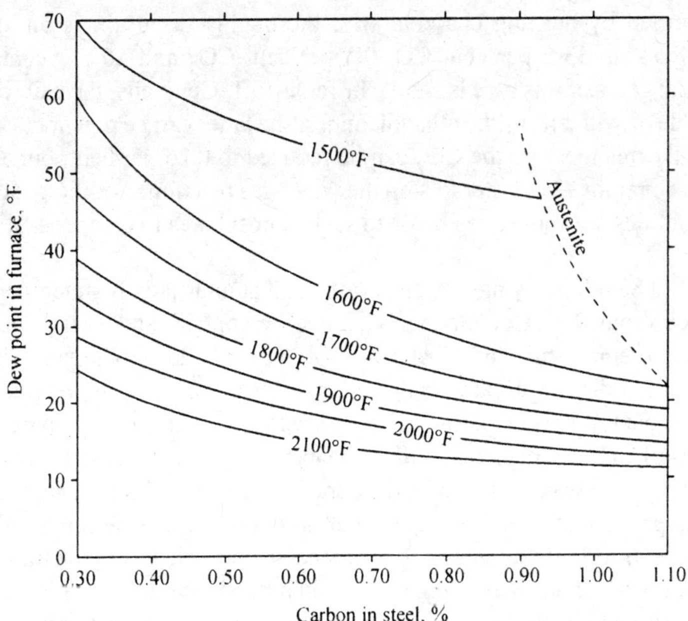

Fig. 16.4. Dew point of endothermic atmosphere in equilibrium with plain carbon steel as a function of carbon content of the steel.

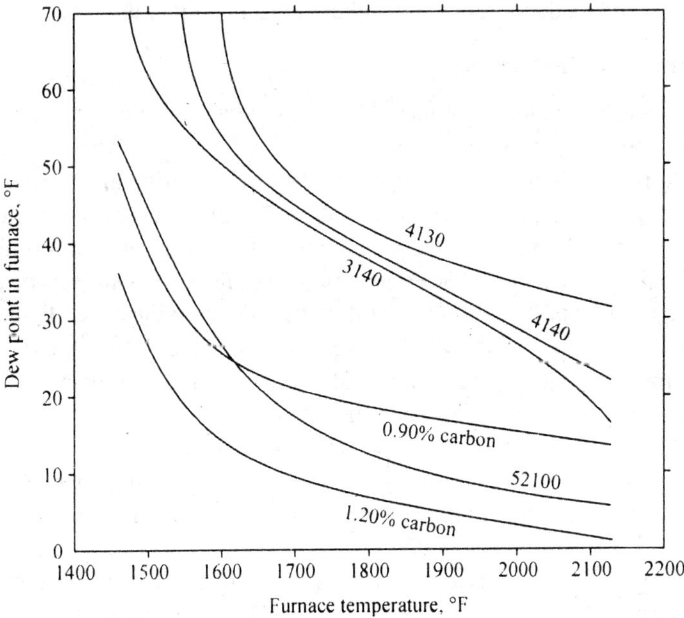

Fig. 16.5. Dew point of endothermic atmosphere in equilibrium with common tool and alloy steels as a function of temperature.

The atmosphere described above (purified exothermic and endothermic) are perhaps the most widely used for the heat treatment of tool steel. Additional types of atmospheres have been employed, however.

Charcoal gas, formed by burning charcoal with thoroughly dried air, is an atmosphere that, at equilibrium, should contain 33.5 per cent CO, 0.5 per cent CO_2 and 66 per cent N_2. The range of analysis of commercial gases of this type is shown in Table 16.1. Generally, the CO_2 can be scrubbed out with sodium hydroxide or with one of the ethanolamines (the latter can be regenerated by boiling, which drives off the CO_2). By this method, the CO_2 can be reduced to 0.25 per cent, but as indicated before, H_2O or H_2 must also be removed in order to stop the water-gas reaction. Additions of CH_4, ammonia, or benzol in minute quantities are sometimes made to such atmospheres to compensate for the presence of H_2, CO_2 and H_2O.

Pure nitrogen would be relatively inert to steel (although at unusually high temperature, even a plain carbon steel can pick up some N_2). Commercial N_2, however, contains at least 0.20 per cent O_2, can hold 2 per cent H_2O at room temperature (65°F) and 0.05 per cent at –20°F. Nitrogen with even such small amounts of moisture has a distinct decarburising action; even if the nitrogen were completely dried and contained only 0.1 per cent O_2, it would decarburise the steel worse than air. An impure nitrogen that would contain small amounts of CH_4 or CO or H_2 to offset the moisture content would be much preferred.

Mixtures of H_2 and N_2 are generally obtained by cracking anhydrous ammonia to obtain a mixture of 75 per cent H_2 and 25 per cent of N_2. To this gas, small amounts of methane can be added as natural gas, or hydrocarbons can be introduced by bubbling the cracked ammonia through a liquid hydrocarbon. By diluting such mixtures with air, or by burning fuel gas in air and steam to form large quantities of CO_2, instead of CO and then chemically removing the CO_2, an atmosphere containing 10 per cent H_2 and 90 per cent N_2 can be obtained that is of somewhat lower cost than the high-hydrogen atmosphere. These gases can be recirculated many times and thus somewhat reduce the high cost of ammonia installations. Typical dissociated ammonia atmosphere are shown in Table 16.1.

Evaluation of Controlled Atmosphere

There are several methods of evaluating controlled atmosphere; however, for valid results, all of them require that the following two precautions be taken:

1. Decarburisation or scale must be absent from the samples originally placed in the furnace. The presence of scale on the surface will lead to decarburisation unless the atmosphere is so highly reducing that such scale is immediately decomposed. If decarburisation is already present, with or without the presence of scale, the concentration gradient will cause carbon to diffuse from the interior to the exterior of the sample and effectively increase the zone of partial decarburisation.
2. Samples must be placed in the furnace so that no surfaces are shielded from the gaseous atmosphere and an adequate circulation of the gases must be maintained.

Koebel has outlined the following evaluation methods:

1. Surface hardness tests may be misleading, as hardness at the surface (especially in high-carbon high-alloy steels) is often not affected by a change in carbon content. However, *Scott* and *Fisher*, in discussion, have pointed out methods that make the pyramid hardness test satisfactory is such instances.
2. Microscopic examination is used extensively in production or troubleshooting to determine neutral, carburising, or decarburising conditions. The microscopic test on any hypereutectoid steel is qualitative and thus is of limited value for research studies.
3. Chemical analysis of surface layers finds wide use and gives reliable results but is time consuming.
4. Change of weight method is applicable when no scaling occurs simultaneously. This method has been used extensively in research investigations on controlled atmospheres.

LIQUID BATHS FOR HEATING

In addition to furnace atmosphere, protection and control of surface chemistry may be achieved by means of various liquid heating mediums. These heating mediums employed in the heat treatment of steel fall into three general classes: metals, salts and oils. The first two are used for both hardening and tempering, while oil baths are restricted to tempering operations only. The working range for the three classes is approximately as follows: for lead, 650 to 1700°F (345 to 925°C); for salts, 350 to 2350°F (175 to 1290°C); and for oils, from room temperature up to approximately 600 to 700°F. Certain alloys of lead and tin that are similar in composition to terne plate or solder find occasional use at temperatures under 650°F and would be used solely for stress-relieving or tempering operations.

Lead Baths

Since lead and salt baths are both used for hardening operations, it will be well to note the characteristics that differentiate these two classes of heating materials. Lead oxidises readily but otherwise is stable and has no appreciable chemical action on the steel with which it comes in contact. Salts, however, tend to decompose and some have a marked chemical effect on the steel. Lead, if clean and unoxidised, does not adhere to the surface, while salts invariably cover the metal with a thin film of the bath material.

Lead is higher in specific gravity, higher in thermal conductivity, lower in specific heat and lower in the latent heat of fusion. The high specific gravity of lead means that steel will not sink in lead but must be immersed with suitable fixtures. Steel, however, sinks in all of the commercial salts that are used for hardening and tempering.

The melting point of lead is lower than that of any of the salts used for hardening, but is higher than that of most of the salts used for tempering. The melting point of a bath material not only determines the minimum temperature of operation but also has a pronounced effect on the heating rate when charging cold stock. The lower the melting point, the less will be the weight of the frozen material formed; the greater the difference between the melting point and the bath temperature, the more rapidly this layer will be removed. Other things being equal, a bath operating at 1500°F (815°C) with a melting point of 700°F (370°C) will heat more rapidly than one operating at the same temperature with a melting point of 1300°F (705°C). The high thermal conductivity of lead is one of its most valuable properties, tending toward rapid heating, uniform temperature throughout the bath and a high thermal efficiency. This occurs despite the low latent heat of fusion of lead. Because of the low thermal conductivity of salts, layers of solidified bath material tend to form more extensively in all salt baths than they do in lead baths when cold stock is placed in the pot. Pre-heating to temperature below the melting point decreases the amount of frozen material. Lead can, therefore, be satisfactorily used at temperatures closer to the melting point. Where intricate parts are inclined to warp or crack in hardening, the retardation of heating caused by the solidified layer is distinctly advantageous and salt baths are preferred.

Lead so greatly excels in heat conductivity that, for any temperature within its range of operation, the other heating mediums do not approach it in the rate of heating. This is illustrated in Fig. 16.6, which shows the heating curves obtained at a temperature of 1500 and 900°F (815 and 480°C) in lead compared with the heating rate of the same size specimen obtained in a cyanide bath held at 1500°F and in an oven furnace at 1500 and at 900°F.

Molten lead is readily oxidised by air at all temperatures and must be protected by a suitable covering to prevent excess formation of dross. Lead oxide is sufficiently insoluble in molten lead to preclude any appreciable contamination of the bath as a whole. This tendency of lead to oxidise is a troublesome feature resulting in the loss of lead and dirty stock. At low temperatures, most coverings are inadequate and special means are often required to clean the steel after removing it from the lead bath. Above 850°F (455°C), however, lead can be well protected by wood charcoal, coal coke, carburising compound, petroleum coke, or other compounds of similar nature and the bad effects of oxidation kept to a minimum; under proper conditions, the steel will come out remarkably clean.

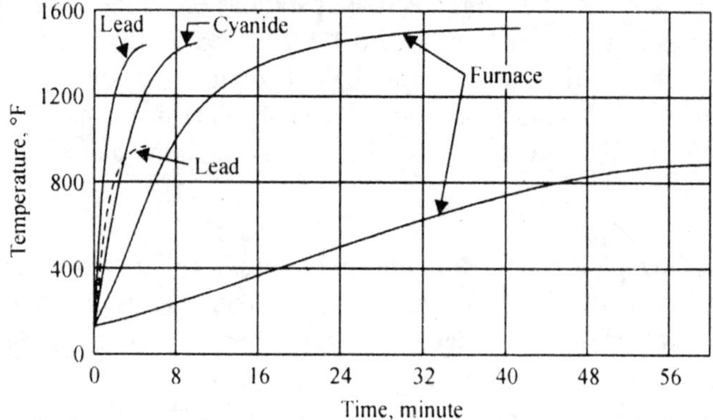

Fig. 16.6. Comparative heating rates of steel in a lead bath, a cyanide bath and an oven furnace. Baths and furnace held at 900 and 1500°F.

Salt Baths

Salt baths can be conveniently classified into three groups: low-temperature salts, medium-temperature salts and high-temperature salts. The chemical compositions of typical heat treating salts are shown in Table 16.3.

Low-temperature salts are used up to 1000 to 1100°F (540 to 595°C) and are, therefore, used for tempering operations only. These salts are composed of nitrates and nitrites of the alkali metals. Sodium nitrate melting at 586°F (310°C) may be used at temperature above 650°F (345°C). The eutectoid mixture of 56 per cent potassium nitrate and 44 per cent sodium nitrite melts at 295°F and may be used above 325°F. A mixture of 51.3 per cent potassium nitrate and 48.7 per cent sodium nitrate melts at 425°F and is suitable for use above 475°F (245°C).

At temperature above 1050 to 1100°F (565 to 595°C), these baths attack the work and the bath containers. Local overheating of the bath container must not be permitted or rapid failure will occur. The work introduced into these baths should be thoroughly cleaned of oil, brine, scale, or soot before immersion. Contamination will result if this is not done, resulting in sludge formation, which increases the dragout and the difficulty of cleaning after tempering. If uncontaminated, any of these nitrate mixtures will clean easily in water.

These baths may also be used for quenching carbon and low-alloy steels. As a note of caution, it should be pointed out that nitrates and cyanides react together violently when hot and the two should always be kept apart. Thus austenitising or quenching salts containing cyanide should never be used with these baths. The bath may be maintained at any temperature between 300 and 1000°F (150 and 540°C), depending on the physical properties required. Some decomposition of the bath constituents may occur because of the local action of the hot work being quenched and some pitting of the work may take place. The intensity of these actions, of course, depends on the size of the work and its temperature before quenching.

Table 16.3. Nominal compositions of salts for heat treating

Weight Percentage									*Approximate melting point, °F*	*Recomm-ended heating range, °F*
NaCl	*KCl*	*$BaCl_2$*	*$NaNO_3$*	*$NaNO_2$*	*KNO_3*	*$CaCl_2$*	*NaCN*	*Na_2CO_3*		
Low-temperature salts										
–	–	–	(a)	40 to 50	50 to 60(a)	–	–	–	290	325 to 1200
–	–	–	40 to 50	–	50 to 60	–	–	–	440	500 to 1200(b)
–	–	–	96 min	–	–	–	–	–	700	750 to 1200(b)
Intermediate-temperature salts										
30 to 40(a)	(a)	–	–	–	–	–	15 to 40	30 to 50	1020	1150 to 1500
15 to 25	–	25 to 35	–	–	–	45 to 55(c)	–	–	900	950 to 1400
45 to 55	45 to 55	–	–	–	–	–	–	–	1250	1350 to 1650
15 to 25	20 to 30	50 to 60	–	–	–	–	–	–	1100	1250 to 1700
20 to 30	–	70 to 80	–	–	–	–	–	–	1300	1400 to 1700
10 to 20	–	80 to 90	–	–	–	–	–	–	1400	1500 to 2000
High-temperature salts										
–	–	98 min	–	–	–	–	–	–	1800	1900 to 2450
4 to 8	–	92 to 96	–	–	–	–	–	–	1600	1750 to 2300

(a) Either of the two materials marked by an (a) in this line, or any mixture of the two materials, may appear in the percentage shown under the one material. (b) The materials should not be operated above 1100°F in a fuel-heated spot. (c) This material is hygroscopic and should be added to molten salt with great care.

Salt baths used for the medium-temperature range vary more widely in composition than those used for the low-temperature range, but in general, they are eutectics of chlorides, or chlorides and carbonates. These bath mixtures can be used from 1100 to 1650°F (540 to 900°C). The main constituents of these baths are sodium carbonate, potassium carbonate, sodium chloride and barium or calcium chloride. Carbonates in a salt bath tend to decarburise steel and it is necessary to make additions of sodium cyanide to prevent the formation of a soft skin. With steels, the desirability of this practice is questionable and baths consisting of chlorides only are generally preferred. The lowest melting chloride mixtures contain calcium chloride, barium chloride and sodium chloride and melt at about 900°F. Calcium chloride is extremely deliquescent and tends to corrode the work. The alkali chloride mixtures are more easily handled and an example might be a mixture of 56 per cent potassium chloride and 44 per cent sodium chloride, which melts at 1220°F (660°C).

Although chloride mixtures are inert to steel when first melted,they gradually oxidise to oxychlorides and then to oxides or carbonates. As a result, through constant use, these baths will decarburise steel. In order to counteract this tendency, rectifiers such as boric acid, fused borax, or boric oxide, which convert the oxides and carbonates to stable metaborates that can be removed and discarded, are often used. Methyl chloride gas is also used as a bath rectifier and has the advantage of producing no sludge. However, these rectifiers are of no value when added to a chloride-carbonate bath and should be used only in straight chloride mixtures.

High-temperature salt baths are generally used in a temperature range from 1800 to 2400°F (980 to 1315°C) and are used primarily for hardening high speed steel. The salts used in these bath mixtures are barium chloride, borax, sodium fluoride, silicates and occasionally magnesia or lime. The borax baths are not eutectic mixtures but are truly glasses; they have no sharply defined melting point but soften gradually when heated, the melting range being between 1600 and 1900°F (870 and 1040°C). Perhaps their greatest disadvantage is that they are not easily cleaned from the work. It is almost imperative that a salt quench be used.

This quench consists of the previously described, medium-temperature chloride-carbonate baths, with or without cyanide, maintained just above their melting point and preferably below 1200°F (650°C). Such baths react chemically, with some sputtering, with a high-temperature salt. In the quench pot, the products of the reaction form a heavy sludge that must be frequently removed and thrown away. The barium chloride salts have almost completely replaced the borax baths for hardening high speed steels because they are water soluble and can be more readily cleaned from the work. Despite this, it is general practice to hot quench in chloride salt mixtures that are even more water soluble. A build-up of barium chloride in the quenching bath raises its melting point and frequent additions of lower-melting chlorides are required to keep the bath fluid at the lowest possible temperature.

The rectification of high-temperature salt baths to prevent them from decarburising can be accomplished with silicon or carbon. Baths in which the electrodes protrude above the surface require daily rectification with either ferrosilicon or silicon carbide. Submerged-electrode baths can be controlled by simply introducing a graphite rod into the molten bath, which will reduce any oxides into metal. The metal adheres to the rod and can be readily scraped off and the graphite rod re-used.

Salts cover the stock with a continuous film, resulting in a large mechanical loss, but because of the solubility of these materials in water, the work is easily cleaned. The adhering film of salt is a major advantage to a fine finish tool, as this film effectively prevents oxidation by air on removal from the bath. The thickness of the film and, consequently, the amount of mechanical loss largely depend on the fluidity of the bath.

The ideal salt bath has not yet been discovered. Virtually all the salts used for heat treating are hygroscopic to some extent; all of them adhere to the work when it is removed; there is always a corrosive action at the surface of the salt bath affecting containers, thermocouple tubes and partly immersed work and this action is more severe with higher temperature.

Several precautions should be observed in operating salt baths. First, no salt bath should be allowed to become contaminated. Pure salts should be used in all mixtures. Sulphates and other impurities may cause excessive corrosion between the bath and the steel. Second, it is necessary to heat salt baths very slowly while they are solid. Solid salts are poor conductors of heat and since no circulation is possible, the rate of heat absorption is limited. Any attempt to heat too quickly only results in overheating the pot, with a consequent shortening of its life. A salt having a low melting point is desirable when large work is to be heated, as the immersion of a large mass of steel lowers the temperature so much that a thick coating of solid salt is formed on the work. If possible, large work should be pre-heated up to the melting point of the bath, then immersed with the bath just above its melting point and the bath and metal slowly heated to the desired temperature together. The melting point of the bath is not so important for small work, provided the melting point is not too near the operating temperature.

SURFACE PROTECTION BY PACKING MATERIALS

The annealing of tool steels and the hardening of many tools are often performed by sealing the steel, completely surrounded by a packing compound, in a suitable container. This procedure ensures a slow uniform rate of heating, thus minimising the danger of warpage and cracking and serves as an alternative method of controlling surface chemistry. The packing materials commonly used include wood charcoal, spent pitch coke, carburising compound, sand, alumina particles, cast iron chips and mica. These materials may carburise, decarburise, or be neutral to steel, depending on the heating temperature, the carbon content of the steel and the density of the packing (also the particle size of the packing material). The principal aim of such materials is to exclude the decarburising gases from contact with the steel. If oxygen is to be excluded, the packing must be very tight and the sealing of the container perfect, for most gases can diffuse through 'ordinary seals' at a surprisingly rapid rate at elevated temperature. Materials that contain carbonaceous substances are, therefore, somewhat better where this is the case. In the annealing operation, where scale is present on the bars as charged in the furnace, decarburisation may occur unless a carbonaceous material is present. Needless to say, the presence of moisture in the packing material is not tolerable as decarburisation will take place very readily.

Figure 16.7 indicates diagrammatically the ranges of usefulness of each of the principal packing compounds. This should be used only as a guide, as the presence of scale, the age of the packing compound, the moisture content and the carbon content of the steel will appreciably alter the range. Furthermore, the temperature at which one material will change from neutral to decarburising is not necessarily as definite as that indicated. It is to be noted that the effect of carbon content is not included. The chart applies particularly to high carbon steels. Pitch coke, not shown in Fig. 16.7, will carburise at almost all temperature if the volatile materials are not driven off and will likewise show inconsistent results since its composition will change with re-use. (This latter statement applies to cast iron chips as well.) To obtain spent pitch coke, an excellent packing material for tool steels, the volatile material should be eliminated by firing above 1800°F (980°C), or at least to a temperature at which it will be used.

The presence of a piece of carbon in a furnace without controlled atmosphere facilities near, but not in contact with, the steel can cause marked decarburisation. Carbonaceous muffles, however, can carburise or decarburise, depending on the amount and type of energiser contained.

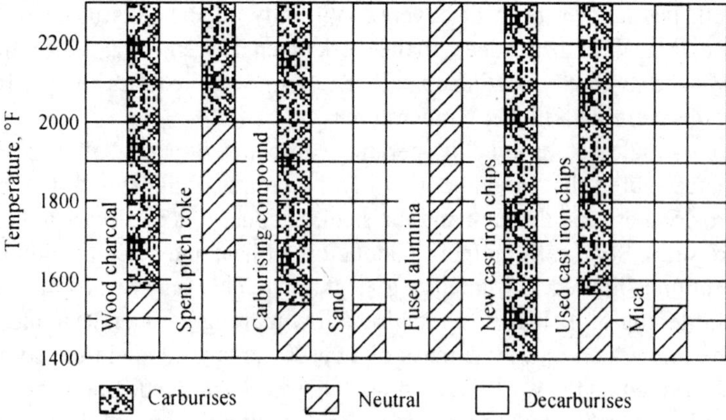

Fig. 16.7. Approximate range of usefulness of each of the principal packing compounds. Temperature of change from one behaviour to another are actually not sharp. For cast iron chips the temperature below which decarburisation will take place depends on the carbon content of the chips.

QUENCHING MEDIUMS FOR STEEL

Since most tool steels are used in a nearly fully hardened condition, a discussion of the quenching of steel must necessarily be restricted to those quenching operations that have for their purpose the formation of martensite or a product possessing a hardness much greater than that of pearlite or spheroidised structures. As already pointed out (concerning the science and principles of heat treatment), in order to form martensite in any steel, it is necessary to heat that steel to an elevated temperature with the attendant formation of an austenitic or partially austenitic structure and to cool that steel at a rate that will be sufficiently rapid to completely suppress the formation of pearlite or spheroidite. The formation of pearlite or spheroidite occurs in the temperature range between 1300 and 950°F (705 and 510°C) at a rate which increases rapidly as the steel approaches the lower portion of this range. It is the speed of this high-temperature transformation that controls the rate of quenching required. Once this temperature range has been passed during the cooling or quenching operation, pearlite cannot form and the decomposition product of the austenite will, in most instances, be martensite. The latter forms continuously as the temperature is lowered below the M_s point, which occurs in most tool steels between 500 and 350°F.

The rate of cooling that is necessary to suppress the formation of pearlite has been called the critical cooling velocity and, of course, is related to the speed of the pearlite transformation. There are three main factors that affect this quantity:

1. Degree of solution of the carbide at heat treating temperature.
2. Grain size of the steel.
3. Composition of the steel.

The first two factors above allow a certain amount of variation in the critical cooling velocity of any one steel. With higher heat treating temperature, almost invariably, more carbides are dissolved and the grain size is increased; both factors permit a slower cooling rate to achieve the desired suppression of the pearlite transformation. However, these first two are seldom of great enough influence to allow a drastic change in the type of quenching medium required. There are, however, exceptions to this in instances where, by increasing the quenching temperature, certain 'border-line' steels can be quenched in a less drastic medium.

The composition of the steel is the most important item influencing the selection of the quenching medium. The presence of alloying elements in tool steels makes it possible to classify the steels into three groupings:

1. Shallow or water-hardening steels.
2. Oil-hardening steels.
3. Air or salt-hardening steels.

This indicates that quenching in either water, oil, salt, or air is sufficient to achieve hardening. This classification, of course, reading from top to bottom, depends on an increase in alloy content. It does not, however, indicate that all steels with high alloy content are air-hardening steels. The alloying element present must be effective in increasing the hardenability of the steel; thus, high-carbon steels containing single large additions of tungsten and cobalt are not air-hardening steels or even oil-hardening steels, for these two elements appear to be ineffective in this respect. Nor does this classification imply that there are three distinct rates of cooling and only three rates available to the heat treater. There are gaps between the three quenching mediums listed with respect to cooling rates, particularly between water- and oil-base quenching mediums, but within each class many variations in effective rate and quenching behaviour are to be found.

A section devoted to the study of quenching should, therefore, include a discussion of the principal characteristics of the three main types of quenching mediums and a discussion of the factors that would lead to variations in these characteristics. Such a discussion is of interest and importance to the steel user and manufacturer only in so far as it indicates those characteristics that affect the hardening of the steel in question and the economy and safety of operation. An evaluation of the efficiency of quenching mediums must necessarily be based on these items, with the former holding the most important position in this respect. A review of the findings of studies of the physics of cooling of hot objects in relatively cool fluid mediums is, however, necessary for even an elementary understanding of these characteristics.

Mechanism of Quenching

Numerous extended studies have been made relating to the mechanism of cooling of hot objects in liquid quenching mediums. It is perhaps noteworthy that all of these investigators are in substantial agreement concerning the various phenomena that occur during the quenching operation. It will be well to indicate here, in brief, a summary of their findings as it bears an important relation to the problem of hardening all types of steel.

Figure 16.8. shows typical cooling curves obtained in quenching 2 inch diam cylinders of austenitic stainless steel in agitated water and agitated oil. Three curves are shown for each quenching medium; the cooling curve obtained at the centre, the cooling curve obtained 0.16 inch from the surface and the estimated course of the surface cooling phenomenon. These curves emphasise the fact that water cooling produces a much faster rate and produces much higher temperature gradients throughout the piece at any one time during the quenching cycle than does oil quenching. The magnitude of these temperature gradients is of importance in determining the dimensional changes, the residual stresses and the tendency toward cracking of the steel.

Under certain conditions, it is possible to show that there are three distinct and separate stages encountered during the cooling of hot metals in liquids. Fig.16.9 is included to illustrate this point. It shows the temperature at the centre of a small cylinder of steel quenched from 1525°F (830°C) into still water maintained at 137°F.

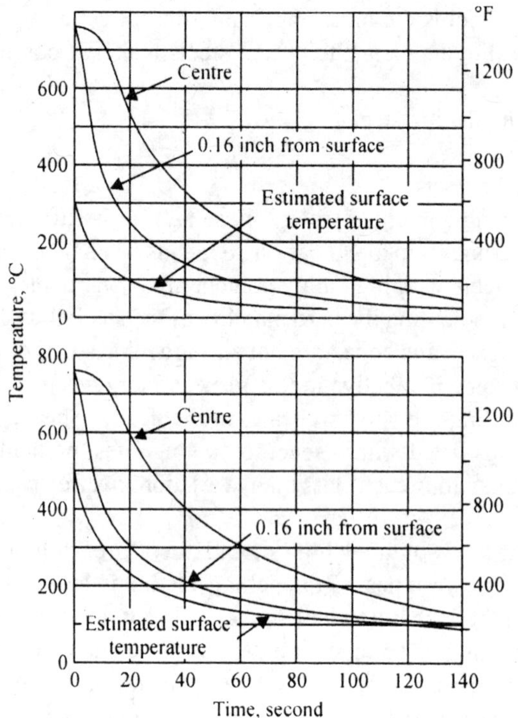

Fig. 16.8. Top: Observed cooling curve of centre and of a point 0.16 inch from the surface of a long austenitic steel cylinder 2 inch in diameter during quenching in agitated water. Bottom: Same, quenched in agitated oil.

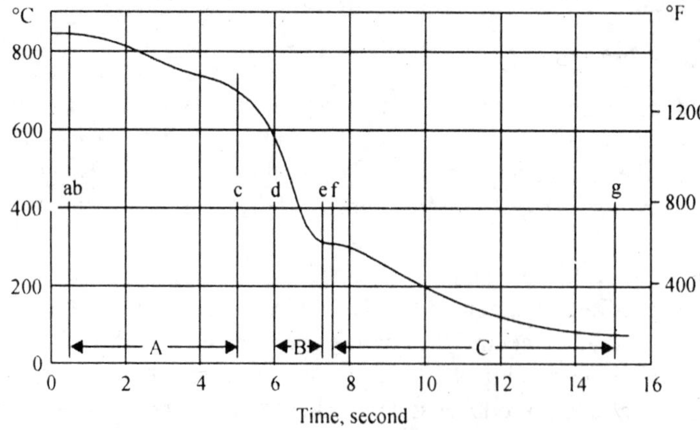

Fig. 16.9. Typical cooling curve for the centre of a small cylinder during quenching in still water at 1370°F, illustrating the three stages of surface cooling.

Under these somewhat exaggerated conditions, the three stages are clearly evident and have been described as follows: Stage A indicated on the curve can be termed the vapour cooling stage. During this first stage, immediately on placing the hot steel in the liquid quenching bath, a certain amount of liquid vapourises adjacent to the steel surface and forms a continuous film or blanket of vapour around the body to be cooled. Cooling can proceed through this film only by radiation and consequently, a slow cooling rate is the result. The second stage or cooling, stage B, indicated on the curve has often been termed the vapour transport stage or the stage of violent boiling. During this period, the continuity of the vapour blanket is lost. A certain amount of the vapour migrates into the liquid and condenses; new liquid rushes in and vapourises at the steel surface, extracting an amount of heat equal to the heat of vapourisation of the liquid and is subsequently condensed in the body of the quenching bath. This action is extremely violent and results in a rapid extraction of heat from the steel body at a rate largely dependent on the amount of liquid material that is vapourised. When the temperature of the surface of the steel becomes lower than the boiling point of the liquid, vapour can no longer form and the second stage of cooling has ended. Cooling from this point, during stage C indicated on the diagram, can be termed simple liquid cooling and occurs only by conduction and convection. This produces a much slower rate of cooling than in the preceding stage. The rate of cooling here depends largely on the specific heat of the liquid, its thermal conductivity and degree of agitation of the quenching bath.

For cooling curves of this type, refer to the centre of a cylindrical bar or the centre of a sphere. Actually, however, this behaviour is apparent at almost any point below the immediate surface zone of the quenched article. A complete discussion of the properties of the quenching mediums that affect these general characteristics of cooling curves will be made at a later time. It appears, however, that in any one quenching medium the most important factor to be controlled is the bath temperature. Certain quenching mediums exhibit a strong tendency to change their cooling power as the bath temperature is altered. Generally, this results in a lowered quenching efficiency as the temperature of the bath is raised. If the temperature coefficient of the bath efficiency is too great, the bath will likely lead to variable results, as the temperature in most quenching installations will normally fluctuate over a relatively wide range.

Typical cooling curves obtained on oil quenching have been shown previously. It was noted that the rate of cooling when quenching in oil is much slower than that obtained by water quenching and that the temperature gradients were materially decreased. The same three stages that have just been described for water quenching are found in quenching in oil. An initial slow rate caused by a relatively stable vapour blanket is followed by a period of rapid heat extraction during the boiling stage and finally by a very slow rate of cooling through the lower-temperature regions. With oil quenching, the most rapid rate of cooling is generally found at a temperature about 80 per cent of the quenching temperature.

Typical curves for cooling rate versus temperature obtained on air cooling steel specimens are shown in Fig. 16.10. Here it is to be noted that the first stage of cooling is not apparent and that the most rapid cooling rate at any location occurs at about 97 per cent of the quenching temperature. However, it is to be noted that the rate of cooling under these conditions is exceedingly small when compared to either oil or water quenching mediums.

Relation of Cooling Power of Quenching Mediums to the Steel

These general characteristics of quenching mediums must be correlated to the behaviour of the steel itself during cooling. No liquid quenching medium is perfect with respect to the hardening of steel, for each one has certain features that prevent it from being universally used. It is, of course, important to suppress the formation of pearlite by cooling rapidly through the high-temperature range, but it is also important that the cooling through the low-temperature range be relatively slow to minimise volume changes and internal stresses. For this reason, we might arbitrarily define an ideal quenching medium as one that would show an initially rapid cooling rate to below 1000°F (540°C) and then a relatively slow rate between 700°F (370°C) and room temperature.

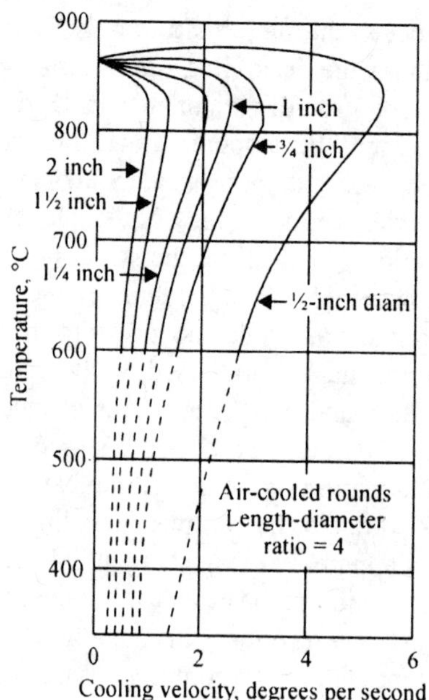

Fig. 16.10. Cooling velocity—temperature curves taken at the centre of 0.89% C steel rounds of different sizes cooled from 875°C (1606°F) in still air at 22°C (72°F). Dashed lines indicate an extrapolation in which the heat effects of the transformation are neglected.

With regard to the suppression of the formation of pearlite, it is obvious that the first two stages encountered in the cooling of steel are of importance. The duration of the vapour stage, maximum rate of cooling during the boiling stage and temperature at which this maximum rate occurs have an important bearing in this respect. The cooling rate is only effective in suppressing the formation of pearlite after the steel has cooled to below the critical temperature and thus, any length of time consumed in cooling to this temperature, no matter at what rate, is of little consequence unless carbide or ferrite precipitates. Once the steel cools to below the critical temperature, if the vapour stage still persists and it usually does, the cooling may be so retarded that pearlite will form. The temperature at which the maximum rate of cooling occurs is likewise of importance. Several liquids might cool a steel from 1500 to 700°F (815 to 370°C) in the same total time, but the liquid that exhibited a maximum cooling rate at the temperature of the maximum rate of formation of pearlite would be the best with respect to the hardening phenomenon.

The third stage of the cooling phenomenon and in some steels what amounts to the end of the second stage, is of vital importance during the formation of martensite. Martensite forms over a temperature range extending from the M_s point to near the room temperature (in some steels it actually extends to sub-atmospheric temperatures). The M_s point varies with composition and the heat treating temperature (degree of solution of the carbide). The formation of martensite results in a volume increase leading to high internal stresses. In addition, large temperature gradients throughout the steel, an invariable result of rapid cooling, cause a non-homogeneous thermal contraction leading to still higher internal stresses.

These stresses appear to be developed to the greatest extent during the latter stages of the quenching operation and may lead to cracking. Slow cooling through this range would, therefore, be desirable. There are wide differences in the cooling rate through this third stage of the quenching operation between water and oil, oil being approximately one-tenth as fast as water. The ideal quenching medium should show a rapid, or at least controlled, cooling rate through the upper transformation zone and a slow cooling rate through the lower transformation zone.

Tests for Rating Liquid Cooling Efficiency

To test and rate the various quenching mediums and to determine how they fit into this ideal picture, there are many parameters that we can use. Of course, the hardening tests are without question the most logical and important tests. However, many other tests and many methods of expressing the results of these tests have been used.

Direct cooling curves, which have been shown previously, have been extensively used. From curves of this type, many investigators have chosen to express graphically that rate of cooling at each different temperature. This method of plotting the rate of cooling versus the temperature is shown in Figs. 16.10 and 16.11. In the latter figure, the cooling curves obtained on 0.787 inch diam silver spheres are shown after quenching in water, salt, a mineral oil, still air and compressed air at a pressure of 14.2 psi. The large differences in the cooling rates at the temperature of the maximum rate are more closely shown by this type of plot and likewise the variation and range of temperatures over which the maximum rate occurs is clearly displayed. In the samples quenched in water and mineral oil, a slight motion of the sample through the bath was maintained during the cooling operation. The vapour stage on cooling in water is not markedly apparent. Motion tends to make the vapour unstable to bath temperatures lower than 70°F.

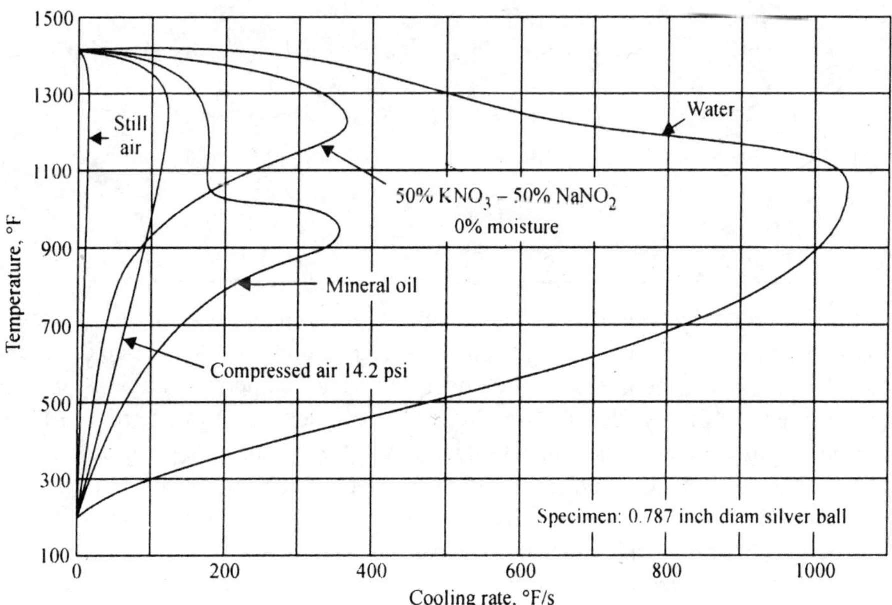

Fig. 16.11. Variation in the rate of cooling with temperature during quenching silver spheres in water, oil, salt and air.

A 'five-second test' can be used to rate the ability of a quenching medium to extract heat from a hot body during the initial stages of the quenching operation and thus rates the ability of the quenching medium to suppress the formation of pearlite. In this test, specimens 1 inch in diameter by 2½ inch long are generally heated to 1500°F (815°C) in a controlled atmosphere furnace and quenched into the liquid medium to be tested, which is contained in a calorimeter. After the first specimen is quenched into the bath and held motionless for exactly 5 seconds, it is removed, the stirring mechanism is started and the maximum rise in bath temperature is noted. A duplicate specimen, heated under identical conditions, is then quenched into the bath after the latter has returned to the original base temperature. Constant agitation is maintained during this second run and the maximum rise in bath temperature is noted when all of the heat has been extracted. The '5-seconds value' is then calculated as the percentage of the total rise in temperature of the bath during the complete cooling obtained in the 5-seconds run. Table 16.4 shows typical results of a series of tests conducted with stainless steel specimens and different quenching mediums. As indicated, quenching oils will vary from 16 to 37 per cent and water will give a value of approximately 53.4 per cent. Latest developments in quenching oils indicate that values above 35 per cent can be obtained. Such rapid quenching oils are of value particularly where 'border-line' hardenability is present.

Table 16.4. 'Five-second test' results on a number of quenching oils.

Quenching medium	*½ heat removal in first 5 s*
Water	53.4
Oil No. 1	36.9
Oil E	30.8
Oil F	31.0
Oil G	20.4
Oil H	25.8
Oil I	19.3
Oil J	27.9
Oil K	15.6
Oil L	18.8
Oil No. 2	17.0
Oil No. 2	19.0

Other methods to quantitatively express the efficiency or cooling power of a quenching medium involve the expression of the rate of cooling at any one temperature, that is, the rate of cooling at 1300 or at 1000°F (705 or 540°C); the rate of cooling over a range of temperatures, that is, the average rate over the range from 1300 to 1000°F (705 to 540°C); or the time to cool to one half of the quenching temperature. The purpose of each of these means of indicating the rate of cooling is to attempt a correlation with the rate of transformation of austenite to pearlite. Since pearlite forms over a range of temperatures, the second and third methods are to be preferred to the first, but as yet only by use of the third method has any direct correlation been achieved with the hardenability.

Two other methods of expressing the efficiency of the quenching medium involve:

1. The expression of the heat transfer coefficient or H value. These values refer only to the time to cool during the upper part of the temperature range and therefore refer to the hardenability of the steel and the suppression of the pearlite transformation. The advantage of this mode of representation lies in the fact that it presents numerical data independent of the conditions of experiment.
2. *Rose* has calculated the heat transfer value from cooling rate curves for many different commercial quenching mediums. These are likewise independent of conditions of experiment and therefore do not entail the drawback that the results cannot be correlated with those obtained by other

investigators because the size, shape and thermal properties of the specimen are included in the cooling rates measured. *Rose* has further presented different values of the heat transfer coefficient for the three stages of cooling.

Trembour and *Scott* have proposed that an accurate evaluation of the ability of quenching liquids to prevent soft spots on shallow-hardening steels can be obtained by quenching relatively large rounds of SAE 1045 in motionless baths maintained at different temperatures. After etching the quenched pieces, the proportion of the surface area that has not hardened is found to increase rapidly with bath temperature. The baths can therefore be rated by stating the temperature at which a certain proportion of soft surface is found.

All of these tests and methods of rating the efficiency actually relate to the cooling power of the liquid. Other tests can be applied to specific quenching mediums, the results of which may prove valuable for standardisation and specific purposes and from which the efficiency in certain phases of the cooling process can be inferred. These tests are largely physical property tests and are to be discussed when the separate liquids are considered.

Water-base Quenching Mediums

Water-base quenching mediums are used for quenching of shallow-hardening steels. The general characteristics have already been given. These include a vapour stage in which a relatively slow rate of cooling occurs, a very violent boiling stage in which an exceedingly rapid cooling rate occurs and finally a liquid cooling stage in which the rate of cooling is relatively high. The maximum cooling rate that occurs in quenching in water is close to the ideal and few other liquids can even approach it. The ideal quench for shallow-hardening steels would be one that could produce the desired surface cooling rate without the presence of agitation. Agitation is extremely useful, but because of the mechanism involved, it is limited to simple short objects. If symmetrical flow of the water relative to the work is not obtained, any agitation or water spray, whether submerged or not, is conductive to warpage. Tap water and even distilled water, is not satisfactory as quenching medium for shallow-hardening steels for two reasons:

1. The vapour stage is too long.
2. The rate of cooling through the liquid cooling stage is too rapid.

In water quenching, the vapour blanket is not too stable below 70°F. Under conditions of moderate circulation, however, if the water temperature rises above 70°F and a sufficient flow is not maintained to destroy the vapour film, the duration of the vapour stage can be appreciable. *Speith* and *Lange* have investigated the cooling power of tap water, distilled water and water to which small amounts of salt have been added and a summary of their results is shown in Fig. 16.12. It will be noted that with distilled water the vapour stage is so persistent that the maximum rate of cooling does not occur until a temperature will below that of the maximum rate of pearlite formation is reached. The freedom of distilled water from salts appears responsible for this phenomenon.

Tap water proved to be better despite the possible presence of undissolved gases, which have been considered of great importance by other investigators. The soft spot test of *Trembour* and *Scott* also failed to indicate that normal contents of atmospheric gases were of consequence. (Highly aerated water is detrimental, however and the use of air as a means of circulation of water is to be avoided.) Boiling normal tap water for 6 hrs. reduced the soft spot area only slightly (from 2.3 to 2.0 per cent). The addition of 0.2 per cent sodium chloride to both tap water and distilled water was found by *Speith* and *Lange* to decrease the duration of the vapour stage and to equalise the maximum cooling rate obtained with the two types of base solutions.

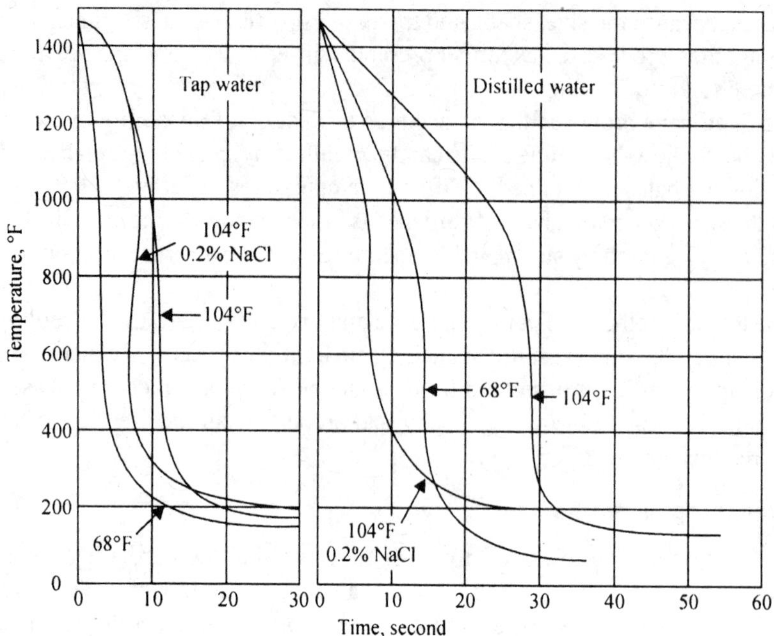

Fig. 16.12. Cooling curves for a 0.787 inch diam silver ball quenched in tap water and distilled water at 68 and 104°F and in tap water and distilled water with additions of 0.2 per cent NaCl at 104°F.

Fig.16.13 shows the effect of bath temperature on the duration of the vapour stage. It will be noted that as the bath temperature increases above 70°F, a rapid increase in the duration of the vapour stage occurs and thus all water-base quenching baths should be kept below 70°F if soft spots are to be avoided. Also shown in this figure is the effect of small amounts of soap on the duration of the vapour stage and the effect of small amounts of sodium chloride. Soap evidently stabilises the vapour film perhaps by a surface tension effect and thereby materially decreases the rate of cooling. Sodium chloride and most inorganic salts act in the reverse direction.

The second objection to water-base quenching mediums is found in the third or liquid cooling stage. Because of the high heat conductivity and high specific heat of water, cooling is much too fast for safety. Fortunately, most steels that are water quenched do not harden throughout the cross-section and only a small proportion of the total structure is subject to the drastic volume changes that occur when martensite forms. The resistance to cracking in this instance is considerably greater than it would be if the steel hardened throughout the cross-section. Invariably, more trouble is found in water quenching deep-hardening steels because of this phenomenon. When deep-hardening steels are hardened, it is necessary to resort to a slow quench, especially during the third stage of the cooling operation. If this is done, the distortion is minimised and the danger of cracking is lessened. The cooling power during the third stage can be somewhat reduced by increasing the water bath temperature, but this is an impractical solution; Fig. 16.13 shows that the duration of the vapour stage is greatly increased and that hardening is not achieved. It has already been pointed out that small additions of inorganic salts to water solutions materially decrease the duration of the vapour stage. *Zavarine* has shown that crystals of salt are precipitated momentarily and explosively thrown away from the steel surface during the initial stages of brine quenching, continuously causing the film of steam to be disrupted and leading to violent agitation.

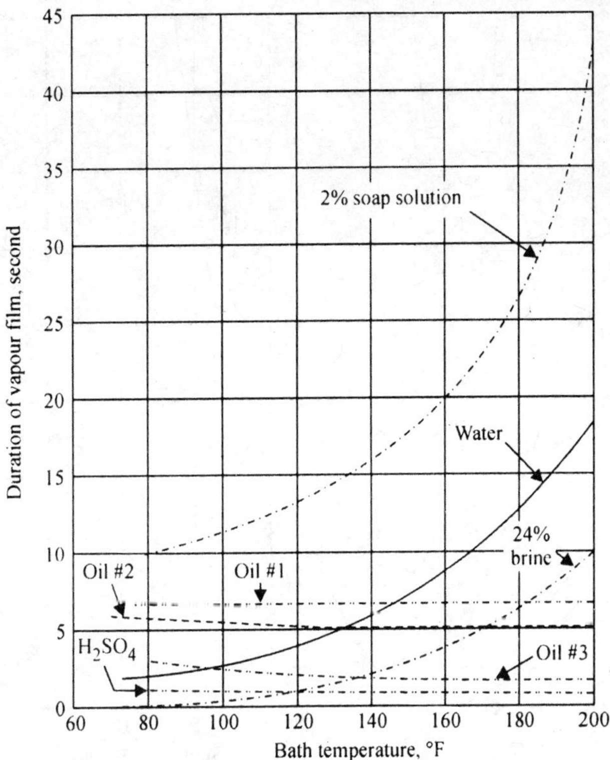

Fig.16.13. Effect of bath temperature on the duration of vapour stage when quenching into the indicated medium.

Brine solutions are also known to remove the scale on specimens much more rapidly than water. Water can dissolve up to 24 per cent of sodium chloride by weight and this saturated solution reduces the duration of the vapour stage to a minimum. However, salt additions appear to reduce somewhat the rate of cooling during the second stage. Fortunately, this effect is small at low salt concentrations and it is more than offset by the contraction of the first stage. A compromise is usually reached in selecting the composition of a brine quenching bath and the usual composition will vary from 5 to 12 per cent sodium chloride. Fig. 16.14 shows the effect of NaCl additions on the percentage of soft spots obtained at different bath temperatures. The optimum value seems to be 9 per cent NaCl. Table16.4 shows the density relationships obtained in making up solutions of different concentrations of sodium chloride and water. Brine solutions are corrosive to equipment but are not dangerous to handle. Solutions of other inorganic salts—especially sodium hydroxide, which has proved equally as good as brine (Fig. 16.14)—are used to achieve the same purpose, but for shallow hardening steels, brine quenching is more widely accepted as a standard quench than any other. Soft spot tests show that 3 per cent sodium hydroxide is optimum (Fig.16.14).

Oil Quenching Mediums

As stated before, the general characteristics of oil quenching that distinguish it from water quenching involve a relatively stable vapour phase and a very slow rate of cooling during the final stages of the quenching operation.

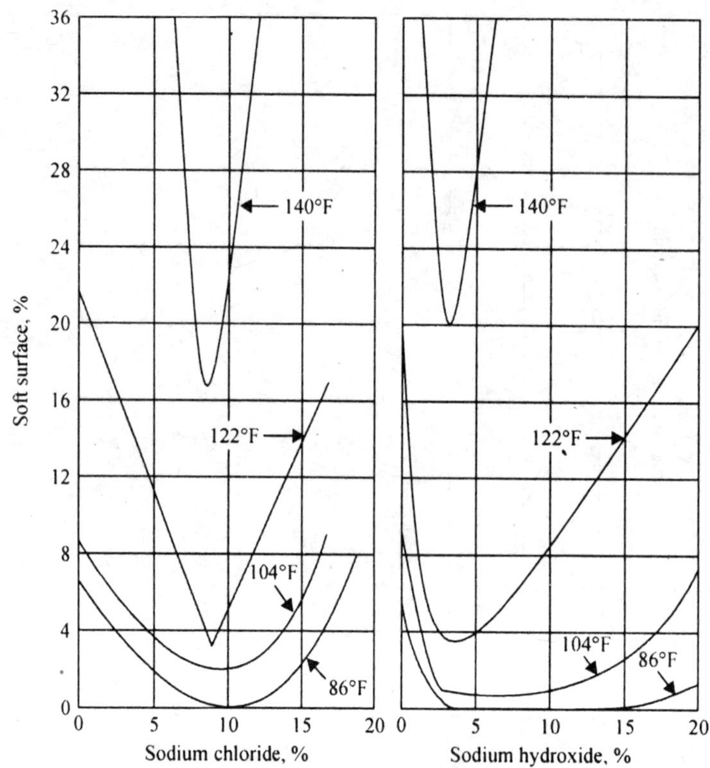

Fig. 16.14. Effect of concentration of aqueous quenching solutions of sodium chloride and sodium hydroxide on the formation of soft spots at a series of quenching bath temperatures.

Table 16.4. Brine: Relation of density to concentration.

% NaCl	*Lb salt/gal of water*	*Density at 68°F*	*Salinometer reading*	*Degrees Baume*
0	0	1.0	0	1.0
5	0.43	1.039	18.9	4.8
10	0.89	1.071	37.8	9.6
15	1.39	1.108	56.6	14.2
20	1.92	1.148	75.6	18.7
25	2.44	1.189	94.5	23.1

When compared with water, mineral oils show a stable vapour blanket that cannot be destroyed even by the most vigorous agitation. Consequently, the influence of agitation of the bath is relatively small. The vapour film is formed not only by oil vapours but also to a large extent by gases that are products of an oil cracking process occurring at high temperatures. It has been found that the duration of the vapour stage in oils is longer, the lower the boiling point of the oil and it has been suggested that a minimum be placed on the boiling point. However, the boiling point of oil is an indefinite quantity and must be defined arbitrarily. Actually, oil is a mixture of many hydrocarbons, each of which may boil at its own peculiar temperature. If the average boiling point of this mixture is high, the duration of the vapour stage will be slightly lessened.

The stage of violent boiling in oil produces a relatively fast cooling rate, but it is not the same as obtained with water quenching. Cooling curves of the usual type do not clearly show the difference in the cooling rate obtained, but when rate versus temperature curves are plotted, as in Fig. 16.11, it is found that the maximum rate of cooling with water is two to three times that obtained with most oil during this stage. The final stage of the cooling process in oils results in a very slow rate of cooling. During this stage, the specific heat, conductivity and temperature of the oil are controlling factors. The specific heat of most oils is one half that of water and the conductivity is approximately one fourth that of water. The ratio between the rate of cooling of oil and the rate of cooling of water through this low-temperature range is approximately 1 to10. In this stage, circulation has a pronounced effect as it does in water solutions and the slope of the temperature-time cooling curve can be materially altered by increasing the rate of circulation. The effect of viscosity of the oil is said to be almost negligible if sufficient stirring is used.

The effect of the properties of different quenching oils on the characteristics of the cooling phenomenon are not completely known. However, the following facts can be stated with safety. The boiling points of most oils are so high that the duration of the vapour stage is little affected by a moderate change in bath temperature. This is contrary to the situation with water quenching in which the low boiling temperature and the close proximity of bath temperature to boiling temperature make it important to maintain a low bath temperature.

Steels actually cool faster during the second stage if the oil is warm, probably because of the lowered viscosity permitting more rapid movement of the vapour away from the steel surface. The main advantage, however, in using hot oil in quenching is the great reduction in the rate of cooling during the third stage, which results in lowered residual stresses. It is usually recommended to keep oil between 90 and 140°F, with the best temperature range at 120 to 130°F if such a close control is at all possible.

The following considerations may be presented relating to the quenching power of mineral oils of different densities. Light low-boiling-point oils show a pronounced formation of vapour because of their relatively high contents of volatile components and their cooling power is low. In general, cooling power increases with increasing density and increasing boiling temperature.

However, it should not be overlooked that at the same time the maximum rates of cooling move toward higher temperatures and in certain steels it is likely that a better hardness penetration will be obtained with a lighter oil. Thus, the variation of tendency of the transformation to pearlite as a function of temperature will play a decisive role.

As the bath temperature is increased to a certain limiting temperature, a slight increase in cooling power is noted, mainly associated with decreased viscosity. Beyond the limiting temperature, however, the rates show a pronounced decrease in all stages of the quenching process and vapour is copiously formed.

Tests commonly used to govern the selection of quenching oil are the specific gravity, viscosity, flash point, fire point, pour test, carbon residue, steam emulsion and saponification number. The specific gravity is important in that it is the prime quantity affecting the specific heat, thermal conductivity and heat of vapourisation, each of which plays an important part in determining the rate of cooling through the last two stages of the cooling process. A low viscosity is important in oils to ensure free drainage of the oil from the specimens and to reduce the loss from 'drag out'. A low viscosity also is maintained to increase the cooling rate during the second stage.

It should perhaps be mentioned that, although oils may differ widely in viscosity at atmospheric temperatures, at the usual recommended quenching bath temperature the differences are usually insignificant. The curves of viscosity versus temperature for almost all commercial quenching oils approach each other rapidly as the temperature is raised above 100°F. The flash point and the fire point are generally desired high in order to reduce the fire hazard resulting from the quenching of hot bodies in the oil. However, minor changes in the flash point and fire point have little effect on this quantity as these points are generally well above the bath operating temperature in any actual case.

The pour test ensures ample viscosity so that the oil can be handled when it is cold. A low carbon residue is desired because it is said to suggest stability of the oil; however, no satisfactory correlation of any laboratory test with the stability of the quenching oil has been made. The steam emulsion test is made on oils to indicate the presence of any amounts of entrained water and the saponification number is an indication of the type of oil.

The addition of fatty oils, such as rapeseed oil, to mineral oils slightly increases the cooling power, largely by diminishing the stability of the vapour stage. Pure rape oils have the well-known characteristic of a greater cooling power than almost any mineral oil, but they show a high degree of sensitivity to quenching cracks because of high cooling rates at low temperatures. Fig. 16.15 shows the comparison of the rate versus temperature cooling curves of a pure mineral oil and a mineral oil to which a small percentage of fatty oil has been added.

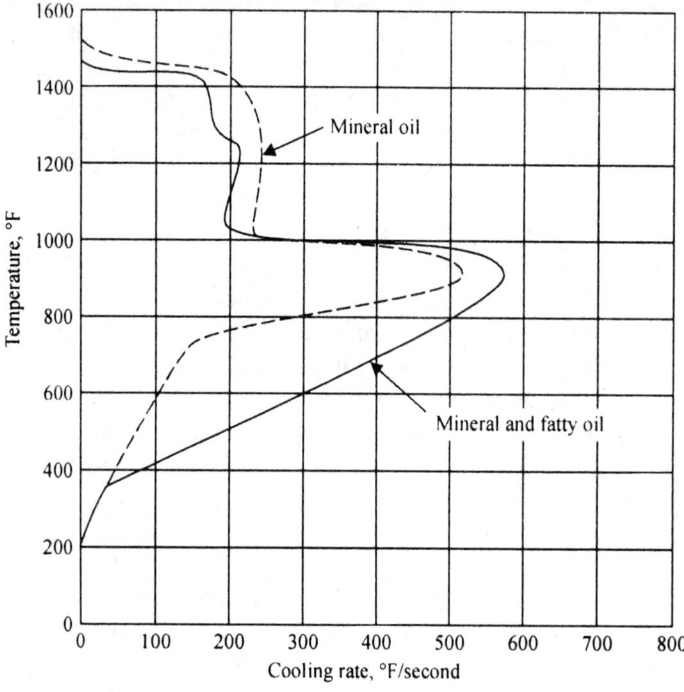

Fig. 6.15. Variation of the cooling power of a mineral oil caused by the addition of small amounts of a fatty oil.

Such mixtures made by compounding different types of oil have long been on the market. It is possible by their use to obtain rates of cooling that are somewhat intermediate between that of mineral oil and that of water. Figure 16.16 into shows that a somewhat greater cooling rate in the low-temperature zone is encountered. It should also be mentioned that straight mineral oils must be used when specimens covered with salt are introduced into the bath, that is, in quenching from fused cyanide or sodium carbonate baths. The alkaline quality of the salt causes saponification of the oil and leads to its rapid deterioration.

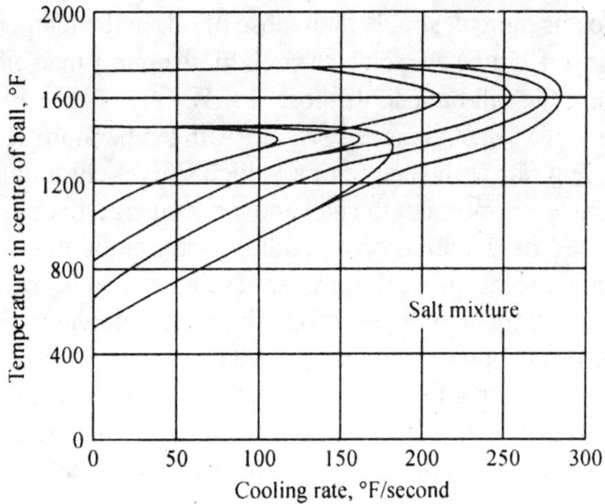

Fig. 16.16. Cooling rates at the centre of a ½-inch Si-Cr-Al steel ball on quenching from 1742°F and 1472°F into a salt bath held at various temperatures.

Intermediate Quenches

Many water-base equivalents of quenching oils have been attempted, but these have not been successful. Although it is possible by making such solutions in different composition ranges to obtain a desired cooling rate during a certain stage of the cooling process, it has not yet been possible to obtain desirable cooling characteristics throughout the whole temperature range. Mixtures of glycerine and water can give the same rate of cooling during the boiling stage as oil but still show the rapid rate of cooling during the liquid cooling stage that is obtained with water. For oil-hardening steels, the glycerine content must be fairly high (60 to 90 per cent), whereas for deep water-hardening steels, the glycerine content is less (10 to 40 per cent). The same comment applies to solutions of ethylene glycol and sodium silicate in water. Another disadvantage of these water-base solutions that attempt to produce a cooling power intermediate between that of water and oil lies in the fact that there is a pronounced effect of the temperature of the bath on the cooling efficiency. *Rose* has made an exhaustive investigation of many such solutions and has studied water glass solutions of many different compositions and at many different temperatures. The summary of his findings is as follows:

'The foregoing experiments, made to explore quenching mediums having a quenching power intermediate between that of water and oil, show clearly that with water-base solutions or emulsions, the increase in the rate of cooling within the range of low temperatures exerts invariably a detrimental effect'.

Molten Salt Quenching Mediums

Unlike most other liquid quenching mediums, molten salts do not vapourise on contact with hot steel and thus, there is no vapour blanket or vapour transport stage. Heat transfer is principally by conduction and convection and cooling rates equivalent to those produced by an oil quench can be readily obtained (Fig. 16.11). It may be noted that although the 50:50 mixture of sodium nitrite and potassium nitrate (Fig. 16.11) provides a maximum cooling rate equal to that of oil, this maximum cooling rate occurs at a somewhat higher temperature (1200°F instead of 925°F). In addition, below about 1050°F, the cooling rate in molten salt is considerably less than that in mineral oil. Thus, the cooling characteristics of the salt mixture are in these two respects more desirable than those of oil. In the temperature range of maximum rate of pearlite formation (above 1050°F), the salt bath cools more rapidly than oil, while in the temperature range of martensite formation, the salt bath cools more slowly.

The maximum cooling rate of a salt bath increases as the bath temperature decreases. This is illustrated in Fig. 16.16, where the cooling rate at the centre of a ½ inch silicon-chromium-aluminium steel ball is shown to depend on the initial temperature of the ball and the temperature of the salt. The cooling power of a salt bath can be increased by agitation or by adding small amounts of water to the molten salt (Fig. 16.17). *Harvey* has found that the quenching power of salt bath can be increased by several hundred per cent by agitation. With a 50:50 mixture of potassium nitrate and sodium nitrite containing 2.6 per cent moisture, maximum cooling rates approaching that of water can be achieved. Water may be added to a molten nitrate-nitrite bath by directing a fine stream of water into the vortex around a pump or impeller shaft. Turbulence carries the water into solution without spattering or hazard to operating personnel. It is necessary to make periodic additions of water since it evaporates quite rapidly (an addition of 0.5 per cent water will evaporate in less than 24 hours).

A convenient method for estimating the water content of the salt bath consists in removing a small sample (200 to 500 cm^3) in a Pyrex beaker and taking temperature readings every 30 seconds as the salt cools. The freezing point may be noted as a thermal arrest and is related to the amount of water present. The freezing point decreases with increasing water content.

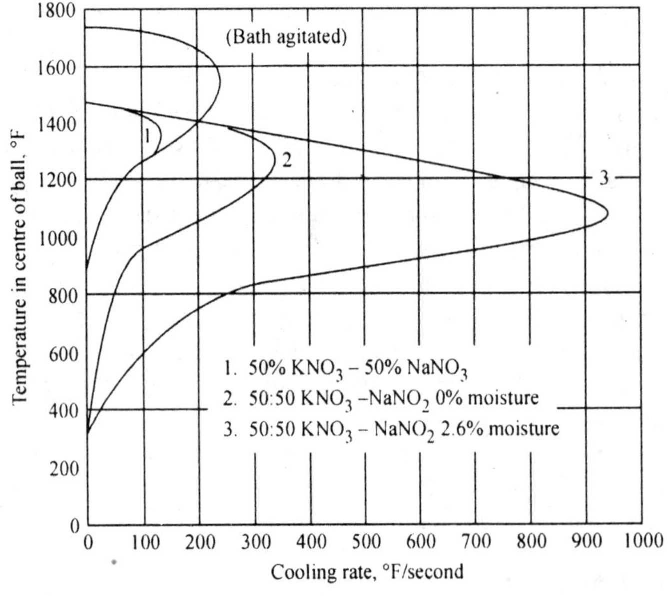

Fig.16.17. Cooling rates on quenching in various salt baths.

Salt bath quenching is applicable to a wide variety of both oil- and air-hardening steels. High speed steels austenitised in salt baths are frequently quenched in mixtures of sodium and potassium chloride operating in the range of 1100 to 1200°F. These salts are more easily washed from the piece than the less-soluble barium chloride used in the high-heat salt bath. Air-hardening steels such as the high-carbon high-chromium types may be quenched in salt to avoid the scaling that occurs on air cooling. The low-temperature nitrate and nitrite salts are often employed for marquenching and austempering.

Air Cooling

Cooling rate versus temperature curves for air cooling are shown in Fig. 16.10. These apply only to cooling in still air. By varying the velocity of the air striking the body to be cooled, large variations in the cooling rate and in the hardness of mildly air-hardening steels, can be obtained (Fig.16.11). With an air blast, it is possible to obtain cooling rates that equal those of the slow oil quenches and thus, the gap that exists between oil and water quenches is not present between oil and air quenches. So far, no quantitative data are available to indicate the change of cooling rate with change in air velocity; thus, it is necessary to use such vague terms as strong air blast, light air blast and still air in describing the intensity of the quench. *Grossmann* has indicated that the H value for still air cooling is approximately 0.02.

Sub-zero Cooling

The addition of alloying elements to steel depresses the M_S temperature and results in large quantities of retained austenite on quenching to room temperature. Sub-zero cooling is one technique whereby the quantity of retained austenite can be reduced.

Mediums for sub-zero quenching include freezer chests or low-temperature liquids. The latter may be a mixture of substances such as acetone and dry ice (–100°F), or liquefied gases such as nitrogen (–321°F), oxygen (–297°F), helium (–453°F), or hydrogen (–423°F). The first three liquefied gases are preferable because of the explosion hazard involved in the use of hydrogen. Sub-zero cooling normally follows the quench to room temperature in the more conventional quenching mediums (air, oil, water). Excessive delay at room temperature or tempering before sub-zero cooling tends to stabilise the austenite and lessen or completely suppress transformation during further cooling.

Effect of Surface Condition

Many of the tests that have been described in developing the science of quenching have been made on stainless steel or silver to eliminate any spurious effects that might conceivably be introduced by scale on the surface. In many instances, the specimens were polished so that as smooth a surface as possible would be obtained. There are many indications that roughness of the surface and the presence of scale on the surface materially affect the rate of cooling of the steel. Roughness of the surface in some instances appears to be an advantage in that it favours wetting of the surface by quenching medium and thus tends to break down the vapour blanket that forms. Scale, however, definitely prevents efficient heat transfer. This is illustrated by the following example.

A disk of plain carbon steel when quenched in oil after treating in a semi-muffle furnace produced a hardness of approximately Rockwell C 45. When treated in a controlled atmosphere furnace so that no appreciable amounts of scale formed and quenched in the same oil maintained at the same temperature, it hardened to Rockwell C 63 to 64. In this instance, the steel was very close to the critical size for hardening and the slight amount of scale that formed in the semi-muffle furnace was sufficient to retard the cooling rate and cause the formation of large quantities of pearlite at the surface of the sample.

It might also be noted that even in stainless steel, although no appreciable amounts of scale will form at a temperature of 1500°F (815°C), the scale that does form causes a change in the percentage of heat abstracted during the first 5 seconds of the quench as revealed by the 5 seconds test results. Freshly polished samples of stainless steel will give a 5 seconds result in a commercial quenching oil of approximately 14. Reheating the same specimens and repeating the test continuously will result in a gradual increase of this value to 17. At this point, the value holds constant for a number of tests and then a sudden rise is noted as the test is continued. This is, of course, just opposite to the effect of scale on plain carbon steels. It must be remembered that scale on plain carbon steels is relatively thick and that, with stainless steel specimens, the effect of scale in promoting wetting of the surface probably was of greater influence than in retarding the heat transfer from the specimen to the liquid.

Chapter 17

Specific Classes of Tool Steels

INTRODUCTION

Heat treating procedures vary significantly among classes of tool steels and with intended application. The preferred heat treating and hardening procedures, as well as mechanical properties and applications are discussed in this chapter with respect to water and air-hardening tool steels, oil-hardening and high-carbon, high-chromium cold work steels, low-alloy and special purpose high-speed tool steels, and shock-resisting tool steels. Specific examples of heat treating procedures for specific applications for hot work tools are also given.

WATER-HARDENING TOOL STEELS

Water-hardening tool steels containing 0.90 to 1.00 per cent carbon are the most widely used. Carbon content affects heat treating temperatures as indicated in Table 17.1, which outlines recommended heat treating practices for these steels.

As a class, water-hardening tool steels are relatively low in hardenability, although they are arbitrarily classified and available as shallow-hardening, medium-hardening and deep-hardening types. Their low hardenability is frequently an advantage, because it allows tough core properties in combination with high surface hardness. Low cost and adaptability to simple heat treatment are additional advantages offered by these steels.

Water-hardening tool steels are so termed because they are most commonly quenched in an aqueous medium. There are exceptions, however; for example, thin sections may be satisfactorily quenched in oil with less distortion and danger of cracking than if quenched in water or brine.

Example 17.1. In one plant, it was desirable to harden small-diameter punches in oil to reduce breakage and consequent downtime of the presses. A study was made to determine the maximum diameters of water-hardening tool steels that could be fully hardened to a minimum of 60 HRC by oil quenching. Results of the study, indicating the relationship between austenitising temperature, type of steel and punch diameter are shown in Fig. 17.1.

Experimentation proved that a greater degree of uniformity was obtained if the punches were normalised prior to hardening. Normalising temperatures applied were: 870°C (1600°F) for punches up to 6.5 mm (¼ inch) in diameter; 900°C (1650°F) for those over 6.5 mm (¼ inch) in diameter. As indicated in Fig.17.1, austenitising temperature varied from 790 to 900°C (1450 to 1650°F), depending on punch diameter. The punches were austenitised by being heated vertically in a neutral salt bath. They were also quenched vertically, in a compounded oil containing additives. The quenching oil was maintained at 50 to 60°C (120 to 140°F) and circulated up and around workpieces at 190 l/minute (50 gal/minute).

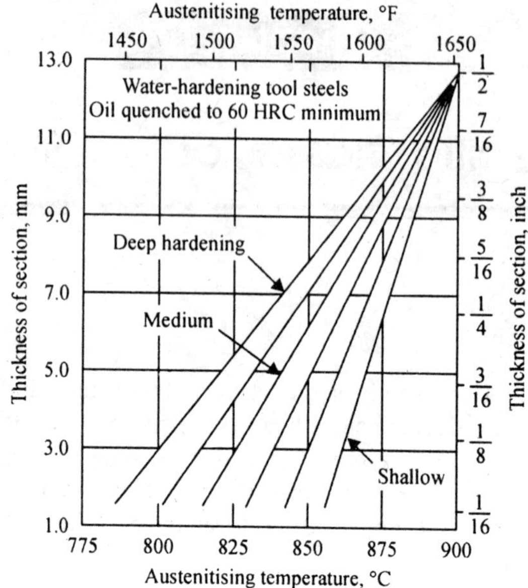

Fig. 17.1. Maximum section thicknesses of three classes of water-hardening tool steel that will develop minimum hardness of 60 HRC when oil quenched from various austenitising temperatures (Example 1).

Table 17.1. Recommended heat treating practice for water-hardening tool steels.

Temperature °C	°F	*Carbon content*	*Hardness after treatment*	*Procedure*
Normalising				
815	1500	0.60 to 0.75	–	Heat through uniformly; hold for 15 minutes (light sections) to 1 hour (heavy sections), then air cool
790	1450	0.75 to 0.90		
870	1600	0.90 to 1.10		
870 to 925	1600 to 1700	1.10 to 1.40		
Annealing				
740 to 760	1360 to 1400	0.60 to 0.90	156 to 201 hp.	Heat through uniformly; hold for 1 to 4 hrs.(a); furnace cool to 510°C (950°F) at 20°C/hr. (40°F/hr.), then air cool
760 to 790	1400 to 1450	0.90 to 1.40		
Hardening(b)				
790 to 845	1450 to 1550	0.60 to 0.80	65 to 68 HRC	Hold at austenitising temperature for 10 to 30 minutes; quench in water or brine (very small pieces may be oil quenched)
775 to 845	1425 to 1550	0.85 to 1.05		
760 to 830	1400 to 1525	1.10 to 1.40		

(a) Holding times vary from about 1 hour, for light sections and small furnace charges, to about 4 hrs., for heavy sections and large furnace charges. (b) For large tools and tools with intricate sections, pre-heating at 565 to 650°C (1050 to 1200°F) is recommended.

Normalising

Except in special instances where experience has proved it beneficial (as in the preceding example), normalising is not recommended for water-hardening tool steels as received from the supplier. Normalising is recommended for these steels after forging or before re-heat treatment, for refining the grain and producing a more uniform structure. Recommended normalising temperatures are given in Table 17.1; as indicated, optimum temperature varies with carbon content.

Decarburisation during air cooling will be minimised if parts are heated in a protective atmosphere or a neutral salt bath. Parts heated in salt are additionally protected during the cooling period by the film of salt that adheres to their surfaces when they are removed from the salt bath. After parts have cooled, the film of salt can be easily removed (except from recesses such as tapped holes) by a water rinse.

Annealing

Tool steels of the W types are received from the supplier in the annealed condition. Thus, annealing by the user is usually unnecessary. Annealing is applied to forged or cold worked carbon tool steel to soften it for easier machining, relieve residual stress, and produce a structure suitable for hardening.

Annealing may be done in an atmosphere furnace (provided the furnace is of a type that can be cooled slowly to below 540°C or 1000°F, in a vacuum, or in an ordinary air furnace after the piece has been protected against surface decarburisation by being packed in a suitable container with an inert material.

Protection against decarburisation (but not against oxidation) may be obtained also by copper plating the surface or by applying a surface-protecting paint. (Not all of these paints are equally effective, and some are difficult to remove; the prospective user should investigate such a paint by trying it under his conditions of operation and then inspecting the treated part for decarburisation.) The workpiece should be heated to the annealing temperature (Table 17.1) and held at temperature for from 1 hour, for thin sections, to about 4 hours for heavy sections. When the steel has been placed in a pack to prevent surface reactions, a general rule of thumb is to allow the assembly to soak at temperature for 1 hour per inch of pack cross-section. Work should then be cooled in the furnace at a rate not exceeding 22 °C/hr. (40 °F/hr.), to 510°C (950°F). Below this temperature, cooling rate is not critical. Hardness after annealing should be in the range of 156 to 201 HB.

Stress Relieving

Stress relieving prior to hardening is sometimes employed to minimise distortion and cracking. The procedure consists of heating the work to 650 to 720 °C (1200 to 1325 °F) and cooling in air. Usually, stress relieving of water-hardening tool steel is limited to complex or severely cold worked parts.

Example 17.2. A piston of W2 steel for a pneumatic clay digger varied in section thickness by as much as 6 to 1. Cracking occurred in the cupped end section when the pistons were hardened by conventional practice. Stress relieving or pre-heating at 675°C (1250°F) prior to hardening eliminated this difficulty. In most instances, stress relieving after hardening and grinding is not employed. Periodic stress relieving of tools that have been in service will reduce the stresses imposed by such service, and is believed to be beneficial in extending service life. Temperatures used for this purpose should never exceed those used for tempering the steel after hardening.

Austenitising

Austenitising temperatures for water-hardening tool steels normally vary from 760 to 845°C (1400 to 1550°F), as indicated in Table 17.1. Higher temperatures are sometimes used for special purposes

(Fig. 17.1). Hardenability increases as austenitising temperature increases. The optimum time at austenitising temperature is from 10 to 30 minutes. Pre-heating is unusual except for very large tools or those with intricate cross-sections (such as the W2 piston cited in Example 2).

If surfaces are to be protected against scaling or decarburisation, an atmosphere furnace, lead bath or salt bath is required. It is particularly important to protect shallow-hardening steels against scaling and decarburisation. Severe scaling can interfere with heat transfer during quenching and slow the required high rate of cooling. Decarburisation will produce a soft surface on any tool steel, but in a deep-hardening steel it can be ground off until the underlying hard high-carbon area is reached. Grinding a shallow-hardening steel will frequently expose the soft core.

Atmospheres

Excellent results are obtained by austenitising water-hardening tool steels in a slightly oxidising atmosphere, as the data in Fig. 17.2, obtained in tests on type W2, indicate. Oxidising atmospheres are inexpensive, and are usually produced by controlled direct-firing burners. The light scale that is produced is removed by the vigorous water or brine quench.

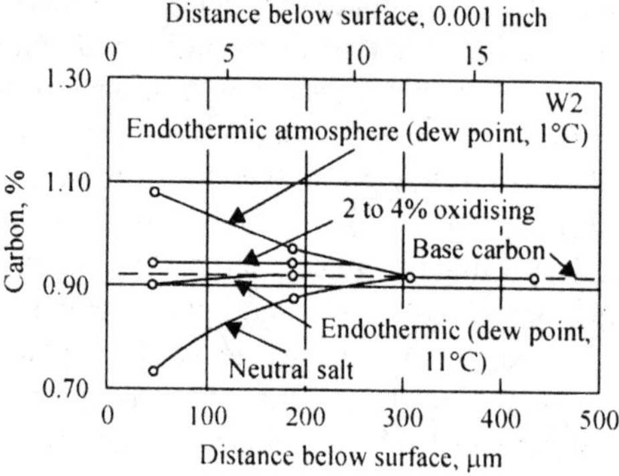

Specimens were heated at 790°C (1450°F) for 1 hr. quenched in brine, annealed in lead at 705° (1300°F), and machined in 0.13 mm (0.005 inch) cuts for analysis.

Fig. 17.2. Effect of furnace atmosphere on the surface carbon content of W2 tool steel.

Endothermic atmospheres also are used, but close control is necessary to match the carbon potential of the atmosphere to the carbon content at the surface (Fig. 17.2). Also, endothermic installations are more expensive than the controlled-burner technique mentioned above.

Salt Baths

Salts baths are widely used and frequently preferred over other heating mediums for hardening type W tool steels. High-temperature salt baths will cause severe decarburisation (Fig. 17.2) if not closely controlled. Other disadvantages of salt baths are that: (a) salt dragout necessitates frequent replenishment of the bath, particularly when many small parts are being treated; and (b) salt is sometimes difficult to remove from parts having complex shapes or tapped holes.

Lead Baths

Lead baths also are used for austenitising water-hardening steels and have advantages and limitations paralleling those of the salt bath, specifically item (b) above.

Quenching

To produce maximum depth of hardness in water-hardening tool steels, it is essential that they be quenched as rapidly as possible. In most instances, water or a brine solution consisting of 10 per cent NaCl (by weight) in water is used. Occasionally, for an even faster quench, an iced brine solution is employed. Cooling rate is a function of size of workpiece as well as of quenching medium; for this reason, small pieces can be quenched in oil (Fig. 17.1). This is particularly useful when heat treating thin-section tools in an atmosphere furnace containing an integral oil quench.

Tempering

Water-hardening tool steels should be tempered immediately after hardening, preferably before they reach room temperature; about 50°C (120°F) is optimum. Salt baths, oil baths and air furnaces are all satisfactory for tempering. However, working temperatures for both oil and salt are limited; the minimum for salt is about 165°C (325°F), and the maximum for oil is usually about 205°C (400°F).

All parts made of these steels should be tempered at temperatures not lower than 175°C (350°F). One hour at temperature is usually adequate; additional soaking time will further lower hardness. Fig. 17.3 shows the effect of tempering temperature on hardness of water-hardening tool steels austenitised at 790, 815 and 845°C (1450, 1500 and 1550°F) and quenched in brine.

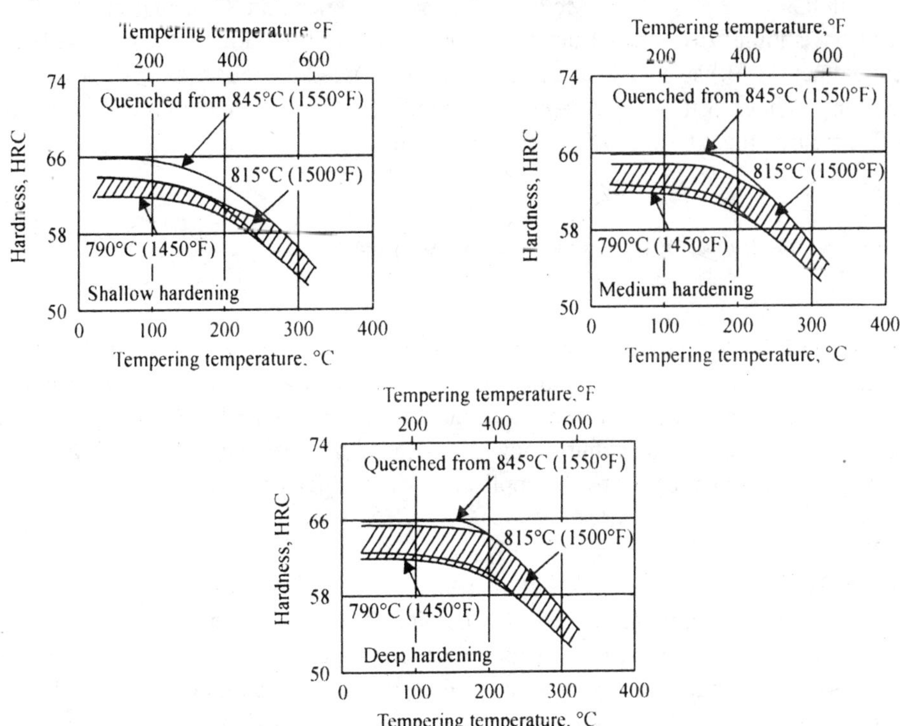

Specimens held for 1 hour at the tempering temperature in a recirculating-air furnace. Cooled in air to room temperature. Data represent 20 25 mm (1 inch) diam specimens for each steel. Compositions of steels: shallow hardening, 0.90 to 1.00 C, 0.18 to 0.22 Mn, 0.20 to 0.22 Si, 0.18 to 0.22 V; medium hardening, 0.90 to 1.00C, 0.25 Mn, 0.25 Si, no alloying elements; deep hardening, 0.90 to 1.00 C, 0.30 to 0.35 Mn, 0.20 to 0.25 Si, 0.23 to 0.27 Cr.

Fig. 17.3. Effect of tempering temperature on surface hardness of water-hardening tool steels austenitised at three different temperatures and quenched in brine.

Tools should be placed in a warm 95 to 120°C (200 to 250°F) furnace immediately after quenching and then be brought to the tempering temperature with the furnace. This is particularly necessary when quenched tools are being accumulated for tempering in a single batch. Allowing quenched tools to stand at room temperature or placing them in a cold furnace will lead to cracking. Except for large pieces, the work will heat at about the same rate as the furnace. The low temperatures used in tempering eliminate the need for atmosphere control. A double temper is frequently used to temper any martensite that may have formed from retained austenite during quenching, and in the first tempering cycle.

The resistance to fracture by impact initially increases with tempering temperature to about 180°C (360°F) but falls off rapidly to a minimum at about 260°C (500°F). This is known as '500°F embrittlement'. For tools subjected to impact loading, tempering temperature should be selected to give an optimum combination of hardness and impact resistance.

SHOCK-RESISTING TOOL STEELS

Recommended heat treating practices for shock-resisting tool steels are outlined in Table 17.2. These steels may be obtained with several variations in composition, for specific applications (for example, S1 steel is available with 0.30 or 0.50 per cent Mo or with up to 10.90 per cent Si). The user of these non-standard compositions should: (a) obtain from the manufacturer information as to the modifications required in heat treatment or (b) select a heat treatment recommended for the shock-resisting tool steel of standard composition that most closely resembles the modified steel. The latter procedure should be followed only after the treatment has been tried on test samples.

Normalising

Normalising is not recommended for the shock-resisting tool steels.

Annealing

The high-silicon types (S2, S4, S5 and S6) are susceptible to graphitisation and decarburisation. Annealing these types at temperatures higher than those indicated in Table 17.2 may produce a softer structure, but it will also increase the danger of graphitisation. The silicon types should not be soaked at temperature. Surfaces should be protected against decarburisation by heating in a protective atmosphere or a vacuum furnace, by the use of pack annealing, or by the application of proprietary paints.

Pack Annealing

Pack annealing consists of surrounding parts with inert material inside a closed container, heating the container to the recommended temperature, and slow cooling. The selection of a packing medium for use with shock-resisting tool steels is difficult; the same general practice has produced different results in different plants. Dry silica sand is usually satisfactory for type S1, and a combination of new and used carburising compound is usually satisfactory for S2, S4 and S5. Burned-off cast iron chips, spent pitch coke, lime and mica are sometimes used, also. Cast iron chips decrease in carbon content and should not be used indefinitely; lime and mica should be used carefully, because they are insulators. Excessive thicknesses of inert material should not be packed around parts, because this complicates handling and lengthens heating time. Wrapping parts tightly in brown paper before surrounding them with packing material helps to keep the surfaces clean.

Table.17.2. Recommended heat treating practice for shock-resisting tool steels.

Steel	Normalising	Annealing: Temperature(a) °C	Annealing: Temperature(a) °F	Annealing: Cooling rate(b) °C/hr.	Annealing: Cooling rate(b) °F/hr.	Annealing: Anealed hardness, HB	Hardening: Temperature, Pre-heat °C	Hardening: Temperature, Pre-heat °F	Hardening: Temperature, Austenitise °C	Hardening: Temperature, Austenitise °F	Holding time, min	Quenching medium	Quenched hardness, HRC
S1	Not rec	790–815	1450–1500	22	40	183–229	–	–	900–955	1650–1750	15–45	O	57–59
S2	Not rec	760–790	1400–1450	22	40	192–217	650(c)	1200(c)	845–900	1550–1650	5–20	B, W	60–62
S4	Not rec	760–790	1400–1450	22	40	192–229	650	1200	870–925	1600–1700	5–20	B, W	61–63
									900–925	1650–1700	5–20	O	57–59
S5	Not rec	775–800	1425–1475	14	25	192–229	760	1400	870–925	1600–1700	5–20	O	58–61
S6	Not rec	800–830	1475–1525	14	25	192–229	760	1400	915–955	1675–1750	10–30	O	56–60
S7	Not rec	815–845	1500–1550	14	25	187–223	650–705	1200–1300	925–955	1700–1750	15–45(d)	A, O	60–61

(a) Lower limit of range should be used for small sections, upper limit for large sections. Holding time varies from about 1 hour for light sections and small furnace charges, to about 4 hours, for heavy sections and large charges; for pack annealing, hold for 1 hour per inch of pack cross-section. (b) Maximum. rate is not critical after work is cooled to about 510°C (950°F). (c) Preferable for large tools, to minimise decarburisation. (d) For open furnace heat treatment. For pack hardening, hold for ½ hour per inch of pack cross-section.

Proprietary Paints

Proprietary paints are available that are intended to protect steel surfaces from decarburisation during annealing. The use of such paints is simpler than the use of a pack anneal, but not all of these paints are effective. Moreover, considerable difficulty may be experiened in removing such paints after heat treatment. The prospective user should test any such paint on a sample of steel prior to adopting it in practice.

Stress Relieving

Stress relieving before hardening is seldom required for shock-resisting tool steel, except for extremely intricate parts of widely varying section thickness (to minimise distortion and cracking) and parts subjected to excessive stock removal (to relieve stresses induced by machining). Treatment of such parts, which involves no microstructural transformation, consists of heating them to 650°C (1200°F) (soaking should be avoided), furnace cooling to about 510°C (950°F), and then removing them from the furnace to cool in air. Stress relieving of tools after tempering is seldom done. In some instances, however, increased tool life has been obtained by removing tools from service and stress relieving them (at a temperature no higher than the original tempering temperature) before returning them to service.

Example 17.3. In one plant, shock-resisting steel tools used for swaging stainless steel would sink a definite amount after a time in service. If kept in service, these tools would crack after swaging about 40000 parts. However, by withdrawing the tools after sinking had ceased and stress relieving them at 230°C (450°F) for 1 hour per inch of cross-section, tool life was more than doubled.

Austenitising

Austenitising temperatures for shock-resisting tool steels vary from 815 to 955°C (1500 to 1750°F). Preheating is not mandatory, but it is sometimes desirable for large tools, to minimise distortion, shorten time at the austenitising temperature and speed up production.

These steels may be austenitised in electric or fuel-fired furnaces or in salt or lead baths. Generally, for austenitising temperatures below 870°C (1600°F), a slightly oxidising atmosphere is best, whereas above 870°C (1600°F) a reducing atmosphere is required. If a semi-muffle fuel-fired furnace is used, the desired atmosphere can be obtained at low cost by adjustment of burners. However, if electrically heated or full-muffle fuel-fired furnaces are used, a prepared atmosphere from an external source is required.

Neutral salt baths are a practical means of heating the type S steels. A salt mixture such as No. 3 in Table 17.2 is satisfactory for types S2, S4 and S5, whereas a mixture such as No. 2 in that table will be more suitable for heating S1. If atmosphere furnaces or neutral salt baths are not available, the shock-resisting steels can be heated in a pack of neutral material such as burned pitch coke or cast iron chips. Packing mediums must be free of oil or other contaminants. Before being placed in the pack, tools should be wrapped with heavy brown paper, to prevent packing material from adhering to them as they are removed for quenching. Types S2, S4 and S5 should be quenched soon after they reach the austenitising temperature; types S1 and S7 are soaked at temperature for 15 to 45 minutes before being quenched (Table 17.2). Types S1 and S7 have the highest hardenability of these steels. The other types, although lower in hardenability than S1 and S7, are higher in hardenability than the W steels.

Tempering

Both the tungsten and the silicon types of shock-resisting tool steel resist softening from tempering to a greater degree than carbon tool steels. Secondary hardening does not occur in these steels, except to a minimal degree in some compositions of the tungsten type.

The effect of tempering temperature on the hardness of various types and compositions of the S steels is shown graphically in Fig. 17.4.

Tools made of shock-resisting steel should be tempered immediately after quenching, or cracking is likely to result, especially if they are quenched in water or brine.

Example 17.4. One plant made an extensive study on how much time could be safely permitted between quenching and tempering of tools made of shock-resisting steels. Results of this study are given in Table 17.3.

Table 17.3. Allowable time between quenching and tempering of shock-resisting tool steels, for prevention of cracking.

Steel	Austenitising temperature °C	°F	Quenching medium	Allowable time prior to tempering, minute
S1	900	1650	Oil	30
S1	980	1800	Oil	15
S2	845	1550	Brine	10
S2	900	1650	Brine	5
S3	815	1500	Brine	10
S3	870	1600	Brine	5
S4	870	1600	Brine	10
S4	925	1700	Brine	5
S4	900	1650	Oil	30
S4	955	1750	Oil	15
S5	870	1600	Oil	30
S5	925	1700	Oil	15

As determined by an extensive study conducted in one plant. Allowable time may vary significantly with size and shape of part.

In this same plant, tool records indicate that double tempering is beneficial for tools made from the S steels. The first tempering operation is done at a temperature 30 to 55°C (50 to 100°F) lower than that of the second and final tempering operation.

Surface Treatments

Surface treatments such as carburising and carbonitriding are often applied to S1-steel. Types S4 and S5 do not take an effective carburised case.

OIL-HARDENING COLD WORK TOOL STEELS

Recommended heat treating practice for oil-hardening cold work tool steels is summarised in Table 17.4.

Normalising

Normalising is desirable and sometimes necessary for parts that have been forged or heated previously to temperatures much higher than the proper austenitising temperature, because it produces a more uniformly refined grain structure. Recommended normalising temperatures are given in Table 17.4.

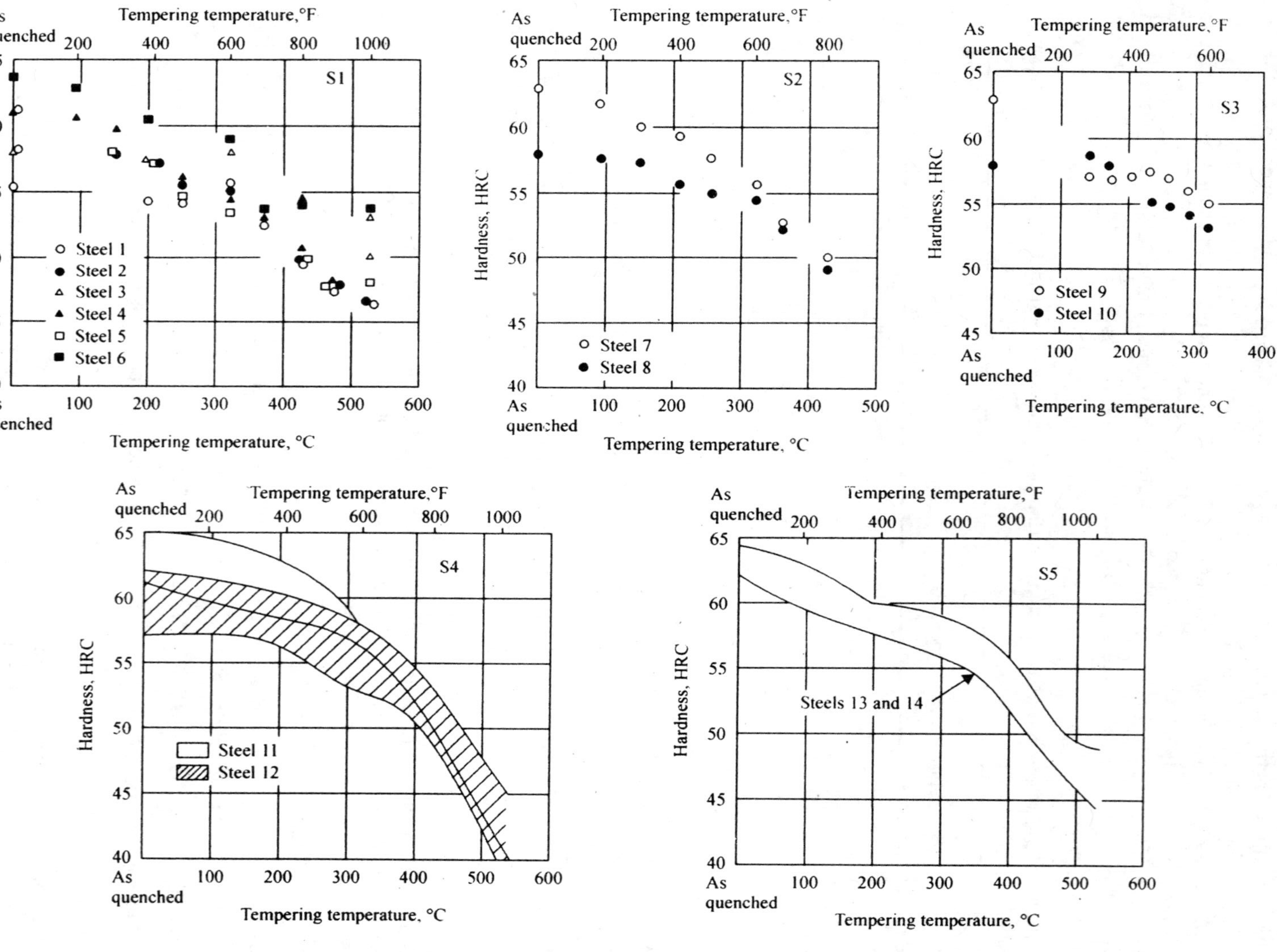

Fig. 17.4. Effect of tempering temperature on hardness of shock-resisting tool steels.

No.	Steel Type	Composition, % C	Si	W or Mo	Cr	V	Quenching Temperature °C	°F	Medium
1	S1	0.43	–	2.00 W	1.30	0.25	955	1750	–
2	S1	0.53	–	2.00 W	1.65	0.25	900	1650	–
3	S1	0.50	–	2.75 W	1.25	0.20	925	1700	–
4	S1	0.55	–	2.50 W	1.50	0.35	925	1700	–
5	S1	0.50	0.75	2.50 W	1.15	0.20	955	1750	Oil
6	S1	0.58	0.95	2.25 W	1.25	0.25	925	1700	Oil
7	S2	0.50	1.10	0.50 Mo	–	0.20	855	1575	Water
8	S2	0.50	1.10	0.50 Mo	–	0.20	900	1650	Oil
9	S3	0.50	–	1.00 W	1.00	–	870	1600	Oil
10	S3	0.50	–	1.00 W	1.00	–	815	1500	Water
11	S4	0.54 to 0.60	1.90 to 2.00	–	0 to 0.34	0 to 0.25	845 to 900	1550 to 1650	Water
12	S4	0.54 to 0.60	1.90 to 2.00	–	–	–	870 to 955	1600 to 1750	Oil
13	S5	0.50	1.60	0.25 Mo	–	–	870	1600	Oil
14	S5	0.60	2.00	0.40 Mo	0.28	0.20	900	1650	Oil

Work should be held at temperature for 15 minutes to 1 hour, depending on section size; prolonged soaking is not desirable. When tools are to be hardened after normalising, precautions must be taken in order to avoid decarburisation during normalising. If tools are to be subsequently machined, annealing is recommended in preference to normalising.

Annealing

Finished or semi-finished tools made from oil-hardening cold work steels should be protected from decarburisation or carburisation during annealing. This can be accomplished by the use of dry exothermic furnace atmospheres. More often, however, it is accomplished by pack annealing, wherein work to be annealed is packed in a box and surrounded with inert protective material, such as clean cast iron chips or 6 to 8 mesh spent pitch coke. Pack annealing permits the use of an open furnace; also, slow heating and cooling occur naturally in the packed box. However, it is important that the work be soaked long enough to permit it to reach the annealing temperature. Recommended annealing temperatures, cooling rates and expected hardness values are given in Table 17.4. Type O1 steel may also be cycle annealed (Table 17.5).

Cycle annealing offers little advantage for large loads, but with individual tools that can be conveniently handled in liquid baths or other conventional furnaces, it enables substantial savings in time.

Stress Relieving

In most instances, stress relieving of finished tools prior to final hardening does not noticeably lessen distortion during hardening. If extreme dimensional accuracy after hardening is required, tools should be stress relieved after rough machining but before final light machining. A recommended stress-relieving treatment consists of heating to 650 to 675°C (1200 to 1250°F), holding at temperature for 1 hour per inch of thickness, and then air cooling.

Table 17.4. Recommended heat treating practice for oil-hardening cold work tool steels.

Steel	Normalising temperature(a) °C	Normalising temperature(a) °F	Annealing Temperature(b) °C	Annealing Temperature(b) °F	Annealing Cooling rate(c) °C/hr.	Annealing Cooling rate(c) °F/hr.	Annealed hardness, HB	Hardening Temperature Pre-heat °C	Hardening Temperature Pre-heat °F	Hardening Temperature Austenitise °C	Hardening Temperature Austenitise °F	Holding time, min	Quenching medium	Quenched hardness, HRC
O1	870	1600	760–790	1400–1450	20	40	183–212	650	1200	790–815	1450–1500	10–30	Oil	63–65
O2	845	1550	745–775	1375–1425	20	40	183–212	650	1200	760–800	1400–1475	5–20	Oil	63–65
O6	870	1600	765–790	1410–1450	10	20	183–217	–	–	790–815	1450–1500	2–5	Oil	63–65
O7	900	1650	790–815	1450–1500	20	40	192–217	650	1200	790–830	1450–1525	10–30	Water	64–66
										845–885	1550–1625	10–30	Oil	64–66(d)

(a) Holding time, after uniform through heating, varies from about 15 minutes, for small sections, to about 1 hour, for large sections. Work is cooled from temperature in still air. (b) Lower limit of range should be used for small sections, upper limit for large sections. Holding time varies from about 1 hour, for light sections and small furnace charges, to about 4 hours, for heavy sections and large charges; for pack annealing, hold for 1 hour per inch of pack cross-section. (c) Maximum. Rate is not critical after cooling to below 540°C (1000°F). (d) Sections larger than 38 mm (1½ inch) will be softer.

Table 17.5. Cycle annealing treatments for four types of tool steel.

Steel	Treatment
O1	Heat to 730°C(1350°F), hold for 4 hours; heats to 780°C (1440°F), hold for 2 hours; cool to 690°C (1275°F), hold for 6 hours; air cool
A2	Heat to 900°C(1650°F), hold for 2 hours; cool to 760°C (1400°F), hold for 6 hours; air cool
A6	Heat to 815°C(1500°F), hold for 2 hours; cool to 650°C (1200°F), hold for 6 hours; air cool
D2	Heat to 900°C(1650°F), hold for 2 hours; cool to 775°C (1425°F), hold for 6 hours; air cool

Pre-heating

Pre-heating of the O steels will minimise distortion during subsequent hardening. It is almost always required for tools that are to be austenitised in liquid baths. Recommended pre-heating temperatures are listed in Table 17.4. Open furnaces can be used for pre-heating, but if scale-free and oxide-free hardening is required, pre-heating must be done with atmosphere control.

Austenitising

Recommended austenitising temperatures for oil-hardening tool steels are given in Table 17.4. Work that has been pre-heated may either be transferred to an austenitising furnace or be heated to the austenitising temperature in the same furnace in which it was pre-heated.

Decarburisation and scaling can be effectively minimised in liquid salt or lead baths, and in furnaces with controlled atmospheres (such as endothermic gas, dissociated ammonia, and argon or other inert gases). However, in all of these there is some danger of decarburisation if conditions are not controlled. Oxides in the molten baths or excess water vapour in the various gases will cause decarburisation. The atmospheres of gas-fired or oil-fired semi-muffle furnaces can be adjusted to contain from 2 to 4 per cent oxygen, a condition that will eliminate decarburisation but not oxidation. Types O1 and O2 can be satisfactorily austenitised in such an atmosphere, but it is not recommended for types O6 and O7. All type O steels may be austenitised in semi-muffle furnaces if packed in inert materials such as spent pitch coke and clean cast iron chips. Adequate time must be allowed to ensure that packed work reaches prescribed temperature. If salt baths are used, a salt mixture such as No. 3 in Table 17.2 is recommended.

Quenching

The optimum temperature range for quenching baths consisting of conventional oils is 40 to 60°C (100 to 140°F); agitation is recommended. Quenching oils that contain additives ('fast' oils) increase the cooling rate of the steel and permit more latitude in the operating temperature of the bath. Tools may be quenched in these oils at 80°C (180°F) without loss of hardness.

Martempering

If control of distortion is particularly important, martempering is sometimes advantageous. In martempering, the work is quenched in a bath of oil or molten salt that is usually held about 15 to 30°C (25 to 50°F) above the M_s temperature, and is held in the bath long enough to allow it to attain substantially equalised temperature throughout. The work is then removed from the bath and air cooled. The slow cooling through the martensitic transformation range permits the transformation of austenite to martensite to take place uniformly throughout the piece, thus minimising distortion. Fig. 17.5 presents a comparison of the dimensional changes in tools made of O1 steel that were oil quenched with those in tools of the same steel martempered at 230°C (450°F) for 10 minute; the martempered tools exhibited markedly less distortion.

Tempering

The O steels should be tempered immediately after quenching (preferably before they quite reach room temperature). These steels usually are not tempered below 120°C (250°F) or above 540°C (1000°F); the most commonly used temperature range is from 175 to 205°C (350 to 400°F). Tempering times vary with section size. Often, a time at temperature of 1 hour per inch of thickness (minimum dimension of heaviest section) or per inch of diameter, with a minimum of 1 hour, is used.

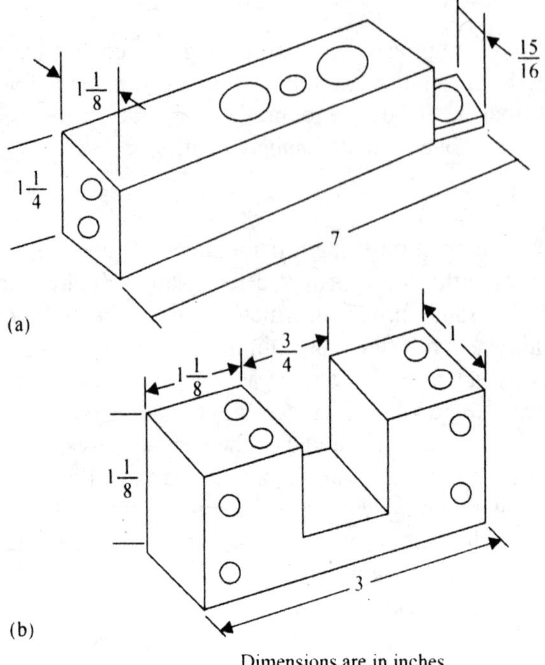

Tools sketched, made of O1 steel, were used for comparison of dimensional changes resulting from martempering at 230°C (450°F) for 10 min and from oil quenching. Five tools of each design, processed by each method, were measured on 5 different days (a) maximum change in flatness along the 180 mm (7 inch) dimension were 0.25 mm (0.010 inch) after martempering. (b) maximum change of the 19 mm (¾ inch) slot width width was 0.1-mm (0.0039 inch) after oil quenching and 0.03-mm (0.0012-inch) after martempering.

Fig. 17.5. Dimensional changes in O1 tools.

Typical hardness values obtained with various tempering temperatures for oil-hardening tool steels are given in Fig. 17.6. The upper curve in each graph represents results from austenitising at the higher side of the range of temperatures indicated, and the lower curve represents results from austenitising at the lower side.

Conventional tools made from the O steels are seldom subjected to multiple tempering or sub-zero treatment. However, for some special tools, such as gages, where dimensional stability is critical, multiple tempering is desirable. In such instances the workpieces should be cooled to below 65°C (150°F) prior to each re-tempering. Sub-zero cooling to –75°C (–100°F) or lower is also helpful in achieving dimensional stability.

MEDIUM-ALLOY AIR-HARDENING, AND HIGH-CARBON HIGH-CHROMIUM, COLD WORK TOOL STEELS

Recommended heat treating practice for medium-alloy air-hardening cold work tool steels (group A) and high-carbon high-chromium cold work tool steels (group D) is summarised in Table 17.6.

Normalising

Except for type A10 (Table 17.6), normalising is not recommended for any of the steels in groups A and D.

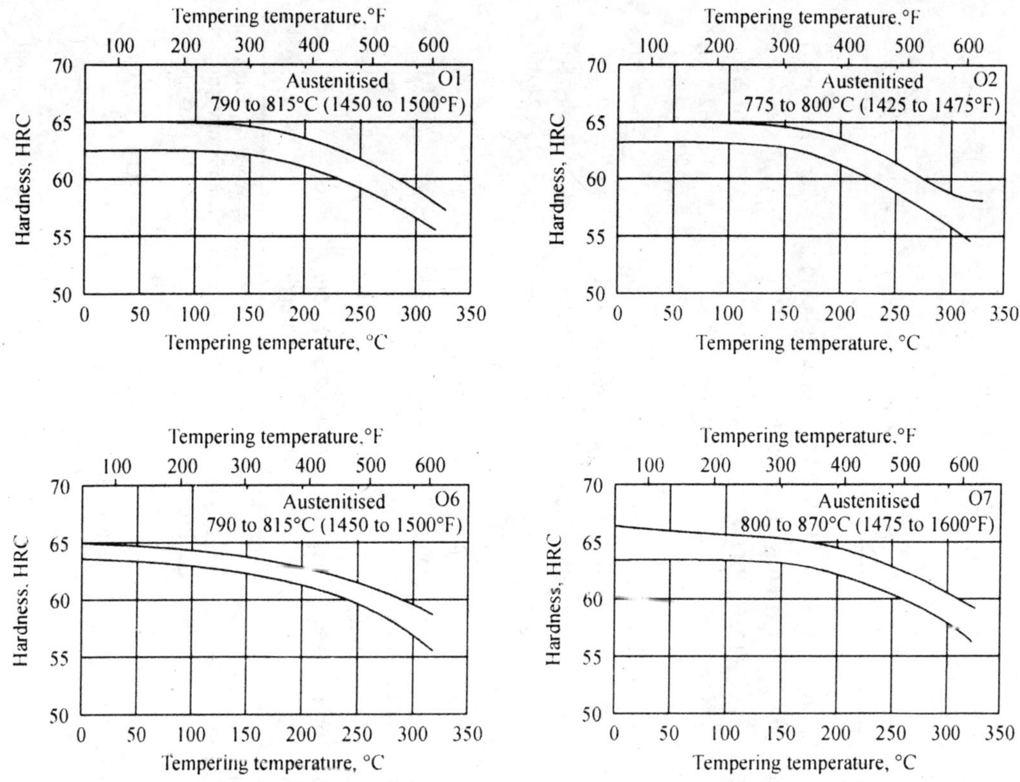

Steels O1, O2 and O6 were austenitised at the temperatures indicated, and then oil quenched. For O7 steel, large, uniform sections were austenitised at 800 to 830°C (1475 to 1525°F) and water quenched, and other sections were austenitised at 830 to 870°C (1525 to 1600°F) and oil quenched. Duration of tempering was 1 hour.

Fig. 17.6. Hardness as a function of tempering temperature, for oil-hardening cold work tool steels.

Annealing

These steels are usually supplied in the annealed condition by the manufacturer. However, they should be annealed after forging and prior to re-hardening. Annealing is required also for previously hardened or welded tools that are to be re-worked.

Recommended annealing temperatures for the various types are given in Table 17.6. Tools should be heated slowly and uniformly to the annealing temperature. Slow heating is particularly important if a hardened tool is being annealed.

Cycle, or isothermal, treatments may be employed for annealing some A and D steels (Table 17.5).

Stress Relieving

Tools made of A and D steels that cannot be ground after hardening are sometimes stress relieved after rough machining. This is particularly advisable for delicate tools and tools that vary markedly in cross-section. Stress relieving is used also on tools that are machined to final shape, if these tools can be straightened after stress relieving and before final heat treatment.

Table 17.6. Recommended heat treating practice for medium-alloy air-hardening, and high-carbon high-chromium, cold work tool steels.

Steel	Normalising temperature(a)	Annealing				Anealed hardness,	Hardening						Quenched hardness,
		Temperature(b)		Cooling rate(c)			Temperature				Holding time,	Quenching	
							Pre-heat		Austenitise				
	°C (°F)	°C	°F	°C/hr.	°F/hr.	HB	°C	°F	°C	°F	min	medium	HRC
Medium-alloy air-hardening cold work tool steels													
A2	Not rec	845–870	1550–1600	22(d)	40(d)	201–229	790	1450	925–980	1700–1800	20-45(e)	A	62-65(f)
A3	Not rec	845–870	1550–1600	22	20	207–229	790	1450	955–1010	1750–1850	25-60(e)	A	–
A4	Not rec	740–760	1360–1400	14(g)	25(g)	200–241	675	1250	815–870	1500–1600	15-90	A	61-64(f)
A5	Not rec	740–760(h)	1360–1400(h)	14	25	229–255	595	1100	790–845	1450–1550	15-45	A	62-63(f)
A6	Not rec	730–745	1350–1375	14	25	217–248	650	1200	830–870	1525–1600	20-45	A	59-63(f)
A7	Not rec	870–900	1600–1650	14(d)	25(d)	235–262	815	1500	955–980	1750–1800	30-60(e)	A	64-67(f)
A8	Not rec	845–870	1550–1600	22	40	192–223	790	1450	980–1010	1800–1850	20-45(e)	A	60-62(f)
A9	Not rec	845–870	1550–1600	14	25	212–248	790	1450	980–1025	1800–1875	20-45(e)	A	56-58(f)
A10	790 (1450)	765–795	1410–1460	8	15	235–269	650	1200	790–815	1450–1500	30-60	A	62-64(f)
High-carbon high-chromium cold work tool steels													
D1	Not rec	870–900	1600–1650	22	40	207–248	815	1500	970–1010	1775–1850	15-45(e)	A	61
D2	Not rec	870–900	1600–1650	22	40	217–255	815	1500	980–1025	1800–1875	15-45(e)	A	64
D3	Not rec	870–900	1600–1650	22	40	217–255	815	1500	925–980	1700–1800	15-45(e)	O	64
D4	Not rec	870–900	1600–1650	22	40	217–255	815	1500	970–1010	1775–1850	15-45(e)	A	64
D5	Not rec	870–900	1600–1650	22	40	223–255	815	1500	980–1025	1800–1875	15-45(e)	A	64
D7	Not rec	870–900	1600–1650	22	40	235–262	815	1500	1010–1065	1850–1950	30-60(e)	A	65

(a) Holding time, after uniform through heating, varies from about 15 minutes, for small sections, to about 1 hour, for large sections. Work is cooled from temperature in still air. (b) Lower limit of range should be used for small sections, upper limit for large sections. Holding time varies from about 1 hour, for light sections and small furnace charges, to about 4 hours, for heavy sections and large charges; for pack annealing, hold for 1 hour per inch of pack cross-section. (c) Maximum rate, 540°C (1000°F) unless footnoted to indicate otherwise. (d)To 705°C (1300°F). (e) For open furnace heat treatment. For pack hardening, hold for ½ hour per inch of pack cross-section. (f) Hardness varies with austenitising temperature. (g) To 650°C (1200°F). One manufacturer recommends cooling from 760 to 540°C (1400 to 1000°F), then re-heating to 730°C (1350°F) and cooling.

There is little advantage in stress relieving completed tools if they cannot be straightened prior to hardening, because a good pre-heat will relieve stresses, and the distortion which occurs in either case will remain uncorrected. Recommended temperatures for stress relieving are:

A2, A7	650 to 675°C (1200 to 1250°F)
A4, A5, A6	675 to 705°C (1250 to 1300°F)
D1 to D7	675 to 705°C (1250 to 1300°F)

Usually, tools can he stress relieved at these temperatures without surface protection. Tools are commonly held at temperatures for 1 hour per inch of cross-section (minimum of 1 hour) and then air cooled.

Pre-heating

Steels of the A and D groups are usually pre-heated before being austenitised for hardening. Pre-heating reduces subsequent distortion in the hardened parts by minimising non-uniform dimensional changes during austenitising. Pre-heating simpler tools made of grades A4, A5, A6 and A10 can often be eliminated if they are austenitised in a furnace instead of a liquid bath, because these steels are austenitised at lower temperatures.

Recommended pre-heating temperatures are listed in Table 17.6. Holding time at temperature is usually 1 hour per inch of maximum cross-section. Pre-heating temperatures of 790 to 815°C (1450 to 1500°F) are used for tools made from A2, A3, A7, A8 or A9, or from any of the D steels. For these higher temperatures, a liquid bath or a protective furnace atmosphere is required in order to prevent scaling and decarburisation.

Austenitising

Steels of groups A and D can be austenitised in molten salt baths or in various types of furnaces using gaseous atmospheres. Because of their lower austenitising temperatures, types A4, A5, A6 and A10 may also be austenitised in molten lead, or in open furnaces with oxidising atmospheres. However, the latter methods are not satisfactory for the other A steels or for the D steels, because of their higher austenitising temperatures.

If salt baths are used, salt mixtures such as No.2 or No. 3 in Table 17.2 are recommended; the choice between the two depends on required working temperature range. These mixtures may be rectified (for the prevention of decarburisation).

Procedures for austenitising two different parts made of D2 steel by salt bath and by endothermic atmosphere furnace processes are shown in Table 17.7.

Table 17.7 Procedures for salt bath and endothermic atmosphere austenitising of D2 inserts

Bending die inserts	
Salt bath(b)	Pre-heat in air furnace at 650°C (1200°F) for 1½ hour(c); austenitise at 1010°C (1850°F) for 35 minute(c); air cool; remove salt
Endothermic atomoshpere(d)	Charge directly into furnace at 705°C (1300°F) and pre-heat for 1½ hour(c); austenitise at 1010°C (1850°F) for 2 hour; air cool(c)

(Contd...)

Trim die inserts	
Salt bath(b)	Pre-heat in air furnace at 650°C (1200°F), 4 hour(c), then in salt bath at 845°C (1550°F), 1 hour(c)(f); austenitise at 1010°C (1850°F) for 1 hour(c); air cool(c); remove salt
Endothermic atmosphere(d)	Charge directly into furnace at 705°C (1300°F) and pre-heat for 4 hour(c); raise furnace temperature, and austenitise at 1010°C (1850°F) for 4 hours; air cool(c)

(a) After austenitising, inserts 200 by 305 by 38 mm (8 by 12 by 1½ inch) were double tempered (2 hours at 510°C or 950°F, air cool; 2 hours at 510°C or 950°F, air cool) and then nitrided for 48 hours at 510°C (950°F). (b) Salt bath furnace was immerse-electrode type, 380 by 760 by 915 mm (15 by 30 by 36 inch) deep. (c) Manual loading requires 1½ minute per piece. (d) Furnace was radian-tube type, 610 by 915 by 455 mm (24 by 26 by 18 inch) high. (e) After austenitising, die inserts were double tempered (4 hours at 190°C or 375°F, air cool; 4 hours at 190°C or 375°F, air cool). (f) Second pre-heat was necessary because of faster heating rate of salt bath.

In some instances, austenitising will cost more with atmosphere furnaces than with salt baths, and in other instances the reverse will be true. Atmospheres that have proved suitable for austenitising the A and D steels are endothermic, dry dissociated ammonia and dry hydrogen.

Endothermic gas produced by catalytic combination of air and fuel gas is the most widely used atmosphere. This relatively inexpensive gas can be adjusted for desired carbon potential and controlled by dew point.

Dry dissociated ammonia (dew point, –50°C or –60°F) and dry hydrogen (dew point, –75°C or –100°F) are used in applications in which complete freedom from discolouration is required. Vacuum, which excludes all atmosphere, can also be used to austenitise the A and D steels; it is particularly suitable for these steels because their air-hardening characteristics permit slow cooling rates.

Like the O steels, the A and D steels may be packed and then austenitised in semi-muffle furnaces. The packing materials and heat treating procedures employed are similar to those described in the previous section on austenitising of the O grades.

Steels of groups A and D must be held at their austenitising temperatures long enough to obtain required carbide solution if they are to attain maximum hardness. However, hardening from excessively high austenitising temperatures will increase the retained austenite. Although retained austenite can be decreased by repeated tempering or sub-zero cooling (or both), it should be avoided.

Quenching

Steels of groups A and D, except D3, will attain maximum hardness by cooling in still air, unless sections are extremely large. However, the hardenability of these steels varies with different types, as indicated in Table 17.8.

Table 17.8. Hardenability in still air of several A and D tool steels.

Steel	*Centre hardening Size of section that fully hardens at centre*	*Hardness, HRC*	*Surface hardening Size of section that fully hardens at Surface*	*Hardness, HRC*
A2, A4	75-mm (3-inch) diam	59 to 61	100-mm (4-inch) diam	59 ot 61

(Contd...)

Steel	*Centre hardening Size of section that fully hardens at centre*	*Hardness, HRC*	*Surface hardening Size of section that fully hardens at Surface*	*Hardness, HRC*
A5	–	–	100-mm (4-inch) diam	62 to 63
A6	180-mm (7-inch) cube	59 to 60	180-mm (7-inch) cube	60 to 61
D1, D2, D5	100-mm (4-inch) cube	60 to 61	125 by 125 by 255-mm (5 by 5 by 10-inch)	61 to 62

Depending on section size, hardenability, and complexity of shape, the following methods are used to obtain increasingly accelerated cooling of nominally air-hardening steels:

1. Cool in 'still air'—that is, atmospheric air undisturbed by artificial circulation.
2. Cool in 'fan air'—that is, the current of air discharged from a fan.
3. Cool in 'air blast'—that is, the discharge from a high-pressure line.
4. 'Oil quench to black' —that is, quench in oil until the steel is below the temperature at which it glows dull red, then cool to room temperature in air.
5. Oil quench by conventional practice.

Tempering

Tempering practices for A and D steels parallel those described for O steels in the preceding section. Tempering is usually begun when the work reaches a temperature of about 50 to 65°C (120 to 150°F). However, these steels retain some austenite at this temperature range. To maximise transformation of austenite to martensite, cooling to room temperature, or to sub-zero temperature, is sometimes applied.

Opinions vary greatly as to the merits of sub-zero cooling, because it increases the probability of cracking during the cooling cycle. The more usual practice is to begin tempering when parts reach about 50 to 65°C (120 to 150°F) and then double or triple temper. Multiple tempering is effective in decreasing the amount of austenite retained in A and D steels and is a common practice in heat treating them. The general precautions and tempering practices outlined for O steels in the preceding section are followed for the A and D steels. However, because most of the steels in groups A and D (except A4, A5 and A6) soften less rapidly than the group O steels with an increase in tempering temperature (Figs. 17.7 and 17.8), higher tempering temperatures can be used for the A and D steels. A minimum tempering temperature of 205°C (400°F) is a common requirement for A2, A7 and D steels. Tempering temperatures as high as 550°C (1025°F) are frequently used, and even higher temperatures are used for special requirements.

It will be noted in Figs. 17.7 and 17.8 that certain steels (notably A2 and D2) exhibit higher hardness after being tempered at about 540°C (1000°F) than after being tempered at temperatures 55 to 110°C (100 to 200°F) lower.

This reversal in the usual relationship is known as secondary hardening, and is caused by transformation of retained austenite during tempering at the higher temperatures, near 540°C (1000°F). When a steel can be tempered to the same hardness at more than one temperature (for instance, D2 to 58 to 59 HRC), it is advisable to select the highest tempering temperature that will produce the desired hardness and may prevent tool breakage in service.

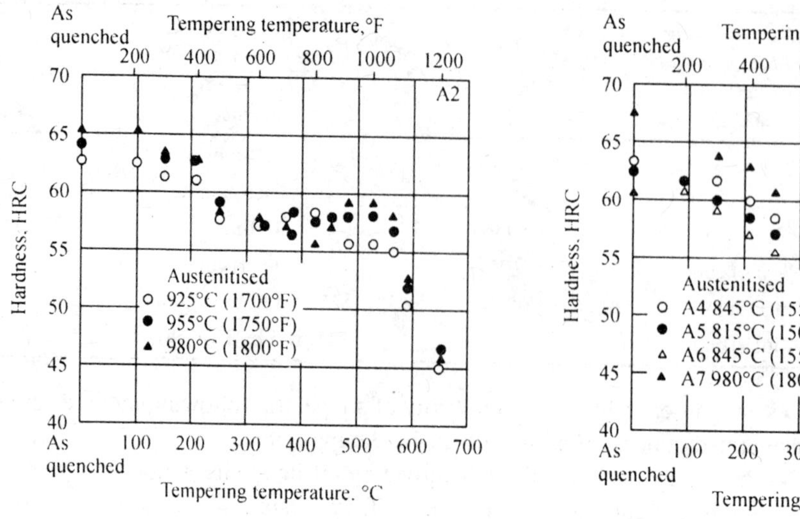

Fig. 17.7. Effect of tempering temperature on hardness of medium-alloy air-hardening steels.

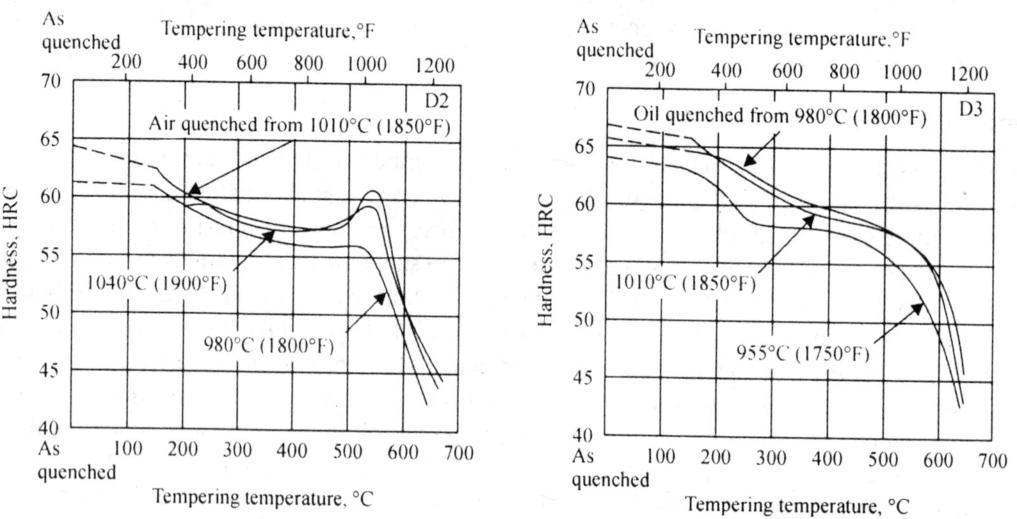

Steels were austenitised in an air furnace; a recirculating pit-type furnace was used for tempering. Curves represent steel from 3 suppliers; average of 5 hardness measurement made on each specimen from each supplier. Specimens were 25 mm (1 inch) in diam and 38 mm (1½ inch) long.

Fig. 17.8. Relation between tempering temperature and hardness for D2 and D3 tool steels.

Nitriding

The A steels (particularly A2 and A7) and the D steels are often nitrided after being hardened and tempered. Nitriding may be done either in a salt bath or in an atmosphere of dissociated ammonia. High tempering temperatures of 510 to 540°C (950 to 1000°F) are used on steels that are to be nitrided. Excessively high nitriding temperatures, recommended range is 510 to 540°C (950 to 1000°F), will reduce hardness of the base metal and should not be used. Austenitising at a higher temperature when hardening prior to nitriding will minimise loss of hardness during nitriding of some D steels (note graph for D2 in Fig. 17.8).

HOT WORK TOOL STEELS

Table 17.9 summarises the heat treating practices commonly employed for this composite group of tool steels.

Normalising

Because these steels as a group are either partially or completely air-hardening, normalising is not recommended except for the high-nickel steel 6F7. After forging or before re-heat treating, 6F7 may be normalised by heating to 845 to 870°C (1550 to 1600°F), preferably in a controlled atmosphere, and cooling in still air.

Annealing

Recommended annealing temperatures, cooling practice and expected hardness values are given in Table 17.9. Heating for annealing should be slow and uniform to prevent cracking, especially when annealing hardened tools. Heat losses from the furnace usually determine the rate of cooling; large furnace loads will cool at a slower rate than light loads. For most of these steels, furnace cooling to 425°C (800°F), at 22°C max. (40°F max.) per hour and then air cooling, will suffice. For types 6F2, 6F3 and 6H1, an isothermal anneal (Table 17.9) may be employed to advantage for small tools that can be handled in salt or lead baths or for small loads in batch-type furnaces; however, isothermal annealing has no advantage over conventional annealing for large die blocks or large furnace loads of these steels.

To minimise scaling and decarburisation, small parts are usually pack annealed, while large and heavy die blocks are more commonly annealed in controlled-atmosphere furnaces.

Packing material should preferably be spent cast iron chips or spent pitch coke-petroleum coke heated to 1205°C (2200°F) in a semi-closed container to drive off gas and moisture. Lime, sand or mica is sometimes used, but under such material if mixed with a small amount of charcoal or other carburising material, the steel may be decarburised. Packing material should be dry and free of all oxidising materials, should separate all metal surfaces, and should fill the container. Containers should be sealed after packing. Holding time at the annealing temperature is 1 hour per inch of container thickness. The H steels must have a neutral packing material, because they are extremely susceptible to both carburisation and decarburisation. In controlled-atmosphere furnaces, the work should be supported so that it does not touch the bottom of the furnace. This will ensure uniform heating and permit free circulation of the atmosphere around the work. Workpieces should be supported so that they will not sag or distort under their own weight.

Grades 6F4 and 6F7 may be annealed without packing or controlled atmosphere if light scaling is not objectionable, because they are annealed at lower temperatures (Table 17.9).

Stress Relieving

It is sometimes advantageous to stress relieve tools made of hot work steel after rough machining but prior to final machining, by heating them to 650 to 730°C (1200 to 1350°F). This treatment minimises distortion during hardening, particularly for dies or tools that have major changes in configuration or deep cavities. However, closer dimensional control can be obtained by hardening and tempering after rough machining and prior to final machining, provided that the final hardness obtained by this method is within the machinable range.

Table. 17.9. Recommended heat treating practice for hot work tool steels.

Steel	Normalising temperature(a) °C	Normalising temperature(a) °F	Annealing Temperature(b) °C	Annealing Temperature(b) °F	Annealing Cooling rate(c) °C/hr.	Annealing Cooling rate(c) °F/hr.	Annealed hardness, HB	Hardening Temperature Pre-heat °C	Hardening Temperature Pre-heat °F	Hardening Temperature Austenitise °C	Hardening Temperature Austenitise °F	Holding time, min.	Quenching medium	Quenched hardness, HRC
Chromium hot work tool steels														
H10	Not recommended		845-900	1550-1650	22	40	192-229	815	1500	1010-1040	1850-1900	15-40(d)	A	56-59
H11	Not recommended		845-900	1550-1650	22	40	192-229	815	1500	995-1025	1825-1875	15-40(d)	A	53-55
H12	Not recommended		845-900	1550-1650	22	40	192-229	815	1500	995-1025	1825-1875	15-40(d)	A	52-55
H13	Not recommended		845-900	1550-1650	22	40	192-229	815	1500	995-1040	1825-1900	15-40(d)	A	49-53
H14	Not recommended		870-900	1600-1650	22	40	207-235	815	1500	1010-1065	1850-1950	15-40(d)	A	55-56
H16	Not recommended		870-900	1600-1650	22	40	212-241	815	1500	1120-1175	2050-2150	2-5	A, O	55-58
H19	Not recommended		870-900	1600-1650	22	40	207-241	815	1500	1095-1205	2000-2200	2-5	A, O	52-55
Tungsten hot work tool steels														
H20	Not recommended		870-900	1600-1650	22	40	207-235	815	1500	1095-1205	2000-2200	2-5	A, O	53-55
H21	Not recommended		870-900	1600-1650	22	40	207-235	815	1500	1095-1205	2000-2200	2-5	A, O	43-52
H22	Not recommended		870-900	1600-1650	22	40	207-235	815	1500	1095-1205	2000-2200	2-5	A, O	48-57
H23	Not recommended		870-900	1600-1650	22	40	212-255	815	1500	1205-1260	2200-2300	2-5	O	33-35(e)
H24	Not recommended		870-900	1600-1650	22	40	217-241	815	1500	1095-1230	2000-2250	2-5	A, O	44-55
H25	Not recommended		870-900	1600-1650	22	40	207-235	815	1500	1150-1260	2100-2300	2-5	A, O	46-53
H26	Not recommended		870-900	1600-1650	22	40	217-241	870	1600	1175-1260	2150-2300	2-5	A, O, S	63-64
Molybdenum hot work tool steels														
H41	Not recommended		815-870	1500-1600	22(f)	40(f)	207-235	730-845	1350-1550	1095-1190	2000-2175	2-5	A, O, S	64-66
H42	Not recommended		845-900	1550-1650	22	40	207-235	730-845	1350-1550	1120-1220	2050-2225	2-5	A, O, S	54-62
H43	Not recommended		815-870	1500-1600	22(g)	40(g)	207-235	730-845	1350-1550	1095-1190	2000-2175	2-5	A, O, S	54-58
Other alloy tool steels														
6G	Not recommended		790-815	1450-1500	22(h)	40(h)	197-229	Not required	Not required	845-855	1550-1575	–	O(j)	63 min(k)
6F2	Not recommended		780-795	1440-1460	22(m)	40(h)	223-235	Not required	Not required	845-870	1550-1600	–	O(j)	63 min(k)
6F3	Not recommended		760-775	1400-1425	22(n)	40(h)	235-248	Not required	Not required	900-925	1650-1700	–	A(p)	63 min(k)
6F4	Not recommended		705	1300	(q)	(q)	262-285	815	1500	1010-1020	1850-1870	–	O, A	38-41(e)
6F5	Not recommended		845	1550	(r)	(r)	262-285	Not required	Not required	870	1600	–	O, A	58-59
6F6	Not recommended		845(pack)	1550(pack)	(s)	(s)	196	650-705(t)	1200-1300(t)	925-955(t)	1700-1750(t)	–	O(u)	(v)
6F7	845-870	1550-1600	670	1240	22	40	260-300	730	1350	915	1675	–	A	54-55

(Contd...)

Steel	Normalising temper-ature(a) °C	Normalising temper-ature(a) °F	Annealing Temperature(b) °C	Annealing Temperature(b) °F	Annealing Cooling rate(c) °C/hr.	Annealing Cooling rate(c) °F/hr.	Annealed hardness, HB	Hardening Temperature Pre-heat °C	Hardening Temperature Pre-heat °F	Hardening Temperature Austenitise °C	Hardening Temperature Austenitise °F	Holding time, min.	Quenching medium	Quenched hardness, HRC
6H1	Not recommended		845	1550	22(w)	40(w)	202-235	760-790	1400-1450	900-940	1650-1725	–	A	48–49
6H2	Not recommended		815-845	1500-1550	22	40	202-235	705-760	1300-1400	980-1065	1800-1950	–	O, A	52-55

Note: A, air; O, oil; S, salt. (a) Holding time, after uniform through heating, varies from about 15 minutes, for small sections, to about 1 hour, for large sections. Work is cooled from temperature in still air. (b) Lower limit of range should be used for small sections, upper limit for large sections. Holding time varies from about 1 hour, for light sections and small furnace charges, to about 4 hours, for heavy sections and large charges; for pack annealing, hold for 1 hour per inch of pack cross-section. (c) Maximum rate, 425°C (800°F) unless footnoted to indicate otherwise. (d) For open furnace heat treatment. For pack hardening, hold for ½ hour per inch of pack cross-section. (e) Temper to precipitation harden. (f) To 540°C (1000°F). (g) To 480°C (900°F). (h) To 370°C (700°F). (i) To 205 to 175°C (400 to 350°F), then air cool. (j) Temper immediately. (k) For isothermal annealing, furnace cool to 650°C (1200°F), hold for 4 hours, furnace cool to 425°C (800°F), then air cool. (l) For isothermal annealing, furnace cool to 670°C (1240°F), hold for 4 hours furnace cool to 425°C (800°F), then air cool. (m) Cool with forced-air blast to 205 to 175°C (400 to 350°F), then cool in still air. (n) Air cool from annealing temperature. (o) Furnace cool, at 20°C (40°F), (max) per hour, to 425°C (800°F), re-heat to 595 ± 14°C (1100±25°F), furnace cool to 425°C (800°F), then air cool. (q) Furnace cool at 17°C (30°F) per hour to 540°C (1000°F), re-heat to 790°C (1450°F), furnace cool at 10°C (20°F) per hour to 540°C (1000°F), then air cool. (r) Heat in pack or in controlled atmosphere. (s) To 50°C (125°F). (t) Pack heating, 59 to 60 HRC; atmosphere heating, 54 to 55 HRC. (u) For isothermal annealing, hold at 845°C (1550°F) for 2 hours, furnace cool to 745°C (1375°F), hold for 4 to 6 hours, then air cool.

Pre-heating

Pre-heating prior to austenitising is nearly always recommended for all hot work steels except 6G, 6F2, 6F3 and 6F5. These four steels may or may not require pre-heating, depending on size and configuration of the workpieces. Recommended pre-heating temperatures for all the other types are given in Table 17.9.

Die blocks or other tools for open furnace treatment should be placed in a furnace that is not over 260°C (500°F). Work that is packed in containers may be safely placed in furnaces at 370 to 540°C (700 to 1000°F). Once the workpieces (or container) have attained furnace temperature, they are heated slowly and uniformly, at 85 to 110°C (150 to 200°F) per hour, to the pre-heating temperature (Table 17.9) and held for 1 hour per inch of thickness (or per inch of container thickness, if packed). Thermocouples should be placed adjacent to the pieces in containers. Controlled atmospheres or other protective means must be used above 650°C (1200°F) to minimise scaling and decarburisation. A slightly reducing atmosphere is especially recommended for pre-heating of H41.

For certain parts—for example, intricate die-casting dies—pre-heating is omitted. Distortion of such parts is sometimes lessened by packing them and heating them slowly and uniformly throughout the entire range to the quenching temperature.

Austenitising

Austenitising temperatures recommended for the hardening of hot work tool steels are given in Table 17.9. Rapid heating from the pre-heating temperature to the austenitising temperature is preferred for types H16 through H43 and for type 6F4.

Except for steels H10 through H14 (Table 17.9), time at the austenitising temperature should only be sufficient to heat the work completely through; prolonged soaking is not recommended. Time cycles for several specific conditions are indicated in the next section entitled 'Examples of Heat Treating Procedure for Hot work Tools'.

The equipment and method employed for austenitising are frequently determined by the size of the workpiece. For tools weighing less than about 227 kg (500 lb), any of the methods would be suitable. However, larger tools or dies would be difficult to handle in either a salt bath or a pack.

Tools or dies made of hot work steel must be protected against carburisation and decarburisation when being heated for austenitising. Carburised surfaces are highly susceptible to heat checking. Decarburisation causes decreased strength, which may result in fatigue failures; and on die-casting dies, the molten casting metal will 'weld on' to decarburised surfaces and may cause 'washout' because of poor wear resistance of the decarburised surface. However, the principal detrimental effect of decarburisation is to mislead the heat treater as to the actual hardness of the die. To obtain specified hardness of the decarburised surface, the die is tempered at too low a temperature. The die then goes into operation at excessive internal hardness and breaks at the first application of load.

An endothermic atmosphere produced by a gas generator is probably the most widely used protective medium. The dew point is normally held from 2 to 7°C (35 to 45°F) in the furnace, depending on carbon content of the steel and operating temperature. A dew point of 3 to 4°C (38 to 40°F) is ideal for most steels of type H11 or H13 when austenitised at 1010°C (1850°F).

The packing of work in spent pitch coke before heating it for austenitising has been used extensively in small shops where it has not been feasible to invest in special equipment. This procedure is generally used for small dies. New pitch coke is generally heated to 1040 to 1205°C (1900 to 2200°F) to burn off any combustibles that may be present as well as to remove any excessive moisture. The spent pitch coke is then sifted to remove the fines (the coke should also be sifted before re-use). Normal procedure for this method is to wrap the workpiece in plain brown wrapping paper and place it in a heat-resistant metal box in the bottom of which is about 50 mm (2 inch) of spent pitch coke. The workpiece should be covered and surrounded with approximately 50 to 100 mm (2 to 4 inch) of spent pitch coke. The cover is then placed on the box and sealed with asbestos paste. The box is then ready to be placed in a furnace, which need not be provided with controlled atmosphere.

Quenching

Hot work steels range from high to extremely high in hardenability. Most of them will achieve full hardness by cooling in still air; however, even with those types having the highest hardenability, sections of die blocks may be so large that insufficient hardening results. In such instances, an air blast or an oil quench is required to achieve full hardness. Hot work steels are never water quenched. Recommended quenching media are listed in Table 17.9.

If blast cooling is used, air should be blasted uniformly on the surface to be hardened. All air must be dry. When being air quenched, dies or other tools should not be placed on concrete floors or in locations where water vapour may strike them.

Some of the hot work steels (especially the tungsten and molybdenum types) will scale considerably during cooling to room temperature in air. An interrupted quench reduces this scaling by eliminating the long period of contact with air at elevated temperature, but it also increases distortion. The procedure is best carried out by quenching from the austenitising temperature in a salt bath held at 595 to 650°C (1100 to 1200°F), holding in the quench until the workpiece reaches the temperature of the bath, and then withdrawing the piece and allowing it to cool in air. An alternative, but less precise, procedure is to quench in oil at room temperature or slightly above and judge by colour (faint red) when the workpiece has reached 595 to 650°C (1100 to 1200°F); the piece is then quickly withdrawn and permitted to cool to room temperature in air. While cooling, the pieces should be placed in a suitable rack, or be supported by wires, in such a manner that air is permitted to come in contact with all surfaces.

Steel H23 requires a different type of interrupted quench, because ferrite precipitates rapidly in this steel at 595°C (1100°F) and M_s is below room temperature. Type H23 should be quenched in molten salt at 165 to 190°C (325 to 375°F) and then air cooled to room temperature. This steel will not harden in quenching but will do so by secondary hardening during the tempering cycle.

Parts quenched in oil should be completely immersed in the oil bath, held until they have reached bath temperature, and then transferred immediately to the tempering furnace. Oil bath temperatures may range from 55 to 150°C (130 to 300°F), but should always be below the flash point of the oil. Oil baths should be circulated and kept free of water.

Tempering

Hot work tool steels should be tempered immediately after quenching, even though sensitivity to cracking in this stage varies considerably among the various types (for example, air-quenched 6F4 may be safely kept at room temperature for several hours before tempering, whereas 6G, 6F2 and 6F3 are susceptible to cracking if they are cooled substantially below 175°C or 350°F before tempering.

Hot work steels are usually tempered in air furnaces of the forced-convection type. Salt baths are used successfully for smaller parts, but for large complex parts salt bath tempering may induce too severe a thermal shock and cause cracking. The effect of tempering temperature on hardness of hot work tool steels is shown in Fig. 17.9.

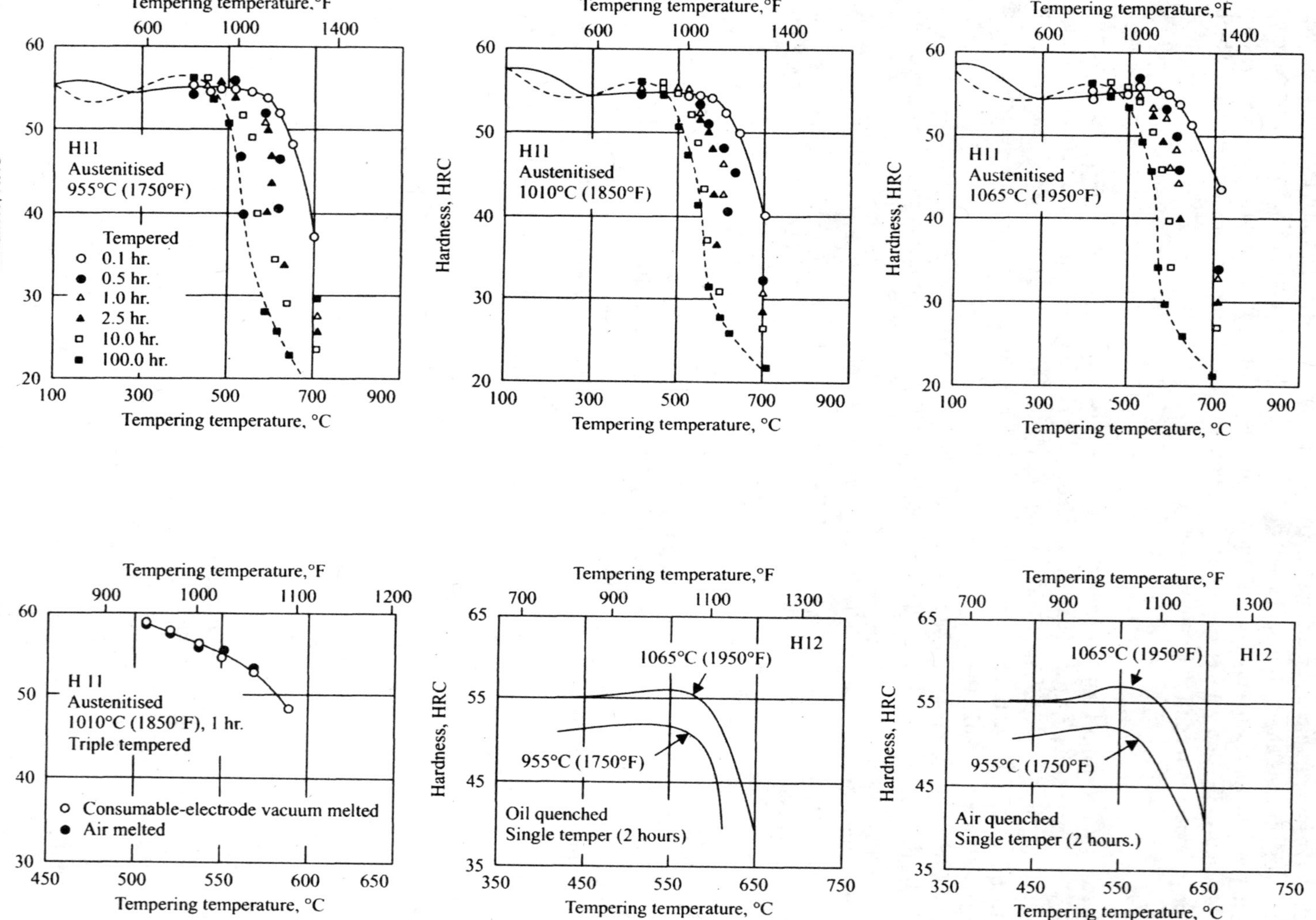

Fig. 17.9(a). Effect of temperature on hardness of chromium hot work tool steels **(Cond...)**.

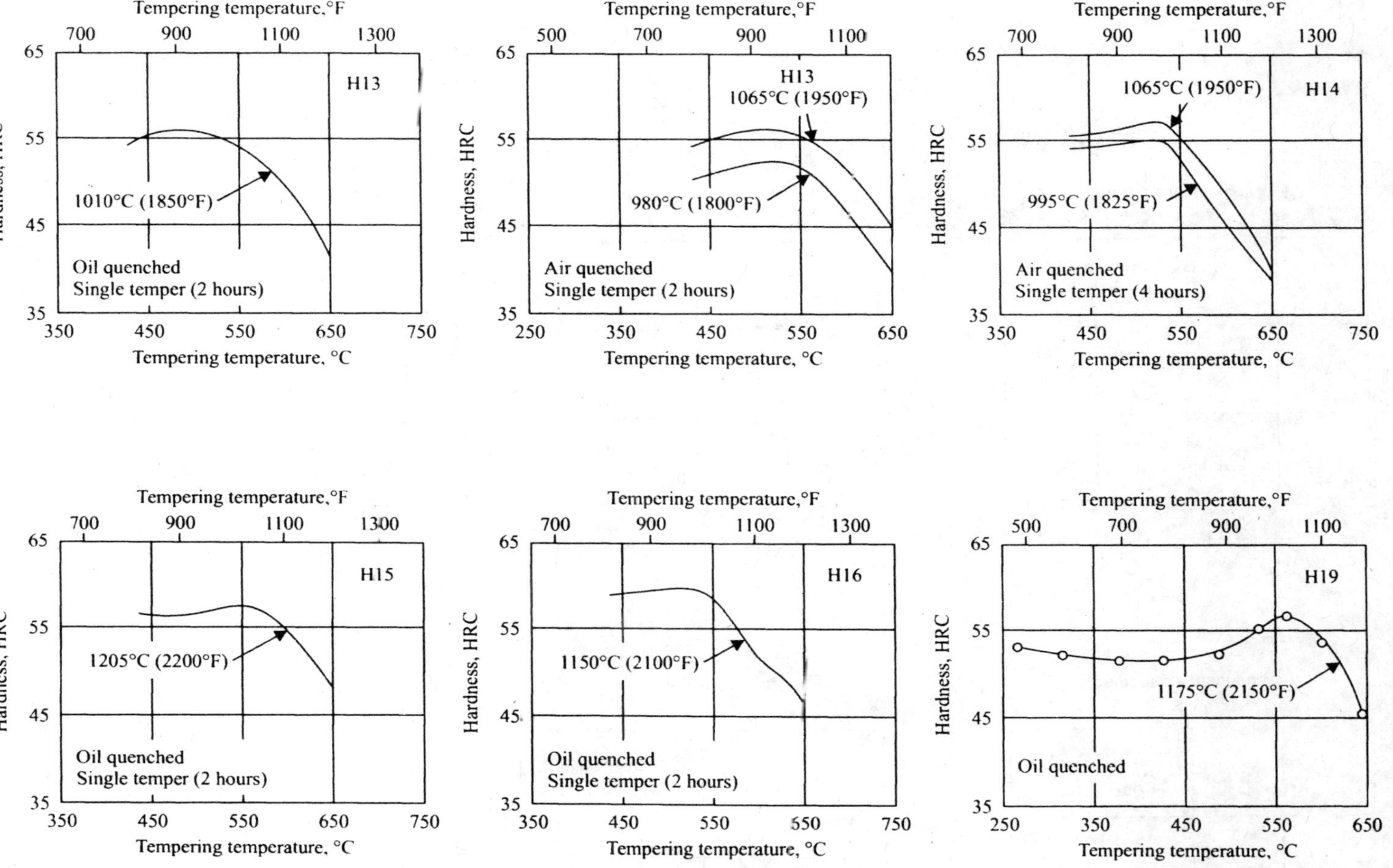

Fig. 17.9(a). Effect of temperature on hardness of chromium hot work tool steels .

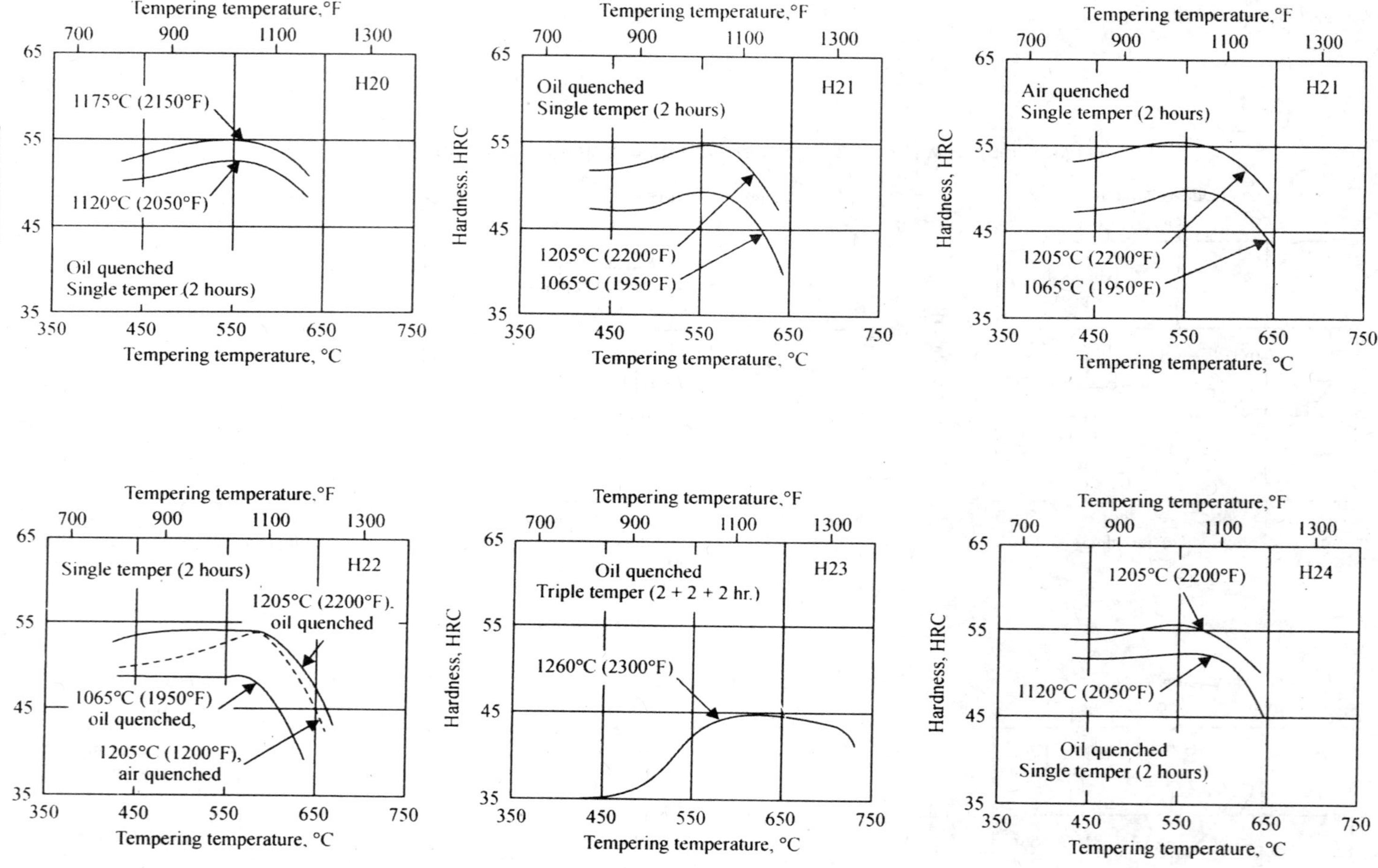

Fig.17.9 (b). Effect of temperature on hardness of tungsten hot work tool steels **(Cond...)**.

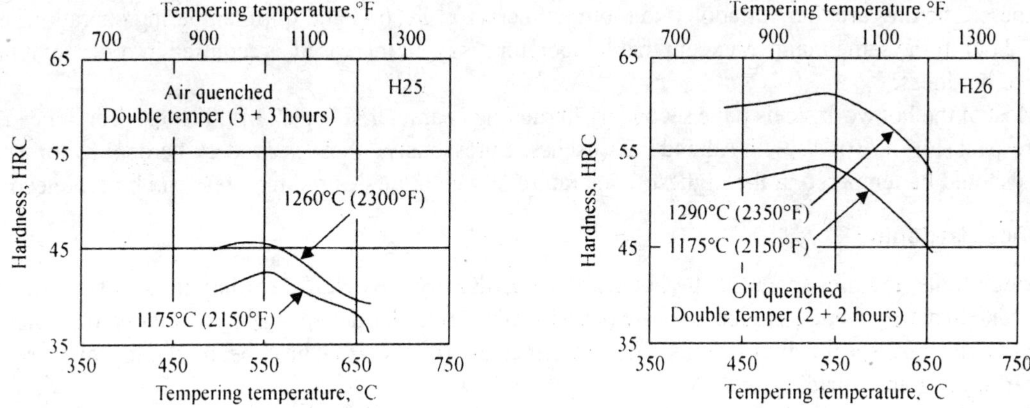

Fig.17.9(b). Effect of temperature on hardness of tungsten hot work tool steels.

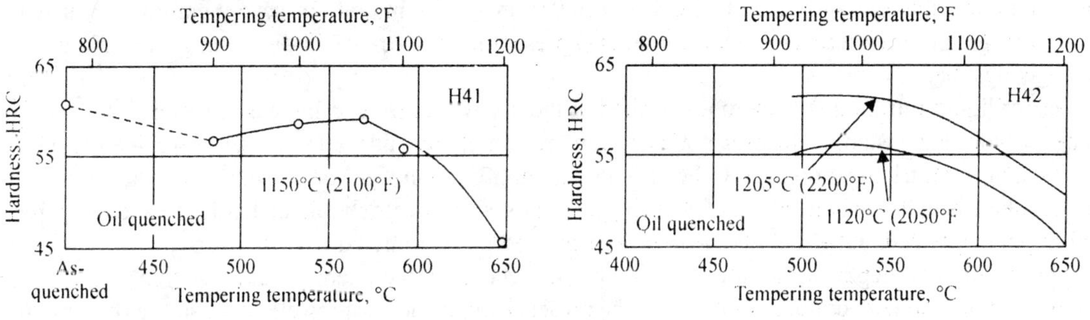

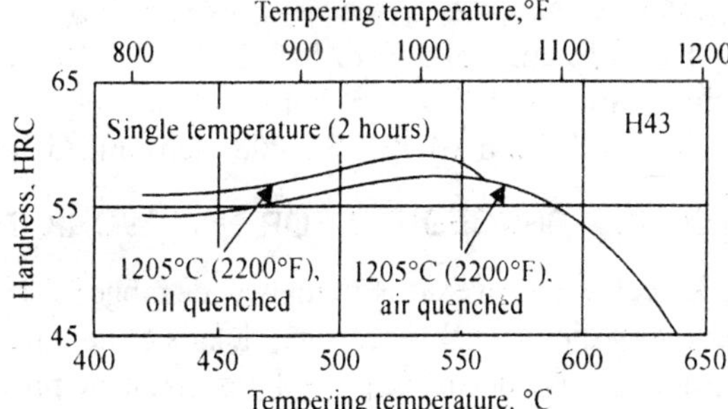

Fig.17.9(c). Effect of temperature on hardness of molybdenum hot work tool steels.

Multiple tempering ensures that any retained austenite that transforms to martensite during the first tempering cycle is tempered before a tool is placed in service. Multiple tempering also minimises cracks due to stress originating from the hardening operation.

Multiple tempering has proved particularly advantageous for large or sharp-cornered die blocks that are not permitted to reach room temperature before the first tempering operation.

Example 17.5. In one plant where many die blocks are heat treated, standard practice is as follows. When the dies have air cooled to 52°C (125°F), they are placed in a tempering furnace maintained at 565°C (1050°F). After the dies have reached furnace temperature, they are soaked for 1 hour per inch of thickness. The dies are then air cooled to room temperature. Second and third tempering operations are carried out in the same manner, except that temperature may be increased as required in order to obtain desired hardness.

Most of the hot work steels have secondary hardening characteristics; H23 is the most pronounced in this respect (Fig. 17.9). As with A2 and D2 (discussed previously), these secondary-hardening hot work steels should be tempered at the highest temperature at which the desired hardness can be produced.

Surface Hardening

Although tools and dies made of the hot work steels usually have sufficient hardness to perform the tasks for which they were designed, they are occasionally surface hardened to acquire improved resistance to wear or heat for special applications. The two principal processes that have been used for this purpose are carburising and nitriding.

Carburising is usually limited to hot work steels having a carbon content of 0.35 per cent or lower. Type H12 has been reported to achieve a carburised surface hardness of 60 to 62 HRC. The carburised case should be shallow—for example, 0.4 mm (0.015 inch) maximum—or severe embrittlement will occur. The greater the thermal shock (or gradient) present in service—as in die casting—the shallower the case must be.

Gas or liquid nitriding is sometimes applied to the hot work steels to increase resistance to heat or wear, or both. For instance, dies for hot extrusion are sometimes nitrided to increase service life. One disadvantage of nitriding, however, is the difficulty it imposes on the re-working of tools or dies. Another disadvantage is that it may accentuate heat checking. Hot work steels should be hardened and tempered before being nitrided, but should be neither decarburised nor carburised.

The quality and depth of the nitrided case are influenced by the chemical composition of the steel and by the time and temperature of nitriding. The presence of nitride-forming elements such as chromium and vanadium is helpful to the attainment of a satisfactory case. The fact that most of the hot work steels reach a secondary hardening peak when tempered in the vicinity of 540°C (1000°F) is beneficial, because nitriding is usually accomplished in a range of 510 to 540°C (950 to 1000°F) over a period of 15 to 24 hours. The nitrided case, in addition to being very hard, may be brittle. Brittleness increases with depth of case; hence, shallow, 0.08 to 0.2 mm (0.003 to 0.008 inch), nitrided cases are usually applied.

EXAMPLES OF HEAT TREATING PROCEDURE FOR HOT WORK TOOLS

Tools and dies made of hot work steel extend over an extremely wide range of sizes and weights (sometimes up to several tonnes, as in the largest die blocks). Therefore, details of heat treating techniques may vary considerably. The following examples give details of procedures that have prove successful in practice:

Example 17.6. (H21 hot extrusion die). A typical method for heat treating a 75 mm (3 inch) thick, 200 mm (8 inch) OD, 75 mm (3 inch) hole, hot extrusion die made of H21 steel comprises the following:

1. Pre-heat at 815 to 845°C (1500 to 1550°F), either in a slightly oxidising atmosphere or in neutral salt.
2. Transfer to furnace (6 to 12 per cent reducing atmosphere or neutral salt bath) operating at 1175°C (2150°F). Hold in furnace for approximately 20 minutes after the die has reached 1175°C (2150°F).
3. Cool in still air to about 65°C (150°F).
4. Temper at 565°C (1050°F) for 4 hours.
5. Cool to near room temperature.

6. Retemper at 650°C (1200°F) for 4 hours.
7. Air cool.

Example 17.7. (H11 Mandrel). Mandrel made of H11, used in conjunction with the H21 die in Example 17.6, above, are heat treated as follows:

1. Pre-heat at 760°C (1400°F) in a slightly oxidising atmosphere.
2. Transfer to atmosphere furnace (1 to 3 per cent excess O_2) operating at 1010°C (1850°F) and hold for 20 minute plus 5 minute for each inch of thickness.
3. Air cool to near room temperature (oil quenching can also be used).
4. Temper (or preferably, double temper) for desired hardness.

Example 17.8. (H13 Die Block). One plant employs the following procedure for heat treating die blocks made of H13 that weigh less than 23 kg (50 lb):

1. Insert eyebolt to facilitate handling.
2. Wrap die block in waxed paper and place in a heat-resistant container on a bed of spent pitch coke 75 to 100 mm (3 to 4 inch) deep.
3. Seal cover on container with asbestos paste.
4. Place container in furnace (not atmosphere-controlled) operating at 760°C (1400°F); bring to furnace temperature and hold for 4 hours.
5. Raise furnace temperature at 30°C (50°F) per hour to 1010°C (1850°F) and hold charge at this temperature for 6 hours.
6. Remove die block from container by use of eyebolt.
7. Cool in still air to 345°C (650°F) (temperature-indicating crayons may be used), then place in furnace operating at 345°C (650°F) and cool in furnace at 30°C (50°F) per hour to 95°C (200°F). (If the die block has no sharp corners or major changes in configuration, the interrupted cooling may be omitted).
8. Remove from furnace and cool in air to 40°C (100°F).
9. Place in tempering furnace operating at 565°C (1050°F), bring to furnace temperature and hold for 8 hours, air cool to room temperature and check hardness.
10. Repeat step 9, except that it may be necessary to increase tempering temperature so that final hardness will be 46 to 49 HRC.

Example 17.9. (H13 die block). The following procedure has proved successful for heat treating large die blocks, 1590 kg (3500 lb), made of H13 steel:

1. Load die block into electrically heated bell-type furnace. The sequence of operations begins when furnace temperature reaches 95°C (200°F).
2. Raise furnace temperature at 30°C (50°F) per hour to 370°C (700°F).
3. Introduce nitrogen atmosphere to furnace and increase furnace temperature at 55°C (100°F) per hour to 790°C (1450°F): hold for 1 hour, then shut off nitrogen, introduce endothermic atmosphere—dew point, 3 to 4°C (38 to 40°F)—and hold for an additional 5 hours.
4. Increase furnace temperature at 55°C (100°F) per hour to 1040°C (1900°F) and hold for 6 hours.
5. Remove die block and air cool to 65°C (150°F).
6. Place die block in tempering furnace operating at 205°C (400°F); bring to furnace temperature and hold for 7 hours.
7. Increase furnace temperature at 40°C (100°F) per hour to 565°C (1050°F) and hold for 16 hour.
8. Air cool to room temperature. Hardness is about 46 to 48 HRC.

9. Temper die block a second time, repeating steps 6, 7 and 8 but increasing final temperature to 580°C (1075°F) because a finished hardness of 42 to 43 HRC is desired.
10. Temper a third time, repeating steps 6, 7 and 8 without modification.

Example 17.10. (6F3 forging die). In one plant, dies, 495 by 215 by 150 mm (19½ by 8½ by 6 inch), used for forging pitman arms are heat treated to a final hardness of 40 to 42 HRC by the following procedure:

1. Pre-heat at 260°C (500°F).
2. When dies have attained furnace temperature, raise temperature at 55 to 85°C (100 to 150°F) per hour to 915°C (1675°F); use controlled atmosphere above 760°C (1400°F).
3. Hold at 915°C (1675°F) for 6 hours.
4. Air-blast cool to 175°C (350°F) (temperature-indicating crayons used).
5. Place in tempering furnace operating at 175 to 205°C (350 to 400°F). When dies have attained furnace temperature, raise temperature at 85°C (150°F) per hour to 595°C (1100°F) and hold for 9 hours.
6. Air cool to room temperature; check hardness.
7. Re-temper, repeating steps 5 and 6 except for final temperature, which will depend on hardness obtained from first tempering.

The following three examples indicate the procedures employed in one plant for heat treating 6F2, 6F4 and H12 components used for hot upset forging of pinions.

Example 17.11. (6F2 heating tool and gripper die). Final hardness of 40 to 42 HRC is obtained on this tool by pre-heating, austenitising, quenching, tempering and re-tempering as follows:

1. Pre-heat at 260°C (500°F).
2. When dies have attained furnace temperature, raise temperature at 55 to 85°C (100 to 150°F) per hour to 855°C (1575°F).
3. Hold at 855°C (1575°F) for 1 hour per inch of thickness.
4. Quench in oil at 55°C (130°F), to 175°C (350°F) (temperature-indicating crayons used); transfer as quickly as possible to tempering furnace.
5. Place in tempering furnace operating at 175 to 205°C (350 to 400°F). When dies have attained furnace temperature, raise temperature at 55 to 85°C (100 to 150°F) per hour to 595°C (1100°F) and hold for 1 hour per inch of thickness.
6. Cool in still air to room temperature; check hardness.
7. Re-temper, repeating 5 and 6 except for final temperature, which depends on hardness obtained from first tempering.

Example 17.12. (6F4 slab insert). These inserts, requiring final hardness of 39 to 41 HRC, are heat treated as follows:

1. Pre-heat at 260°C (500°F).
2. When inserts have attained furnace temperature, raise temperature at 55 to 85°C (100 to 150°F) per hour to 815°C (1500°F) (use controlled atmosphere above 760°C or 1400°F).
3. Solution treat at 1020°C (1870°F) for 1 hr. per inch of thickness.
4. Cool in still air to room temperature.
   (Note: When these inserts were quenched to only 95°C (200°F), threads broke out of the die during tapping.)
5. Precipitation harden by heating at 260°C (500°F) until temperature of insert equals furnace temperature, raising furnace temperature at 55 to 85°C (100 to 150°F) per hour to 450°C (840°F)

for 3 hours plus 1 hour per inch of thickness (minimum time, 4 hours) and air cooling to room temperature.

Example 17.13. (H12 punch insert). Heat treating to a desired final hardness of 40 to 42 HRC comprises the following procedure:

1. Pre-heat at 260°C (500°F).
2. When inserts have attained furnace temperature, raise temperature at 55 to 85°C (100 to 150°F) per hour to 870°C (1600°F) (use controlled atmosphere above 760°C or 1400°F).
3. Raise furnace temperature to 1010°C (1850°F); hold inserts at 1010°C (1850°F) for 1 hour per inch of thickness.
4. Quench inserts in still air until they are cool enough to be hand-held. Transfer immediately to the tempering furnace.
5. Place in tempering furnace operating at 260°C (500°F). When inserts have attained furnace temperature, raise temperature at 55 to 85°C (100 to 150°F) per hour to 595°C (1100°F). Hold at 595°C (1100°F) for 1 hour per inch of thickness.
6. Cool in still air and check hardness.
7. Re-temper, repeating steps 5 and 6 except for final temperature, which will depend on the hardness that was obtained from the first tempering cycle.
8. Temper for a third time, if time permits.

HIGH SPEED TOOL STEELS

Table 17.10 do not recommend normalising of high speed tool steels.

Steels in the intermediate group, M50 and M52, are less expensive than standard high speed steels and may occasionally be used in place of standard high speed steels.

The intermediate high speed steels do not resist tempering to the same extent as M2 and therefore, they cannot be expected to perform as well as high speed steels in applications where red hardness is important. For example, in such applications as lathe tools and drills, where the tool is in continuous contact with the workpiece and high surface temperatures are the rule, M50 and M52 steels would not be expected to perform as well as standard high speed steels. When contact with the workpiece is intermittent or surface temperatures are low, in such applications as hack and band saw blades, blanking dies and some special woodworking tools, M50 and M52 steels may perform adequately. M50 steel is also used in ball and roller bearing races for elevated temperatures. Other applications include woodworking tools, hydraulic pump assemblies, pump pistons and pump vanes. If greater abrasion resistance is required, but not as much as afforded by standard high speed steels, then M52 may be a logical choice.

Table 17.10. Recommended heat treating practice for high speed tool steels.

		Annealing			*Hardening*				
			Cooling	*Annealed*	*Temperature*		*Holding*		*Quenched*
Steel	*Normalising*	*Temperature(a), °C*	*rate(b) °C/h*	*hardness, HB*	*Pre-heat, °C*	*Austenitise(c), °C*	*time, minute*	*Quenching medium*	*hardness, HRC*
Tungsten high speed tool steels, standard group									
T1	Not recommended	870–900	20	217–255	815–870	1260–1300	2–5	O, A, S	63–65
T2	Not recommended	870–900	20	223–255	815–870	1260–1300	2–5	O, A, S	64–66
T4	Not recommended	870–900	20	229–269	815–870	1260–1300	2–5	O, A, S	64–66

(Contd...)

Steel	Normalising	Annealing			Hardening					
		Temperature(a), °C	Cooling rate(b) °C/h	Annealed hardness, HB	Temperature Pre-heat, °C	Austenitise(c), °C	Holding time, minute	Quenching medium	Quenched hardness, HRC
T5	Not recommended	870–900	20	235–285	815–870	1275–1300	2–5	O, A, S	64–66
T6	Not recommended	870–900	20	248–302	815–870	1275–1300	2–5	O, A, S	64–66
T8	Not recommended	870–900	20	229–255	815–870	1260–1300	2–5	O, A, S	64–66
T15	Not recommended	870–900	20	241–277	815–870	1205–1260	2–5	O, A, S	65–67
Molybdenum high speed tool steels, standard group									
M1	Not recommended	815–870	20	207–235	730–845	1175–1220	2–5	O, A, S	64–66
M2	Not recommended	870–900	20	212–241	730–845	1190–1230	2–5	O, A, S	65–66
M3	Not recommended	870–900	20	223–255	730–845	1205–1230	2–5	O, A, S	64–66
M4	Not recommended	870–900	20	223–255	730–845	1205–1230	2–5	O, A, S	64–66
M6	Not recommended	870	20	248–277	790	1175–1205	2–5	O, A, S	63–66
M7	Not recommended	815–870	20	217–255	730–845	1175–1220	2–5	O, A, S	64–65
M10	Not recommended	815–870	20	207–255	730–845	1175–1220	2–5	O, A, S	64–66
M30	Not recommended	870–900	20	235–269	730–845	1205–1230	2–5	O, A, S	64–66
M33	Not recommended	870–900	20	235–269	730–845	1205–1230	2–5	O, A, S	64–66
M34	Not recommended	870–900	20	235–269	730–845	1205–1230	2–5	O, A, S	64–66
M36	Not recommended	870–900	20	235–269	730–845	1220–1245	2–5	O, A, S	64–66
M41	Not recommended	870–900	20	235–269	730–845	1190–1215	2–5	O, A, S	63–66
M42	Not recommended	870–900	20	235–269	730–845	1165–1190	2–5	O, A, S	63–66
M43	Not recommended	870–900	20	248–269	730–845	1150–1175	2–5	O, A, S	63–66
M44	Not recommended	870–900	20	248–285	730–845	1200–1225	2–5	O, A, S	63–66
M46	Not recommended	870–900	20	235–269	730–845	1190–1220	2–5	O, A, S	63–66
M47	Not recommended	870–900	20	235–269	730–845	1175–1205	2–5	O, A, S	63–66
High speed tool steels, intermediate group									
M50	Not recommended	830–845	20	197–235	730–845	1095–1120	2–5	O, A, S	63–65
M52	Not recommended	830–845	20	197–235	730–845	1120–1175	2–5	O, A, S	63–65

Note: O, oil; A, air; S, salt. (a) Pack annealing is recommended, for minimum decarburisation. Steels should be held at temperature for 1 hour per inch of thickness of the container. (b) Maximum. Rate is not critical after work (in pack, if employed) has been furnace cooled to 650°C (1200°F). (c) If steels are austenitised in a salt bath, austenitising temperatures should be 14°C (25°F) lower than those in the ranges given.

Annealing

High speed steel must be fully annealed after forging or when re-hardening is required. To minimise decarburisation, pack annealing in tightly closed containers is recommended. The packing material can be dry sand or lime to which a small amount of charcoal has been added; burned cast iron chips also are satisfactory. Because the packing material acts to insulate the container and thereby slow down heating, the container should be filled in such a way with the steel to be annealed that a minimum amount of packing material is required.

After the steel has reached the annealing temperature range (Table 17.10), it should be held at temperature for 1 hour per inch of thickness of the container and should then be slowly cooled in the furnace (at a rate not exceeding 20°C or 40°F per hour) until it reaches a temperature of 650°C (1200°F), when a faster rate of cooling is permissible.

Pre-heating

Austenite begins to form at about 760°C (1400°F), and pre-heating for hardening to slightly above this temperature will minimise stresses that might be set up because of the transformation. If the prevention of partial decarburisation is important, a pre-heating temperature of 705 to 790°C (1300 to 1450°F) generally will be used. When this is not a problem, pre-heating at 815 to 900°C (1500 to 1650°F) is satisfactory.

Double pre-heating—in one furnace at 540 to 650°C (1000 to 1200°F) and in another at 845 to 870°C (1550 to 1600°F)—is often recommended, to minimise thermal shock.

If a single pre-heat is used, the T types of high speed steels are preferably pre-heated at 815 to 870°C (1500 to 1600°F), and the remaining M types at 730 to 845°C (1350 to 1550°F). It is common practice to pre-heat for twice the length of time required at the austenitising temperature. Accordingly, to ensure a uniform flow of work, the capacity of the pre-heating installation is generally twice that of the austenitising installation.

Although pre-heating is recommended for all high speed steels, small tools and those that do not incorporate sharp notches or abrupt changes in section, such as small tool bits and solid drill rod blanks, may be placed directly into the austenitising furnace with reasonable safety. If consumable carbonaceous muffles are used, the pre-heating temperature must not exceed about 650°C (1200°F), because the type of atmosphere they provide is ineffective in preventing decarburisation at higher temperatures.

Austenitising

High speed steels depend on the solution of various complex alloy carbides during austenitising to develop their heat-resisting qualities and cutting ability. These carbides do not dissolve to an appreciable extent unless the steel is heated to temperatures near the melting point (Table 17.10). Therefore, exceedingly accurate temperature control is required in austenitising high speed steel. Steels containing about 3 per cent or more vanadium may be held at the austenitising temperature approximately 50 per cent longer than the lower-vanadium types. The relatively pure vanadium carbide phase inherent in the microstructure of the high-vanadium steels is virtually insoluble at temperatures below the melting point and acts to restrict grain growth, thus permitting longer soaking times without detriment. However, the recommended austenitising temperatures for these steels should not be exceeded.

Single-point tools intended for heavy-duty cutting often can be effectively austenitised at 10 to 15°C (15 to 30°F) above the nominal austenitising temperature. The higher temperature increases alloy solution, temper resistance and hot hardness, but it also results in some sacrifice in toughness. To impart added toughness, fine-edged tools, such as taps and chasers, may be hardened at temperature 15 to 30°C (25 to 50°F) below the nominal austenitising temperature. Punches and dies that do not require maximum hardness may be austenitised for maximum toughness at temperatures 55 to 110°C (100 to 200°F) below the nominal temperature. Other adjustments in austenitising temperature depend on the type of heating equipment employed. Fullmuffle furnaces employing a controlled atmosphere rich in carbon monoxide are usually operated at the higher temperature of the recommended range. Salt baths usually are operated 15 to 30°C (30 to 50°F) below the top of the range.

The effect of austenitising temperature on the as-quenched hardness of M2 steel is shown in Fig.17.10. Below 1175°C (2150°F), M2 cannot develop full hardness on quenching, because of insufficient carbide solution. At temperatures above approximately 1230°C (2250°F), the as-quenched hardness of M2 decreases because of too much carbon and alloy solution and an excess of retained austenite in the as-quenched steel. Fig. 17.11 illustrates the improved toughness of M2, as measured by the Izod unnotched impact test, that results from the use of lower-than-normal austenitising temperatures. Numerous investigators have shown that the optimum means for attaining maximum toughness in high speed steel is through reduced austenitising temperatures rather than by full austenitising and overtempering to an equivalent hardness level.

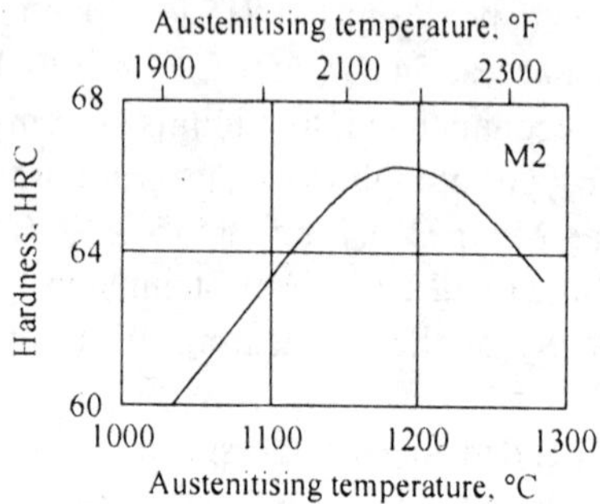

Fig. 17.10. Effect of austenitising temperatures on the as-quenched hardness of M2 steel.

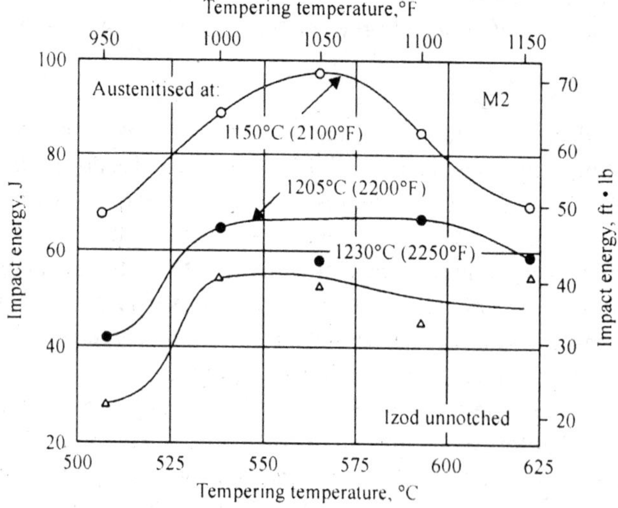

Fig. 17.11. Effect of austenitising and tempering temperatures on impact strength of M2.

Figure 17.12 illustrates the sacrifice in high-temperature hardness of M2 that results from the use of reduced austenitising temperatures.

Quenching

High speed steels can be quenched in air, oil or molten salt. However, except for thin tools, which are air quenched between plates to keep them straight, it is customary to quench in oil from muffle or semi-muffle furnaces and in molten salt from a high-temperature salt bath. After its temperature has been equalised in the salt quench, the tool is air cooled. For large cutters heated in a furnace, an interrupted oil quench is often used to minimise quenching strains and prevent cracking. This consists of cooling the cutters in the oil only until they lose colour (about 540°C or 1000°F) and then cooling in air.

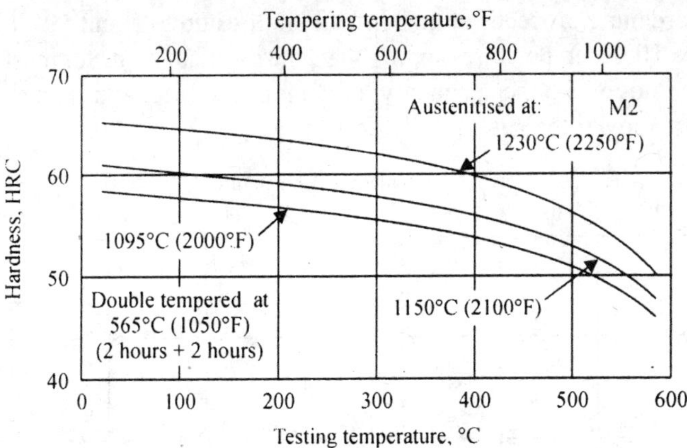

Fig.17.12. Effect of austenitising temperature on high-temperature hardness of M2 steel.

After quenching, high speed steel tools usually possess high residual stress, and to prevent cracking, it is good practice to transfer them from the quenchant to a tempering furnace before they have cooled to below 65°C (150°F). This is particularly important for large or intricate tools, for which a delay between quenching and tempering or permitting the work to cool to too low a temperature will usually induce cracking. If the work cannot be transferred to a tempering furnace at once, it should be put in a holding furnace maintained at 120 to 205°C (250 to 400°F) until a tempering furnace is available.

Bainitic Hardening

Bainitic hardening has been used in a few applications. To produce a 'primary bainitic' structure, this treatment is performed by arresting the quench from the austenitising temperature at approximately 260°C (500°F), holding for 4 hours, then cooling to room temperature. This produces a structure with about 55 per cent bainite and the remainder retained austenite. Subsequent tempering at normal tempering temperature transforms the retained austenite and tempers the bainite to a Rockwell C hardness 1 to 3 points lower than normal for the selected tempering temperature.

Partial Hardening to Improve Machinability

Annealed high speed steel may be partially hardened to approximately 270 to 300 HB to improve machinability. At these hardnesses, high speed steels, including the sulphurised types, are less likely to tear in shaving or back-off operations. Typical heat treating to achieve this result consists of heating to 855 to 870°C (1575 to 1600°F), holding for at least 1 hour, quenching in oil, and tempering at 635 to 665°C (1175 to 1225°F) to obtain the desired hardness. If the austenitising temperature does not exceed 870°C (1600°F), this treatment will not cause grain coarsening in the final hardening operation.

Certain machining operations, such as drilling and rough milling, should be performed in the annealed condition to obtain maximum tool life.

Tempering

As shown in Fig.17.13 for an M2 steel austenitised at 1220°C (2225°F), the hardness of high speed steel is directly affected by tempering temperature and time. From the slope of the curves in Fig.17.13, it can be seen that M2 undergoes secondary hardening at temperatures above approximately 370°C (700°F) and that secondary hardening proceeds at higher temperatures up to about 595°C (1100°F), depending on time at temperature. These temperatures approximate the practical limits for most tempering operations; lower temperatures do not evoke the secondary hardening response, and higher temperatures produce hardnesses considerably lower than are usually desired.

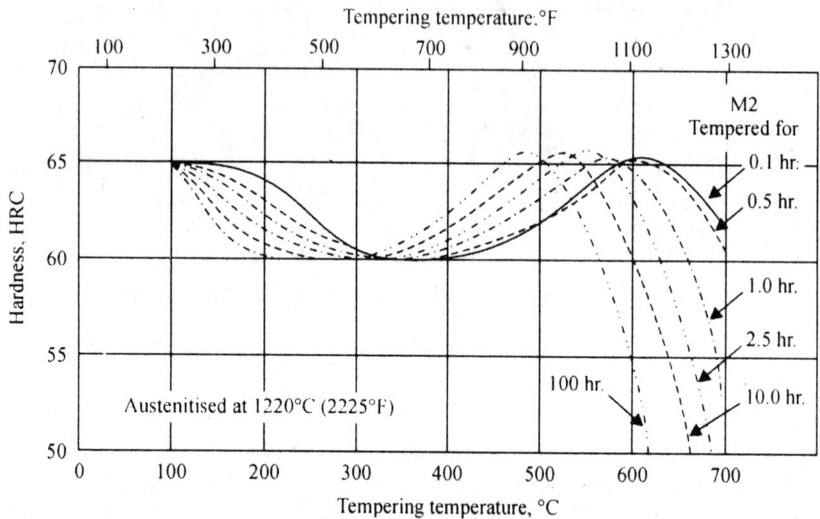

Fig. 17.13. Effect of tempering temperature and time on hardness of M2 high speed steel.

Emphasising the practical time temperature range, the response of several M and T types of high speed steel to tempering at 425 to 705°C (800 to 1300°F) for periods ranging from ½ to 10 hours is indicated in Fig.17.14.

The effect of austenitising temperature on the tempering characteristics of several high speed steels tempered from 480 to 675°C (900 to 1250°F) is shown graphically in Fig.17.15. For all of the steels for which data are plotted, the highest austenitising temperature results in maximum solution of alloy carbides—which, during subsequent tempering, produces the maximum response to secondary hardening.

High speed steels normally are subjected to a minimum of two separate tempering treatments within the range of 540 to 595°C (1000 to 1100°F). The duration of each treatment is usually 2 hours or more at temperature. This process ensures attaining consistent martensitic structures, because the amount of retained austenite in the as-quenched condition will vary significantly because of variations in heat chemistry, prior thermal history, hardening temperature and quenching conditions. It is essential that the time-temperature combination of the first tempering operation be adequate to condition the retained austenite. Consequently, the first tempering treatment is sometimes longer and at a slightly higher temperature than the second, because the latter is used to temper the freshly formed martensite that develops on cooling from the first temper. Moreover, multiple tempering gains in importance in attaining an acceptable structure if short tempering times are used. The hardness of single and double tempered M2 steel austenitised at various temperatures, as affected by tempering temperature, is shown in Fig. 17.16.

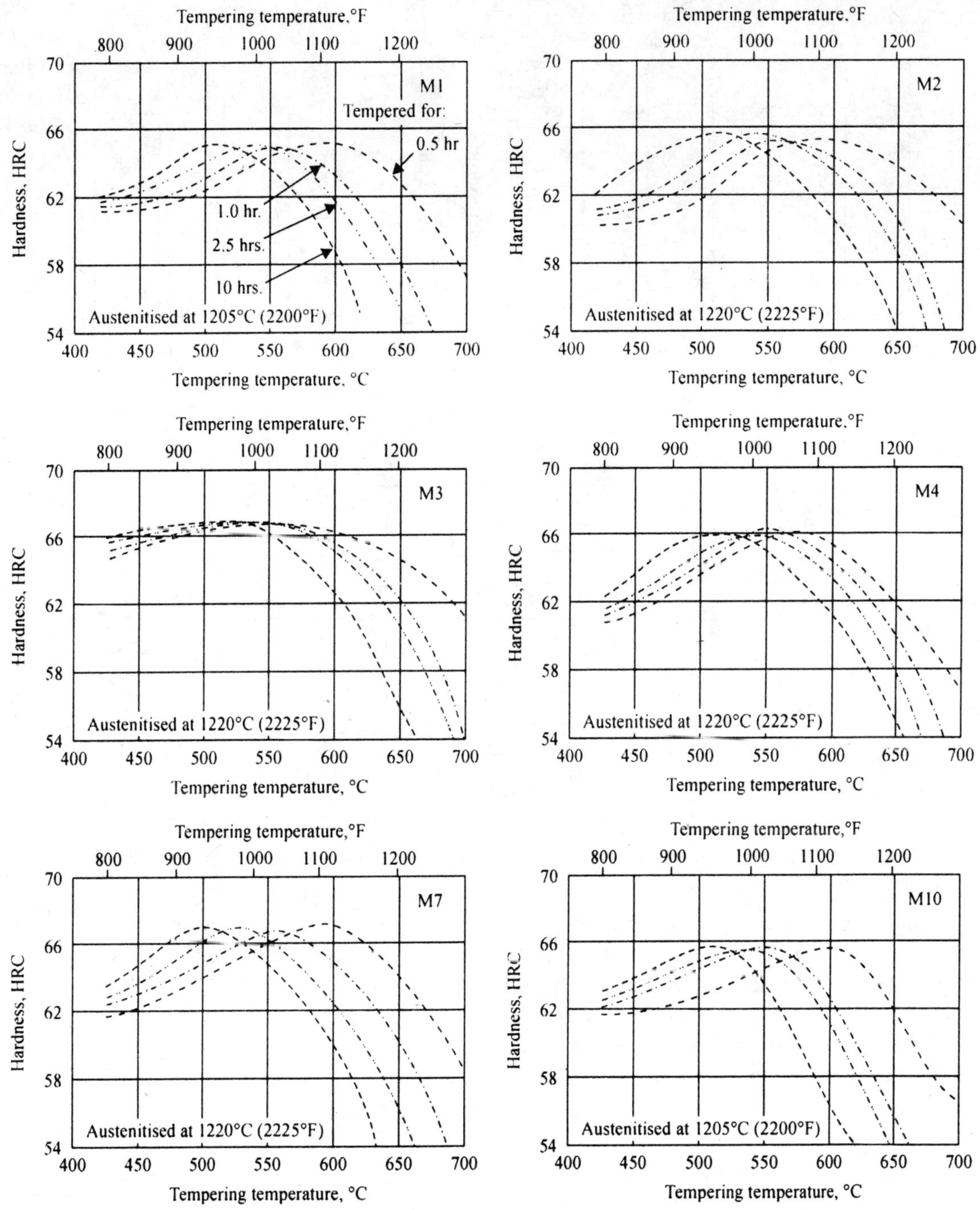

Fig. 17.14. Effect of tempering at temperatures from 425 to 705°C (800 to 1300°F) for periods of ½ to 10 hours on hardness of high speed steels **(Contd...).**

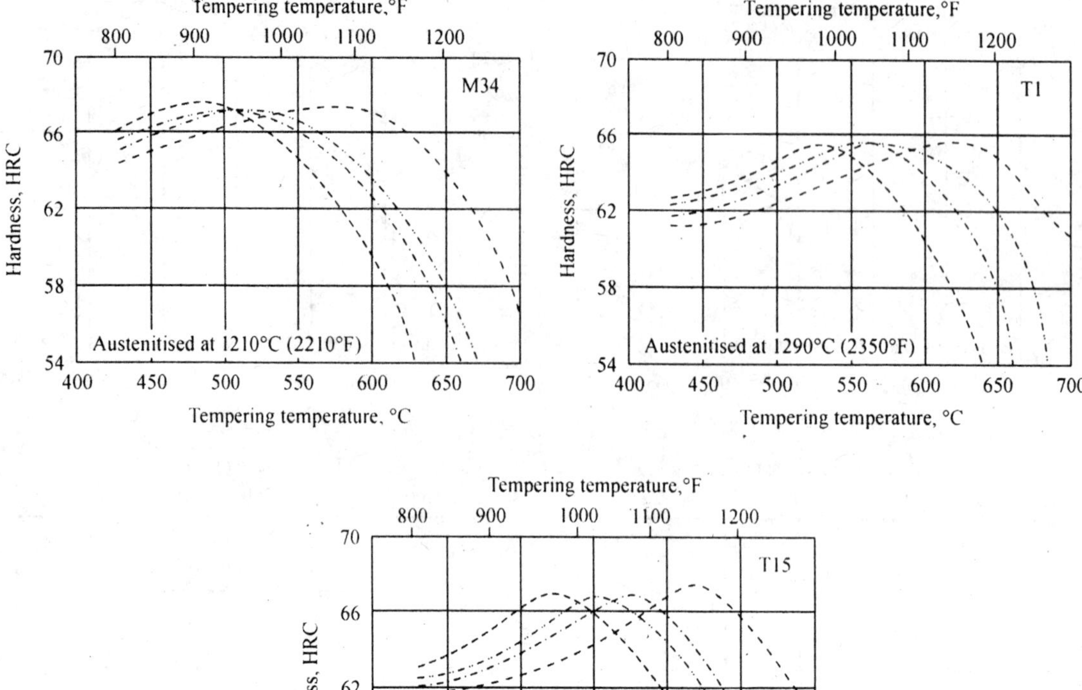

Fig. 17.14. Effect of tempering at temperatures from 425 to 705°C (800 to 1300°F) for periods of ½ to 10 hours on hardness of high speed steels.

Tempering at too low a temperature or for too short a time, or both, may not adequately condition the 20 to 30 per cent retained austenite present after initial quenching, and the steel will still retain abnormally large quantities of austenite after cooling from the initial temper. This austenite will not transform until the steel is cooled from the second temper, and a third temper is then required to temper the martensite so formed. In order to carry these reactions as near to completion as possible, high speed steel should always be cooled to room temperature between tempers. The beneficial effect of multiple tempering on mechanical properties of T1 high speed steel is shown in Table 17.11.

Forced-air furnaces are generally conceded to be the most desirable for tempering high speed steel, because the heat is transmitted from the heating elements to the work by convection; consequently, the transfer of heat is gradual, and there is little danger of the work cracking as the result of thermal shock. It is advisable to place the work in a tempering chamber maintained in the temperature range of 205 to 260°C (400 to 500°F) and to bring the work up to the tempering temperature slowly with the furnace. This is particularly important for large or intricate tools, because too rapid a heating rate may lead to cracking.

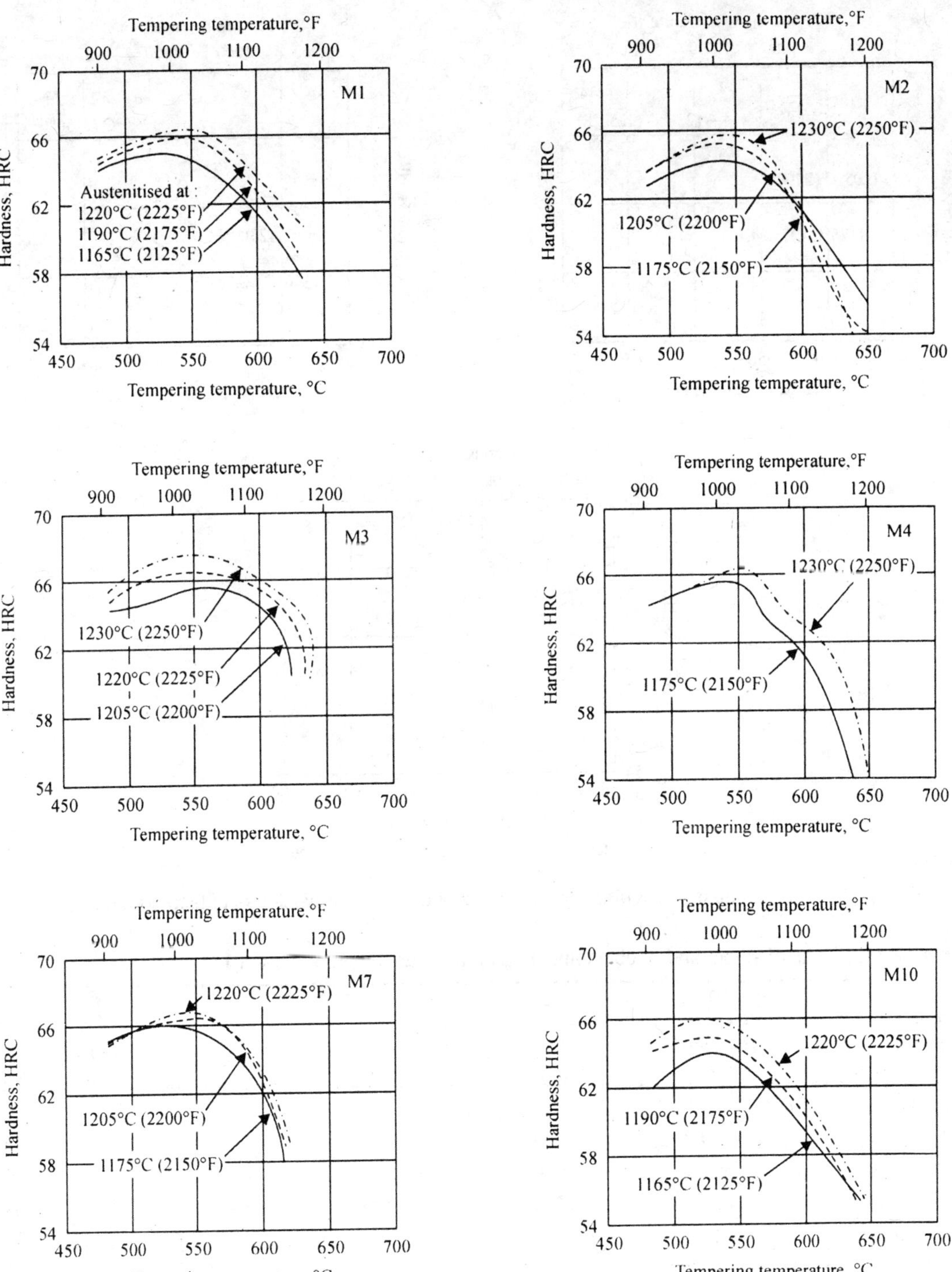

Fig. 17.15. Effect of austenitising and tempering temperatures on hardness of high speed steels **(Contd...).**

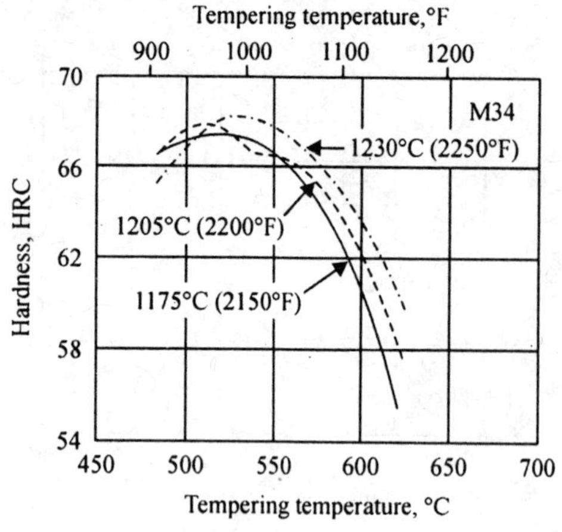

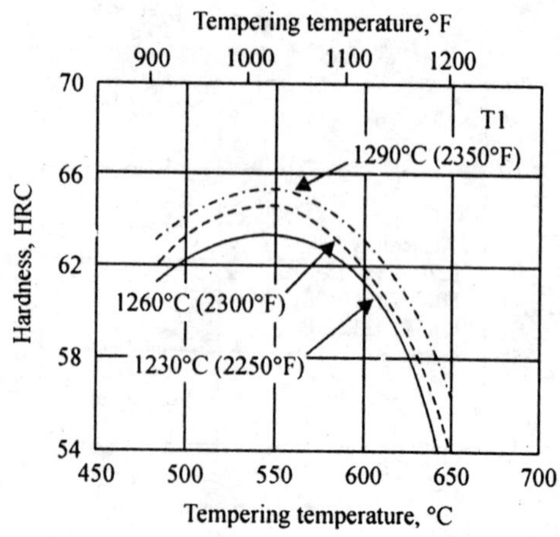

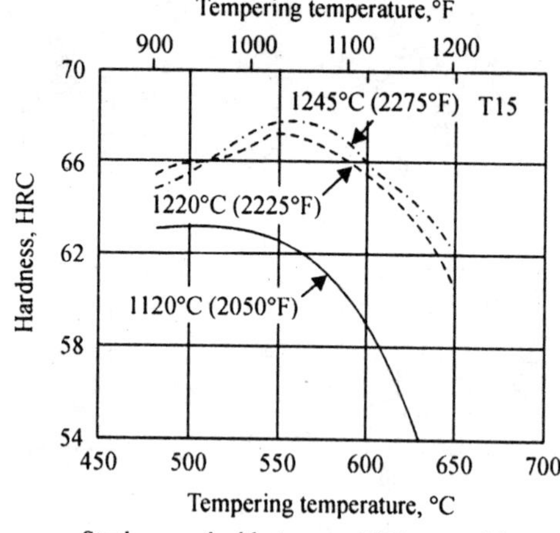

Steels were double tempered (2 hours + 2 hours).

Fig. 17.15. Effect of austenitising and tempering temperatures on hardness of high speed steels.

Table 17.11. Effects of single and double tempering on mechanical properties of T1.

Time at tempering temperature	*Hardness, HRC*	*Bend strength*		*Torsion-impact strength*	
		MPa	*ksi*	*J*	*ft.lb*
Single tempering at 565°C (1050°F)					
6 minutes	65.1	2150	312	22	16
1 hours	65.7	1860	270	41	30
2½ hours	65.0	2810	408	65	48
5 hours	64.5	2590	376	65	48
Double tempering at 565°C (1050°F)					
2½ hours + 2½ hours	64.5	3130	454	85	63

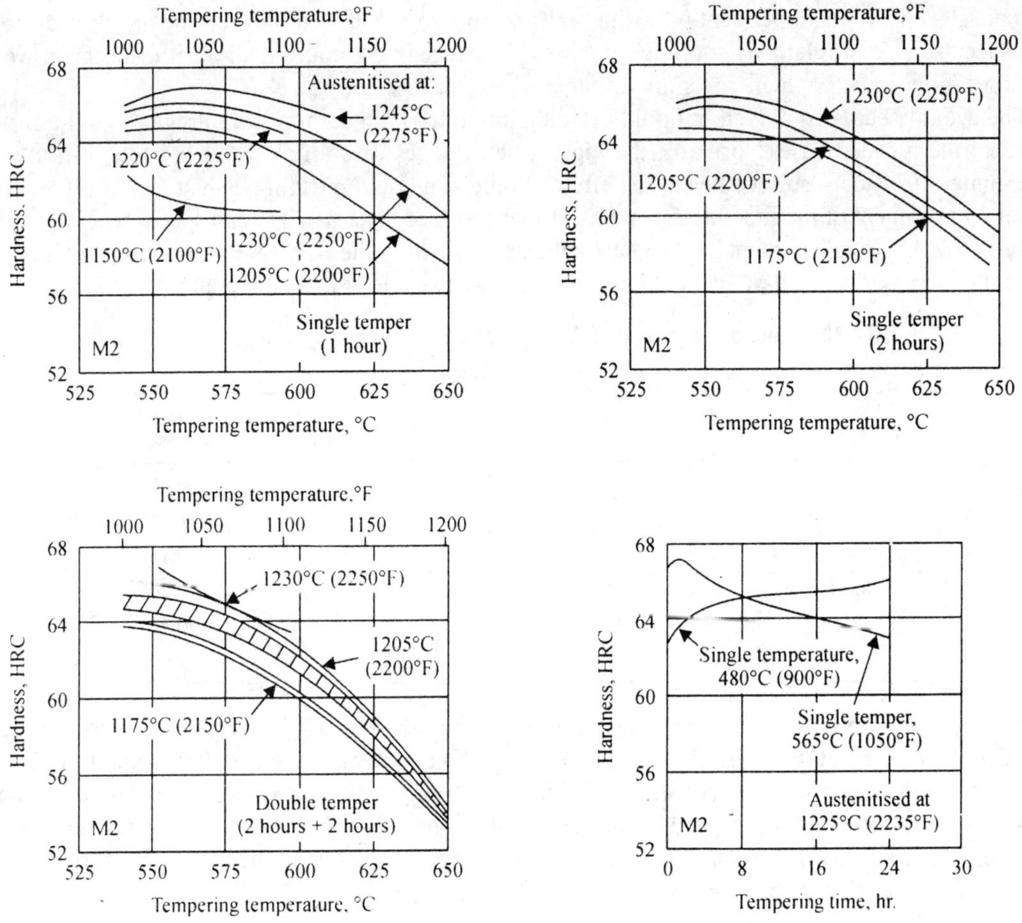

Fig. 17.16. Effect of austenitising temperature and tempering conditions on hardness of M2 speed steel.

The very rapid heating rates of molten lead or salt baths, and the attendant thermal shock, usually militate against their successful use for tempering high speed steel tools of other than simple shape and design, unless they are pre-heated to about 315°C (600°F) before being introduced into the bath.

Refrigeration

Refrigeration treatment may be employed to transform retained austenite. The hardened and tempered tool is cooled to at least –85°C (–120°F) and then tempered or re-tempered at normal tempering temperatures. Carburised surfaces will respond satisfactorily to the –85°C (–120°F) treatment, even when they have been tempered prior to refrigeration.

Nitriding

Liquid nitriding is preferred to gas nitriding for high speed steel cutting tools because it is capable of producing a more ductile case with a lower nitrogen content. Although any of the liquid nitriding baths or processes may be used to nitride high speed steel, the commercial bath consisting of 60 to 70 per cent sodium salts and 30 to 40 per cent potassium salts is most commonly employed. The nitriding cycle for high speed steel is of relatively short duration, seldom exceeding 1 hour; in all other respects, however, the procedures and equipment are similar to those used for low-alloy steels.

The cyanide baths employed in liquid nitriding introduce both carbon and nitrogen into the surface layers of the nitrided case. Normally, the highest percentages of both elements are found in the first 0.025 mm (0.001 inch) surface layer. The effect of time in a liquid nitriding bath at 565°C (1050°F) on the nitrogen content of the first 0.025 mm (0.001 inch) surface layer of a T1 high speed steel is shown in Table 17.12. A nitrogen content of 0.06 per cent was obtained in the first 3 minute at temperature and it gradually increased to 1.09 per cent at the end of a 6 hours cycle at this temperature.

Table 17.12. Effect of nitriding time on surface nitrogen content of T1.

Nitrogen content of first 0.025 mm (0.001 inch) layer.

Time at 565°C (1050°F)	*Nitrogen, %*
3 minutes	0.06
10 minutes	0.093
30 minutes	0.15
90 minutes	0.26
3 hours	0.58
6 hours	1.09

As shown in Table 17.13, carbon also was absorbed by the steel, at nitriding temperatures as low as 455°C (850°F). In a 30 minutes nitriding cycle, the carbon content of the first 0.025 mm (0.001 inch) surface layer increased with an increase in the nitriding temperature. However, it was reported that only a portion of the carbon was absorbed by the steel, most of the carbon being mechanically attached to the surface, filling microscopic pits. (This pitting is not dangerous under normal conditions, because the pits are shallower than ordinary grinding or machining marks).

Table 17.13. Carbon content of nitrided T1.

Carbon content of the first 0.025-mm (0.001-inch) surface layer of steel originally containing 0.705 per cent C. Some of the carbon was in pits on the surface, rather than diffused into the steel.

Nitriding			*Surface carbon, %*
Temperature		*Time, minute*	
°C	*°F*		
455	850	30	0.85
510	950	30	0.99
565	1050	30	1.18
565	1050	360	1.63

High speed steel tools that are nitrided in fresh baths or for short times show steep nitrogen and hardness gradients. To avoid these steep gradients, which are believed responsible for the brittleness of the case after such treatments, the use of longer immersion time, higher temperature, or a thoroughly aged bath is recommended. To avoid brittleness of case when relatively short immersion times are used, the cyanate content of the bath should exceed 6 per cent. These conditions often will lower the surface hardness as well as the hardness gradient.

Figure 17.17 compares the hardness gradients obtained on specimens of T1 high speed steel nitrided at 565°C (1050°F) for 90 minutes in a new bath and for various lengths of time in an aged bath.

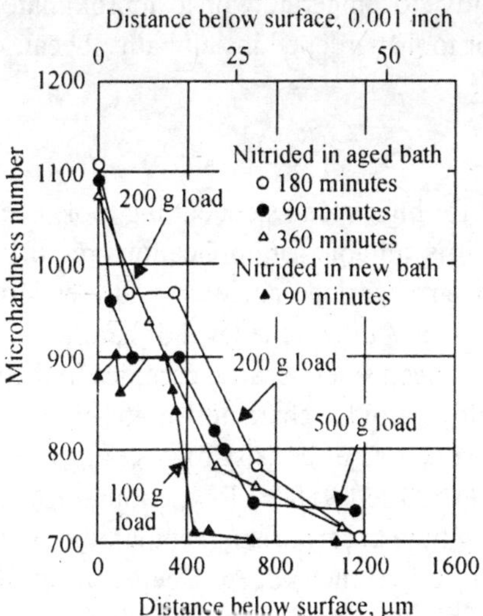

Fig. 17.17. Effect of bath condition and immersion time on hardness gradients in type T1 high speed steel specimens nitrided at 565°C (1050°F).

Nitriding of decarburised high speed steel tools should be avoided, because it results in a brittle surface condition. For those surfaces that have been softened from grinding, nitriding is frequently employed as an offsetting corrective measure.

Liquid nitriding provides high speed steel tools with high hardness and wear resistance and a low coefficient of friction. These properties enhance tool life in two somewhat related ways. The high hardness and wear resistance lower the abrading action of chips and work on the tool, and the low frictional characteristics serve to create less heat at and behind the tool point, in addition to assistant in the prevention of 'chip pick-up'.

Steam Treating

Steam treating produces a uniform, soft layer of iron oxide on the surface of finished high speed steel tools. This layer, approximately 0.005 mm (0.0002 inch) thick, has lubricant-retaining and antigalling properties, and in some applications will improve tool life by reducing too-edge build-up. The oxide layer is removed from the tool after a short interval of operation; during this interval, the cutting surfaces of the tool develop a burnished surface that adds further to antigalling characteristics.

Steam treatment requires a special furnace with a sealed retort from which all air can be displaced by steam, which is admitted at controlled rates. The presence of any moisture in the furnace prior to the admission of the steam will cause rusting and an unsatisfactory surface finish.

A typical processing cycle involves placing the work in the special furnace, heating to approximately 370°C (700°F) and equalising. After a suitable equalising time, which depends on the load, the steam is admitted at controlled rates for approximately ½ hour. The furnace is then partly sealed to develop positive steam pressure, and the temperature is raised to 525°C (975°F). The steam can then be shut off and the work removed from the furnace and cooled normally.

The treatment produces a blue-black film whose appearance is improved by subsequent dipping in oil. This treatment may sometimes be combined with normal tempering treatments, because the type of film produced is relatively insensitive to temperature up to approximately 580°C (1075°F). Steam treating offers an additional advantage for tools hardened in salt baths, because it effectively reduces the pitting that can result from adhering salt.

Carburising

Carburising is not recommended for high speed steel cutting tools because of the extreme brittleness of the case so produced. However, it is suitable for applications requiring extreme wear resistance in the absence of impact or highly concentrated loading, such as are encountered with certain types of cold work dies made from high speed steel. At the same level of hardness, the carburised layer does not have the heat resistance of normal high speed steel because carbides in the microstructure are predominantly Fe_3C, rather than the complex alloy carbides characteristic of high speed steel.

Carburising cycles for high speed steel consist of packing in a carburising medium, heating to approximately 1040 to 1065°C (1900 to 1950°F) long enough to develop the depth of case desired, and air cooling. The usual holding time at carburising temperature is from 10 to 60 minutes, to produce a case 0.05 to 0.25 mm (0.002 to 0.010 inch) deep. Deeper cases should be avoided because of the extreme brittleness developed. This treatment carburises the surface and serves as the austenitising treatment for hardening the entire piece. The carburised layer will harden to 65 to 70 HRC at the surface.

HARDENING OF SPECIFIC HIGH SPEED STEEL TOOLS

Broaches

Broaches require maximum edge hardness because of the continuous cutting action and light chip load to which they are subjected. This indicates a minimum hardness of 65 HRC for the standard grades and 66 HRC for the premium grades of high speed steel.

Broaches should be suspended vertically in the hardening furnace to avoid undue distortion, and should be quenched under controlled and uniform cooling conditions. Broaches should be straightened while still warm from the hardening operation, and should be cooled to at least 65°C (150°F) before tempering. These precautions are particularly important for large diameters.

Chasers

Chasers, because they usually are quite small, present no particular problem in hardening with regard to straightness or residual stress. Hardness recommendations for chasers depend largely on the type of application and the pitch of the thread. For cutting steel, the following hardnesses are commonly used:

Fine-pitch threads	61 to 63 HRC
Coarse-pitch threads	64 to 65 HRC
Acme threads	60 to 62 HRC

For cutting cast iron or plastics, chasers should be heat treated to the maximum attainable hardness, because these materials are cut without any significant cutting force but require maximum abrasion resistance. For Acme threads, however, it is sometimes advisable to underharden.

Milling Cutters

Fine-tooth cutters and those with fragile forms should be hardened to 63 to 64 HRC. Heavy-duty milling cutters and cutters for use on soft, abrasive materials should be hardened to the maximum hardness obtainable for the particular type of steel.

Drills

Hardening techniques for drills vary, depending on the diameter of the drill. Straightness of these tools is extremely important. Various jigging methods are employed, but it is usually advisable to heat treat drills vertically suspended by their shanks in order to reduce distortion in the hardening operation. Straightening is best accomplished in the as-hardened condition before tempering. In tempering, the tempering furnace must not be overloaded, and all drills must receive the correct tempering temperature and time at temperature.

Specific recommendations for the hardness of drills for cutting steel are as follows:

1. Most drills 5 mm (3/16 inch) in diameter and smaller are usually hardened to 63 to 65 HRC. (Drills of this size used for plastics, aluminium or magnesium may have hardness as high as 65 HRC.)
2. Drills over 5 mm (3/16 inch) in diameter, to 63 to 65 HRC.
3. Heavy-duty drills normally use grades of high speed steel providing hardnesses equal to or higher than those noted above. (These drills generally arc designed for maximum rigidity and require maximum abrasion resistance.)

Taps

Taps, like drills, are slender in section and require hardening techniques that minimise distortion. This generally means hardening in thc vertical position suspended in suitable jigs. Taps should be straightened in the as-hardened condition before tempering. Tempering of these tools must be carefully controlled to allow adequate heating time. Specific hardness recommendations for taps that are to be used to cut steel are as follows:

Fine-pitch taps	63 to 65 HRC
Coarse-pitch taps	63 to 65 HRC
Acme taps	62 to 64 HRC
Pipe taps	62 to 64 HRC

Reamers

Reamers encounter a minimum chip load but require maximum wear resistance. For this reason, they are always hardened to the maximum hardness attainable for each grade of steel.

Form Tools

Form tools of all types also should have maximum hardness. In general, a minimum of 65 HRC is necessary, and for the premium grades hardnesses ranging from 68 to 70 HRC are frequently desirable.

Hobs

Because of their shaving action, hobs require maximum edge hardness. They may become oval in shape if they are not placed in the hardening furnace in the vertical position. Such placement may require special fixtures. Techniques and temperatures in both hardening and tempering must be accurately controlled if tools of this type are to be produced successfully and economically.

The hardness of fragile tooth forms may have to be reduced to 62 to 64 HRC to avoid breakage, although the lower hardness results in a shorter production life.

Thread Rolling Dies

Thread rolling dies are usually made of A2 or D2 steel, although dies made of high speed steel frequently afford superior results, particularly in rolling the harder materials. For fragile thread forms, thread rolls should be hardened to 60 to 62 HRC. For heavier thread forms and those used to roll high-strength materials, hardnesses of 63 to 65 HRC are recommended; however, at these higher hardnesses, dies are more susceptible to breakage.

Threading Dies

Most threading dies are made of carbon steel; however, button and acorn dies justify the use of high speed steel. The relation between hardness and thread form for threading dies is the same as that recommended for taps and chasers.

Tool Bits

Standard tool bits, as well as cheeking tools, offset-heat bits and other special types, all require maximum hardness. Standard-duty tool bits should be hardened to 65 to 66 HRC, whereas tool bits made from the higher-alloy high speed steels should be hardened to 67 to 69 HRC when possible.

Bearing Components

The heat treatment of M50 high speed steel bearing components for aerospace applications must be capable of producing a part with high hardness, uniformly fine grain size and dimensional stability over a wide temperature range.

M50 steel has a nominal composition of 0.83C-4.0Cr-4.0Mo-1.0V with a M_s temperature of approximately 163 to 165°C (325 to 330°F). The time-temperature transformation (TTT) diagram for M50 is illustrated in Fig.17.18.

Virtually any cooling rate capable of cooling the austenitised part to 205°C (400°F) or below in 15 minutes will produce high hardness. To minimise distortion, residual stress and crack susceptibility, a cooling similar to the 'idealised' rate shown in Fig.17.18. is desirable.

The following practices and procedures are recommended for heat treating M50 bearing components to provide optimum bearing properties:

1. M50 can be satisfactorily heat treated in vacuum or protective atmosphere furnace. However, most bearing manufacturers prefer to heat treat these bearing components in a neutral molten salt bath or baths.
2. Parts should be pre-heated prior to the austenitising cycle to minimise the required soak time at the high austenitising temperature. If a single pre-heat is employed, a bath temperature of 815 to 875°C(1500 to 1600°F) with a cycle of 5 to 15 minutes is recommended.

If multiple pre-heat baths are available, recommended bath temperatures and cycles are as follows:

Cycle	Temperature °C	°F	Time(a), minute
Two pre-heat baths			
1	675–730	1250–1350	10–15
2	815–870	1500–1600	5–15
Three pre-heat baths			
1	675–730	1250–1350	10–15
2	815–870	1500–1600	5–15
3	955–1010	1750–1850	5–10

(a) Time predicated on relative load size/bath capacity.

3. The high-temperature bath cycle is the most critical operation in heat treating M50 steel. Following pre-heating, parts should be austenitised at 1105 to 1120°C (2025 to 2050°F) for 3 to 10 minutes, depending on cross-section and gross load weight. Optimum cycles in the austenitising bath may be established empirically by varying the soak cycle in the high temperature bath in ½ minute increments and evaluating resultant grain size and hardness. Grain size is more easily measured on as-quenched samples; however, hardness should be checked on parts subsequent to final tempering operations. Ideally, the cycle will be as short as possible to minimise grain growth while producing desired hardness.
4. Following austenitising, parts should be quenched in 540 to 595°C (1000 to 1100°F) molten salt for 5 to 10 minutes. The quench minimises internal stresses and the core-to-surface thermal differential prior to subsequent air cooling and martempering operations.
5. Parts should be subjected to a 175 to 190°C (350 to 375°F) martemper bath for 5 to 15 minutes following quench or quench/air cool operations. The martemper bath, which operates between 15 and 30°C (25 and 50°F) above the M_s temperature for M50, equalises core-to-surface thermal differentials and facilitates subsequent transformation of austenite into martensite with minimal residual stress, distortion or cracking potential. To avoid undesirable intermediate transformation products, the interval between austenitising and martempering should not exceed 15 minutes.
6. Following martempering, parts should be air cooled to room temperature prior to washing, tempering or sub-zero treatment. The air-cooling equipment and conditions should provide uniform cooling of parts from the 175 to 190°C (350 to 375°F) martempering bath to room temperature within 30 to 60 minutes. Shorter cooling rates may result in increased residual stress, distortion or susceptibility to stress cracking.
7. M50 steel requires multiple tempers to provide maximum toughness and dimensional stability. Parts should be subjected to a minimum of three tempers of 540 to 550°C (1000 to 1025°F) for 2 to 4 hours, with cooling to room temperature between each temper. Failure to cool to below 55°C (100°F) between tempers may result in retained austenite. Tempering may be performed either in neutral molten salts or in atmosphere or air furnaces.
8. Subjection to subzero temperatures prior to and/or after initial tempering enhances transformation of retained austenite to martensite. Common deep-freeze cycles for M50 are –70 to –85°C (–90 to –120°F) for 2 to 4 hours. Use of lower temperatures provides little if any added benefit. The deep-freeze cycle provides maximum benefit when employed before tempering; however, it is

not recommended for parts not subjected to martempering or parts susceptible to cracking. When parts are sub-zero treated before tempering, caution should be exercised to ensure that the total elapsed time between martempering and tempering does not exceed 5 hours. Use of prior stress-relief cycles reduces effectiveness of deep-freeze operation. When equipment, time constraints or part design are unfavourable for performing deep freezing prior to tempering, the parts should be subjected to deep freeze between the first and second tempering operations.

9. Parts requiring re-treating should be annealed prior to re-hardening to minimise susceptibility to developing duplex/non-uniform grain.

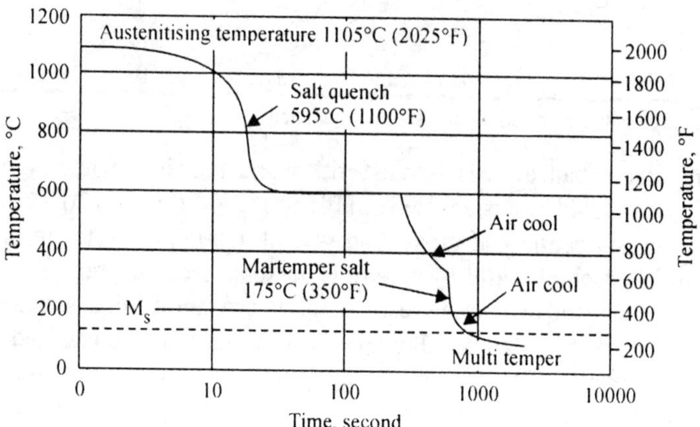

Fig. 17.18. TTT diagram for M50 steel.

LOW-ALLOY SPECIAL-PURPOSE TOOL STEELS

These steels are similar in composition to the water-hardening tool steels, except that the addition of chromium and other elements provides the L steels with greater wear resistance and hardenability. Types L1, L3, L4 and L7 are similar to the production steel 52100 and are used for similar applications.

Because of their relatively low austenitising temperatures, the L steels are easily heat treated. Recommended heat treating practice is summarised in Table17.14.

Normalising

Normalising should follow forging or any other operation in which the steel has been exposed to temperatures substantially above the transformation range. For the L steels, normalising consists of through heating to 870 to 900°C (1600 to 1650°F) (Table 17.14) and cooling in still air. The use of a protective atmosphere is recommended.

Annealing

Annealing must follow normalising and precede any re-hardening operation. Recommended annealing temperatures and cooling rates, as well as expected as-annealed hardness values, are given in Table 17.14.

Stress Relieving

Stress relieving prior to hardening may be advantageous for complex tools, to minimise distortion during hardening. A common practice for complex tools is to rough machine, heat to 620 to 650°C (1150 to 1200°F) for 1 hour per inch of cross-section, cool in air, and then finish machine prior to hardening.

Table 17.14. Recommended heat treating practice for low-alloy special-purpose tool steels.

	Normalising temperature(a)		Annealing					Hardening					
			Temperature(b)		Cooling rate(c)		Annealed hardness, HB	Austenitising temperature(d)		Holding time, min	Quenching medium	Quenched hardness, HRC(e)
Steel	°C	°F	°C	°F	°C/hr.	°F/hr.		°C	°F			
L1	900	1650	775–800	1425–1475	22	40	179–207	790–845	1450–1550	10–30	O,W	64
L2	870–900	1600–1650	760–790	1400–1450	22	40	163–197	790–845	1450–1550	10–30	W	63
L3	900	1650	790–815	1450–1500	22	40	174–201	845–925	1550–1700	10–30	O	63
								775–815	1425–1500	10–30	W	64
								815–870	1500–1600	10–30	O	64
L6	870	1600	760–790	1400–1450	22	40	183–212	790–845	1450–1550	10–30	O	62
L7	900	1650	790–815	1450–1500	22	40	183–212	815–870	1500–1600	10–30	O	64

Note: O, oil; W, water. (a) Holding time, after uniform through heating, varies from about 15 minutes, for small sections, to about 1 hour, for large sections. Work is cooled from temperature in still air. (b) Lower limit of range should be used for small sections, upper limit for large sections. Holding time varies from about 1 hour, for light sections and small furnace charges, to about 4 hours, for heavy sections and large charges; for pack annealing, hold for 1 hour per inch of pack cross section. (c) Maximum. Rate is not critical after cooling to below 540°C (1000°F). (d) These steels are seldom pre-heated. (e) Typical average values; subject to variations depending on austenitising temperature and quenching medium.

Austenitising

Austenitising temperatures recommended for hardening the L steels are listed in Table 17.14; pre-heating is seldom employed for steels in this group. Salt or lead baths and atmosphere furnaces are all satisfactory for austenitising these steels. A neutral salt, such as No. 3 in Table 17.2 is recommended. This salt may be deoxidised, for control of decarburisation, by the method indicated earlier in this chapter in the section on rectification of salt baths.

Quenching

Oil is the quenching medium most commonly used for the L steels. Water or brine may be used for simple shapes, or for large sections that do not attain full hardness by oil quenching. Rolling-mill rolls made of L7 are an example of parts for which water or brine quenching is used. These steels respond well to martempering.

Tempering

Tools made of the L steels should be quenched only to a temperature at which they can be handled with bare hands, about 50°C (125°F), and should be tempered immediately thereafter; otherwise, cracking is likely to occur.

The tempering characteristics of these steels are plotted in Fig. 17.19. For most applications, the S steels are used at near-maximum hardness. It is recommended that tools made of any of these low-alloy steels be tempered at a minimum of 120°C (250°F), even though maximum hardness is desired. Double tempering also is recommended.

CARBON-TUNGSTEN SPECIAL-PURPOSE TOOL STEELS

Recommended heat treating practice for these steels is summarised in Table 17.15.

As a group, these steels are shallow hardening and usually are quenched in water or brine. Steel F3, because of the chromium addition, is the highest in hardenability.

Normalising and Annealing

These steels should be normalised after they have been forged or otherwise subjected to temperatures above their hardening temperatures. Normalising and annealing practices are essentially the same as those recommended in the preceding section on the heat treating of the low-alloy special-purpose steels. Recommendations for normalising and annealing the F steels are given in Table 17.15.

Stress Relieving

Stress relieving as outlined previously for the low-alloy special-purpose steels may be advantageously applied also to the F steels. The same procedure as that described for the L steels would be used.

Austenitising

Pre-heating and austenitising temperatures recommended for the carbon-tungsten special-purpose tool steels are given in Table 17.15. Equipment and practice are generally the same as those previously described for the low-alloy special-purpose steels.

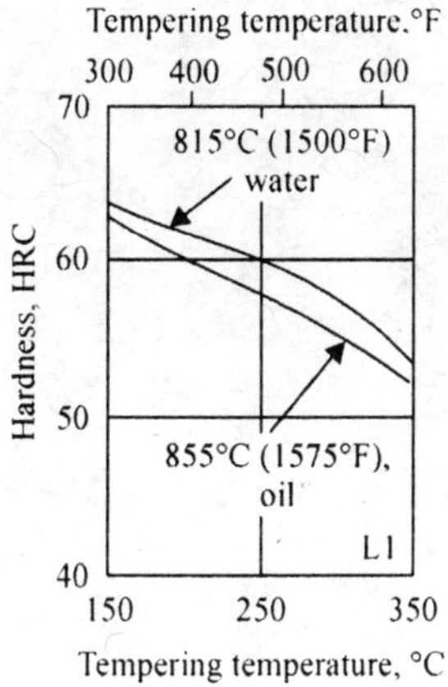

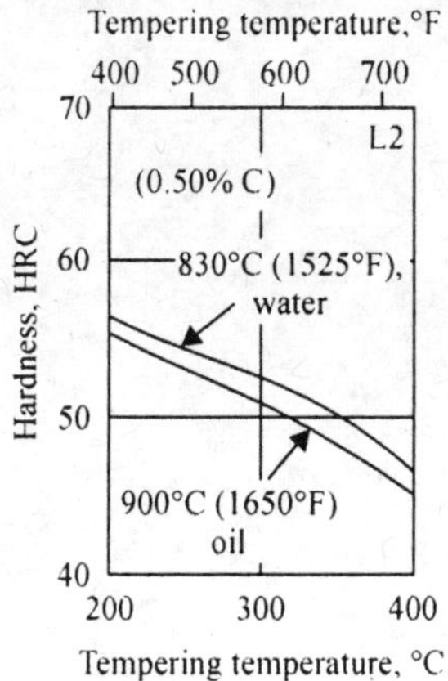

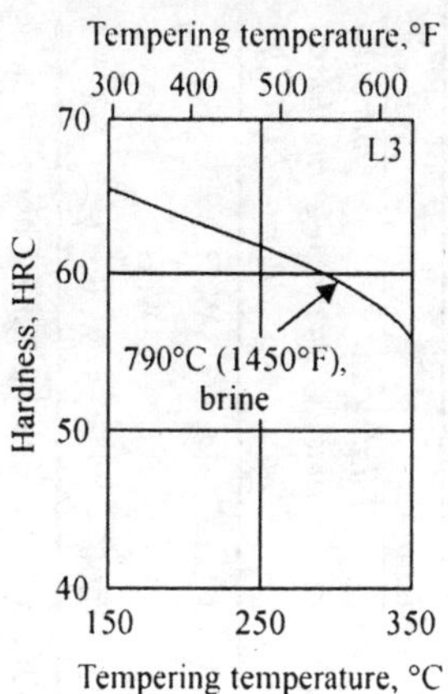

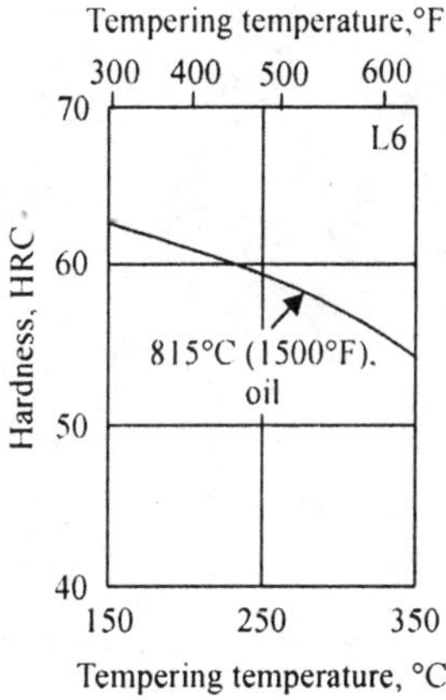

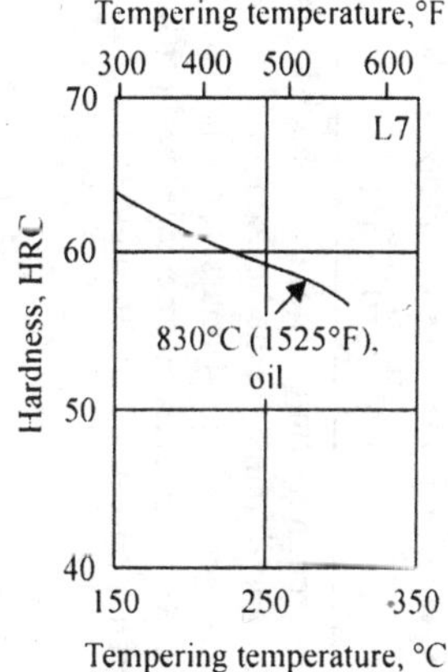

Fig. 17.19. Hardness of low-alloy special-purpose tool steels after tempering for 2 hour

Quenching

Water or brine quenching causes high distortion in parts made of the F steels. This is often used to advantage in the re-hardening of worn dies that have been used for cold drawing of bars and tubes. Such dies are 'flush quenched'—that is, a spout of water is directed into the bore, thus causing shrinkage and allowing further use of dies for the same product size.

Tempering

Because tools made of the F steels (cold drawing dies, for example) are used mainly for applications requiring wear resistance, they are usually placed in service at or near their maximum hardness. Therefore, tempering temperatures higher than 205°C (400°F) are seldom used. Tempering temperature versus hardness for the F steels is shown graphically in Fig. 17.20.

Table 17.15. Recommended heat treating practice for carbon-tungsten special-purpose tool steels.

Steel	Normalising temperature(a)		Annealing					Hardening							
			Temperature(b)		Cooling rate(c)		Annealed hardness,	Temperature				Holding time,	Quenching	Quenched hardness,
								Pre-heat		Austenitise				
	°C	°F	°C	°F	°C/hr.	°F/hr.	HB	°C	°F	°C	°F	min	medium	HRC(e)
F1	900	1650	760–800	1400–1475	20	40	183–207	650	1200	790–870	1450–1600	15	W, B	64
F2	900	1650	790–815	1450–1500	20	40	207–235	650	1200	790–870	1450–1600	15	W, B	66
F3	900	1650	790–815	1450–1500	20	40	212–248	650	1200	790—870	1450–1600	15	W, B, O	66

Note: W, water; B, brine; O,oil. (a) Holding time, after uniform through heating, varies from about 15 minutes, for small sections, to about 1 hour, for large sections. Work is cooled from temperature in still air. (b) Lower limit of range should be used for small sections, upper limit for large sections. Holding time varies from about 1 hour, for light sections and small furnace charges, to about 4 hours, for heavy sections and large charges; for pack annealing, hold for 1 hour per inch of pack cross-section. (c) Maximum cooling rate. Rate is not critical after steel has been cooled to below 540°C (1000°F). (d) Typical average values; subject to variations depending on austenitising temperature and quenching medium employed.

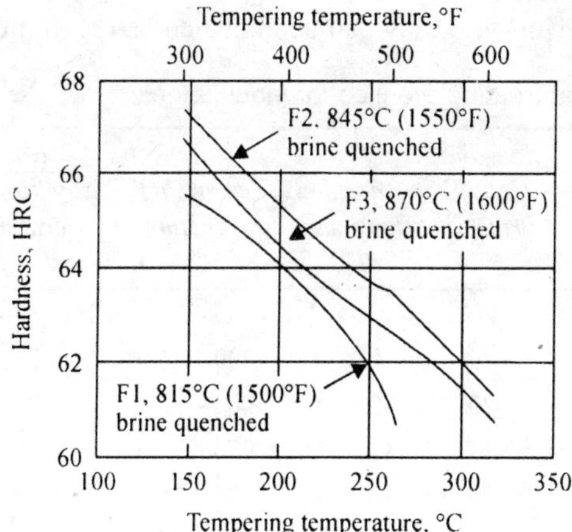

Fig. 17.20. Tempering characteristics of carbon-tungsten special-purpose tool steels tempered 2 hour after being brine quenched.

MOULD STEELS

The wide variations in composition, method of forming the mould cavity, moulding method, and material to be moulded are major influences on choice of mould material as well as method of heat treating. The two most common methods of heat treating the mould steels are: (i) preharden the steel (or partially machined mould or die) to about 30 to 36 HRC, finish machine, and use at this hardness level; and (ii) case harden by carburising. Nitrided moulds have proved successful in some instances, but nitriding is not used extensively.

When moulds are carburised or nitrided, the same procedures are used as for production steels.

Heat treating practice for the mould steels is summarised in Table 17.16. P21 is a special type heat treated by the manufacturer and delivered ready for the user to machine and place in operation without further treatment. As noted in Table 17.16, this steel is hardened by solution treating and ageing.

Annealing

Annealing temperatures and expected resulting hardness values are indicated in Table 17.16. For some types, such as P1, the annealing temperature is not critical. A more important factor is surface protection, especially if the mould cavities will be formed by hubbing. If surfaces are allowed to carburise, even slightly, during annealing, subsequent hubbing will be impaired.

Usually, parts are packed in an inert material such as spent pitch coke and are held at annealing temperature only long enough to become heated through; they are then cooled in the pack to below 540°C (1000°F), after which they may be removed from the pack. If hubbing is to follow, it is usually preferable to use the lower side of the annealing temperature range to minimise the danger of carburising, even though annealing at the higher side of the range will result in slightly lower hardness. Atmosphere-controlled furnaces that can be programmed for slow cooling can also be used for annealing. For hubbing deep cavities, two or more in-process anneals are sometimes required.

When cavities will be formed entirely by machining (sometimes a combination of hubbing and machining is used), annealing usually is neither necessary nor desirable, because slightly harder structures can be machined more easily. Steels as received from the manufacturer are usually suitable for machining.

If hardened moulds require re-working, they can be annealed as recommended in Table 17.16.

Table 17.16. Recommended heat treating practice for mould steels.

		Annealing			*Hardening (after carburising)*				
Steel	*Normalising temperature(a), °C*	*Temper-ature(b) °C*	*Cooling rate(c) °C/hr.*	*Annealed hardness, HB*	*Carburising temperature, °C*	*Austenitising temperature, °C*	*Holding time, min.*	*Quenching medium*	*Quenched hardness, HRC*
P1	Not required	730–900	20	81–101	900–925	790–800	15	W, B	62–64
P2	Not required	730–815	20	103–123	900–925	830–845	15	O	62–65
P3	Not required	730–815	20	109–137	900–925	800–830	15	O	62–64
P4	Not recommended	870–900	15	116–128	970–995	970–995	15	A	62–65
P5	Not required	845–870	20	105–116	900–925	845–870	15	O, W	62–65
P6	Not required	845	8	183–217	900–925	790–815	15	A, O	60–62
P20	900	760–790	20	149–179	870–900(d)	815–870	15	O	58–64
P21	900	Not recommended				Hardened by solution treating and ageing(e)			

Note: W, water; B, brine; O,oil. (a) Holding time, after uniform through heating, varies from about 15 minutes, for small sections, to about 1 hour, for large sections. Work is cooled from temperature in still air. (b) Lower limit of range should be used for small sections, upper limit for large sections. Holding time varies from about 1 hour, for light sections and small furnace charges, to about 4 hours, for heavy sections and large charges; for pack annealing, hold for 1 hour per inch of pack cross-section. (c) Maximum. Rate is not critical after cooling to below 540°C (1000°F). (d) When applicable. (e) Solution treatment: Hold at 705 to 730°C (1300 to 1350°F) for 1 to 3 hours, quench in air or oil; approximate solution treated hardness, 24 to 28 HRC. Ageing treatment: Re-heat to 510 to 550°C (950 to 1025°F); approximate aged hardness, 40 to 30 HRC.

Variations in heat treatment, as necessitated by differences in composition, properties and intended use, are discussed in the following sections for steels P1 to P20.

P1 steel

P1 steel can be carburised by any of the regular practices. Whether the steel is re-heated to the austenitising temperature or quenched from a programmed furnace depends on equipment used. Full hardness (Table 17.16) can be achieved only by water or brine quenching. Practice varies as to working hardness range. This steel usually is used only for hubbed moulds for injection moulding of general-purpose plastics.

A minimum tempering temperature of 175°C (350°F) is recommended. This will retain a finished surface hardness of 60 HRC or slightly higher. However, a more commonly desired hardness range is 54 to 58 HRC, which is obtained by tempering at 260 to 315°C (500 to 600°F). If the distortion encountered from water quenching cannot be tolerated for a particular mould design, a type of mould steel that can be hardened by oil quenching must be used instead of P1.

P2 steel

P2 steel also is a hubbing steel, although it is less easily hubbed than P1. Carburising and hardening practice and the working hardness range are the same as for P1, except that the alloy content of P2 increases hardenability so that full hardness can usually be obtained by oil quenching, thus minimising distortion.

P3 steel

P3 steel is also hubbed, but it is less easily hubbed than P1 or P2. Except that P3 is usually oil quenched, the carburising and hardening practice for it is essentially the same as that outlined above for P1. The operating hardness range may vary from 54 to 64 HRC, but common practice is to temper at about 315°C (600°F) to achieve a final hardness of 54 to 58 HRC.

P4 steel

P4 steel is sometimes used hubbed, but because of its resistance to cold deformation it is more often used for machined moulds or dies. Of all the steels in this group, P4 is the most resistant to wear and to softening by tempering. Because of these properties, it is commonly used for injection moulding of plastics that require high curing temperatures and for dies used for die casting low-melting alloys. For the latter application, a common practice is to carburise P4 in cast iron chips to obtain a slight increase in carbon content at the surface. The effect of carburising practice, as well as case and core hardness values after tempering, is shown in Fig. 17.21.

Because of its high alloy content, P4 steel can be hardened by air cooling. However, it is sometimes quenched in oil to minimise scaling during cooling. For use in plastic moulds, the most common working range is 56 to 60 HRC, which may be obtained by tempering the carburised and hardened moulds at 205 to 315°C (400 to 600°F) (Fig. 17.21).

P5 steel

P5 steel, in which chromium is the major alloying element, approaches P1 in ease of hubbing and has a core strength equivalent to that P3. After carburising, a surface hardness of 65 HRC can be achieved by water quenching, or slightly lower values by oil quenching. Choice of quenching medium depends on mould configuration, allowable distortion, and required hardness. A common working range is 54 to 58 HRC; this can be obtained by tempering at about 260°C(500°F).

P6 steel

P6 steel, because it can seldom be annealed to a hardness of less than 183 HB (Table 17.16), is difficult to hub, and hence it is usually used for machine-cut cavities. It can be carburised by conventional practice. Because of its hardenability, heavy sections of P6 can be oil quenched to full hardness from 790 to 815°(1450 to 1500°F). The as-quenched surface hardness is not quite so high as for some other types, because the high nickel content of P6 promotes retention of austenite. Some of this retained austenite is transformed in tempering, with the result that after tempering, with the result that after tempering up to about 260°C (500°F) the hardness will be little or no lower than that obtained after quenching. By tempering at 315°C (600°F), the most common working hardness range (54 to 58 HRC) is obtained. In some plants, a working hardness range of 58 to 61 HRC, obtained by tempering at 260°C (500°F), is considered preferable.

P20 steel

P20 steel is a popular mould material for either injection or compression moulding, and also for die casting low-melting alloys.

For injection moulding of the general-purpose plastics or die casting of low-melting alloys, P20 is usually used in the pre-hardened condition. It is available at hardness levels of about 300 HB or slightly higher. In this condition, cavities are machined and the dies or moulds placed in service without further heat treatment. Annealed moulds or dies can be austenitised at 845 to 870°C (1550 to 1600°F), oil quenched, and tempered at 540°C (1000°F), to obtain a hardness of about 300 HB.

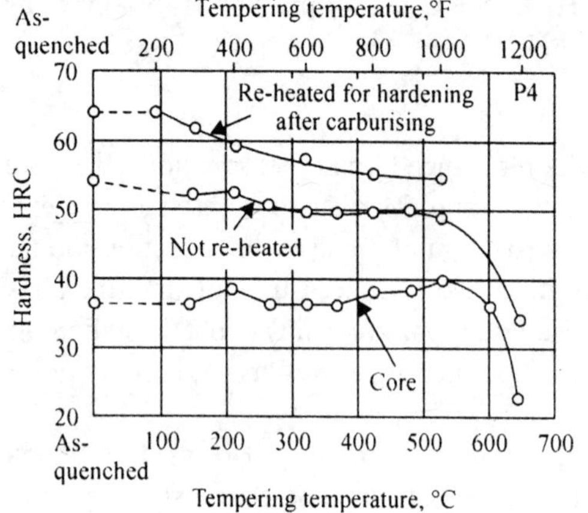

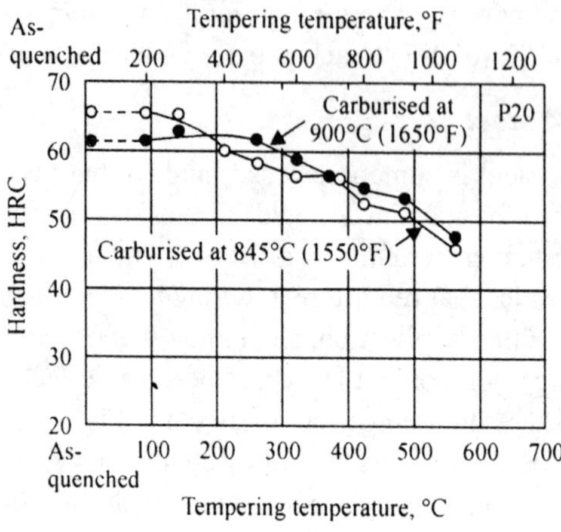

(a) Upper curve represent steel carburised in hardwood charcoal at 915 to 925°C (1675 to 1700°F) for 8 hours air cooled in pack, re-heated at 940 to 955°C (1725 to 1750°F), cooled in air and tempered. Middle curve represents steel carburised in cast iron chips. at 940 to 955°C (1725 to 1750°F), removed from pack, cooled in air and tempered.
(b) Surface hardness after heating at temperature for 2 hours in carburising compound, oil quenching, and tempering.

Fig. 17.21. Tempering characteristics of carburised mould steels.

Type P20 is often carburised for moulds used in compression moulding, particularly for moulding the more abrasive plastics. Carburising temperatures no higher than 900°C (1650°F) are recommended for this steel, because higher temperatures may impair polishability; otherwise, conventional carburising practice is used, and moulds may be quenched in oil directly from the carburising temperature. A common working range is 54 to 58 HRC.

Tempering characteristics for P20 carburised at two different temperatures are given in Fig. 17.21.

This steel is sometimes nitrided for special applications. Conventional nitriding practice is employed. Before being nitrided, P20 should first be quenched and tempered to about 300 HB as outlined above, and cavities should be machined; following this sequence will ensure freedom from carburisation or decarburisation.

Chapter 18

Control of Distortion in Tool Steels

INTRODUCTION

Distortion in tool steel parts includes all irreversible changes in size and shape that result from processing, from heat treatment and from temperature variations and loading in service. A basic understanding of distortion is important for two reasons. First, most tool steel parts must interact with other parts in service and excessive distortion may prevent them from interacting in the intended manner. Second, finishing operations for correcting distortion not only are expensive but also may destroy some desirable properties and introduce others that are undesirable.

Changes in size or shape of tool steel parts may be either reversible or irreversible. Reversible changes are those caused by stressing in the elastic range or by temperature variations that neither cause changes in the metallurgical structure nor induce stresses that exceed the elastic range. Under such conditions, the initial dimensional values can be restored by a return to the original state of stress or temperature.

The upper limit of reversible dimensional change in a tool steel is determined by the stress required to initiate deformation (that is, the elastic limit corresponding to a pre-selected value of plastic strain), the elastic deformation per unit stress (modulus of elasticity), the effect of temperature on these properties, the coefficient of thermal expansion and the temperature-time combinations at which stress relief and phase changes occur.

For practical purposes the modulus of elasticity of all tool steels, regardless of composition or heat treatment, is 210 GPa (30×10^6 psi) at room temperature. Therefore, if a tool steel part deforms excessively under service loading but returns to its original dimensions when the load is removed, a change in grade or type of tool steel, or in heat treatment will not be useful. To counteract excessive elastic distortion it is necessary to: (i) reduce the applied stress by increasing the section size; and (ii) use a tool material with a higher modulus of elasticity (such as cemented tungsten carbide).

Irreversible changes in size or shape of tool steel parts are those caused by stresses that exceed the elastic limit or by changes in metallurgical structure (most notably, phase changes). Such irreversible changes sometimes can be corrected by thermal processing (annealing, tempering or cold treating) or by mechanical processing to remove excess material or to re-distribute residual stresses.

NATURE AND CAUSES OF DISTORTION

Distortion is a general term encompassing all irreversible dimensional changes. There are two main types: size distortion, which involves expansion or contraction in volume or linear dimensions without changes in geometrical form; and shape distortion, which entails changes in curvature or angular relations, as in twisting, bending, and/or non-symmetrical changes in dimensions. Frequently, both size distortion and shape distortion (Fig. 18.1) occur during a heat treating operation.

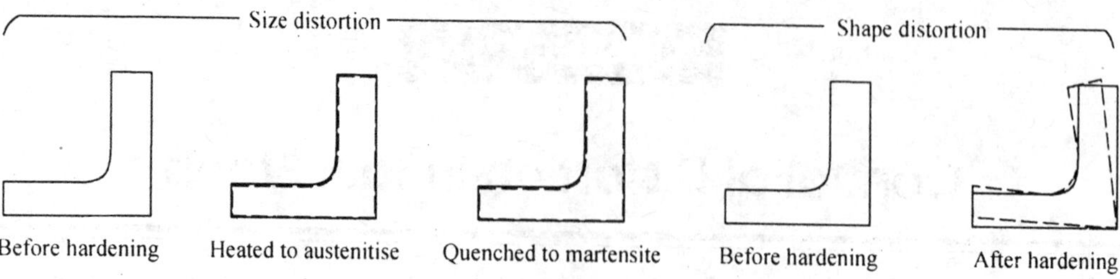

Fig. 18.1. Size and shape distortion in hardening.

Size distortion is the result of a change in volume produced by a change in metallurgical structure during heat treatment. Shape distortion results from either residual or applied stresses. Residual stresses developed during heat treatment are caused by thermal gradients within the metal (producing differing amounts of expansion or contraction) by non-uniform changes in metallurgical structure and by non-uniformity in the composition of the metal itself, such as that due to segregation.

Steps Involved During Changes in Metallurgical Structure While Heat Treatment of Tool Steel

Changes in metallurgical structure during heat treatment of tool steels are produced by the three steps described below.

The first step involves heating an annealed structure (usually consisting of ferrite and spheroidal carbides, commonly called spheroidite) to about 800°C (1450°F) or higher to change the ferrite to austenite and to dissolve all or most of the spheroidal carbides to the austenite. For plain carbon or low-alloy tool steels, austenitising results in a contraction in volume. The extent of volumetric contraction decreases with increasing amounts of carbon present in the composition. This can be approximated as follows:

$$V_{SA} = -4.64 + 2.21\,(\%\,C) \qquad \text{... (18.1)}$$

where V_{SA} is the volume change in per cent that occurs when spheroidite transforms to austenite. By use of this equation, it can be estimated that, if heated to a temperature high enough to dissolve all of the carbon in the austenite, a 0.50 per cent carbon tool steel would exhibit a volume change of –3.53 per cent, a common type containing 1 per cent carbon would exhibit a change of –2.43 per cent and a very-high-carbon type containing 1.5 per cent carbon would exhibit a change of –1.33 per cent.

However, tool steels having carbon contents higher than that of the eutectoid composition are normally austenitised at temperatures only high enough to dissolve the eutectoid amount of carbon. Under these circumstances, 1 per cent carbon and 1.5 per cent carbon tool steels would exhibit changes in volume of –2.77 and –2.53 per cent, respectively, after austenitising. These percentages are less than that calculated directly from Eq. 18.1 because an allowance must be made for the volume occupied by undissolved carbides, which is about 3.5 per cent for the 1.0 per cent carbon steel and about 12 per cent for the 1.5 per cent carbon steel. The second step involves cooling fast enough to cause the austenite to transform to martensite. The steel expands on transformation, the amount of expansion being in inverse proportion to the amount of carbon in solution in the austenite:

$$V_{AM} = 4.64 - 0.53\,(\%C) \qquad \text{... (18.2)}$$

where V_{AM} is the per cent volume change that occurs when austenite transforms to martensite. By use of Eq. 18.2, it can be estimated that a 0.5 per cent carbon tool steel would exhibit a volume increase for this transformation of 4.37 per cent and that 1.0 and 1.5 per cent carbon steels would exhibit increases of 4.07 per cent and 3.71 per cent, respectively, if austenitised at the normal austenitising temperature (only 0.8 per cent carbon, the eutectoid amount of carbon, in solution and again allowing for the volume occupied by undissolved carbides).

Table 18.1. Microconstituents in various tool steels after hardening.

Steel	*Hardening treatment*	*As-quenched hardness, HRC*	*Martensite, vol %*	*Retained austenite vol %*	*Undissolved carbides, vol %*
W1	790°C (1450°F), 30 minute; WQ	67.0	88.5	9	2.5
L3	845°C (1550°F), 30 minute; OQ	66.5	90	7	3.0
M2	1225°C (2235°F), 6 minute; OQ	64	71.5	20	8.5
D2	1040°C (1900°F), 30 minute, AC	62	45	40	15

Equations 18.1 and 18.2 can be used to calculate the net change in dimensions in a tool steel when it is heat treated to transform it from an annealed to a fully hardened (martensitic) state. For the examples referred to above, normal heat treatment would produce net volume increase of $-3.53 + 4.37 = 0.84$ per cent in the –0.5 per cent carbon tool steel, $-2.77 + 4.07 = 1.30$ per cent in the 1.0 per cent carbon steel and $-2.53 + 3.71 - 1.18$ per cent in the 1.5 per cent carbon steel. Net changes in linear dimensions would be about one third the corresponding net changes in volume.

The third step involves re-heating the freshly formed martensite to relatively low temperatures (tempering) to increase toughness and reduce lattice stress. Tempering produces various changes in metallurgical structure, depending on temperature and time at temperature.

After very long times at room temperature or shorter times at temperatures up to 200°C (400°F), the high carbon martensite in plain carbon and low-alloy tool steels decomposes into low-carbon martensite (about 0.25 per cent carbon) plus epsilon carbide, with an accompanying contraction in volume. At higher tempering temperatures, 200 to 430°C (400 to 800°F), the martensite decomposes into ferrite plus cementite.

Transformation of the maximum amount of austenite to martensite on quenching usually requires continuous cooling to below the martensite-finish temperature (M_f), which for a eutectoid tool steel is about –50°C (–60°F). To prevent cracking of very large or very intricate pieces, it is common practice to remove the tool from the quenching medium and begin tempering it while it is still slightly too warm to hold comfortably in the bare hands (about 60°C, or 140°F). Under these conditions, a substantial proportion of the structure (10 per cent or more) may still be austenite. Most alloying elements lower the M_f temperature. Consequently, more austenite is retained at room temperature in the more highly alloyed tool steels. On tempering at increasing temperatures in the range 120 to 260°C (250 to 500°F), increasing amounts of this retained austenite transform to bainite for some tool steel compositions with an accompanying expansion in volume.

Table 18.2. Typical dimensional changes in hardening and tempering.

Tool steel	Hardening treatment			Total change in linear dimensions, % after quenching	Total change in linear dimensions, %, after tempering at:											
	Temperature		Quenching medium		°C	150	205	260	315	370	425	480	510	540	565	595
	°C	°F			°F	300	400	500	600	700	800	900	950	1000	1050	1100
O1	815	1500	Oil	0.22		0.17	0.16	0.18	–	–	–	–	–	–	–	–
O1	790	1450	Oil	0.18		0.09	0.12	0.13	–	–	–	–	–	–	–	–
O6	790	1450	Oil	0.12		0.07	0.10	0.14	0.10	0.00	–0.05	–0.06	–	–0.07	–	–
A2	955	1750	Air	0.09		0.06	0.06	0.08	0.07	–	0.05	0.04	–	0.06	–	–
A10	790	1450	Air	0.04		0.00	0.00	0.08	0.08	0.01	0.01	0.02	–	0.01	–	0.02
D2	1010	1850	Air	0.06		0.03	0.03	0.02	0.00	–	–0.01	–0.02	–	0.06	–	–
D3	955	1750	Oil	0.07		0.04	0.02	0.01	–0.02	–	–	–	–	–	–	–
D4	1040	1900	Air	0.07		0.03	0.01	–0.01	–0.03	–	–0.4	–0.03	–	0.05	–	–
D5	1010	1850	Air	0.07		0.03	0.02	0.01	0.00	–	0.3	0.03	–	0.05	–	–
H11	1010	1850	Air	0.11		0.06	0.07	0.08	0.08	–	0.3	0.01	–	0.12	–	–
H13	1010	1850	Air	–0.01		–	–	–	–	–	–	0.00	–	0.06	–	–
M2	1210	2210	Oil	–0.02		–	–	–	–	–	–	–	–0.06	0.10	0.14	0.16
M41	1210	2210	Oil	–0.16		–	–	–	–	–	–	–	–0.17	0.08	0.21	0.23

Depending on the alloy content of the tool steel, all, some, or none of the retained austenite will transform during tempering. In some highly alloyed tool steel compositions, cementite redissolves at tempering temperature of 540 to 595°C (1000 to 1100°F) to form alloy carbides, which induces an additional expansion in volume. The formation of alloy carbides during tempering is characteristic of tool steels containing large amounts of carbide-forming elements such as chromium, molybdenum and tungsten, which are found in high-speed tool steels.

SIZE DISTORTION IN TOOL STEELS

Typical volume percentages of martensite, retained austenite and undissolved carbides are given in Table 18.1 for four different tool steels is quenched from their recommended austenitising temperatures.

Typical changes in linear dimensions for several tool steels are given in Table 18.2. As shown in this table, some tool steels such as A10 show very little size change when hardened and tempered over the entire range from 150 to 600°C (300 to 1100°F).

Other types, such as the M2 and M41 high speed steels, expand about 0.2 per cent (2 mm/m, or 0.002 inch/inch) when hardened and tempered in the temperature range of 540 to 595°C (1000 to 1100°F) to develop full secondary hardness. Although the information in Table 18.2 is useful in comparing size distortion in several tool steels, the factor of shape distortion makes it impossible to use these data alone to predict dimensional changes of a particular tool made from any of these steels. Densities and thermal expansion characteristics for several classes of tool steels are presented in Table 18.3.

SHAPE DISTORTION IN TOOL STEELS

The strength of any tool steel decreases rapidly above about 600°C (1100°F). At the austenitising temperature, the yield strength is so low that plastic deformation often occurs simply from the stresses induced in the part by gravity. Therefore, long parts, large parts and parts of complex shape must be properly supported at critical locations to prevent sagging at the hardening temperature.

Rapid heating increases shape distortion, especially in large tools and in complex tools containing both light and heavy sections. If the rate of heating is high, light sections will increase in temperature much faster than heavy sections. Likewise, the outer surfaces in heavy sections will increase in temperature much faster than the interior. Differences in thermal expansion due to the differences in temperature between light and heavy sections or between surface and interior in heavy sections will be enough to set up large stresses in the material. Under these stresses, the hotter regions will deform plastically, to relieve the thermally induced stress.

Eventually, the hotter portions will reach the furnace temperature, while the cooler portions will continue to increase in temperature. At this point, a decrease in thermal differential begins, which will cause a partial reversal in thermal stress which produced plastic deformation when the temperature differential was high. This may cause the part to undergo further plastic deformation, but to a lesser extent than the deformation caused by the initial high temperature differential. Such deformation will occur in a different direction. Slow heating minimises distortion by keeping temperature differentials low and thermal stresses within the elastic range of the material throughout the heating cycle. Ideally, all heat treatment of tool steel parts should start from a cold furnace to provide the greatest freedom from shape distortion during heating. Starting from a cold furnace is neither very practical nor energy efficient unless heat treating is being done in a vacuum furnace. When heat treating in fused salt or an atmosphere furnace, preheating the parts at an intermediate temperature prior to heating them to the austenitising temperature provides the best compromise.

Table 18.3. Density and thermal expansion of selected tool steels.

Type	Density		Thermal expansion									
			μm/m.K from 20°C to:					μin./in. °F from 68°F to:				
	Mg/m³	lb/in.³	100°C	205°C	425°C	540°C	650°C	200°F	400°F	800°F	1000°F	1200°F
W1	7.84	0.283	10.4	11.0	13.1	13.8(a)	14.2(b)	5.76	6.13	7.28	7.64(a)	7.90(b)
W2	7.85	0.283	–	–	14.4	14.8	14.9	–	–	8.0	8.2	8.3
S1	7.88	0.255	12.4	12.6	13.5	13.9	14.2	6.9	7.0	7.5	7.7	7.9
S2	7.79	0.281	10.9	11.9	13.5	14.0	14.2	6.0	6.6	7.5	7.8	7.9
S5	7.76	0.280	–	–	12.6	13.3	13.7	–	–	7.0	7.4	7.6
S6	7.75	0.280	–	–	12.6	13.3	–	–	–	7.0	7.4	–
S7	7.76	0.280	–	12.6	13.3	13.7(a)	13.3	–	7.0	7.4	7.6(a)	7.4
O1	7.85	0.283	–	10.6(c)	12.8	14.0(d)	14.4(d)	–	5.9(c)	7.1	7.8(d)	8.0(d)
O2	7.66	0.277	11.2	12.6	13.9	14.6	15.1	6.2	7.0	7.7	8.1	8.4
O7	7.8	0.283	–	–	–	–	–	–	–	–	–	–
A2	7.86	0.284	10.7	10.6(c)	12.9	14.0	14.2	5.96	5.91(c)	7.2	7.8	7.9
A6	7.84	0.283	11.5	12.4	13.5	13.9	14.2	6.4	6.9	7.5	7.7	7.9
A7	7.66	0.277	–	–	12.4	12.9	13.5	–	–	6.9	7.2	7.5
A8	7.87	0.284	–	–	12.0	12.4	12.6	–	–	6.7	6.9	7.0
A9	7.78	0.281	–	–	12.0	12.4	12.6	–	–	6.7	6.9	7.0
D2	7.70	0.278	10.4	10.3	11.9	12.2	12.2	5.8	5.7	6.6	6.8	6.8
D3	7.70	0.278	12.0	11.7	12.9	13.1	13.5	6.7	6.5	7.2	7.3	7.5
D4	7.70	0.278	–	–	12.4	–	–	–	–	6.9	–	–
D5	–	–	–	–	–	12.0	–	–	–	–	6.7	–
H10	7.81	0.281	–	–	12.2	13.3	13.7	–	–	6.8	7.4	7.6
H11	7.75	0.280	11.9	12.4	12.8	12.9	13.3	6.6	6.9	7.1	7.2	7.4
H13	7.76	0.280	10.4	11.5	12.2	12.4	13.1	5.8	6.4	6.8	6.9	7.3
H14	7.89	0.285	11.0	–	–	–	–	6.1	–	–	–	–
H19	7.98	0.288	11.0	11.0	12.0	12.4	12.9	6.1	6.1	6.7	6.9	7.2
H21	8.28	0.299	12.4	12.6	12.9	13.5	13.9	6.9	7.0	7.2	7.5	7.7
H22	8.36	0.302	11.0	–	11.5	12.0	12.4	6.1	–	6.4	6.7	6.9
H26	8.67	0.313	–	–	–	12.4	–	–	–	–	6.9	–
H42	8.15	0.295	–	–	–	11.9	–	–	–	–	6.6	–
T1	8.67	0.313	–	9.7	11.2	11.7	11.9	–	5.4	6.2	6.5	6.6
T2	8.67	0.313	–	–	–	......	–	–	–	–	–	–
T4	8.68	0.313	–	–	–	11.9	–	–	–	–	6.6	–
T5	8.75	0.316	11.2	–	–	11.5	–	6.2	–	–	6.4	–
T6	8.89	0.321	–	–	–	–	–	–	–	–	–	–

(Contd...)

Type	Density		Thermal expansion									
			µm/m.K from 20°C to:					µin./in. °F from 68°F to:				
	Mg/m³	lb/in.³	100°C	205°C	425°C	540°C	650°C	200°F	400°F	800°F	1000°F	1200°F
T8	8.43	0.305	–	–	–	–	–	–	–	–	–	–
T15	8.19	0.296	–	9.9	11.0	11.5	–	–	5.5(c)	6.1	6.4	–
M1	7.89	0.285	–	10.6(c)	11.3	12.0	12.4	–	5.9(c)	6.3	6.7	6.9
M2	8.16	0.295	10.1	9.4(c)	11.2	11.9	12.2	5.6	5.2(c)	6.2	6.6	6.8
M3, class 1	8.15	0.295	–	–	11.5	12.0	12.2	–	–	6.4	6.7	6.8
M3, class 2	8.16	0.295	–	–	11.5	12.0	12.8	–	–	6.4	6.7	7.1
M4	7.97	0.288	–	9.5(c)	11.2	12.0	12.2	–	5.3(c)	6.2	6.7	6.8
M7	7.95	0.287	–	9.5(c)	11.5	12.2	12.4	–	5.3(c)	6.4	6.8	6.9
M10	7.88	0.255	–	–	11.0	11.9	12.4	–	–	6.1	6.6	6.9
M30	8.01	0.289	–	–	11.2	11.7	12.2	–	–	6.2	6.5	6.8
M33	8.03	0.290	–	–	11.0	11.7	12.0	–	–	6.1	6.5	6.7
M36	8.18	0.296	–	–	–	–	–	–	–	–	–	–
M41	8.17	0.295	–	9.7	10.4	11.2	–	–	5.4	5.8	6.2	–
M42	7.98	0.288	–	–	–	–	–	–	–	–	–	–
M 46	7.83	0.283	–	–	–	–	–	–	–	–	–	–
M47	7.96	0.288	10.6	11.0	11.9	–	12.6	5.9	6.1	6.6	–	7.0
L2	7.86	0.284	–	–	14.4	14.6	14.8	–	–	8.0	8.1	8.2
L6	7.86	0.284	11.3	12.6	12.6	13.5	13.7	6.3	7.0	7.0	7.5	7.6
P2	7.86	0.284	–	–	13.7	–	–	–	–	7.6	–	–
P5	7.80	0.282	–	–	–	–	–	–	–	–	–	–
P6	7.85	0.283	–	–	–	–	–	–	–	–	–	–
P20	7.85	0.283	–	–	12.8	13.7	14.2	–	–	7.1	7.6	7.9

(a) From 20 to 500°C (68 to 930°F). (b) From 20 to 600°C (68 to 1110°F). (c) From 20 to 260°C (68 to 500°F). (d) From 38°C (100°F).

During quenching, large temperature differences between surface and interior and between light and heavy sections, can cause severe shape distortion, because of thermal stress and mechanical stress produced by a martensitic transformation. This problem is most severe if the hardenability of the steel is so low that a fast cooling rate is required to obtain full hardness. In such a situation, especially when making a large or complex part, it may be best to substitute a high-hardenability, air-hardening tool steel, which requires only a slow cooling rate to fully harden.

However, if lower-hardenability steels requiring liquid quenching are used, fixturing and pressure die quenching can help minimise distortion. Long symmetrical parts should be fixtured and should be quenched in the vertical position with vertical agitation of the quench mediums.

SPECIAL TECHNIQUES FOR CONTROLLING SHAPE DISTORTION

Special quenching procedures such as martempering and austempering may also be useful for controlling distortion in parts which have an appropriate configuration and been made of material having appropriate hardenability. In martempering, parts are quenched in hot molten salt fast enough to avoid transformation to high-temperature transformation products such as ferrite or pearlite. The parts are held at a bath temperature in the range from slightly above to slightly below the M_s just long enough to equalise the interior and surface temperatures. The parts are then removed from the bath and allowed to air cool to room temperature. Slow cooling through the martensitic transformation range reduces distortion as compared with rapid quenching. Martempered tools must be given the usual tempering treatment.

Austempering can be used to reduce distortion if a hardness no higher than 57 HRC is acceptable for the application. In austempering, parts are also quenched in hot molten salt but by temperature selection are forced to transform into bainite rather than martensite. Bainite forms at temperature above those at which martensite forms. The parts must be held long enough at a temperature above the M_s (usually about 230°C or 450°F) to permit the austenite to transform to lower bainite. When air cooled to room temperature, austempered tools exhibit less shape distortion and generally require no subsequent tempering.

Besides being reduced through control of rates of heating and cooling, shape distortion can be reduced by employing a localised method of heating and quenching such as flame hardening, induction hardening, electron beam or laser hardening to treat only that portion of the tool that must be hardened.

Controlling Out-of-Roundness

Controlling out-of-roundness is important for certain precision applications, such as class C and D cutting hobs made of high speed steels. Class C and D hobs must be held to close size limits because they are not ground to size after heat treatment, but rather are used-in the unground condition.

Normal size distortion in hardening and tempering can be accommodated by making the tool slightly oversize or slightly undersize, as required, before heat treating. High speed steel bars, however, have been observed to go out-of-round as much as 0.05 mm (0.002 inch) during heat treatment. The pattern of size distortion shown in Fig. 18.2a can occur. It appears to be related to the initial shape of the cast ingot and to the specific primary-mill processing used to reduce the ingot into bars. By changing steelmaking, forging and rolling procedures, out-of-roundness has been reduced to the smaller differential pattern shown in Fig. 18.2b, where the difference between high and low points is only 0.005 mm (0.0002 inch). High speed steel bars made this way are marketed by a few tool steel producers as 'close tolerance hob stock'. An even better method of combating out-of-roundness is to use high speed tool steel bars made from hot isostatically pressed powders, which maintain the best possible symmetry during conventional heat treatment.

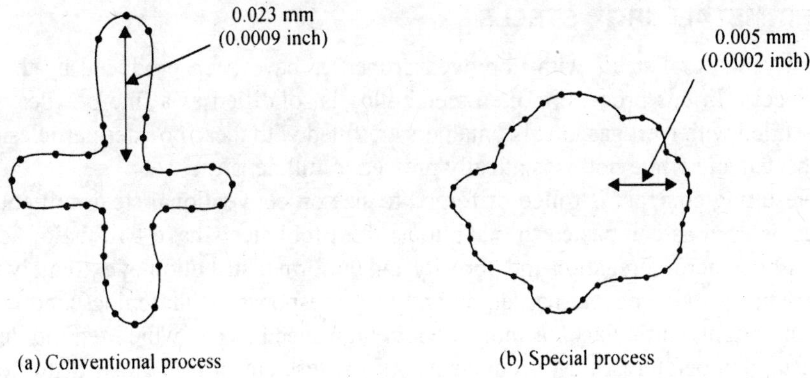

Drawings produced by calculation from precision measurements of diameter. Charts are plots on polar co-ordinates depicting variations in diameter after heat treatment for a bar that was round within ±1.25 μm (±0.00005 inch) before heat treatment.

Fig. 18.2. Typical diameter changes during heat treatment for high speed steel bars.

Stabilisation

Stabilisation involves reducing the amount of retained austenite in heat treated material. Retained austenite that can slowly transform and produce distortion if the material is later heated or subjected to stress. Stabilisation also reduces internal (residual) stress, which makes distortion in service due to stress relaxation less likely to occur. Stabilisation is most important for tools that must retain their exact size and shape over long periods of time (i.e., gauges, blocks).

If the tool steel chosen provides the required hardness after tempering at a relatively high temperature, it is possible to reduce the amount of retained austenite and the internal stress by multiple tempering. Initial tempering reduces internal stress and conditions the retained austenite so that it can transform to martensite on cooling from the tempering temperature. Usually, a second or third retempering is necessary to reduce the internal stress set up by the tranformation of retained austenite.

Single or repeated cold treatment to a temperature below M_f will cause most of the retained austenite to transform to martensite in plain carbon or low alloy tool steels that must be tempered at low temperatures to achieve the hardness required. Cold treatment may be applied either before or after the first temper. If, however, the tools tend to crack because of the additional stress induced by dimensional expansion during cold treatment, it is generally prudent to apply cold treatment after the tools have been tempered the first time. When cold treatment is applied after the first temper, the amount of retained austenite that transforms during the cold treatment may be considerably less than desired because some of the austenite may have been stabilised by tempering prior to cold treating. Cold treatment is usually done in a commercial refrigeration unit capable of attaining –70 to –95°C (–100 to –140°F). Tools must be retempered promptly after return to room temperature following cold treatment to reduce internal stress and increase the toughness of the newly formed martensite.

For some tools, a small percentage of retained austenited is desirable to improve toughness and provide a favourable internal stress pattern that will help the tool withstand service stresses. For these tools, a full stabilising treatment may actually result in tools that are unfit to perform their required functions.

POWDER METALLURGY STEELS

In recent years, tool steels with improved properties have been produced by the powder metallurgy (P/M) process. In this process, molten metal alloy is solidified as a fine powder by spraying it into a chamber filled with inert gas. Steel containers are filled with these powder particles, evacuated of gases, sealed, heated and white hot, isostatically pressed to full density.

The resulting compact is rolled or forged to size on conventional steel-mill equipment or, in some instances, is used as compacted to make tools. P/M tool steels have two major advantages: complete freedom from macrosegregation and porosity and uniform distribution of extremely fine carbides. These characteristics provide deeper hardening and faster response to hardening conditions (Fig. 18.3). The latter is important, particulary for molybdenum high speed steels, which tend to decarburise rapidly at austenitising temperatures. P/M products also show less out-of-roundness distortion in large-diameter bars (Table 18.4).

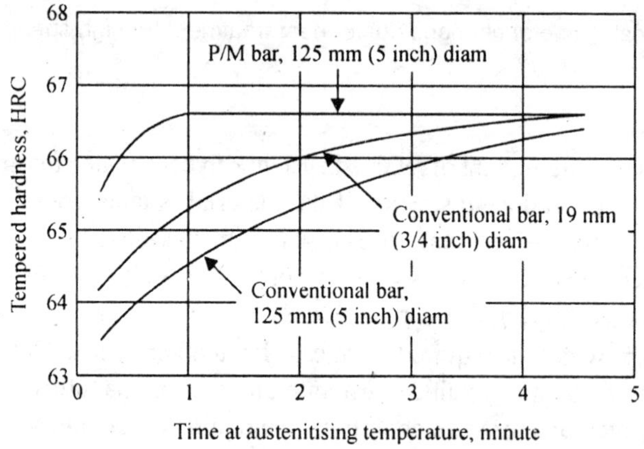

Hardness at mid-radius was evaluated for bars oil quenched from 1200°C (2200°F) and tempered 2 + 2 + 2 hours at 550°C (1025°F).

Fig. 18.3. Comparison of response to hardening for P/M and conventionally produced bars of M2S (HC) tool steel.

When sulphur is added to P/M tool steels, they exhibit a very fine homogeneous distribution of sulphides. This uniform sulphide distribution promotes better machinability. After heat treating, the refined hardened and tempered P/M tool steels grindability and greater toughness than conventionally processed (cast and wrought) tool steels.

The following AISI compositions of high speed steels were available in P/M form: M2, M2 with high sulphur and high carbon, M3 class 2 with sulphur, M4, M35 with sulphur, M42 and T15. P/M steels can be substituted for their conventional counterparts in all applications and are particularly advantageous when heavy sections are required.

The freedom from gross segregation provided by the P/M process makes it possible to readily fabricate new higher-alloy tool steels compositions. One type now available, which contains 1.50 C, 3.75 Cr, 3.00 V, 10 W, 5.25 Mo and 9.00 Co, is reported to have the highest hot hardness of any high speed steel.

Table 18.4. Out-of-roundness distortion in large-diameter bars of M2S tool steel.

Bar diameter		*Production*	*Typical out-of-roundness(a)*	
mm	*inch*	*method*	*mm*	*inch*
75	3	P/M	0.008	0.0003
		Conventional	0.020	0.0008
125	5	P/M	0.013	0.0005
		Conventional	0.033	0.0013
190	7.5	P/M	0.015	0.0006
		Conventional	0.051	0.0020

(a) Maximum diameter minus minimum diameter after normal hardening treatment.

SURFACE TREATMENTS

In many applications, service life of tool steels can be increased by surface treatments.

Oxide coatings

Oxide coatings, provided by treatment of the finish-ground tool in an alkali-nitrate bath or by steam oxidation, prevent or reduce adhesion of the tool to the workpiece. Oxide coatings have doubled tool life—particularly in machining of gummy materials such as soft copper and non-free-cutting low-carbon steels.

Plating

Plating of finished high speed steel tools with 0.0025 to 0.0125 mm (0.1 to 0.5 mil) of chromium also prolongs tool life by reducing adhesion of the tool to the workpiece. Chromium plating is relatively expensive and precautions must be taken to prevent tool failure in service due to hydrogen embrittlement.

Electroless nickel plating has been used successfully as a replacement for chromium plating, both in routine production and for salvage plating operations on tool steel parts. Because plating by this method is accomplished by means of chemical reduction, it does not depend on any galvanic coupling between dissimilar metals and there is no electrolysis involved. Therefore, there is no danger of hydrogen embrittlement. Plated hardness is in the high Rockwell 50's range, with good, uniform plated thickness on all surfaces and the plated surfaces have a low coefficient of friction.

Carburising

Carburising is not recommended for high speed steel cutting tools because the cases on such tools are extremely brittle. However, carburising is useful for applications such as cold work dies that require extreme wear resistance and that are not subjected to impact or highly concentrated loading. Carburising is done at 1040 to 1065°C (1900 to 1950°F) for short periods of time (10 to 60 minute) to produce a case 0.05 to 0.25 mm (0.002 to 0.010 inch) deep. The carburising treatment also serves as an austenitising treatment for the whole tool. A carburised case on high speed steels has a hardness of 65 to 70 HRC, but does not have the high resistance to softening at elevated temperatures exhibited by normally hardened high speed steel.

Nitriding

Nitriding successfully increases the life of all types of high speed steel cutting tools. However, gas nitriding in dissociated ammonia produces a case that is too brittle for most applications. Liquid nitriding for about 1 hr. at 565°C (1050°F) provides a light case, increasing both surface hardness and resistance to adhesion. For nitrided high speed steel taps, drills and reamers used in machining annealed steel, five fold increases in life have been reported, with average increases of 100 to 200 per cent. Obviously, if this nitrided case is removed when the tool is reground, the tool must then be retreated, which reduces the cost advantage of the process.

In addition, special surface-treatment processes, such as aerated nitriding baths, improve resistance to adhesive wear without producing excessive brittleness. Sulphur-containing nitriding baths provide a high-sulphur surface layer for additional resistance to seizing.

Sulphide Treatment

Sulphide treatment is a low temperature (190°C, or 375°F) electrolytic process using sodium and potassium thiocyanate provides a seizing-resistant iron sulphide layer. This process can be used as a final treatment for all types of hardened tool steels without much danger of overtempering.

MARAGING STEELS

Certain high-nickel maraging steels are being used for special noncutting tool applications; 18Ni(250) is the type most frequently used. For more demanding applications, the higher-strength 18Ni(300) is often preferred. Where maximum abrasion resistance is required, any of the maraging steels can be nitrided.

Maraging steels achieve full hardness—nominally 50 HRC for 18Ni(250), 54 HRC for 18Ni(300) and 58 HRC for 18Ni(350)—by means of a simple ageing treatment, usually 3 hr. at about 480°C (900°F). Because the development of hardness does not depend on cooling rate, full hardness can be developed uniformly in massive sections, with almost no distortion. Decarburisation is of no concern in these alloys, because they contain very little carbon and because their ageing temperature is relatively low. However, if the long-time service temperature exceeds the ageing temperature, maraging steels overage and undergo significant drop in hardness.

The 18Ni(300) grade is used for aluminum die-casting dies and cores, aluminum hot forging dies, dies for moulding plastics and various support tooling used in extrusion of aluminum. In die casting of aluminum, maraging steel dies can be used at higher hardness than is possible with dies made of 1113 tool steel because maraging steel is not as prone to heat checking. Because the ageing process results in very little size change, it is possible to machine the intricate impressions for plastic moulding dies to final size prior to final hardening.

For moulding extremely abrasive types of plastics, the higher surface hardness provides by 18Ni(350) maraging steel is desirable.

Chapter 19

Heat Treating of Stainless Steels

INTRODUCTION

Heat treating of stainless steel serves to produce changes in physical condition, mechanical properties and residual stress level and to restore maximum corrosion resistance when that property has been adversely affected by previous fabrication or heating. Frequently, a combination of satisfactory corrosion resistance and optimum mechanical properties is obtained in the same heat treatment.

AUSTENITIC STAINLESS STEELS

The austenitic stainless steels may be divided into five groups:

1. The conventional austenitics, such as types 301, 302, 303, 304, 305, 308, 309, 310, 316 and 317.
2. The stabilised compositions, primarily types 321, 347 and 348.
3. The low-carbon grades, such as types 304L, 316L and 317L.
4. The high-nitrogen grades, such as AISI types 201, 202, 304N and 316N and thc nitronic series of alloys.
5. The highly alloyed austenitics, such as 317LM, 317LX, JS700, JS777, 904L, AL-4X, 2RK65, Carpenter 20Cb-3, Sanicro 28, AL-6X and 254 SMO.

In furnace loading, the high thermal expansion of austenitic stainless steels (about 50 per cent higher than that of a mild carbon steel) should be considered. The spacing between parts should be adequate to accommodate this expansion. Stacking, when necessary, should be employed judiciously, to avoid deformation of parts at elevated temperature.

Conventional Austenitics

Conventional austenitics cannot be hardened by heat treatment but will harden as a result of cold working. These steels are usually purchased in an annealed or cold worked state. Following welding or thermal processing, a subsequent re-anneal may be required for optimum corrosion resistance, softness and ductility. During annealing, chromium carbides, which markedly decrease resistance to intergranular corrosion, are dissolved. Annealing temperatures, vary somewhat with the composition of the steel.

Because carbide precipitation can occur at temperatures between 425 and 900°C (800 and 1650°F), it obviously is desirable that the annealing temperature should be safely above this limit. Moreover, because all carbides should be in solution before cooling begins and because the chromium carbide dissolves slowly, the highest practical temperature consistent with limited grain growth is selected. This temperature is in the vicinity of 1095°C (2000°F).

Cooling from the annealing temperature must be rapid, but it must also be consistent with limitations of distortion. Whenever considerations of distortion permit, water quenching is used, thus ensuring that dissolved carbides remain in solution (because it precipitates carbides more rapidly, type 310 invariably requires water quenching). Where practical considerations of distortion rule out such a fast cooling rate, cooling in an air blast is used. With some thin-section parts, even this intermediate rate of cooling produces excessive distortion and parts must be cooled in still air. If cooling in still air does not provide a rate sufficient to prevent carbide precipitation, maximum corrosion resistance will not be obtained. A solution to this dilemma is the use of a stabilised grade or the low-carbon alloys.

Stabilised Austenitic Alloys

Stabilised austenitic alloys, namely types 321, 347, 348 and Carpenter 20Cb-3, contain controlled amounts of titanium or of niobium, which render the steel nearly immune to intergranular precipitation of chromium carbide and its adverse effects on corrosion resistance. Nevertheless, these alloys may require annealing to relieve stresses, increase softness and ductility, or provide additional stabilisation.

To obtain maximum softness and ductility, the stabilised grades are annealed at the temperatures shown in Table 19.1. Unlike the unstabilised grades, these steels do not require water quenching or other acceleration of cooling from the annealing temperature to prevent subsequent intergranular corrosion; air cooling is generally adequate.

Table 19.1. Recommended annealing temperatures for austenitic stainless steels.

		Temperature(a)	
UNS No.	*Designation*	*°C*	*°F*
Conventional grades			
S30100, S30200, S30215	301, 302, 302B	1010 to 1120	1850 to 2050
S30300, S30323	303, 303Se	1010 to 1120	1850 to 2050
S30400, S30500, S30800	304, 305, 308	1010 to 1120	1850 to 2050
S30900, S30908	309, 309S	1040 to 1120	1900 to 2050
S31000, S31008	310, 310S	1040 to 1065	1900 to 1950
S31600	316	1040 to 1120	1900 to 2050
S31700	317	1065 to 1120	1950 to 2050
Stabilised grades			
S32100	321	955 to 1065	1750 to 1950
S34700, S34800	347, 348	980 to 1065	1800 to 1950
N08020	Carpenter 20Cb-3	925 to 955	1700 to 1750
Low-carbon grades			
S30403	304L, 304LN	1010 to 1120	1850 to 2050
S31603, S31703	316L, 316LN, 317L	1040 to 1110	1900 to 2025
High-nitrogen grades			
S20100, S20200	201, 202	1010 to 1120	1850 to 2050
S30451	304N	1010 to 1120	1850 to 2050

(Contd...)

UNS No.	Designation	Temperature(a) °C	°F
S31651	316N	1010 to 1120	1850 to 2050
S24100	Nitronic 32, Carpenter 18Cr-2Ni-12Mn	1010 to 1065	1850 to 1950
S24000	Nitronic 33	1040 to 1095	1900 to 2000
S21904	Nitronic 40, Carpenter 21Cr-6Ni-9Mn	980 to 1175	1800 to 2150
S20910	Nitronic 50, Carpenter 22Cr-13Ni-5Mn	1065 to 1120	1950 to 2050
S21800	Nitronic 60	1040 to 1095	1900 to 2000
S28200	Carpenter 18-18 PLUS	1040 to 1095	1900 to 2000
Highly alloyed grades			
	317LM, 317LX, 317L PLUS, 317LMO, 7L4	1120 to 1150	2050 to 2100
	JS700, JS777	1065 to 1150	1950 to 2100
N08904	904L, AL-4X, 2RK65	1075 to 1125	1965 to 2055
N08028	Sanicro 28	–	
N08366	AL-6X	1205 to 1230	2200 to 2250
S31254	254 SMO	1150 to 1205	2100 to 2200

(a) Temperatures given are for annealing a composite structure. Time at temperature and method of cooling depend on thickness. Light sections may be held at temperature for 3 to 5 minutes per 2.5 mm (0.10 inch) of thickness, followed by rapid air cooling. Thicker sections are water quenched. For many of these grades, a post-weld heat treatment is not necessary. For proprietary alloys, alloy producers may be consulted for details. Although cooling from the annealing temperature must be rapid, it must also be consistent with limitations of distortion.

When maximum corrosion resistance of the stabilised austenitic grades is required, it may be necessary to employ a heat treatment known as a stabilising anneal. The treatment consists of holding at 845 to 900°C (1550 to 1650°F) for up to 5 hrs., depending on section thickness. It may be applied either prior to or in the course of fabrication and it may be followed by short-time stress relieving at 705°C (1300°F) without danger of harmful carbide precipitation.

Carpenter 20Cb-3 stainless steel is unlike the conventional stabilised austenitics (types 321, 347 and 348) because of its higher alloy content and improved corrosion resistance. This alloy normally is stabilised and annealed at 925 to 955°C (1700 to 1750°F). For special applications, the alloy can be annealed at higher temperatures (up to 1150°C or 2100°F), but this is premissible only if the alloy will not be subject to welding or heating temperatures over 540°C (1000°F). Certain restrictions on furnace atmosphere are mandatory. Furnace combustion must be carefully controlled to eliminate carburising or excessively oxidising conditions. Because the properties of the stabilised steels are based on their original carbon content, carbon absorption cannot be tolerated. Excessively oxidising conditions cause the formation of a scale that is difficult to remove in subsequent descaling operations. Direct impingement of flame on the work must be prevented. The sulphur content of the furnace atmosphere, particularly in oil-fired furnaces, must be kept low; natural gas should be used, not producer gas.

Low-carbon Austenitics

Low-carbon austenitics are intermediate in tendency to precipitate chromium carbides to the stabilised and unstabilised grades. Carbon content (0.03 per cent maximum) is low enough to reduce precipitation of intergranular carbides. This characteristic of limited sensitisation is of particular value in welding, flame cutting and other hot working operations. They do not require the quenching treatment that unstabilised grades require to retain carbon in solid solution. Nevertheless, the low-carbon alloys are not satisfactory for long-time service in the sensitising temperature range of 540 to 760°C (1000 to 1400°F) because they are not completely immune to the formation of carbides deleterious to corrosion resistance. Recommended annealing temperatures for the low-carbon alloys are given in Table 19.1.

The effects of sensitisation and susceptibility to general corrosion vary among the low-carbon alloys, depending on their chemical composition. Because they contain molybdenum, types 316L and 317L are susceptible to sigma phase formation as a result of long-time exposure at 650 to 870°C (1200 to 1600°F). However, the corrosion resistance of these grades can be improved by employing a stabilising treatment (ASTM A262C), consisting of holding at 885°C (1625°F) for 2 hrs., prior to stress relieving at 675°C (1250°F). After receiving the stabilising heat treatment, these alloys pass the sulphuric acid-copper sulphate test (ASTM A262, Practice E) for freedom from intergranular carbide precipitation.

Magnetic Permeability

The low-carbon alloys are frequently used in the production of articles requiring low magnetic permeability. These materials are non-magnetic in the fully annealed condition, with permeabilities below 1.02 maximum at 200 gausses, but may develop ferromagnetic qualities as a result of cold working during fabrication. Cold working may generate some low-carbon martensite, which is strongly magnetic. Fusion welding with a low-nickel filler rod is another possible cause of magnetism. Magnetism due to any of these causes can be eliminated by a full anneal to restore the alloy to its fully austenitic condition.

Bright Annealing

All grades of austenitic stainless steel can be bright annealed in either pure hydrogen or dissociated ammonia, provided the dew point of the atmosphere is less than –50°C (–60°F) and the workpieces, upon entering the furnace, are dry and scrupulously clean. The furnaces used in bright annealing must be clean, moisture-free and tight if low dew points are to be maintained.

If a low dew point is not maintained, a thin greenish oxide will form on the work. This oxide is very difficult to remove in subsequent descaling operations.

To maintain close control of dew point, atmosphere samples should be withdrawn from the furnace at frequent intervals and tested or continuously monitored, as in most commercial operations. Traces of oxygen in hydrogen gas can be removed before the gas enters the furnace by passing the gas through a catalytic tower that causes excess oxygen to combine with the hydrogen to form water vapour. The gas is then passed through activated alumina to remove moisture.

In using dissociated ammonia, it is important that maximum dissociation be obtained before the gas enters the furnace. The presence of any undissociated ammonia will result in objectionable nitriding action. Because the undissociated gas is entirely soluble in water, its removal can be easily accomplished. However, the remaining fully dissociated product must be processed through drying towers to restore the required low dew point.

High-nitrogen Austenitic Stainless Steels

High-nitrogen austenitic stainless steels are heat treated in the same manner and are subject to the same problems (carbide precipitation and distortion) as conventional austenitics. They cannot be hardened by heat treatment but will harden by cold working. High-nitrogen austenitics are annealed to ensure maximum corrosion resistance, softness and ductility. Rapid cooling is preferred. Annealing temperature ranges are listed in Table 19.1.

Highly Alloyed Austenitic Stainless Steels

Highly alloyed austenitic stainless steels contain large amounts of molybdenum to provide very good resistance to chloride corrosion. They usually are produced with low carbon to avoid sensitisation and may contain copper for increased acid resistance. These alloys are austenitic in the mill-annealed condition but may form sigma or delta ferrite phases under certain conditions of heat treatment or service. Those phases may be detrimental to corrosion resistance and mechanical properties. Annealing temperatures are confined to a narrow range to avoid sigma phase formation at lower temperature, or delta ferrite at higher temperature. Recommended annealing temperatures are given in Table 19.1. Rapid cooling following annealing is usually advisable, especially in heavy sections. Stress relief treatments may be used below the annealing temperature range, but holding times should be held to a minimum to avoid sigma phase and sensitisation problems.

FERRITIC STAINLESS STEELS

The ferritic stainless steels may be divided into two groups, the conventional ferritics such as types 409, 430, 434 and 446 and the low-interstitial ferritics such as types 439, 444, E-BRITE, SEA-CURE, AL 29-4C and AL 29-4-2. The ferritic stainless steels are not hardened by quenching but rather develop minimum hardness and maximum ductility, toughness and corrosion resistance in the annealed and quenched condition. Therefore, the only heat treatment applied to the ferritics is annealing. This treatment relieves stresses developed during welding or cold working and provides a more homogeneous structure by dissolving transformation products formed during welding. Post-weld heat treatment of the low-interstitial ferritic stainless steels is generally unnecessary and is frequently undesirable. Table 19.2 summarises current annealing practices for the ferritic grades.

Table 19.2. Recommended annealing treatments for ferritic stainless steels.

		Treatment temperature	
UNS No.	*Designation*	°C	°F
Conventional ferritic grades			
S40500	405	650 to 815	1200 to 1500
S40900	409	870 to 900	1600 to 1650
S43000	430	705 to 790	1300 to 1450
S43020	430F	705 to 790	1300 to 1450
S43400	434	705 to 790	1300 to 1450
S44600	446	760 to 830	1400 to 1525
Low-interstitial ferritic grades			
S43035	439	870 to 925	1600 to 1700

(Contd...)

UNS No.	Designation	Treatment temperature °C	Treatment temperature °F
S44400	444	955 to 1010	1750 to 1850
S44626	E-BRITE	760 to 955	1400 to 1750
S44660	SEA-CURE, SC-1	1010 to 1065	1850 to 1950
	AL 29-4C	1010 to 1065	1850 to 1950
S44800	AL 29-4-2	1010 to 1065	1850 to 1950
S44635	MONIT	1010 to 1065	1850 to 1950

Post-weld heat treating of the low interstitial ferritic stainless steels is generally unnecessary and frequently undesirable. Any annealing of these grades should be followed by water quenching or very rapid cooling.

Austenite-Martensite Embrittlement

When grades such as 430 and 434 are cooled rapidly from above 925°C (1700°F), they may become brittle from austenite transforming to as much as 30 per cent martensite. This may be corrected by a tempering treatment such as 650 to 790°C (1200 to 1450°F), which softens the alloy.

475°C (885°F) Embrittlement

A potentially harmful form of embrittlement common to the ferritic grades can develop from prolonged exposure to, or slow cooling within, the temperature range from about 370 to 540°C (700 to 1000°F) with maximum rate of embrittlement occurring at about 475°C (885°F). The embrittlement is caused by precipitation of the α' phase and the effects of embrittlement increase rapidly with chromium content. Certain heat treatments must be controlled to avoid embrittlement. The brittle condition can be eliminated by the treatments listed in Table 19.2, using temperatures clearly above the upper boundary of embrittlement, followed by rapid cooling to prevent a recurrence.

Intermetallic Phase Embrittlement

Intermetallic phases, such as sigma, chi and Laves, may form at elevated temperatures in ferritic stainless steels containing more than about 14 per cent chromium. These intermetallic phases increase hardness (sometimes usefully) and decrease ductility, notch toughness and corrosion resistance. The temperature range over which these phases form is approximately 600 to 1000°C (1100 to 1830°F).

Intermetallic phase embrittlement is primarily a service problem where long exposures at elevated temperatures are involved. These phases can be dissolved by heating to above 1000°C (1830°F).

Duplex Stainless Steels

Duplex stainless steels consist of a mixed microstructure of austenite and ferrite. Some duplex stainless steels are rich in ferrite, others in austenite and others are equally balanced. Compared with type 316, the annealed duplex alloys provide improved resistance to chloride stress corrosion cracking. Although the duplex grades are generally not as resistant as the low-interstitial ferritics, they are normally available in heavier section thicknesses. Another useful characteristic of the duplex grades is that they typically have yield strengths more than twice that of the conventional austenitic stainless steels.

Duplex stainless steels, such as SAF 2205, AF 22, DP 3 and Ferralium alloy 225, are alloyed with 0.15 to 0.20 per cent nitrogen. This minimises alloy element segregation between the ferrite and austenite, thereby improving the as-welded corrosion resistance compared with the type 329 alloy. Recommended annealing temperatures for duplex stainless steels are listed below:

UNS No.	*Designation*	*Temperature °C (°F)*
S32900	329, 7-Mo	925–955 (1700–1750)
S31500	3RE60	975–1025 (1785–1875)
–	SAF 2205, AF 22	1020–1100 (1870–2010)
–	DP 3	1065–1175 (1950–2150)
S32550	Ferralium alloy 225	1065–1175 (1950–2150)

Note: Cooling from the annealing temperature must be rapid, but it also must be consistent with limitations of distortion.

MARTENSITIC STAINLESS STEELS

The heat treating of martensitic stainless steel is essentially the same as for plain carbon or low-alloy steels, in that maximum strength and hardness depend chiefly on carbon content. The principal metallurgical difference is that the high alloy content of the stainless grades causes the transformation to be so sluggish and the hardenability to be so high, that maximum hardness is produced by air cooling in the centre of sections up to approximately 30.5 cm (12 inch) thick. Surface hardness ranges for the various heat treated conditions from fully annealed to fully hardened are given in Fig. 19.1.

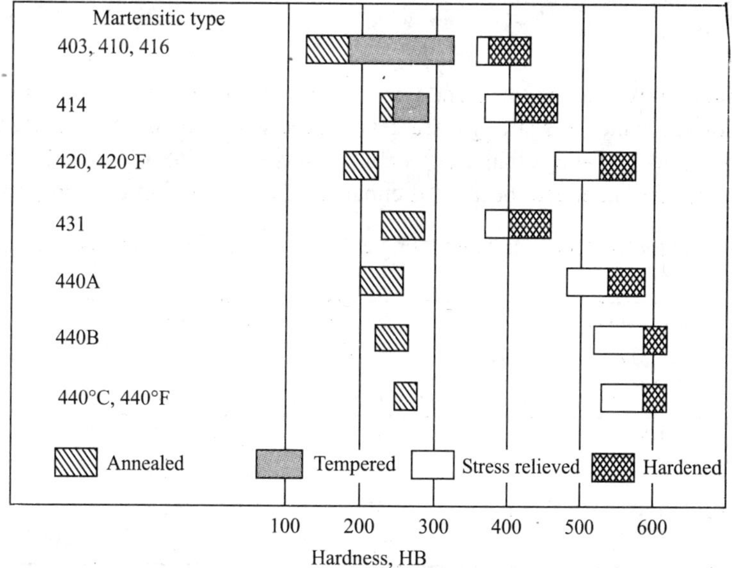

Fig. 19.1. Effect of heat treatments on the hardness of martensitic stainless steels.

The martensitic stainless steels are more sensitive to heat treating variables than are carbon and low-alloy steels; rejection rates due to faults in heat treating are correspondingly high.

Prior Cleaning

To avoid contamination, all parts and heat treating fixtures must be cleaned thoroughly before they are placed in the furnace. Proper cleaning is particularly important when the heat treatment is to be performed in a protective atmosphere. Grease, oil and even location lines made by an ordinary lead pencil, can cause carburisation. Perspiration stains from fingerprints are a source of chloride contamination and may cause severe scaling in oxidising atmospheres. Furthermore, a protective atmosphere cannot be effective unless it is permitted to make unobstructed contact with metal surfaces.

Pre-heating

Martensitic stainless steels normally are hardened by being heated to the austenitising range of 925 to 1065°C (1700 to 1950°F) and then cooled in air or oil.

The thermal conductivity of stainless steels is characteristically lower than that of carbon and alloy steels. Accordingly, high thermal gradients and high stresses during rapid heating may cause warpage and cracking in some parts. To avoid these problems, pre-heating is usually recommended in the treatment of martensitic stainless steels. In annealing or hardening, the following parts should be pre-heated: (i) heavy section parts; (ii) parts with both thin and thick sections; (iii) parts with sharp corners and re-entrant angles; (iv) heavily ground parts; (v) parts machined with heavy deep cuts; (vi) parts that have been cold formed or straightened; and (vii) previously hardened parts that are being re-heat treated.

Pre-heating is normally accomplished at 760 to 790°C (1400 to 1450°F) and heating need be continued only long enough to ensure that all portions of each part have reached the pre-heating temperature. Large heavy parts are sometimes pre-heated at approximately 540°C (1000°F) prior to the 790°C (1450°F) pre-heat. Types 403, 410 and 416 require less pre-heating than the higher-carbon types 414, 431, 420 and the 440 grades.

Austenitising

Austenitising temperatures, soaking times and quenching mediums are summarised in Table 19.3. When maximum corrosion resistance and strength are desired, the steel should be austenitised at the high end of the temperature range. For alloys that are to be tempered above 565°C (1050°F), the low side of the austenitising range is recommended, because it enhances ductility and impact properties.

Table 19.3. Procedures for hardening and tempering wrought martensitic stainless steels to specific strength and hardness levels.

	Austenitising(a)			*Tempering temperature(d)*						
	Temperature(b)		*Quenching*	*°C*		*°F*		*Tensile strength*		*Hardness,*
Type	*°C*	*°F*	*medium(c)*	*Min.*	*Max.*	*Min.*	*Max.*	*MPa*	*ksi*	*HRC*
403, 410	925–1010	1700–1850	Air or oil	565	605	1050	1125	760–965	110–140	25–31
				205	370	400	700	1105–1515	160–220	38–47
414	925–1050	1700–1925	Air or oil	595	650	1100	1200	760–965	110–140	25–31
				230	370	450	700	1105–1515	160–220	38–49

(Contd...)

Type	Austenitising(a) Temperature(b) °C	°F	Quenching medium(c)	Tempering temperature(d) °C Min.	Max.	°F Min.	Max.	Tensile strength MPa	ksi	Hardness, HRC
416, 416(Se)	925–1010	1700–1850	Oil	565	605	1050	1125	760–965	110–140	25–31
				230	370	450	700	1105–1515	160–220	35–45
420	985–1065	1800–1950	Air or oil(e)	205	370	400	700	1550–1930	225–280	48–56
431	985–1065	1800–1950	Air or oil(e)	565	605	1050	1125	860–1035	125–150	26–34
				230	370	450	700	1210–1515	175–220	40–47
440A	1010–1065	1850–1950	Air or oil(e)	150	370	300	700	–	–	49–57
440B	1010–1065	1850–1950	Air or oil(e)	150	370	300	700	–	–	53–59
440C, 440F	1010–1065	1850–1950	Air or oil(e)	–	160	–	325	–	–	60 min
				–	190	–	375	–	–	58 min
				–	230	–	450	–	–	57 min
				–	355	–	675	–	–	52–56

(a) Pre-heating to a temperature within the process annealing range (Table 19.5) is recommended for thin-gauge parts, heavy sections, previously hardened parts, parts with extreme variations in section or with sharp re-entrant angles and parts that have been straightened or heavily ground or machined, to avoid cracking and minimise distortion, particularly for types 420, 431 and 440A, B, C and F. (b) Usual time at temperature ranges from 30 to 90 minutes. The low side of the austenitising range is recommended for all types subsequently tempered to 25 to 31 HRC; generally, however, corrosion resistance is enhanced by quenching from the upper limit of the austenitising range. (c) Where air or oil is indicated, oil quenching should be used for parts more than 6.4 mm (¼ inch) thick; martempering baths at 150 to 400°C (300 to 750°F) may be substituted for an oil quench. (d) Generally, the low end of the tempering range of 150 to 370°C (300 to 700°F) is recommended for maximum hardness, the middle for maximum toughness and the high end for maximum yield strength. Tempering in the range of 370 to 565°C (700 to 1050°F) is not recommended, because it results in low and erratic impact properties and poor resistance to corrosion and stress corrosion. (e) For minimum retained austenite and maximum dimensional stability, a sub-zero treatment –75°C ± 10°C (–100°F ± 20°F) is recommended; this should incorporate continuous cooling from the austenitising temperature to the cold transformation temperature.

The effect of austenitising temperature on the as-quenched hardness of three martensitic grades is shown in Fig. 19.2. The hardness increases with increasing austenitising temperature to about 980°C (1800°F), then decreases because of austenite retention and (occasionally) the formation of delta ferrite.

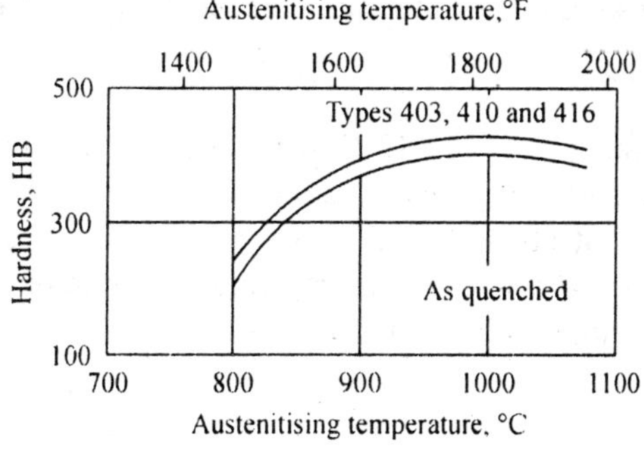

Specimens were martensitic stainless steels containing 0.15 per cent maximum carbon.

Fig. 19.2. Effect of austenitising temperature on as-quenched hardness.

Certain anomalies in these steels that should be considered before specifying a heat treating procedure are exemplified in the opposing injurious effects of the high and low extremes of austenitising temperature, depending on the subsequent tempering temperature. For example, type 431 showed the following Izod impact properties that are caused by retained austenite:

Temperature, °C		
Austenitising	*Tempering*	*Izod impact, J*
980	315	20.3–33.9
1065	315	40.7–81.3
980	595	74.6–108.5
1065	595	61.0–74.6

Soaking Times

Soaking times employed in the hardening of martensitic stainless steels represent a compromise between: (i) achieving maximum solution of chromium-iron carbides for maximum strength and corrosion resistance; and (ii) avoiding decarburisation, excessive grain growth, retained austenite, brittleness and quench cracking. For sections 13 mm (½ inch) thick and under, a soaking time of 30 to 60 minutes is sometimes recommended. For most parts, adding 30 minutes for each additional inch of thickness or fraction thereof has proven adequate. However, soaking times should be doubled if parts to be hardened have been fully annealed or isothermally annealed.

The effect of soaking time at austenitising temperature and of other variables, on the impact strength and room temperature hardness of types 403, 420 and 431 is plotted in Figs. 19.3, 19.4 and 19.5.

Quenching

Because of their high hardenability, martensitic stainless steels can be quenched in either oil or air. Some decrease in corrosion resistance and ductility, resulting from air quenching, may occur in these grades. These steels may precipitate carbides in grain-boundary areas if heavy sections are cooled slowly through the temperature range of about 870 to 540°C (1600 to 1000°F). Too slow a cooling rate in bright annealing these alloys may impair their corrosion resistance. Although oil quenching is preferred, air cooling may be required for large or complex sections, to prevent distortion or quench cracking. Martempering is particularly easy with these steels because of their high hardenability.

Retained Austenite

The higher-carbon martensitic grades, such as 440°C and the higher-nickel type 431, are likely to retain large amounts of untransformed austenite in the as-quenched structure, frequently as much as 30 per cent by volume. Stress relieving at about 150°C (300°F) has little effect. Delayed transformation, particularly in type 440°C, may occur as a result of temperature fluctuations in service, thus resulting in embrittlement and unacceptable dimensional changes.

Sub-Zero Cooling

A portion of the austenite retained in quenching may be transformed by sub-zero cooling to about –75°C (–100°F) immediately after quenching. To obtain maximum transformation of retained austenite, double tempering may be necessary. Parts should be air cooled to room temperature between the tempering cycles.

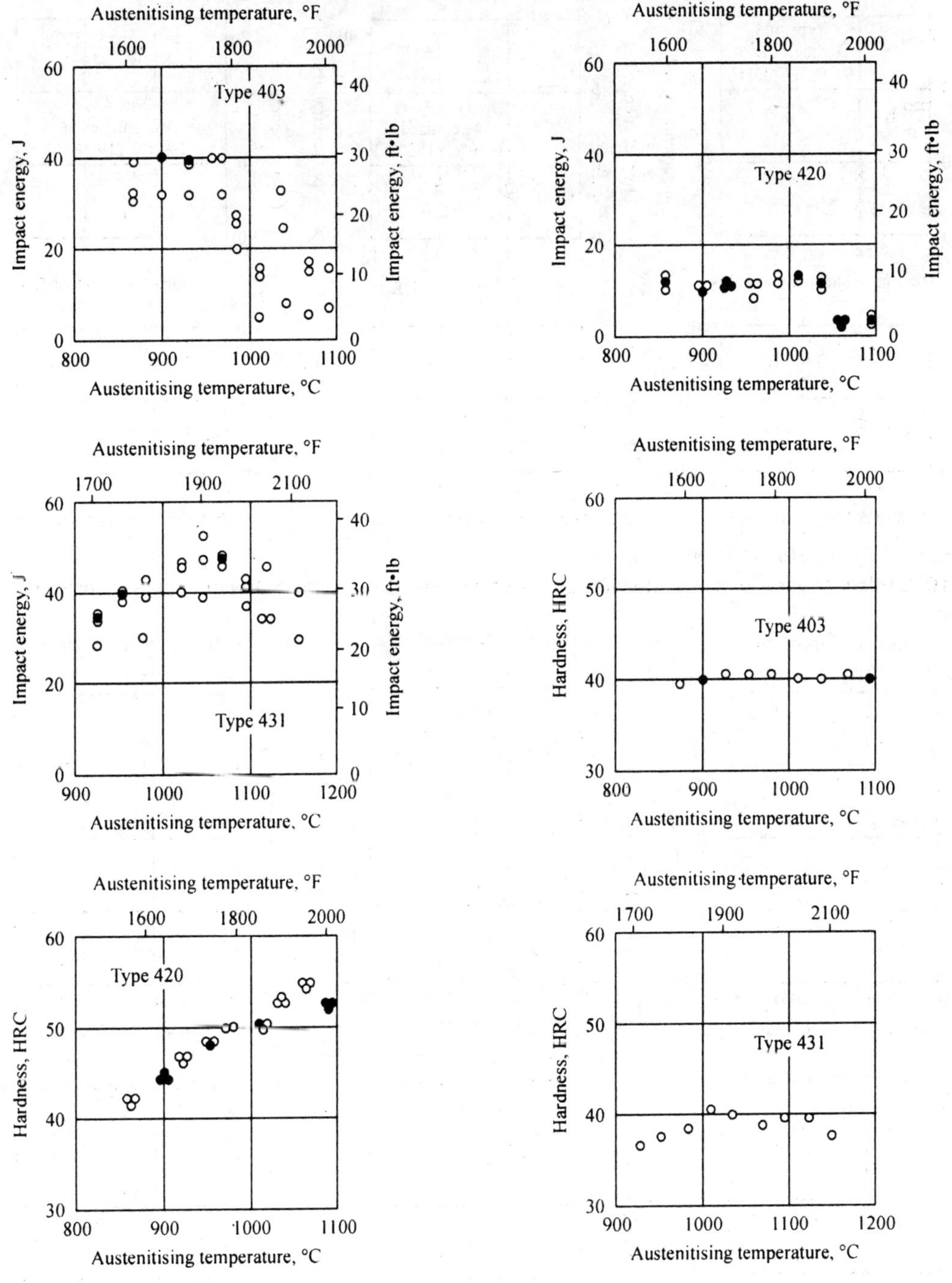

Fig. 19.3. Effect of variations in austenitising temperature on hardness and impact strengths of martensitic stainless steels.

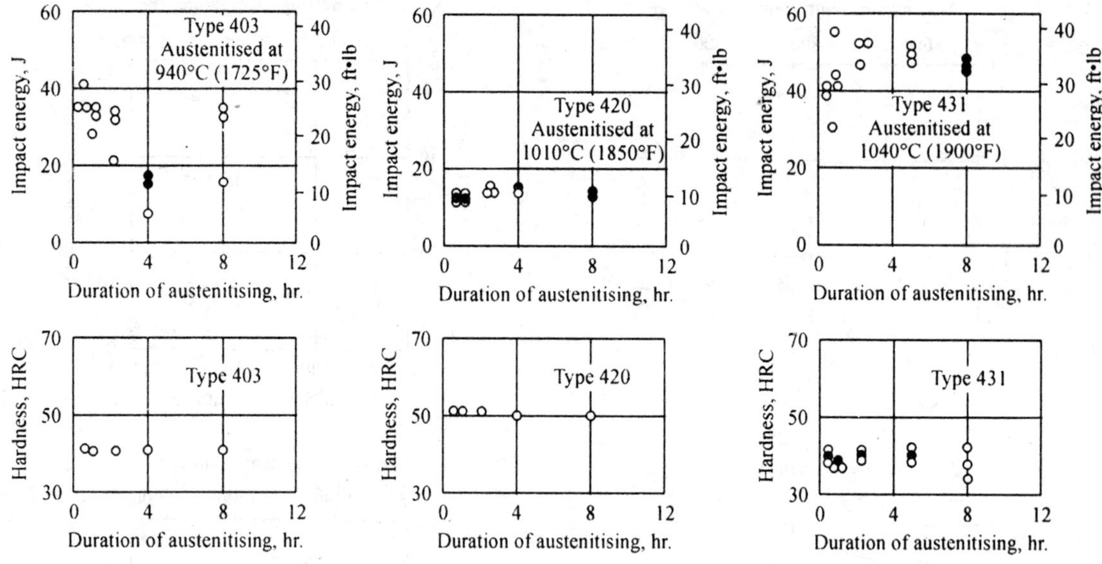

Specimens were tempered at 480°C (900°F) for 4 hrs.

Fig. 19.4. Effect of variations in austenitising time on hardness and impact strengths of martensitic stainless steels.

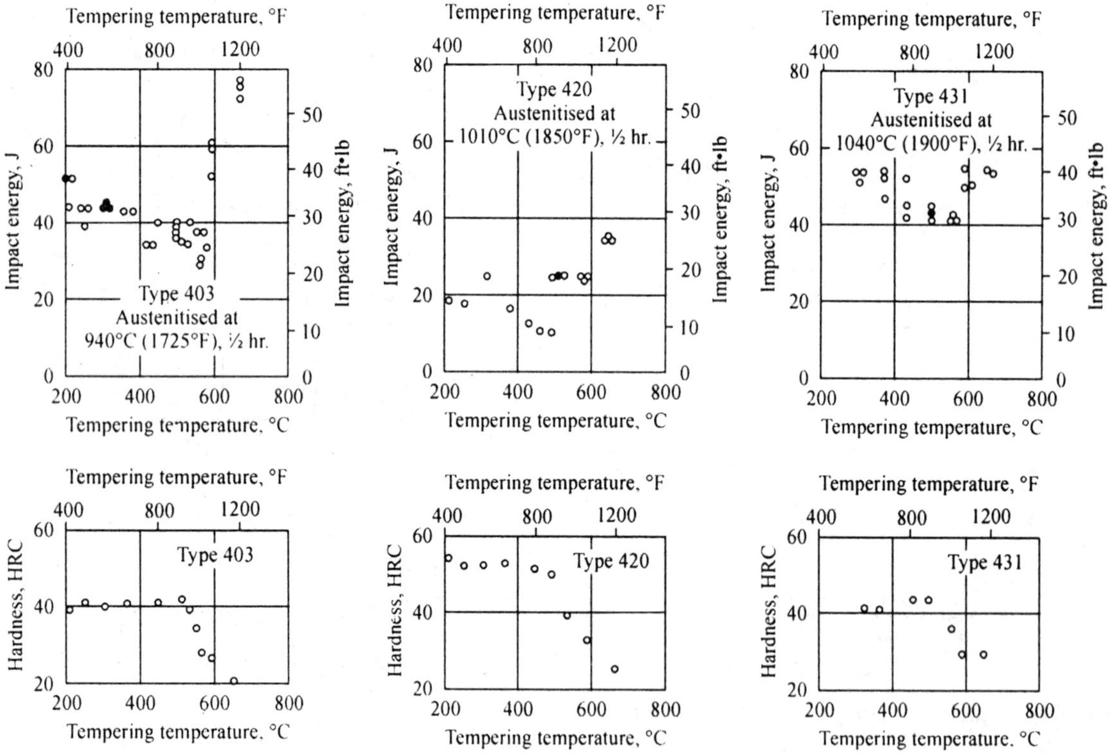

Fig. 19.5. Effect of variations in tempering temperature on hardness and impact strengths of martensitic stainless steels.

Sub-zero cooling is frequently included in the hardening treatment of parts such as the slides and sleeves of slide valves and bearings requiring maximum dimensional stability.

Re-heating

For fully hardened steels, increasing degrees of recovery are achieved by:

Stress relieving

Stress relieving at 150 to 370°C (300 to 700°F) after hardening, to reduce transformation stresses without significantly affecting microconstituents or mechanical properties.

Tempering

Tempering at intermediate temperatures, to modify properties.

Sub-critical annealing

Sub-critical annealing (variously called process, mill, or low annealing) in the upper portion of the ferritic range, just below the lower critical Ac_1 temperature, to achieve maximum softening without the complications of re-entering the gamma or austenitic field.

Full annealing

Full annealing for maximum softening by a return to the austenitic range, followed by slow cooling.

Although the strength, elongation and hardness curves shown in Figs. 19.6 to 19.11 appear to have the same general form as those of low-alloy steel, the increase in tensile strength and hardness between 400 and 510°C (750 and 950°F) may be associated with a serious decrease in notch toughness, while tempering on the high side of the temperature range generally coincides with a decrease in corrosion resistance. The tempering temperatures most frequently employed to achieve desired hardness and other mechanical properties are included in Table 19.3.

The Izod impact curves in Figs. 19.6 to 19.11 reveal a loss of impact strength when parts are tempered within the range of 370 to 650°C (700 to 1200°F). Tempering within this range also results in decreased corrosion resistance, particularly resistance to stress-corrosion cracking (Fig. 19.12). Double tempering (cooling parts to room temperature after the first tempering treatment) also is beneficial for resistance to stress corrosion.

Annealing

Temperatures and resulting hardnesses for process (sub-critical) annealing, full annealing and isothermal annealing are given in Table 19.4. Full annealing is an expensive and time-consuming treatment; it should be used only when required for subsequent severe forming. Types 414 and 431 do not respond to full or isothermal annealing procedures within a reasonable soaking period.

Isothermal annealing is recommended where maximum softening is required and adequate facilities for controlled slow cooling are not available. Full annealing, isothermal annealing and especially repeated process annealing promote the formation of coarse carbides that take longer to dissolve at austenitising temperatures. Sub-critical annealing is recommended for all applications that do not require maximum softness.

Table 19.4. Annealing temperatures and procedures for wrought martensitic stainless steels.

Type	*Process (subcritical) annealing Temperature(a), °C*	*Hardness*	*Full annealing Temperature(b)(c), °C*	*Hardness*	*Isothermal annealing(c) Procedure (d)*	*Hardness*
403, 410	650–760	86–92 HRB	830–885	75–85 HRB	Heat to 830 to 885°C; hold 6 hrs. at 705°C	85 HRB
414	650–730	99 HRB–24 HRC	Not recommended		Not recommended	
416, 416(Se)	650–760	86–92 HRB	830–885	75–85 HRB	Heat to 830 to 885°C; hold 2 hrs. at 720°C	85 HRB
420	675–760	94–97 HRB	830–885	86–95 HRB	Heat to 830 to 885°C; hold 2 hrs. at 705°C	95 HRB
431	620–705	99 HRB-30 HRC	Not recommended		Not recommended	
440A	675–760	90 HRB-22 HRC	845–900	94–98 HRB	Heat to 845 to 900°C; hold 4 hrs. at 690°C	98 HRB
440B	675–760	98 HRB-23 HRC	845–900	95 HRB–20 HRC	Same as 440A	20 HRC
440C, 440F	675–760	98 HRB-23 HRC	845–900	98 HRB–25 HRC	Same as 440A	25 HRC

(a) Air cool from temperature; maximum softness is obtained by heating to temperature at high end of range; (b) Soak thoroughly at temperature within range indicated; furnace cool to 790°C; continue cooling at 15 to 25°C/hr. to 595°C; air cool to room temperature; (c) Recommended for applications in which full advantage may be taken of the rapid cooling to the transformation temperature and from it to room temperature; and (d) Pre-heating to a temperature within the process annealing range is recommended for thin-gauge parts, heavy sections, previously hardened parts, parts with extreme variations in section or with sharp re-entrant angles and parts that have been straightened or heavily ground or machined to avoid cracking and minimise distortion, particularly for types 420 and 431 and 440A, B, C and F.

Salt Baths

Many stainless steel parts are heat treated in molten salt, with excellent results. The baths usually employed consist of barium chloride with 5 to 35 per cent sodium or potassium chloride. Alkaline-earth and other metallic oxides build up in these baths through use, but these oxides are not harmful to low-carbon stainless steels. However, if these salt baths are to be used also for hardening other alloy steels, to avoid surface decarburisation it is necessary to rectify the baths with graphite to remove the metallic oxides and with methyl chloride gas to convert the alkaline-earth oxides back to chlorides. A bath treated with methyl chloride will carburise low-carbon stainless steels unless it is aged for at least 24 hrs. before the stainless is treated in it. To avoid this problem, stainless steel parts should be heat treated in a salt bath reserved exclusively for stainless steels.

Protective Atmospheres

Argon or helium, if used as protective atmospheres, should be exceptionally dry (with a dew point below –50°C or –60°F). Because they are expensive and cannot be generated, they are rarely used. Exothermic and endothermically generated gas can be used with excellent results.

These require dew point or infrared control so as not to carburise or decarburise the stainless grade being heat treated. Endothermic gas containing approximately 40 per cent hydrogen can embrittle martensitic stainless steels that are oil quenched.

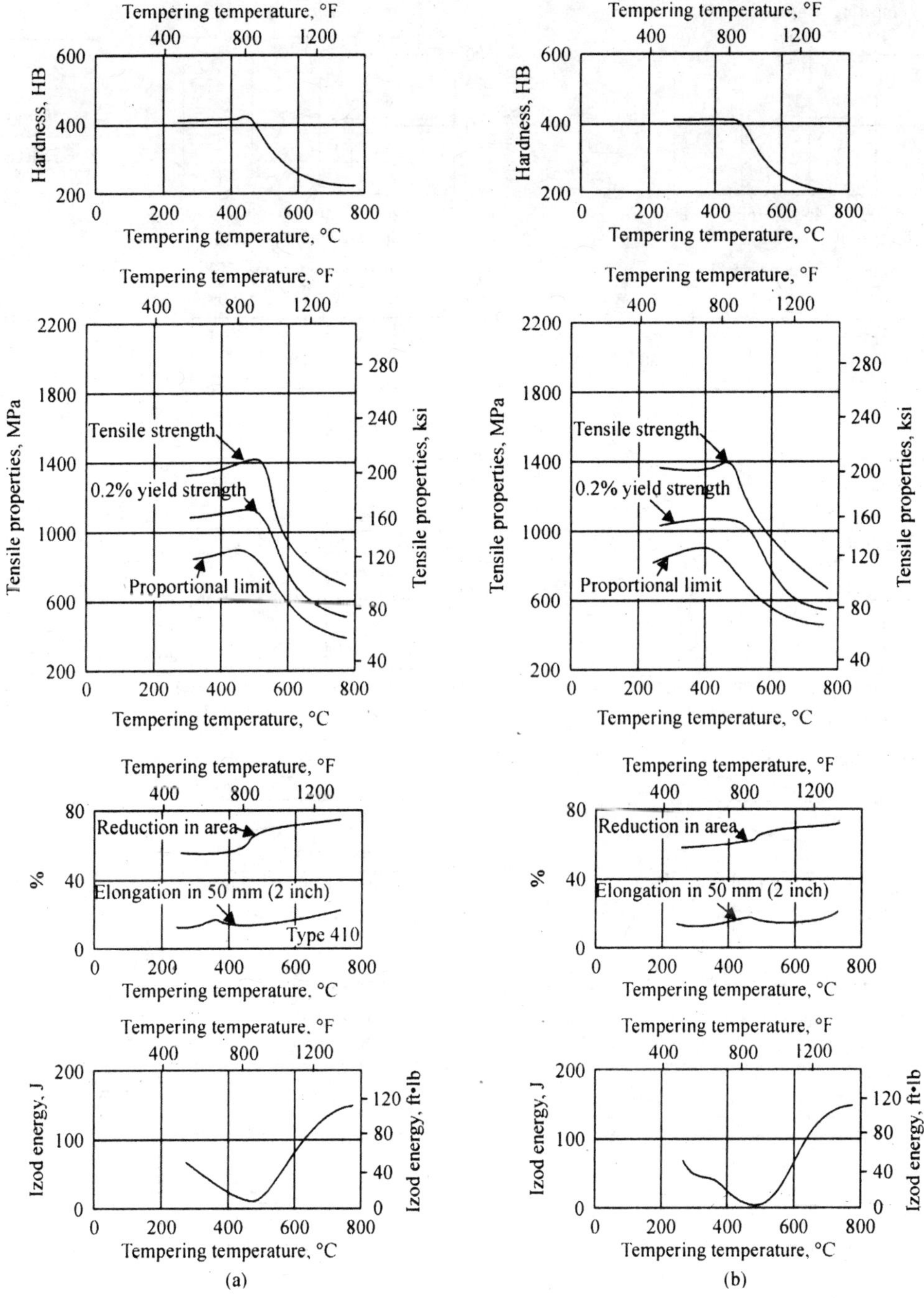

Austenitised 30 minutes; oil quenched to 65° to 95°C (150 to 200°F); double stress relieved at 175°C (350°F) for 15 minute and water quenched; tempered 2 hrs. (a) Quenched from 925° C (1700°F). (b) Quenched from 1010°C (1850°F)

Fig. 19.6. Effect of austenitising and tempering temperatures on mechanical properties of type 410 martensitic stainless steel.

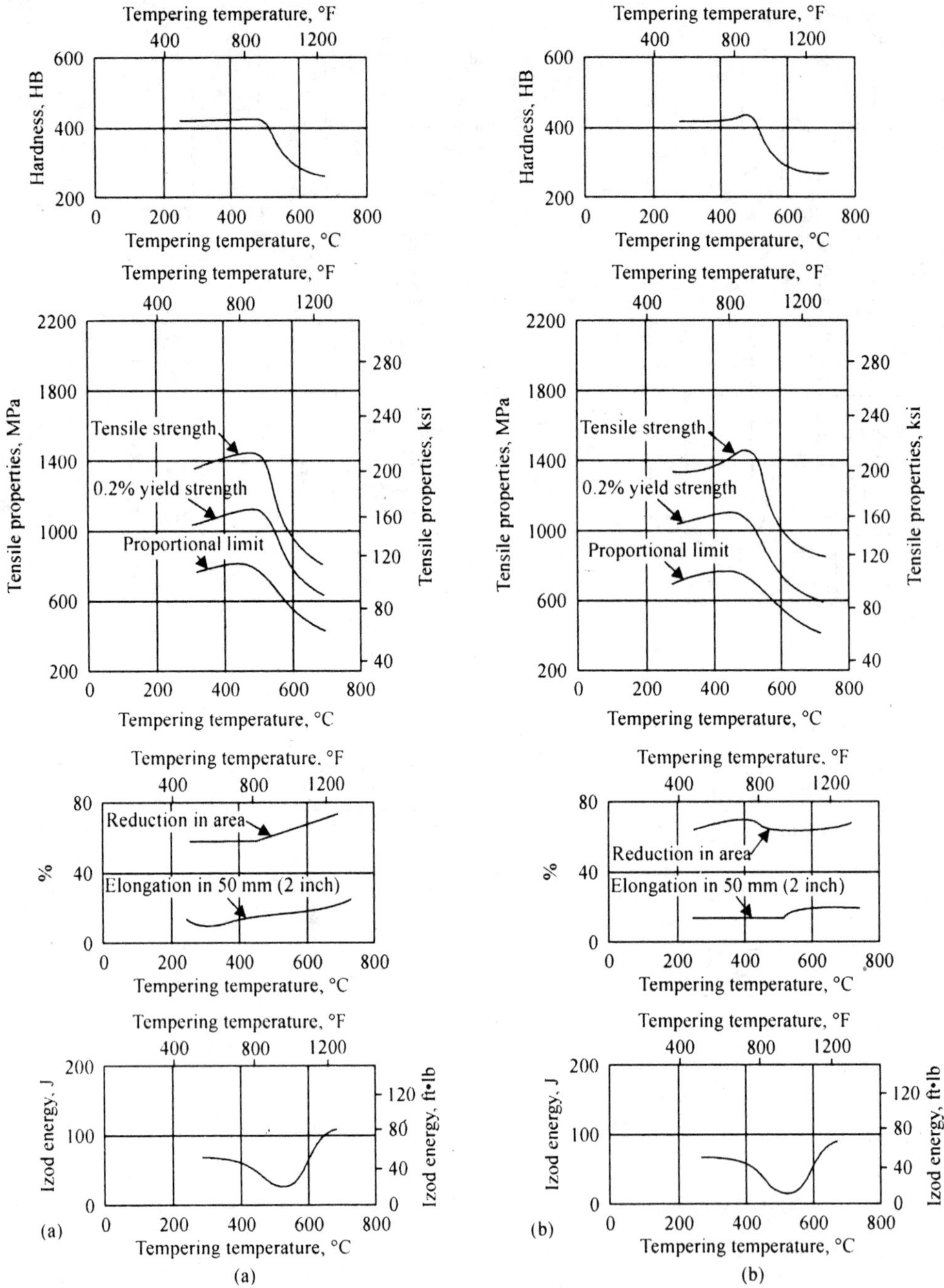

Austenitised 30 minutes; oil quenched to 65 to 95°C (150 to 200°F); double stress relieved at 175°C (350°F) for 15 minute and water quenched; tempered 2 hr. (a) Quenched from 925° C (1700°F); (b) Quenched from 1035°C (1900°F)

Fig. 19.7. Effect of austenitising and tempering temperatures on typical mechanical properties of type 414 martensitic stainless steel.

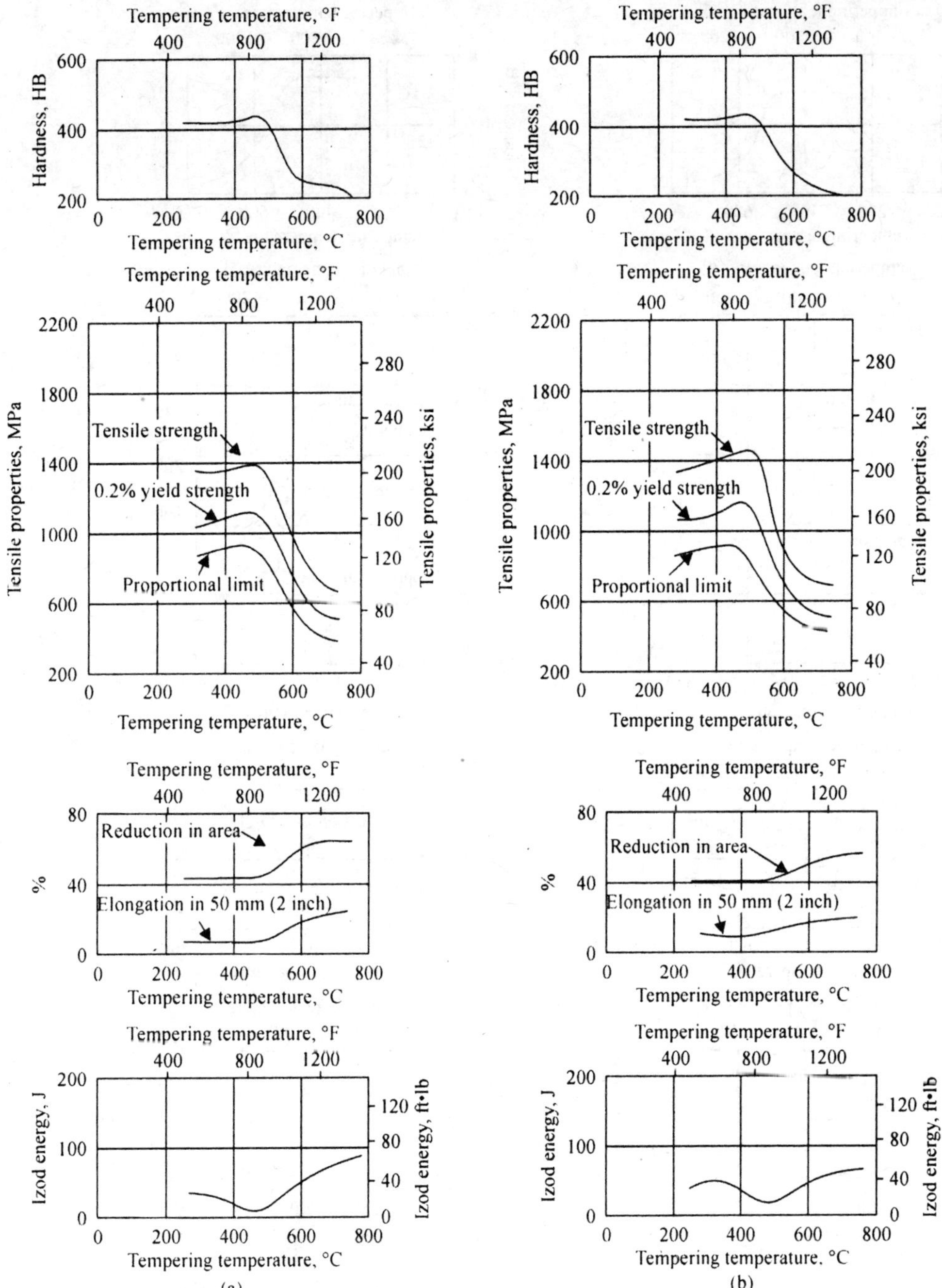

Austenitised 30 minutes; oil quenched to 65 to 95°C (150 to 200°F); double stress relieved at 175°C (350°F) for 15 minutes and water quenched; tempered 2 hrs. (a) Quenched from 925° C (1700°F); (b) Quenched from 980°C (1800°F)

Fig. 19.8. Effect of austenitising and tempering temperatures on typical mechanical properties of type 416 martensitic stainless steel.

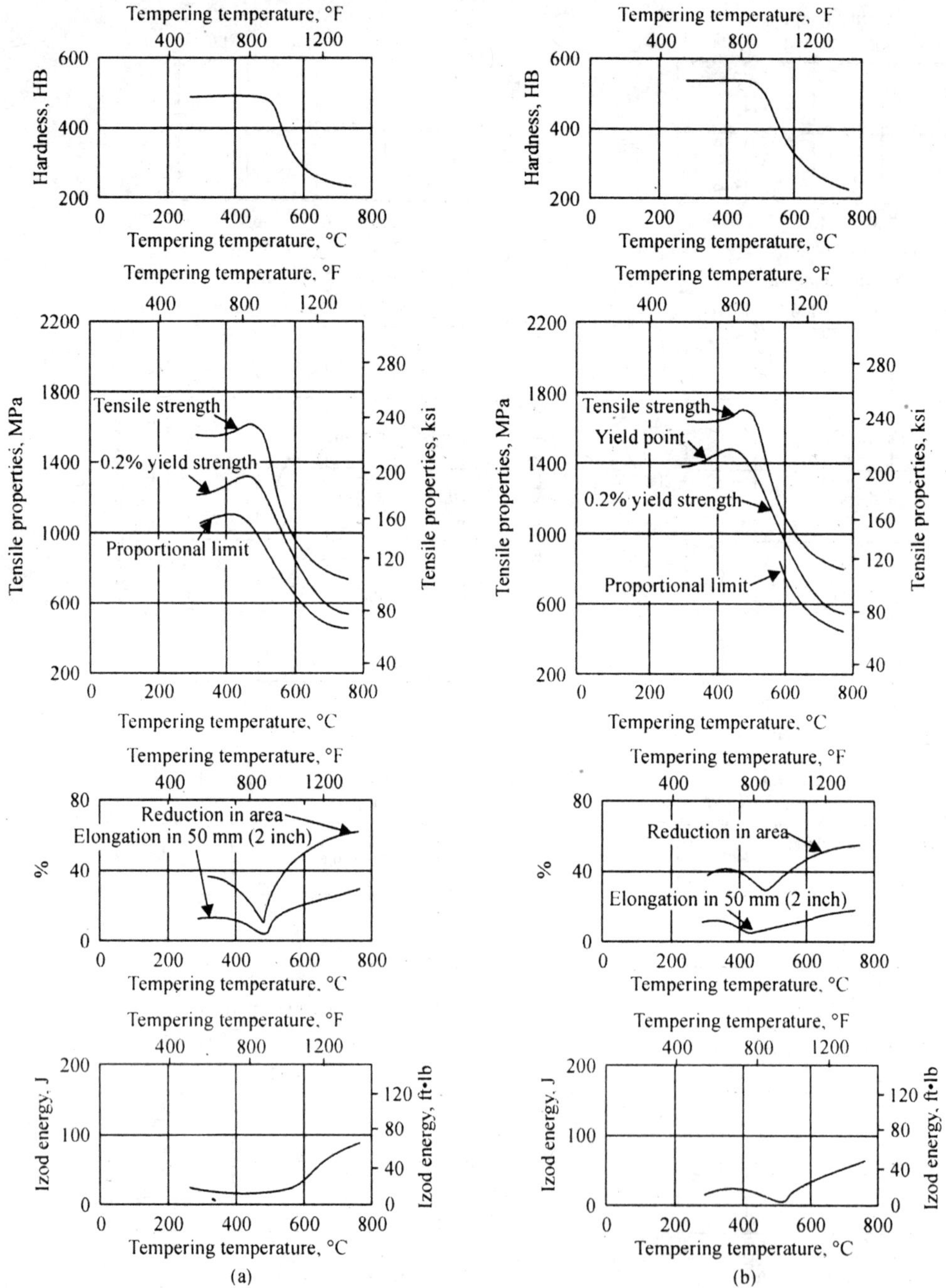

Austenitised 30 minutes; oil quenched to 65 to 95°C (150 to 200°F); double stress relieved at 175°C (350°F) for 15 minutes and water quenched; tempered 2 hrs. (a) Quenched from 925° C (1700°F); (b) Quenched from 1025°C (1875°F)

Fig. 19.9. Effect of austenitising and tempering temperatures on typical mechanical properties of type 420 martensitic stainless steel.

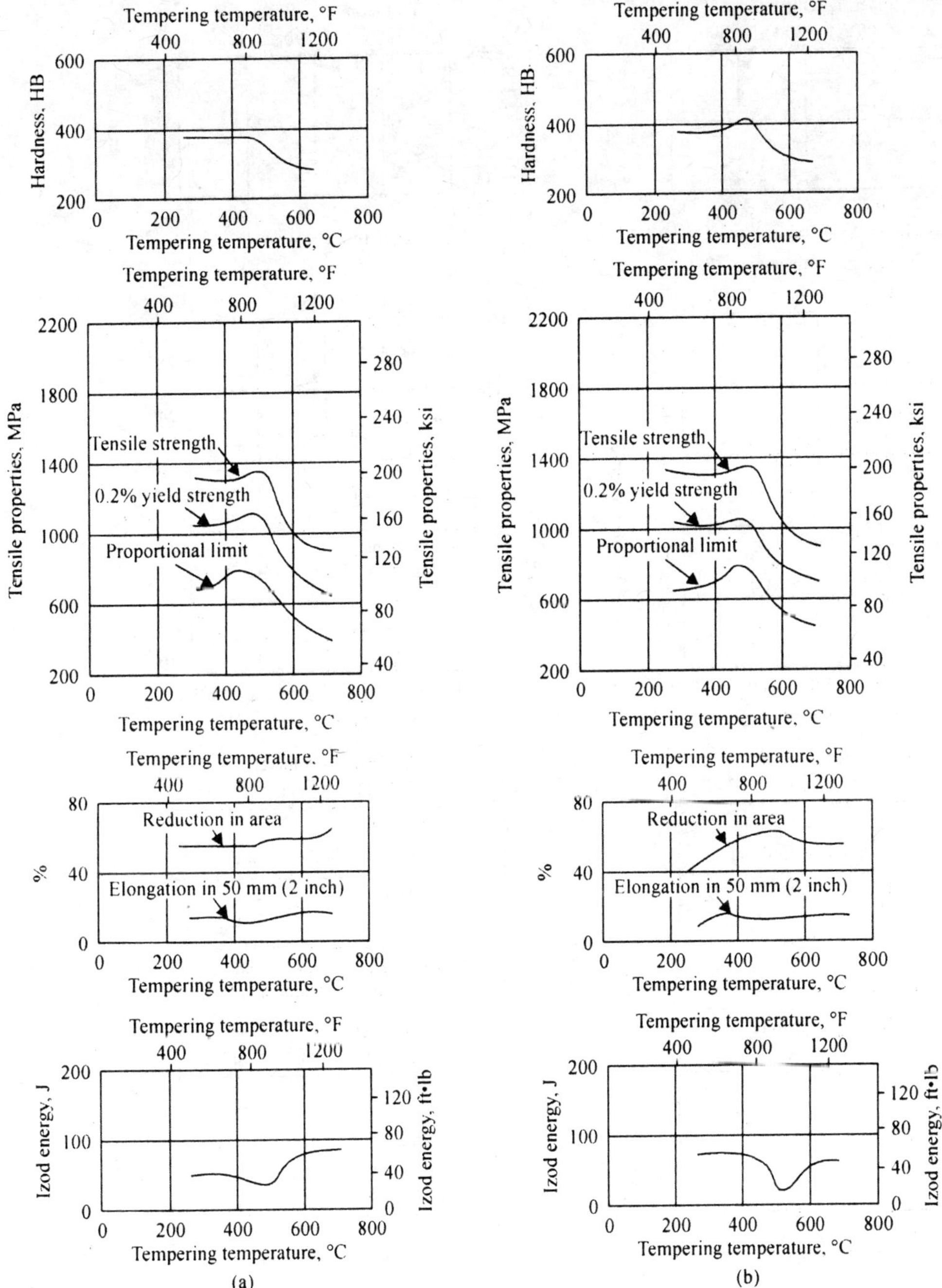

Austenitised 30 minutes; oil quenched to 65 to 95°C (150 to 200°F); double stress relieved at 175°C (350°F) for 15 minutes and water quenched; tempered 2 hrs. (a) Quenched from 925° C (1700°F). (b) Quenched from 1040°C (1900°F)

Fig. 19.10. Effect of austenitising and tempering temperatures on typical mechanical properties of type 431 martensitic stainless steel.

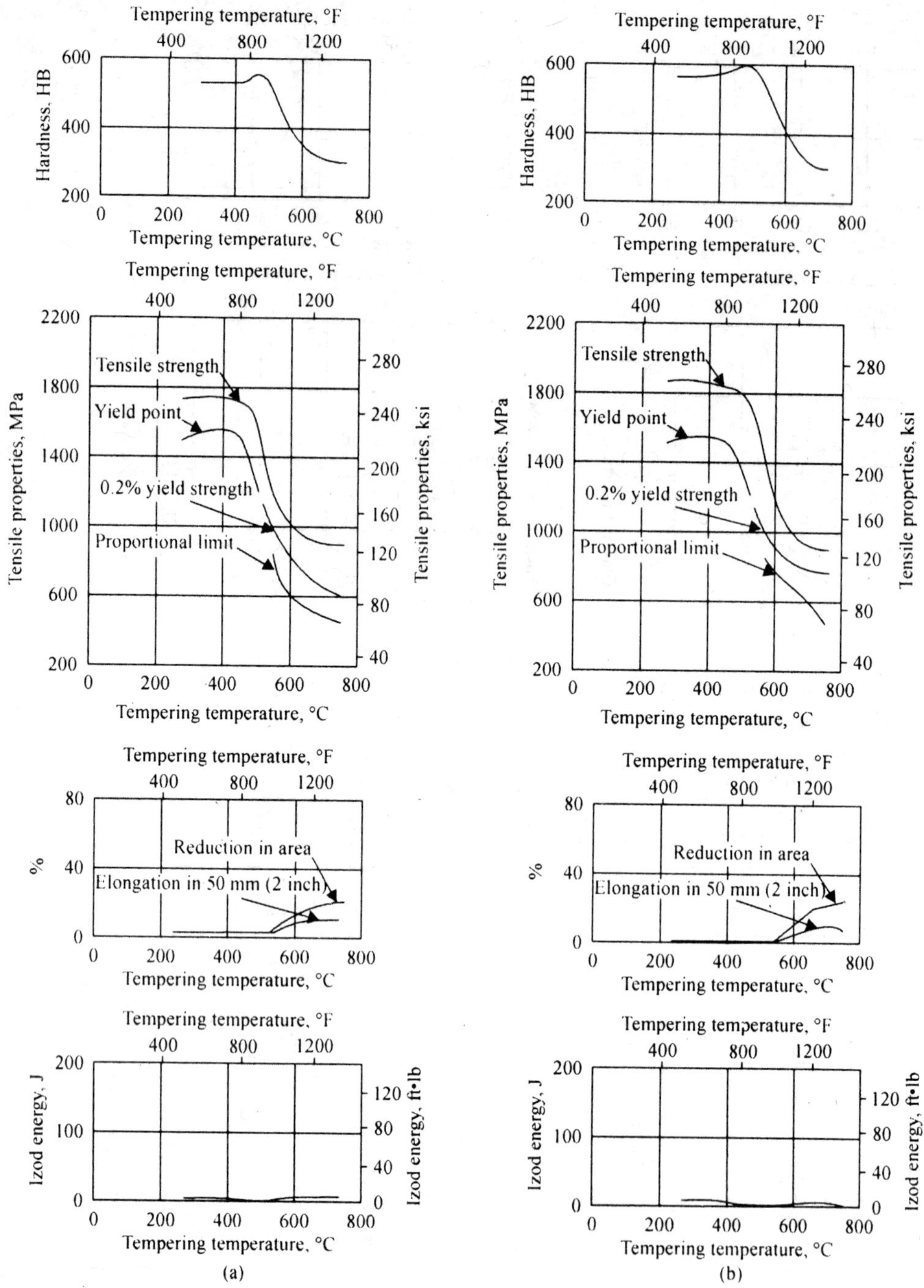

Austenitised for 1 hr. at 925°C (1700°F) and 2 hrs. at 1040°C (1900°F); oil quenched to 65 to 95°C (150 to 200°F); double stress relieved at 175°C (350°F) for 15 minutes and water quenched; tempered 2 hr.s (a) Quenched from 925°C (1700°F); (b) Quenched from 1040°C (1900°F)

Fig. 19.11. Effect of austenitising and tempering temperatures on typical mechanical properties of type 440C martensitic stainless steel.

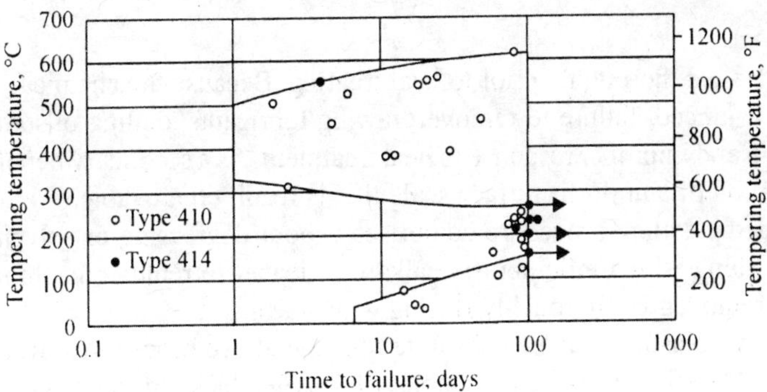

Data apply to a stress level of 550 MPa (80 ksi) for tests in a salt fog cabinet.

Fig. 19.12. Effect of tempering temperature on the stress-corrosion characteristics of two martensitic stainless steels at high stress.

An exothermic gas ratio of 6.5 or 7 to 1 is satisfactory for grades of stainless containing not more than 0.15 per cent carbon. When an endothermic atmosphere is used, dew points for specific steels and austenitising temperatures should be:

Alloy	*Austenitising temperature, °C*	*Dew point, °C*
420	1010	10–12
403, 410, 414, 416, 431	980	16–18
440C	1040	2–4

Hydrogen Embrittlement

Hydrogen embrittlement can become an important concern in the martensitic grades, generally increasing with hardness and carbon content. (It is variable and less acute in ferritic steels and is virtually unknown in the austenitic grades).

The embrittling hydrogen may be acquired as result of the melting process, a heat treating atmosphere, or chemical and electrochemical processes such as pickling and electroplating.

Most heat treating atmospheres contain hydrogen in the form of: (i) moisture; (ii) hydrocarbons; and (iii) elemental hydrogen as an atmosphere or a dissociation product. The use of pure hydrogen or dissociated ammonia for bright annealing in one plant was associated with cracking of wire coils of types 431 and 440C, although other plants have reported no similar difficulty. Nevertheless, it is possible that some loss in ductility may result from the bright annealing of any of the martensitic stainless steels.

Embrittlement has been found in oil-quenched types 403, 410, 414 and 431. Air quenching or the subsequent tempering of oil-quenched material releases the entrapped hydrogen and ductility is restored. If a green oxide appears on the oil-quenched alloy, this must be removed prior to tempering or it will interfere with the release of entrapped hydrogen.

PRECIPITATION-HARDENING STAINLESS STEELS

In the heat treating of precipitation-hardening stainless steels, areas of primary interest include: (i) cleaning prior to heat treatment; (ii) furnace atmospheres; (iii) time-temperature cycles; (iv) effect of variations in cycles; and (v) scale removal after heat treating.

Prior Cleaning

All parts must be cleaned thoroughly prior to heat treating. Because the chemical composition of these steels is delicately balanced, failure to remove drawing lubricants, cutting oils and grease can lead to surface carburisation and improper response to heat treatment. As a secondary benefit, thorough cleaning promotes the formation of a uniform surface scale that is readily removable.

The recommended cleaning procedure comprises vapour degreasing or solvent cleaning, followed by mechanical scrubbing with a mild abrasive alkaline cleaner to remove insoluble soils. All traces of cleaners should be removed by thoroughly rinsing with warm water.

Wet or dry abrasive blasting may be substituted for the above procedures. Recommended grits and operating details for blasting are given in Table 19.5. After blasting, all traces of abrasive must be removed from the work by scrubbing thoroughly.

Table 19.5. Recommended conditions for abrasive blast cleaning of precipitation-hardening stainless steels prior to heat treatment(a).

Abrasive		*Nozzle*			*Air pressure*		*Cleaning speed*	
		Size						
Material	*Grit No.*	*mm*	*inch*	*Angle, °*	*KPa*	*psi*	*mm²/s*	*inch²/min*
Alumina (dry)	30	6.4	1/4	45 to 60	170 to 655(b)	25 to 95(b)	130 to 215	12 to 20
Garnet or alumina (dry)	36	9.5	3/8	60	240	35	645	60
Wet blasting	220	6.4	1/4	45 to 60	170 to 655(b)	25 to 95(b)	65 to 110	6 to 10

(a) All abrasive must be removed by thorough scrubbing. (b) Depending on metal thickness.

In some applications, cleaning prior to heat treating may be accomplished by closely controlled pickling in a 10 per cent HNO_3-2 per cent HF aqueous solution at 45 to 60°C (110 to 140°F). Time should be limited to 2 to 3 minutes. This method is not recommended for cleaning severely formed or previously heat treated parts.

Furnaces

Furnaces fired with oil or natural gas are not entirely satisfactory for the heat treatment of these steels where finished surfaces are not to be subsequently machined. In such units, it is difficult to control combustion contaminants and to eliminate flame impingement on the parts being treated. Electric furnaces or gas-fired radiant-tube furnaces are generally used for heat treating precipitation-hardening stainless steels.

Furnace Atmospheres

Air is a satisfactory furnace atmosphere for austenite-conditioning and annealing operations. Controlled reducing atmospheres, such as dissociated ammonia or bright-annealing gas, introduce the potential hazard of nitriding or carburising, either of which has a deleterious effect on mechanical properties.

Bright annealing may be done in hydrogen, argon or helium atmospheres, provided a dew point of –55°C (–65°F) or lower is maintained. The cooling rate from the annealing temperature must be approximately equal to that of cooling in still air. Austenite-conditioning treatments at temperatures as high as 925 to 955°C (1700 to 1750°F) may also be performed in dry hydrogen, argon or helium, maintaining the same low dew point. A scale-free surface will be obtained.

The lower austenite-conditioning temperatures, such as 760°C (1400°F), present difficulties in achieving scale-free surfaces in dry hydrogen, argon or helium. An air atmosphere is generally used at these temperatures. For complete freedom from scale or discolouration at the lower temperatures, a vacuum furnace is required. Final hardening of these steels is performed at relatively low temperatures and an air atmosphere is acceptable for these treatments.

Heat Treating Procedures

Recommended procedures for full annealing, austenite conditioning, transformation cooling and age-tempering (precipitation hardening) are given in Table 19.6.

Table 19.6. Recommended heat treating procedures for semi-austenitic precipitation-hardenable stainless steels.

UNS S17400

Homogenisation. 1175 ± 15°C (2150 ± 25°F), 2 hrs. + 30 minutes per 25 mm (1 inch)(a)

Austenite conditioning (solution treatment): 1040 ± 15°C (1900 ± 25°F), 30 minutes + 30 minutes per 25 mm (1 inch)(a)

Transformation cooling. To below + 30°C (+90°F)

Precipitation hardening. To obtain minimum tensile strengths shown, use the following treatments for wrought alloys(b):

MPa	*ksi*		
1310	190	480 ± 5°C (900 ± 10°F)	1 hr.
1170	170	495 ± 5°C (925 ± 10°F)	4 hr.
1070	155	550 ± 5°C (1030 ± 10°F)	4 hr.
1030	150	565 ± 5°C (1050 ± 10°F)	4 hr.
1000	145	580 ± 5°C (1075 ± 10°F)	4 hr.
930	135	620 ± 5°C (1150 ± 10°F)	4 hr.

Comparable treatments for cast materials(b):

MPa	*ksi*		
1240	180	480 ± 5°C (900 ± 10°F)	4 hr.
1170	170	495 ± 5°C (925 ± 10°F)	4 hr.
1030	150	540 ± 5°C (1000 ± 10°F)	4 hr.
900	130	595 ± 5°C (1100 ± 10°F)	4 hr.

UNS S17700

Full annealing. 1065 ± 15°C (1950 ± 25°F), 3 minutes + 1 minute per 0.25 mm (0.01 inch); air cool

RH treatments

Austenite conditioning. 955 ± 15°C (1750 ± 25°F), 10 minutes + 1 minute per 0.25 mm (0.01 inch); air cool(c)

Transformation cooling. To below – 70°C (–90°F), 8 hrs.

Precipitation hardening. To obtain minimum tensile strengths shown, the following treatments are recommended(b):

MPa	*ksi*		
1450	210	510 ± 5°C (950 ± 10°F)	1 hr.
1240	180	565 ± 5°C (1050 ± 10°F)	1 hr.
1170	170	580 ± 5°C (1075 ± 10°F)	1 hr.
1030	150	595 ± 5°C (1100 ± 10°F)	1 hr.

(Contd...)

TH treatments

Austenite conditioning. 760 ± 15°C (1400 ± 25°F), 1½ hr; cool within 1 hr. to below 15°C (60°F) but above 0°C (32°F) and hold at least ½ hr. before precipitation hardening

Precipitation hardening. To obtain minimum tensile strengths shown, the following treatments are recommended(b):

MPa	*ksi*		
1240	180	565 ± 5°C (1050 ± 10°F)	1½ hr.
1170	170	580 ± 5°C (1075 ± 10°F)	1½ hr.
1030	150	595 ± 5°C (1100 ± 10°F)	1½ hr.

UNS S15700

Full annealing. Same as for 17–7 PH

RH treatments

Austenite conditioning. Same as for 17–7 PH

Transformation cooling. Same as for 17–7 PH

Precipitation hardening. To obtain minimum tensile strengths shown, the following treatments are recommended(b):

MPa	*ksi*		
1550	225	510 ± 5°C (950 ± 10 °F)	1hr.
	565 ± 5 (1050 ± 10°F)		1hr.
1310	190°C		

TH treatments

Austenite conditioning. 760 ± 15°C (1400 ± 25°F). 1½ hr. cool within 1 hr. to below 15°C (60°F) but above 0°C (32°F) and hold at least ½ hr. before precipitation hardening

Precipitation hardening. For minimum tensile strength of 1310 MPa (190 ksi)(b): 565 ± 5 °C (1050 ± 10°F), 1½ hr.

UNS S35000

Full annealing. Wrought materials only: 1065 ± 15°C (1950 ± 25°F), 3 minutes + 1 minute per 0.25 mm (0.01 inch); air cool

Austenite conditioning. 930 ± 5°C (1710 ± 10°F), 10 minutes + 1 minute per 0.25 mm (0.01 inch); air cool(c)

Transformation cooling. To –75 ± 5°C (–100 ± 10°F), 3 hrs. (minimum)

Precipitation hardening. To obtain minimum tensile strengths shown, the following treatments are recommended(b):

MPa	*ksi*		
1270	185	455 ± 5°C (850 ± 10°F)	3 hr.
1170	170	510 ± 5°C (950 ± 10°F)	3 hr.
1140	165	540 ± 5°C (1000 ± 10°F)	3 hr.

UNS S35500

Homogenisation. Castings only: 1095 ± 15°C (2000 ± 25°F), 2 hrs.; air cool (water quench sections over 50 mm, or 2 inch). Bar and forgings: 1050 ± 15°C (1925 ± 25°F), 1 to 3 hrs.; water quench. Cool all forms to below –70°C (–90°F), hold 3hrs. minimum.

Full annealing. 1025 ± 15°C (1875 ± 25°F), 1 hr. per 25 mm (1 inch); water quench

Machinability treatment. 760 ± 15 °C (1400 ± 25°F), 3hrs.; air cool. Refrigerate to –70°C (–90°F), hold for 3 hrs. Reheat to 565 ± 15°C (1050 ± 25°F), 3 hrs.

(Contd...)

For –70°C (–90°F) transformation:

Austenite conditioning. Castings: 980 ± 15 °C (1800 ± 25°F), 2 hrs; air cool (oil quench sections over 3 mm, or 0.125 inch). Wrought materials: 930 ± 15 °C (1710 ± 25°F), 15 minutes per inch; air cool (oil quench sections over 3 mm or 0.125 inch)

Transformation cooling. To –75 ± 5°C (–100 ± 10°F), 3 hrs.

Precipitation hardening. To obtain minimum tensile strengths shown, use the following treatments for wrought alloys(b):

MPa	*ksi*		
1310	190	455 ± 5°C (850 ± 10°F)	3hrs.
1170	170	540 ± 5°C (1000 ± 10°F)	3hrs.

For castings, to obtain minimum tensile strength of 1240 MPa (180 ksi)(b): 455 ± 5°C (850 ± 10°F), 2 hr.

UNS S45000

Solution annealing. 1040 ± 15°C (1900 ± 25°F), 1 hr. at heat (d), water quench

Precipitation hardening. Typical tensile strengths shown may be obtained by the following treatments:

MPa	*ksi*		
1345	195	480 ± 5°C (900 ± 10°F)	4 hrs., air cool
1170	170	540 ± 5°C (1000 ± 10°F)	4 hrs., air cool
1105	160	565 ± 5°C (1050 ± 10°F)	4 hrs., air cool
965	140	620 ± 5°C (1150 ± 10°F)	4 hrs., air cool

UNS S45500

Solution annealing. 830 ± 15°C (1525 ± 25°F), 1 hr. at heat (d), water quench.

Precipitation hardening. Typical tensile strengths shown may be obtained by the following treatments:

MPa	*ksi*		
1725	250	480 ± 5°C (900 ± 10°F)	4 hrs., air cool
1620	235	510 ± 5°C (950 ± 10°F)	4 hrs., air cool
1450	210	540 ± 5°C (1000 ± 10°F)	4 hrs., air cool
1310	190	565 ± 5°C (1050 ± 10°F)	4 hrs., air cool

UNS S15500

Solution annealing. 1040 ± 15°C (900 ± 25°F), 1 hr. at heat (d), water quench.

Precipitation hardening. Typical tensile strengths shown may be obtained by the following treatments:

MPa	*ksi*		
1380	200	480 ± 5°C (900 ± 10°F)	1 hr., air cool
1170	170	550 ± 5°C (1025 ± 10°F)	4 hr., air cool
1000	145	620 ± 5°C (1150 ± 10°F)	4 hr., air cool

H1150M condition (after annealing). 760 ± 8°C (1400 ± 15°F), 2 hrs., air cool + 620 ± 5°C (1150 ± 10°F), 4 hrs., air cool

Typical tensile strength. 860 MPa (125 ksi)

UNS S13800

Annealing. 925 ± 8°C (1700 ± 15°F), 1 hr. (d), air cool or oil quench.

(Contd...)

Precipitation hardening. Typical tensile strengths shown may be obtained by the following treatments:			
MPa	*ksi*		
1550	225	510 ± 5°C (950 ± 10°F)	4 hrs., air cool
1310	190	565 ± 5°C (1050 ± 10°F)	4 hrs., air cool
1000	145	620 ± 5°C (1150 ± 10°F)	4 hrs., air cool
H1150M condition (after annealing). 760 ± 8°C (1400 ± 15°F), 2 hrs., air cool + 620 ± 5°C (1150 ± 10°F), 4 hrs., air cool			
Typical tensile strength. 895 MPa (130 ksi).			

(a) To prevent cracking and ensure uniform properties, cool as follows: 75 mm (3 inch) and less, oil quench or air cool; 75 to 150 mm (3 to 6 inch), air cool; 150 mm (6 inch) and over, air cool under cover. All parts must be cooled to below + 30°C (+ 90°F) prior to the precipitation-hardening cycle; (b) If hardness exceeds maximum specified, re-heat treat at a slightly higher temperature for a minimum of 30 minutes; (c) Air cool to room temperature; do not re-heat before transformation cooling; and (d) Time at heat is dependent upon section size. Normally, a 1hr. hold at temperature is suggested.

17–4 PH (UNS S17400)

17–4 PH (UNS S17400) is a precipitation hardening steel that has an essentially martensitic structure and limited formability when supplied in the solution treated condition. Fabrication is followed by hardening in the range of 480 to 620°C (900 to 1150°F) (Table 19.6).

This alloy should not be put into service in any application in the solution treated condition, because in this condition its ductility can be relatively low and its resistance to stress-corrosion cracking is poor. Hardening to any of the strength levels shown in Table 19.6 improves both toughness and resistance to stress corrosion.

17–7 PH (UNS S17700)

17–7 PH (UNS S17700) is normally supplied in the annealed condition (condition A), in which it is soft and formable. Heat treatment is accomplished through the use of the TH or RH procedures indicated in Table 19.6. The choice of method of heat treatment is in most instances dictated by the ease with which the particular sequence fits into the user's production techniques.

This alloy is also supplied in the cold rolled condition (condition C). Here, transformation has been achieved by cold rolling and heat treatment is reduced to a single step: 480°C (900°F) for 1 hour. Although strength and stress-corrosion resistance are greatly increased by this treatment, ductility is reduced and formability is limited.

pH 15-7 Mo (UNS S15700)

pH 15–7 Mo (UNS S15700) is a high-strength modification of 17–7 PH. It is supplied in the same condition as 17–7 PH and requires identical heat treating procedures. Table 19.6 illustrates the strength levels obtainable with PH 15–7 Mo.

15–5 Ni Alloys (UNS S15500)

15–5 Ni alloys (UNS S15500) are normally supplied in the annealed condition. They should not be put into service in the annealed condition because of susceptibility to stress corrosion cracking. The alloy can be hardened by heating to a temperature in the range of 480 to 620°C (900 to 1150°F) for 1 to 4 hours depending on the temperature, then air cooling. Table 19.6 illustrates typical strength levels obtainable.

13–8 Mo Alloys (UNS S13800)

13–8 Mo Alloys (UNS S13800) are normally supplied in the annealed condition. They can be hardened to high strength levels by a single low temperature treatment. Table 19.6 illustrates typical strength levels versus hardening procedures.

AM-350 and PYRO MET 350 (UNS S35000)

AM-350 and PYRO MET 350 (UNS S35000) are normally purchased in the fully annealed condition; however, after severe forming or cold working, it may require a second annealing treatment. The annealing temperature limits, indicated in Table 19.6, are critical. High temperatures reduce strength; lower temperatures adversely affect formability.

After annealing and conditioning at 930°C (1710°F), AM-350 is usually sub-zero cooled and then aged at 455°C (850°F) for 3 hrs.; this treatment produces maximum strength. Maximum toughness is achieved by ageing in the range of 480 to 540°C (900 to 1000°F). The recommended temperature for sub-zero cooling must be carefully observed. Cooling to much lower temperatures, such as –195°C (–320°F), results in incomplete transformation, as does failure to hold at the recommended temperature for at least 3 hrs.

AM-355 and PYRO MET 355 (UNS S35500)

AM-355 and PYRO MET 355 (UNS S35500) flat products are supplied in either the solution treated or solution treated and cold rolled condition, whereas bar products are usually supplied in the equalised and overtempered condition for best machinability. Most castings are supplied in the as-cast condition.

Although, as indicated in Table 19.6, the homogenising and austenite-conditioning treatments applied to castings differ from those applied to wrought materials (higher temperatures being used for castings in both treatments), sub-zero cooling is required for all forms, to obtain maximum toughness and corrosion resistance. The full annealing treatment shown in Table 19.6 would normally be applicable only to flat products. The machinability treatment is required for obtaining good machining characteristics in this alloy. Wrought materials should be aged at 455°C (850°F) for maximum strength and at 540°C (1000°F) for maximum ductility and toughness. The usual ageing treatment for castings consists of holding at 455°C (850°F) for 2 hrs.

Custom 450 (UNS S45000)

Custom 450 (UNS S 45000) stainless is normally supplied in the annealed condition, requiring no further heat treatment for many applications. It is easily fabricated in the annealed condition. A single-step hardening treatment develops higher strength with good ductility and toughness.

The recommended minimum hardening temperature of 480°C (900°F) produces the optimum combination of strength, ductility and toughness. Hardening at temperatures up to 620°C (1150°F) increases the ductility and decreases strength (Table 19.6).

Custom 455 (UNS S45500)

Custom 455 (UNS S45500) stainless is normally in the annealed condition. It is relatively soft and easily formable in the annealed condition. A single-step hardening treatment develops exceptionally high yield strengths with good ductility and toughness. Harden by heating in the range of 480 to 565°C (900 to 1050°F). (Table 19.6).

Variations in Heat Treating Cycles

One of the principal advantages of these materials is their versatility. Although certain heat treatments have been classified as standard, there are applications where deviations from these standards are desirable. The series of curves in Figs. 19.13 through 19.17 illustrates how these deviations affect mechanical properties.

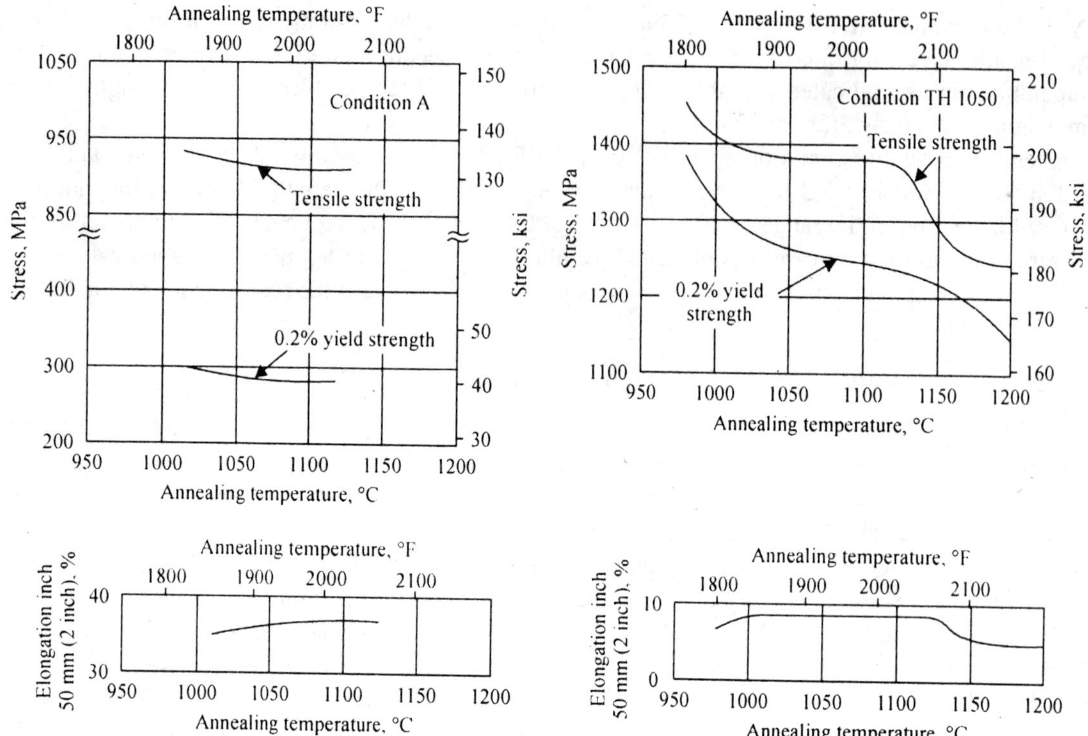

Fig. 19.13. Effect of variations in annealing temperature on typical mechanical properties of 17–7 PH sheet, strip and plate.

Scale Removal after Heat Treating

The amount and nature of scale vary with the degree of cleanness of the work being treated, the furnace atmosphere and the temperature and duration of heat treatment. In the following discussion, it will be assumed that all heat treating operations are performed in an air atmosphere. A variety of descaling methods may be employed; the choice depends on the type of steel and the facilities available.

In removing scale formed during homogenisation or full annealing, the use of 10 per cent HNO_3 2 per cent HF aqueous solution at 45 to 60°C (110 to 140°F) has been effective. Exposure to the acid solution should be limited to a period of 3 minutes. Removal of loosened scale may be facilitated by the use of high-pressure water or steam. A uniform surface is evidence of a well-cleaned part.

The austenite-conditioning treatments produce a scale that is best removed by mechanical means. Acids should be avoided, because they are a possible source of intergranular attack. Wet grit blasting processes have been widely used to remove these scales and have been found to be highly satisfactory.

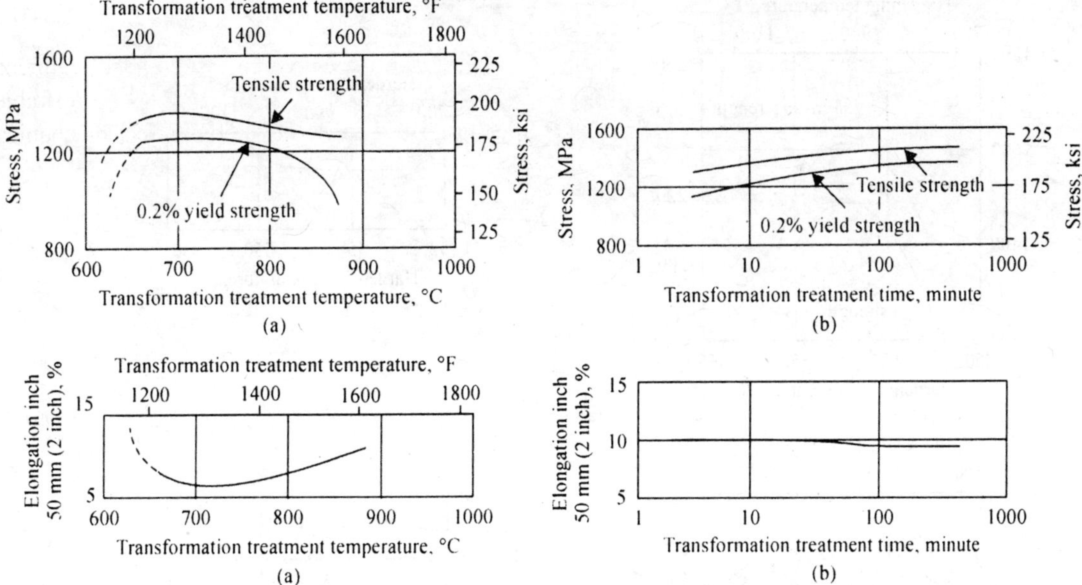

Fig. 19.14. Effect of variations in transformation treatment temperature and time on typical mechanical properties of 17–7 PH sheet, strip and plate.

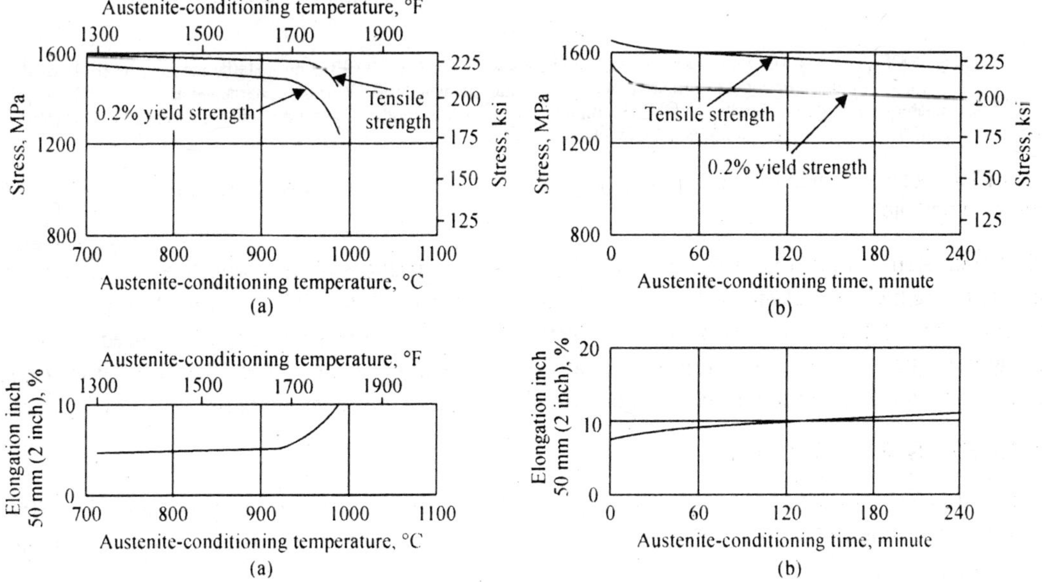

Fig. 19.15. Effect of variations in austenite-conditioning temperature and time on typical mechanical properties of 17–7 PH sheet, strip and plate.

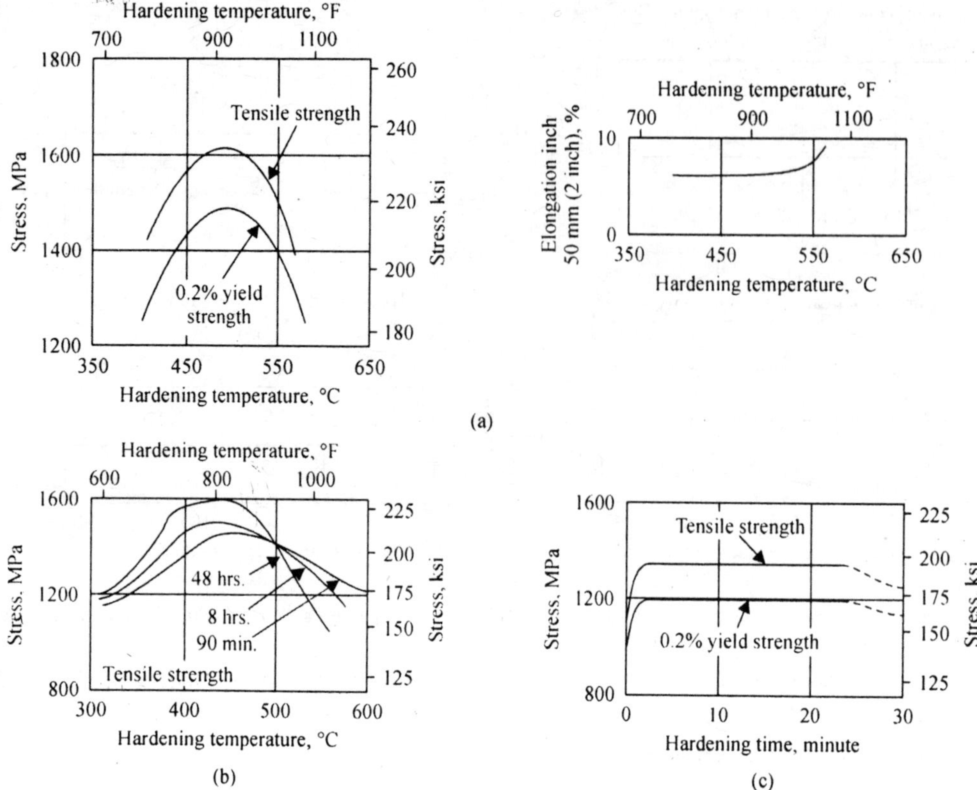

(a) Heated at 955°C (1750°F), 10 minutes air cooled, liquid cooled to −75°C (−100°F), 8 hrs.; Hardened for 1 hr. (b) Heated at 760°C (1400°F), 90 minutes; air cooled to room temperature, water quenched to 15°C (60°F); hardened as indicated. Elongation data not available; (c) Heated at 760°C (1400°F), 90 minutes; air cooled to room temperature, water quenched to 15°C (60°F); hardened at 565°C (1050°F). Elongation data not available.

Fig. 19.16. Effect of variations in hardening temperature and time on typical mechanical properties of 17–7 PH sheet, strip and plate.

The final step in heat treating (precipitation hardening) produces a discolouration of heat tint. It is desirable to use mechanical means to remove this oxide from 17–7 PH, PH 15–7 Mo, AM-350 and AM-355. The HNO_3-HFsolution has been used on these steels, but extreme care is required to prevent intergranular attack. The acid solution may be used satisfactorily with 17–4 PH. To a lesser extent, electropolishing has also been used to remove the final heat tint resulting from precipitation hardening.

STAINLESS STEEL CASTINGS

The heat treatment of stainless steel castings follows closely in purpose and procedure the thermal processing of comparable wrought materials. However, the differences in detail warrant separate consideration here.

Because they are not cold worked or cold formed, castings of the older, conventional martensitic grades CA-15 and CA-40 (UNS J91150 and J91153) do not require sub-critical annealing to remove the effects of cold working. However, in work-hardenable ferritic alloys, machining and grinding stresses are relieved at temperatures from about 260 to 540°C (500 to 1000°F). Casting stresses in the martensitic castings noted above should be relieved by sub-critical annealing prior to further heat treatment. When these hardened martensitic castings are stress relieved, the stress-relieving temperature must be kept below the final tempering or ageing temperature.

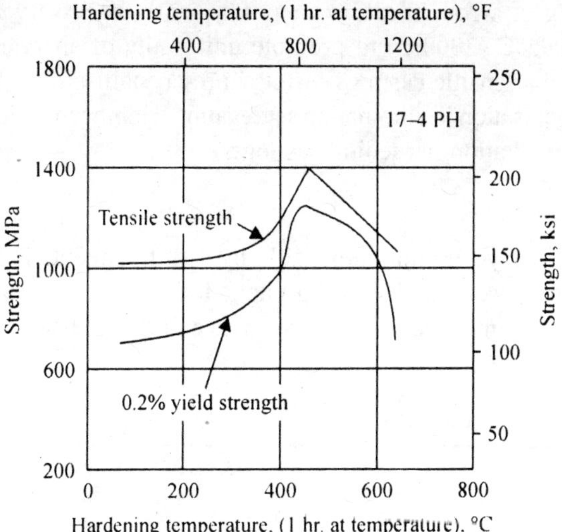

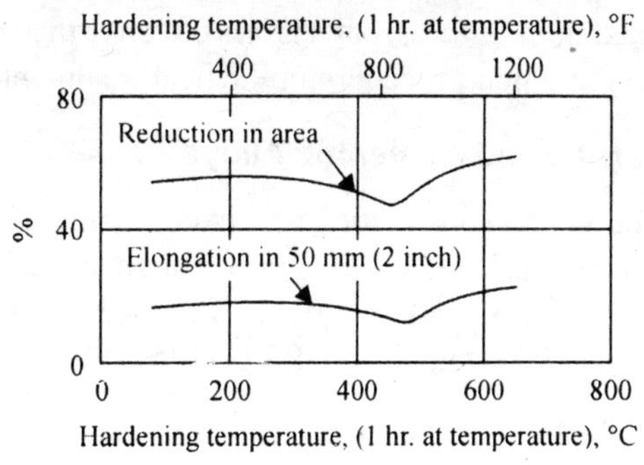

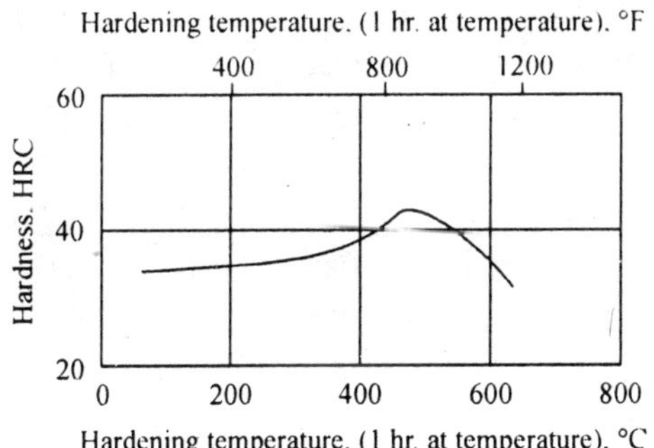

Data are average values for bars 25 to 89 mm (1 to 3½ inch) in diameter from four heats.

Fig. 19.17. Effect of hardening temperature on typical room-temperature properties of 17–4 PH.

An improved, cast martensitic alloy, CA-6NM (UNS-J91540), possesses better casting behaviour, improved weldability and equals or exceeds all of the mechanical, corrosion and cavitation resistance properties of CA-15 and, as a result, has largely replaced the older alloy. Both CA-6NM and CA-15 castings are normally supplied in the normalised condition at 955°C (1750°F) minimum and tempered at 595°C (1100°F) minimum.

However, when it is necessary or desirable to anneal CA-6NM castings, a temperature of 790 to 815°C (1450 to 1500°F) should be used. The alloy should be furnace cooled or otherwise slow cooled to 595°C (1100°F), after which it may be cooled in air. When stress relieving is required, CA-6NM may be heated to 620°C (1150°F) maximum, followed by slow cooling to prevent the formation of martensite.

Homogenisation

Alloy segregation and dendritic structures may occur in castings and may be particularly pronounced in heavy sections. Because castings are not subjected to the high-temperature mechanical reduction and soaking treatments entailed in the mill processing of wrought alloys, it is frequently necessary to homogenise some alloys at temperatures above 1095°C (2000°F) to promote uniformity of chemical composition and microstructure. Full annealing of martensitic castings results in re-crystallisation and maximum softness, but it is less effective than homogenisation in eliminating segregation. Homogenisation is a common procedure in the heat treatment of precipitation-hardening castings.

Ferritic and Austenitic Alloys

The ferritic, austenitic and mixed ferritic-austenitic alloys are not hardenable by heat treatment. They can be heat treated to improve their corrosion resistance and machining characteristics. The ferritic alloys CB-30 and CC-50 (UNS J91803 and J92615) are annealed to relieve stresses and reduce hardness by being heated above 790°C (1450°F) (Table 19.7).

Table 19.7. Annealing of ferritic and austenitic stainless steel castings.

	Minimum temperature			*Typical ultimate tensile strength(b)*	
Type	*°C*	*°F*	*Quench(a)*	*MPa*	*ksi*
For full softness					
CB-30	790	1450	FC + A(c)	660	95
CC-50	790	1450	A	670	97
For maximum corrosion resistance					
CE-30	1095	2000	W, O, A	670	97
CF-3, CF-3M	1040	1900	W, O, A	530	77
CF-8, CF-8C(d)	1040	1900	W, O, A	530	77
CF-8M, CF-12M(e)	1040	1900	W, O, A	550	80
CF-16F, CF-20	1040	1900	W, O, A	530	77
CH-20	1095	2000	W, O, A	610	88
CK-20	1095	2000	W, O, A	520	76
CN-7M	1120	2050	W, O, A	480	69

(a) FC, furnace cool; W, water; O, oil; A, air. (b) Approximate. (c) Furnace cool to 540°C (1000°F), then air cool. (d) CF-8C may be re-heated to 870 to 925°C (1600 to 1700°F), then air cooled, for precipitation of niobium carbides. (e) CF-12M should be quenched from a temperature above 1095°C (2000°F).

The austenitic alloys achieve maximum resistance to intergranular corrosion by the high-temperature heating and quenching procedure known as solution annealing (Table 19.7). As-cast structures, or castings exposed to temperatures in the range from 425 to 870°C (800 to 1600°F), may contain complex chromium carbides precipitated preferentially along grain boundaries in wholly austenitic alloys. This microstructure is susceptible to intergranular corrosion, especially in oxidising solutions. (In partially ferritic alloys, carbides tend to precipitate in the discontinuous ferrite pools; thus, these alloys are less susceptible to intergranular attack.) The purpose of solution annealing is to ensure complete solution of carbides in the matrix and to retain these carbides in solid solution.

Solution annealing procedures for all austenitic alloys are similar and consist of heating to a temperature of about 1095°C (2000°F), holding for a time sufficient to accomplish complete solution of carbides and quenching at a rate fast enough to prevent re-precipitation of the carbides—particularly while cooling through the range from 870 to 540°C (1600 to 1000°F). Temperatures to which castings should be heated prior to quenching vary somewhat, depending on the alloy (Table 19.7).

As shown in Table 19.7, a two-step heat treating procedure may be applied to the niobium-containing CF-8C (UNS J92710) alloy. The first treatment consists of solution annealing. This is followed by a stabilising treatment at 870°C to 925°C (1600 to 1700°F), which precipitates niobium carbides, prevents formation of the damaging chromium carbides and provides maximum resistance to intergranular attack.

Because of their low carbon contents, CF-3 and CF-3M (UNS J92700 and J92800) as cast do not contain enough chromium carbides to cause selective intergranular attack and hence may be used in some corrodents in this condition; for maximum corrosion resistance, however, these grades require solution annealing.

Martensitic Alloys

Castings of the CA-6NM composition should be hardened by air cooling or oil quenching from a temperature of 1010 to 1065°C (1850 to 1950°F). Even though the carbon content of this alloy is lower than that of CA-15, this fact in itself and the addition of molybdenum and nickel enable the alloy to harden completely without significant austenite retention when cooled as suggested.

The choice of cooling medium is determined primarily by the maximum section size. Section sizes in excess of 125 mm (5 inch) will harden completely when cooled in air. CA-6NM is not prone to cracking during cooling from elevated temperatures. For this reason, no problem should arise in the air cooling or oil quenching of configurations that include thick as well as thin sections.

A wide choice of mechanical properties is available through the choice of tempering temperature. Castings of CA-6NM are normally supplied normalised and tempered at 595 to 620°C (1100 to 1150°F). Re-austenitising occurs upon tempering above 620°C (1150°F), the amount of re-austenitisation increasing with increasing temperature. Depending on the amount of this transformation, cooling from such tempering temperatures may adversely affect both ductility and toughness through the transformation to untempered martensite. Even though the alloy is characterised by a decrease in impact strength when tempered in the range of 370 to 595°C (700 to 1100°F), the minimum reached is significantly higher than that of CA-15. This improvement in impact toughness results from the presence of molybdenum and nickel in the composition and from the lower carbon content. The best combination of strength with toughness is obtained when the alloy is tempered above 510°C (950°F).

Fig. 19.18 describes the effect of tempering temperature on the hardness, strength, ductility and toughness properties of CA-6NM and illustrates that strengths even higher than those considered typical can be obtained by tempering at lower temperatures without a disturbing loss of ductility or toughness.

The minor loss of toughness and ductility that does occur is associated with the lesser degree of tempering which takes place at the lower temperature and not with embrittlement, as might be the situation with other 12 per cent chromium steels that contain no molybdenum. The addition of molybdenum to 12 per cent chromium steels makes them unusually stable thermally and normally not susceptible to embrittlement in the annealed or annealed and cold worked conditions, even when exposed for long periods of time at 370 to 480°C (700 to 900°F).

Another significant practical advantage of CA-6NM is its relative freedom from the rapid drop in hardness when tempered above about 510°C (950°F). Fig. 19.19 shows clearly that a given increase in tempering temperature produces a much more gradual decrease in hardness as compared with CA-15. This makes heat treating much easier and cheaper and decreases the frequency of rejects and/or the necessity for re-heat treatment.

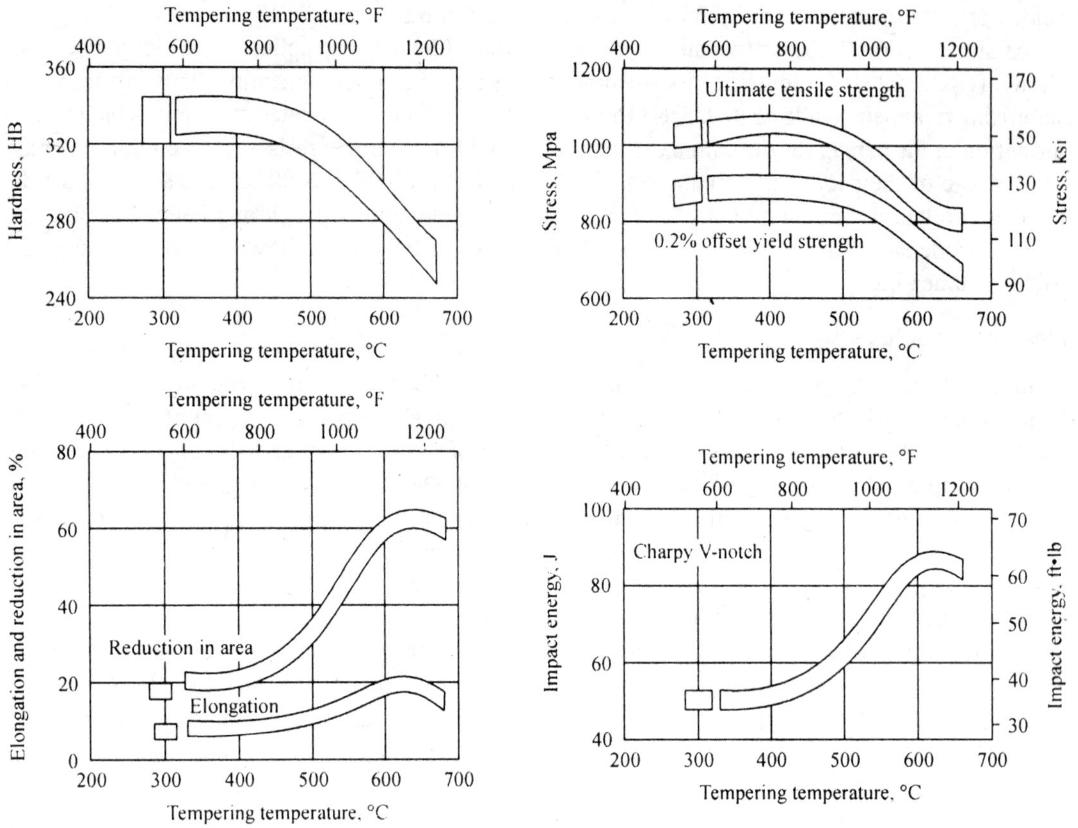

Fig. 19.18. Effect of tempering temperature on the mechanical properties of a CA-6NM standard keel block.

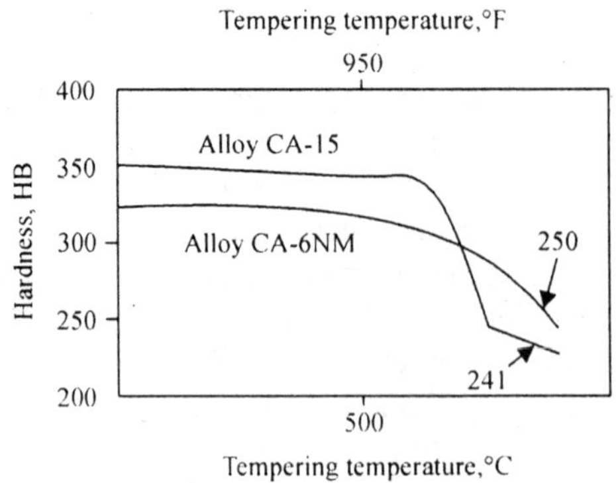

Fig. 19.19. Influence of tempering temperature on the hardness of CA-6NM and CA-15.

The hardening procedures for CA-15 castings are similar to those used for the comparable wrought alloy (type 410). Austenitising consists of heating to 955 to 1010°C (1750 to 1850°F) and soaking for a minimum of 30 minutes; the high side of this temperature range is normally employed. Parts are then cooled in air or quenched in oil. To reduce the probability of cracking in the brittle, untempered martensitic condition, tempering should take place immediately after quenching.

Tempering is performed in two temperature ranges: (i) up to 370°C (700°F) for maximum strength and corrosion resistance and (ii) from 595 to 760°C (1100 to 1400°F) for improved ductility at lower strength levels. Tempering in the range of 370 to 595°C (700 to 1100°F) is normally avoided, because of the resultant low impact strength. Fig. 19.20 shows the nominal mechanical properties obtained in CA-15 castings as a function of tempering temperature. Additional data on mechanical properties are given in Table 19.8.

Table 19.8. Effects of four methods of heat treatment on typical mechanical properties of cast CA-15 (a).

	Ultimate tensile strength		*Yield strength*		*Elongation in 50 mm (2 inch),%*	*Reduction area, %*
Heat treatment(b)	*MPa*	*ksi*	*MPa*	*ksi*		
Treatment 1	1230	178	1010	146	9.0	13.0
Homogenise: 1 hr. at 1040°C (1900°F), AC	1250	181	970	141	12.5	28.0
Solution anneal: ⅓ hr at 955°C (1750°F), OQ	1280	185	990	143	7.0	14.0
Temper: 3 hrs. at 300°C (575°F), AC	1320	191	1020	148	8.0	12.5
Treatment 2	1260	183	1120	162	6.5	9.5
Anneal: 1 hr. at 900°C (1650 °F), FC	1300	188	1130	164	5.5	16.0
Solution anneal: 1¼ hr. at 1010°C (1850°F), OQ	1340	194	1070	155	9.0	23.0
Temper: 3 hrs. at 370°C (700°F), OQ	1380	200	1050	152	12.0	42.0
Treatment 3(c)	790	115	480	70	15.5	60.0
Anneal: 1 hr. at 900°C (1650 °F),FC	810	117	630	91	16.5	37.0
Solution anneal: 1¼ hr. at 1010°C (1850°F), OQ	830	120	680	98	9.5	23.0
Temper: 2 hrs. at 620°C (1150°F), AC	860	125	590	85	12.5	32.0
Treatment 4(d)	680	99	520	76	21.0	65.0
Anneal: 1 hr. at 900°C (1650 °F), FC	710	103	540	79	20.5	56.0
Solution anneal: 1¼ hr. at 995°C (1825°F), OQ	710	103	540	79	18.5	61.5
Temper: 2 hrs. at 705°C (1300°F), AC	720	104	550	80	20.5	60.0

(a) Specimens taken from shell mould cast keel blocks; data indicate results obtained on four specimens treated by each method. (b) Each treatment comprised three processes as listed. AC, air cool; OQ, oil quench; FC, furnace cool; FAC forced-air cool. (c) AMS 5351-B. (d) MIL-S-16993.

These data are based on several heats of shell cast CA-15 alloy. The standard heat treating procedures for CA-15, CA-40 and CA-6NM are given in Table 19.9. In the hardened and tempered condition, CA-40 provides higher tensile strength and lower ductility than CA-15 tempered at the same temperature. Both alloys can be annealed by cooling slowly from the range 845 to 900°C (1550 to 1650°F).

Precipitation-Hardening Alloys

It is desirable to subject precipitation-hardenable castings to a high-temperature homogenisation treatment to reduce alloy segregation and to obtain more uniform response to subsequent heat treatment.

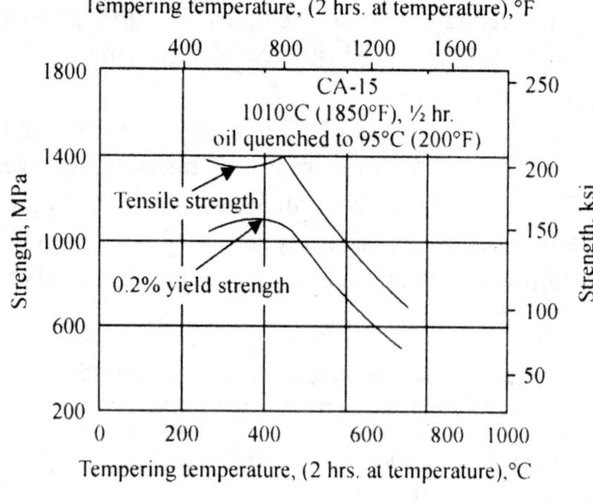

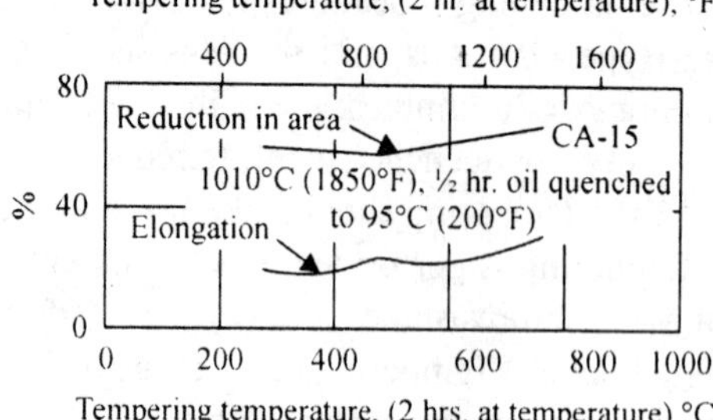

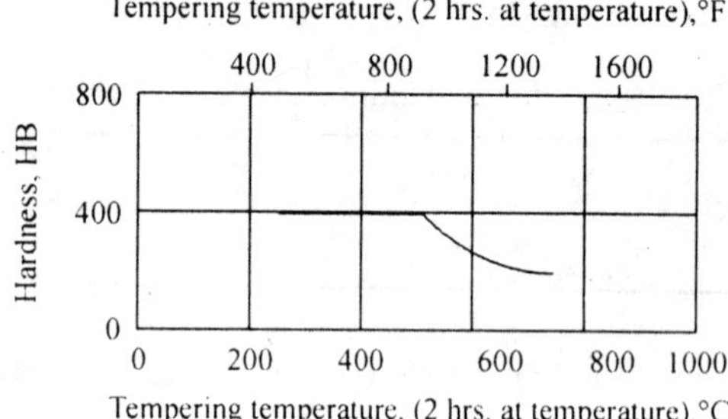

Fig. 19.20. Effect of tempering temperature on typical room temperature mechanical properties of CA-15 castings.

Table 19.9. Heat treatment of martensitic stainless steel castings.

Annealing temperature(a)		*Hardening treatment* *Austenitising temperature(b)*		*Tempering temperature*		*Typical ultimate tensile strengths(c)*	
°C	*°F*	*°C*	*°F*	*°C*	*°F*	*MPa*	*ksi*
CA-15 alloy							
845–900	1550–1650	–	–	–	–	550	80
	–	925–1010(d)	1700–1850(d)	370 max(e)	700 max(e)	1380	200
	–	925–1010(d)	1700–1850(d)	595–760	1100–1400	690–930	100–135
CA-40 alloy							
845–900	1550–1650	–	–	–	–	620	90
	–	980–1010	1800–1850	315 max(e)	600 max(e)	1520	220
	–	980–1010	1800–1850	595	1100	1030	150
	–	980–1010	1800–1850	650	1200	970	140
	–	980–1010	1800–1850	760	1400	760	110
CA-6NM alloy							
760–815	1450–1500	–	–	–	–	550	80
	–	950–980	1750–1800	595–620	1100–1150	830	120

(a) Annealing for maximum softness; slow furnace cool from temperature; (b) Quench in oil or air; (c) Approximate; (d) Hold at temperature for a minimum of 30 minutes; and (e) Tempering at 370 to 595°C (700 to 1100°F) is not recommended because low impact ductility results.

Even investment castings that are cooled slowly from the pouring temperature exhibit more nearly uniform properties when they have been homogenised. Recommended homogenising treatments for precipitation-hardening alloys 17–4 PH and AM-350 are included in Table 19.6.

17–4 PH Castings

When 17–4 PH (ASTM CB-7Cu-1 and CB-7Cu-2) is cast in plastic-bonded shell moulds, the surface is carburised by decomposition of the binder. The added carbon prevents proper heat treating response of the casting surface. Satisfactory response is obtained when surface carbon is removed prior to the homogenisation treatment.

In addition to homogenisation, other heat treating procedures for 17–4 PH castings include solution annealing and precipitation hardening. Details of these procedures are given in Table 19.7. The preferred temperature range for precipitation hardening is 480 to 595°C (900 to 1100°F). The mechanical properties obtained at different ageing temperatures are given in Table 19.10.

Table 19.10. Effect of temperature of 1-hr. ageing treatment on typical properties of investment cast 17–4 PH stainless steels(a).

Ageing temperature		*Ultimate tensile strength*		*Yield strength, 0.2 %*		*Elongation in 50 mm (2 inch),%*	*Hardness, HRC*
°C	*°F*	*MPa*	*ksi*	*MPa*	*ksi*		
Alloy with 0.15 to 0.40% niobium							
As cast		1070	153	770	112	3.5	–
480	900	1380	200	1070	153	15	44
510	950	1360	197	1080	157	13	42
540	1000	1130	164	970	141	14	39
565	1050	1120	163	1040	151	16	35
595	1100	1120	162	990	143	16	34
650	1200	1010	147	860	125	15	30
Alloy without niobium							
As cast		1120	162	990	143	2.7	38
480	900	1360	198	1140	166	12	43
510	950	1250	182	1110	161	13	42
540	1000	1280	186	1100	159	14	38
565	1050	980	142	910	132	16	35
595	1100	1080	157	840	122	16	34
650	1200	1050	153	900	130	12	32

(a) As determined on cast test bars. Before ageing, specimens were homogenised (1½ hr. at 1150°C or 2100°F, air cool) and solution annealed (½ hr. at 1040°C or 1900°F, oil quench); sub-zero transformation not employed. After 1 hr. at ageing temperature, specimens were air cooled.

The tendency of 17–4 PH castings to overage is reduced by the addition of about 0.25 per cent combined niobium plus tantalum to the alloy. The effect of time at ageing temperature on the mechanical properties of niobium-free and niobium-containing 17–4 PH investment castings is shown in Table 19.11.

Table 19.11. Effect of ageing time at 480°C (900°F) on typical properties of investment cast 17–4 PH stainless steels (a).

	Ultimate tensile strength		*Yield strength*		*Elongation in 25 mm (1 inch), %*	*Hardness, HRC*
Ageing time, hr.	*MPa*	*ksi*	*MPa*	*ksi*		
Alloy with 0.15 to 0.40% niobium						
1/2	1380	201	1280	185	7	45
1	1380	200	1070	155	15	44
2	1340	194	1060	153	13	45
4	1300	188	1080	156	9	43
Alloy without niobium						
½	1380	201	1080	156	10	43
1	1360	198	1130	164	12	43
2	1390	202	1080	157	12	44
4	1180	171	980	142	16	38
8	1130	164	970	141	14	37

(a) Treatment prior to ageing: 1½ hr. at 1150°C (2100°F), air cool; 1 hr. at 1040°C (1900°F), oil quench.

AM-350 and AM-355

Although investment castings made of these alloys do not necessarily require a homogenising treatment, this treatment provides a more uniform response to subsequent heat treatment. Shell mould and sand castings made of AM-355 that were extremely brittle without homogenisation re-gained ductility after homogenising at 1095°C (2000°F) for 2 hr. minimum. Heat treating procedures and effects of tempering temperatures up to 650°C (1200 °F) on mechanical properties of AM-355 shell mould castings are given in Table 19.12.

Table 19.12. Effects of tempering temperature on typical properties of shell mould cast AM-355(a).

	Ultimate tensile strength		*Yield strength(b)*		*Elongation in 50 mm (2 inch), %*	*Reduction in, area, %*
Condition	*MPa*	*ksi*	*MPa*	*ksi*		
Annealed	1290	187	480	70	6	3.5
Sub-zero transformed	1400	203	960	140	6	2.5
Tempered 3 hrs. at:						
480°C (900°F)	1440	209	1170	170	20	9
540°C (1000°F)	1320	192	1100	159	34	13
595°C (1100°F)	1190	173	940	136	35	14
650°C (1200°F)	1010	147	590	86	33	15

(a) Treatment prior to tempering: 1½ hr. at 1095°C (2000°F), furnace cool to 980°C (1800°F); soak at 980°C (1800°F) for 1½ hr., water quench; sub-zero cool at –85°C (–120°F) for 6 hrs. (b) 0.2 per cent offset.

When AM-355 castings are welded, maximum mechanical properties are obtained when the castings are fully heat treated after welding (Table 19.13). Heat treatments prior to welding have little effect on properties when a complete heat treatment follows welding.

Table 19.13. Effect of heat treating and welding sequence on typical properties of AM-355 castings.

	Ultimate tensile strength		*Yield strength, 0.2%*		*Elongation in 50 mm (2inch), %*
Sequence	*MPa*	*ksi*	*MPa*	*ksi*	
Heat treated after welding					
A	1450	210	1100	160	15
B	1380	200	1120	162	12
C	1410	205	1070	155	12
D	1410	205	1100	160	8
Not heat treated after welding					
E	1070	155	830	120	11

Note: Heat treating and welding procedures and sequences were as follows: A-2 hr. at 1095°C (2000°F), air cool; 2 hrs. at 1010°C (1850°F), water quench; 3 hrs. at –75°C (–100°F), 3 hrs. at 455°C (850°F), air cool; 3 hrs. at 510°C (950°F), air cool; helium-shielded arc welding; repeat heat treatment; B-2 hrs. at 1095°C (2000°F), air cool; 2 hrs. at 1010°C (1850°F), water quench; 3 hrs. at –75°C (–100°F), 3 hrs. at 455°C (850°F), air cool; 3 hrs. at 510°C (950°F), air cool; 2hrs. at 1010°C (1850°F), water quench; helium-shielded arc welding; repeat heat treatment through 510°C (950°F) tempering. C–Helium-shielded arc welding; 2 hrs. at 1095°C (2000°F), air cool; 2 hrs. at 1010°C (1850°F), water quench; 3 hrs. at –75°C (–100°F), 3 hrs. at 455°C (850°F), air cool; 3 hrs. at 510°C (950°F), air cool; D-2 hrs. at 1095°C (2000°F), air cool; 2 hrs. at 1010°C (1850°F), water quench; helium-shielded arc welding; 2 hrs. at 1095°C (2000°F), air cool; 2 hrs. at 1010°C (1850°F), water quench; 3 hrs. at –75°C (–100°F), 3 hrs. at 455°C (850°F), air cool; 3 hrs. at 510°C (950°F), air cool; E-2 hrs. at 1095°C (2000°F), air cool; 2 hrs. at 1010°C (1850°F), water quench; 3 hrs. at –75°C (–100°F), 3 hrs. at 455°C (850°F), air cool; 3 hrs. at 510°C (950°F), air cool; helium-shielded arc welding.

SECTION V

Heat Treatment of Non Ferrous Metals and Alloys

Chapter 20

Heat Treating of Aluminium Alloys

INTRODUCTION

Heat treating, in its broadest sense, refers to any of the heating and cooling operations that are performed for the purpose of changing the mechanical properties, the metallurgical structure or the residual stress state of a metal product. When the term is applied to aluminium alloys, however, its use frequently is restricted to the specific operations employed to increase strength and hardness of the precipitation-hardenable wrought and cast alloys. These usually are referred to as the 'heat treatable' alloys, to distinguish them from those alloys in which no significant strengthening can be achieved by heating and cooling. The latter, generally referred to as 'non-heat-treatable' alloys, when in wrought form depend primarily on cold work to increase strength. Heating to decrease strength and increase ductility (annealing) is used with alloys of both types; metallurgical reactions may vary with type of alloy and with degree of softening desired. Except for the low-temperature stabilisation treatment sometimes given *5xxx* series alloys (which is a mill treatment and not discussed in this chapter), complete or partial annealing treatments are the only ones used for non-heat-treatable alloys.

PRECIPITATION HARDENING

One essential attribute of a precipitation-hardening alloy system is a temperature-dependent equilibrium solid solubility characterised by increasing solubility with increasing temperature. Although this condition is met by most of the binary aluminium alloy systems, many exhibit very little precipitation hardening and these alloys ordinarily are not considered heat treatable. Alloys of the binary Al-Si and Al-Mn systems, for example, exhibit relatively insignificant changes in mechanical properties as a result of heat treatments that produce considerable precipitation.

The solubility-temperature relationship required for precipitation hardening is illustrated by the Al-Cu system (Fig. 20.1). The equilibrium solid solubility of copper in aluminium increases as temperature increases–from about 0.20 per cent at 250°C (480°F) to a maximum of 5.65 per cent at the eutectic melting temperature of 548°C (1018°F). (It is considerably lower than 0.20 per cent at temperatures below 250°C). For Al-Cu alloys containing from 0.2 to 5.6 per cent Cu, two distinct equilibrium solid states are possible. At temperatures above the lower curve in Fig. 20.1 (solvus), the copper is completely soluble and when the alloy is held at such temperatures for sufficient time to permit needed diffusion, the copper will be taken completely into solid solution. At temperatures below the solvus, the equilibrium state consists of two solid phases: solid solution, α, plus an intermetallic-compound phase, θ (Al_2Cu). When such an alloy is converted to all solid solution by holding above the solvus temperature and then the temperature is decreased to below the solvus, the solid solution becomes supersaturated and the alloy seeks the equilibrium two-phase condition; the second phase tends to form by solid-state precipitation.

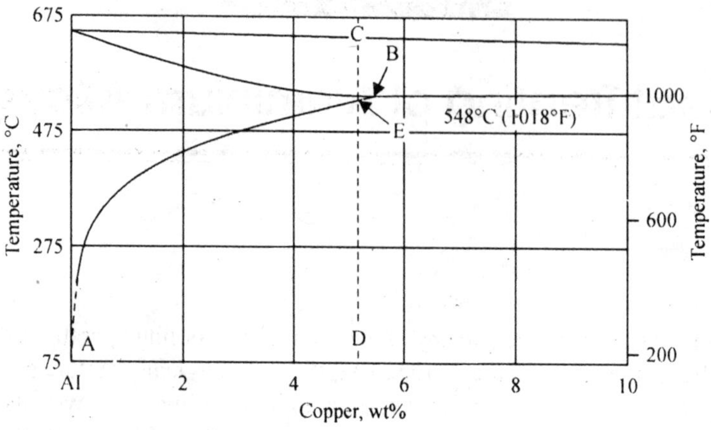

Line AB represents the increase in solubility of copper in solid aluminium with increasing temperature.

Fig. 20.1. Aluminium-rich end of the aluminium-copper equilibrium diagram.

The preceding description is a gross oversimplification of the actual changes that occur under different conditions even in simple binary Al-Cu alloys. A variety of different nonequilibrium precipitate structures is formed at temperatures below the solvus. In alloys of the Al-Cu system, a succession of precipitates is developed from a rapidly cooled supersaturated solid solution (SSS). These precipitates develop sequentially either with increasing temperature or with increasing time at temperatures between room temperature and the solvus. The several stages are identified by the following notation:

$$\text{SSS} \longrightarrow \text{GP zones} \longrightarrow \theta'' \longrightarrow \theta' \longrightarrow \theta\ (Al_2Cu)$$

At temperatures in the natural ageing range (about –20 to +60°C, or 0 to 140°F), the distribution of copper atoms changes with time from random to the disklike planar aggregates termed GP zones, which form on particular crystallographic planes of the aluminium matrix. These aggregates create coherency strain fields that increase resistance to deformation and their formation is responsible for the changes in mechanical properties that occur during natural ageing. At higher temperatures, transition forms of approximate composition Al_2Cu develop and further increase strength. In the highest strength condition, both the θ″ and θ′ transition precipitates may be present. When time and temperature are increased sufficiently to form high proportions of the equilibrium, the alloy softens and is said to be 'overaged'.

Commercial Alloys

Commercial alloys whose strength and hardness can be significantly increased by heat treatment include 2xxx, 6xxx and 7xxx series wrought alloys (except 7072) and 2xx.0, 3xx.0 and 7xx.0 series casting alloy. Some of these contain only copper, or copper and silicon, as the primary strengthening alloy addition(s). Most of the heat treatable alloys, however, contain combinations of magnesium with one or more of the elements copper, silicon and zinc. Characteristically, even small amounts of magnesium in concert with these elements accelerate and accentuate the strength changes attributable to precipitation hardening.

Heat Treatable Wrought Alloys

In the heat treatable wrought alloys, with some notable exceptions (2024, 2219 and 7128), such solute elements are present in amounts that are within the limits of mutual solid solubility at temperatures below the eutectic temperature (lowest melting temperature). In contrast, some of the casting alloys of the 2xx.0 series and all of the 3xx.0 series alloys contain amounts of soluble elements that far exceed solid-solubility limits. In these alloys, the phase formed by combination of the excess soluble elements with the aluminium will never be dissolved, although the shapes of the undissolved particles may be changed by partial solution.

Heat Treatable Aluminium Alloy

Most of the heat treatable aluminium alloy systems exhibit multistage precipitation and undergo accompanying strength changes analogous to those of the Al-Cu system. The structural changes from the solid solution to the high-strength tempers involve precipitate particles that are too small to be resolved by light microscopy. In some instances, the presence of precipitation can be detected by etching, but definitive determination of the details of precipitate structure responsible for the changes in mechanical and other properties generally requires electron microscopy and special x-ray techniques.

In some alloys, sufficient precipitation occurs in a few days at room temperature to yield stable products with properties that are adequate for many applications. These alloys sometimes are precipitation heat treated to provide increased strength and hardness in wrought or cast products. Other alloys with slow precipitation reactions at room temperature are always precipitation heat treated before being used.

Precipitation heat treatments generally are low-temperature, long-term processes. Temperatures range from 115 to 190°C (240 to 375°F); times vary from 5 to 48 hours.

Time-Temperature Cycles for Precipitation Hardening

Choice of time-temperature cycles for precipitation heat treatment should receive careful consideration. Larger particles of precipitate result from longer times and higher temperatures; however, the larger particles must, of necessity, be fewer in number with greater distances between them. The objective is to select the cycle that produces optimum precipitate size and distribution pattern. Unfortunately, the cycle required to maximise one property, such as tensile strength, is usually different from that required to maximise others, such as yield strength and corrosion resistance. Consequently, the cycles used represent compromises that provide the best combinations of properties.

To recap, heat treatment to increase strength of aluminium alloys is a three-step process:

1. Solution heat treatment: dissolution of soluble phases.
2. Quenching: development of supersaturation.
3. Ageing: precipitation of solute atoms either at room temperature (natural ageing) or elevated temperature (artificial ageing or precipitation heat treatment).

Typical solution and precipitation heat treatments for mill products are given in Tables 20.1 and 20.2 and treatments for castings are given in Table 20.3.

SOLUTION HEAT TREATING

To take advantage of the precipitation-hardening reaction, it is necessary first to produce a solid solution. The process by which this is accomplished is called solution heat treating and its objective is to take into solid solution the maximum practical amounts of the soluble hardening elements in the alloy. The process consists of soaking the alloy at a temperature sufficiently high and for a time long enough to achieve a nearly homogeneous solid solution.

Table 20.1. Typical solution and precipitation heat treatments for aluminium alloy mill products.
These times and temperatures are typical for various forms, sizes and methods of manufacture and may not exactly describe optimum treatments for specific items.

		Solution heat treatment(a)			*Precipitation heat treatment*			
		Metal temperature(b)			*Metal temperature(b)*			
Alloy	*Product form*	°C	°F	*Temper designation*	°C	°F	*Time(c), hr.*	*Temper designation*
2011	Rolled or cold finished rod and bar	525	975	T3(d)	160	320	14	T8(d)
				T4	–	–	–	–
				T451(e)	–	–	–	–
2014(f)	Flat sheet	500	935	T3(d)	160	320	18	T6
				T42	160	320	18	T62
	Coiled sheet	500	935	T4	160	320	18	T6
				T42	160	320	18	T62
	Plate	500	935	T42	160	320	18	T62
				T451(e)	160	320	18	T651(e)
	Rolled or cold finished wire, rod and bar	500	935	T4	160(g)	320(g)	18	T6
				T42	160(g)	320(g)	18	T62
				T451(e)	160(g)	320(g)	18	T651(e)
	Extruded rod, bar, shapes and tube	500	935	T4	160(g)	320(g)	18	T6
				T42	160(g)	320(g)	18	T62
				T4510(e)	160(g)	320(g)	18	T6510(e)
				T4511(e)	160(g)	320(g)	18	T6511(e)
	Drawn tube	500	935	T4	160(g)	320(g)	18	T6
				T42	160(g)	320(g)	18	T62
	Die forgings	500(h)	935(h)	T4	170	340	10	T6
	Hand forgings and rolled rings	500(h)	935(h)	T4	170	340	10	T6
				T452(j)	170	340	10	T652(j)
2017	Rolled or cold finished wire, rod and bar	500	935	T4	–	–	–	–
				T42	–	–	–	–
				T451(e)	–	–	–	–
2018	Die forgings	510(k)	950(k)	T4	170	340	10	T61
2024(f)	Flat sheet	495	920	T3(d)	190	375	12	T81(d)
				T361(d)	190	375	8	T861(d)

(Contd...)

		Solution heat treatment(a)			*Precipitation heat treatment*			
		Metal temperature(b)			*Metal temperature(b)*			
Alloy	*Product form*	*°C*	*°F*	*Temper designation*	*°C*	*°F*	*Time(c), hr.*	*Temper designation*
				T42	190	375	9	T62
					190	375	16	T72
2024(f)	Coiled sheet	495	920	T4	–	–	–	–
				T42	190	375	9	T62
					190	375	16	T72
	Plate	495	920	T351(e)	190	375	12	T851(e)
				T361(d)	190	375	8	T861(d)
				T42	190	375	9	T62
	Rolled or cold finished wire, rod and bar	495	920	T4	190	375	12	T6
				T351(e)	190	375	12	T851(e)
				T36(d)	190	375	8	T86(d)
				T42	190	375	16	T62
	Extruded rod, bar, shapes and tube	495	920	T3	190	375	12	T81
				T3510(e)	190	375	12	T8510(e)
				T3511(e)	190	375	12	T8511(e)
				T42	190	375	16	T62
	Drawn tube	495	920	T3(d)	–	–	–	–
				T42	–	–	–	–
2025	Die forgings	515	960	T4	170	340	10	T6
2036	Sheet	500	930	T4	–	–	–	–
2117	Rolled or cold finished wire and rod	500	935	T4	–	–	–	–
				T42	–	–	–	–
2218	Die forgings	510(k)	950(k)	T4	170	340	10	T61
		510(m)	950(m)	T41	240	460	6	T72
2219(f)	Flat sheet	535	995	T31(d)	175	350	18	T81(d)
				T37(d)	165	325	24	T87(d)
				T42	190	375	36	T62
	Plate	535	995	T31(d)	175	350	18	T81(d)
				T37(d)	175	350	18	T87(d)
				T351(e)	175	350	18	T851(e)
				T42	190	375	36	T62
	Rolled or cold finished wire, rod and bar	535	995	T351(e)	190	375	18	T851(e)

(Contd...)

Alloy	Product form	Solution heat treatment(a) Metal temperature(b) °C	°F	Temper designation	Precipitation heat treatment Metal temperature(b) °C	°F	Time(c), hr.	Temper designation
2219(f)	Extruded rod, bar, shapes and tube	535	995	T31(d)	190	375	18	T81(d)
				T3510(e)	190	375	18	T8510(e)
				T3511(e)	190	375	18	T8511(e)
				T42	190	375	36	T62
	Die forgings and rolled rings	535	995	T4	190	375	26	T6
	Hand forgings	535	995	T4	190	375	26	T6
				T352(j)	175	350	18	T852(j)
2618	Forgings and rolled rings	530	985	T4	200	390	20	T61
4032	Die forgings	510(h)	950(h)	T4	170	340	10	T6
6005	Extruded rod, bar, shapes and tube	530(n)	985(n)	T1	175	350	8	T5
6009(p)	Coiled sheet	555	1030	T4	175	350	8	T6
6010(p)	Coiled sheet	565	1050	T4	175	350	8	T6
6053	Die forgings	520	970	T4	170	340	10	T6
6061(f)	Sheet	530	985	T4	160	320	18	T6
				T42	160	320	18	T62
	Plate	530	985	T4(q)	160	320	18	T6(q)
				T42	160	320	18	T62
				T451(e)	160	320	18	T651(e)
	Rolled or cold finished wire, rod and bar	530	985	T4	160(r)	320(r)	18	T6
					160(r)	320(r)	18	T89(d)
					160(r)	320(r)	18	T93(s)
					160(r)	320(r)	18	T913(s)
					160(r)	320(r)	18	T94(s)
				T42	160(r)	320(r)	18	T62
				T451(e)	160(r)	320(r)	18	T651(e)
	Extruded rod, bar, shapes and tube	530(n)	985(n)	T4	175	350	8	T6
				T4510(e)	175	350	8	T6510(e)
				T4511(e)	175	350	8	T6511(e)
		530	985	T42	175	350	8	T62
6061(f)	Drawn tube	530	985	T4	160(r)	320(r)	18	T6
				T42	160(r)	320(r)	18	T62

(Contd...)

Alloy	Product form	*Solution heat treatment(a)* Metal temperature(b) °C	°F	Temper designation	*Precipitation heat treatment* Metal temperature(b) °C	°F	Time(c), hr.	Temper designation
	Die and hand forgings	530	985	T4	175	350	8	T6
	Rolled rings	530	985	T4	175	350	8	T6
				T452(j)	175	350	8	T652(j)
6063	Extruded rod, bar, shapes and tube	(n)	(n)	T1	205(t)	400(t)	1	T5
		520(n)	970(n)	T4	175(u)	350(u)	8	T6
		520	970	T42	175(u)	350(u)	8	T62
	Drawn tube	520	970	T4	175	350	8	T6
					175	350	8	T83(d)(n)
					175	350	8	T831(d)(n)
					175	350	8	T832(d)(n)
				T42	175	350	8	T62
6066	Extruded rod, bar, shapes and tube	530	990	T4	175	350	8	T6
				T42	175	350	8	T62
				T4510(e)	175	350	8	T6510(e)
				T4511(e)	175	350	8	T6511(e)
	Drawn tube	530	990	T4	175	350	8	T6
				T42	175	350	8	T62
	Die forgings	530	990	T4	175	350	8	T6
6070	Extruded rod, bar, shapes and tube	545(n)	1015(n)	T4	160	320	18	T6
				T42	160	320	18	T62
6151	Die forgings	515	960	T4	170	340	10	T6
	Rolled rings	515	960	T4	170	340	10	T6
				T452(j)	170	340	10	T652(j)
6262	Rolled or cold finished wire, rod and bar	540	1000	T4	170	340	8	T6
					170	340	12	T9(s)
				T451	170	340	8	T651(e)
				T42	170	340	8	T62
6262	Extruded rod, bar, shapes and tube	540(n)	1000(n)	T4	175	350	12	T6
				T4510(e)	175	350	12	T6510(e)
				T4511(e)	175	350	12	T6511(e)
		540	1000	T42	175	350	12	T62

(Contd...)

Alloy	Product form	Solution heat treatment(a) Metal temperature(b) °C	°F	Temper designation	Precipitation heat treatment Metal temperature(b) °C	°F	Time(c), hr.	Temper designation
	Drawn tube	540	1000	T4	170	340	8	T6
					170	340	8	T9(s)
				T42	170	340	8	T62
6463	Extruded rod, bar, shapes and tube	(n)	(n)	T1	205(t)	400(t)	1	T5
		520(n)	970(n)	T4	175(u)	350(u)	8	T6
		520	970	T42	175(u)	350(u)	8	T62
6951	Sheet	530	985	T4	160	320	18	T6
				T42	160	320	18	T62
7001	Extruded rod, bar, shapes and tube	465	870	W	120	250	24	T6
					120	250	24	T62
				W510(e)	120	250	24	T6510(e)
				W511(e)	120	250	24	T6511(e)
7005	Extruded rod, bar and shapes	–	–	–	–	–	–	T53(v)
7050	Plate	475	890	W51(e)	(w)	(w)	(w)	T7651(x)
					(y)	(y)	(y)	T73651(x)
	Extrusions	475	890	W510(e)	(w)	(w)	(w)	T76510(x)
				W511(e)	(w)	(w)	(w)	T76511(x)
	Die and hand forgings	475	890	W	(y)	(y)	(y)	T736(x)
				W52(e)	(y)	(y)	(y)	T73652(x)
7075(f)	Sheet	480	900	W	120(z)	250(z)	24	T6
					120(z)	250(z)	24	T62
					(w)	(w)	(w)	T76(x)
					(y)(aa)	(y)(aa)	(y)(aa)	T73(x)
	Plate	480	900	W	120(z)	250(z)	24	T62
				W51(e)	(y)(aa)	(y)(aa)	(y)(aa)	T7351(e)(x)
					120(z)	250(z)	24	T651(e)
					(w)	(w)	(w)	T7651(x)
7075(f)	Rolled or cold finished wire, rod and bar	490	915	W	120	250	24	T6
					120	250	24	T62
					(y)(aa)	(y)(aa)	(y)(aa)	T73(x)
				W51(e)	120	250	24	T651(e)
					(y)(aa)	(y)(aa)	(y)(aa)	T7351(e)(x)
	Extruded rod, bar, shapes and tube	465	870	W	120(bb)	250(bb)	24	T6

(Contd...)

Alloy	Product form	*Solution heat treatment(a)* *Metal temperature(b)* °C	°F	*Temper designation*	*Precipitation heat treatment* *Metal temperature(b)* °C	°F	*Time(c), hr.*	*Temper designation*
					120(bb)	250(bb)	24	T62
					(y)(aa)	(y)(aa)	(y)(aa)	T73(x)
					(w)	(w)	(w)	T76(x)
				W510(e)	120(bb)	250(bb)	24	T6510(e)
					(y)(aa)	(y)(aa)	(y)(aa)	T73510(e)(x)
					(w)	(w)	(w)	T76510(x)
				W511(e)	120(bb)	250(bb)	24	T6511(e)
					(y)(aa)	(y)(aa)	(y)(aa)	T73511(e)(x)
					(w)	(w)	(w)	T76511(x)
	Drawn tube	465	870	W	120	250	24	T6
					120	250	24	T62
					(y)(aa)	(y)(aa)	(y)(aa)	T73(x)
	Die forgings	470(h)	880(h)	W	120	250	24	T6
					(y)	(y)	(y)	T73(x)
				W52(j)	(y)	(y)	(y)	T7352(j)(x)
	Hand forgings	470(h)	880(h)	W	120	250	24	T6
					(y)	(y)	(y)	T73(x)
				W52(j)	120	250	24	T652(j)
					(y)	(y)	(y)	T7352(j)(x)
	Rolled rings	470	880	W	120	250	24	T6
7175	Die forgings	(cc)	(cc)	W	(cc)	(cc)	(cc)	T66(cc)
		(cc)	(cc)	W	(cc)	(cc)	(cc)	T736(x)(cc)
		(cc)	(cc)	W52(j)	(cc)	(cc)	(cc)	
T73652(j)(x)(cc)								
	Hand forgings	(cc)	(cc)	W	(cc)	(cc)	(cc)	T736(x)(cc)
		(cc)	(cc)	W52(j)	(cc)	(cc)	(cc)	T7365(j)(x)(cc)
7475	Sheet	515(dd)	960(dd)	W	120	250	3	
					plus 155	315	3	T61(dd)
					(w)		(w)	(w)
T761(x)(dd)								
	Plate	510(dd)	950(dd)	W51(e)	120	250	24	T651(dd)
					(w)	(w)	(w)	T7651(x)(dd)
					(y)	(y)	(y)	T7351(x)(dd)
Alclad 7475	Sheet	495	920	W	120	250	3	
					plus 155	315	3	T61(dd)
					(w)	(w)	(w)	T761(x)(dd)

(a) Material should be quenched from the solution-treating temperature as rapidly as possible and with minimum delay after removal from the furnace. When material is quenched by total immersion in water, unless otherwise indicated, the water should be at room temperature and should be suitably cooled so that it remains below 38°C (100°F) during the quenching cycle. Use of high-velocity, high-volume jets of cold water also is effective for some materials; (b) The nominal temperatures listed should be attained as rapidly as possible and maintained within ±6°C (±10°F) of nominal during the time at temperature; (c) Approximate time at temperature. The specific time will depend on the time required for the load to reach temperature. The times shown are based on rapid heating, with soak time measured from the time the load reaches a temperature within 6°C (10°F) of the applicable temperature; (d) Cold working subsequent to solution heat treatment and prior to any precipitation heat treatment is necessary to attain the specified properties for this temper; (e) Stress relieved by stretching to produce a specified amount of permanent set subsequent to solution heat treatment and prior to any precipitation heat treatment; (f) These heat treatments also apply to alclad sheet and plate in these alloys; (g) An alternative treatment of 8 hr. at 177°C (350°F) also may be used; (h) Solution heat treatment is followed by quenching in water 60 to 82°C (140 to 180°F); (j) Stress relieved by 1 to 5 per cent cold reduction subsequent to solution heat treatment and prior to precipitation heat treatment; (k) Solution heat treatment is followed by quenching in water at 100°C (212°F); (m) Solution heat treatment is followed by quenching in room-temperature air blast; (n) By suitable control of extrusion temperature, product may be quenched directly from extrusion press to provide specified properties for this temper. Some products may be adequately quenched in room-temperature air blast; (p) See U.S. Patent 4 082 578; (q) Applicable to tread plate only; (r) An alternative treatment of 8 hr. at 171°C(340°F) also may be used; (s) Cold working subsequent to precipitation heat treatment is necessary to attain the specified properties for this temper; (t) An alternative treatment of 3 hr. at 182°C (360°F) also may be used; (u) An alternative treatment of 6 hr. at 182°C (360°F) also may be used; (v) No solution heat treatment; 72 hr. at room temperature following press quench, followed by two-stage precipitation heat treatment comprised of 8 hr. at 107°C (225°F) plus 16 hr. at 149°C (300°F); (w) Ageing practice varies with product, size, nature of equipment, loading procedures and furnace-control capabilities. The optimum practice for a specific item can be ascertained only by actual trial treatment of the item under specific conditions. Typical procedures involve a two-stage treatment comprised of 3 to 30 hr. at 121°C(250°F) followed by 15 to 18 hr. at 163°C (325°F) for extrusions. An alternative two-stage treatment of 8 hr. at 99°C(210°F) followed by 24 to 28 hr. at 163°C (325°F) also may be used; (x) Ageing of aluminium alloys 7050, 7075, 7175 and 7475 from any temper to the T73 or T76 temper series requires closer-than-normal controls on ageing variables such as time, temperature, heatup rate, etc., for any given item. In addition, when material in a T6-type temper is reaged to a T73- or T76-type temper, the specific condition of the T6 material (such as property levels and other effects of processing variables) is extremely important and will affect the capability of the reaged material to conform to the requirements specified for the applicable T73- or T76-type temper; (y) Two-stage treatment comprised of 6 to 8 hr. at 107°C (225°F) followed by: 24 to 30 hr. at 163°C (325°F) for sheet and plate; 8 to 10 hr. at 177°C (350°F) for rolled or cold finished rod and bar; 6 to 8 hr. at 177°C (350°F) for extrusions and tube; 8 to 10 hr. at 177°C (350°F) for forgings in the T73 temper and 6 to 8 hr. at 177°C (350°F) for forgings in the T7352 temper; and (z) An alternative two-stage treatment comprised of 4 hr. at 96°C (205°F) followed by 8 hr. at 157°C (315°F) also may be used. (aa) For sheet, plate, tube and extrusions, an alternative two-stage treatment comprised of 6 to 8 hr. at 107°C (225°F) followed by 14 to 18 hr. at 168°C(335°F) may be used, provided that a heatup rate of approximately 14°C/hr. (25°F/hr.) is employed. For rolled or cold finished rod and bar, the alternative treatment is 10 hr. at 177°C (350°F). (bb) An alternative three-stage treatment comprised of 5 hr. at 99°C (210°F), 4 hr. at 121°C (250°F) and then 4 hr. at 149°C (300°F) also may be used. (cc) 7175-T736 and -T73652 heat treatments are directed to specific results, may vary from supplier to supplier and are either proprietary or patented. (dd) Must be preceded by soak at 466 to 477°C (870 to 890°F).

Determination of Solution Heat Treating Temperature

Nominal commercial solution heat treating temperature is determined by the composition limits of the alloy and an allowance for unintentional temperature variations. Although ranges normally listed allow variations of ±6°C (±10°F) from the nominal, some highly alloyed, controlled-toughness, high-strength alloys require that temperature be controlled within more restrictive limits. Broader ranges may be allowable for alloys with greater intervals of temperature between their solvus and eutectic melting temperatures.

Overheating

Care must be exercised to avoid exceeding the initial eutectic melting temperature. If appreciable eutectic melting occurs as a result of overheating, properties such as tensile strength, ductility and fracture toughness may be degraded. Specifications generally categorise as unacceptable those materials that exhibit microstructural evidence of overheating. The grain-boundary melting that occurs when the eutectic melting temperature of the alloy is exceeded. The material becomes brittle and nonsalvageable. This dangerous condition usually is not detectable by either visual examination or nondestructive testing.

Although maximum temperature must be restricted to avoid melting, the lower limit should, when possible, be above the temperature at which complete solution occurs (solvus). In the 5.25 per cent Cu alloy represented by line CD in Fig. 20.1, these temperatures would be 548 and 536°C (1018 and 996°F), respectively. However, under production conditions, a range of 537 to 546°C (999 to 1014°F) would probably be used, to provide a 2.2°C (4°F) margin to safeguard against eutectic melting and a 1.7°C (3°F) cushion on the low side for increased solution and diffusion rates.

For certain alloys such as 2024 and 2219, which can be represented on Fig. 20.1 as alloys containing more than 5.65 per cent Cu, complete solution can never occur. For these alloys, the minimum solution heat treating temperature is established so that it is as close as practical to the eutectic temperature while providing a margin of safety commensurate with the capability of the equipment. The proximity of typical solution-treating temperature ranges to eutectic melting temperatures for three common alloys is shown in the Table 20.2.

Table 20.2. Proximity of typical solution temperature ranges to eutectic melting temperature for three common alloys.

	Solution treating temperature		*Eutectic melting temperature*	
Alloy	*°C*	*°F*	*°C*	*°F*
2014	496–507	925–945	510	950
2017	496–507	925–945	513	955
2024	488–499	910–930	502	935

Nonequilibrium Melting

When high heating rates are employed, the phenomenon of nonequilibrium melting must be considered. This phenomenon can also be explained with the help of the Al-Cu phase diagram (Fig. 20.1). The room-temperature microstructure of an F-temper product containing 4 per cent Cu consists of a solid solution of copper in aluminium and particles of Al_2Cu. When this product is heated slowly, the Al_2Cu begins to dissolve and if heating is slow enough, all of the Al_2Cu is dissolved when temperatures above the solvus (500°C or 932°F) are reached.

When the heating rate is high, however, much of the Al_2Cu remains undissolved. If a material with this microstructure is heated at or above the eutectic temperature of 548°C (1018°F), melting will begin at the interface between the Al_2Cu and the matrix. With sufficient time above the eutectic temperature , this metastable liquid will dissolve to form a solid solution and will leave no trace provided that hydrogen gas has not condensed at the interface to form a void. If the product is quenched before the liquid has time to equilibrate, however, it will solidify and form fine eutectic rosettes. This nonequilibrium melting should not be confused with true equilibrium melting, which would occur in an alloy containing more than 5.65 per cent Cu.

In such an alloy, eutectic melting is equilibrium melting. No matter how long such an alloy is held above the eutectic temperature, the liquid will never solidify. In commercial alloys, which usually are ternaries or quaternaries of the major alloying elements, the situation is more complex. Different phases have different solvus temperatures and nonequilibrium melting may occur at different temperatures depending on composition, size of precipitates and rate of heating. When new solution heat treating equipment (which provides higher heating rates) is employed, careful examination of alloy microstructures should be included as part of the certification process.

Underheating

When the temperatures attained by the parts or pieces being heat treated are appreciably below the normal range, solution is incomplete and strength somewhat lower than normal is expected. The shallow slope of the solvus at its intersection with the composition line (typified by point E on line AB in Fig. 20.1) indicates that a slight decrease in temperature below point E will result in a large reduction in the concentration of the solid solution and a correspondingly significant decrease in final strength. The effect of solution-treating temperature on the strength of two aluminium alloys is illustrated by the following data:

Solution treating temperature		*Tensile strength*		*Yield strength*	
°C	°F	*MPa*	*ksi*	*MPa*	*ksi*
6061-T6 sheet 1.6 mm (0.064 inch) thick					
493	920	301	43.7	272	39.4
504	940	316	45.8	288	41.7
516	960	333	48.3	305	44.3
527	980	348	50.5	315	45.7
2024-T4 sheet 0.8 mm (0.032 inch) thick					
488	910	419	60.8	255	37.0
491	915	422	61.2	259	37.5
493	920	433	62.8	269	39.0
496	925	441	63.9	271	39.3

In the tabulation above, note especially the effects of small increments of temperature, within the normal range, on the properties of 0.8-mm (0.032-inch) 2024-T4 sheet.

Solution Treating Time

The time at the nominal solution heat treating temperature ('soak time') required to effect a satisfactory degree of solution of the undissolved or precipitated soluble phase constituents and to achieve good homogeneity of the solid solution is a function of microstructure before heat treatment. This time requirement can vary from less than a minute for thin sheet to as much as 20 hour for large sand or plaster-mould castings.

Guideline information for soak times required for wrought products of various section thicknesses is given in Table 20.3. Similar guidelines for castings are presented in Table 20.4. The time required to heat a load to the treatment temperature in furnace heat treatment also increases with section thickness and furnace loading and thus total cycle time increases with these factors.

Table 20.3. Soak times and maximum quench delays for solution treatment of wrought aluminium alloys. (See Table 20.1 for solution-treating temperatures).

	Soak time, minutes				*Maximum*
	Air furnace(b)		*Salt bath(c)*		*quench*
Thickness(a), mm (inch)	*min.*	*max(d)*	*min.*	*max(d)*	*delay, second*
Thru 0.41 (0.016)	20	25	10	15	5
0.51 (0.020)	20	30	10	20	7
0.64 (0.025)	25	35	15	25	7
0.81 (0.032)	25	35	15	25	7
1.02 (0.040)	30	40	20	30	10
1.27 (0.050)	30	40	20	30	10
1.35 (0.053)	30	40	20	30	10
1.80 (0.071)	35	45	25	35	10
2.03 (0.080)	35	45	25	35	10
2.29 (0.090)	35	45	25	35	10
2.54 (0.100)	40	55	30	45	15
3.18 (0.125)	40	55	30	45	15
4.06 (0.160)	50	60	35	45	15
4.57 (0.180)	50	60	35	45	15
6.35 (0.250)	55	65	35	45	15
Over 6.35 (0.250) thru 12.7 (0.500)	65	75	45	55	15
For each additional 12.7 (½) or fraction	+30	+30	+20	+20	(e)
Rivets (all)	60	–	30	–	5

(a) Minimum dimension of thickest section. (b) Soak time begins when all pyrometer instruments recover to original operating temperature. (c) Soak time begins at time of immersion except when a heavy charge causes bath temperature to drop below specified minimum, in which case soak time begins when bath regains minimum temperature. (d) Applicable to alclad materials only. (e) Increases in thickness above 12.7 mm (½ inch) do not affect maximum quench delay, which remains constant at 15 second.

Table 20.4. Typical heat treatments for aluminium alloy sand and permanent mould castings.

		Type	*Solution heat treatment(b)*			*Ageing treatment*		
		of	*Temperature(c)*		*Time,*	*Temperature(c)*		*Time,*
Alloy	*Temper*	*casting(a)*	*°C*	*°F*	*hr.*	*°C*	*°F*	*hr.*
201.0	T6	S	510–515;	950–960;	2			
			525–530	980–990	14–20	155	310	20
	T7	S	510–515;	950–960;	2			
			525–530	980–990	14–20	190	370	5
204.0	T4	S or P	520	970	10	–	–	–
208.0	T55	S	–	–	–	155	310	16
222.0	O(d)	S	–	–	–	315	600	3
	T61	S	510	950	12	155	310	11

(Contd...)

Alloy	Temper	Type of casting(a)	Solution heat treatment(b) Temperature(c) °C	°F	Time, hr.	Ageing treatment Temperature(c) °C	°F	Time, hr.
	T551	P	–	–	–	170	340	16–22
	T65	–	510	950	4–12	170	340	7–9
242.0	O(e)	S	–	–	–	345	650	3
	T571	S	–	–	–	205	400	8
		P	–	–	–	165–170	330–340	22–26
	T77	S	515	960	5(f)	330–355	625–675	2 (min)
	T61	S or P	515	960	4–12(f)	205–230	400–450	3–5
295.0	T4	S	515	960	12	–	–	–
	T6	S	515	960	12	155	310	3–6
	T62	S	515	960	12	155	310	12–24
	T7	S	515	960	12	260	500	4–6
296.0	T4	P	510	950	8	–	–	–
	T6	P	510	950	8	155	310	1–8
	T7	P	510	950	8	260	500	4–6
319.0	T5	S	–	–	–	205	400	8
	T6	S	505	940	12	155	310	2–5
		P	505	940	4–12	155	310	2–5
328.0	T6	S	515	960	12	155	310	2–5
332.0	T5	P	–	–	–	205	400	7–9
333.0	T5	P	–	–	–	205	400	7–9
	T6	P	505	940	6–12	155	310	2–5
	T7	P	505	940	6–12	260	500	4–6
336.0	T551	P	–	–	–	205	400	7–9
	T65	P	515	960	8	205	400	7–9
354.0	–	(g)	525–535	980–995	10–12	(h)	(h)	(h)
355.0	T51	S or P	–	–	–	225	440	7–9
	T6	S	525	980	12	155	310	3–5
		P	525	980	4–12	155	310	2–5
	T62	P	525	980	4–12	170	340	14–18
	T7	S	525	980	12	225	440	3–5
		P	525	980	4–12	225	440	3–9
	T71	S	525	980	12	245	475	4–6
		P	525	980	4–12	245	475	3–6
C355.0	T6	S	525	980	12	155	310	3–5
	T61	P	525	980	6–12	Room temperature		8 (min.)
						155	310	10–12
356.0	T51	S or P	–	–	–	225	440	7–9

(Contd...)

Alloy	Temper	Type of casting(a)	Solution heat treatment(b) Temperature(c) °C	°F	Time, hr.	Ageing treatment Temperature(c) °C	°F	Time, hr.
	T6	S	540	1000	12	155	310	3–5
		P	540	1000	4–12	155	310	2–5
	T7	S	540	1000	12	205	400	3–5
		P	540	1000	4–12	225	440	7–9
	T71	S	540	1000	10–12	245	475	3
		P	540	1000	4–12	245	475	3–6
A356.0	T6	S	540	1000	12	155	310	3–5
	T61	P	540	1000	6–12	Room temperature		8 (min.)
						155	310	6–12
357.0	T6	P	540	1000	8	175	350	6
	T61	S	540	1000	10–12	155	310	10–12
A357.0	–	(g)	540	1000	8–12	(h)	(h)	(h)
359.0	–	(g)	540	1000	10–14	(h)	(h)	(h)
A444.0	T4	P	540	1000	8–12	–	–	–
520.0	T4	S	430	810	18(j)	–	–	–
535.0	T5(d)	S	400	750	5	–	–	
705.0	T5	S	–	–	–	Room temperature or		21 days
						100	210	8
		P	–	–	–	Room temperature or		21 days
						100	210	10
707.0	T5	S	–	–	–	155	310	3–5
		P	–	–	–	Room temperature or		21 days
						100	210	8
		T7	S	530	990	8–16	175	350
4–10								
		P	530	990	4–8	175	350	4–10
710.0	T5	S	–	–	–	Room temperature		21 days
711.0	T1	P	–	–	–	Room temperature		21 days
712.0	T5	S	–	–	–	Room temperature or		21 days
						155	315	6–8
713.0	T5	S or P	–	–	–	Room temperature or		21 days
						120	250	16
771.0	T53(d)	S	415(k)	775(k)	5(k)	180(k)	360(k)	4(k)
	T5	S	–	–	–	180(k)	355(k)	3–5(k)
	T51	S	–	–	–	205	405	6
	T52	S	–	–	–	(d)	(d)	(d)
	T6	S	590(k)	1090(k)	6(k)	130	265	3
	T71	S	590(e)	1090(e)	6(e)	140	285	15

(Contd...)

Alloy	Temper	Type of casting(a)	Solution heat treatment(b) Temperature(c) °C	°F	Time, hr.	Ageing treatment Temperature(c) °C	°F	Time, hr.
850.0	T5	S or P	–	–	–	220	430	7–9
851.0	T5	S or P	–	–	–	220	430	7–9
	T6	P	480	900	6	220	430	4
852.0	T5	S or P	–	–	–	220	430	7–9

(a) S, sand; P, permanent mould; (b) Unless otherwise indicated, solution treating is followed by quenching in water at 65 to 100°C (150 to 212°F); (c) Except where ranges are given, listed temperatures are ±6°C or ±10°F; (d) Stress relieve for dimensional stability as follows: hold 5 hr. at 413 ± 14°C (775 ± 25°F); furnace cool to 345°C (650°F) over a period of 2 hr. or more; furnace cool to 230°C (450°F) over a period of not more than ½ hr.; furnace cool to 120°C (250°F) over a period of approximately 2 hr.' cool to room temperature in still air outside the furnace; (e) No quench required; cool in still air outside furnace; (f) Air-blast quench from solution-treating temperature; (g) Casting process varies (sand, permanent mould or composite) depending on desired mechanical properties; (h) Solution heat treat as indicated, then artificially age by heating uniformly at the temperature and for the time necessary to develop the desired mechanical properties; (j) Quench in water at 65 to 100°C (150 to 212°F) for 10 to 20 second only; (k) Cool to room temperature in still air outside furnace.

Soak time for alclad sheet and for parts made from alclad sheet must be held to a minimum, because excessive diffusion of alloying elements from the core into the cladding reduces corrosion protection. For the same reason, reheat treatment of alclad sheet less than 0.76 mm (0.030 inch) thick generally is prohibited and the number of reheat treatments permitted for thicker alclad sheet is limited.

The soak times for wrought alloys take into account the normal thermal lag between furnace and part and the difference between surface and centre temperatures for commercial equipment qualified to the standards of MIL-H-6088. The rapid heating rates of salt baths permit all immersion time to be counted as soak time unless the bath temperature drops below the minimum of the range. Even then, soak time begins as soon as the bath temperature returns to the minimum. In air furnaces, soak time does not begin until all furnace instruments return to their original set temperature–that is, the temperature reading before insertion of the load.

In air furnaces, thermocouples may also be attached to or buried in, parts located in the load in such a manner as to represent the hottest and coldest temperatures in each zone. In this way, it is possible to ensure that adequate soaking is obtained.

Special consideration is given also to establishing soak times for hand and die forgings; soak time in some specifications is extended to complete solution and homogenisation in areas that received marginal reduction during forging. Considerable variation exists in the amount of soak time added; some specifications call for an arbitrary addition, such as one hour and others require one hour per inch of thickness of the original forging.

In air furnaces, careful attention should be given to arrangement of the load. Air flow and natural temperature distribution within the furnace should be arranged (a) to offer minimum resistance to air flow, (b) to produce the least disturbance in the natural temperature distribution and (c) to afford constant replenishment of the envelope of air around each part. It is common practice to specify a minimum spacing of 50 mm (2 inch) between parts, but large complex shapes may require considerably greater spacing. Many operators have found conservative loading practices to be more economical in the long run than heavier loading, because with lighter loads heating rates are higher and fewer rejections and service failures are encountered.

High-Temperature Oxidation

There is a condition, commonly but erroneously known as HTO or high-temperature oxidation, which can lead to deterioration of properties in aluminium alloys.

Moisture in contact with aluminium at high temperature serves as a source of nascent hydrogen, which diffuses into the metal. Foreign materials, such as sulphur compounds, function as decomposers of the natural oxide surface film, eliminating it as a barrier either between the moisture and the aluminium or between the nascent hydrogen and the aluminium. The most common manifestation of high-temperature oxidation is surface blistering, but occasionally the only manifestations are internal discontinuities or voids, which can be detected only by careful ultrasonic inspection or by metallographic techniques.

It is important to recognise that the symptoms of high-temperature oxidation are identical to those of unsoundness or high gas content in the original ingot or of other improper mill practice. Blisters resulting from ingot defects, improper extrusion or improper rolling may be lined up in the direction of working. However, it usually is impossible to distinguish among defect sources and therefore the possibility that a contaminated atmosphere is the cause of the defects must be checked.

Not all alloys and product forms are equally vulnerable to this type of attack. The 7*xxx* series alloys are most susceptible, followed by the 2*xxx* alloys. Extrusions undoubtedly are the most susceptible, form; forgings are probably second. Low-strength alloys and alclad sheet and plate are relatively immune to high-temperature oxidation. (Blistering of alclad material as a result of inadequate bonding is not the same as the blistering caused by high-temperature oxidation).

If the protective oxide film formed during mill operations is removed from the mill product by a subsequent mechanical conditioning operation, such as sanding, the conditioned surface will be more susceptible to high-temperature oxidation than those from which the film was not removed.

Moisture can be minimised by thoroughly drying parts and racks before they are charged. Drain holes often are needed in racks of tubular construction, to avoid entrapment of water. Another common requirement is adjustment of the position of the quench tank with respect to furnace doors and air intake. Because it is unlikely that all moisture can be eliminated from the atmosphere in a production heat treating furnace, it is extremely important to eliminate all traces of other contaminants from both the parts and the furnace atmosphere.

The most virulent contaminants in attacking aluminium are sulphur compounds. Residues from forming or machining lubricants, or from a sulphur dioxide protective atmosphere used in prior heat treatment of magnesium, are potential sources of sulphur contamination. In one plant, surface contamination resulted from sulphur-containing materials in tote boxes used to transport parts. In another, an epidemic of blistering was cured by rectifying a 'sour' degreaser. In a third instance, it was found that a vapour-degreasing operation was not completely removing a thin, hard waxy residue and an alkaline cleaning operation was added.

Very often, the source of contamination is obscure and difficult to detect and the problem must be combated in another way. The most common of the alternative methods is use of a protective fluoborate compound in the furnace. Such a compound usually is effective in minimising the harmful effects of moisture and other undesirable contaminants because it forms a barrier layer or film on the aluminium surface. The additive is not a universal solution; in some applications, high-temperature oxidation has occurred even though a fluoborate compound was employed. Also, the use of such compounds, particularly ammonium fluoborate, may present a hazard to personnel if used in poorly sealed furnaces or in furnaces that discharge their atmosphere into enclosed areas.

Protective fluoborate compounds accentuate staining or darkening of the parts being heat treated. (At times, this attack, particularly on parts located near the protective-compound container during heat treatment, has been severe enough to be termed 'corrosion'). Although this minor nuisance might be considered a small price to pay for solution of a problem of high-temperature oxidation, the residual compound in the furnace dissipates slowly. Therefore, subsequent loads of alloys and product forms whose end uses require bright surfaces and that are not susceptible to high-temperature oxidation, may be detrimentally affected.

Successful use of fluoborate protective compounds appears to depend on specifying the right amount for each furnace; this must be established on a trial-and-error basis. One aircraft manufacturer adds 4 gram/m^3 (0.004 oz/ft^3) of furnace chamber to each load. Another adds 0.45 kg (1 lb) per shift to a metal container hung on the furnace chamber wall, thus avoiding loss of the compound during quenching.

A second method of combating high-temperature oxidation is to anodise the work before it is heat treated. The resultant aluminium oxide film prevents attack by contaminants in the furnace atmosphere. The only deterrents to the use of anodising are its cost (in money and time) and the slight surface frostiness which results from the subsequent stripping operation.

The usual objection to the blistered surface produced by high-temperature oxidation is its unsightly appearance. This often can be improved (for salvage purposes) by applying local pressure to flatten each blister and then finishing by a mechanical process such as polishing, buffing, sanding or abrasive blasting. In general, the effect of HTO on static properties and fatigue strength is slight. However, if a void resulting from HTO is located close to another stress concentration, such as a hole, much greater degradation of fatigue strength is likely. In critical aluminium alloy forgings, any blistering must be evaluated carefully for its effect on the integrity of the part. Any 'cosmetic' salvage should be performed only after it has been established that the blisters are superficial and will not remain in the finished product.

QUENCHING

In most instances, to avoid those types of precipitation that are detrimental to mechanical properties or to corrosion resistance, the solid solution formed during solution heat treatment must be quenched rapidly enough (and without interruption) to produce a supersaturated solution at room temperature–the optimum condition for precipitation hardening. The resistance to stress-corrosion cracking of certain Cu-free Al-Zn-Mg alloys, however, is improved by slow quenching. Most frequently, parts are quenched by immersion in cold water or, in continuous heat treating of sheet, plate or extrusions in primary fabricating mills, by progressive flooding or high-velocity spraying with cold water. However, parts of complex shape, often with both thin and thick sections (such as die forgings, most castings, impact extrusions and components formed from sheet) are commonly quenched in a medium that provides somewhat slower cooling. This medium may be water at 65 to 80°C (150 to 180°F), boiling water, an aqueous solution of polyalkaline glycol, or some other fluid medium such as forced air or mist.

If appreciable precipitation during cooling is to be avoided, two requirements must be satisfied. First, the time required for transfer of the load from the furnace to the quenching medium must be short enough to preclude slow precooling into the temperature range where very rapid precipitation takes place. For alloy 7075, this range was determined to be 400 to 290°C (750 to 550°F) and some sources quote this range (or a slightly different range) as the most critical range for quenching of any aluminium alloy. Later work has shown that the most critical range is alloy-dependent and as will be discussed in detail under 'Quench-Factor Analysis', significant errors can result from the assumption that precipitation in negligible outside of a so-called 'critical range'.

The second requirement for avoidance of appreciable precipitation during quenching is that the volume, heat-absorption capacity and rate of flow of the quenching medium be such that little or no precipitation occurs during cooling. Any interruption of the quench that might allow reheating into a temperature range where rapid precipitation can occur must be prohibited.

For maximum dimensional stability, some forgings and castings are fan cooled or still-air cooled. In such instances, precipitation-hardening response is limited, but satisfactory values of strength and hardness are obtained. Extrusions produced without separate solution heat treatment can be air or mist quenched, but thicker sections may require water quenching by immersion or spraying. Alloys that are relatively dilute, such as 6063 and 7005, are particularly well suited to air quenching and their mechanical properties are not greatly affected by its low cooling rate.

Lower quenching rates are employed for forgings, castings and complex shapes to minimise warpage or other distortion and the magnitude of residual stresses developed as a consequence of temperature nonuniformity from surface to interior. The effects of these low rates on mechanical properties vary with alloy composition and temper. The effect on yield strength of average quenching rates through the range from 400 to 290°C (750 to 550°F) are shown for four alloys in Fig. 20.2. For alloys relatively high in sensitivity to quenching rate, such as 7075, rates of about 300°C/second (540°F/second) or higher are required in order to obtain near-maximum strength after precipitation heat treatment. The other alloys in Fig. 20.2 maintain their strengths at cooling rates as low as about 100°C/second (180°F/second). As cautioned previously, significant precipitation can occur outside this temperature range and thus curves such as those in Fig. 20.2 are useful mainly for comparison of relative sensitivities to quench rate. Moreover, for high strength alloys, toughness and corrosion resistance may be impaired without significant loss of tensile strength. Where such properties are important, further restrictions on minimum quench rate may be necessary.

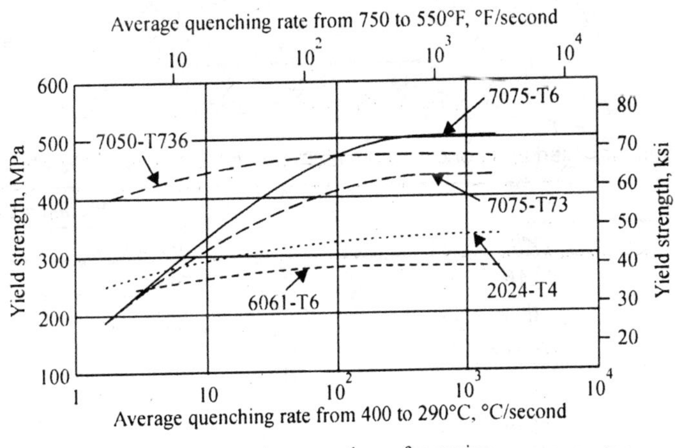

Fig. 20.2. Effect of quenching rate through critical temperature range on final yield strength of four wrought aluminium alloys.

Delay in Quenching

Whether the transfer of parts from the furnace to the quench is performed manually or mechanically, it must be completed in less than the specified maximum time. The maximum allowable transfer time or 'quench delay' varies with the temperature and velocity of the ambient air and the mass and emissivity of the parts. From cooling curves such as those illustrated in Fig. 20.3, maximum quench delays (see table accompanying Fig. 20.3) can be determined that will ensure complete immersion before the parts cool below 400°C (750°F). MIL-H-6088 specifies maximum quench delays for high-strength alloys of 5, 7, 10 and 15 second for thickness ranges of (a) up to 0.016 inch (0.41 mm); (b) 0.017 to 0.031 inch (0.43 to 0.79 mm); (c) 0.032 to 0.090 inch (0.81 to 2.29 mm); and (d) over 0.090 inch, respectively. Quench delay is conservatively defined as commencing 'when the furnace door begins to open or the first corner of a load emerges from a salt bath' and ending 'when the last corner of the load is immersed in the water quench tank'. Recommended maximum quench-delay times are listed in Table 20.3. However, exceeding the maximum delay time is permitted if temperature measurements of the load prove that all parts are above 415°C (775°F) when quenched.

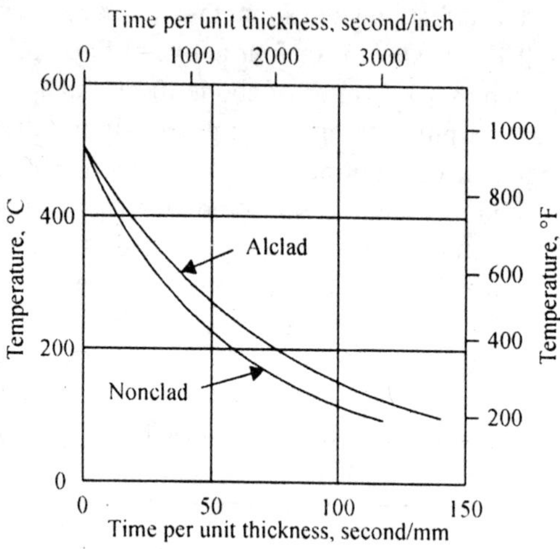

Fig. 20.3. Cooling curves for alclad and nonclad aluminium products cooled from 495°C (920°F) in forced air.

Thickness		*Maximum quench delay, s*	
mm	*inch*	*Alclad*	*Nonclad*
0.41	0.016	6.4	4.4
0.51	0.020	8.0	5.5
0.64	0.025	10.0	6.8
0.81	0.032	12.8	8.8
1.02	0.040	20.0	11.0

Air temperature, 25°C (80°F); air velocity, 2.3 m/s (450 ft/min.). Tabulated values of quench delay (maximum delay before the material being quenched has cooled below 400°C, or 750°F) were determined from cooling curves shown.

It is relatively easy to control quench delay in day-to-day operations by using a stopwatch or, if necessary, by attaching thermocouples to parts. However, although the cooling rate between 400 and 260°C (750 and 500°F) is most critical and must be extremely high for many high-strength alloys, it cannot be directly measured in production operations. It is usual to rely on standardised practices, augmented by results of tension tests and tests of susceptibility to intergranular corrosion.

Water-Immersion Quenching

Water-immersion quenching normally is controlled in practice by stipulating maximum quench-delay time and maximum water temperature. The first requirement controls the cooling rate during transfer and for high-strength alloys, often is based on the criterion of complete immersion before the metal cools below 415°C (775°F). The second requirement controls the cooling rate during immersion. MIL-H-6088 specifies that for water-immersion quenching, except quenching of forgings and castings, the temperature of the water shall not exceed 38°C (100°F) upon completion of quenching. This requirement controls both the temperature of the quench water prior to immersion and the ratio of the combined mass of load and rack to the volume of water. However, to ensure adequate quenching effectiveness, it is necessary also that the cooling fluid flow past all surfaces of each part during the first few seconds after immersion. Before parts enter the furnace, their placement in racks or baskets should be compatible with this requirement. During the first few seconds of quenching, agitation of the parts or the water should be sufficient to prevent local increases in temperature due to the formation of steam pockets.

In one application, it was found that 2024-T4 plates 13 by 760 by 760 mm (½ by 30 by 30 inch), quenched singly into a large volume of still water, were quite susceptible to intergranular corrosion. This susceptibility disappeared completely when the quenching practice was modified by adding sufficient agitation to break up the insulating blanket of steam that formed on the surface of the hot metal. Quenching practices for small parts such as fasteners and hydraulic fittings have been modified for the same reason. Dumping in bulk from baskets has been replaced by methods, such as the use of shaker hearth furnaces or special racking, which permit parts to be quenched singly.

Spray Quenching

For spray quenching, the quench rate is controlled by the velocity of the water and by volume of water per unit area per unit time of impingement of the water on the workpiece. Rate of travel of the workpiece through the sprays is an important variable.

Local increases in temperature that occur within the first few seconds of quenching, caused by a phenomenon such as plugged spray nozzles, are particularly deleterious. The remaining 'internal heat' may be sufficient to reheat the surface region. When this happens, a large loss in strength occurs at the previously quenched surface. The loss of strength in the affected area of a heavy part is much more severe than that caused by an inadequate quenching rate alone. This is illustrated for 75 mm (3 inch) thick 7075-T62 plate in Fig. 20.4, which compares, at various depths, the properties of a plate for which quenching was interrupted on one side after 3 second with those of a plate that was quenched from one side only.

Quench-Factor Analysis

Although useful as first approximations, average quenching rates and critical temperature ranges are too qualitative to permit accurate prediction of the effects of quenching rate when the rate of cooling does not change smoothly. For such instances, a procedure known as 'quench-factor' analysis' has been developed. To understand this procedure, one must recognise that the rate of precipitation during quenching depends on two competing factors: supersaturation and diffusion. At high temperatures, supersaturation is low and so precipitation rate is low despite the high diffusion rate. At low temperatures, diffusion rate is low and thus precipitation rate is low despite the high degree of supersaturation. At intermediate temperatures, precipitation rate is highest. Consequently, times to produce equal amounts of precipitation follow a C-shape pattern.

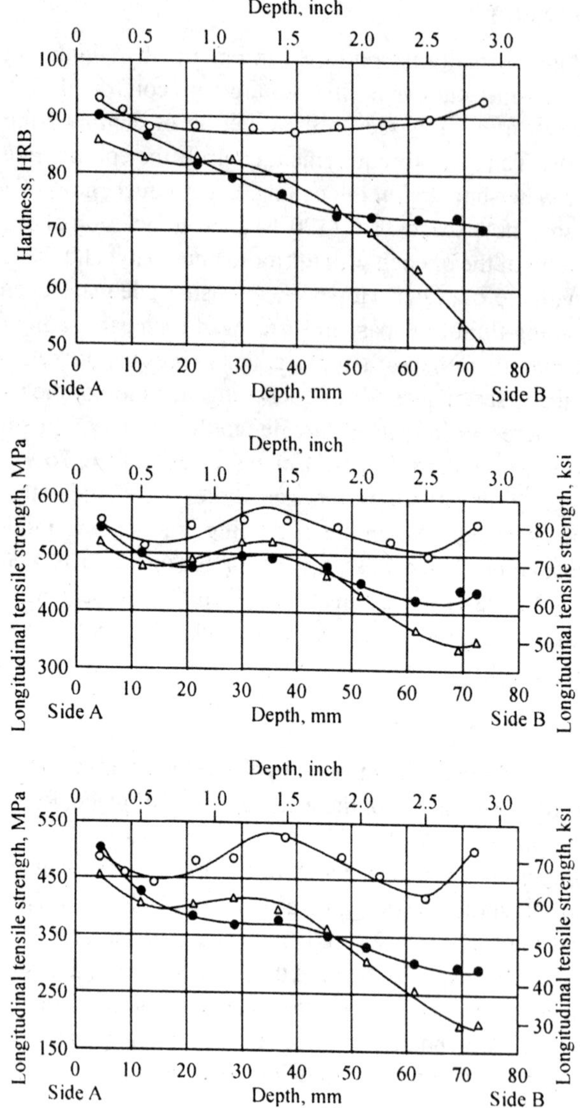

Fig. 20.4. Through-thickness property variations due to quench rate and temperature-rise effects in 7075-T62 plate 75 mm (3 inch) thick.

Using this C_t-curve, a quench curve can be analysed to predict quantitatively how certain properties will be affected. For example, it has been shown that the strengths of several alloys and the type of corrosion that will attack alloy 2024-T3, can be predicted by means of this method.

Precipitation kinetics for continuous cooling can be described by the equation:

$$\zeta = 1 - \exp(k\tau) \quad \text{... (Eq. 20.1)}$$

where ζ is the fraction transformed, k is a constant, and:

$$\tau = \int \frac{dt}{C_t}$$

where t is time and C_t is critical time as a function of temperature; the locus of critical times is the C-curve. When $\tau = 1$, the fraction transformed, ζ, equals the fraction transformed value designated by the C-curve. This integral can be graphically integrated using the method illustrated in Fig. 20.5 and discussed below.

Predicting Corrosion Behaviour

Alloy 2024-T4, for example, is susceptible to intergranular corrosion when a critical amount of solute is precipitated during quenching, but will corrode in the less severe pitting mode when lesser amounts are precipitated. For predicting the effects of proposed quenching conditions on the corrosion characteristics of 2024-T4, the postulated quench curve is drawn and the quench factor is calculated using the C-curve in Fig. 20.6. Corrosion characteristics are predicted from the plot in Fig. 20.7. When the quench factor, τ, is less than 1.0, continuously quenched 2024-T4 will corrode by pitting.

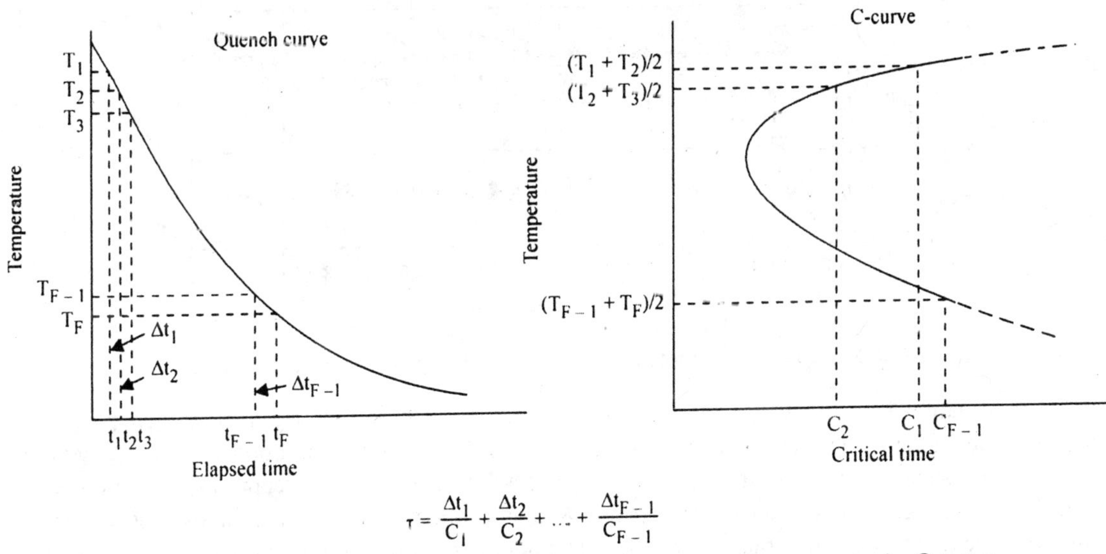

$$\tau = \frac{\Delta t_1}{C_1} + \frac{\Delta t_2}{C_2} + \ldots + \frac{\Delta t_{F-1}}{C_{F-1}}$$

Fig. 20.5. Method of determining quench factor, τ, using a cooling curve and a C-curve.

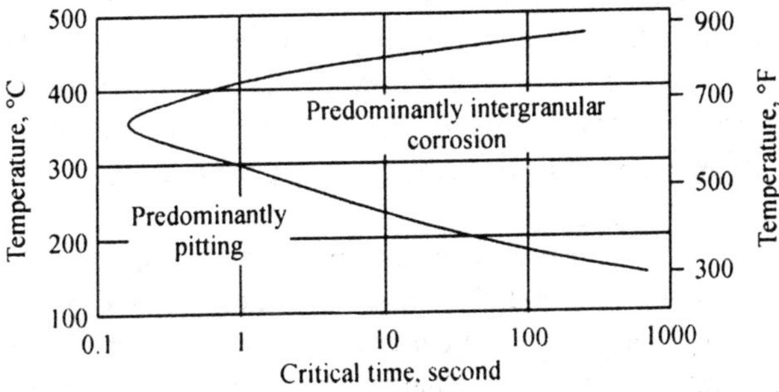

Fig. 20.6. C-curve indicating type of corrosion attack on 2024-T4 sheet.

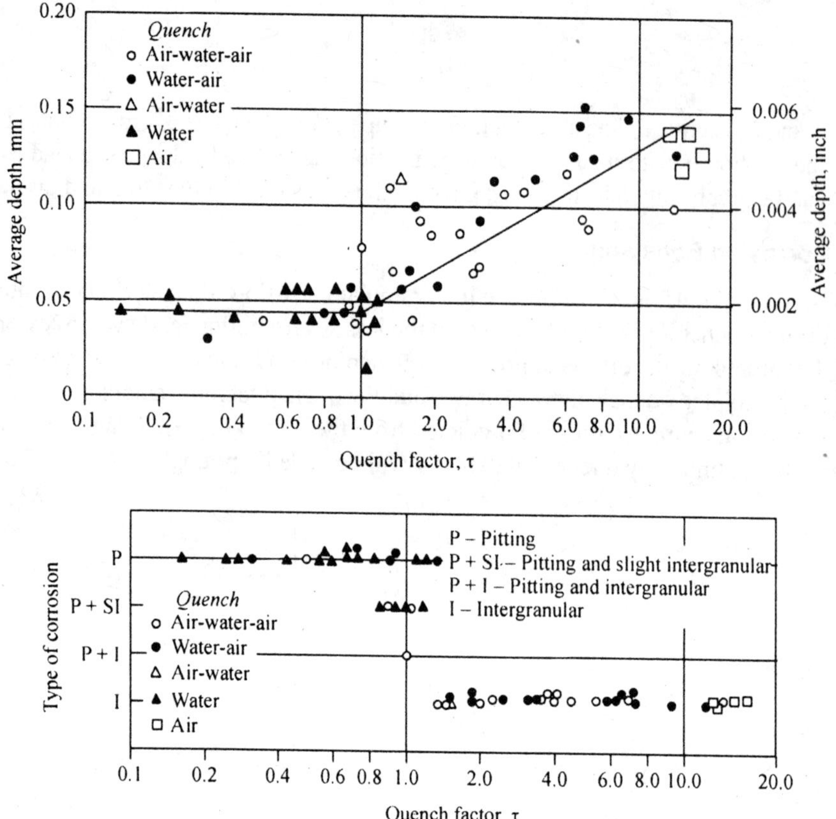

Fig. 20.7. Type and depth of attack on 2024-T4 sheet vs quench factor.

These relationships are applied to studies of effects of proposed changes in quench practice on design of new quenching systems. For example, consider that the goal of a proposed quenching system for 2024-T4 sheet products is to minimise warpage while preventing susceptibility to intergranular corrosion. Warpage occurs when the stresses imposed by temperature differences across the parts exceed the flow stress. As quenching rate decreases, the tendency for large differences in temperature to occur decreases but the tendency for intergranular corrosion to occur increases.

The C-curve in Fig. 20.6 indicates that quenching rate can be decreased near the solution heat treating temperature and near room temperature without greatly sacrificing corrosion characteristics, but this information does not provide a quantitative answer. Simple calculations, however, can reveal a multitude of hypothetical cooling curves that provide slow quenching during a large portion of the quench cycle but sufficiently rapid quenching where critical times are short so that desirable corrosion characteristics are obtained.

As an example, one-, two- and three-step quench curves that would ensure acceptable corrosion behaviour in 2024-T4 sheet (quench factor, 0.99) were calculated. Some of these curves are plotted in Fig. 20.8. This illustration shows that 2024 can be quenched at a rate of 470°C/second (850°F/second) or higher and still develop acceptable corrosion characteristics if the quenching rate is linear from the solution temperature to 150°C (300°F). If sheet 3.2 mm (0.125 inch) thick is air-blast quenched (rate of heat removal, 5.68 W/m^{2}·°C) to 395°C (740°F), however, the quenching rate from 395 to 150°C must be at least 945°C/second (1700°F/second) to maintain the acceptable corrosion behaviour. It may also be air-blast quenched to 395°C, spray quenched at 3300°C/second (6000°F/second) to 250°C (480°F), then air-blast quenched to 150°C.

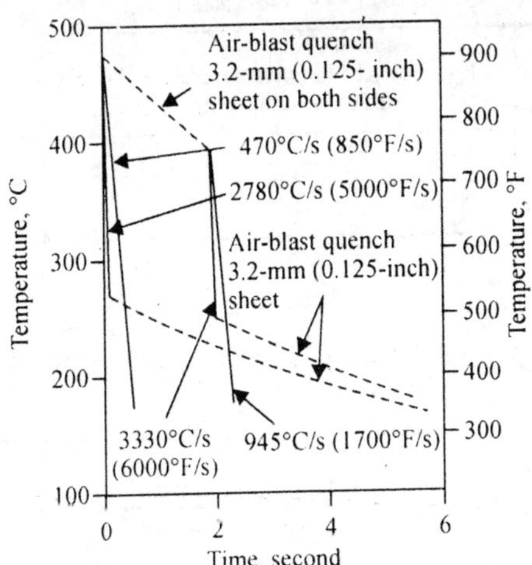

Fig. 20.8. Quench curves for 2024-T4 sheet, to eliminate susceptibility to intergranular corrosion.

Other curves could be drawn, of course, but the important points are that air-blast quenching cannot be continued to more than a few degrees below 395°C (740°F) and cannot be initiated at more than few degrees above 270°C (520°F) even if infinite quenching rates are attained from 395 to 270°C.

Predicting Yield Strength

Predicting yield strength is more complex than predicting corrosion behaviour and requires some knowledge of the relationship between extent of precipitation and loss in ability to develop property. Because attainable strength of precipitation-hardening aluminium alloys is a function of the amount of solute remaining in solid solution after quenching, relationships between strength, σ, attainable after continuous cooling and quench factor, τ, can be expressed as follows:

$$\sigma = \sigma_{max} \exp (k_1 \tau) \quad \text{... (Eq. 20.2)}$$

where σ_{max} is the strength attainable with an infinite quenching rate, k_1 is 0.005013 and:

$$\tau = \frac{dt}{C_{t99.5}}$$

where t is time and $C_{t99.5}$ is the C-curve for $\sigma_{99.5}$—that is, critical time as a function of temperature to reduce attainable strength to 99.5 per cent of σ_{max}.

The advantage of predicting yield strength from quench factor instead of from average quenching rate is illustrated by the following comparison. Four specimens of alloy 7075-T6 quenched by various means (Fig. 20.9) were selected. Yield strengths were predicted both from average quenching rate between 400 and 290°C (750 and 550°F) and from quench factor. Quench factor was calculated using the C-curve for 99.5 per cent maximum yield strength for 7075-T6 (Fig. 20.10) and yield strength was estimated from the above equation defining the quench factor, τ (Fig. 20.11).

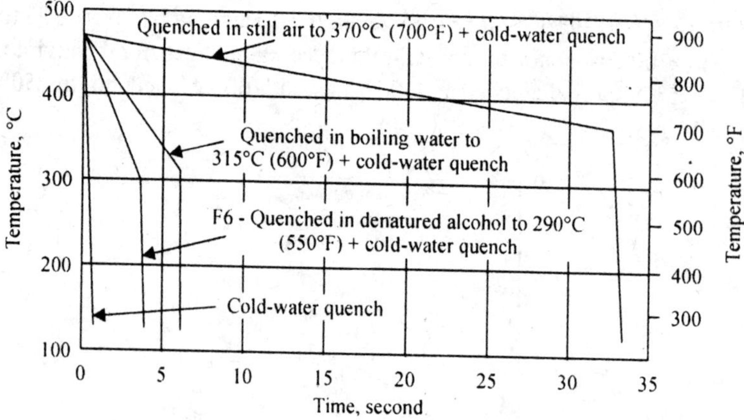

Fig. 20.9. Cooling curves for 7075-T6 sheet.

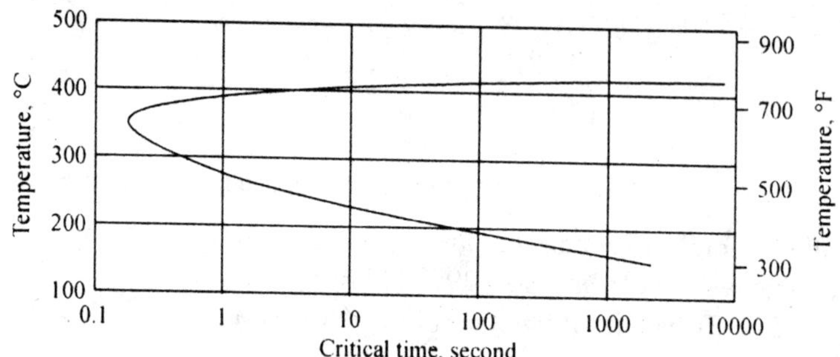

Fig. 20.10. C-curve for 99.5% maximum yield strength of 7075-T6 sheet.

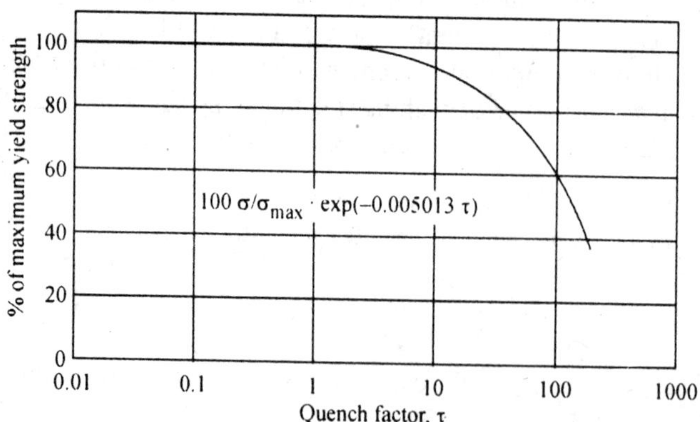

Fig. 20.11. Yield strength vs quench factor.

A comparison of predicted yield strength with actual yield strength is given in Table 20.5. Yield strengths predicted from quench factor agree very well with measured yield strengths for all specimens, the maximum error being 19.3 MPa (2.8 ksi). Yield strengths predicted from average quenching rates, however, differ from measured values by as much as 226 MPa (32.8 ksi).

Table 20.5. Yield-strength values for 7075-T6 sheet predicted from cooling curves using average quench rate and quench factor.

Quench	*Average quench rate from 400 to 290°C (750 to 550°F)*		*Quench factor,*	*Measured yield strength*		*Yield strength predicted from average quench rate*		*Yield strength predicted from quench factor*	
	°C/s	*°F/s*	*τ*	*MPa*	*ksi*	*MPa*	*ksi*	*MPa*	*ksi*
Cold water	935	1680	0.464	506	73.4	499	72.4	498	72.3
Denatured alcohol to 290°C (550°F), then cold water	50	90	8.539	476	69.1	463	67.2	478	69.4
Boiling water to 315°C (600°F, then cold water	30	55	15.327	458	66.4	443	64.2	463	67.1
Still air to 370°C (700°F), then cold water	5	9	21.334	468	67.9	242	35.1	449	65.1

The advantage of using the quench factor for predicting yield strength from cooling curves is apparent. Cooling curves that have long holding times either above or below the critical temperature range from 400 to 290°C (750 to 550°F) cannot be used to predict yield strength from average quenching rate. In such instances, prediction yield strength on the basis of quench factor is particularly advantageous.

An underlying assumption of both quench-factor analysis and average-cooling-rate estimation is that the only effect of temperature is on the kinetics of precipitation. Recent evidence indicates that this assumption is not valid, however, when portions of the metal are quenched locally but reheated significantly before quenching is complete.

Quenching to Minimise Residual Stress and Warpage

Although cold-water immersion or flushing is most common, because it produces the most effective quench (and has been required by MIL-H-6088 for 2014, 2017, 2024, 2117, 7075 and 7178 alloys except forgings), it presents problems involving residual stress and warpage.

Residual stresses in heavy sections of aluminium alloys originate from differential thermal expansion during quenching–that is, the still-warm central material contracts, pulling in the already cooled outer shell. The magnitude of stresses increases with section size, as shown in Fig. 20.12.

The distribution pattern of residual stresses in as-quenched parts (compression in the outer layers and tension in the central portion) is usually desirable in service. Compressive stresses inhibit failure by fatigue and stress corrosion–two mechanisms that initiate in the outer fibres. Unfortunately, metal-removal operations required after heat treating often expose material that is stressed in tension. Also, metal-removal operations that are asymmetrical (with respect to residual stresses) cause distortion by redistributing residual stresses. When close-tolerance parts are being fabricated, the resulting warpage can be costly and difficult to correct.

Although service performance is sometimes a factor, the major incentive for reducing residual stress differentials has been a reduction in warpage during machining or an improvement in shape before machining.

One approach to reducing the cooling-rate differential between surface and centre is the use of a milder quenching medium–water that is hotter than that normally used or water-glycol solutions. Boiling water, which is the slowest quenching medium used for thick sections, is sometimes employed for quenching wrought products even though it lowers mechanical properties and corrosion resistance. Quenching of castings in boiling water, however, is standard practice and is reflected in design allowables.

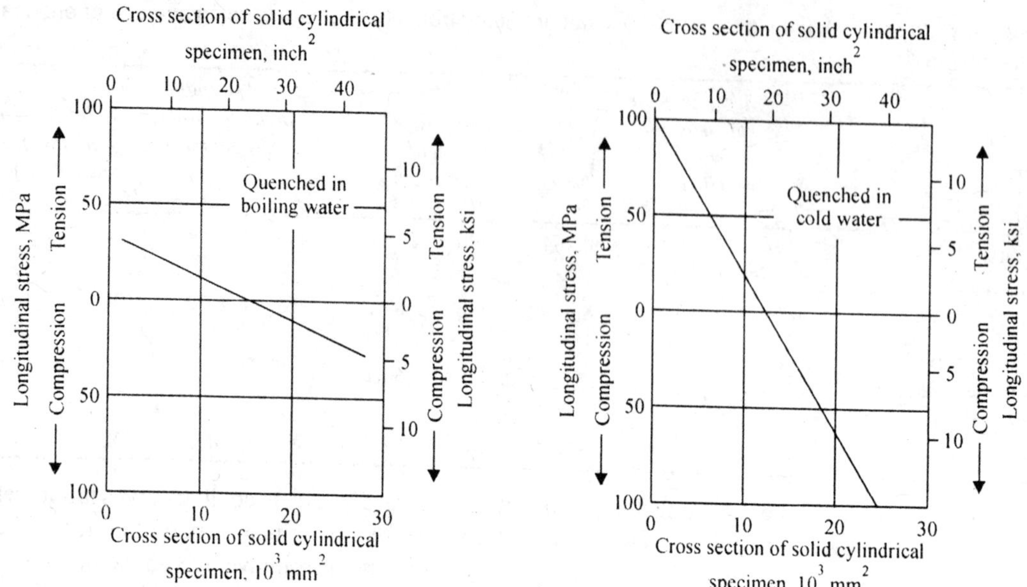

Fig. 20.12. Effects of quenching from 540°C (1000°F) on residual stresses in solid cylinders of alloy 6151.

Another approach to the minimisation of residual stresses that is generally successful consists of rough machining to within 3.2 mm (0.125 inch) or less of finish dimensions, heat treating and then finish machining. This procedure is intended to reduce the cooling-rate differential between surface and centre by reducing thickness; other benefits that accrue if this technique is used to reduce or reverse surface tension stresses in finished parts are improvements in strength, fatigue life, corrosion resistance and reduced probability of stress-corrosion cracking.

Several factors (especially quenching warpage) sometimes preclude general use of this procedure. The thinner and less symmetrical a section, the more it will warp during quenching and the residual stresses resulting from straightening of warped parts (plus straightening costs) often are less desirable than the quenching stresses. Holding fixtures and die quenching may be helpful, but precautions must be taken to ensure that they do not retard quenching rates excessively. Other factors that must be considered are the availability of heat treating facilities and whether or not the advantages of such a manufacturing sequence offset the delay and cost entailed in a double-machining setup.

Warpage of thin sections during quenching is also a problem. Even in the same load, symmetry of cooling usually varies significantly among identical parts and the resultant inconsistent warpage usually requires costly hand straightening. Consequently, a significant amount of effort has been devoted to reducing or eliminating warpage by changing racking positions to achieve symmetry of cooling.

For sheet-metal parts, one manufacturer uses a double screen floor in the quenching rack to reduce the force of initial contact between water and parts. Others allow parts to 'free fall' from rack to quench tank. Spacing and positioning on the rack are carefully controlled so that parts will enter the water with minimum impact. With this technique, water turbulence must be avoided, because it will often cause parts to float for a few seconds, greatly reducing their cooling rate.

Because of the difficulties encountered with quenching in cold water, milder quenchants have been employed. Indiscriminate use of milder quenchants can have catastrophic effects; however, when their use is based on sound engineering judgement and a metallurgical knowledge of the effects on the specific alloy, significant cost savings or performance improvements can be realised. The most frequent advantage is the reduction in costly straightening operations and in resultant uncontrolled residual stresses. For example, one aircraft manufacturer utilises water-spray and air-blast quenching for weldments and complex formed parts made from 6061, an alloy whose corrosion resistance is insensitive to quenching rate. Straightening requirements are negligible and through careful control of racking and coolant flow, the decrease in mechanical properties is minimised, as shown by the data in Fig. 20.13.

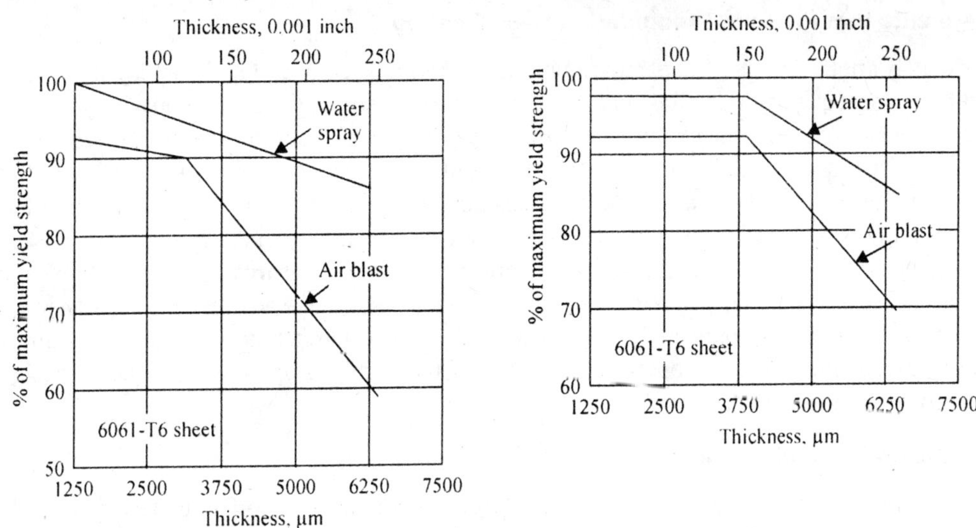

Water-immersion quench equals 100 per cent. Control of coolant flow will minimise decrease in mechanical properties.

Fig. 20.13. Effect of quenching medium on strength of 6061-T6 sheet.

Another development for reducing straightening costs is quenching in water-polymer solutions. Quenching of formed sheet-metal parts in aqueous solutions of polyalkylene glycol or in similar inversely soluble media has significantly reduced the cost of straightening these parts after quenching. The Aerospace Metals Engineering Committee has recommended, for several alloys, maximum thicknesses that can be quenched in solutions of specific concentrations while maintaining acceptable property levels. Typical parameters for quenching wrought products (other than forgings) in glycol-water solutions are presented in Table 20.6.

Table 20.6. Limits for quenching in glycol-water solutions.

Data are for wrought aluminium alloy products other than forgings.

Glycol concentration, vol %	*Alloys*	*Maximum thickness*	
		mm	*inch*
12 to 16	2014, 2017, 2117, 2024, 2219	2.03	0.080
	7075, 7175	25.4	1.000
17 to 22	2014, 2017, 2117, 2024, 2219	1.80	0.071
	7075, 7079, 7175, 7178, 6061	12.7	0.500
23 to 28	2014, 2017, 2117, 2024, 2219	1.60	0.063
	7075, 7079, 7175, 7178, 6061	9.53	0.375

(Contd...)

Glycol concentration, vol %	Alloys	Maximum thickness mm	inch
29 to 34	2014, 2017, 2117, 2024, 2219	1.02	0.040
	7075, 7079, 7175, 7178, 6061	6.35	0.250
35 to 40	7075, 7079, 7175, 7178, 6061	2.03	0.080

Treatments that Precede Precipitation Heat Treating

Immediately after being quenched, most aluminium alloys are nearly as ductile as they are in the annealed condition. Consequently, it is often advantageous to form or straighten parts in this temper. Moreover, at the mill level, controlled mechanical deformation is the most common method of reducing residual quenching stresses. Because precipitation hardening will occur at room temperature, forming or straightening usually follows as soon after quenching as possible. In addition, maximum effectiveness in stress relief is obtained by working the metal immediately after quenching.

In some alloys, notably those of the 2xxx series, cold working of freshly quenched material greatly increases its response to later precipitation heat treatment. Mills take advantage of this phenomenon by applying a controlled amount of rolling (sheet and plate) or stretching (extrusion, bar and plate) to produce higher mechanical properties. However, if the higher properties are used in design, reheat treatment must be avoided.

Forming and Straightening

These operations vary in degree from minor corrections of warpage to complete forming of complex parts from solution-treated flat blanks. Particular value is gained when enough forming can be done at this stage of processing to eliminate the distortion caused by quenching. However, production operation must be adjusted so that most of the plastic deformation is accomplished before an appreciable amount of precipitation hardening takes place.

Although the most severe forming operations may have to be arranged to avoid natural ageing, it often is desirable to allow some natural ageing to occur and thus avoid formation of Lüders lines. This condition of nonuniform deformation is most likely to occur shortly after quenching and diminishes significantly after a few hours of natural ageing.. Complete freedom from Lüders lines, however, may require one or two days of natural ageing prior to forming. Thus, the forming operation may have to be timed so as to obtain the most appropriate trade-off of these characteristics for the specific parts involved. Lüders lines also can be reduced by employing low strain rates or by forming at temperatures of 150 to 175°C (300 to 350°F).

Residual stresses in sheet-metal parts formed in the quenched condition are higher than those in parts formed in the annealed condition. Consequently, forming in the quenched condition should be selected judiciously for parts that are critical in fatigue (Fig. 20.14) or stress corrosion.

Re-solution heat treatment of parts formed after quenching often causes excessive grain growth in critically strained regions and thus is not recommended.

Mechanical Stress Relief

Deformation consists of stretching (bar, extrusions and plate) or compressing (forgings) the product sufficiently to achieve a small but controlled amount (1 to 3 per cent) of plastic deformation. If the benefits of mechanical stress relieving are needed, the user should refrain from reheat treating.

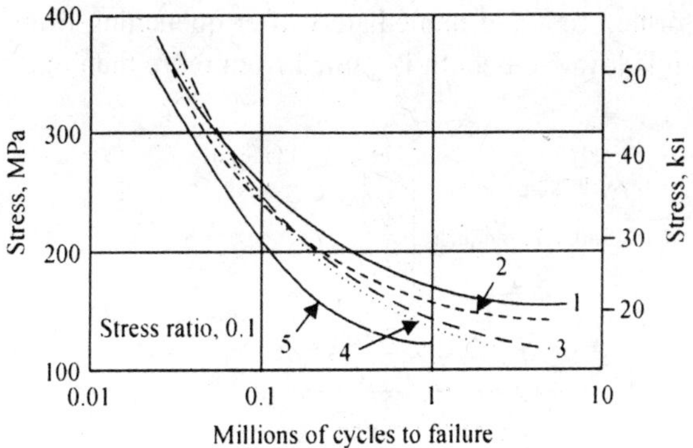

Sheet 1.02 mm (0.040 inch) thick was annealed, solution heat treated and quenched and then fatigue tested. The sheet represented by curve 1 was not bent. All other sheet was bent 90° in the annealed condition. Flattening (unbending) was done in either the annealed condition (curve 2) or the solution heat treated and quenched condition (curves 3, 4 and 5). Details of bending and flattening were as follows: (1) Not bent. (2) Bend radius, 3.18 mm (1/8 inch); flattened in annealed condition. (3) Bend radius, 3.18 mm (1/8 inch); flattened in quenched condition after 3 days of storage at –18 to –12°C (0 to 10°F). (5) Bend radius, 1.59 mm (1/16 inch); flattened in quenched condition after 3 days of storage at –18 to –12°C (0 to 10°F).

Fig. 20.14. Effects of flattening on fatigue characteristics of alclad 2024-T4 sheet.

Fig. 20.15 illustrate the beneficial effect of 3 per cent permanent deformation in compression on a large forging. These methods are most readily adaptable to mill and forge shop products and require equipment of greater capacity than that found in most manufacturing plants. Application of these methods to die forgings and extrusions usually requires construction of special dies and jaws. Stretching generally is limited to material of uniform cross section; however, it has been applied successfully to stepped extrusions and to a 3-by-14-m (10-by-47-ft) aircraft wing skin roll-tapered to a thickness range of 7.1 to 3.2 mm (0.280 to 0.125 inch).

Specific combinations of the supplemental digits are used to denote the tempers produced when mechanical deformation is used primarily to relieve residual stresses induced during the quenching operation. For products stress relieved by stretching, the digits 51 follow the basic T_x designation (T451, for example). For products stress relieved by compressive deformation, the supplementary digits are 52. An additional digit is added to designations for extrusions: an added zero specifies that the product has not been straightened after final stretching; an added one indicates that straightening may have been performed after final stretching.

Thermal Treatments for Relieving Stresses

Numerous attempts have been made to develop a thermal treatment that will remove, or appreciably reduce, quenching stresses. Normal precipitation heat treating temperatures are generally too low to provide appreciable stress relief. Exposure to higher temperatures (at which stresses are relieved more effectively) results in lower properties. However, such treatments sometimes are utilised when even moderate reduction of residual stress levels are important enough so that some sacrifice in mechanical properties can be accepted. The T7 temper for castings is a typical example of this kind of treatment and is discussed later.

Other thermal stress-relief treatments, known as 'subzero treatment' and 'cold stabilisation', involve cycling of parts above and below room temperature. The temperatures chosen are those that can be readily obtained with boiling water and mixtures of dry ice and alcohol–namely, +100 and –73°C (+212 and –100°F)–and the number of cycles ranges from one to five. The maximum reduction in residual stress that can be effected by these techniques is about 25 per cent. The maximum effects can be obtained only if the subzero step is performed first and immediately after quenching from the solution-treating temperature while yield strength is low. No benefit is gained from more than one cycle.

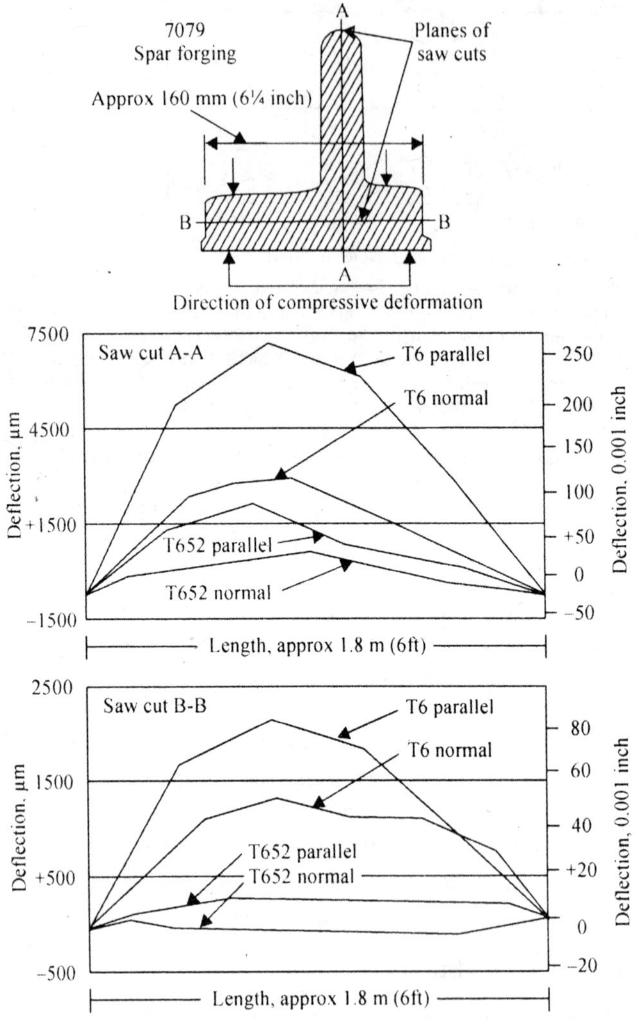

Fig. 20.15. Effect of 3 per cent permanent deformation in compression (T652 treatment) on distribution of stress in a large forging.

A 25 per cent reduction in residual stress is sometimes sufficient to permit fabrication of a part that could not be made without this reduction. However, if a general reduction is needed, as much as 83 per cent relief of residual stress is possible by increasing the severity of the 'uphill quench'–that is, more closely approximating the reverse of the cooling-rate differential during the original quench. This may be accomplished by a patented process that involves extending the subzero step to –195°C (–320°F) and then very rapidly 'uphill quenching' in a blast of live steam (Fig. 20.16). The rate of reheating is extremely critical and therefore, to ensure proper application of the steam blast, a special fixture usually is required for each part.

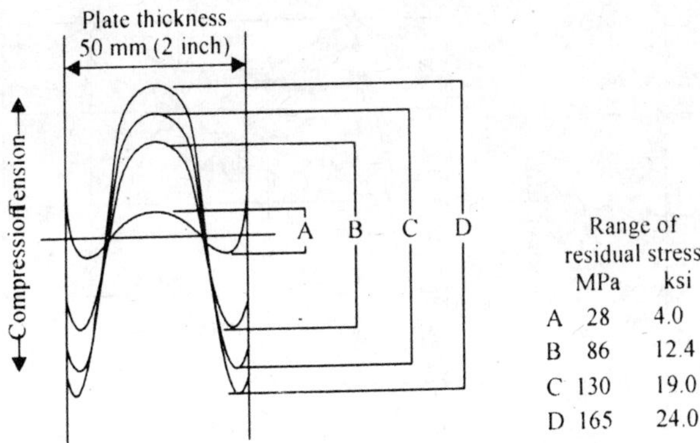

A: cooled to −195°C (−320°F), then uphill quenched in a steam blast. B: cooled to −75°C (−100°F), then uphill quenched in a steam blast. C: cooled to −75 or −195°C (−100 or −320°F), then uphill quenched in boiling water. D: standard specimen, quenched and aged to T6 temper in conventional manner with no further treatment. Note: uphill quenching treatments (single cycle only) were applied from ½ to 1½ hour after quenching from the recommended solution-treating temperature. All specimens were aged to the T6 temper after uphill quenching.

Fig. 20.16. Effectiveness of various uphill quenching treatments in reducing residual quenching stresses in 2014 plate.

This process will not solve all problems of warpage in machining. It may reduce warpage internally but increase warpage of the extreme outer layers, although in the opposite direction (Fig. 20.17). Also, the effect of the altered residual-stress pattern on performance must be evaluated carefully for each part. This is particularly important for parts subjected to cyclic loading or exposed to corrosive environments such as marine atmospheres, especially if the process is introduced after the start of production and original performance tests are not repeated. Further disadvantages are the cost and hazard involved in handling liquid nitrogen and live steam.

Natural Ageing

The more highly alloyed members of the 6xxx wrought series, the copper-containing alloys of the 7xxx group and all of the 2xxx alloys are almost always solution heat treated and quenched. For some of these alloys–particularly the 2xxx alloys–the precipitation hardening that results from natural ageing alone produces useful tempers (T3 and T4 types) that are characterised by high ratios of tensile to yield strength and high fracture toughness and resistance to fatigue. For the alloys that are used in these tempers, the relatively high supersaturation of atoms and vacancies retained by rapid quenching causes rapid formation of GP zones and strength increases rapidly, attaining nearly maximum stable values in four or five days. Tensile-property specifications for products in T3 and T4- type tempers are based on a nominal natural ageing time of four days. In alloys for which T3 or T4-type tempers are standard, the changes that occur on further natural ageing are of relatively minor magnitude and products of these combinations of alloy and temper are regarded as essentially stable after about one week.

In contrast to the relatively stable condition reached in a few days by 2xxx alloys that are used in T3 and T4-type tempers, the 6xxx alloys and to an even greater degree the 7xxx alloys are considerably less stable at room temperature and continue to exhibit significant changes in mechanical properties for many years. The differences in rate and duration of changes in tensile yield strength of representative alloys of the three types are illustrated in Fig. 20.18. Because of the relative instability of the 7xxx alloys, the naturally aged temper (after solution heat treatment and quenching) is designated by the suffix letter W. For specific description of this condition, the time of natural ageing should be included.

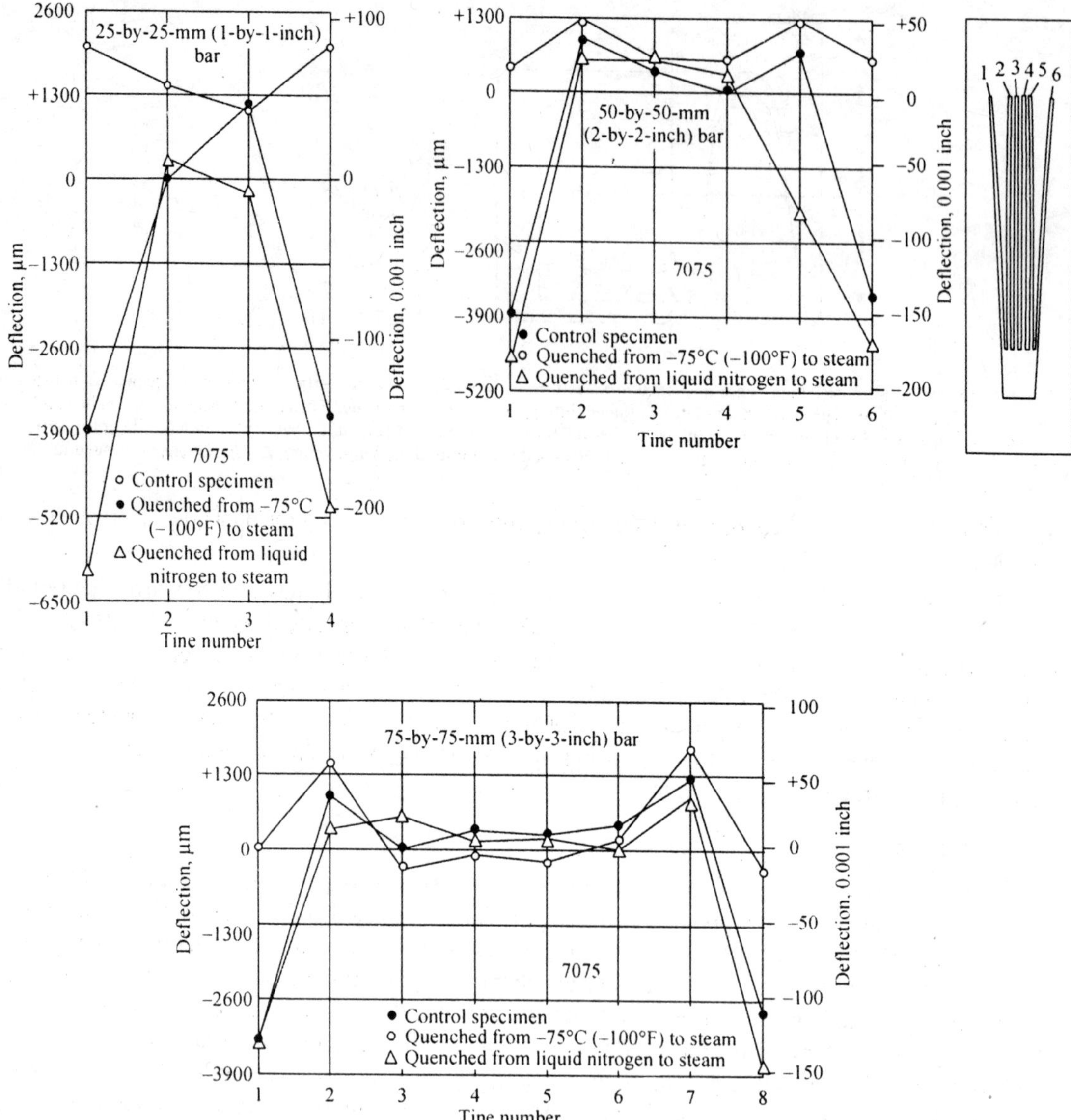

Six-tine specimen shown was machined from 50-by-50-mm (2-by-2-inch) bar. Similar specimens machined from 25-by-25-mm (1-by-1-inch) and 75-by-75-mm (3-by-3-inch) bars had four and eight tines, respectively.

Fig. 20.17. Effect of uphill quenching on deflection of tines.

Ageing characteristic vary from alloy to alloy with respect to both time to initial change in mechanical properties and rate of change, but ageing effects always are lessened by reductions in ageing temperature (Fig. 20.18). With some alloys, ageing can be suppressed or delayed for several days by holding at a temperature of −18°C (0°F) or lower. It is usual practice to complete forming and straightening before ageing changes mechanical properties appreciably. When scheduling makes this impractical, ageing may be avoided in some alloys by refrigerating prior to forming.

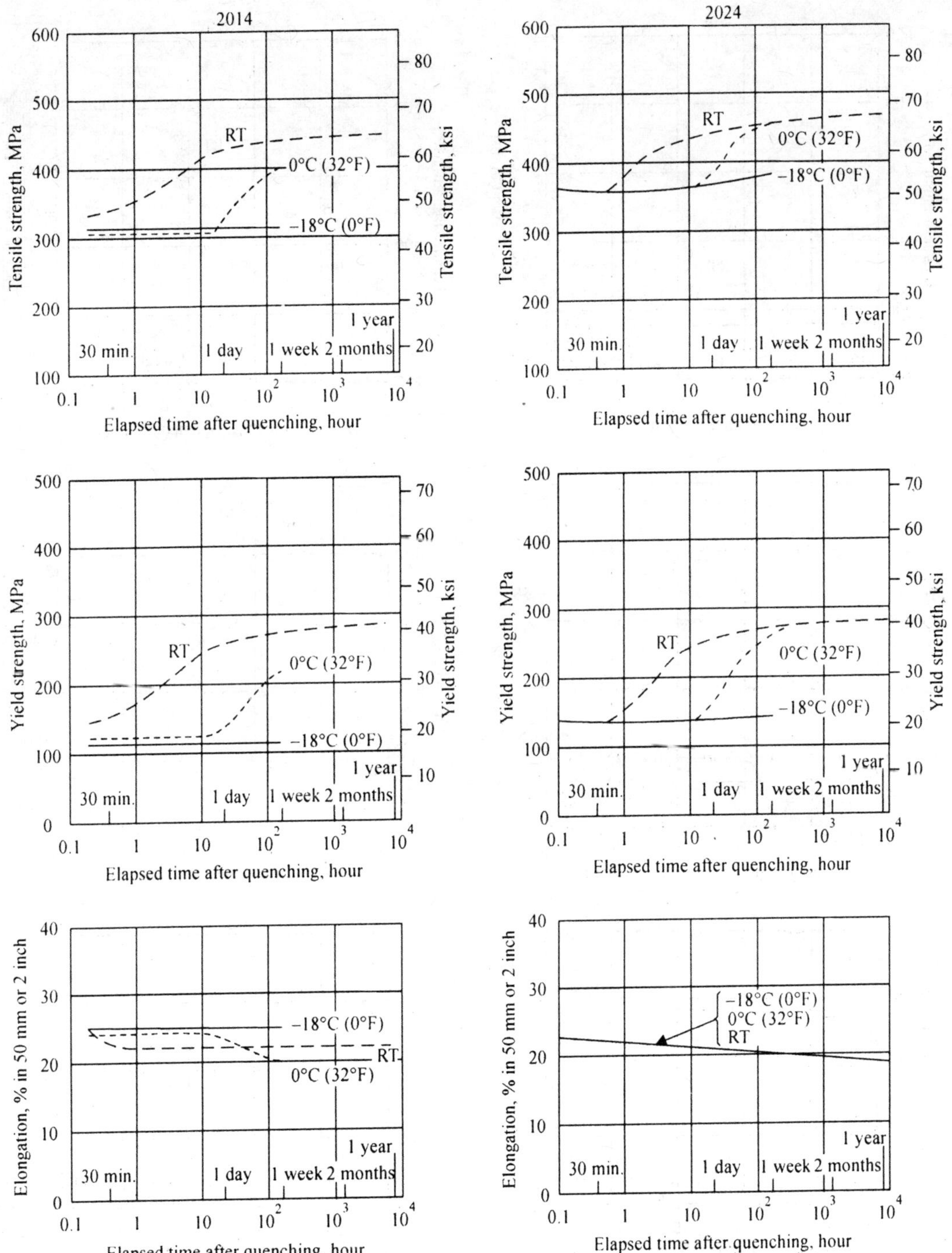

Fig. 20.18. Ageing characteristics of aluminium sheet alloys at room temperature, at 0°C (32°F) and at –18°C (0°F) **(Contd...)**.

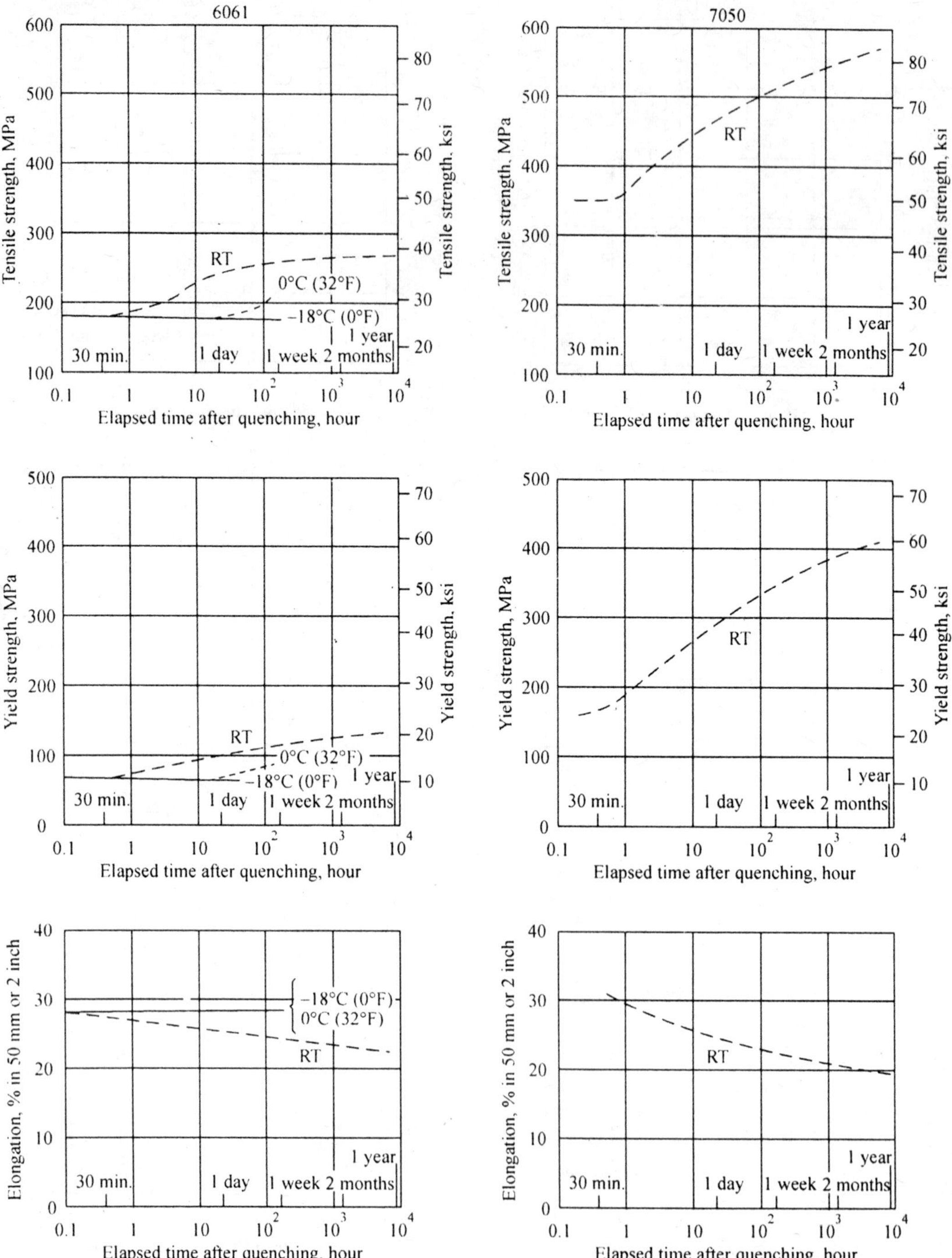

Fig. 20.18. Ageing characteristics of aluminium sheet alloys at room temperature, at 0°C (32°F) and at –18°C (0°F) **(Contd...).**

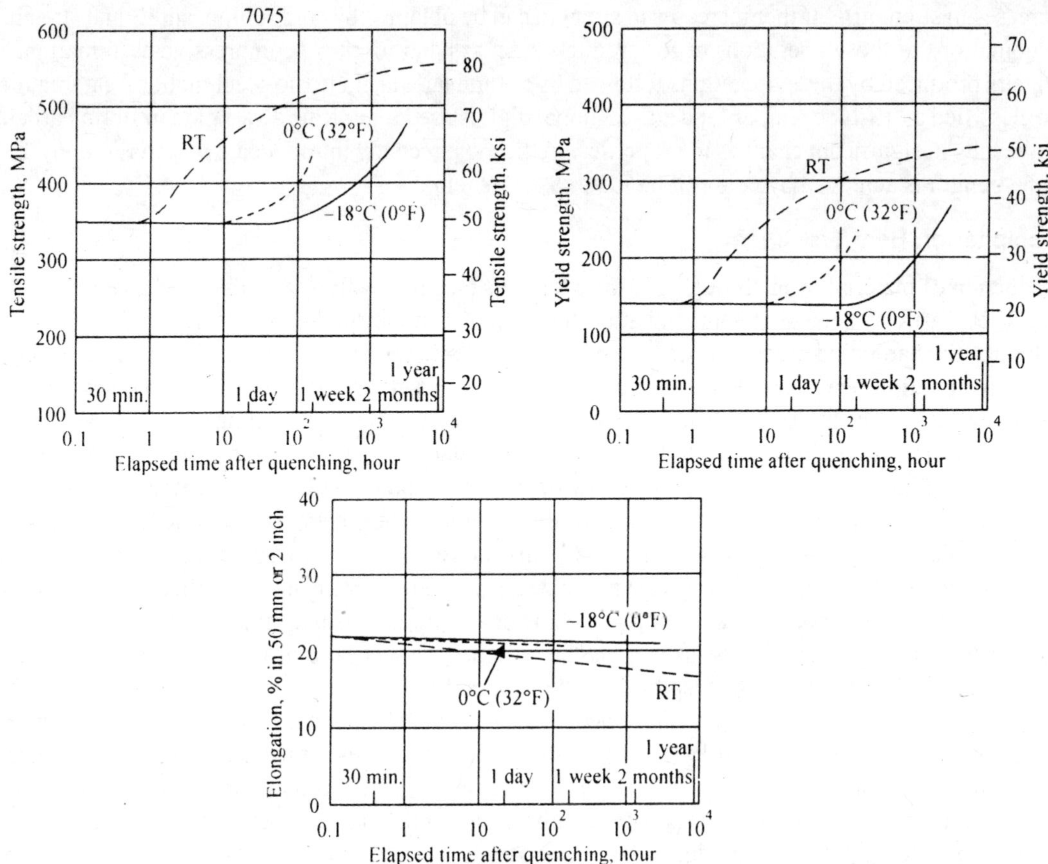

Fig. 20.18. Ageing characteristics of aluminium sheet alloys at room temperature, at 0°C (32°F) and at −18°C (0°F).

It is conventional practice to refrigerate alloy 2024-T4 rivets to maintain good driving characteristics. Full-size wing plates for current-generation jet aircraft have been solution heat treated and quenched at the primary fabricating mill, packed in dry ice in specially designed insulated shipping containers and transported by rail about 2000 miles to the aircraft manufacturer's plant for forming.

Unanticipated difficulties may arise as a result of failure to control refrigerator or part temperature closely enough. If opening of the cold box to insert or remove parts is done too frequently, the cooling capacity of the refrigerator may be exceeded. At times, the rate at which heavy-gage parts can be cooled in a still-air cold box has been found to be insufficient. This problem has been solved in one plant by immersing parts in a solvent at −40°C (−40°F) before placing them in the refrigerator.

Cold Work Strain Hardening

The T3-type tempers are distinguished from T4-type tempers by significant mechanical-property differences resulting from cold work strain hardening associated with certain mechanical operations performed after quenching. Roller or stretcher leveling to achieve flatness or straightness introduces modest strains (on the order of 1 to 4 per cent) that cause changes in mechanical properties (primarily, increases in strength). Further increases in strength can be obtained by cold rolling, additional stretching, combinations of these operations or for products such as hand forgings, compressive deformation. The tempers produced by these operations followed by natural ageing alone (no precipitation heat treatment) are classified as T3-type tempers and an additional digit is used to indicate a variation in strain hardening that results in significant changes in properties. In the most recently introduced 2*xxx* aircraft alloy, 2324, high strength is achieved by cold rolling plate to a T39 temper.

Precipitation Heat Treatment

Production of material in T5 through T10-type tempers (see the section on temper designations near the end of this article) necessitates precipitation heat-treating at elevated temperatures (artificial ageing). Although the hardening precipitate developed by this operation is submicroscopic, structures before and after precipitation heat treatment often can be distinguished by etching metallographic specimens. In aluminium alloys in the solution heat treated and quenched conditions, colouration contrast between grains of differing orientation is relatively high, particularly in 2*xxx* series wrought alloys and 2*xx*.0 series casting alloy. This contrast is noticeably decreased by precipitation heat treatment.

Differences in type, volume fraction, size and distribution of the precipitated particles govern properties as well as the changes observed with time and temperature and these are all affected by the initial state of the structure. The initial structure may vary in wrought products from unrecrystallised to recrystallised and may exhibit only modest strain from quenching or additional strain from cold working after solution heat treatment. These conditions, as well as the time and temperature of precipitation heat treatment, affect the final structure and the resulting mechanical properties.

Because mechanical properties and other characteristics change continuously with time and with temperature, as shown in Fig. 20.19 (a), (b) and (c) by typical curves for three wrought alloys, treatment to produce a combination of properties corresponding to a specific alloy-temper combination requires one or more rather specific and co-ordinated combinations of time and temperature, with both parameters being subject to practical limitations. Recommended commercial treatments often are compromises between (a) time and cost factors; and (b) the probability of obtaining the intended properties, with consideration of allowances for variables such as composition within specified range and temperature variations within the furnace and load. Use of higher temperatures may reduce treatment time; but if the temperature is too high, characteristic features of the precipitation-hardening process reduce the probability of obtaining the required properties.

T6 and T7 Tempers

Precipitation heat treatment following solution heat treatment and quenching produces T6- and T7-type tempers. Alloys in T6-type tempers generally have the highest strengths practical without sacrifice of the other properties and characteristics found by experience to be satisfactory and useful for engineering applications. Alloys in T7 tempers are 'over-aged', which means that some degree of strength has been sacrificed or 'traded off' to improve one or more other characteristics. Strength may be sacrificed to improve dimensional stability, particularly in products intended for service at elevated temperatures, or to lower residual stresses in order to reduce warpage or distortion in machining. T7-type tempers frequently are specified for cast or forged engine parts. Precipitation heat treating temperatures used to produce these tempers generally are higher than those used to produce T6-type tempers in the same alloys.

Two important groups of T7-type tempers—the T73 and T76 types—have been developed for the wrought alloys of the 7*xxx* series, which contain more than about 1.25 per cent copper.

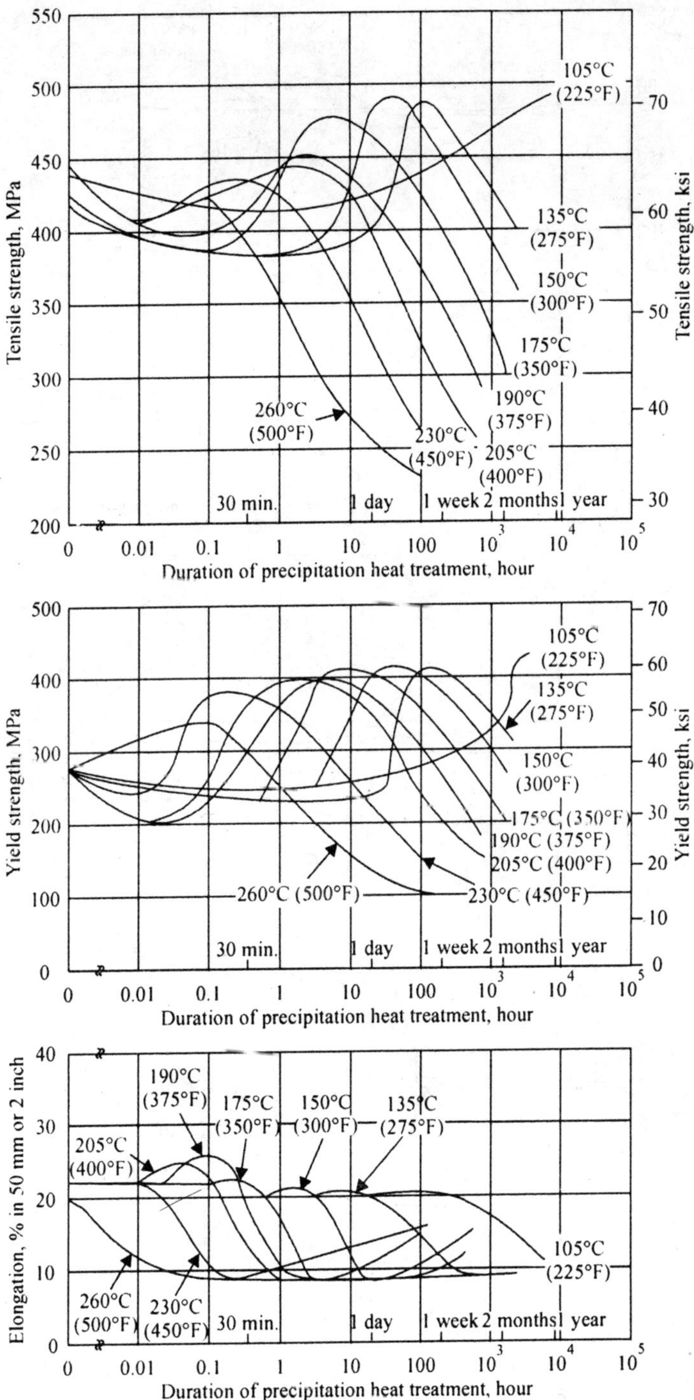

Fig. 20.19a. Ageing characteristics of alloy 2014 sheet.

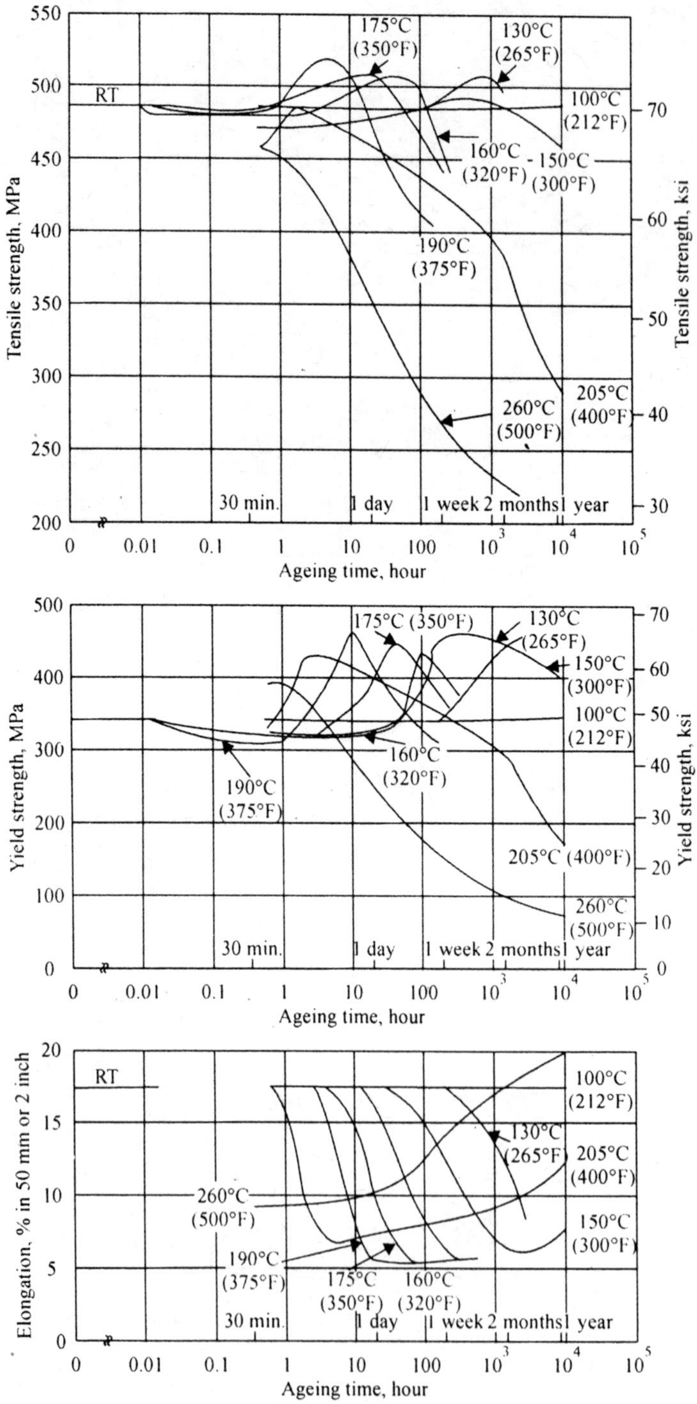

Fig. 20.19b. Ageing characteristics of alloy 2024 sheet.

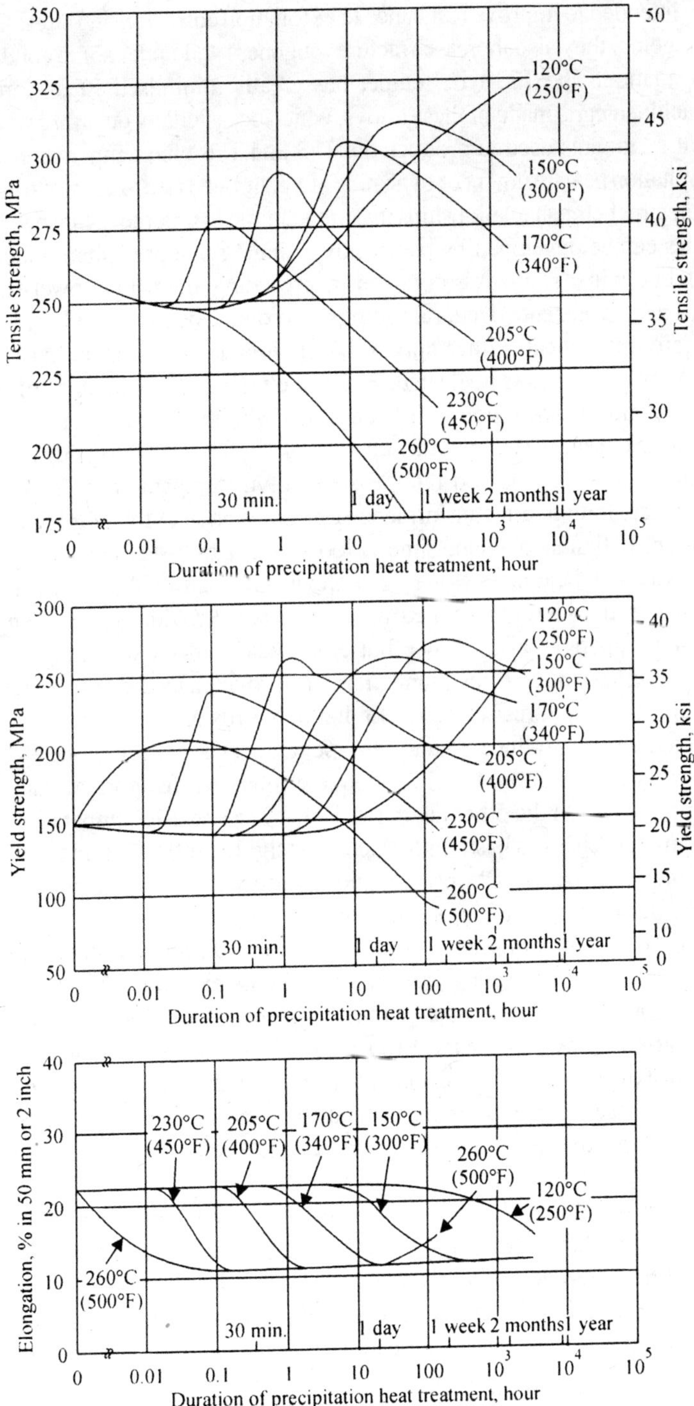

Fig. 20.19c. Ageing characteristics of alloy 6061 sheet.

These tempers are intended to improve resistance to exfoliation corrosion and stress-corrosion cracking, but as a result of overageing, they also increase fracture toughness and under some conditions, reduce rates of fatigue crack propagation. The T73-type temper has greatly minimised stress-corrosion cracking of large and complex machined parts made of these alloys, which occasionally occurred with T6-type tempers. The precipitation heat treatments used to produce the T73 and T76-type tempers consist either of a two-stage isothermal precipitation heat treatment or of heating at a controlled rate to a single treatment temperature. The microstructural/electrochemical relationships that are required in order to achieve the desired corrosion-resisting characteristics can be developed by using only a single-stage precipitation heat treatment above about 150°C (300°F), but higher strength is obtained by preceding this with a lower-temperature stage or with a slow, controlled heatup. Extended natural ageing can provide the same results, but the times required at room temperature are impractical. Either during the preliminary stage or during slow heatup, a fine, high-density dispersion of GP zones is nucleated. Either the time and temperature of the first step or the rate of heating must be controlled to produce GP zones that will not dissolve but will transform to the η' precipitate when heated to the ageing temperature above 150°C (300°F). The ageing practice that produces the results in the shortest time depends on the GP zone solvus temperature. This temperature, in turn, depends on vacancy concentration, a factor influenced by solution heat treating temperature and quench rate and on composition. If first-step ageing time is too short, if first-step ageing temperature is too far below the GP-zone solvus, or if heating rates are too high, the GP zones will dissolve above 150°C (300°F) and the resultant coarse and widely distributed precipitate will provide lower strength. The T76-type treatments have the same operational sequence but employ second-stage heating only long enough to develop a resistance to exfoliation corrosion higher than that provided by the T6-type tempers. Materials in the T73-type temper also have high resistance to exfoliation corrosion.

Recommended treatments to produce T5 and T6-type tempers and those of the T7-type employed for dimensional and property stabilisation, provide adequate tolerance for normal variations encountered with good operating practices. On the other hand, the T73, T736 and T76 tempers for alloys 7049, 7050, 7075, 7175 and 7475 involve changes in strength that occur significantly more rapidly at the temperatures employed in the second stage of the T7x precipitation heat treatment cycle compared to the changes occurring at the temperatures employed to produce the T6 temper.

As illustrated in Fig. 20.20, variations in soak time of several hours and variations in soak temperature of up to 11°C (20°F) from the nominal ageing practice of 24 hours at 120°C (250°F) affect the strength of 7075-T6 by as much as 28 MPa (4 ksi). In contrast, similar variations in second-step soak time and temperature for 7075-T73–that is, variations from 24 hours at 165°C (325°F)–affect strength by up to 150 MPa (22 ksi).

Consequently, control of both temperature and time to achieve the mechanical properties and corrosion resistance specified for these tempers is more critical than the control required in producing the T6 temper. Moreover, rate of heating from the first to the second ageing step must be considered, because precipitation occurs during this period.

Heat treaters attempt to adjust to these new problems by empirically modifying soak times to compensate for precipitation during heating and for effects of soaking at temperatures above or below the nominal. A method has been developed recently, however, that permits quantitative compensation for the effects of precipitation during heating and of soaking either above or below the recommended temperature. For overageing, these effects can be described by the following equation:

$$YS = Y \exp-\left(\frac{t_c}{F_{YS}} + \theta\right) \qquad \text{... (Eq. 20.3)}$$

where YS is yield strength; Y is a term having units of strength that is dependent on alloy, fabrication and test direction; t_c is time at soak temperature; F_{YS} is a temperature-dependent term; and

$$\theta = \frac{dt}{F_{YS}},$$

where t is time during heating.

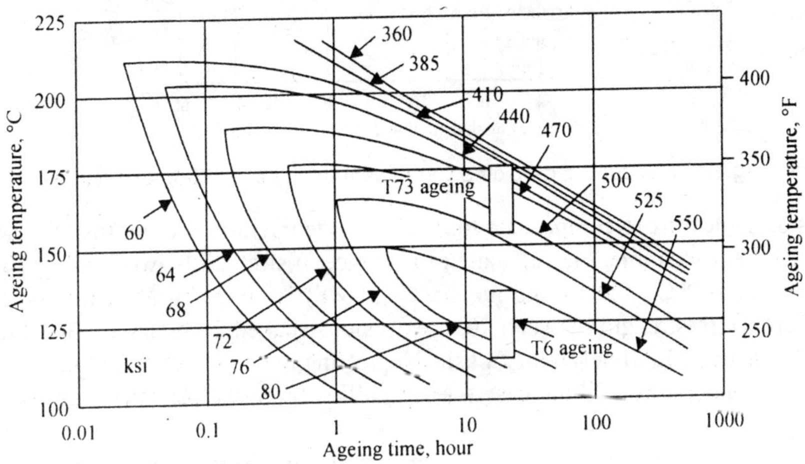

Fig. 20.20. Iso-yield-strength curves for alloy 7075.

Equation 20.3 provides the basis for selection of a nominal ageing time that will result in the desired yield strength and gives the furnace operator a method of compensating for heating rate and for differences between desired and attained soak temperatures.

Specifics will be illustrated using data for alloy 7050. The value of F_{YS} for 7050 can be calculated by use of the following equation:

$$F_{YS} = 1.45 \times 10^{-6} \exp\left(\frac{32562}{T_F + 460}\right) \text{hour} \qquad \text{... (Eq. 20.4)}$$

where T_F is temperature in °F or

$$F_{YS} = 1.45 \times 10^{-16} \exp\left(\frac{18090}{T_K}\right) \text{hour}$$

where TK is temperature in K.

In one experiment, lengths of 7050-W (4 days) extrusions were aged 24 hours at 120°C (250°F) plus the equivalent of 3 to 42 hours at 165°C (325°F). For the second step, a logarithmic heatup was used in which 10 hours were required for the load to reach 155°C (315°F) and nominal soak temperature was 165°C (325°F).

Fig. 20.21 indicates that yield strength generally agreed with values predicted using Equation 20.3. The deviation of the curve for short-transverse strength at the short ageing times indicates that the method is inadequate for predicting strength on the underageing side of the ageing curve.

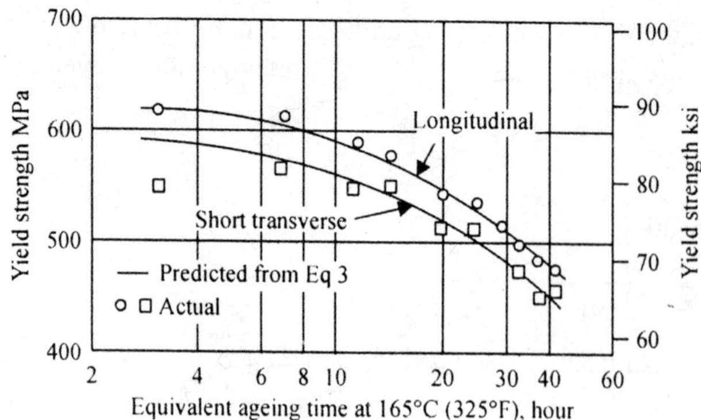

Fig.20.21. Actual vs predicted yield strengths for alloy 7050 extrusions.

The effects of neglecting to compensate for soaking at temperatures other than the nominal can be large (Fig. 20.22). For example, the calculated difference in strength between alloy 7050 extrusions soaked 29 hours at 160°C (320°F) and at 165°C (325°F) is about 50 MPa (about 7 ksi) and the calculated difference in strength between 7050 extrusions soaked 29 hours at 155°C (315°F) and at 170°C (335°F) is about 100 MPa (about 14 ksi). Neglecting to compensate for time spent heating the work to the soak temperature will increase the variability. Strength loss attributed to heatup was 14 MPa (2 ksi).

$$t_{350} = 29/\exp(1.28) = 29/3.6 = 8 \text{ hr.}$$

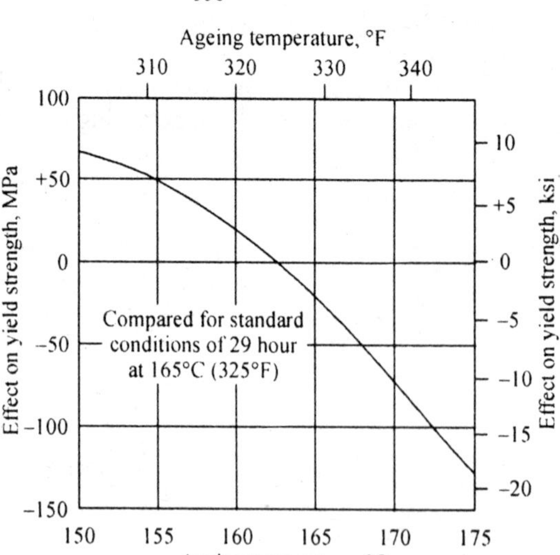

Fig. 20.22. Effect of ageing temperature on yield strength of alloy 7050-T736.

These kinetic relationships also can assist in selection of equivalent ageing times for alternate second-step ageing temperatures. Equations 20.3 and 20.4 can be rearranged to yield the following equations:

$$t_{T_2} = t_{T_1} \exp \frac{32562}{T_1 + 460} - \frac{32562}{T_2 + 460} \qquad \text{... (20.5)}$$

where t_{T_1} is ageing time at temperature T_1, t_{T_2} is ageing time at temperature T_2 that will provide equivalent yield strength and T_1 and T_2 are in °F. For example, the time at 175°C (350°F) equivalent to ageing alloy 7050 for 29 hours at 165°C (325°F) is calculated as follows:

T8-Type Tempers

In some alloys—particularly certain alloys of the 2*xxx* series—strain introduced by cold working after solution heat treatment and quenching nucleates a finer, denser precipitate dispersion that considerably increases strength, as illustrated for alloy 2024 in Fig. 20.23. This effect is the basis for the higher-strength T8-type tempers of alloys 2011, 2024, 2124, 2219 and 2419, which are produced by applying controlled amounts of cold rolling, stretching, or combinations of these operations.

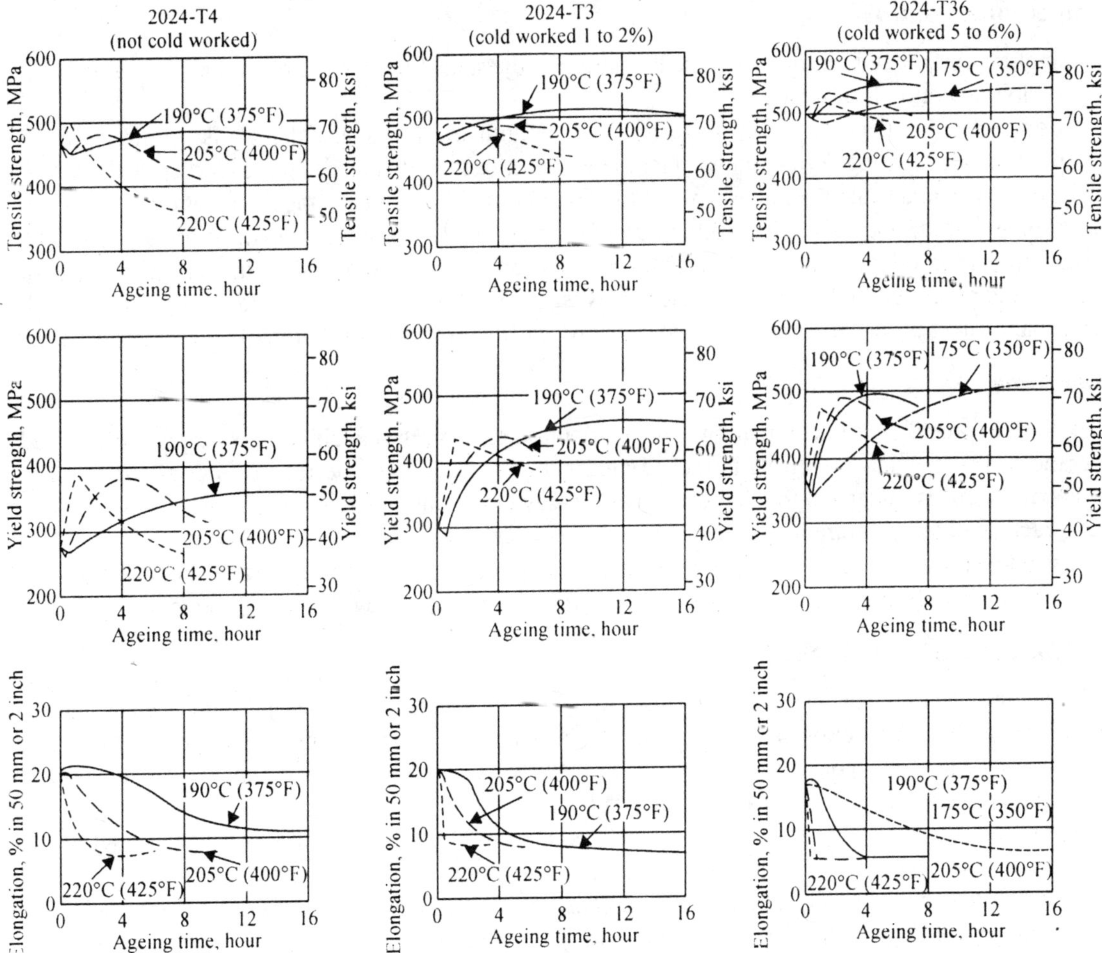

Fig. 20.23. Effects of cold work after quenching and before ageing on tensile properties of alloy 2024 sheet.

Alloys 2024, 2124 and 2219 in T8-type tempers are particularly well suited for supersonic and military aircraft; alloy 2219 in such tempers and alloy 2014-T65, were the principal materials for the fuel and oxidiser tanks (which also served as the primary structure) of the Saturn V space vehicles. Re-solution heat treatment of mill products supplied in these tempers can result in grain growth and in substantially lower strength than is normal for the original temper. Such reheat treatment is not recommended.

Alloys of the 7xxx series do not respond favourably to the sequence of operations used to produce T8-type tempers and no such tempers are standard for these alloys. The strains associated with stretching or compressive stress 999rel of *7xxx* alloys have relatively little effect on the mechanical properties of material precipitation heat treated to T6-type tempers. On the other hand, these operations have measurable detrimental effects on final strength when T73, T736 or T76-type tempers are produced, particularly in the direction opposite the direction of cold work. Accordingly, specification properties are somewhat lower for the stress-relieved versions of these tempers. Decreasing the overageing time to compensate for the loss in strength is not advisable, because this would impair development of the desired corrosion characteristics.

Temperature Control

Temperature control and uniformity present essentially the same problems in precipitation heat treating as they do in solution heat treating.

Good temperature control and uniformity throughout the furnace and load is required for all precipitation heat treating. Recommended temperatures are generally those that are least critical and that can be used with practical time cycles. Except for *7xxx* alloys in T7*x* tempers, these temperatures generally allow some latitude and should have a high probability of meeting property specification requirements. Furnace radiation effects seldom are troublesome except in those few furnaces that are used for both solution and precipitation heat treating. Generally, such situations should be avoided, because the high heat capacity needed for the higher temperatures may be difficult to control at normal ageing temperatures.

Soak Time

Soak time in precipitation heat treating is not difficult to control; the specified times carry rather broad tolerances. Heavier loads with parts racked closer together and even nested, are not abnormal. The principal hazard is undersoaking due to gross excesses in loading practices. Some regions of the load may reach soak temperature long after soak time has been called. Placement of load thermocouples is critical and limiting the size and spacing of a load may be necessary for ageing to the T73 and T76 tempers. As discussed above, soak time is not as critical for peak aged (T6 and T8) tempers.

HARDENING OF CAST ALLOYS

In hardening of cast alloys, selection of specific cycle that will produce the optimum combination of properties in all configurations is not possible. Foundry practice (chills, gating, type of mould) plays an important role in the response of a casting, or a portion of a casting, to heat treatment. Hence, the heat treating practice used for each casting usually is based on preliminary testing.

Premium-quality casting specifications such as MIL-A-21180 can require different strengths and ductility levels in the same castings. Yield strength is largely controlled by the limiting hardening-element level and thus is not controllable inside a given casting. This leaves the heat treater only the option of hardening to the highest specified yield strength. Ductility, however, is controlled for a given yield strength by soundness and microstructural fineness and is thus determined in the foundry and not by the heat treater. Tensile strength is not a fundamental property in this sense, but rather the result of the ductility at a given yield strength.

In general, the principles and procedures for heat treating wrought and cast alloys are similar. The major differences between solution-treating conditions for castings and those for wrought products are found in soak times and quenching media. Solution of the relatively large microconstituents present in castings requires longer soaking periods than those used for wrought products (Table 20.4). When heat treatment of castings must be repeated, solution times become similar to those for wrought products, because the gross solution and homogenisation has been accomplished and is irreversible under normal conditions. Reduction of quenching stresses and distortion dictate that quenching be done in boiling water and sometimes in milder media. A commercially important variety is a mixture of polyalkylene glycol and water, which has no detrimental effect on properties for thicknesses under approximately 3.2 mm (0.125 inch).

Cast products of heat treatable aluminium alloys have the highest combinations of strength, ductility and toughness when produced in T6-type tempers. Developing T6-type tempers in cast products requires the same sequence of operations employed in developing tempers of the same type in wrought products–solution heat treating, quenching and precipitation heat treating.

A comparison of the precipitation-hardening characteristics of sand and permanent mould castings of alloy 356.0-T4 is provided in Fig. 20.24. Because of the finer cast structure and higher supersaturation of the more rapidly solidified permanent mould castings, their tensile properties are superior to those of sand castings of the same composition similarly heat treated.

Among precipitation treatments unique to castings are those resulting in the T5 and T7 tempers. The T5 temper is produced merely by applying a precipitation treatment to the as-cast casting, without previous solution treatment. A moderate increase in strength is achieved without warpage and subsequent straightening. High hardness and dimensional and strength stability at elevated temperatures account for the almost universal use of materials in T5 tempers for pistons and other engine parts. Some applications demand combinations of strength, toughness and dimensional stability that cannot be met by heat treating to T5, T6 or T8-type tempers. For these applications, T7-type tempers are developed by solution heat treating, quenching in a medium that provides a moderate cooling rate and then precipitation heat treating at a temperature higher than those used to develop T5, T6 and T8-type tempers. Heat treating to T7-type tempers results in lower strength than that of material in T6- or T8-type tempers, develops high ductility and toughness and carries precipitation far enough to minimise further precipitation during service.

EFFECTS OF REHEATING

The precipitation characteristics of aluminium alloys must be considered frequently during evaluation of the effects of reheating on mechanical properties and corrosion resistance. Such evaluations are necessary for determining standard practices for manufacturing operations , such as hot forming and straightening, adhesive bonding and paint and dry-film lubricant curing and for evaluating the effects of both short-term and long-term exposure to elevated temperatures in service.

The stage of precipitation that exists in an alloy at the time of reheating plays a significant role in the effects of reheating. Consequently, it is extremely dangerous to reheat material in a solution heat treated temper without first carefully testing the effects of such reheating. In one such test, 2024-T4 sheet was found to be very susceptible to intergranular corrosion when subjected to a 15-minute drying operation at 150°C (300°F) during the first 8 hours after quenching; no susceptibility was evident when the same drying operation was performed more than 16 hr. after quenching. In another test, 7075-W (0.2 to 600 hours) bar and plate were reheated for hot forming at 175°C (350°F) for 20 minute. Strengths after ageing to the T6 temper were 10 to 15 per cent lower than those for standard 7075-T6. In contrast, similar reheating of T6 material for up to 1 hour at 175°C (350°F) produce no detrimental effect.

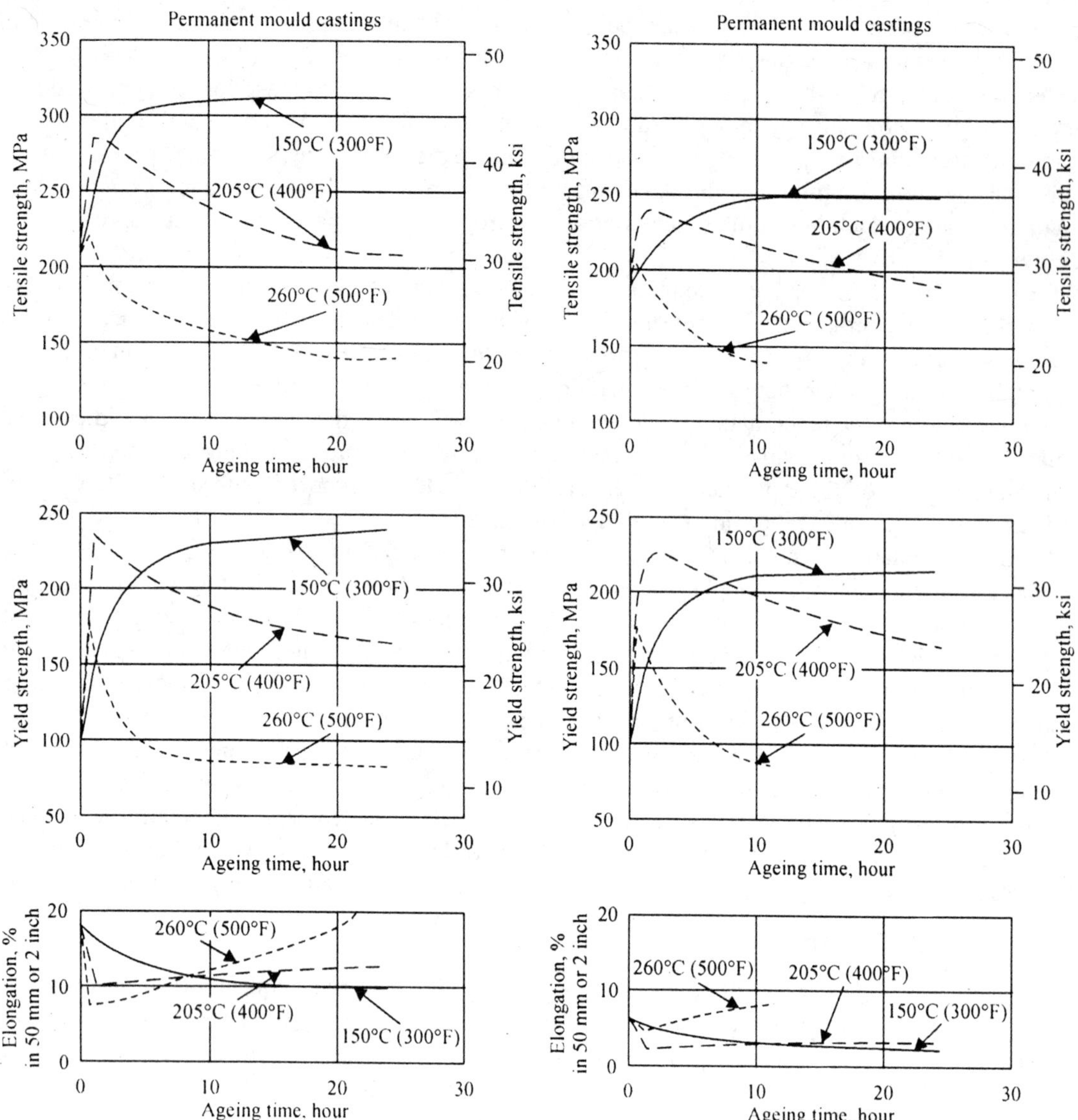

Fig. 20.24. Comparison of the precipitation-hardening characteristics of 356.0-T4 sand and permanent mould castings.

If reheating is performed on material in the W or T4 condition, its effect can be estimated from families of precipitation heat treating curves such as those presented in Fig. 20.19. Such curves can also be used for reheating of precipitation heat treated material at the precipitation heat treating temperature. For reheating at other temperatures, other data may be needed (Fig. 20.25). The heat treating and reheating curves may be used as the bases for limitations on reheating (Table 20.7).

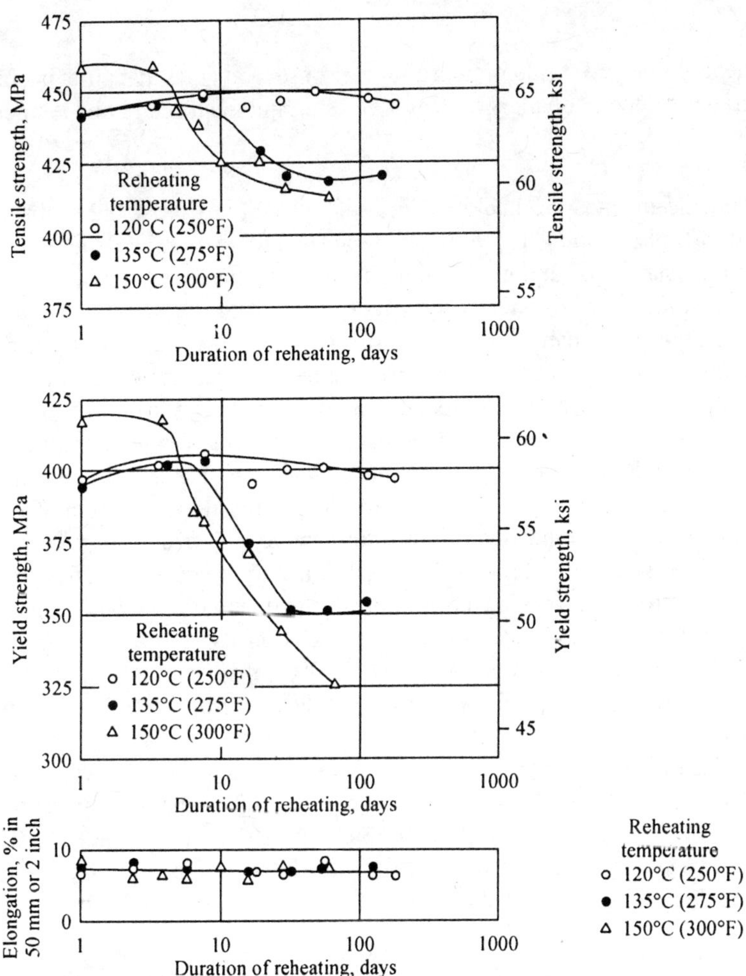

Fig. 20.25. Effects of reheating on tensile properties of alclad 2024-T81 sheet.

Table 20.7. Reheating schedules for wrought aluminium alloys.

The schedules given in this table normally will not decrease strength more than 5 per cent.

	Reheating time at a temperature of:						
Alloy and temper	*150°C (300°F)*	*165°C (325°F)*	*175°C (350°F)*	*190°C (375°F)*	*205°C (400°F)*	*220°C (425°F)*	*230°C (450°F)*
2014-T4	(a)	(a)	(a)	(a)	(a)	(a)	(a)
2014-T6	20–50 hr.	8–10 hr.	2–4 hr.	½–1 hr.	5–15 min.	(b)	(b)
2024-T3, 2024-T4	(a)	(a)	(a)	(a)	(a)	(a)	(a)
2024-T81, 2024-T86	20–40 hr.	–	2–4 hr.	1 hr.	½ hr.	15 min.	5 min.
6061-T6, 6062-T6, 6063-T6	100–200 hr.	50–100 hr.	8–10 hr.	1–2 hr.	½ hr.	15 min.	5 min.
7075-T6, 7178-T6	10–12 hr.	1–2 hr.	1–2 hr.	½–1 hr.	5–10 min.	(b)	(a)

(a) Reheating not recommended (b) Bring to temperature.

ANNEALING

Annealing treatments employed for aluminium alloys are of several types that differ in objective. Annealing times and temperatures depend on alloy type as well as on initial structure and temper.

Full Annealing

The softest, most ductile and most workable condition of both non-heat-treatable and heat treatable wrought alloys is produced by 'full annealing' to the temper designated 'O'. Strain-hardened products in this temper normally become recrystallised, but hot worked products may remain unrecrystallised. In the case of heat treatable alloys, the solutes are sufficiently thoroughly precipitated to prevent natural age hardening. A higher maximum temperature than that used for stress-relief annealing, controlled cooling to a lower temperature and additional holding time at the lower temperature generally are employed.

For both heat treatable and non-heat-treatable aluminium alloys, reduction or elimination of the strengthening effects of cold working is accomplished by heating at a temperature from about 260 to about 440°C (500 to 825°F). The rate of softening is strongly temperature-dependent; the time required to soften a given material by a given amount can vary from hours at low temperatures to seconds at high temperatures. If the purpose of annealing is merely to remove the effects of strain hardening, heating to about 345°C (650°F) will usually suffice. If it is necessary to remove the hardening effects of a heat treatment or of cooling from hot working temperatures, a treatment designed to produce a coarse, widely spaced precipitate is employed. This usually consists of soaking at 415 to 440°C (775 to 825°F) followed by slow cooling (28°C/hr. or 50°F/hr., max) to about 260°C (500°F). The high diffusion rates that exist during soaking and slow cooling permit maximum coalescence of precipitate particles and result in minimum hardness.

As a result of this treatment, only partial precipitation occurs in 7*xxx* alloys and a second treatment (soaking at 230 ±6°C or 450 ±10°F, for 2 hours) is required. When the need arises for small additional improvements in formability, cooling at 28°C/hr. (50°F/hr.) should be extended to 230°C (450°F) and the material should be soaked at 230°C for 6 hours. The effects of eliminating or prolonging the 230°C second step on the ductility of 7075-O sheet are compared with the standard treatment in Table 20.8.

Table 20.8. Effects of annealing treatments on ductility of 7075-O sheet.

Annealing treatment	*Elongation in tension(a), % in 50 mm or 2 inch, for thickness of:*			*Bend angle(b), degrees, for thickness of:*		*Elongation in bending(c), % in 50 mm or 2 inch, for thickness of:*	
	0.5 mm (0.020 inch)	*1.6 mm (0.064 inch)*	*2.6 mm (0.102 inch)*	*1.6 mm (0.064 inch)*	*2.6 mm (0.102 inch)*	*1.6 mm (0.064 inch)*	*2.6 mm (0.102 inch)*
Treatment 1(d)	12	12	12	82	73	48	50
Treatment 2(e)	14	14	14	91	76	58	57
Treatment 3(f)	16	16	–	92.5	84	56	60

(a) Uniform elongation of gridded tension specimens. (b) Bend angle at first fracture. (c) Elongation in bend test for 1.3 mm (0.05 inch) gage spanning fracture. (d) Soak 2 hr. at 415 ± 14°C (775 ± 25°F); furnace cool to 260°C (500°F) at 30°C/hr. (50°F/hr.); air cool. (e) Soak 2 hrs. at 425°C (800°F), air cool; soak 2 hrs. at 230°C (450°F), air cool; (f) Soak 1 hr. at 425°C (800°F); furnace cool to 230°C (450°F) at 30°C/hr. (50°F/hr.); soak 6 hrs. at 230°C (450°F), air cool.

In annealing, it is important to ensure that the proper temperature is reached in all portions of the load; therefore, it is common to specify a soaking period of at least 1 hour. The maximum annealing temperature is moderately critical; it is advisable not to exceed 415°C (775°F), because of oxidation and grain growth. The heating rate can be critical, especially for alloy 3003, which usually requires rapid heating for prevention of grain growth. Relatively slow cooling, in still air or in the furnace, is recommended for all alloys to minimise distortion. Typical annealing conditions used for some alloys in common use are listed in Table 20.9.

Table 20.9. Typical full annealing treatments for some common wrought aluminium alloys.

These treatments, which anneal the material to the 'O' temper, are typical for various sizes and methods of manufacture and may not exactly describe optimum treatments for specific items.

Alloy	*Metal temperature* °C	°F	*Approximate time at temperature, hr.*	*Alloy*	*Metal temperature* °C	°F	*Approximate time at temperature, hr.*
1060	345	650	(a)	5457	345	650	(a)
1100	345	650	(a)	5652	345	650	(a)
1350	345	650	(a)	6005	415(b)	775(b)	2–3
2014	415(b)	775(b)	2–3	6009	415(b)	775(b)	2–3
2017	415(b)	775(b)	2–3	6010	415(b)	775(b)	2–3
2024	415(b)	775(b)	2–3	6053	415(b)	775(b)	2–3
2036	385(b)	725(b)	2–3	6061	415(b)	775(b)	2–3
2117	415(b)	775(b)	2–3	6063	415(b)	775(b)	2–3
2124	415(b)	775(b)	2–3	6066	415(b)	775(b)	2–3
2219	415(b)	775(b)	2–3	7001	415(c)	775(c)	2–3
3003	415	775	(a)	7005	345(d)	650(d)	2–3
3004	345	650	(a)	7049	415(c)	775(c)	2–3
3105	345	650	(a)	7050	415(c)	775(c)	2–3
5005	345	650	(a)	7075	415(c)	775(c)	2–3
5050	345	650	(a)	7079	415(c)	775(c)	2–3
5052	345	650	(a)	7178	415(c)	775(c)	2–3
5056	345	650	(a)	7475	415(c)	775(c)	2–3
5083	345	650	(a)	Brazing sheet			
5086	345	650	(a)	No.11			
5154	345	650	(a)	and 12	345	650	(a)
5182	345	650	(a)	No. 21			
5254	345	650	(a)	and 22	345	650	(a)
5454	345	650	(a)	No. 23			
5456	345	650	(a)	and 24	345	650	(a)

(a) Time in the furnace need not be longer than necessary to bring all parts of the load to annealing temperature. Cooling rate is unimportant. (b) These treatments are intended to remove the effects of solution treatment and include cooling at a rate of about 30°C/hr. (50°F/hr.) from the annealing temperature to 260°C (500°F). Rate of subsequent cooling is unimportant. Treatment at 345°C (650°F), followed by uncontrolled cooling, may be used to remove the effects of cold work, or to partly remove the effects of heat treatment. (c) These treatments are intended to remove the effects of solution treatment and include cooling at an uncontrolled rate to 205°C (400°F) or less, followed by reheating to 230°C (450°F) for 4 hr. Treatment at 345°C (650°F), followed by uncontrolled cooling, may be used to remove the effects of cold work or to partly remove the effects of heat treatment. (d) Cooling rate to 205°C(400°F) or below is less than or equal to 30°C/hr. (50°F/hr.).

Products that can be heated and cooled very rapidly, such as wire, are annealed by continuous processes that require a total heating and cooling time of only a few seconds. Continuous annealing of coiled sheet is accomplished in a total time of a few minutes. For these extremely rapid operations, maximum temperature may exceed 440°C (825°F).

Although material annealed from the precipitation-hardened condition usually has sufficient ductility for most forming operations, this ductility often is slightly lower than that of material that has not been subjected to prior heat treatment–that is, material annealed at the producing source. Therefore, when maximum ductility is required, annealing of a previously heat treated product is sometimes unsuccessful.

Partial Annealing

Annealing of cold worked non-heat-treatable wrought alloys to obtain intermediate mechanical properties (H2-type tempers) is referred to as 'partial annealing' or 'recovery annealing'. Temperatures used are below those that produce extensive recrystallisation and incomplete softening is accomplished by substructural changes in dislocation density and rearrangement into cellular patterns (polygonisation). Bendability and formability of an alloy annealed to an H2-type temper generally are significantly higher than those of the same alloy in which an equal strength level is developed by a final cold working operation (H1-type temper). Treatments to produce H2-type tempers require close control of temperature to achieve uniform and consistent mechanical properties.

Fig. 20.26. shows changes in yield strength as functions of temperature and time for sheet of two non-heat-treatable alloys (1100 and 5052) initially in the highly cold worked condition (H18 temper). From these curves, it is apparent that, by selection of appropriate combinations of time and temperature, mechanical properties intermediate to those of cold worked and fully annealed material can be obtained. It is also evident that yield strength depends much more strongly on temperature than on time of heating.

Stress-Relief Annealing

For cold worked wrought alloys, annealing merely to remove the effects of strain hardening is referred to as 'stress-relief annealing'. Such treatments employ temperatures up to about 345°C (650°F), or up to 400 ± 8°C (750 ± 15°F) for 3003 alloy and cooling to room temperature. No appreciable holding time is required. Such treatment may result in simple recovery, partial recrystallisation or full recrystallisation. Age hardening may follow stress-relief annealing of heat treatable alloys, however, because a concentration of soluble alloying elements sufficient to cause natural ageing remains in solid solution after such treatments.

A special form of stress-relief temper is used for heat treatable alloy products that subsequently will be inspected ultrasonically. The product is heated to its normal solution heat treating temperature, then cooled in still air to room temperature. This temper is referred to as the O1 temper.

Controlled-Atmosphere Annealing and Stabilising

Aluminium alloys that contain even very small amounts of magnesium will form a surface magnesium oxide unless the atmosphere in the annealing furnace is free of moisture and oxygen. Examples include alloy 3004, which is used for cooking utensils and alloys of the 5xxx series.

Another problem that control of the annealing atmosphere helps to overcome or avoid is 'oil staining' by oil-base roll lubricants that do not burn off at lower annealing temperatures. If the oxygen content of the furnace atmosphere is kept very low during such annealing, the oil will not oxidise and stain the work.

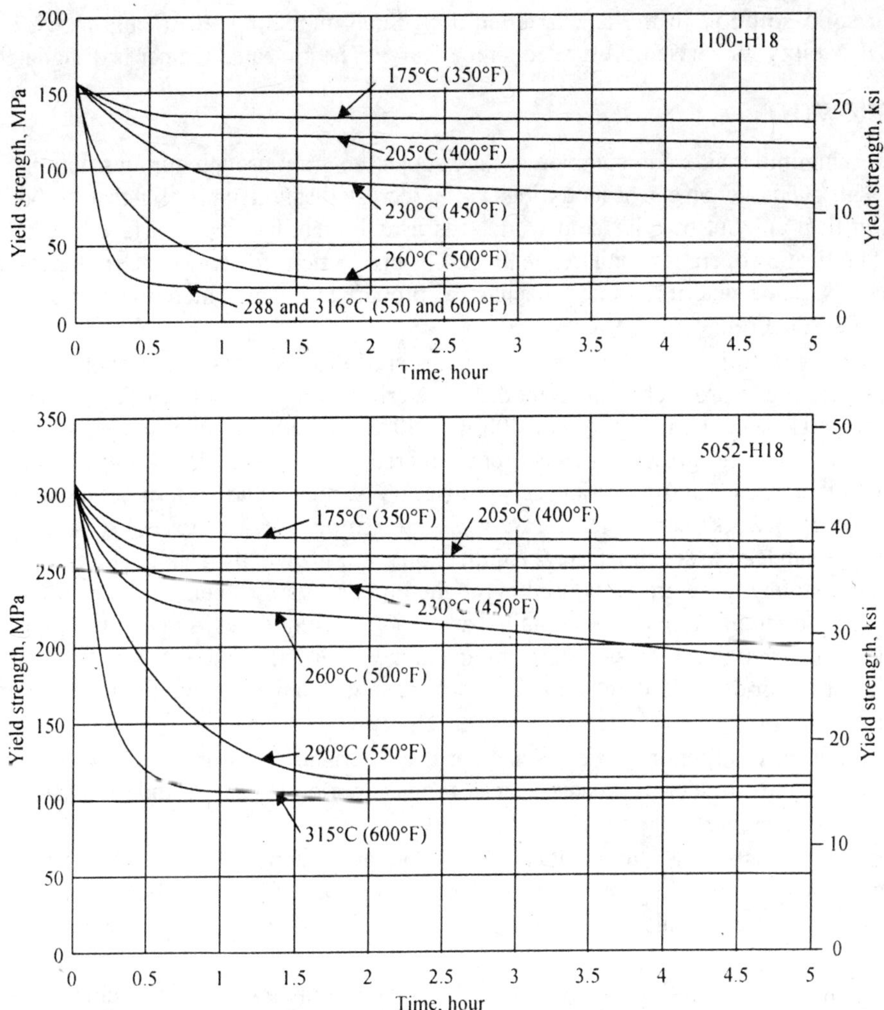

Fig. 20.26. Representative isothermal annealing curves for alloys 1100-H18 and 5052-H18.

Temperature Control

Temperature control for full and partial annealing is somewhat more critical than for stress-relief annealing; the temperatures and times specified are selected to produce recrystallisation and in the case of heat treatable alloys, a precipitate of maximum size; for this the cooling rate must be closely controlled. Even allowing the load to cool in the furnace may result in an excessively high rate. Similarly, lowering the furnace-control instrument by 28°C (50°F) each hour may produce stepped cooling, which is not satisfactory for severe forming operations. For maximum softening, a continuous cooling rate of not more than 28°C/hr. is recommended.

Annealing of Castings

Annealing of castings for 2 to 4 hours at temperatures from 315 to 345°C (600 to 650°F) provides the most complete relief of residual stresses and precipitation of the phases formed by the excess solute retained in solid solution in the as-cast condition. Such annealing treatments provide maximum dimensional stability for service at elevated temperatures. The annealed temper is designated 'O'.

GRAIN GROWTH

Many of the aluminium alloys in common use are subject to grain growth during solution treatment or annealing. This phenomenon can occur during or after recrystallisation of material that has been subjected to a small critical amount of prior cold work. It is usually manifested by surface roughening during subsequent fabrication operations and frequently results in rejections for appearance or functional reasons. Less frequently, some deterioration of mechanical properties is encountered and this is undesirable regardless of surface-roughening effects.

Degree of susceptibility to grain growth varies with alloy, structure and chemical-composition variation and from one product form to another. The critical range of cold work is ordinarily about 5 to 15 per cent. Usually, temperatures of 400°C (750°F) and above must be reached before grain growth occurs, but some growth has been encountered at temperatures as low as 345°C (650°F). Grain growth that occurs during initial recrystallisation is more a function of composition, structure and degree of cold work than of temperature *per se*; temperatures in excess of 455°C (850°F) in common alloys can lead to secondary-recrystallisation grain-growth problems. The common symptom indicating moderately large-grain material is roughening or 'orange peel' on the external surfaces of bends. Severe growth of grains to fingernail size and larger sometimes is evident in parts made from annealed (O temper) material by stretch forming and then thermal treating or similar operations. This type of grain growth often is detected during subsequent anodising, etching and chemical milling operations.

Cracking during welding or brazing is another characteristic which may indicate that severe grain growth has occurred. In such instances, cracks propagate along grain boundaries that provide little obstruction to their progress.

If the surface roughening is objectionable from either an appearance or a functional aspect, the desirability of surface-smoothing operations, such as sanding or buffing, must be evaluated. If reductions in mechanical properties are suspected, these must be established by test and evaluated in relation to the anticipated service.

In one application, a part that had been made by stretch forming O-temper 2-mm (0.080-inch) sheet and heat treating exhibited significantly lower tensile and yield strengths in portions where severe grain growth had occurred than in portions having normal grain size:

		Tensile strength		*Yield strength*	
Test	*Grain structure*	*MPa*	*ksi*	*MPa*	*ksi*
Transverse					
1	Coarse	265	38.5	247	35.8
2	Coarse	263	38.2	241	35.0
3	Fine	311	45.1	261	37.8
Longitudinal					
1	Coarse	259	37.6	243	35.3
2	Coarse	269	39.0	245	35.6
3	Fine	305	44.2	270	39.1

In other similar investigations, no detrimental effects have been discovered and in many cases such parts have served satisfactorily in critical applications.

When a grain-growth problem is discovered, it is too late to change the condition of the parts in question, but several possible methods are available for preventing recurrence of the difficulty. The simplest of these is relieving the causative stress by interjecting a stress-relief anneal into the manufacturing sequence immediately prior to the solution-treating or full-annealing cycle in which the grain growth occurred. This approach is usually successful and practical. Another possibility is to adjust the amount of stress present in the part immediately prior to the critical heat treatment so that the stress level is outside the critical range. This may be done by adding a cold working operation before forming, such as prestretching of blanks, or by forming in multiple stages with a stress-relief anneal before each stage.

A third method that is sometimes successful consists of increasing the heating rate during the critical heat treatment by reducing the size of furnace loads or by changing from an air furnace to a salt bath. In one application, severe grain growth was found during bending of alloy 1100 rectangular tubing. The roughening of the inside surfaces of the parts, which occurred during forming of the large-grain material, impaired their functioning as radar waveguides. Investigation disclosed that, to minimise handling marks, the material was procured in the strain -hardened (H14) temper and was stress-relief annealed at 345°C (650°F) immediately prior to forming. Grain growth occurred during annealing as a result of the moderate amount of cold work introduced at the mill. The problem was eliminated by changing the stress-relieving operation to a 5 minute heating period in an air furnace operating at 540°C (1000°F). The explanation advanced for the success of this treatment was that, due to the rapid heating rate, the temperature of the material was raised through the recrystallisation range for the less severely cold worked grains before the critically cold worked grains had time to grow appreciably.

FURNACE EQUIPMENT AND ACCESSORIES

Both molten salt baths and air-chamber furnaces are suitable for solution heat treating of aluminium alloys. The choice of furnace equipment depends largely on the alloy and the configuration of the parts to be processed. Both heating media have advantages and disadvantages. Oil- and gas-fired furnaces, in designs that allow the products of combustion to come in contact with the work, are usually unsatisfactory because they promote high-temperature oxidation.

Salt Baths

Salt baths heat the work faster (Table 20.3) than air furnaces, provided that the amount of work introduced at any one time is controlled to prevent the temperature of the bath from falling below the desired range. If the temperature is permitted to fall below the minimum limit, much of the advantage of the salt bath is lost, because of the necessity for reheating the large mass of salt.

Salt baths are also more readily adapted to the introduction, at any time, of small amounts of work requiring different soaking periods. (Economical utilisation of air furnaces usually dictates accumulation of a large load of parts of similar thickness before charging). Also, the buoyant effect of the salt reduces distortion during heating and the large reservoir of heat facilitates temperature control and uniformity.

Salt bath operation entails special housekeeping requirements. Dragout is costly and unsightly. Because residual salt on parts may result in corrosion, all salt must be completely removed, including that from crevices and blind holes. In addition, salt residue from the quench water must be kept to a minimum by a constant water overflow or by providing a fresh-water rinse for all parts after quenching. When these provisions are impractical, corrosion can be inhibited by adding 14 g (½ oz) of sodium or potassium dichromate to each 45 kg (100 lb) of the molten salt.

Precautions

Molten salt baths are potentially hazardous and require special precautions. Operators must be protected from splashing and dripping of the hot salt. Because heated nitrates are powerful oxidising agents, they must never be allowed to come in contact with combustibles and reducing agents, such as magnesium and cyanides. Most authorities advise against inserting aluminium alloys containing more than a few per cent of magnesium into molten nitrate. To avoid exposure of personnel to nitrous fumes produced during decomposition of nitrates, good ventilation is essential.

When molten nitrates are being used, the possibilities of explosions resulting from both physical and chemical reactions must be avoided. The former result from rapid expansion of gases entrapped beneath the surface of the bath. Hence, parts entering the bath must be clean and dry; they must also be free of pockets or cavities that contain air or other gases. Chemical-reaction explosions result from rapid breakdown of the nitrates due to overheating or reaction with the pot material. Stainless steel pots (preferably of type 321 or 347) are more resistant to scaling than those made of carbon steel or cast iron and therefore present a lower probability of local overheating. Sludge or sediment accumulations in bottom-heated pots can also lead to local overheating. Overheat controls are essential to ensure against temperatures exceeding 595°C (1100°F).

It is vitally important that water be kept away from a nitrate tank. In controlling a nitrate fire, do not use water or any fire extinguisher containing water. The best extinguisher is dry sand, a supply of which should be kept near the tank.

Extra sacks of salt should be stored in a dry place, distant from the tank. If the fresh salt being added to the bath is even slightly damp, it should be added very slowly or when the bath is frozen.

Air Furnaces

Air furnaces are used more widely than salt baths because they permit greater flexibility in operation temperature. When production schedules and the variety of alloys requiring heat treatment necessitate frequent changes in temperature, the time and cost of adjusting the temperature of a large mass of salt makes the use of an air furnace almost mandatory. However, waiting periods are often required to allow the walls of air furnaces to stabilise at the new temperature before parts are introduced. Otherwise, parts may radiate heat to colder walls or absorb radiant heat from hotter walls and the temperature indicated by the control instrument will not reflect actual metal temperature in the usual manner. Air furnaces are also more economical when the product mix includes a few rather large parts; holding the temperature of a large volume of salt in readiness for an occasional large part is far more expensive than heating an equal volume air.

TEMPERATURE CONTROL

The importance of close temperature control in solution treating has been noted in a previous section of this chapter.

Each control zone of each furnace should contain at least two thermocouples. One thermocouple, with its instrument, should act as a controller, regulating the heat input; the other should act independently as a safety cutoff, requiring manual reset if its set temperature (usually the maximum of the specified range) is exceeded during the solution-treating cycle.

Safety cutoffs are mandatory for salt baths, to guard against explosions and often have paid for themselves in air furnaces by saving a load of parts or even the furnace itself. It is important, however, that they be tested periodically (by deliberately overshooting the empty furnace) to guard against 'frozen' or corroded contacts resulting from prolonged periods of idleness.

At least one of the instruments for each zone should be of the recording type and both instruments should have restricted scales–for instance, 400 to 600°C, rather than 0 to 600°C. This is required for maximum accuracy because manufacturers' guarantees are specified in terms of per cent of scale.

In the placement of instruments, exposure to extremes in ambient temperature, humidity, vibration, dust and corrosive fumes should be avoided. Ambient temperatures between 5 to 50°C (40 to 120°F) are satisfactory, but temperature changes of 6°C/hr. (10°F/hr.) or more should be avoided. It is also essential that instruments and thermocouple circuits be shielded from electromagnetic fields commonly associated with the leads of high-amperage furnace heating elements.

Temperature-sensing elements must be capable of responding more rapidly to temperature changes than the materials being processed. Therefore, thermocouple wire diameter should not exceed 1½ times the thickness of the minimum-gauge material to be heat treated and should in no case exceed 14 gauge. Thermocouples for salt baths should be enclosed in suitable protection tubes. Air-furnace thermocouples should be installed in open-end protection tubes, with the thermocouple junction extending sufficiently beyond the tube to prevent any loss in sensitivity.

Temperature-sensing elements should be located in the furnace work chamber, not in ducts and plenums and should be as close as possible to the working zone. Specification MIL-H-6088C restricts distance between the sensing element and the working zone to a maximum of 102 mm (4 inch). The safety-cutoff thermocouple should be located to reflect the highest temperature in the working zone. The control thermocouple should be located in a position where it will read a temperature approximately halfway between the hottest and coldest temperatures.

Probe Checks

After the temperature-measurement equipment is properly installed, it must be checked frequently for accuracy. This is accomplished by inserting a calibrated probe thermocouple into the furnace adjacent to each furnace thermocouple and comparing its reading on a calibrated test potentiometer with that indicated by the furnace instrument. Correction factors should be applied to the furnace instruments after each probe check, but if the correction required exceeds 3°C (5°F), the source of the deviation should be corrected. MIL-H-6088C recommends that this check be made weekly, but many operators make the check as frequently as once each shift.

Temperature-Uniformity Surveys

In controlling the temperature of parts that are being heat treated it must first be determined that the temperature indicated by the furnace instruments truly represents the temperature of the nearby air or salt. Second, the uniformity of temperature within the working zone must be shown to be within a range of 11°C or 20°F (6°C or 10°F, for precipitation heat treatment of alloy 2024). This is accomplished by measuring the temperature at several test locations, using calibrated test thermocouples and a calibrated test potentiometer and reading furnace instruments nearly simultaneously. MIL-H-6088 recommends monthly surveys with one test location per 1.1 m^3, or 40 ft^3 (0.7 m^3, or 25 ft^3, for air furnaces on initial survey), but with a minimum of nine test locations distributed as shown in Fig. 20.27. Despite the large size of some furnaces, rather surprising temperature uniformities have been reported. In one instance, the initial survey of an air furnace measuring 12.5 by 1.2 by 3.0 m (41 by 4 by 10 ft) showed maximum temperature variations of +1.7, –1.1°C (+3, –2°F). When a partition 0.3 m (1 ft) thick was lowered, converting the furnace to two chambers 6.1 by 1.2 by 3.0 m (20 by 4 by 10 ft) each, the spread was +1.1, –0.6°C (+2, –1°F) in one section and +0.6, –1.1°C (+1, –2°F) in the other.

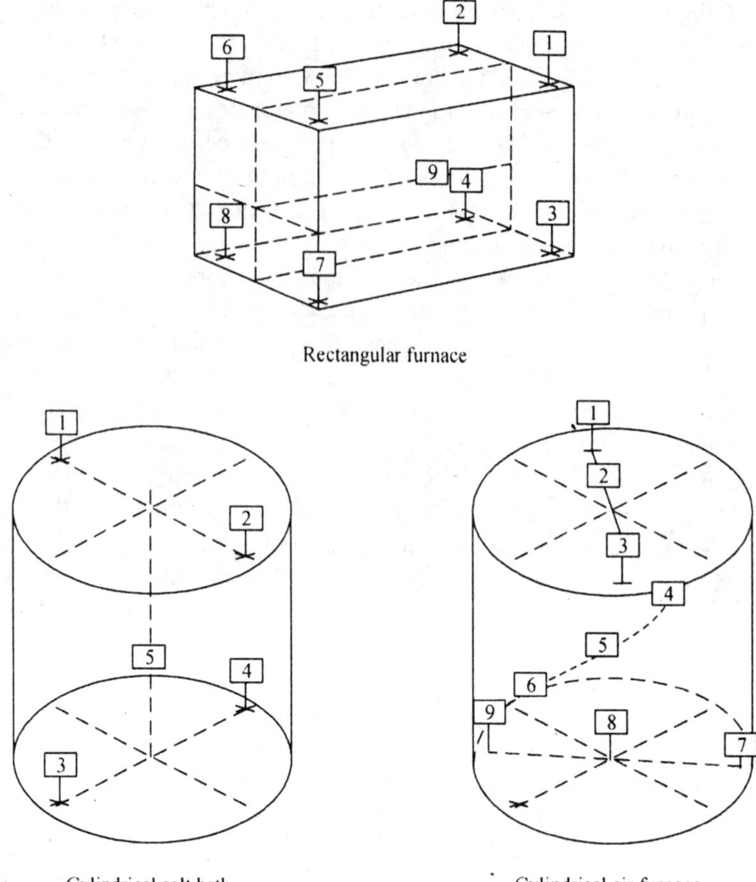

Fig. 20.27. Location of thermocouples for surveying temperature uniformity in the working zones of air furnaces and salt baths.

For each furnace load, one thermocouple (the 'cold' couple) should be placed in the coldest area of the furnace and another (the 'hot' couple) in the hottest area. In addition to these two thermocouples, a load thermocouple should be installed. The load couple should be of approximately the same gauge as the sheet or other product being heat treated. If heavy plate, forgings or castings are being heat treated, a similar discarded item should be used at the controlling load couple. The thermocouple should be placed in a drilled hole and packed to hold it firmly in place during the heat treating cycle. In some instances, the items being heat treated can be used as the load couples. The thermocouples can be placed in holes drilled in areas that will be removed in making the finished article.

It is important that items of different thicknesses–1-mm (0.040-inch) sheet and 25-mm (1-inch) plate, for example–not be heat treated in the same furnace load.

In salt baths, uniformity surveys usually are made by holding a probe thermocouple in each location until thermal equilibrium is reached; in air furnaces, a mock heat treating cycle is required. First, the air furnace is stabilised at the test temperature. Then a rack containing the test thermocouples is inserted into the furnace. By using multiple switches or a multipoint recording instrument, all test thermocouples and furnace instruments are read every 5 minute. As the temperature approaches the test range, it is advisable to increase the frequency of readings to detect possible overshooting. After thermal equilibrium is reached, readings should be continued until the recurrent temperature pattern is established.

Surveys of salt baths generally are considered acceptable whether they are made while the bath is empty or filled with work. It is controversial whether surveys of air furnaces should be made with or without a load. Undoubtedly, recovery overshoots are most likely to occur with a very light load and would not be detected if a heavier load were used. Certainly, if all loads are essentially alike, surveys should be made with typical loads. With widely varying loads, the optimum approach is to make several surveys initially, including one with an empty furnace and then to make succeeding surveyes with an empty furnace to ensure against changes in furnace characteristics. If any changes are made in the furnace that might affect temperature distribution, such as repair of vanes or louvers, several surveys should be repeated.

Another aspect of the problem of temperature control in air furnaces is the necessity of ensuring that the temperature of the parts is the same as that of the surrounding air. Furnace components whose temperature differs from the air temperature must be suitably shielded to prevent radiation to or from the parts being heat treated. In a furnace used for solution heat treating of rivets, unshielded heating elements have been known to produce part temperatures as much as 19°C (35°F) higher than the control temperature, resulting in eutectic melting and cracking. In two other instances, reradiation through inadequate shielding produced a radiation effect of as much as 11°C (20°F). One of these problems was solved by painting the shield with reflective aluminium paint and the other by adding a 12.7-mm- (½-inch-) thick layer of asbestos to the 1.6-mm (1/16- inch) stainless steel shield.

Furnace-wall temperatures that differ appreciably from the temperature of the parts also must be avoided. Consequently, when the operating temperature of an air furnace is changed, waiting periods are required after the furnace instrument indicates stability, to allow the furnace walls to stabilise at the new temperature. The magnitude of this limitation is directly proportional to the efficiency of the furnace as an insulated chamber, but possibilities of such radiation should be recognised even in thin-wall furnaces.

Radiation Effects

Radiation effects are potentially dangerous because they often cannot be detected by ordinary thermocouples. Specially prepared radiation panels with thermocouples attached are used and their readings are compared with adjacent free thermocouples. These panels normally are made of material of the same gauge as the thinnest parts to be heat treated and should have a single surface area of about 650 cm^2 (100 $inch^2$). A thermocouple is attached to the centre of the panel by welding or peening. In order to detect the maximum effect, panel surfaces should be darkened so that their emissivity is at least as high as that of any material to be processed. During the test, the panel surfaces should be parallel to the suspected source or recipient of radiation. As an example of the number of panels required, several aerospace companies specify one panel for every 1.5 linear metres (5 linear feet) of furnace wall.

Instrument Calibration

All instruments and thermocouples must be accurately calibrated and it is essential that the calibrations be traceable directly to the National Bureau of Standards. The chain of traceability should consist of not more than four links for sensing elements and three links for measuring elements. To illustrate, if the article calibrated by the National Bureau of Standards is called a primary standard, then the chain of traceability of measuring elements should consist of primary standard, test potentiometer and furnace instrument. Similarly, the chain for sensing elements should consist of primary standard, secondary standard, test thermocouple and furnace thermocouple. Every effort should be made to ensure that the temperature indicated by the furnace instruments is as close as possible to the actual temperature. To achieve this, it is necessary to apply correction factors obtained during calibration to the next lower echelon of accuracy. Even then, if all errors inherent in the chain are in the same direction, a considerable difference will exist between the measured and actual temperatures. Therefore, it is advisable to operate as close to the mean of the desired range as possible.

DIMENSIONAL CHANGES DURING HEAT TREATMENT

In addition to the completely reversible changes in dimensions that are simple functions of temperature change and are caused by thermal expansion and contraction, dimensional changes of more permanent character are encountered during heat treatment. These changes are of several types, some of mechanical origin and others caused by changes in metallurgical structure. Changes of mechanical origin include those arising from stresses developed by gravitational or other applied forces, from thermally induced stresses or from relaxation of residual stresses. Dimensional changes also accompany recrystallisation, solution and precipitation of alloying elements.

Solution Heat Treatment

Distortions as a result of creep during solution heat treatment should be avoided by proper loading of parts in baskets, racks or fixtures or by provision of adequate support for long pieces of plate, rod, bar and extrusions heat treated in horizontal roller hearth furnaces. Sheet is provided with air-pressure support in continuous heat treating furnaces to avoid scratching, gouging and distortion. If parts are to be solution heat treated in fixtures or racks made of materials (such as steel) with coefficients of thermal expansion lower than that of the aluminium being treated, allowance should be made for this differential expansion to ensure that expansion of the aluminium is not restricted. Straightening immediately after solution heat treating may be preferable to fixturing.

Solution of phases formed by major alloying elements causes volumetric expansion or contraction, depending on the alloy system and this may have to be taken into account in heat treatment of long pieces. For example, solution heat treatment and quenching of alloy 2219 causes lengthwise contraction of about 22 mm/m (0.002 inch/inch). Solution heat treatment and quenching of alloys of the 7xxx series is accompanied by lengthwise expansion—about 0.6 mm/m (0.0006 inch/inch) for alloy 7075 rod or plate.

Quenching

The most troublesome changes in dimensions and shape are those that occur during quenching or that result from stresses induced by quenching. Due to its nonuniform cooling, quenching may produce warpage or distortion, particularly in thin material and in thin sections of parts that contain variations in thickness. For thick-section products or parts, changes in external shape may be small because of rigidity, but the interior-to-surface temperature gradients that form with rapid cooling create residual stresses; these stresses normally are compressive at the surfaces and tensile in the interior.

As previously discussed, warpage or distortion of thin-section material can be reduced by using a quenching medium that provides slower cooling; however, cooling must be sufficient to produce the required properties. Slower quenching can also reduce the magnitude of residual stresses in thicker parts or pieces, as shown in Fig. 20.12 for cylindrical specimens of alloy 6151 quenched in cold or boiling water. Stress range (maximum tensile stress plus maximum compressive stress) for a cylinder with a radius of 89 mm (3.5 inch) is about 205 MPa (30 ksi) when the cylinder is quenched in cold water but less than 70 MPa (10 ksi) when it is quenched in boiling water. The effects of average cooling rate through the temperature range from 400 to 290°C (750 to 550°F) on longitudinal stress ranges developed in alloy 2014 cylinders 75 mm (3 inch) in diameter are shown in Fig. 20.28.

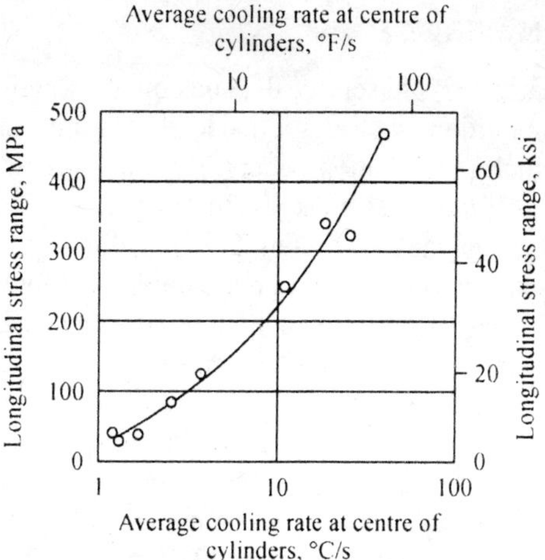

Cylinders were 75 mm (3 inch) in diameter by 230 mm (9 inch) long. Cooling rate was measured from 400 to 290°C (750 to 555°F). Stress range is maximum tensile stress plus maximum compressive stress.

Fig. 20.28. Effect of quenching rate on longitudinal stress ranges in alloy 2014-T4 cylinders quenched in various media.

High stresses induced by rapid quenching generally are reduced only modestly by the precipitation heat treatments used to produce T6- or T8-type tempers. Consequently, for the alloys that require rapid cooling to develop the properties of these tempers, those incorporating mechanical stress relief (Tx51, Tx52) usually are specified when substantial metal must be removed to produce final shapes. Other T8-type tempers, such as T86 and T87, also have low residual stress as a result of the stretching required to produce them.

Heat Treatments for Precipitation and Stabilisation

The most significant dimensional changes associated with precipitation heat treatments and stabilising heat treatments arise from concurrent dilution of the solid solution (which changes lattices parameter) and formation of precipitate. Changes in density and specific volume resulting from these changes in metallurgical structure are the reverse of those caused by solution of the alloy phases. However, because the strongest tempers are those in which the precipitate is present in nonequilibrium transition forms, the amount of change during precipitation heat treatment does not totally compensate for the previous (and opposite) change that occurred during solution heat treatment. Most of the heat treatable alloys expand (grow) during precipitation heat treatment. Exceptions are alloys of the 7*xxx* wrought series and the 7*xx*.0 casting series, which exhibit contraction.

In alloys of the 2*xxx* series, the amount of growth decreases with increasing magnesium content. Thus, growth of about 1.5 mm/m (0.0015 inch/inch) can be expected during precipitation heat treatment of alloy 2219-T87, about 0.5 mm/m (0.005 inch/inch) for treatment of alloy 2014-T6 and less than 0.1 mm/m (0.0001 inch/inch) for treatment of alloy 2024-T851. Alloys 7050 and 7075, on the other hand, contract about 0.3 mm/m (0.0003 inch/inch) on precipitation heat treating from the W temper to the T6 temper and about 0.7 mm/m (0.0007 inch/inch) on treating from the W temper to the T73 temper.

Stabilising T7-type treatments cause greater amounts of growth than the T5, T6 or T8-type treatments for the same alloys. This increased growth is associated either with formation of increased amounts of transition precipitates or with transformation of transition precipitates to equilibrium phases.

DIMENSIONAL STABILITY IN SERVICE

Dimensional stability of heat treated parts in service depends on alloy, temper and service conditions. Of the latter, excluding mechanical conditions such as applied loads, the most important is service temperature range relative to the range in which precipitation occurs. Residual stresses constitute another source of dimensional changes. Stress relief minimises changes due to residual stresses and most mill products usually are supplied in tempers that include stress relief. Potential dimensional change as a result of further precipitation in parts that operate at elevated temperatures is minimised for wrought products by use of T7-type stabilising treatments and for castings by use of T5-type treatments. However, components of high-precision equipment, such as instruments for aerospace guidance systems and optical and telescopic devices, may require special supplementary treatments during manufacture to further reduce stresses or subsequent precipitation. (These treatments are discussed below, under 'Stability of Precision Equipment').

The T3- and T4-type tempers are the least stable dimensionally because of possible precipitation in service. Alloys 2024 and its variants have the smallest dimensional change in ageing; the total change from the quenched to the average state is of the order of 0.06 mm/m (0.00006 inch/inch), less than the change due to a temperature variation of 3°C (5°F). These alloys therefore can be used in the T3- and T4-type tempers, except for precision equipment. For all other alloys, T6- or T8-type tempers should be used, because in these tempers all the alloys have good dimensional stability.

Stability of Precision Equipment

Proper maintenance of high-precision devices, such as gyros, accelerometers and optical systems, requires use of materials in which dimensional changes from metallurgical instability are limited to 10 µm/m (10 µinch/inch). Several laboratory investigations and considerable practical experience have shown that wrought alloys 2024 and 6061 and casting alloy 356.0 are well suited and generally preferred for such applications. Dimensional changes were not greater than 10 µm/m when alloys 2024-T851 and T62, 6061-T651 and T62 and 356.0-T51, T6 and T7 were tested for more than a year at room temperature and for several months at 70°C (160°F) and when the same alloys were tested with repeated thermal cycling between +20 and –70°C (+68 and – 94°F).

Because stresses applied or induced by acceleration in such devices generally are not high, strength levels lower than those of the highest-strength tempers frequently are satisfactory. To increase precision of machining to intended dimensions, as well as to promote maximum stability, it is common practice to apply additional thermal treatments for stress relief and precipitation of 1 to 2 hours at temperatures of 175 to 205°C (350 to 400°F) after rough machining. These additional treatments sometimes are repeated at successive stages of processing and even after final machining. In addition, it has been claimed that one or two cyclic treatments consisting of cooling to –100°C (–150°F), holding for 2 hours, heating to 232 to 240°C (450 to 465°F) and again holding for 2 hours can improve dimensional stability of 356-T6 castings.

QUALITY ASSURANCE

Quality-assurance criteria that heat treated materials must meet always include minimum tensile properties and for certain alloy and tempers, adequate fracture toughness and resistance to detrimental forms of corrosion (such as intergranular or exfoliation attack) or to stress-corrosion cracking. All processing steps through heat treatment must be carefully controlled to ensure high and reliable performance.

Tensile Tests

In general, the relatively constant relationships among various properties allow the use of tensile properties alone as acceptance criteria. The minimum guaranteed strength is ordinarily that value above which it has been statistically predicted with 95 per cent probability that 99 per cent or more of the material will pass. The inherent variability within lots and among specimens from a given piece is shown in Fig. 20.29. Testing provides a check for evidence of conformance; process capability and process control are the foundations for guaranteed values.

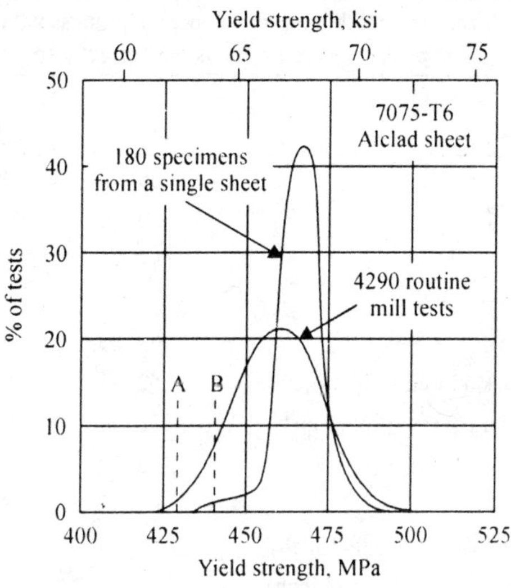

A is 95 per cent probability that not more than 1 per cent of all material will fall below this value; B is 95 per cent probability that not more than 10 per cent of all material will fall below this value. (A and B refer only to curve representing 4290 routine mill tests)

Fig. 20.29. Comparison of distribution of yield strength in heat treated 7075-T6 clad sheet product with distribution in a single sheet.

Published minimum guaranteed values are applicable only to specimens cut from a specific location in the product, with their axes oriented at a specific angle to the direction of working as defined in the applicable procurement specification. In thick plate, for example, the guaranteed values apply to specimens taken from a plane midway between the centre and the surface and with their axes parallel to the width dimension (long transverse). Different properties should be expected in specimens taken from other locations, or in specimens whose axes were parallel to the thickness dimension (short transverse). However, the specified 'referee' locations and orientations do provide a useful basis for lot-to-lot comparisons and constitute a valuable adjunct to other process-control measures.

Tensile tests can be used to evaluate the effects of changes in the process, provided specimens are carefully selected. A variation in process that produces above-minimum properties on test specimens, however, is not necessarily satisfactory. Its acceptability can be judged only by comparing the resulting properties with those developed by the standard process on similarly located specimens. Finally, variations in heat treating procedure are likely to affect the relationships among tensile properties and other mechanical properties. In applications where other properties are more important than tensile properties, the other properties should be checked also.

Hardness Tests

Hardness tests are less valuable for acceptance or rejection of heat treated aluminium alloys than they are for steel. Nevertheless, hardness tests have some utility for process control. Typical hardness values for various alloys and tempers are given in Table 20.10. Fig. 20.30 shows the general relationship between longitudinal tensile strength and hardness for aluminium alloys.

Table 20.10. Typical acceptable hardness values for wrought aluminium alloys.

Acceptable hardness does not guarantee acceptable properties; acceptance should be based on acceptable hardness plus written evidence of compliance with specified heat treating procedures. Hardness values higher than the listed maximums are acceptable provided that the material is positively identified as the correct alloy.

		Hardness			
Alloy and temper	*Product form(a)*	*HRB*	*HRE*	*HRH*	*HR15T*
2014-T3, -T4, -T42	All	65–70	87–95	–	–
5014-T6, -T62, -T65	Sheet(b)	80–90	103–110	–	–
	All others	81–90	104–110	–	–
2014-T61	All	–	100–109	–	–
2024-T3	Not clad(c)	69–83	97–106	111–118	82.5–87.5
	Clad, thru 1.60 mm (0.063 inch)	52–71	91–100	109–116	80–84.5
	Clad, over 1.60 mm (0.063 inch)	52–71	93–102	109–116	–
2024-T36	All	76–90	100–110	–	85–90
2024-T4, -T42(d)	Not clad	69–83	97–106	111–118	82.5–87.5
	Clad, thru 1.60 mm (0.063 inch)	52–71	91–100	109–116	80–84.5
	Clad, over 1.60 mm (0.063 inch)	52–71	93–102	109–116	–
2024-T6, -T62	All	74.5–83.5	99–106	–	84–88
2024–T81	Not clad	74.5–83.5	99–106	–	84–88
	Clad	–	99–106	–	–
2024-T86	All	83–90	105–110	–	87.5–90
6053–T6	All	–	79–87	–	74.5–78.5
6061-T4(d)	Sheet	–	60–75	88–100	64–75
	Extrusions; bar	–	70–81	82–103	67–78
6061-T6	Not clad, 0.41 mm (0.016 inch)	–	–	–	75–84
	Not clad, 0.51 mm (0.020 inch) and over	47–72	85–97	–	78–84
	Clad	–	84–96	–	–
6063-T5	All	–	55–70	89–97	62.5–70
6063-T6	All	–	70–85	–	–
6151-T6	All	–	91–102	–	–
7075-T6, -T65	Not clad(e)	85–94	106–114	–	87.5–92
	Clad:				
	Thru 0.91 mm (0.036 inch)	–	102–110	–	86–90
	Over 0.91 thru 1.27 mm				

(Contd...)

Alloy and temper	Product form(a)	Hardness HRB	HRE	HRH	HR15T
	(over 0.036 thru 0.050 inch)	78–90	104–110	–	–
	Over 1.27 thru 1.57 mm				
	(over 0.050 thru 0.062 inch)	76–90	104–110	–	–
	Over 1.57 thru 1.78 mm				
	(over 0.062 thru 0.070 inch)	76–90	102–110	–	–
	Over 1.78 mm (0.070 inch)	73–90	102–110	–	–
7079-T6, -T65	All(e)	81–93	104–114	–	87.5–92
7178-T6	Not clad(f)	85 min.	105 min.	–	88 min.
	Clad:				
	Thru 0.91 mm (0.036 inch)	–	102 min.	–	86 min.
	Over 0.91 thru 1.57 mm				
	(over 0.036 thru 0.062 inch)	85 min.	–	–	–
	Over 1.57 mm (0.062 inch)	88 min.	–	–	–

(a) Minimum hardness values shown for clad products are valid for thicknesses up to and including 2.31 mm (0.091 inch); for heavier-gauge material; cladding should be locally removed for hardness testing or test should be performed on edge of sheet. (b) 126 to 158-HB (10 mm ball, 500 kg load). (c) 100 to 130 HB (10 mm ball, 500 kg load). (d) Alloys 2024-T4, 2024-T42 and 6061-T4 should not be rejected for low hardness until they have remained at room temperature for at least three days following solution treatment. (e) 136 to 164 HB (10-mm ball, 500-kg load). (f) 136 HB minute (10 mm ball, 500 kg load).

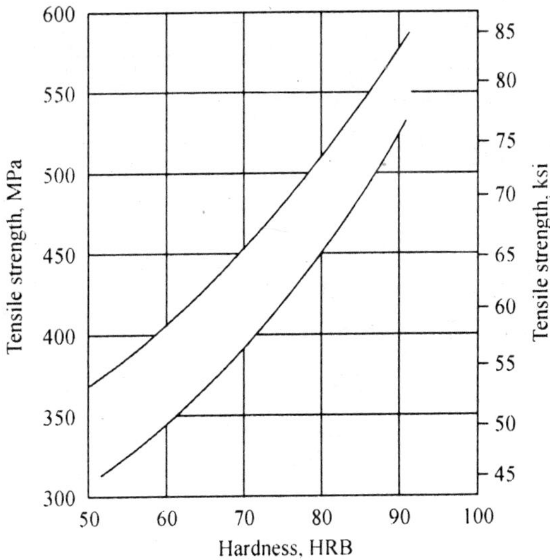

Fig. 20.30. Tensile strength vs hardness for various aluminium alloys and tempers.

Intergranular-Corrosion Test

The most common test for susceptibility to intergranular corrosion is carried out as follows:

1. Use a specimen that has at least 19 cm^2 (3 inch^2) of surface area.

2. Remove any cladding by filling or etching.
3. Clean the specimen by immersing it for 1 minute in a solution containing 5 per cent concentrated nitric acid and 0.5 per cent hydrofluoric acid at a temperature of 95°C (200°F); rinse in distilled water. Immerse for 1 minute in concentrated nitric acid at room temperature; rinse in distilled water.
4. Immerse the specimen for 6 hours in a freshly prepared solution containing 57 gram of sodium chloride and 10 ml of 30 per cent hydrogen peroxide per litre of water at a temperature of 30 ± 5°C (86 ± 9°C). More than one specimen may be corroded in the same container provided that at least 4.6 ml of solution is used for each square centimetre (30 ml/inch^2) of specimen surface and that the specimens are electrically insulated from each other.
5. After the immersion period, wash the specimen with a soft-bristle brush to remove any loose corrosion product. Cut a cross-sectional specimen at least 19 mm ($\frac{3}{4}$ inch) long through the most severely corroded area; mount and metallographically polish this specimen.
6. Examine the cross-sectional specimen microscopically at magnifications of 100× and 500× both before and after etching with Keller's reagent.
7. Describe the results of the microscopic examination in terms of the five degrees of severity of intergranular attack illustrated in Fig. 20.31, as follows: x areas of severity (a), y areas of severity (b), z areas of severity (c) and so on.

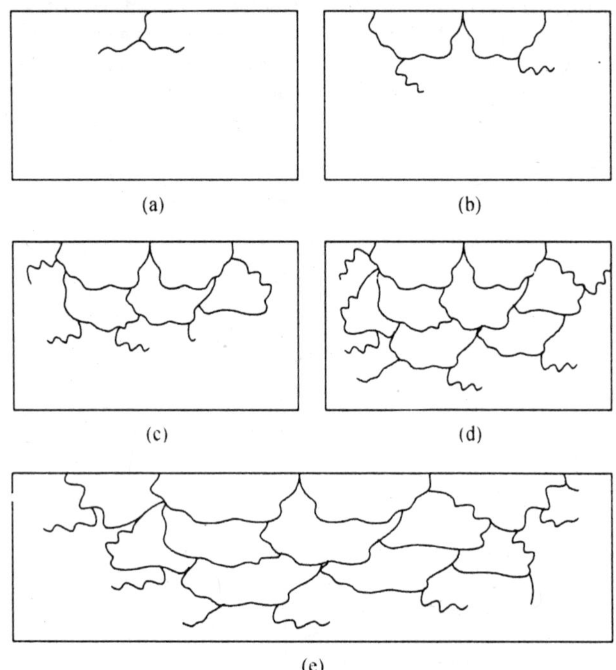

Severity of intergranular attack (schematic), as observed microscopically in transverse sections after test for susceptibility to intergranular corrosion. Top of each area shown in surface exposed to corrosive solution.

Fig. 20.31. Five degrees of severity of intergranular attack.

Electrical Conductivity

For control of the corrosion and stress-corrosion characteristics of certain tempers, notably the T73 and T76 types, the materials must meet combination criteria of yield strength plus electrical conductivity. Although these criteria are based on indirect measurements of properties, their validity for ensuring the intended corrosion and stress-corrosion resistance has been firmly established by extensive correlation and testing.

Low tensile strengths may be accompanied by high levels of electrical conductivity, so electrical conductivity is sometimes used as a quality-assurance diagnostic tool. However, because the correlation between strength and electrical conductivity is strongly a function of chemical composition and fabricating practice, use of electrical conductivity is not recommended except for rough screening. This screening must be followed by hardness testing and then by tensile testing if the hardness tests indicate that the heat treatment was suspect.

Indices of Fracture Toughness

In products of the newer high-strength alloys 2124, 2224, 2324, 7050, 7149, 7150, 7175 and 7475, which were developed to provide high fracture toughness, minimum values of the applicable indices, K_{Ic} or K_c, are being established by accumulation of statistical data from production lots as a basis for guaranteed minimum values. These indices and their attendant values indicate the resistance that a material possesses in regard to crack propagation. Existing specifications for certain products, alloys and tempers contain minimum values of these fracture-toughness indices.

Some specifications allow use of the less-expensive notch tensile test as a basis for release of high-toughness alloy products. In these instances, correlations between K_{Ic} and the ratio of the notch tensile strength and the yield strength of a companion smooth specimen were used to establish the appropriate notch-yield ratio as a lot-release criterion.

TEMPER DESIGNATIONS FOR HEAT TREATABLE ALUMINIUM ALLOYS

The temper designations used in the United States for heat treatable aluminium alloys are part of the system that has been adopted as an American National Standard (ANSI H35.1). Used for all wrought and cast product forms except ingot, the system is based on the sequences of mechanical or thermal treatments, or both, used to produce the various tempers. The temper designation follows the alloy designation and is separated from it by a hyphen. Basic temper designations consist of individual capital letters. Major subdivisions of basic tempers, where required, are indicated by one or more digits following the letter. These digits designate specific sequences of treatments that produce specific combinations of characteristics in the product. Variations in treatment conditions within major subdivisions are identified by additional digits. The conditions during heat treatment (such as time, temperature and quenching rate) used to produce a given temper in one alloy may differ from those employed to produce the same temper in another alloy. Designations for the common heat treated tempers and descriptions of the sequences of operations used to produce those tempers, are given below:

Basic Temper Designations

O *annealed*

Applies to wrought products that are annealed to obtain lowest strength temper and to cast products that are annealed to improve ductility and dimensional stability. The O may be followed by a digit other than zero.

W *Solution heat treated*

An unstable temper applicable to any alloy that naturally ages (spontaneously ages at room temperature) after solution heat treatment. This designation is specific only when the period of natural ageing is indicated–for example, W ½ hour (See also the discussion of the Tx51, Tx52 and Tx54 tempers, in the section below on subdivision of the T temper).

T *Heat treated to produce stable tempers other than O*

Applies to products that are thermally treated, with or without supplementary strain hardening, to produce stable tempers. The T is always followed by one or more digits, as discussed below.

Subdivision of T temper

In T-typedesignations, the T is followed by a number from 1 to 10; each number denotes a specific sequence of basic treatments, as follows:

T1 *Cooled from an elevated-temperature shaping process and naturally aged to a substantially stable condition*

Applies to products that are not cold worked after an elevated-temperature shaping process such as casting or extrusion and for which mechanical properties have been stabilised by room-temperature ageing. If the products are flattened or straightened after cooling from the shaping process, the effects of the cold work imparted by flattening or straightening are not recognised in specified property limits.

T2 *Cooled from an elevated-temperature shaping process, cold worked and naturally aged to a substantially stable condition*

Applies to products that are cold worked specifically to improve strength after cooling from a hot working process such as rolling or extrusion and for which mechanical properties have been stabilised by room-temperature ageing. The effects of cold work, including any cold work imparted by flattening or straightening, are recognised in specified property limits.

T3 *Solution heat treated, cold worked and naturally aged to a substantially stable condition*

Applies to products that are cold worked specifically to improve strength after solution heat treatment and for which mechanical properties have been stabilised by room-temperature ageing. The effects of cold work, including any cold work imparted by flattening or straightening, are recognised in specified property limits.

T4 *Solution heat treated and naturally aged to a substantially stable condition*

Applies to products that are not cold worked after solution heat treatment and for which mechanical properties have been stabilised by room-temperature ageing. If the products are flattened or straightened, the effects of cold work imparted by flattening or straightening are not recognised in specified property limits.

T5 *Cooled from an elevated temperature shaping process and artificially aged*

Applies to products that are not cold worked after an elevated-temperature shaping process such as casting or extrusion and for which mechanical properties or dimensional stability, or both, have been substantially improved by precipitation heat treatment. If the products are flattened or straightened after cooling from the shaping process, the effects of the cold work imparted by flattening or straightening are not recognised in specified property limits.

T6 *Solution heat treated and artificially aged*

Applies to products that are not cold worked after solution heat treatment and for which mechanical properties or dimensional stability, or both, have been substantially improved by precipitation heat treatment. If the products are flattened or straightened, the effects of the cold work imparted by flattening or straightening are not recognised in specified property limits.

T7 *Solution heat treated and stabilised*

Applies to products that have been precipitation heat treated to the extent that they are overaged. Stabilisation heat treatment carries the mechanical properties beyond the point of maximum strength to provide some special characteristic, such as enhanced resistance to stress-corrosion cracking or to exfoliation corrosion.

T8 *Solution heat treated, cold worked and artificially aged*

Applies to products that are cold worked specifically to improve strength after solution heat treatment and for which mechanical properties or dimensional stability, or both, have been substantially improved by precipitation heat treatment. The effects of cold work, including any cold work imparted by flattening or straightening, are recognised in specified property limits.

T9 *Solution heat treated, artificially aged and cold worked*

Applies to products that are cold worked specifically to improve strength after they have been precipitation heat treated.

T10 *Cooled from an elevated-temperature shaping process, cold worked and artificially aged*

Applies to products that are cold worked specifically to improve strength after cooling from a hot working process such as rolling or extrusion and for which mechanical properties or dimensional stability, or both, have been substantially improved by precipitation heat treatment. The effects of cold work, including any cold work imparted by flattening or straightening, are recognised in specified property limits.

When it is desirable to identify a variation of one of the ten major T tempers described above, additional digits, the first of which cannot be zero, may be added to the designation.

The following specific sets of additional digits have been assigned to stress-relieved wrought products:

Tx51 Stress relieved by stretching

Applies to the following products when stretched to the indicated amounts after solution heat treatment or after cooling from an elevated-temperature shaping process:

Product form	*Permanent set, %*
Plate	1½ to 3
Rod, bar, shapes, extruded tube	1 to 3
Drawn tube	½ to 3

Applies directly to plate and to rolled or cold finished rod and bar. These products receive no further straightening after stretching. Applies to extruded rod, bar, shapes and tubing and to drawn tubing, when designated as follows:

Tx510

Products that receive no further straightening after stretching.

Tx511

Products that may receive minor straightening after stretching to comply with standard tolerances.

Tx52

Stress relieved by compressing. Applies to products that are stress relieved by compressing after solution heat treatment, or after cooling from a hot working, or after cooling from a hot working process to produce a permanent set of 1 to 5 per cent.

Tx54

Stress relieved by combining stretching and compressing. Applies to die forgings that are stress relieved by restriking cold in the finish die. (These same digits—and 51, 52 and 54—may be added to the designation W to indicate unstable solution heat treated and stress-relieved tempers.)

The following temper designations have been assigned to wrought products heat treated from the O or the F temper to demonstrate response to heat treatment:

T42 *Solution heat treated from the O or the F temper to demonstrate response to heat treatment and naturally aged to a substantially stable condition*

T62 *Solution heat treated from the O or the F temper to demonstrate response to heat treatment and artificially aged*

Temper designations T42 and T62 also may be applied to wrought products heat treated from any temper by the user when such heat treatment results in the mechanical properties applicable to these tempers.

Subdivision of O Temper

In temper designations for annealed products, a digit following the O indicates special characteristics. For example, O1 denotes that a products has been heat treated according to a time temperature schedule approximately the same as that used for solution heat treatment and then air cooled to room temperature, to accentuate ultrasonic response and provide dimensional stability; this designation applies to products that are to be machined prior to solution heat treatment by the user.

Chapter 21

Heat Treating of Copper Alloys

INTRODUCTION

Heat treating processes that are applied to copper and copper alloys include homogenising, annealing, stress relieving, solution treating, precipitation (age) hardening and quench hardening and tempering.

HOMOGENISING

Homogenising is a process wherein prolonged high-temperature soaking is used to reduce the extent of chemical segregation or coring, which occurs as a natural result of solidification in some alloys. It is applied to copper alloys to improve the hot and cold ductility of cast billets for mill processing and occasionally to castings to meet specified hardness, ductility or toughness requirements.

Homogenisation is required most frequently for alloys with wide freezing ranges, such as tin (phosphor) bronzes, copper, nickels and silicon bronzes. Although coring occurs to some extent in alpha brasses, alpha aluminium bronzes and copper-beryllium alloys, these alloys survive primary mill processing and become homogenised during normal process working and annealing. There is rarely a necessity to apply homogenisation to finished or semi-finished mill products.

The time and temperature required for the process varies with the alloy, the cast grain size and the desired degree of homogenisation. Typical soak times vary from 3 to over 10 hours. Temperatures normally are above the upper annealing range, to within 50°C (90°F) of the solidus temperature.

The normal precautions that apply to annealing should be used for homogenisation of any particular alloy. The furnace atmosphere should be selected so as to control both surface and internal oxidation. Where there is appreciable danger of liquefying segregated phases, the materials, particularly castings, should be well supported and heated slowly through the final 100°C (180°F).

Typical applications of homogenisation are:

1. Alloy C71900 (Cu-Ni-Cr) billets: 1040 to 1065°C (1900 to 1950°F) for 4 to 9 hours, to prevent cracks, seams and excessive wood-fibre structure in extrusions.
2. Alloy C52100 and C52400 (phosphor bronzes, 8 and 10 per cent tin): 775°C (1425°F) for 5 hours, to reduce embrittlement in billets and slabs that are to be cold rolled.

ANNEALING

Annealing is a heat treatment intended to soften and increase the ductility and/or toughness of metals and alloys. It is applied to wrought products during and after mill processing and to castings. The process includes heating, holding and cooling and a proper process description should include heating rate, temperature, time at temperature, atmosphere and cooling rate where each may affect results.

Wrought Products

Annealing of cold worked metal is accomplished by heating to a temperature that produces re-crystallisation and if desirable, by heating beyond the re-crystallisation temperature to produce grain growth. Temperatures commonly used for annealing cold worked coppers and copper alloys are given in Table 21.1.

Table 21.1. Annealing temperatures for cold worked coppers and copper alloys.

Alloy	*Common name*	*Annealing temperature* °C	*Annealing temperature* °F
Wrought coppers			
C10200	Oxygen-free copper	425–650	800–1200
C11000	Electrolytic tough pitch copper	250–650	500–1200
C11300, C11400, C11500, C11600	Silver-bearing tough pitch copper	400–475	750–900
C12000	Phosphorus-deoxidised copper, low residual phosphorus	325–650	600–1200
C12200	Phosphorus-deoxidised copper, high residual phosphorus	375–650	700–1200
C14500	Phosphorus-deoxidised , tellurium-bearing copper	425–650	800–1200
Wrought copper alloys			
C17000, C17200, C17500	Beryllium copper	775–925(a)	1425–1700(a)
C21000	Gilding metal	425–800	800–1450
C22000	Commercial bronze	425–800	800–1450
C22600	Jewelry bronze	425–750	800–1400
C23000	Red brass	425–725	800–1350
C24000	Low brass	425–700	800–1300
C26000	Cartridge brass	425–750	800–1400
C26800, C27000, C27400	Yellow brass	425–700	800–1300
C28000	Muntz metal	425–600	800–1100
C31400	Leaded commercial bronze	425–650	800–1200
C33000, C33500	Low-leaded brass	425–650	800–1200
C33200, C34200, C35300	High-leaded brass	425–650	800–1200
C34000, C35000	Medium-leaded brass	425–650	800–1200
C35600	Extra-high-leaded brass	425–650	800–1200
C36000	Free-cutting brass	425–600	800–1100
C36500, C36600, C36700, C36800	Leaded Muntz metal	425–600	800–1100
C37000	Free-cutting Muntz metal	425–650	800–1200
C37700	Forging brass	425–600	800–1100
C38500	Architectural bronze	425–600	800–1100

(Contd...)

Alloy	*Common name*	*Annealing temperature* °C	°F
C44300, C44400, C44500	Inhibited admiralty	425–600	800–1100
C46200	Naval brass	425–600	800–1100
C48200, C48500	Leaded naval brass	425–600	800–1100
C50500	Phosphor bronze	475–650	900–1200
C51000, C52100, C54200	Phosphor bronze	475–675	900–1250
C53200, C53400, C54400	Free-cutting phosphor bronze	475–675	900–1250
C60600, C60800	Aluminium bronze	550–650	1000–1200
C61000	Aluminium bronze	600–675	1100–1250
C61300, C61400	Aluminium bronze	750–875	1400–1600
C61800, C61900, C62400	Aluminium bronze	600–650(b)	1100–1200(b)
C63000	Aluminium bronze	650–700(c)	1200–1300(c)
C63200	Aluminium bronze	675–725(c)	1250–1350(c)
C64200	Aluminium bronze	Above 650	Above 1200
C65100	Low-silicon bronze	475–675	900–1250
C65500	High-silicon bronze	475–700	900–1300
C67000, C67500	Manganese bronze	425–600	800–1100
C68700	Aluminium brass	425–600	800–1100
C70600	Copper nickel, 10 per cent	600–825	1100–1500
C71500	Copper nickel, 30 per cent	650–825	1200–1500
C75200, C75700, C77000	Nickel silver	600–825	1100–1500

(a) Solution-treating temperature; See Table 21.5 for temperatures for specific alloys. (b) Cool rapidly (cooling method important in determining result of annealing). (c) Air cool (cooling method important in determining result of annealing).

Annealing is primarily a function of metal temperature and time at temperature. Except for multiphase alloys and certain precipitation-hardening alloys or alloys susceptible to fire cracking, rates of heating and cooling are relatively unimportant. Source and application of heat, furnace design, furnace atmosphere and shape of workpiece are important, because they affect finish, cost of annealing and uniformity of results obtained. Because of the multiplicity of influential variables, it is difficult to tabulate a definite annealing schedule that will result in completely re-crystallised metal of specific grain size. The effects of annealing temperature on the tensile strength, elongation and grain size of hard drawn (63 per cent) C27000 (yellow brass) wire annealed for 1 hour and the effect of annealing time on the grain size of C27000 strip are illustrated in Fig. 21.1.

The annealing response of alloy C26000 (cartridge brass) strip after a reduction of 40.6 per cent by cold rolling is shown in Fig. 21.2. Time at temperature was 1 hour. The actual increases in hardness and tensile properties shown at temperatures below the re-crystallisation range are typical of alloys such as brasses, nickel silvers, phosphor bronzes and alpha aluminium bronzes. Depending on the individual alloy, these increases are attributable to phenomenon of the strain-ageing and/or lattice-ordering type.

An increased amount of cold work prior to annealing lowers the re-crystallisation temperature. The lower the degree of prior deformation, the larger the grain size after annealing. For fixed temperature and duration of annealing, the larger the original grain size before working, the larger the grain size after re-crystallisation.

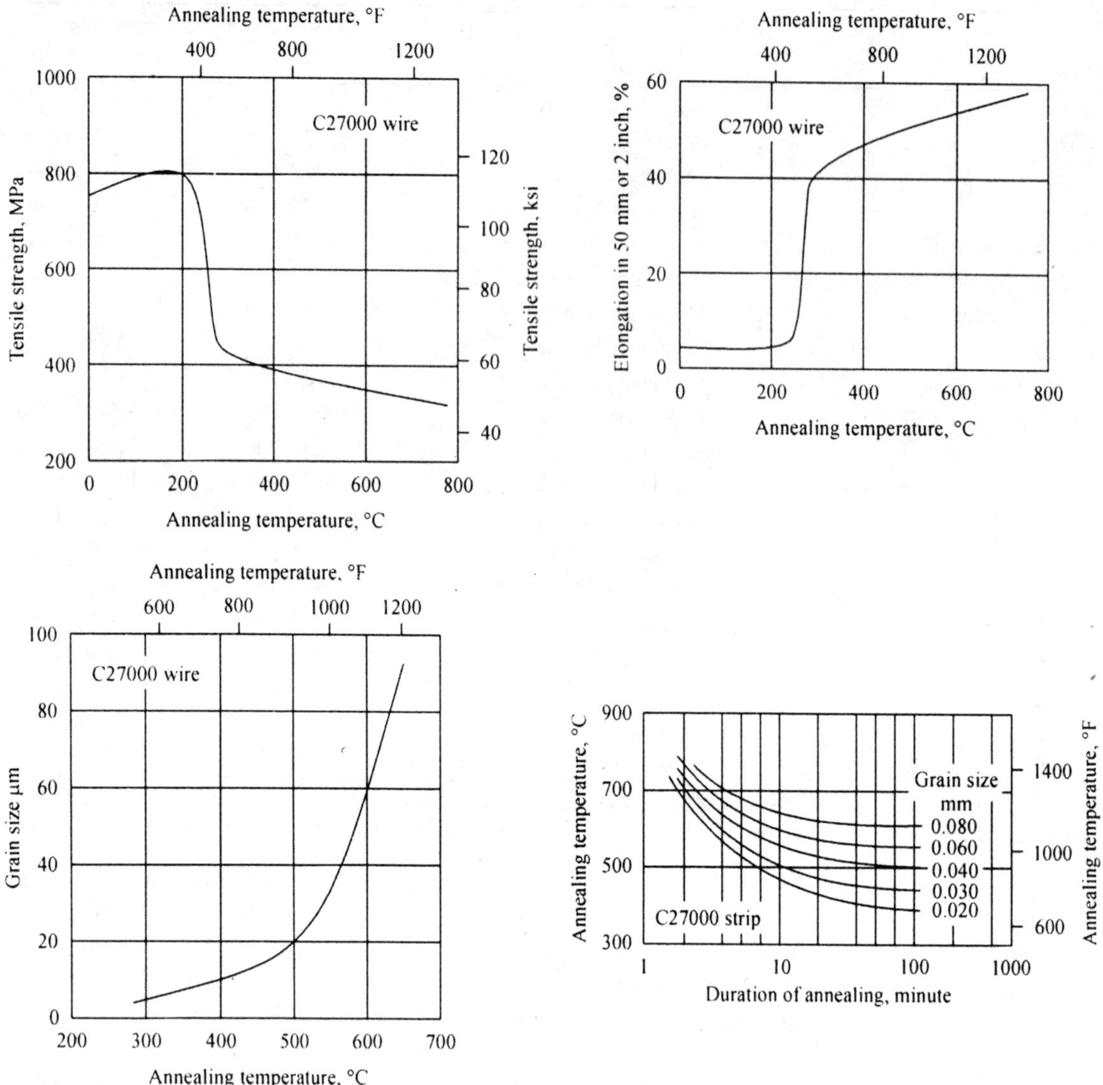

Effects of annealing temperature (annealing time: 1 hour) on tensile strength, elongation and grain size of C27000 wire hard drawn 63 per cent; effect of annealing time on grain size of C27000 strip 1.3 mm (0.050 inch) thick.

Fig. 21.1. Effects of annealing temperature and time on characteristics of C27000 wire and strip.

In commercial mill practice, copper alloys usually are annealed at successively lower temperatures as the material approaches the final anneal, with intermediate cold reductions of at least 35 per cent and as high as 50 to 60 per cent in single or multiple passes wherever practicable. The higher temperatures initially used accelerate homogenisation and the resulting large grains permit more economical reduction during the early working operation. During subsequent anneals, the grain size should be decreased gradually to approximate the final grain size required. This point usually is reached one or two anneals before the final anneal. With such a sequence and with sufficiently severe intermediate reductions, it is then possible to produce a uniform final grain size within a lot and from lot to lot.

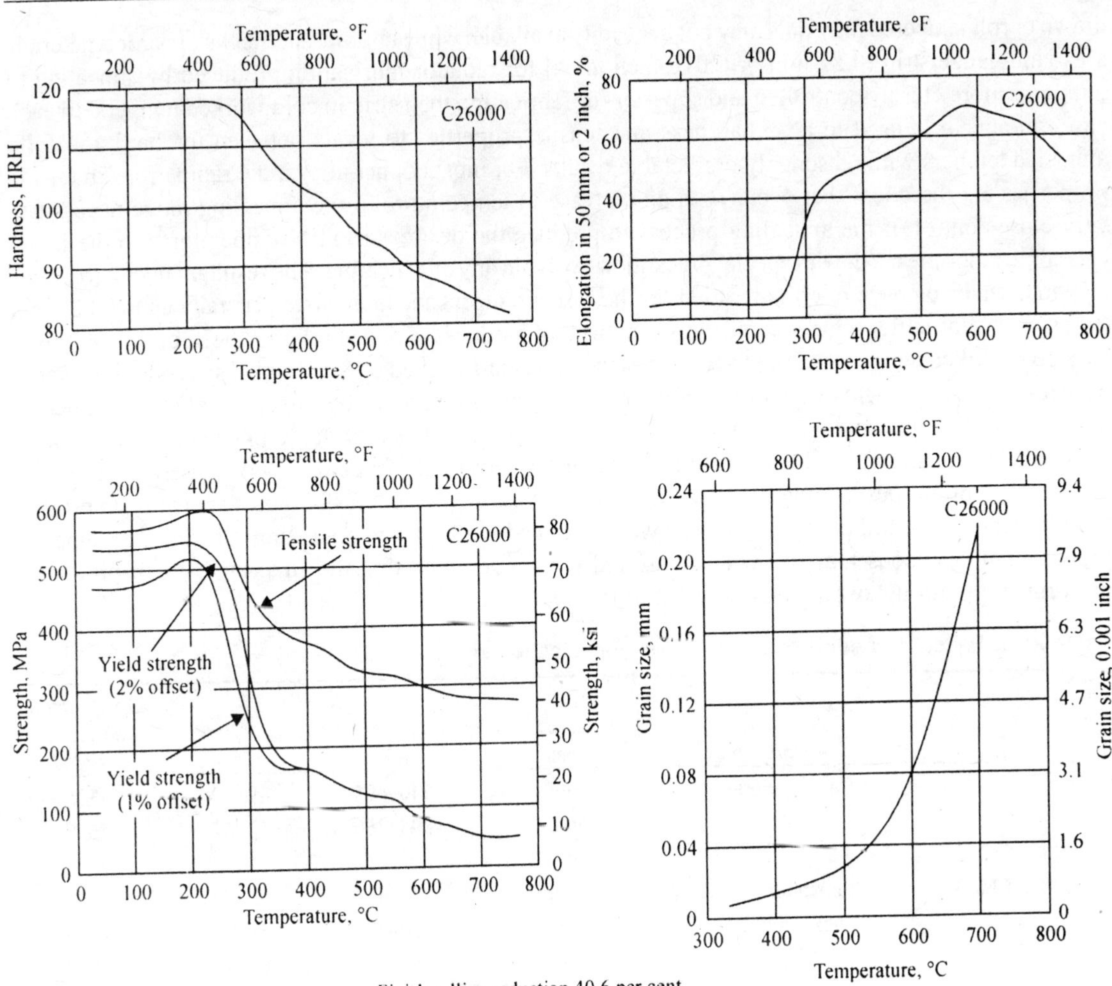

Fig. 21.2. Annealing data on alloy C26000.

The grain size and mechanical properties required for further cold working vary considerably with the alloy and with the amount and kind of further cold work to be done. The object of annealing for cold working is to obtain the maximum combination of ductility and strength. However, when press drawn parts are to be finished by polishing and buffing, the grain size should be as fine as practicable to keep the surface texture smooth and thus to avoid the need for excessive buffing and the attendant excessive finishing costs. The anneal must be governed by definite specifications and co-ordinated with cold working operations, to yield the desired finished properties.

Annealing to Specific Properties

Although specific properties are most frequently produced by controlled cold working of annealed material, there are occasions in which annealing to temper is necessary or advantageous. In hot rolling of cooper alloy plate—particularly large patterns—the finishing temperature may not be consistent or controllable and varying degrees of work hardening may occur. Also, small quantities and/or odd sizes of required drawn or roll tempered materials may not be readily available, while appropriate stocks of harder materials are. Thin-gauge strip (0.25 mm or 0.010 inch thick) for radiator fabrication produced by annealing to temper is more closely controlled and superior for fabrication than strip in cold worked tempers. In each case, an anneal is used to alter hardness and tensile properties to levels between the hard and fully annealed tempers with reasonably predictable results. For most copper alloys, the rapid drops in tensile properties and hardness that occur with an increase in temperature in the annealing range necessitate very close control of the annealing process to produce the desired results. Temperatures in the lower annealing range are used, with special precautions to avoid any over heating. The resultant microstructures may indicate incomplete re-crystallisation for the harder tempers and grain sizes generally up to 0.025 mm (0.001 inch) for softer tempers. Tensile strengths and hardness levels similar to those of ⅛, ¼ and ½ hard cold worked tempers can be produced by annealing hard worked brasses, nickel silvers and phosphor bronzes. While the yield strength for a given final hardness tends to be lower for alloys annealed to temper than for those cold worked to temper, the fatigue resistance of some phosphor bronze spring materials in annealed ½ hard tempers appears to be superior to that of cold worked material. Table 21.2 gives typical properties of annealed-to-temper mill materials. The successful use of annealing to provide specific tempers in mill products requires well-regulated working and annealing schedules designed to produce homogeneous material with controlled grain size, such that the final anneal can produce a uniform result throughout a material lot.

Table 21.2. Typical properties of copper alloys annealed to temper.

Alloy	*Common name*	*Annealed temper*	*Tensile strength* *MPa*	*ksi*	*Approximate hardness, HR30T*
C26000	Cartridge brass	¼ hard	340–405	49–59	43–51
		½ hard	395–460	57–67	56–66
C51100, C53200, C53400, C54400	Phosphor bronze	½ hard	380–485	55–70	57–73
C75200	Nickel silver	¼ hard	400–495	58–72	49–67
		½ hard	455–550	66–80	62–72

General Precautions

For best results in annealing copper and copper alloys, the following precautions should be observed.

Sampling and testing

Test specimens must represent the extreme conditions of the furnace load. For copper alloys that do not contain grain-growth inhibitors, the best and most accurate test for the extent of annealing is the size of the average grain. Grain size is usually the basis for acceptance or rejection of the material. This determination requires special equipment not always available in the plants of consumers or fabricators. For convenience in testing, Rockwell-type hardness testers are used to approximate the grain size; ASTM specifications correlate Rockwell hardness with grain-size values for many copper alloys.

Effect of pre-treatment

Because the amount of cold working and the anneal prior to cold working greatly affect the results of annealing after cold working, any schedule that is set up must take this pre-treatment into account. Once a schedule has been established, both the anneal and the pre-treatment must be adhered to for consistent results.

Effect of time

In most furnaces, there is an appreciable difference between the temperature of the metal and that of the furnace; consequently, time in the furnace greatly affects the final temperature of the metal. For a fixed anneal and furnace temperature, time must vary with type of work load.

Oxidation

Oxidation should be held to a minimum, to reduce loss of metal and costs of pickling, and to improve finish. In some instances, specially prepared atmospheres are used to produce a bright or clean annealed material. Control of furnace atmosphere usually also results in better furnace economy.

Effect of lubricants

Lubricants on metal to be annealed may cause staining that is difficult to remove. Regardless of the type of furnace or the article to be annealed, it is advisable to eliminate as much of the lubricant as possible before the metal is heated.

Hydrogen embrittlement

When copper that contains oxygen is to be annealed, the hydrogen in the furnace atmosphere must be kept to a minimum. This reduces the embrittlement caused by the combination of the hydrogen in the atmosphere with the oxygen in the copper, forming water vapour under pressure and resulting in minute porosity in the metal. For temperatures lower than about 480°C (900°F), the hydrogen content of the atmosphere preferably should not exceed 1 per cent and as the temperature is increased, hydrogen content should approach zero.

Impurities

Occasionally, it is difficult to obtain proper grain growth by annealing under standard conditions that previously have resulted in the desired grain size. This difficulty some times may be traced to impurities in the alloy.

Loading

It usually is inadvisable to anneal a variety of different sizes or kinds of material in the same charge, because of the different rates of heating and the resulting final metal temperatures.

Fire cracking

Fire cracking occurs when some alloys that contain residual stresses are heated too rapidly. Leaded alloys are particularly susceptible to fire cracking. The remedy is to heat slowly until the stresses are relieved. Special types of cold deformation, such as 'springing' (flexing or reeling through a straightener), aid considerably in preventing fire cracking by inducing countervailing mechanical stresses.

Thermal shock

Thermal shock or fatigue occurs when rapid and extreme changes in temperatures occur. Stresses that result in thermal shock are influenced by thermal expansion, thermal conductivity, strength, toughness and the rate of change of the temperature and the condition of the material. Brasses containing lead, lead and tin or lead and certain impurities including bismuth or tellurium may be hot short. If they are repeatedly subject to extreme temperature changes, they may be subject to thermal shock, especially if highly stressed in tension on the surface.

Cooling

Alpha brasses containing less than 70 per cent copper may contain beta that is formed during casting or during heat treatment above 600°C (1110°F), especially if the metal section is massive. Quenching rapidly will entrap beta in the brass. Slow cooling will permit the time and temperature to convert the beta to alpha brass.

Sulphur stains

Excessive sulphur in the fuel or lubricant will cause discolouration of the metal; red stains appear on yellow brass and black or reddish-brown stains on copper-rich alloys.

Castings

Annealing is applied to castings of some duplex alloys, such as manganese bronzes and aluminium bronzes, where it is intended to correct the effects of mould cooling. The extremely slow cooling of sand and plaster castings or the rapid cooling of permanent mould or die castings, can produce microstructures resulting in high hardness and/or low ductility and occasionally inferior corrosion resistance. Typical annealing treatments for castings are in the range of 580 to 700°C (1075 to 1300°F) for 1 hour at temperature. For aluminium bronzes, rapid cooling by water quenching or high-velocity air is advisable.

STRESS RELIEVING

Stress relieving is a process intended to relieve internal stress in materials or parts without appreciably affecting their properties. Stress-relieving heat treatments are applied to copper alloys as one means of accomplishing this objective.

During processing or fabrication of copper alloys by cold working, strength and hardness increase due to plastic strain. Because plastic strain is accompanied by elastic strain, residual stresses remain in the resultant product. If allowed to remain in sufficient magnitude, residual surface tensile stresses can result in stress-corrosion cracking of material in storage or service, unpredictable distortion of material during cutting or machining and hot cracking of materials during processing, brazing or welding. In brasses that contain more than 15 per cent zinc, stress-corrosion cracking or 'season cracking', can occur if both sufficient amounts of residual tensile stress and small amounts of atmospheric ammonia are present. Other copper alloys, such as cold worked aluminium bronzes and silicon bronzes, may also suffer stress corrosion cracking under more severe environments.

Although mill practice for stress relief frequently involves mechanical means such as flexing, cross-roll straightening or shot peening, stress-relief heat treatments are employed for some tubular products and odd shapes. Thermal stress relief is also used for formed parts and fabrications made by material users. It is important to recognise that thermal stress relief reduces residual stress by eliminating part of the residual elastic strain, whereas mechanical stress relief merely re-distributes residual stress into a less detrimental pattern. Stress-relief heat treatments are carried out at temperatures below those normally used for annealing. Typical process stress-relieving temperatures for 19 wrought copper alloys are given in Table 21.3. Temperatures for treatment of cold formed or welded structures are generally 50 to 110°C (90 to 200°F) higher.

Table 21.3. Typical stress-relieving temperatures for 19 wrought copper alloys.

Alloy	*Common name*	*Stress- relieving temperature(a)* °C	°F
C21000	Gilding metal	190	375
C22000	Commercial bronze	205	400
C23000	Red brass	230	450
C24000	Low brass	260	500
C26000	Cartridge brass	260	500
C27000	Yellow brass	260	500
C28000	Muntz metal	205	400
C36000	Free-cutting brass	245	475
C44300, C44400, C44500	Inhibited admiralty	290	550
C51000, C52100	Phosphor bronze	205	400
C61300, C61400	Aluminium bronze	345	650
C65500	High-silicon bronze	345	650
C70600, C71500	Copper nickel	260	500
C75200	Nickel silver	260	500

(a) time at temperature, 1 hour.

From a practical standpoint, higher-temperature/shorter-time treatments are preferable. However, to guarantee the preservation of mechanical properties, lower temperatures and longer times are sometimes necessary. The optimum cycle produces adequate stress relief without adversely affecting properties. As illustrated in Fig. 21.2, some alloys may undergo slight increases in properties during stress relief heat treatment.

To detect the presence of significant residual stress and to evaluate the effectiveness of stress-relieving treatments, samples of material may be tested with mercurous nitrate solutions as described in ASTM B154. Because of the hazards of mercurous salts, tests in high concentrations of moist ammonia have also been used. Warping of rod or tube during longitudinal saw slitting has also been used as a crude field test for residual stress.

HARDENING

Copper alloys that are hardened through heat treatment are of two general types: those that are softened by high-temperature quenching and hardened by lower-temperature treatments and those that are hardened by quenching from high temperatures through martensitic-type reactions. Alloys that harden during low-to-intermediate-temperature treatments following solution quenching include precipitation hardening, spinodal-hardening and order-hardening types. Quench-hardening alloys comprise aluminium bronzes, nickel-aluminium bronzes and a few copper-zinc alloys. Quench-hardened alloys normally are tempered to improve toughness and ductility and reduce hardness in a manner similar to that for alloy steels.

Low-Temperature-Hardening Alloys

For purposes of comparison, Table 21.4 lists examples of the various types of low-temperature-hardening alloys, as well as with typical heat treatments and attainable property levels for these alloys.

Additional details are given in the three sub-sections that follow.

Table 21.4. Typical heat treatments and resulting properties for several low-temperature-hardening alloys.

Alloy	Solution-treating temperature(a) °C	Solution-treating temperature(a) °F	Ageing treatment Temperature °C	Ageing treatment Temperature °F	Time hr.	Hardness	Electrical conductivity % IACS
Precipitation hardening							
C15000	980	1795	500–550	930–1025	3	30 HRB	87–95
C17000, C17200, C17300	760–800	1400–1475	300–350	575–660	1–3	35–44 HRC	22
C17500, C17600	900–950	1650–1740	455–490	850–915	1–4	95–98 HRB	48
C18000(b), C81540	900–930	1650–1705	425–540	800–1000	2–3	92–96 HRB	42–48
C18200, C18400, C18500, C81500	980–1000	1795–1830	425–500	800–930	2–4	68 HRB	80
C94700	775–800	1425–1475	305–325	580–620	5	180 HB	15
C99400	885	1625	482	900	1	170 HB	17
Spinodal hardening							
C71900	900–950	1650–1740	425–760	800–1400	1–2	86 HRC	4–4
C72800	815–845	1500–1550	350–360	660–680	4	32 HRC	–

(a) Solution treating is followed by water quenching. (b) Alloy C18000 (81540) must be double aged—typically, 3 hours at 540°C (1000°F) followed by 3 hours at 425°C (800°F)—to develop the higher levels of electrical conductivity and hardness.

Precipitation-hardening alloys

Most copper alloys of the precipitation-hardening type find use in electrical and heat-conduction applications. The heat treatment must therefore be designed to develop the necessary mechanical strength and electrical conductivity.

The resulting hardness and strength depend on both the effectiveness of the solution quench and the control of the precipitation (ageing) treatment. Note: 'age hardening' or 'ageing' is used in heat treating practice as substitutes for the terms 'precipitation' or 'spinodal hardening'. Copper alloys harden by elevated temperature treatment rather than ambient temperature (natural) ageing as in the case of some aluminium alloys. As solute atoms proceed through the coagulation, coherency, and precipitation cycle in the quenched alloy lattice, the hardness increases, reaches a peak and then decreases with time. Electrical conductivity increases continuously with time until some maximum is reached, normally in the fully precipitated condition.

The optimum condition generally preferred results from precipitation treatment of temperature and duration just beyond those that correspond to the hardness ageing peak. Cold working prior to precipitation ageing tends to improve the heat treated hardness. In the case of lower-strength wrought alloys such as C18200 (Cu-Cr) and C15000 (Cu-Zr), some heat treated hardness may be sacrificed to attain increased conductivity, with final hardness and strength being enhanced by cold working. Two precipitation treatments are necessary in order to develop maximum electrical conductivity and hardness in alloy C18000 (Cu-Ni-Si-Cr) because of two distinct precipitation mechanisms.

Problems in producing anticipated properties in precipitation-hardening alloys may be diagnosed in accordance with the following guidelines:

Problem	*Diagnosis*
Low hardness	Solution temperature too low; solution quench delayed or cooling rate too low; ageing temperature too low and/or time too short (underageing) or temperature too high and/or time too long (overageing)
Low hardness; low conductivity	Inadequate solution treatment and/or underageing
Low hardness; high conductivity	Inadequate solution treatment and/or overageing
High hardness; low conductivity	Underageing; contaminated material

When precipitation hardening is performed at the mill, further treatment following fabrication of parts is not required. However, it may be desirable to stress relieve parts to remove stress induced during fabrication, particularly for highly formed cantilever-type springs and intricate machined shapes that require maximum resistance to relaxation at moderately elevated temperatures.

Spinodal-hardening alloys

Alloys that harden by spinodal decomposition are hardened by a treatment similar to that used for precipitation hardening alloys. The soft and ductile spinodal structure is generated by a high-temperature solution treatment followed by quenching. The material can be cold worked or formed in this condition. A lower-temperature spinodal-decomposition treatment, commonly referred to as ageing, is then used to increase the hardness and strength of the alloy. Spinodal-hardening alloys are basically copper-nickel alloys with chromium or tin additions. The hardening mechanism is related to a miscibility gap in the solid solution and does not result in precipitation. The spinodal-hardening mechanism results in chemical segregation of the alpha crystal matrix on a very fine (Angstrom) scale and requires the use of the electron microscope to discern the metallographic effects. Since no crystallographic changes take place, spinodal-hardening alloys retain excellent dimensional stability during hardening.

Order-hardening alloys

Certain alloys, generally those that are nearly saturated with an alloying element dissolved in the alpha phase, will undergo an ordering reaction when highly cold worked material is annealed at a relatively low temperature. Alloys C61500, C63800, C68800 and C69000 are examples of copper alloys that exhibit this behaviour. Strengthening is attributed to short-range ordering of the solute atoms within the copper matrix, which greatly impedes the motion of dislocations through the crystals.

The low-temperature order-annealing treatment also acts as a stress-relieving treatment, which raises yield strength by reducing stress concentrations in the lattice at the focuses of dislocation pile-ups. As a result, order-annealed alloys exhibit improved stress-relaxation characteristics.

Order annealing is done for relatively short times at relatively low temperatures, generally in the range from 150 to 400°C (300 to 750°F). Because of the low temperature, no special protective atmosphere is required. Order hardening is frequently done after the final fabrication step to take full advantage of the stress-relieving aspect of the treatment, especially where resistance to stress relaxation is desired.

Quench Hardening and Tempering

Quench hardening and tempering (also referred to as 'quench-and-temper hardening') is used primarily for aluminium bronze and nickel aluminium bronze alloys and occasionally for some cast manganese bronze alloys with zinc equivalents of 37 to 41 per cent. Aluminium bronzes with 9 to 11.5 per cent

Al and nickel aluminium bronzes with 8.5 to 11.5 per cent Al, respond in a practical way to quench hardening by a martensitic type reaction. Alloys higher in aluminium content generally are too susceptible to quench cracking, whereas those with lower aluminium contents do not contain enough high-temperature beta phase to respond to quench treatments.

HEAT TREATING EQUIPMENT

Although basic furnace design is similar for all copper alloys, consideration must be given to the annealing temperature range and method of cooling. Solid-solution alloys that do not precipitation harden usually are annealed at temperatures below 760°C (1400°F) and may be cooled at any convenient rate. Precipitation or spinodal hardenable alloys are solution treated at temperatures up to 1040°C (1900°F) and require rapid quenching to ambient temperatures.

Batch-type Atmosphere Furnaces

Batch-type atmosphere furnaces may be heated electrically or by oil or gas. When non-explosive atmospheres are used, electrically heated furnaces permit the atmosphere to be introduced directly into the work chamber.

Furnaces that are heated by gas or oil and that employ protective atmospheres sometimes use a muffle to contain the atmosphere and protect the work from the direct fire of the burners. A properly constructed and safely operated muffle that prevents infiltration of air by maintaining positive pressure is always required when explosive atmospheres, such as hydrogen, are used.

When protective atmospheres are used during annealing, the work must be cooled in the atmosphere nearly to room temperature, to prevent surface scale or discolouration. Metal temperatures above 65°C (150°F) in air may result in light tarnishing. If some degree of surface oxidation and discolouration can be tolerated, direct-natural-gas-fired furnaces may be used. The products of combustion from the gas-air burners are controlled to yield reducing combustion products similar in composition to manufactured protective atmospheres. Parts annealed in reducing atmospheres developed by control of the furnace air/gas ratio require cleaning to restore lustre.

Continuous Atmosphere Furnaces

Continuous atmosphere furnaces offer versatility for solution heat treating of a wide variety of products. Usually, the furnace consists of a vestibule that provides a seal for the atmosphere and in some instances pre-heats the work, a heating chamber of sufficient length to ensure complete solution treating and a cooling or quenching chamber that also serves as an atmosphere seal.

Because the work usually is conveyed at a fixed rate through the furnace, moderate temperature gradients are less harmful than in batch furnaces. When long heating chambers are required, the furnace may be divided into individual temperature-controlled heating zones. Thus, it is practical to develop a high temperature in the entrance zone to facilitate heating of the work to the desired temperature. The cooling chamber may be either a long tunnel through which cool protective atmosphere is circulated or a water-quench zone supplied with a protective atmosphere.

Products such as stampings, machined shapes, castings and small assemblies are conveyed through the furnace on an endless belt or conveyor chain. Long sections such as tubing, bar and flat products or heavy sections that permit stacking or trays, may be conveyed on roller hearth. In rolling-mill operations, the product is uncoiled at the entrance of the furnace and pulled through the furnace by terminal equipment at the exit end; thus, there are no moving parts within the furnace. For wire products, annealing is carried out in bell furnaces, with the wire reel-wound or in-line resistance annealing is performed upon exit of the product from the drawing machine prior to reel-winding.

Salt baths

Molten neutral salts may be used for annealing, stress relieving, solution heat treating or ageing of copper alloys. The composition of the salt mixture depends on the temperature range required. For heating between 705 and 870°C (1300 and 1600°F), mixtures of sodium chloride and potassium chloride are commonly used. Various mixtures of barium chloride with sodium and potassium chlorides are used for a wider temperature range (595 to over 1095°C or 1100 to over 2000°F). The latter mixtures are compatible with each other and are commonly used in multiple-furnace operations when it is advantageous to pre-heat the work in one mixture at a low temperature and then transfer the work to a high-temperature bath. The least common neutral salts are mixtures of calcium chloride, sodium chloride and barium chloride. They have an operating temperature range of 540 to 870°C (1000 to 1600°F) but usually are operated between 540 and 650°C (1100 and 1200°F).

The sodium chloride-carbonate mixtures (not true neutral salts) are used between 595 and 925°C (1100 and 1700°F), primarily for annealing.

For operating temperatures below 540°C (1000°F), the only practical mixtures are the nitrate-nitrite salts. Cyanide-base salts have limited application for heating of copper alloys. Although copper is soluble in cyanide, these salts can be used, with precaution, when a very bright finish is required.

None of the above salt mixtures is applicable for solution treating of standard beryllium copper alloys, because of intergranular attack, pitting or discolouration.

Ageing and stress-relleving operations

Ageing and stress-relieving operations require furnace equipment that can be controlled to within 3°C (5°F) throughout the work zone. Unless cleaning after heating is permissible, it may be necessary to use controlled atmosphere or vacuum equipment.

Because of the necessity for close temperature control, forced-convection (recirculating-air) and salt-bath furnaces are commonly used for ageing and stress relieving. Forced-convection furnaces may be of box, bell or pit type. Each is equipped with a fan that recirculates the constant-temperature atmosphere over the work. When forced convection furnaces are fired by gas or oil and protective atmosphere or vacuum is used, the work must be contained in a properly operating muffle chamber or retort to seal off all products of combustion and to prevent infiltration or air. Temperature variations and heating and cooling times are compared in the following example.

Example 21.1. A comparison was made of temperature variations in a bell furnace and in a pit retort furnace during heat treating of small flat springs made of beryllium copper (Fig. 21.3). Both furnaces were rated at 30 kW. The load in each furnace weighed 90 kg (200 lb) and contained 55000 to 60000 springs. An exothermic gas, produced by a generator using an air/gas ratio of 6.75 to 1 (capacity, 10 m^3/hr., or 350 ft^3/hr.), was used as the protective atmosphere. The composition of the atmosphere was 6.5 per cent CO, 6 per cent CO_2, 10 per cent H_2, rem N_2; dew point was 2°C (35°F) after refrigeration (18 to 21°C or 65 to 70°F, as generated).

Salt baths can reduce total furnace time by up to 30 per cent, as compared to that required with atmosphere furnaces (Fig. 21.4). Salt baths are particularly valuable when the age-hardening time is of short duration and precise control of time at the ageing temperature is required.

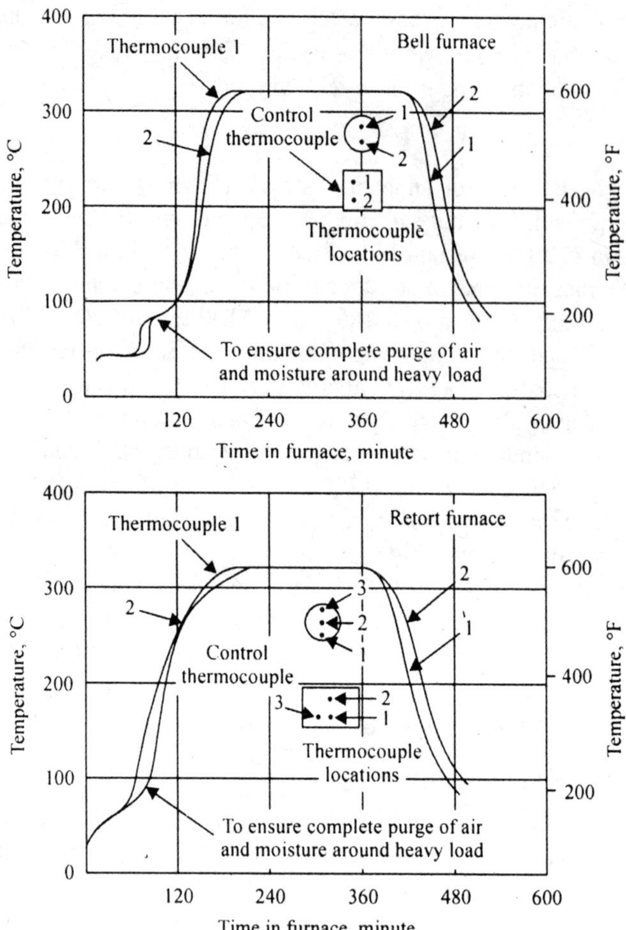

Fig. 21.3. Temperature variations in two types of furnaces (Example 1).

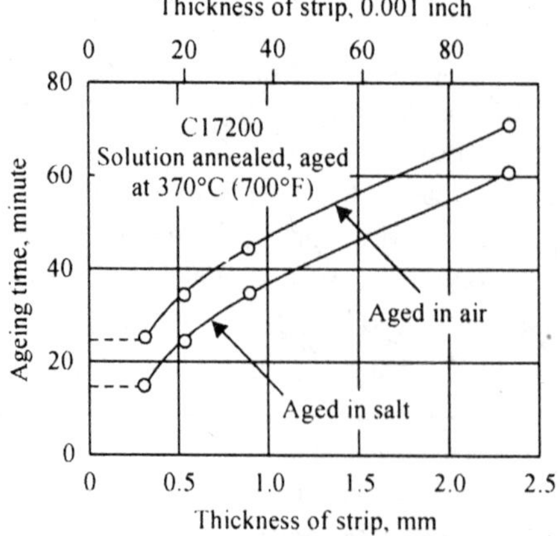

Fig. 21.4. Effects of metal thickness and heating medium on ageing time required to develop maximum strength in C17200 strip.

Commercially available nitrate-nitrite salt mixtures (40 to 50 per cent sodium nitrate, remainder sodium or potassium nitrite) that melt at 143°C (290°F) are used for ageing and stress relieving. All material to be heated in salt should be properly cleaned and dried before being immersed in the molten salt; any organic substance (such as oil or grease) will react violently with the nitrate-nitrite salt.

PROTECTIVE ATMOSPHERES

Selection of protective atmospheres for heat treating of copper and copper alloys is influenced by the temperature used in the heat treating process.

Heating Above 705°C (1300°F)

An exothermic atmosphere is the least expensive protective atmosphere for heat treating of copper alloys. The air/gas ratio is adjusted to produce a combusted gas that contains 2 to 7 per cent hydrogen for use in muffle furnaces operating at 705 to 995°C (1300 to 1825°F). This atmosphere is used successfully for solution treating of alloys such as beryllium coppers, chromium coppers, zirconium coppers and copper-nickel-silicon alloys.

Usually, combusted gases are dried with a surface cooler, using tap water to keep the water/hydrogen ratio reducing throughout the heating and cooling cycle. It may be necessary to lower the dew point further by refrigerating the gas. If the furnace atmosphere is not sufficiently reducing or if the muffle leaks air, a 'sub-scale' or internal oxidation of the hardening elements below the surface of the metal, results. Sub-scale formation can occur rapidly above 845°C (1550°F) if the atmosphere becomes oxidising.

Dissociated ammonia is used primarily for annealing and brazing operations. The gas is very flammable, and can explode if air enters the furnace while at an elevated temperature or if the furnace is improperly purged before reaching the elevated temperature.

Dissociated ammonia can be partly or completely burned with air to reduce cost and flammability. The hydrogen content can be controlled within a range of 1 to 24 per cent, the remainder being nitrogen saturated with water vapour. Water must be removed to maintain a reducing atmosphere.

Hydrogen is highly reducing to copper oxide at elevated temperatures and is recommended for elevated-temperature bright annealing and brazing.

Commercial hydrogen contains about 0.2 per cent oxygen, which if not removed may cause internal oxidation of the reactive alloying elements in the copper.

When mixed with air, hydrogen is explosive at elevated temperature; therefore, the furnace must be purged before being heated to high temperature and air must not enter the furnace.

Heating Below 705°C (1300°F)

Combusted gas (lean exothermic atmosphere) is the most widely used protective atmosphere for annealing of copper and copper alloys. Because of its low sulphur content, natural gas is the preferred fuel for production of combusted gas. The air/gas ratio is adjusted to produce a hydrogen content of 0.5 to 1 per cent. Combusted gas is dried before entering the furnace, to prevent discolouration and staining of the metal by water vapour during the cooling cycle. Steam is the cheapest atmosphere for protecting copper alloys during annealing. Although the annealed metal is not as bright as when heated in a combusted-fuel-gas atmosphere, it is satisfactory for some applications. For products such as tightly wound coils of strip, steam can be used during the heating cycle and combusted fuel gas can be used during cooling.

Inert gases, dissociated ammonia burned with air, and vacuum are more expensive and are not in common use for annealing of copper alloys. A major disadvantage of vacuum is that heating and cooling are slow because heat is transferred by radiation only.

COPPER-BERYLLIUM ALLOYS

Solution Treating

Wrought copper-beryllium alloy mill products generally are supplied solution treated or solution treated and cold worked. Material in these conditions can be fabricated without further heat treatment. Thus, solution treating typically is not a part of the fabricating process unless it is necessitated by a special requirement such as softening of the material for additional forming or is employed as a salvage operation for parts that have been incorrectly heated for precipitation hardening.

Solution treating must be carefully controlled to produce desired grain size, dimensional tolerances and mechanical properties and to prevent surface oxidation. Table 21.5 gives recommended schedules for solution treating and precipitation hardening of the five major copper-beryllium alloys that are produced in wrought form. Optimum mechanical and physical properties for specific applications can be attained by varying these schedules, but the temperatures and times given in this table constitute the most conventional practice and typically provide maximum tensile strength. In addition to the wrought copper-beryllium alloys, there is a wide variety of copper-base casting alloys (C81300 through C82800) that contain beryllium. Appropriate solution-treating and ageing schedules for these alloys are dictated by levels of beryllium and other additives.

Table 21.5. Solution treating and precipitation hardening of copper-beryllium alloys.

	Solution treatment(a)			*Ageing treatment*		
	Temperature		*Time(b), hr.*	*Temperature*		*Time, hr.*
Alloy	*°C*	*°F*		*°C*	*°F*	
C17000	775–800	1425–1475	½–3	300–330	575–625	1–3
C17200	775–800	1425–1475	½–3	300–330	575–625	1–3
C17300	775–800	1425–1475	½–3	300–330	575–625	1–3
C17500	900–925	1650–1700	½–3	455–480	850–900	1–3
C17510	900–925	1650–1700	½–3	455–480	850–900	1–3

(a) All alloys are cooled immediately and rapidly from the solution-treating temperature. Thin sections such as strip can be cooled in circulating atmosphere; heavier sections require water quenching. (b) Shorter times may be desirable to minimise grain growth, particularly for thin sections.

Solution-treating temperature limits must be adhered to if optimum properties are to be obtained from the precipitation-hardening treatment. Exceeding the upper limit causes grain coarsening in wrought material and overheating in wrought and cast materials. Coarse grain size impairs formability; overheating results in a brittel material that does not respond fully to precipitation hardening.

Solution treating below the specified minimum temperature results in insufficient solution of the beryllium-rich phase. This results in lower hardness after precipitation hardening (Fig. 21.5).

Effect of time

The time at the solution-treating temperature depends on the amount of beryllium-rich phase that must be dissolved. Solution of this phase must be complete to produce maximum strength after precipitation hardening.

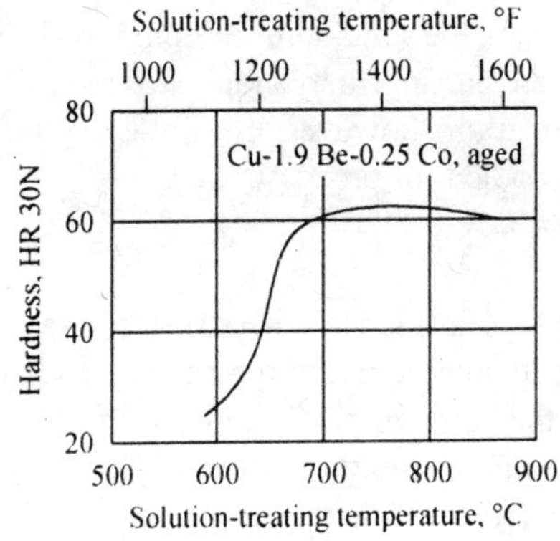

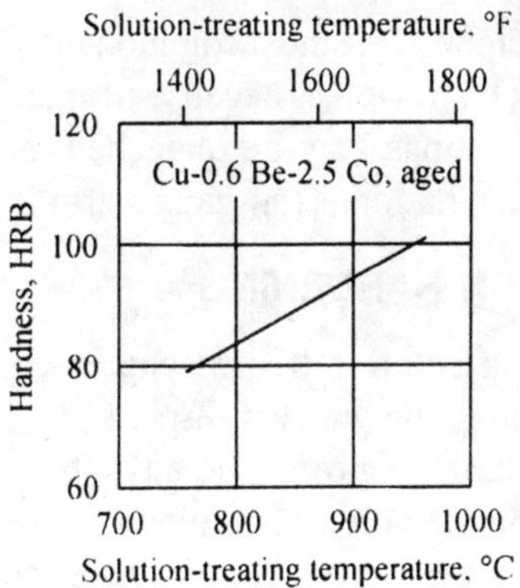

Fig. 21.5. Effect of solution-treating temperature on hardness of C17200 and C17500 after ageing.

In cast products, the as-cast structure usually contains a large amount of microsegregation within the dendritic pattern. Therefore, castings must be heated for a time sufficient to homogenise the structure. A minimum of 3 hours at temperature is recommended for this purpose.

Solution treating of wrought material also removes the effects of cold working and permits additional forming. Some grain growth will occur during softening for additional forming, because the solution-treating temperature is above the re-crystallisation temperature. Therefore, to minimise grain growth, excessive time at temperature must be avoided. It is recommended that wrought alloys be held at temperature 1 hour for each inch or fraction of an inch of section thickness. The optimum time for a specific application must be determined by mechanical testing and microscopic examination of the alloy.

Effect of oxidation

When copper-beryllium alloys are solution treated in air or in an oxidising atmosphere, two types of oxidation are encountered. A continuous and tenacious oxide surface layer forms on alloys with high beryllium contents. Low-beryllium alloys form a loosely adhering scale and are subject to internal oxidation.

The oxide layer on high-beryllium alloys does not significantly affect the mechanical properties of the precipitation-hardened material, but it is abrasive and causes severe wear of tools and dies if not removed. In addition to the abrasive effect of the oxide, oxidation of low-beryllium alloys decreases mechanical properties. This is caused by the surface layer of internal oxidation, which reduces the effective section thickness of the material.

The oxides on both types of alloys may be removed by chemical or abrasive cleaning methods.

Quenching

Quenching is critical phase of the solution-treating process. Successful treatment requires that the material be quenched immediately, and at the highest possible rate, after being removed from the furnace. Any time lapse during transfer from the furnace to the quenching medium will permit some cooling and will cause precipitation. Precipitation is rapid at elevated temperatures and its occurrence significantly affects the properties obtained during subsequent precipitation hardening. The maximum allowable delay before quenching depends on the mass of the load, the size of the parts and the transfer equipment. Mechanical testing and microscopic examination of the structure should be used to evaluate the effectiveness of the quenching operation.

Water quenching is the most common method of retaining the solid-solution condition in both wrought and cast products; however, because of their shape, some castings may crack as the result of the rapid cooling. Such castings may be quenched in oil or forced air; however, the slower cooling rates may cause some precipitation. Thin-gauge strip is typically cooled in forced air.

PRECIPITATION HARDENING

Cold working of solution-treated copper-beryllium alloys influences the strength attainable through subsequent ageing; the greatest response to ageing occurs in material in the cold rolled hard temper. In general, work hardening offers no advantages beyond the hard temper, because formability is poor and control of the precipitation hardening treatment for maximum strength is critical. For some application, however, wire is drawn to higher levels of cold work prior to precipitation hardening.

Table 21.6 lists the properties usually specified for mill products of the common copper-beryllium alloys and Fig. 21.6 shows the time required to develop maximum tensile strength in one of these alloys aged at various temperatures. The ageing times in Fig. 21.6 vary slightly from those given in Table 21.6 for the same alloy; the latter are primarily for acceptance-test purposes.

Table 21.6. Properties and precipitation treatments usually specified for copper-beryllium alloys.

Initial condition	*Standard ageing treatment Time, hr.*	*Temperature °C*	*°F*	*Tensile strength MPa*	*ksi*	*Yield strength(a) MPa*	*ksi*	*Elongation(b), %*	*Hardness(b)*	*Electrical conductivity, % IACS*
C17200										
Flat products:										
Annealed	None	–	–	415–540	60–78	195–380	28–55	35–60	45–78 HRB	17–19
¼ hard	None	–	–	515–605	75–88	415–550	60–80	10–40	68–90 HRB	16–18
½ hard	None	–	–	585–690	85–100	515–655	75–95	10–25	88–96 HRB	15–17
Hard	None	–	–	690–825	100–120	620–770	90–112	2–8	96–102 HRB	15–17
Annealed(d)	3	315	600	1140–1345	165–195	965–1205	140–175	4–10	35–40 HRC	22–25
Annealed	½	370	700	1105–1310	160–190	895–1205	130–175	3–10	34–40 HRC	22–25
¼ hard(d)	2	315	600	1205–1415	175–205	1035–1275	150–185	3–6	37–42 HRC	22–25
¼ hard	1/3	370	700	1170–1380	170–200	965–1275	140–185	2–6	36–42 HRC	22–25
½ hard(d)	2	315	600	1275–1485	185–215	1105–1345	160–195	2–5	39–44 HRC	22–25
½ hard	¼	370	700	1240–1450	180–210	1070–1345	155–195	2–5	38–44 HRC	22–25
Hard(d)	2	315	600	1310–1575	190–220	1140–1415	165–205	1–4	40–45 HRC	22–25
Hard	¼	370	700	1275–1480	185–215	1105–1415	160–205	1–4	39–45 HRC	22–25
Rod, bar and plate:										
Annealed	None	–	–	415–585	60–85	185–205	20–30	35–60	45–85 HRB	17–19
Hard	None	–	–	585–895	85–130	515–725	75–105	10–20	88–103 HRB	15–17
Annealed(d)	3	315	600	1140–1345	165–200	1000–1205	145–175	3–10	36–41 HRC	22–25
Hard(d)	2	315	600	1205–1550	175–225	1035–1380	150–200	2–5	39–45 HRC	22–25
Wire(e):										
Annealed	None	–	–	450–590	65–85	185–240	20–35	35–55	–	17–19

(Contd...)

Initial condition	*Standard ageing treatment Time, hr.*	*Temperature °C*	*°F*	*Tensile strength MPa*	*ksi*	*Yield strength(a) MPa*	*ksi*	*Elongation(b), %*	*Hardness(b)*	*Electrical conductivity, % IACS*
¼ hard	None	–	–	620–795	90–115	485–655	70–95	10–35	–	15–17
½ hard	None	–	–	760–930	110–135	620–760	90–110	4–10	–	15–17
¾ hard	None	–	–	895–1070	130–155	760–930	110–135	2–8	–	15–17
Annealed(d)	3	315	600	1140–1310	165–190	1000–1205	145–175	3–8	–	22–25
Annealed	½	370	700	1105–1310	160–190	930–1205	135–175	3–8	–	22–25
¼ hard(d)	2	315	600	1205–1415	175–205	1105–1310	160–190	2–5	–	22–25
¼ hard	¼	370	700	1170–1415	170–205	1035–1310	150–190	2–5	–	22–25
½ hard(d)	1½	315	600	1310–1480	190–215	1205–1380	175–200	1–3	–	22–25
½ hard	¼	370	700	1275–1480	185–215	1170–1380	170–200	1–3	–	22–25
¾ hard(d)	1	315	600	1345–1585	195–230	1245–1415	180–205	1–3	–	22–25
¾ hard	¼	370	700	1310–1585	190–230	1205–1415	175–205	1–3	–	22–25
C17000										
Flat products:										
Annealed	None	–	–	415–540	60–78	170–365	25–55	35–60	45–78 HRB	17–19
¼ hard	None	–	–	515–605	75–88	310–515	45–75	10–40	68–90 HRB	16–18
½ hard	None	–	–	585–690	85–100	450–620	65–90	10–25	88–96 HRB	15–17
Hard	None	–	–	690–825	100–120	550–760	80–110	2–8	96–102 HRB	15–17
Annealed	3	315	600	1035–1240	150–180	895–1105	130–165	4–10	33–39 HRC	22–25
Annealed(d)	3	345	650	1105–1275	160–185	860–1140	125–165	4–10	34–40 HRC	22–25
¼ hard	2	315	600	1105–1310	160–190	860–1140	135–170	3–6	34–40 HRC	22–25
¼ hard(d)	3	330	625	1170–1345	170–195	895–1170	130–170	3–6	36–41 HRC	22–25
½ hard	2	315	600	1170–1380	170–200	895–1170	145–175	2–5	36–41 HRC	22–25
½ hard(d)	2	330	625	1240–1380	180–200	965–1240	140–180	2–5	38–42 HRC	22–25
Hard	2	315	600	1240–1450	180–210	965–1240	155–180	2–5	38–42 HRC	22–25
Hard(d)	2	330	625	1275–1415	185–205	1070–1345	155–195	2–5	39–43 HRC	22–25
Rod and bar:										
Annealed	None	–	–	415–585	60–85	185–205	20–30	35–60	45–85 HRB	17–19
Hard	None	–	–	585–895	85–130	515–725	75–105	10–20	88–103 HRB	15–17
Annealed	3	315	600	1035–1240	150–180	860–1070	125–155	4–10	32–39 HRC	22–25
Annealed(d)	3	345	650	1105–1275	160–185	930–1140	135–165	4–10	34–40 HRC	22–25
Hard	2	315	600	1140–1380	165–200	930–1140	135–165	2–5	36–41 HRC	22–25
Hard(d)	2	345	650	1205–1415	175–205	965–1170	140–170	2–5	38–42 HRC	22–25
C17500, C17510										
Rod, bar, plate and flat products:										
Annealed	None	–	–	240–380	35–55	185–205	20–30	20–35	20–43 HRB	25–30
Hard	None	–	–	515–585	75–85	380–550	55–80	3–10	78–88 HRB	20–30
Annealed	3	480	900	690–760	100–120	550–690	80–100	10–20	92–100 HRB	45–60

(Contd...)

Initial condition	Standard ageing treatment Time, hr.	Temperature °C	°F	Tensile strength MPa	ksi	Yield strength(a) MPa	ksi	Elongation(b), %	Hardness(b)	Electrical conductivity, % IACS
Annealed(d)	3	455	850	725–825	105–120	550–725	80–105	8–12	93–100 HRB	45–52
Hard	2	480	900	760–860	110–130	690–825	100–120	8–15	95–103 HRB	45–60
Hard(d)	2	455	850	795–930	115–135	725–860	105–125	5–8	97–104 HRB	45–52

(a) At 0.2 per cent offset; (b) In 50 mm or 2 inch; (c) Rockwell B and C hardness values are accurate only if metal is at least 1 mm (0.040 inch) thick; (d) Heat treatment that provides optimum strength; and (e) For wire diameters greater than 1.3 mm (0..050 inch).

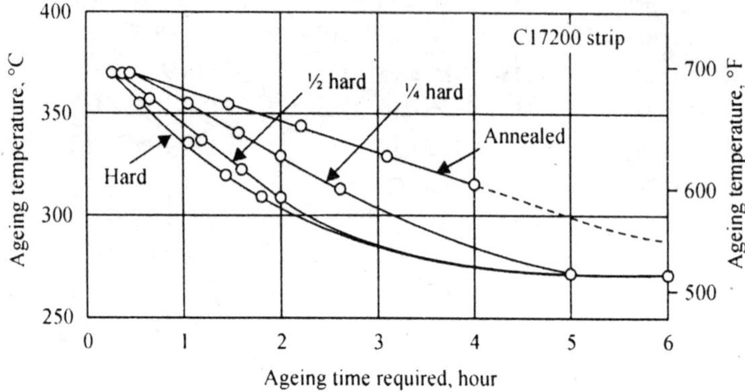

Ageing time required for development of maximum strength in annealed, ¼ hard, ½ -hard and hard C17200 strip aged at various temperatures in a recirculating-air furnace.

Fig. 21.6. Time/temperature relationships in ageing of C17200 strip.

Special combinations of properties can be obtained by varying either the ageing time or the ageing temperature. Table 21.7 shows the age-hardening response from underageing to overageing for cold rolled material in various tempers. Note that as tensile strength increases, elongation decreases and does not recover substantially with overageing but electrical conductivity continues to increase. The response of alloy C17200 (1.9 Be-0.2 Co/Ni) at other ageing temperatures within the hardening range (290 to 400°C, or 550 to 750°F) will be similar to the response at 370°C (700°F), but the corresponding time cycles will vary.

Table 21.7. Effects of special precipitation-hardening treatments on mechanical properties and electrical conductivity of Cu-1.9Be strip.

Initial condition	Ageing treatment Time, min	Temperature °C	°F	Tensile strength MPa	ksi	Yield strength(a) MPa	ksi	Elongation(b), %	Electrical conductivity, % IACS	Fatigue strength(c) MPa	ksi	Modulus of elasticity GPa	10^6 psi
Alloys C17200													
Annealed	None	–	–	465	67.5	250	36	49	18.0	205	30	115	16.5
	5	370	700	855	124	695	101	18	19.5	–	–	120	17.5
	15	370	700	1195	173	1055	153	10	22.0	–	–	125	18.0

(Contd...)

Initial condition	*Ageing treatment Time, min*	*Temperature °C*	*°F*	*Tensile strength MPa*	*ksi*	*Yield strength(a) MPa*	*ksi*	*Elongation(b), %*	*Electrical conductivity, % IACS*	*Fatigue strength(c) MPa*	*ksi*	*Modulus of elasticity GPa*	*10^6 psi*
	30	370	700	1260	182.5	1060	153.5	6	23.0	–	–	125	18.0
	60	370	700	1240	180	1055	153	5	25.5	255	37	130	18.5
	120	370	700	1195	173.5	1040	151	6	26.0	–	–	130	18.5
	240	370	700	1150	167	980	142	6	26.5	–	–	130	19.0
¼ hard	None	–	–	570	82.5	485	70.5	21	17.0	220	32	115	17.0
	5	370	700	1115	162	945	137	9	18.5	–	–	125	18.0
	15	370	700	1250	181	1115	162	6	20.5	–	–	130	18.5
	30	370	700	1290	187	1125	163.5	4	23.5	290	42	130	18.5
	60	370	700	1230	178.5	1060	154	3	25.5	–	–	130	18.5
	120	370	700	1185	172	1000	145	4	26.5	–	–	130	19.0
	240	370	700	1155	167.5	970	141	6	27.0	–	–	130	19.0
½ hard	None	–	–	605	87.5	555	80.5	17	16.0	230	33	115	17.0
	3	370	700	1010	146.5	885	128	11	18.0	230	33	125	18.0
	5	370	700	1280	186	1110	161	3	21.0	295	43	125	18.0
	15	370	700	1310	190	1175	170.5	2	23.0	305	44	130	18.5
	30	370	700	1325	192.5	1180	171	2	24.5	305	44	130	18.5
	60	370	700	1280	185.5	1105	160	2	25.0	295	43	130	18.5
	120	370	700	1200	174	1040	150.5	3	26.0	275	40	130	18.5
	240	370	700	1185	172	1035	150	3	27.0	275	40	130	19.0
	420	370	700	1010	146.5	860	125	10	27.0	200	29	130	19.0
Hard	None	–	–	730	106	690	100	5	15.0	270	39	120	17.5
	5	370	700	1300	188.5	1125	163	3	18.0	–	–	125	18.0
	15	370	700	1360	197	1195	173	2	21.0	–	–	130	18.5
	30	370	700	1310	190	1170	170	1	24.5	315	46	130	19.0
	60	370	700	1295	188	1105	160	1	26.5	–	–	130	19.0
	120	370	700	1240	180	1090	158	2	27.5	–	–	130	19.0
	240	370	700	1215	176	1055	153	2	27.5	–	–	130	19.0
Alloys C17500													
Annealed	None	–	–	350	51	170	25	30	25	–	–	110	16.3
	120	425	800	805	117	625	91	14	44	–	–	135	19.3
	120	455	850	835	121	675	98	14	48	–	–	140	20.0
	120	480	900	805	116.5	625	91	14	48	215	31	140	20.0
	120	510	950	795	115	600	87	16	48.5	–	–	140	20.0
Hard	None	–	–	440	63.5	425	61.5	2	27.8	–	–	125	18.3
	120	425	800	985	142.5	860	125	11	44.0	–	–	140	20.0
	120	455	850	915	133	800	116	13	45.0	–	–	140	20.0
	120	480	900	850	123	760	110.5	13	47.5	250	36	140	20.0
	120	510	950	800	116	705	102	12	49.0	–	–	140	20.0

(a) At 0.2 per cent offset. (b) In 50 mm or 2 inch (c) 10^7 cycles.

Recommended precipitation-hardening cycles for solution-treated copper-beryllium castings are presented in Table 21.8.

Effect of Temperature

Close control of temperature is critical in the conventional ageing of copper-beryllium alloys. As indicated by the data in Tables 21.6 and 21.7, a change in temperature affects the time required for development of maximum properties. Also, the higher temperatures can result in lower property values. Normal commercial control of ±6°C (±10°F) is adequate for temperatures in the range from 315 to 370°C (600 to 700°F).

Table 21.8. Recommended precipitation-hardening schedules and resulting properties for solution-treated copper-beryllium castings.

	Solution treatment			*Ageing treatment*			*Tensile*		*Yield*		*Elonga-*		*Electrical*
	Temperature		*Time,*	*Temperature*		*Time,*	*strength*		*strength(a)*		*tion(b),*		*conductivity*
Alloy	°C	°F	*min*	°C	°F	*min*	*MPa*	*ksi*	*MPa*	*ksi*	%	*Hardness*	*% IACS*
C81300	980–1010	1800–1850	60	480	900	120	365	53	250	36	11	89 HB(c)	60
C81700	900–925	1650–1700	60	455	850	180	635	92	470	68	8	217 HB(d)	48
C81800	900–925	1650–1700	60	480	900	180	705	102	515	75	8	92 HRB	45
C82000	900–925	1650–1700	180	480	900	180	690	100	515	75	8	195 HB(d)	45
C82100	900–925	1650–1700	60	455	850	180	635	92	470	68	8	217 HB(d)	48
C82200	900–925	1650–1700	60	445–455	835–850	120	655	95	515	75	8	96 HRB	45
C82400	785–850	1450–1560	60	345	650	180	1035	150	965	140	1	34 HRC	25
C82500	785–800	1450–1475	60	345	650	180	1105	160	795	115	1	40 HRC	20
C82600	785–800	1450–1475	60	345	650	180	1105	160	1035	150	1	40 HRC	19
C82700	785–800	1450–1475	80	345	650	180	1070	155	895	130	0	39 HRC	20
C82800	785–800	1450–1475	60	345	650	180	1140	165	1000	145	1	42 HRC	18

(a) At 0.2 per cent extension under load. (b) In 50 mm or 2 inch. (c) 500 kg load. (d) 3000 kg load.

Problems involving temperature usually arise when test data are translated into production-control data. A strip specimen of the material tested at the mill may just meet the minimum specification requirements; however, when the fabricator heat treats a large mass of parts made of the same material, tests may indicate properties below the minimum requirements. The low properties may result from heat treatment of too large a mass of parts relative to the capacity of the furnace, from inadequate time at the proper temperature or from use of a higher temperature to gain production speed or to obtain fixture conformity when the parts are in fixtures.

Effect of Grain Size

The effect of grain size on the properties of heat treated material is less significant for copper-beryllium alloys than for solid-solution alloys such as brass and phosphor bronze. The relatively high temperatures required for solution treating of copper-beryllium alloys usually override the effects of cold work and time at temperature. Low solution-treating temperatures will result in fine grain size, but, if the temperature is too low to dissolve the beryllium-rich phase, response to ageing will be affected adversely and the benefits obtained from the fine grain size will be nullified. For this reason, grain sizes below about 0.015 mm (0.0006 inch) are not practical for most beryllium-copper products, regardless of dimensions. With normal commercial practice and depending on the product, the grain size of solution treated material will range from about 0.015 to 0.060 mm.

Fixturing for Close Tolerances

Excellent dimensional accuracy can be achieved by properly supporting beryllium copper parts during ageing. Usually, overageing is necessary to hold close tolerances. Fixture design should be based on the following principles:

1. Fixtures should be of minimum weight.
2. Excessive clamping pressure should be avoided, to prevent stripping of clamping-screw threads and warping of fixtures.
3. Parts should be held only at critical locations.
4. A maximum number of parts should be held by a minimum number of clamping screws.
5. Design should minimise warping of the fixture, thus maximising its service life.
6. When configuration of the part permits, the fixture should be designed so that parts can be stacked. Parts having no more than two planes often can be stacked (Fig. 21.7), provided that no burrs are present.

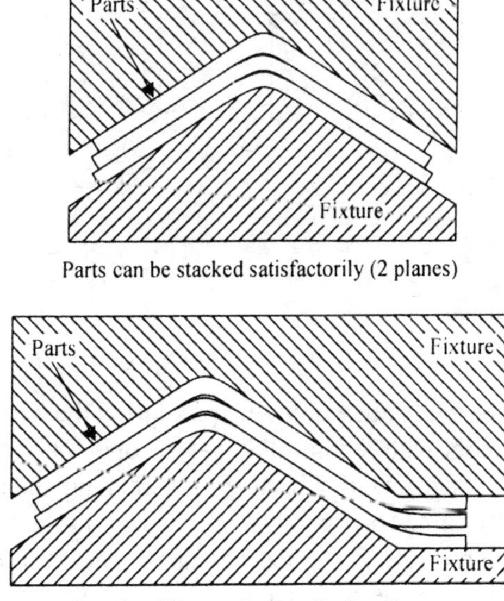

Fig. 21.7. Parts that can and cannot be stacked in fixtures for control of dimensional tolerances during ageing.

Example 21.2: A comparison was made of the dimensional variations that occurred during ageing of solenoid guides heated as loose pieces and as fixtured pieces. The fixture that was used supported, with adequate pressure, all inside and outside surfaces of the part. Although the total cost of ageing the guides as fixtured pieces was nearly 2½ times the cost of ageing them as loose pieces (total cost per piece, based on ageing of 1,00,000 pieces per year), a significant improvement (reduction) in dimensional variations was achieved by the use of fixtures (Table 21.9).

Table 21.9. Dimensional variations in beryllium-copper solenoid guides aged with and without fixtures (Example 2).

	Dimensional variation(a)	
Dimension	*Loose pieces*	*Fixtured pieces*
A(b)	Up to 3/16 inch	1/32 inch max.
B	–1/8 to + 1/16 inch	±0.010 inch
C	–4° to + 2°	± ½°
D	–4° to + 2°	± ½°

(a) Assuming no deviations from the die operation. (b) Twist in 3 inch.

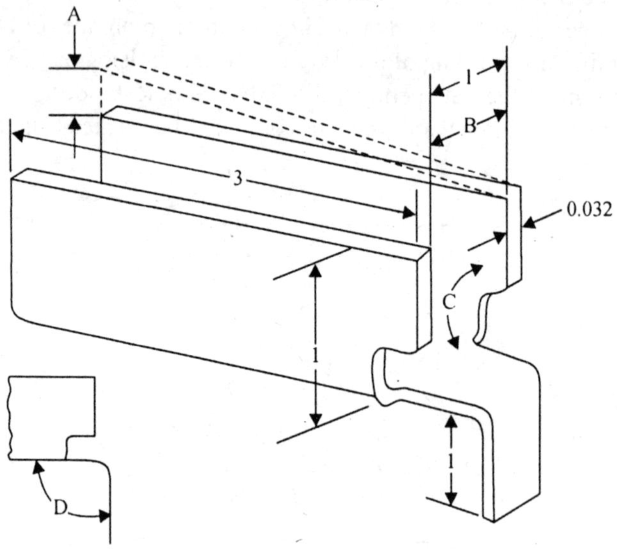

Linear dimensions are in inches.

An understanding of the behaviour of beryllium-copper during the hardening treatment is helpful in design and use of fixtures for parts with a controlled gap or opening. These may be U-shaped parts, circular parts or clips of various designs. During heating for hardening, the material will move toward the direction in which it was plastically formed or elastically deflected. If both conditions occur before hardening, movement due to the elastic effect prevails. These phenomena may be illustrated by the following example.

Example 21.3: Spring clips like the one shown in Fig. 21.8. were press formed to a slightly undersize ID (24.9 to 25.2 mm or 0.980 to 0.990 inch) so that they would fit snugly on a mandrel 25.4 ± 0.03 mm (1 ± 0.001 inch) in diameter. However, during ageing for ½ hour at 350°C (660°F), the metal moved in the direction of elastic deflection, which caused the clips to be loose on the mandrel. Because precipitation hardening of beryllium-copper is a cumulative time/temperature reaction, the hardening treatment can be interrupted. Therefore, the clips (without being placed on the mandrel) were partially aged for 5 minutes at 350°C (this decreased the original dimension to 24.8 to 25.0 mm or 0.975 to 0.985 inch) and then were placed on the mandrel and aged at 350°C for remaining 25 minutes. Following this ageing treatment, none of the clips was larger than the diameter of the mandrel; the actual final dimension was governed by the time and temperature of ageing—that is, by the amount of stress relief that occurred.

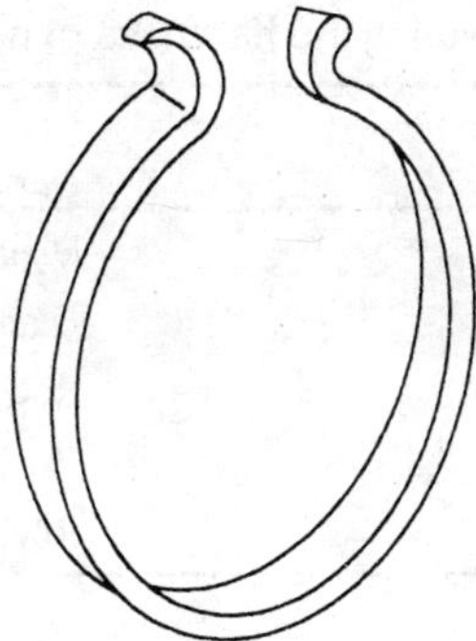

Spring clip representing a type of Beryllium-copper part that should be partially aged without a fixture and then placed on a mandrel for final ageing, for control of diameter.

Fig. 21.8. Problem in dimensional control (Example 3).

Selection of the correct time and temperature for the first step of the interrupted ageing sequence is important but not critical. If the time is inadequate, the direction of movement will not be established; if excessive, there will not be sufficient time for heating the parts on the fixture to allow enough stress relief for close conformity. Duration of the preliminary treatment usually should be about 15 per cent of the total ageing time. Allowance must be made also for the mass effect of the load on the time required for it to reach furnace temperature. Interrupted ageing procedures also can be used to restrict increases in the outside diameters of various parts.

Inspection and Quality Control

In most instances, the completeness of ageing can be verified by hardness testing. Exceptions are tension testing of specimens taken from large parts and simulated service testing to determine elastic performance.

Hardness measurements should always be made with the method and load most suitable for the thickness of the material and the normal level of hardness expected. Table 21.10 indicates suggested methods for testing various thicknesses of a hardened beryllium-copper alloy. Variations obtained with heat treated parts and with strip are presented in the following examples.

Example 21.4: Terminal springs made of a beryllium-copper alloy (0.8 to 1 per cent Be, 0.85 per cent max Ni + Co + Fe, 3 per cent max Zn + Sn) were solution annealed, then aged for 5 hours at 343 ± 6°C (650 ± 10°F) in a batch-type recirculating-air furnace (Fig. 21.9). The furnace load consisted of 2,80,000 pieces. Hardness tests were conducted on specimens of strip material representative of the parts. Specifications required a hardness of 70 to 75 HR15N. Results of hardness tests on 41 production lots within a three-month period are presented in Fig. 21.9.

Example 21.5: Spring contact receptacles stamped from ¼-hard beryllium-copper strip (alloy C17200) were aged in a salt bath at 300°C (575°F) for 30 minutes. The salt pot, 915 by 840 by 760 mm (36 by 33 by 30 inch), contained 1090 kg (2400 lb) of nitrate-nitrite salt (Fig. 21.10).

A strip of about 1000 of these parts was wound on a 'birdcage' reel 265 mm (10½ inch) in diameter and 305 mm (12 inch) wide. Each receptacle weighed about 0.6 gram (0.02 oz); therefore, the total weight of the receptacles on each reel was about 565 gram (20 oz). Each reel weighed about 0.9 kg (2 lb) and one furnace load consisted of eight reels.

Table 21.10. Suggested methods for measuring the hardness of heat treated C17200.

Thickness			*Load,*
mm	*inch*	*Hardness test method*	*kg*
0.03–0.08	0.001–0.003	Diamond pyramid	0.2
0.08–0.38	0.003–0.015	Diamond pyramid	0.5
0.38–0.51	0.015–0.020	Rockwell superficial 15N	15
0.51–1.02	0.020–0.040	Rockwell superficial 15N	30
1.02 and up	0.040 and up	Rockwell B or C	Std

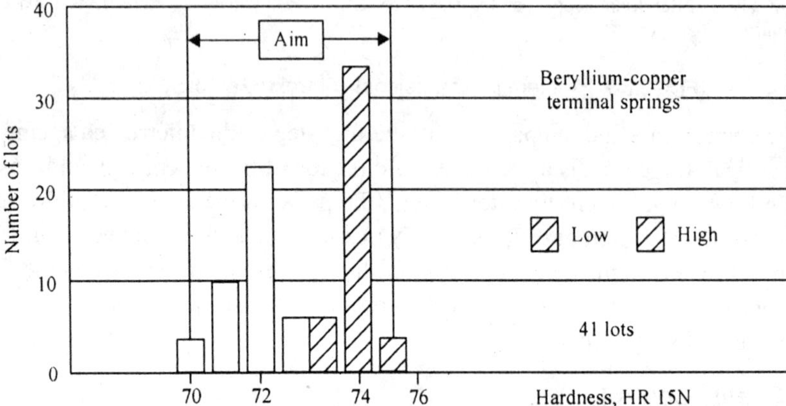

Variation in hardness for 41 lots of heat treated beryllium-copper terminal springs tested over a three-month period; high and low values in each lot are plotted.

Fig. 21.9. Hardness variation in Beryllium-copper springs (Example 4).

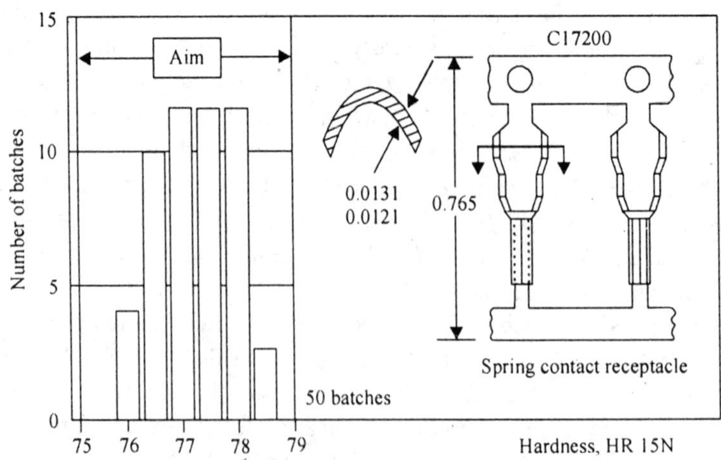

Variation in hardness for 50 batches of aged spring contact receptacles stamped from ¼-hard C17200 strip.

Fig. 21.10. Hardness variation in C17200 spring contact receptacles (Example 5).

Specified hardness for these parts was 75 to 79 HR15N. Fig. 21.11 shows the variation in hardness for 50 batches.

Example 21.6: The variation in tensile strength obtained after ageing hard-temper strip (alloy C17200) is illustrated in Fig. 21.11. The strip, 0.17 to 0.19 mm (0.0065 to 0.0075 inch) thick and 15.9 mm (0.625 inch) wide, was aged 3 hours at 315°C (600°F).

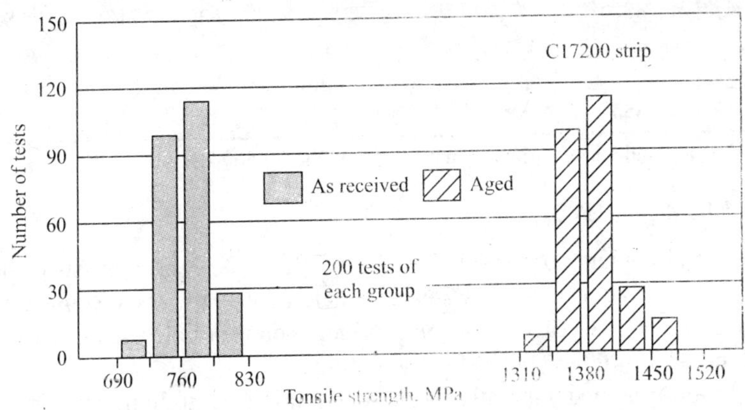

Fig. 21.11. Variation in tensile strength of hard C17200 strip after ageing (Example 6).

COPPER-CHROMIUM ALLOYS

Copper-chromium alloys of 0.5 to 1.0 per cent Cr are solution treated, in molten salt or in controlled-atmosphere furnaces to avoid scaling, at 980 to 1010°C (1800 to 1850°F) and rapidly quenched. Solution-treated chromium-copper is soft and ductile; therefore, it can be cold worked in a manner similar to that used for unalloyed copper.

After being solution treated, the material may be aged for several hours at 400 to 500°C (750 to 930°F) to produce special mechanical and physical properties. A typical ageing cycle is 4 hours or more at 455°C (850°F). Typical effects of heat treatment and cold work on the properties of chromium-copper are shown in Table 21.11. The hard drawn specimens were obtained by reducing the cross-sectional area of solution-annealed specimens by approximately 40 per cent.

Table 21.11. Typical effects of heat treatment and cold work on properties of copper-1 per cent chromium alloys.

Condition	*Ultimate tensile strength* MPa	ksi	*Yield strength(a)* MPa	ksi	*Elongation(b), %*	*Hardness*	*Electrical conductivity, % IACS*
Alloy C18200							
Solution treated	240	35	105	15	42	50 HRF	35–42
Solution treated and aged	350	51	275	40	15	90 HB(c)	75–82
Solution treated and drawn 40 per cent	415	60	310	45	15	65 HRB	40

(Contd...)

Condition	Ultimate tensile strength MPa	ksi	Yield strength(a) MPa	ksi	Elongation(b), %	Hardness	Electrical conductivity, % IACS
Solution treated, hard drawn and aged	435	63	385	56	18	68–75 HRB	80
Solution treated, aged and drawn 30 per cent	480	70	425	62	18	75–80 HRB	80
Alloy C81500							
Cast, solution treated and aged	350	51	275	40	17	105 HB(c)	75–80

(a) At 0.5 per cent extension under load; (b) In 50 mm or 2 inch; and (c) 500 kg load.

COPPER-ZIRCONIUM ALLOYS

Solution treatment of zirconium-copper (99.7 min Cu, 0.13 to 0.30 Zr) consists of heating to 900 to 980°C (1650 to 1795°F) and quenching in water. The material may then be precipitation hardened for 1 to 4 hours at 500 to 550°C (930 to 1020°F). If cold working is done prior to ageing, the ageing temperature is reduced to 370 to 480°C (700 to 900°F) for 1 to 4 hours.

Time at the solution-treating temperature should be minimised, to limit grain growth and possible internal oxidation by reaction of the zirconium with the furnace atmosphere. Because solution and diffusion of the zirconium occur rapidly at the solution-treating temperature, holding at temperature is not required.

Maximum mechanical properties and resistance to softening are developed with maximum solution of zirconium. If material containing 0.15 per cent Zr or more is heated above 980°C (1795°F), the Cu_3Zr phase will begin to melt. A slight amount of melting will not affect mechanical properties; but if excessive melting occurs, ductility of the alloy will decrease.

Normally, as the solution temperature is increased from 900 to 980°C (1650 to 1795°F), the ageing temperature should also be increased to maintain high electrical conductivity. The following ageing treatments will produce the best combination of mechanical properties and electrical conductivity:

Condition	Ageing treatment: Temperature	Time, hr.
Solution treated at 900°C (1650°F)	500°C (930°F)	3
Solution treated at 900°C and cold worked	400° (750°F)	3
Solution treated at 980°C (1795°F)	550°C (1020°F)	3
Solution treated at 980°C and cold worked	450°C (840°F)	3

The increase in strength of zirconium copper-depends primarily on cold work. Although ageing results in some increase in strength, its chief effect is to increase electrical conductivity. The properties developed by various combinations of heat treatment and cold work are given in Table 21.12.

Table 21.12. Effect of heat treatment and cold work on properties of copper-zirconium alloy C15000.

Solution treating temperature(a)		*Amount of cold*	*Ageing Temperature*		*Time,*	*Tensile strength*		*Yield strength*		*Elonga-tion(b),*	*Hard-ness,*	*Electrical conduc-tivity, %*
°C	*°F*	*work, %*	*°C*	*°F*	*hr.*	*MPa*	*ksi*	*MPa*	*ksi*	*%*	*HRB*	*IACS*
900	1650	20	475	885	1	310	45	260	38	25	48	85 min
900	1650	80	425	795	1	425	62	380	55	12	64	85 min
980	1795	None	–	–	–	200	29	41(c)	6(c)	54	–	64
980	1795	20	–	–	–	270	39	250(c)	36(c)	26	37	64
980	1795	80	–	–	–	440	64	420(c)	61(c)	19	73	64
980	1795	None	500	930	3	205	30	90	13	51	–	87
980	1795	None	550	1025	3	205	30	90	13	49	–	95
980	1795	20	400	750	3	330	48	260	38	31	50	80
980	1795	20	450	840	3	330	48	275	40	28	57	92
980	1795	85	400	750	3	495	72	440	64	24	79	85
980	1795	85	450	840	3	470	68	425	62	23	74	91

(a) Hold 30 minute, water quench. (b) In mm or 2 inch. (c) 0.5 per cent extension under load.

MISCELLANEOUS PRECIPITATION-HARDENING ALLOYS

Other alloys which can be age hardened are the nickel-tin bronze alloys C94700 and C94800, copper-nickel-beryllium alloy C96600, and the complex special alloys C99400 and 99500. The solution treating and precipitation hardening treatments for these alloys are shown in Table 21.13.

Table 21.13. Typical heat treating schedules and resulting properties for precipitation hardened miscellaneous alloys.

	Solution treatment			*Tempering treatment*			*Tensile strength*		*Yield strength(a)*		*Elonga-tion(b),*	*Hard-ness*
	Temperature		*Time,*	*Temperature*		*Time,*						
Alloy	*°C*	*°F*	*min*	*°C*	*°F*	*min*	*MPa*	*ksi*	*MPa*	*ksi*	*%*	*HB(c)*
C94700	775–800	1425–1475	120	305–325	580–620	300	585	85	415	60	10	180
C94800	–	–	–	305–325	580–620	360–1000	415	60	205	30	8	120
C96600	995	1825	60	510	950	180	760	110	485	70	7	230
C99400	885	1625	60	480	900	60	545	79	370	54	–	170
C99500	885	1625	60	480	900	60	595	86	425	62	8	196

(a) At 0.2 per cent extension under load for C96600; at 0.5 per cent extension under load for all other alloys; (b) In 50 mm or 2 inch; and (c) 3000 kg load.

A protective atmosphere of exothermic gas or dissociated ammonia is recommended during ageing of these alloys to produce bright surfaces. Parts heated in gas-fired or oil-fired furnaces in which the products of combustion are used as a protective atmosphere may tarnish slightly and require cleaning. Ageing in an oxidising atmosphere results in scaling.

Alloys C19000 and C19100 (Cu-4 per cent Ni-0.25 per cent P) also respond to precipitation hardening. The alloys are solution treated at 705 to 790°C (1300 to 1450°F). A reducing or neutral atmosphere should be used to prevent internal oxidation, especially on thin sections. Water quenching is preferred, while rapid air cooling may be adequate for separate small parts. Precipitation hardening is accomplished by ageing the alloy at 425 to 480°C (800 to 900°F) for 1 to 3 hours.

While annealing is required to facilitate cold working prior to ageing, temperatures as low as 620°C (1150°F) followed by normal air cooling are adequate.

SPINODAL-HARDENING ALLOYS

Table 21.14 lists the recommended solution treatment for the Cu-Ni-Cr and Cu-Ni-Sn spinodal hardening alloys. It is important to maintain control within the limits shown in order to obtain proper heat treating response in the subsequent spinodal ageing treatment. Exceeding the upper limit may result in excessive grain growth in wrought materials, which could impair formability in the solution annealed condition. Overheating cast material may cause incipient melting, resulting in brittle material that does not respond to spinodal hardening, particularly in the Cu-Ni-Sn alloys. Solution treating below the minimum temperature results in incomplete solution and failure of the material to fully harden during the spinodal ageing treatment.

Table 21.14. Recommended solution heat treating temperatures and times for spinodal alloys.

Alloy	*CDA No.*	*Solution heat treating temperature* °C	°F	*Time at temperature, hr.*
Cu-30Ni-3Cr	C71900	900 to 950	1650 to 1740	½ to 2
Cu-4Ni-4Sn	C72600	700 to 760	1300 to 1400	½ to 2
Cu-9Ni-6Sn	C72700	730 to 790	1350 to 1450	½ to 2
Cu-10Ni + 8Sn + 0.2Nb	C72800	805 to 845	1480 to 1550	½ to 2
Cu-15Ni-8Sn	C72900	815 to 860	1500 to 1575	½ to 2

Wrought materials of some of these alloys can be extensively cold worked, with up to 90 per cent reduction, after an effective solution treatment and quench. Solution anneals used between working schedules must be controlled toward the lower end of the temperature range and minimum time to minimise grain growth, because the solution temperature is above the recrystallistion temperature. However, wrought alloys should be solution treated for approximately ½ hour for each inch of section thickness (or fraction thereof).

Homogenisation

Cast microstructures of spinodal hardening alloys generally require homogenisation to encourage uniform distribution of hardening elements and adequate response to the hardening treatments. The C71900 alloy may be homogenised by a prolonged solution treating temperature. The Cu-Ni-Sn alloys have a tendency to develop porosity at higher temperatures and therefore homogenisation is best accomplished by first heating the alloys to 725°C (1335°F) and holding long enough to spheroidise the gamma phase (3 to 12 hours). The temperature is then increased to the regular solution temperature in preparation for the quench.

Oxidation

When spinodol Cu-Ni-Sn alloys are solution treated in air or oxidising atmospheres, oxidation is encountered which can be extremely abrasive to tools although it will not substantially affect the mechanical properties of the spinodal hardened material. The oxide generally does not penetrate far below the surface of these alloys and it can be removed by mechanical, chemical or abrasive cleaning methods.

Quenching

Quenching is a critical step of the solution process for Cu-Ni-Sn alloys. Successful heat treating requires that the material be quenched at the highest possible rate from the solution treating temperature. It is therefore necessary to provide both rapid transfer from furnace to quench tank and an efficient quenching medium. Delays sufficient to cause loss of temperature before the quench could allow the Cu-Ni-Sn intermetallic gamma phase to form, reducing the effectiveness of the entire heat treatment. The quenching media, which is generally water, must be sufficiently cold and agitated to maintain a rapid cooling rate to below the 200°C (400°F) level to ensure that no premature spinodal hardening occurs.

In some circumstances, particularly with the low tin alloys (C72600 and C72700), oil, air or cold gaseous medium quenching may be used for small parts or very thin sections, but a careful evaluation of these methods should be conducted to insure the adequacy of the quench. Both mechanical testing of fully treated materials and metallographic examination of the microstructure should be employed in evaluating the quenching practice. The quenching of spinodal hardening C71900 (Cu-Ni-Cr) alloy is somewhat less critical as to the over all cooling rate, but it is equally important to begin the quench at a temperature above the 950°C (1650°F) temperature.

Spinodal Hardening (Ageing)

Table 21.15 illustrates typical ageing treatments and some resultant properties of various spinodal hardening alloys. The effect of cold working between the solution quench and ageing steps is also illustrated for the Cu-Ni-Sn alloys. The Cu-Ni-Cr (C71900) alloy can be hardened by slow cooling from the solution treating (full annealing) range through 760°C (1400°F) or by a spinodal ageing treatment in the 425 to 760°C (800 to 1400°F) range after a solution treatment. The slow cool from solution temperature tends to produce slightly greater ductility.

The Cu-Ni-Sn alloys are hardened by treating in a rather narrow temperature range of 350 to 360°C (660 to 680°F). Development of the maximum properties requires careful control of temperature and time at temperature. The use of hardness alone to evaluate results may not be adequate, since high hardness may be maintained where excessive ageing causes a decrease in elastic properties. Variations in tensile properties of 68.9 to 103.35 MPa (10 to 15 ksi) are possible without significant hardness change.

Microduplexing

The Cu-Ni-Sn spinodal alloys can be treated using a combination of cold working and heat treatment called microduplexing. The alloys are cold worked to significant reductions (40 to 60 per cent is typical) and given a partial solution treatment below the single phase boundary, typically at 725°C (1335°F). The alloy is then aged at a higher spinodal hardening temperature level of 425°C (800°F) for an extended time. The higher ageing temperature is used to keep the ageing time within reason, as microduplex age hardening takes place much more slowly than full spinodal hardening. This treatment results in moderately high tensile properties and significantly greater ductility. The curves in Fig. 21.12 illustrate the effect of the microduplex treatment compared to conventional spinodal treatment using a 425°C (800°F) age.

Effect of Grain Size

The temperatures required for solution treatment are well into the recrystallisation/growth range for spinodal hardening alloys. The normal grain size obtained in properly treated wrought products will be in the 0.010 to 0.090 mm range, but coarser grain sizes have been used without adverse results. Maintaining low solution temperatures in an attempt to minimise grain growth is good practice, but is secondary to ensuring a complete solution treatment and adequate quench.

Table 21.15. Typical strength and recommended ageing times for various spinodal alloys (a).

Alloy	CDA No.	Solution treated and cold worked temper	Ageing cycle	Tensile strength		Yield strength		Elongation, %
				MPa	ksi	MPa	ksi	
Cu-4Ni-4Sn	C72600	TD 02(½H)	90 min at 350°C (660°F)	635–690	92–100	495–570	72–83 (0.05)	12
Cu-4Ni-4Sn	C72600	TD 06(XH)	90 min at 350°C (660°F)	690–725	100–105	565–620	82–90 (0.05)	9
Cu-4Ni-4Sn	C72600	TD 08(S)	90 min at 350°C (660°F)	705–795	102–115	565–655	82–95 (0.05)	7
Cu-9Ni-6Sn	C72700	TD 04(H)	90 min at 350°C (660°F)	860–1035	125–150	760–895	110–130 (0.05)	8
Cu-9Ni-6Sn	C72700	TD 14(SS)	90 min at 350°C (660°F)	1055–1145	153–166	930–985	135–143 (0.05)	–
Cu-10Ni-8Sn-0.2Nb	C72800	TB 00 cast and solution treated	4–6 hr. at 350°C (660°F)	830–965	120–140	550–690	80–100 (0.01)	3
Cu-10Ni-8Sn-0.2Nb	C72800	TB 00 hot work and solution treated	3–5 hr. at 350°C (660°F)	965–1070	140–155	690–825	100–120 (0.01)	6–14
Cu-10Ni-8Sn-0.2Nb	C72800	TD 01(¼H)	3 hr. at 350°C (660°F)	1140–1240	165–180	895–930	130–135 (0.01)	7
Cu-10Ni-8Sn-0.2Nb	C72800	TD 04(H)	3 hr. at 350°C (660°F)	1205–1380	175–200	930–1000	135–145 (0.01)	7
Cu-10Ni-8Sn-0.2Nb	C72800	TD 06(XH)	3 hr. at 350°C (660°F)	1205–1380	175–200	965–1035	140–150 (0.01)	5
Cu-10Ni-8Sn-0.2Nb	C72800	TD 08(H)	3 hr. at 350°C (660°F)	1240–1380	180–200	1000–1070	145–155 (0.01)	4
Cu-10Ni-8Sn-0.2Nb	C72800	TD 14(SS)	90 min at 350°C (660°F)	1240–1380	180–200	1070–1140	155–165 (0.01)	2.5
Cu-15Ni-8Sn	C72900	TD 14(SS)	90 min at 350°C (660°F)	1140–1380	165–200	1035–1170	150–170 (0.05)	3
Cu-30Ni-3Cr	C71900	Hot extruded	90 min at 760°C (405°F)	550	80	345	50 (0.20)	25

(a) At 350°C (660°F).

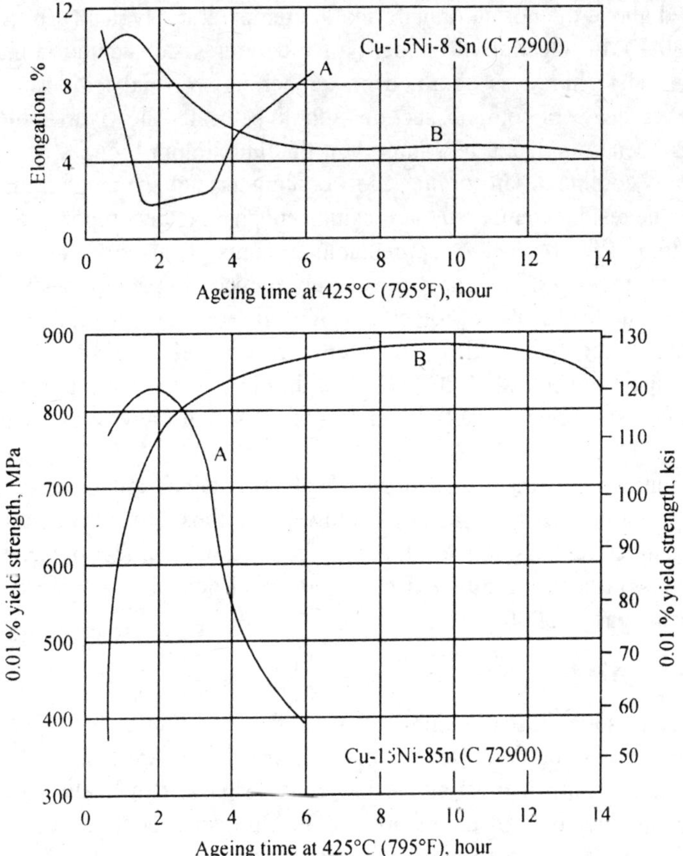

A, water quench for 30 minute at 825°C (1520°F); B, water quench for 1 hour at 725°C (1340°F), microduplexed.

Fig. 21.12. Ageing response of Cu-15Ni-8Sn (C72900) solution treated above and below (microduplexed) the single phase boundary.

An extremely fine grain size (2 to 4 μm) may be produced in the microduplexing treatment or Cu-Ni-Sn alloys because of the much lower solution treating temperature and nucleating effects of undissolved gamma phase particles.

Fixturing in Close Tolerances

Excellent dimensional reproducibility can be achieved without fixturing during heat treatment. This is because the spinodal alloys do not undergo a crystallographic change during hardening. The chemical segregation (spinodal decomposition) is not accompanied by a second phase which can cause distortion during the ageing process. Therefore, stability is greatly increased over precipitation hardening alloys. Fixturing may still be necessary for extremely tight dimensional control but is not generally necessary.

COPPER-ALUMINIUM (ALUMINIUM BRONZE) ALLOYS

The microstructures and consequent heat treatabilities of aluminium bronzes vary with aluminium content much the same as these characteristics vary with carbon content in steels. Unlike steels, aluminium bronzes are tempered above the normal transformation temperature, typically in the range from 565 to 675°C (1050 to 1250°F). In selection of tempering temperatures, consideration must be given both to required properties and to hardness obtained on quenching. Normal tempering time is 2 hours at temperature. Moreover, heavy or complex sections should be heated slowly to avoid cracking. After the tempering cycle has been completed, it is important that aluminium bronzes be cooled rapidly, using water quenching, spray cooling or fan cooling. Slow cooling through the range from 565 to 275°C (1050 to 530°F) can cause the residual tempered martensitic beta phase to decompose, forming the embrittling alpha-gamma eutectoid. The presence of appreciable amounts of this eutectoid structure can result in low tensile elongation, low energy of rupture, severely reduced impact values and reduced corrosion resistance in some media. For adequate protection against detrimental eutectoid transformation, cooling after tempering should bring the alloy to a temperature below 370°C (700°F) within about 5 minutes and to a temperature below 275°C (530°F) within 15 minutes. Normally, the danger of eutectoid transformation is much smaller in nickel-aluminium bronzes and these alloys can be air cooled after tempering.

The normal precautions used in heat treating of steel have been found to be applicable to aluminium bronze, with critical cooling rates being somewhat lower than those for steel. Soaking time should be at least 45 minutes at temperature and longer times may be used without fear of excess surface oxidation. Oil quenching is used on heavy, complex sections to avoid quench cracking, particularly in nickel-aluminium bronzes such as C63000.

Alpha Aluminium Bronzes

Alpha aluminium bronzes are those aluminium bronzes that contain less than 9 per cent aluminium, or less than 8.5 per cent aluminium with up to 3 per cent iron. They are essentially single-phase alloys, except for fine iron-rich particles in those alloys that contain iron. For alpha aluminium bronzes, effective strengthening can be attained only by cold work and annealing and/or stress relieving are the only heat treatments of practical use. The most prevalent alloys of this group are C60600, C61000, C61300 and C61400. In addition, alloys containing up to 9.6 per cent Al, with microstructures containing small amounts of beta phase at high temperatures, have such limited heat treatability that they also can be hardened only by cold work.

Annealing of alpha aluminium bronzes is carried out at temperatures from about 540 to about 870°C (1000 to 1600°F), with the iron-containing alloys requiring temperatures nearer the high end of this range. Alloys of intermediate composition (containing small amounts of beta phase), such as C61900, normally are annealed at 595 to 650°C (1100 to 1200°F). Fig. 21.13 presents annealing curves for typical alpha aluminium bronze alloys C61300 and C61400.

Complex (Alpha-Beta) Aluminium Bronzes

Complex (alpha-beta) aluminium bronzes are those aluminium bronzes whose normal microsturctures contain more than one phase to the extent that beneficial quench-and-temper treatments are possible. These copper-aluminium alloys, with and without iron, are heat treated by procedures somewhat similar to those used for heat treatment of steel and have isothermal transformation diagrams that resemble those of carbon steels. For these alloys, the quench-hardening treatment is essentially a high-temperature soak intended to dissolve all of the alpha phase into the beta phase. Quenching results in a hard room-temperature beta martensite structure and subsequent tempering re-precipitates fine alpha needles in the structure, forming a tempered beta martensite.

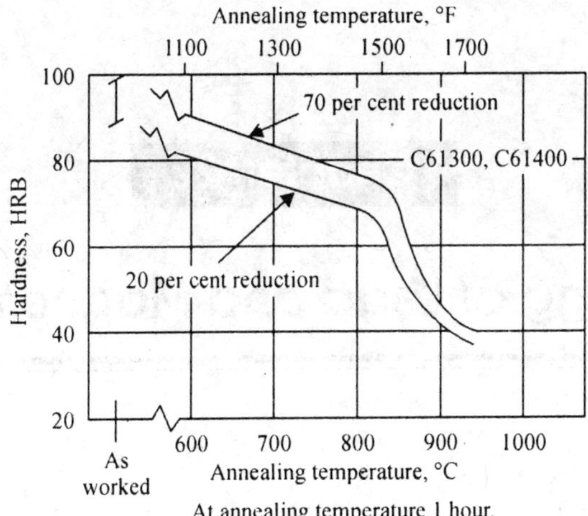

Fig. 21.13. Annealing curves for cold reduced alpha aluminium bronzes.

Table 21.16 gives typical tensile properties and hardnesses of alpha-beta aluminium bronzes after various stages of heat treatment.

Table 21.16. Typical heat treatments and resulting properties for complex (alpha-beta) aluminium bronzes.

Alloy	Typical conditions(a)	Tensile strength MPa	Tensile strength ksi	Yield strength(b) MPa	Yield strength(b) ksi	Elonga-tion(c) %	Hard-ness, HB
C62400	As forged or extruded	620–690	90–100	240–260	35–38	14–16	163–183
	Solution treated at 870°C (1600°F) and quenched, tempered 2 hrs. at 620°C (1150°F)	675–725	98–105	345–385	50–56	8–14	187–202
C63000	As forged or extruded	730	106	365	53	13	187
	Solution treated at 855°C (1575°F) and quenched, tempered 2 hrs. at 650°C (1200°F)	760	110	425	62	13	212
C95300	As cast	495–530	72–77	185–205	27–30	27–30	137–140
	Solution treated at 855°C (1575°F) and quenched, tempered 2 hrs. at 620°C (1150°F)	585	85	290	42	14–16	159–179
C95400	As cast	585–690	85--100	240–260	35–38	14–18	156–179
	Solution treated at 870°C (1600°F) and quenched, tempered 2 hrs. at 620°C (1150°F)	655–725	95–105	330–370	48–54	8–14	187–202
C95500	As cast	640–710	93–103	290–310	42–45	10–14	183–192
	Solution treated at 855°C (1575°F) and quenched, tempered 2 hrs. at 650°C (1200°F)	775–800	112–116	440–470	64–68	10–14	217–234

(a) As-cast condition is typical for moderate sections shaken out at temperatures above 540°C (1000°F) and fan cooled; or mould cooled, annealed at 620°C (1150°F) and fan (rapid) cooled; (b) At 0.5 per cent extension under load; and (c) In 50 mm or 2 inch.

Chapter 22

Heat Treating of Lead and Magnesium Alloys

INTRODUCTION

Lead normally is considered to be unresponsive to heat treatment. Yet, some means of strengthening lead and lead alloys may be required for some applications. Lead alloys for battery components, for example, can benefit from improved creep resistance in order to retain dimensional tolerances for the full service life. Battery grids also require improved hardness to with stand industrial handling.

The absolute melting point of lead is 327.4°C (621.3°F). Therefore, in applications in which lead is used, recovery and re-crystallisation processes and creep properties have great significance. Attempts to strengthen the metal by reducing the grain size or by cold working (strain hardening) have proved unsuccessful. Lead-tin alloys, for example, may re-crystallise immediately and completely at room temperature. Lead-silver alloys respond in the same manner within two weeks.

Transformations that are induced in steel by heat treatment do not occur in lead alloys and strengthening by ordering phenomena, such as in the formation of lattice superstructures, has no practical significance. Despite these obstacles, however, attempts to strengthen lead have met with some success.

Magnesium alloys usually are heat treated either to improve mechanical properties or as a means of conditioning for specific fabricating operations. The type of heat treatment selected depends on alloy composition and form (cast or wrought) and on anticipated service conditions.

Solution heat treatment improves strength and results in maximum toughness and shock resistance. Artificial ageing (precipitation heat treatment) subsequent to solution treatment gives maximum hardness and yield strength, but with some sacrifice of toughness. As applied to castings, artificial ageing without prior solution treatment or annealing is a stress-relieving treatment that also somewhat increases tensile properties. Annealing of wrought products lowers tensile properties considerably and increases ductility, thereby facilitating some types of fabrication. Modifications of these basic treatments have been developed for specific alloys, to obtain the most desirable combinations of properties. For example, increasing the ageing time for some magnesium alloy castings considerably increases yield strength (although with some sacrifice of ductility). Also, combinations of solution treating, strain hardening and artificial ageing are applied to alloy HM21A sheet to improve mechanical properties over those attainable by solution treating and artificial ageing alone.

SOLID-SOLUTION HARDENING

In solid-solution hardening of lead alloys, the rate of increase in hardness generally improves as the difference between the atomic radius of the solute and the atomic radius of lead increases.

Specifically, in one study of possible binary lead alloys, it was found that the following elements, in the order listed, provided successively greater amounts of solid-solution hardening: thallium, bismuth, tin, cadmium, antimony, lithium, arsenic, calcium, zinc, copper and barium. Unfortunately, these elements have successively decreasing solid-solution solubilities and thus the most potent solutes have the most limited solid-solution hardening effects. Within the mid-range of this series, however, are elements that, when alloyed with lead, produce useful strengthening. A useful level of strengthening normally requires solute additions in excess of the room-temperature solubility limit. In most lead alloys, homogenisation and rapid cooling result in breakdown of the supersaturated solution during storage. Although this breakdown produces coarse structures in certain alloys (lead-tin alloys, for example), it produces fine structures in others (lead-antimony alloys, for example). In alloys of the lead-tin system, the initial hardening produced by alloying is quickly followed by softening as the coarse structure is formed.

In lead-antimony alloys, at suitable solute concentrations, the structure may remain single phase with hardening by Guinier-Preston (GP) zones formed during ageing. At higher concentrations and in certain other systems, ageing may produce precipitation hardening as discrete second-phase particles are formed.

Alloys that exhibit precipitation hardening typically are less susceptible to overageing and thus are more stable with time than alloys hardened by GP zones. Lead-calcium and lead-strontium alloys have been observed to age harden through discontinuous precipitation of a second phase—Pb_3Ca in lead-calcium alloys and Pb_3Sr in lead-strontium alloys—as grain boundaries move through the structure.

Solution Treating and Ageing

Useful strengthening of lead can be attained by adding sufficient quantities of antimony to produce hypoeutectic lead-antimony alloys. Small amounts of arsenic have particularly strong effects on the age-hardening response of such alloys and these effects are enhanced by solution treating and rapid quenching prior to ageing.

An investigation was conducted on the effects of additions of 0.15 per cent arsenic on the age-hardening behaviour of five hypoeutectic lead-antimony alloys. Accurately weighed quantities of commercially pure lead, antimony and arsenic were melted under a nitrogen atmosphere to produce the alloys listed below:

Alloy	*Antimony, %*	*Arsenic, %*
200	2.0	–
215	2.0	0.15
400	4.0	–
415	4.0	0.15
600	6.0	–
615	6.0	0.15
800	8.0	–
815	8.0	0.15
1000	10.0	–
1015	10.0	0.15

To minimise segregation during solidification, the melts, each weighing about 100 gram (3.5 oz), were chilled rapidly by casting into a shallow horizontal steel boat. Pieces weighing about 15 gram (0.05 oz) were cut from the castings and were used as hardness-test specimens after being subjected to one of two heat treatments. One group of specimens, designated group A, was air cooled from the liquid state and quenched in ice water as soon as solidification was complete. Another group, designated group B, was solution treated for 4 hours at 250°C (480°F) and then quenched in ice water.

In view of the rapid age hardening of the arsenical alloys, it was desired that the hardness tests be commenced as soon after quenching as possible and one minute was chosen as a reasonable period of time. For group B specimens, this was easily achieved by solution treating the specimens, which had previously been ground with 600 grade paper, in a stream of nitrogen in a vertical tube furnace. The specimens then were dropped into ice water and transferred after 20 seconds to water at room temperature. Hardness of these specimens was measured with a Vickers hardness tester using a 2.5 kg load.

Because the specimens in group A could not be prepared for hardness testing before they were quenched, a different procedure was necessary to permit testing to commence after one minute. These specimens were re-melted in small flat-base cylindrical cups punched from thin aluminium foil. After quenching, the foil was peeled from the specimen and the hardness measurement was made on the flat surface (after the top surface had been quickly smoothed on a file to ensure proper seating in the test machine.

Cooling curves were obtained for the specimens in group A by means of Chromel-Alumel thermocouples, which were thinly sheathed with mild steel. After re-melting, the thermocouple was removed before the beginning of solidification. Quenching then was timed to follow complete solidification on the basis of the previous cooling curves. The group A specimens were cooled at a rate of 1°C/s (1.8°F/s) to solidification at 232°C (450°F). These specimens showed evidence of surface segregation.

Hardness tests were continued for up to two years, with the specimens being stored at ambient temperature, which for the majority of the time was between 20 and 22°C (68 and 72°F). To investigate variations in hardness, tests were made on some of the specimens using a Reichert microhardness tester (5.3 gram load, applied for 10 seconds).

Test results

Results of hardness testing showed that lead-antimony alloys of commercial purity demonstrate significant age hardening, particularly after solution treating, as shown in Figs. 22.1 and 22.2. In the quench-cast alloys of group A, the hardness at any time increased with antimony content. For the solution-treated specimens of group B, ageing was more effective for alloys with lower antimony contents.

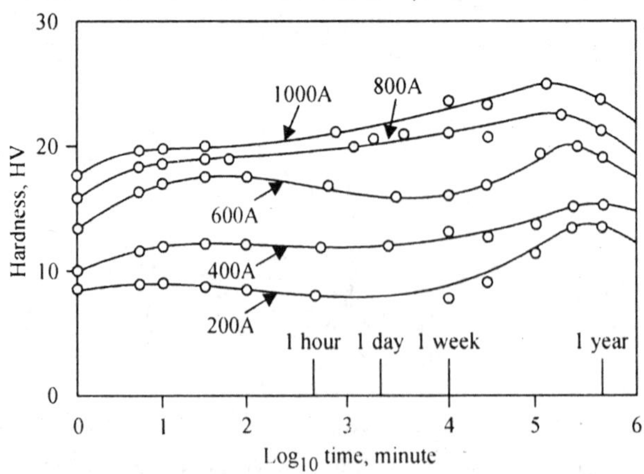

Fig. 22.1. Age hardening of Pb-Sb alloys, solidified and water quenched.

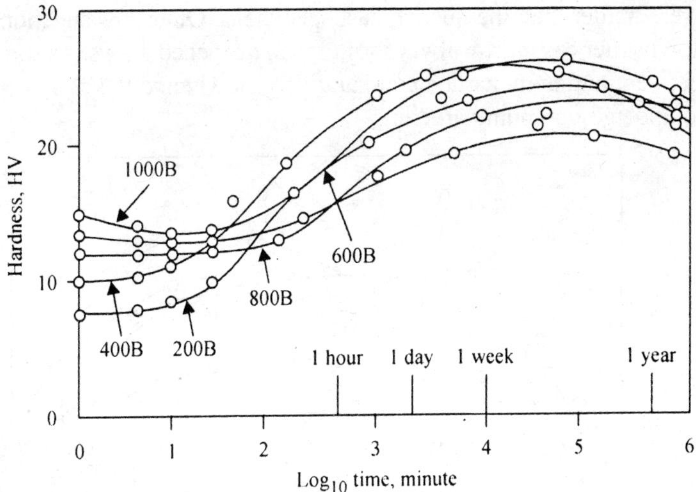

Fig. 22.2. Age hardening of Pb-Sb alloys, solution treated 4 hours at 250°C (480°F) and water quenched.

Figs 22.3 and 22.4 show that both the rate and the extent of hardening are increased by addition of 0.15 per cent arsenic. As shown in Fig. 22.4, increases in hardness of the solution-treated hypoeutectic alloys are pronounced in the first 10 minutes.

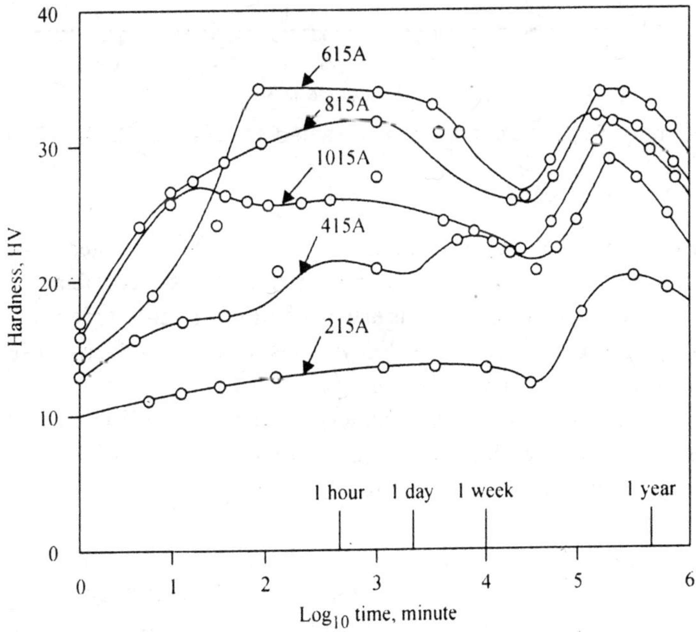

Fig. 22.3. Age hardening of Pb-Sb as alloys, solidified and water quenched.

Hardness stability

For any given alloy, both heat treatments result in hardnesses after 1 minute and after 2 years that are somewhat comparable, as indicated in Figs. 22.1. to 22.4. For most of the two-year period, the solution-treated specimens were harder than the quench-cast specimens. Other investigations have also shown that alloys cooled slowly after casting are always softer than quenched alloys. As shown in Fig. 22.3, the alloys with 2 to 4 per cent antimony harden comparatively slowly and the alloy containing 6 per cent antimony appears to undergo optimum hardening.

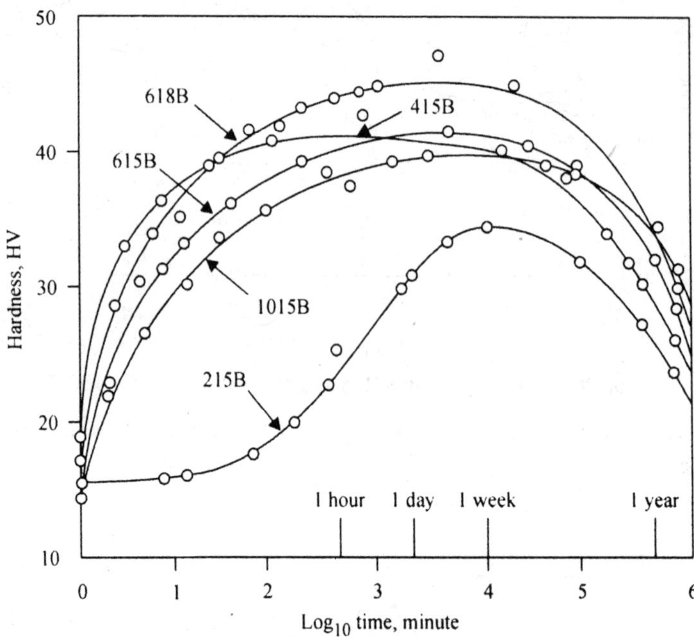

Fig. 22.4. Age hardening of Pb-Sb as alloys, solution treated 4 hour at 250°C (480°F) and water quenched.

Application

To reduce the antimony contents of the positive plates in lead-acid storage batteries, because of antimony's detrimental effect on charge retention, there has been a trend toward replacing eutectic alloys with a Pb-6Sb-0.15As alloy. Battery grids made of this arsenical alloy will age harden slowly after casting and air cooling. However, storing grids for several days constitutes unproductive use of floor space and results in undesirable interruptions in manufacturing sequences.

Although large-scale solution treatment of battery grids might be difficult to justify economically or to achieve without some distortion, quenching of grids cast from arsenical lead-antimony alloys offers an attractive alternative method of effecting improvements in strength. The suitability of quench-cast grids can be assessed by comparing the values given in Fig. 22.3 with the hardness level that battery grids require in order to withstand industrial handling (above 18 HV, the hardness of the eutectic alloy). The alloy containing 2 per cent antimony clearly does not respond sufficiently well to be considered as a possible alternative. The 4 per cent antimony alloy, however, attains a hardness of 18 HV after 30 minutes and the alloys that contain 6, 8 and 10 per cent antimony could be handled almost immediately. Furthermore, the values given in Fig. 22.3 for hardness after two years are superior in all instances to those for air-cooled alloys of similar composition. Hardness curves decline steadily after two years and full evaluation of these alloys for use in the battery industry would require battery performance tests.

Dispersion Hardening

Another mechanism for strengthening of lead alloys involves elements that have low solubilities in solid lead, such as copper and nickel. Alloys that contain these elements can be processed so that no homogenisation results; most of the strengthening that occurs is developed through dispersion hardening, with some solid-solution hardening taking place as a secondary effect. The resulting structure is more stable than those developed by other hardening processes. Dispersion strengthening also has been achieved through powder metallurgy methods in which lead oxide, alumina or similar materials are dispersed in pure lead.

Fabrication

Although alloy selection is important, care must be taken in fabrication as well. Castings should be cooled rapidly to a temperature below that at which the structure breaks down or a coarse structure will be obtained. Age-hardening alloys should be extruded at a temperature and extrusions should not be allowed to cool slowly. Rolled alloys often are processed at insufficient temperatures; when this occurs, homogenisation after rolling is required if age hardening is to produce a beneficial response.

Cold Storage

Cold storage has been shown to improve the response of lead-antimony alloys to age hardening. Cooling a homogenised Pb-2Sb alloy to –10°C (+15°F) and holding for one or two days prior to room-temperature ageing results increases in both the rate of age hardening and the maximum hardness attained. This behaviour has been explained as the result of a reduction in the mobility of quenched-in free vacancies and a consequent reduction in their annihilation. The process allows the vacancies to form complexes with solute atoms and these complexes improve the efficiency of nucleation during ageing.

Service Temperatures

Service temperatures for lead alloys must be kept low to prevent overageing. Some cable-sheathing alloys, for example, have retained most of their creep resistance for up to 20 years, but exposure to elevated temperatures could have reduced this performance substantially. Even the normally stable age-hardened lead-antimony and lead-calcium alloys can be altered detrimentally by high service temperatures or excessive working.

MAGNESIUM ALLOYS

For certain magnesium alloys, development of properties depends almost entirely on heat treatment. In magnesium-zirconium alloys, however, the extremely pronounced grain-refining effect of the zirconium also plays a very important role in improving mechanical properties.

The basic temper designations outlined in Table 22.1 for magnesium alloys are used throughout this section to indicate the various types of heat treatment.

Table 22.1. Basic temper designations.

F	As fabricated
O	Annealed, re-crystallised (wrought products only)
H	Strain hardened (wrought products only):
	H1 Strain hardened only
	H2 Strain hardened and partially annealed

(Contd..)

	H3 Strain hardened and stabilised
W	Solution heat treated; unstable temper
T	Heat treated to produce stable tempers other than F, O or H:
	T2 Annealed (cast products only)
	T3 Solution heat treated and cold worked
	T4 Solution heat treated
	T5 Artificially aged only
	T6 Solution heat treated and artificially aged
	T7 Solution heat treated and stabilised
	T8 Solution heat treated, cold worked and artificially aged
	T9 Solution heat treated, artificially aged and cold worked
	T10 Artificially aged and cold worked

The mechanical properties of most magnesium casting alloys can be improved by heat treatment. Casting alloys can be grouped into six general classes of commercial importance on the basis of composition, as follows:

1. Magnesium-aluminium-manganese (example: AM100A).
2. Magnesium-aluminium-zinc (examples: AZ63A, AZ81A, AZ91C, AZ92A).
3. Magnesium-zinc-zirconium (examples: ZK51A, ZK61A).
4. Magnesium-rare earth metal-zinc-zirconium (examples: EZ33A, ZE41A).
5. Magnesium-rare earth metal-silver-zirconium, with or without thorium (examples: QE22A, QH21A).
6. Magnesium-thorium-zirconium, with or without zinc (examples: HK31A, ZH62A, HZ32A).

In most wrought alloys, maximum mechanical properties are developed through strain hardening, and these alloys generally are either used without subsequent heat treatment or merely aged to a T5 temper. Occasionally, however, solution treatment or a combination of solution treatment with strain hardening and artificial ageing, will substantially improve mechanical properties. Wrought alloys that can be strengthened by heat treatment are grouped into four general classes according to composition:

1. Magnesium-aluminium-zinc (example: AZ80A).
2. Magnesium-thorium-zirconium (example: HK31A).
3. Magnesium-thorium-manganese (examples: HM21A, HM31A).
4. Magnesium-zinc-zirconium (example: ZK60A).

Types of Heat Treatment

The heat treatments commonly used for various magnesium alloys, both cast and wrought, are indicated by temper designations in Table 22.2.

Table 22.2. Heat treatments commonly applied to magnesium alloys.

Alloy	*Heat treatment(a)*	*Alloy*	*Heat treatment(a)*
Casting alloys			
AM100A	T4, T5, T6,	ZE41A	T5
	T61(b)	ZE63A	T6(c)

(Contd..)

Alloy	Heat treatment(a)	Alloy	Heat treatment(a)
AZ63A	T4, T5, T6	ZH62A	T5
AZ81A	T4	ZK51A	T5
AZ91C	T4, T6	ZK61A	T4, T6
AZ92A	T4, T6	Wrought alloys	
EZ33A	T5	AZ80A	T5
HK31A	T6	HM21A	T5, T8, T81(d)
HZ32A	T5		
QE22A	T6	HM31A	T5
QH21A	T6	ZK60A	T5

(a) Indicated by temper designations (see Table 22.1). (b) Same as T6 except aged for longer time to increase yield strength. (c) Thermal treatment must include hydriding. (d) Mill modification of T8 to improve mechanical properties.

Annealing

Wrought magnesium alloys in various conditions of strain hardening or temper can be annealed by being heated at 290 to 455°C (550 to 850°F), depending on alloy, for one or more hours (Table 22.3). This procedure usually will provide a product with the maximum anneal that is practical. Because most forming operations on magnesium are done at elevated temperature, the need for fully annealed wrought material is less than with many other metals.

Table 22.3. Annealing temperatures for wrought magnesium alloys.

Alloy	Original temper	Annealing temperature(a) °C	°F
AZ31B	F,H10, H11, H23, H24, H26	345	650
AZ31C	F	345	650
AZ61A	F	345	650
AZ80A	F, T5, T6	385	725
HK31A	H24	400	750
HM21A	T5, T8, T81	455	850
HM31A	T5	455	850
ZK60A	F, T5, T6	290	550

(a) Time at temperature, 1 hour or more.

Stress relieving of wrought alloys

Stress relieving is used to remove or reduce residual stresses induced in wrought magnesium products by cold and hot working, shaping and forming, straightening and welding.

Table 22.4 gives the recommended stress-relieving times and temperatures for wrought magnesium alloys to obtain assemblies with maximum freedom from stress. When extrusions are welded to hard rolled sheet, the lower stress-relieving temperature and the longer time should be used to minimise distortion—for example, 150°C (300°F) for 60 minutes, rather than 260°C (500°F) for 15 minutes.

Table 22.4. Recommended stress-relieving treatments for wrought magnesium alloys.

	Sheet						*Extrusions and forgings*		
	Annealed			*Hard rolled*					
	Temperature		*Time,*	*Temperature*		*Time,*	*Temperature*		*Time,*
Alloy	°*C*	°*F*	*min*	°*C*	°*F*	*min*	°*C*	°*F*	*min.*
AZ31B	345	650	120	150	300	60	–	–	–
AZ31B-F	–	–	–	–	–	–	260	500	15
AZ61A	345	650	120	205	400	60	–	–	–
AZ61A-F	–	–	–	–	–	–	260	500	15
AZ80A-F	–	–	–	–	–	–	260	500	15
AZ80A-T5	–	–	–	–	–	–	205	400	60
HK31A	345	650	60	290	550	30	–	–	–
HM21A-T5	–	–	–	–	–	–	370	700	30
HM21A-T8	–	–	–	370	700	30	–	–	–
HM21A-T81	–	–	–	400	750	30	–	–	–
HM31A-T5	–	–	–	–	–	–	425	800	60
ZK60A-F	230	450	180	–	–	–	260	500	15
ZK60A-T5	–	–	–	–	–	–	150	300	60

Note: Stress relieving after welding, to prevent stress-corrosion cracking, is necessary only for alloys that contain more than 1.5 per cent aluminium.

Stress relieving of castings

The precision machining of castings to close dimensional limits, the necessity of avoiding warpage and distortion and the desirability of preventing stress-corrosion cracking in welded magnesium- aluminium casting alloys make it mandatory that cast components be substantially free from residual stresses. Although magnesium castings do not normally contain high residual stresses, the low modulus of elasticity of magnesium alloys means that comparatively low stresses can produce appreciable elastic strains.

Residual stresses may arise from contraction due to mould restraint during solidification, from non-uniform cooling after heat treatment or from quenching. Machining operations also can result in residual stress and require intermediate stress relieving prior to final machining.

Weld repairs may introduce severe stresses and should be followed by some type of heat treatment to prevent subsequent movement and cracking, as discussed in this section concerning heat treatment of repair-welded castings.

The following heat treatments for castings will provide stress relief without significantly affecting mechanical properties:

Alloy	*Temper*	*Heat treatment*
Mg-Al-Mn	All	1 hour at 260°C (500°F)
Mg-Al-Zn	All	1 hour at 260°C (500°F)
ZK61A	T5	2 hours at 330°C (625°F) then 48 hours at 130°C (265°F)
ZE41A	All	2 hours at 330°C (625°F)

Solution treating and ageing

Schedules for solution treating and ageing of magnesium alloy castings are summarised in Table 22.5. In solution treating of magnesium-aluminium-zinc alloys, parts should be loaded into the furnace at approximately 260°C(500°F) and then raised to the appropriate solution-treating temperature slowly, to avoid fusion of eutectic compounds and resultant formation of voids. The time required to bring the load from 260°C to the solution-treating temperature is determined by the sizes of the load and by the composition, size, weight and section thickness of the parts, but 2 hours is a typical time. All other heat treatable magnesium alloys can be loaded into furnace at the solution-treating temperature. For alloy HK31A, it is important to bring the load to temperature as rapidly as possible, to avoid grain coarsening.

Table 22.5. Recommended solution-treating and ageing schedules for magnesium alloy castings.

(For castings up to 51 mm (2 inch) in section thickness; heavier sections may require longer times at temperature).

		Ageing(a)			*Solution treating(c)*					*Ageing after solution treating*		
		Temperature			*Temperature*			*Maximum temperature*		*Temperature*		
Alloy	*Final temper*	*°C, ± 6 (b)*	*°F, ± 10(b)*	*Time, hr.*	*°C, ± 6 (b)*	*°F, ± 10(b)*	*Time, hr.*	*°C*	*°F*	*°C, ± 6 (b)*	*°F, ± 10(b)*	*Time, hr.*
Magnesium-aluminium-zinc alloys(d)												
AM100A	T5	232	450	5	–	–	–	–	–	–	–	–
	T4	–	–	–	424(e)	795(e)	16–24(e)	432	810	–	–	–
	T6	–	–	–	424(e)	795(e)	16–24(e)	432	810	232	450	5
	T61	–	–	–	424(e)	795(e)	16–24(e)	432	810	218	425	25
AZ63A	T5	260(f)	500(f)	4f	–	–	–	–	–	–	–	–
	T4	–	–	–	385	725	10–14	391	735	–	–	–
	T6	–	–	–	385	725	10–14	391	735	218(f)	425(f)	5(f)
AZ81A	T4	–	–	–	413(e)	775(e)	16–24(e)	418	785	–	–	
AZ91C	T5	168(g)	335(g)	16(g)	–	–	–	–	–	–	–	–
	T4	–	–	–	413(e)	775(e)	16–24(e)	418	785	–	–	–
	T6	–	–	–	413(e)	775(e)	16–24(e)	418	785	168(h)	335(h)	16(h)
AZ92A	T5	260	500	4	–	–	–	–	–	–	–	–
	T4	–	–	–	407(j)	765(j)	16–24(j)	413	775	–	–	–
	T6	–	–	–	407(j)	765(j)	16–24(j)	413	775	218	425	5
Magnesium-zirconium alloys												
EZ33A	T5	216(k)	420(k)	5(k)	–	–	–	–	–	–	–	–
HK31A(m)	T6	–	–	–	566	1050	2	571	1060	204	400	16
HZ32A	T5	316	600	16	–	–	–	–	–	–	–	–
QE22A(n)	T6	–	–	–	527	980	4–8	538	1000	204	400	8
QH21A(n)	T6	–	–	–	527	980	4–8	538	1000	204	400	8
ZE41A	T5	329(p)	625(p)	2(p)	–	–	–	–	–	–	–	–
ZE63A(q)	T6	–	–	–	480	895	10–72	491	915	141	285	48
ZH62A	T5	329	625	2	–	–	–	–	–	–	–	–
		plus: 177	350	16	–	–	–	–	–	–	–	–

(Contd..)

Alloy	Final temper	Ageing(a) Temperature °C, ± 6 (b)	°F, ± 10(b)	Time, hr.	Solution treating(c) Temperature °C, ± 6 (b)	°F, ± 10(b)	Time, hr.	Maximum temperature °C	°F	Ageing after solution treating Temperature °C, ± 6 (b)	°F, ± 10(b)	Time, hr.
ZK51A	T5	177(r)	350(r)	12(r)	–	–	–	–	–	–	–	–
ZK61A	T5	149	300	48	–	–	–	–	–	–	–	–
	T6	–	–	–	499(s)	930(s)	2(s)	502	935	129	265	48

(a) Ageing of castings to the T5 temper is done from the as-cast condition. (b) Except where quoted differently. (c). After solution treatment and before subsequent ageing, castings are cooled to room temperature by fast fan cooling, except where otherwise indicated. Use carbon dioxide or sulphur dioxide atmosphere above 400°C(750°F). (d) For solution treating, Mg-Al-Zn alloys are loaded into the furnace at 260°C(500°F) and brought to temperature over a 2 hours period at a uniform rate of temperature increase. (e) Alternative treatment, to prevent germination (excessive grain growth): 6 hours at 413 ± 6°C (775 ± 10°F), 2 hours at 352 ± 6°C (665 ± 10°F), 10 hours at 413 ± 6°C (775 ± 10°F). (f) Alternative treatment: 5 hour at 232 ± 6°C (450 ± 10°F).(g) Alternative treatment: 4 hours at 216 ± 6°C (420 ± 10°F). (h) Alternative treatment: 5–6 hours at 216 ± 6°C (420 ± 10°F). (j) Alternative treatment , to prevent germination (excessive grain growth): 6 hours at 407 ± 6°C (765 ± 10°F), 2 hours at 352 ± 6°C (665 ± 10°F), 10 hours at 407 ± 6°C (765 ± 10°F). (k) Alternative treatment, which can be used where maximum resistance to creep at elevated temperature is not of prime importance: 2 hours at 343 ± 6°C (650 ± 10°F). (m) Alloy HK31A castings must be loaded into the furnace already at temperature and brought back to temperature as quickly as possible. (n) Quench from solution-treating temperature either in water at 65°C (150°F) or in other suitable quenching medium. (p) This treatment is adequate for development of satisfactory properties; it may be followed by 16 hours at 177 ± 6°C (350 ± 10°F), to provide very slight improvements in mechanical properties. (q) Alloy ZE63A must be solution treated in a special hydrogen atmosphere, because its mechanical properties are developed through hydriding of some of its alloying elements. Hydriding time depends on section thickness; as a guide, 6.4 mm (¼ inch) sections require approximately 10 hours and 19 mm (¾ inch) sections require about 72 hours. Following solution treatment, ZE63A should be quenched in oil, water spray or air blast. (r) Alternative treatment: 8 hours at 218 ± 6°C (425 ± 10°F) (s) Alternative treatment: 10 hours at 482 ± 6°C (900 ± 10°F).

During ageing, magnesium alloy parts should be loaded into the furnace at the treatment temperature, held for the appropriate period of time, and then cooled in still air. As indicated in Table 22.5, there is a choice of artificial ageing treatments for some alloys; results are closely similar for the alternative treatments given.

Re-heat treating

Under normal circumstances, when mechanical properties are within expected ranges and the prescribed heat treatment has been carried out, re-heat treating is seldom necessary. However, if the microstructures of heat treated castings indicate too high a compound rating or if the castings have been aged excessively by slow cooling after solution treating, re-heat treating is called for. Most magnesium alloys can be re-heat treated with little danger of germination (excessive grain growth). When re-heat treating of alloy HK31A is necessary, however, the castings should be checked carefully for evidence of germination. To prevent germination in Mg-Al-Zn alloys, solution re-heat treating time should be limited to 30 minutes (assuming proper solution treatment of thick sections during prior heat treatment).

Effects of Major Variables

Casting size and section thickness, relation of casting size to volume capacity of the furnace and arrangement of castings in the furnace are mechanical considerations that can affect heat treating schedules for all metals.

Section size and heating time

There is no general rule for estimating time of heating per unit of thickness for magnesium alloys. However, because of the high thermal conductivity of these alloys combined with their low specific heat per unit volume, parts reach soaking temperature quite rapidly. The usual procedure is to load the furnace and then begin the soaking period when the loaded furnace reaches the desired temperature.

The heat treating times given in Table 22.5 have been found to be satisfactory for normal furnace loads and for castings of moderate section thickness. In the heat treating of magnesium alloy castings with thick sections (occasionally as low as 25 mm or 1 inch but usually over 50 mm or 2 inch), a good rule is to double the time at the solution treating temperature. For example, the usual solution treatment for AZ63A castings is 12 hours at about 385°C (725°F), whereas 25 hours at about 385°C is suggested for castings with section thicknesses greater than 50 mm (2 inch). Similarly, the suggested solution treating schedule for preventing excessive grain growth in AZ92A castings is 6 hours at about 405°C (765°F), 2 hours at about 350°C (665°F) and 10 hours at about 405°C; but for castings with sections more than 50 mm (2 inch) thick, it is recommended that the last soak at 405°C be extended from 10 hours to 19 hours. The best way to determine whether or not additional solution-treating time is required is to cut a section through the thickest portion of a scrap casting and examine the centre of the section microscopically: if heat treatment will reveal a low compound rating.

Heat treating time and temperature

As demonstrated by the data in Figs. 22.5, 22.6, 22.7, the mechanical properties of magnesium alloys can be varied within wide limits by varying the heat treating times and temperatures recommended in Table 22.5. (Although, as illustrated in Fig. 22.5, the highest mechanical properties in test bars of QE22A-T6 were obtained by solution treating for 4 hours at 540°C or 1005°F; less distortion due to sagging is experienced in production castings solution treated for 8 hours at 527°C, or 980°F). The risk of incipient melting can also occur at the higher temperature.

Protective atmospheres

Although magnesium alloys can be heat treated in air, protective atmospheres are almost always used for solution treating. Government specification MIL-M-6857, for heat treating of magnesium castings, requires a protective atmosphere for solution treating above 400°C (750°F). Protective atmospheres serve the dual purpose of preventing surface oxidation (which, if severe, can decrease strength) and of preventing active burning should the furnace exceed proper temperature.

The two gases normally used are sulphur dioxide and carbon dioxide. Inert gases also may be used; however, in most instances, these gases are not practical because of higher cost. Sulphur dioxide is available bottled, while carbon dioxide may be obtained either bottled or as the product of recirculated combustion gases from a gas-fired furnace. A concentration of 0.7 per cent (0.5 per cent min.) sulphur dioxide will prevent active burning to a temperature of 565°C (1050°F), provided that melting of the alloy has not occurred. Carbon dioxide in a concentration of 3 per cent will prevent active burning to 510°C (950°F) and a carbon dioxide concentration of 5 per cent will provide protection to about 540°C (1000°F). Although sulphur dioxide is more expensive per unit volume, the volume of sulphur dioxide required for a protective atmosphere is only $\frac{1}{6}$ the volume of carbon dioxide required. Thus, the cost of producing a protective atmosphere with bottled gases is less using sulphur dioxide. Where gas-fired furnaces are used and the atmosphere is obtained by recirculating combustion gases, a carbon dioxide atmosphere has the lower cost.

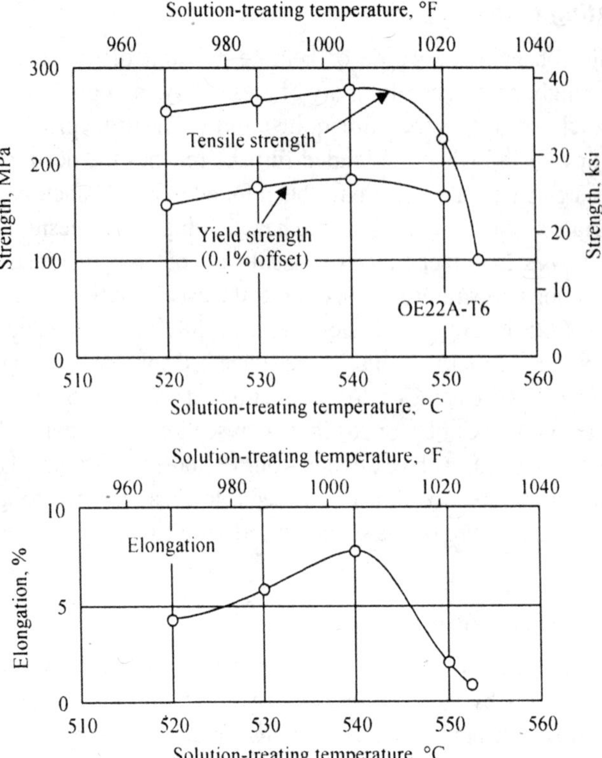

Data were obtained from test bars of casting alloy QE22A-T6 machined from 25 mm (1 inch) diam cast specimens. The bars were held at temperature for 4 hours.

Fig. 22.5. Tensile properties of alloy QE22A-T6 as functions of solution-treating temperature.

The use of sulphur dioxide requires frequent cleaning of furnace controls and fixtures and replacement of furnace parts, because of the formation of corrosive sulphuric acid in the furnace system. When it is desired to heat treat both magnesium and aluminium alloy castings in the same furnace, a carbon dioxide atmosphere is required, because sulphur dioxide is harmful to aluminium. Government regulations regarding the use of sulphur dioxide also suggest the use of carbon dioxide.

Equipment and Processing

In solution treating and artificial ageing of magnesium alloys, it is standard practice to use an electrically heated or gas-fired furnace equipped with high-velocity fan or comparable means for circulating the atmosphere for solution treating sometimes contains sulphur dioxide, only furnaces that are gastight and that provide an inlet for introducing protective atmosphere are suitable.

The atmosphere is circulated past the heating elements and through the load in the basket. A rapid rate of circulation is necessary for maintaining an even distribution of temperature throughout the load. The minimum rate of circulation varies with furnace design and loading practice; one manufacturer recommends a rate equivalent to about 45 changes of atmosphere per minute.

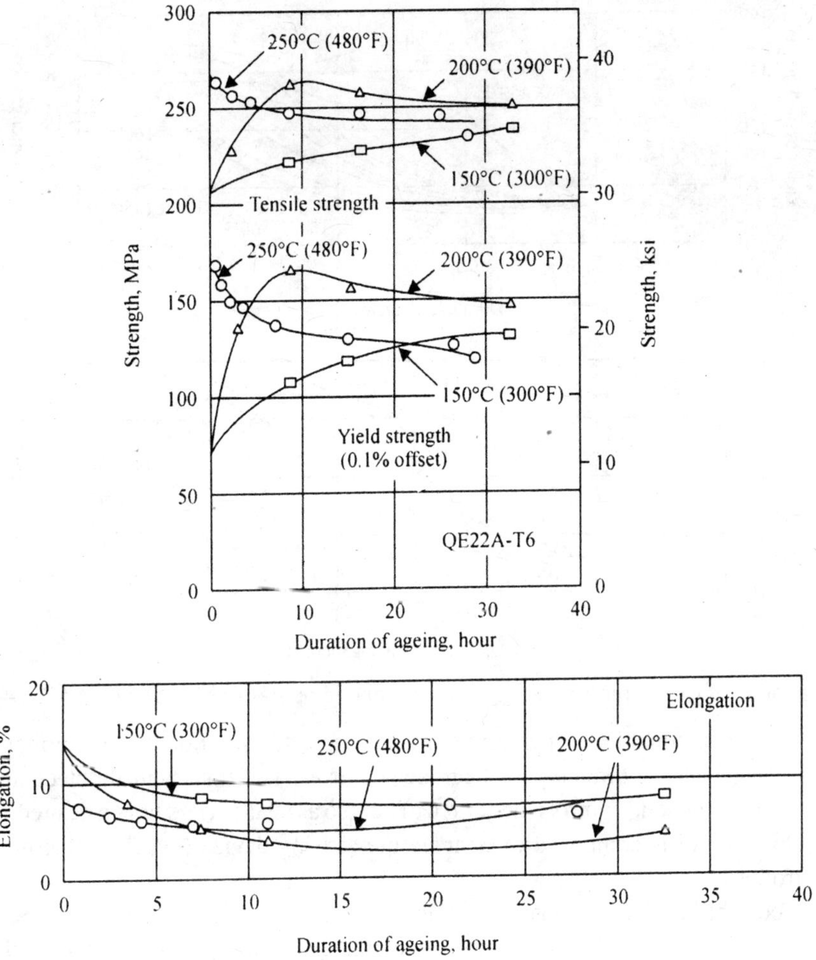

Data were obtained from test bars of casting alloy QE22A-T6 machined from 25 mm (1 inch) diam cast specimens.

Fig. 22.6. Tensile properties of alloy QE22A-T6 as functions of ageing time and temperature.

Furnace loading

Loading of furnaces used for heat treatment of magnesium parts is an important consideration. Parts must be clean and free from grinding dust, shavings, chips, savings and other fines; this is particularly important at the higher temperatures used for solution treatment. A furnace load should be comprised of only one alloy composition because of the variation in non-equilibrium fusion points among various alloys. Furnaces should be loaded in an orderly manner that will not interfere with air circulation and cause uneven heating.

Temperature control

A high degree of temperature control is required for heat treating of magnesium alloys. As indicated in Table 22.5, the maximum allowable temperature variation is ± 6°C (± 10°F) for the solution treating operation.

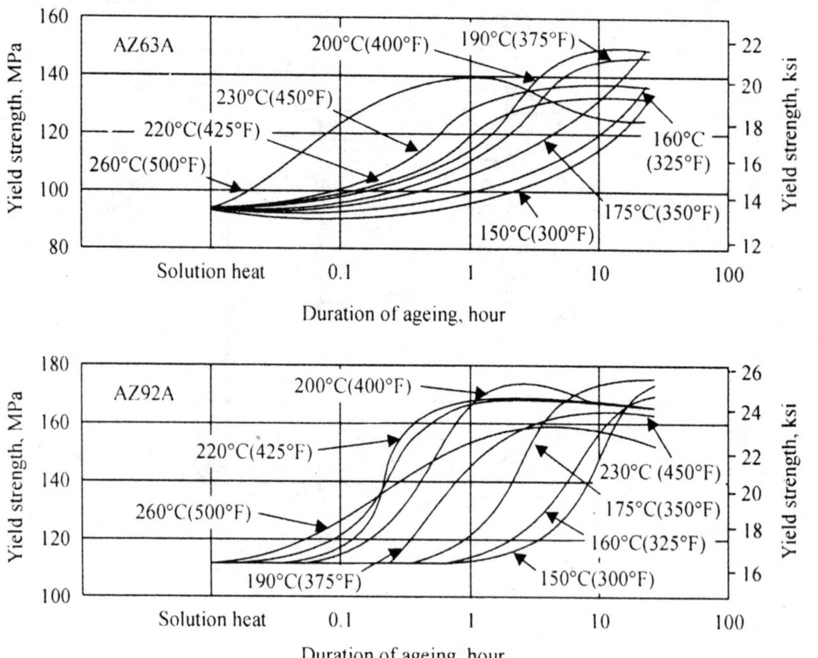

Fig. 22.7. Variation of yield strength with ageing time and temperature for sand cast AZ63A and AZ92A.

The safest and most suitable temperature-control system for solution and precipitation treating of magnesium alloys consists of three types of furnace controls. The first control detects the temperature within the heating chamber and controls the source of heat so as to attain and hold a desired pre-determined temperature. This control is composed of temperature-sensing device (usually a thermocouple) and a recording controller.

The second control is a furnace-charge-temperature sensor and is used to determine when the load has reached temperature. It consists of a thermocouple strategically located in the charge and a temperature-indicating device; an indicator-controller may be used to actuate an alarm, signal or timing device to indicate that the charge has reached temperature. This control is useful for preventing overshooting of the desired maximum temperature because of high heat input or because the load is small in comparison with the size of the furnace.

The third control is a safety device and is used to prevent serious overheating of the load or a possible magnesium fire. It consists of a thermocouple, usually located in the roof of the furnace and an indicating controller that will turn off the source of heat to the furnace. This controller should require re-setting before the furnace will heat again.

Quenching media

Magnesium alloy products normally are quenched in air following solution treatment. Still air usually is sufficient; forced-air cooling is recommended for dense loads or for parts that have very thick sections. The only common exceptions to air quenching are alloys QE22A and QH21A, for which water quenching at 60 to 95°C (140 to 200°F) is used to develop the best mechanical properties. Glycol or oil quenchants will produce similar properties with reduced distortion. For QE22A products subject to distortion as a result of the severity of water quenching, air cooling may be used provided that the cooling rate exceeds 3°C/s (5°F/s). The effects of quenching in still air, in forced air and in water at 65°C (150°F) on the tensile properties of separately cast test bars of QE22A-T6 are indicated by the data in Table 22.6.

Table 22.6. Effect of quenching medium on average tensile properties of QE22A-T6.

	Tensile strength		*Yield strength(a)*		*Elongation(b),*
Quenching medium	*MPa*	*ksi*	*MPa*	*ksi*	*%*
Still air(c)	232	33.6	158	22.9	3.8
Air blast(c)	250	36.2	182	26.4	3.5
Water at 65°C (150°F)(c)	270	39.2	190	27.5	3.0
30% glycol at room temperature(d)	269	39.0	190	27.5	3.0

(a) At 0.2 per cent offset. (b) In 50 mm or 2 inch. (c) Properties determined on bars machined from 25 mm (1 inch) diam separately cast specimens. (d) Properties determined on bars machined from castings.

Control of distortion

The strength of cast magnesium alloys decreases at elevated temperature to such an extent that it is often necessary during solution treating to prevent intricate castings from sagging from their own weight and to keep flat castings from warping as a result of the relief of casting stresses. To accomplish these ends, tie bars are made an integral part of the casting, simple fixturing is used or complicated cast or machined fixtures are produced. The method used depends on the complexity and quantity of castings to be made, and on the degree of dimensional control required. Whatever the method, it should not interfere unnecessarily with the free circulation of heat about the castings. Although fixturing decreases warpage of castings, some castings still require straightening after solution treating. Straightening is most readily done after solution treating, prior to ageing.

Prevention of Heat Treating Problems

Six common problems that may be encountered in heat treating magnesium alloys are oxidation, fusion voids, warpage, grain coarsening, germination and inconsistent properties. Causes and prevention are discussed in Table 22.7.

Table 22.7. Causes and prevention of problems commonly encountered in heat treatment of magnesium alloys.

Oxidation

Cause: Heat treating without use of protective or inert atmosphere; can lead to local weakening of the metal part, and even to burning of the metal in the furnace

Prevention: Heat treat in a controlled atmosphere containing about 0.5 to 1.5 per cent SO_2 or 3 to 5 per cent CO_2, or (less practical, because more costly) in an inert-gas atmosphere. Ensure that furnace is clean and completely dry

Grain coarsening

Cause: Occurs in HK31A as a result of delay in attaining solution temperature or of holding at solution temperature for an excessive period of time

Prevention: Prior to solution treating of HK31A, furnace should be at temperature; castings should be loaded quickly, and the loaded furnace should be closed and brought to temperature as rapidly as possible. Time at temperature should be controlled

(Contd..)

Fusion voids

Cause: Use of improper rate of heating from 260 to 370°C (500 to 700°F) for Mg-Al-Zn alloys or exceeding recommended temperature in solution heat treating of these alloys or of the alloys that contain zinc, thorium and rare earth metals as major alloying elements.

Prevention: Charge furnace with Mg-Al-Zn alloys at 260°C (500°F), then heat gradually to solution-treating temperature over a period of 2 hours. Control solution temperature so as not to exceed designated temperature by more than 6°C (10°F).

Warpage

Cause: Lack of support of castings during heat treatment; uneven distribution of heat.

Prevention: Support long spans of thin cross sections; use jigs for intricate shapes. Distribute load in furnace so as to obtain good circulation of atmosphere.

Germination

Cause: Grain growth, which occurs in AM100A, AZ81A, AZ91C and AZ92A toward the end of the solution-treating cycle

Prevention: Use antigermination heat treating schedules (presented as alternative treatments in footnotes e and j in Table 22.5)

Inconsistent properties

Cause: Insufficient or excessive furnace temperature, inadequate circulation of heat in the furnace, faulty temperature control, very slow cooling from the solution-treating temperature or inadequatesolution-treating time for heavy sections.

Prevention: Check temperature at various positions in furnace with standardised thermocouple. Distribute castings in furnace so as to provide adequate circulation of heat. Check temperature controls often and ensure that controls are located so as to provide uniformity of temperature. Increase solution-treating time to allow complete homogenisation.

Dimensional Stability

In normal service up to approximately 95°C (200°F), all magnesium casting alloys exhibit good dimensional stability and changes.

Some cast magnesium-aluminium-manganese and magnesium-aluminium-zinc alloys in certain tempers exhibit slight permanent growth after relatively long exposure to temperatures exceeding 95°C (Figs. 22.8 and 22.9). This growth, although slight, can give rise to problems.

Example. An aircraft-engine cover plate, sand cast of alloy AZ63A, was used in the as-cast condition (F temper) and the operating temperature of the engine was high enough to cause growth. As a result of this growth, it was necessary to pry off the cover plate when the engine was taken down for overhauling, and the plate could not be replaced because the holes did not line up properly with mating studs. This problem could have been eliminated by proper choice of temper (T5 or T6) prior to placing the cover in service.

In contrast to the growth characteristics of the magnesium-aluminium-zinc alloys are those of the magnesium alloys containing thorium, rare earth metals and zirconium as major alloying elements. These alloys normally are used in the T5 or T6 temper, and they shrink, rather than grow, on exposure to elevated temperature (Table 22.8).

Table 22.8. Contraction of magnesium castings alloys at elevated temperatures.

Temperature		*Unit contraction, 100μm/m (0.0001 inch/inch), after exposure time, hr., of:*			
°C	*°F*	*10*	*100*	*1000*	*5000*
EZ33A-T5					
205	400	1.1	1.3	1.3	1.3
260	500	1.3	1.6	1.8	1.9

(Contd..)

Temperature		Unit contraction, 100μm/m (0.0001 inch/inch), after exposure time, hr., of:			
°C	°F	10	100	1000	5000
315	600	1.2	1.5	1.7	1.8
370	700	1.0	1.2	1.3	1.4
HK31A-T6					
205	400	0.3	0.3	0.3	0.3
260	500	0.3	0.5	0.7	0.7
315	600	0.2	0.5	1.3	1.3
370	700	0.6	1.1	1.1	1.1
HZ32A-T5					
205	400	1.1	1.3	1.4	1.4
260	500	0.8	1.0	1.1	1.2
315	600	0.8	1.0	1.1	1.2
370	700	0.6	0.8	0.9	1.0

Heat Treatment of Weld-Repaired Castings

Magnesium sand castings that have been reclaimed or repaired by welding may be subsequently heat treated to relieve residual stresses. Heat treatment also may be necessary for restoring the mechanical properties of the casting when these have been impaired or modified by pre-heating prior to welding or for heat treating the weld zone as such.

Table 22.9 shows the heat treating schedules recommended for magnesium castings after welding. The heat treating procedures shown are based on both the temper of the casting before the welding operation and the temper desired after welding. These post-welding heat treatments are normally all that is required for adequate stress relief of castings and for optimum mechanical properties in weld areas. The solution treatments here require the use of a protective atmosphere to prevent oxidation or burning. Only the minimum time for complete solution (½ hour) is used for the welded casting alloys AZ81A, AZ91C, and AZ92A if the castings were in the T4 or the T6 condition prior to welding.

Table 22.9. Postweld heat treatmonts for magnesium alloy castings.

Alloy	*Welding rod*	*Temper before welding*	*Desired temper after welding*	*Postweld heat treatment*
AZ63A	AZ63A or	F	T4	12 hours at 385 ± 6°C (725 ± 10°F)(b)
	AZ92A(a)	F	T6	12 hours at 385 ± 6°C (725 ± 10°F)(b), plus 5 hours at 220°C (425°F)
		T4	T4	½ hour at 385 ± 6°C (725 ± 10°F)
		T4 or T6	T6	½ hour at 385 ± 6°C (725 ± 10°F), plus 5 hours at 220°C (425°F)
AZ81A	AZ92A or AZ101	T4	T4	½ hour at 413 ± 6°C (775 ± 10°F(c)
AZ91C	AZ95A or	T4	T4	½ hour at 413 ± 6°C (775 ± 10°F(c)

(Contd..)

Alloy	*Welding rod*	*Temper before welding*	*Desired temper after welding*	*Postweld heat treatment*
	AZ101	T4 or T6	T6	½ hour at 413 ± 6°C (775 ± 10°F)(c), plus 4 hours at 215°C (420°F) or 16 hours at 170°C (335°F)
AZ92A	AZ92A	T4	T4	½ hour at 407 ± 6°C (765 ± 10°F(c)
		T4 or T6	T6	½ hour at 407 ± 6°C (765 ± 10°F(c), 4 hours at 260°C (500°F) or hour at 220°C (425°F)
EZ33A	EZ33A	F or T5	T5	2 hours at 345°C (650°F)(d) or 5 hours at or 220°C (420 or 425°F)
HK31A	HK31A(g)	T4 or T6	T6	16 hours at 205°C (400°F)(e)
HZ32A	HZ32A(g)	F or T5	T5	16 hours at 315°C (600°F)
QE22A	QE22A	F or T6	T6	4 hours at 524 ± 6°C (975 ± 10°F)(c); quench; 8 hours at 205°C (400°F)
QH21A	QH21A	F or T6	T6	4 hours at 524 ± 6°C (975 ± 10°F)(c); quench; 8 hours at 205°C (400°F)
ZE41A	ZE41A(g)	F or T5	T5	2 hours at 330°C (625°F)(f)
ZH62A	ZH62A(g)	F or T5	T5	12 hours at 250°C (480°F)(f)
ZK51A	ZK51A(g)	F or T5	T5	2 hours at 330°C (625°F), plus 16 hours at 175°C (350°F)

(a) AZ63A rod must be used for welding AZ63A in the F temper, because 12 hours at 385°C(725°F) causes germination in welds made with AZ92A rod; AZ92A rod normally is used for welding AZ63A in the T4 or T6 condition unless AZ63A rod is required by specifications. (b) Pre-heat to 260°C (500°F); heat to specified temperature at no more than 83°C/hour (150°F/hour). (c) Use carbon dioxide or sulphur dioxide atmosphere. (d) Heating for 2 hours at 345°C (650°F) results in slight loss of creep strength. (e) Alternative treatment; 1 hour at 315°C (600°F), plus 16 hours at 205°C (400°F). (f) Alternative treatment: 2 hours at 330°C (625°F), plus 16 hours at 175°C (350°F). (g) Or EZ33A.

Welded Mg-Al-Zn alloy castings that do not require solution treatment after welding should be stress relieved 1 hour at 260°C (500°F), to eliminate the possibility of stress- corrosion cracking.

Evaluation of Heat Treated Parts

The effectiveness of heat treating procedures can be determined by hardness and tensile tests and by microscopic examination.

Indentation hardness tests

Indentation hardness tests are rapid and usually can be performed on the parts heat treated, without the necessity for a special test specimen. Brinell and Rockwell E hardness tests are normally used, but the Rockwell 15T superficial test may be required for thin sections. Soft material with large grains should be tested by the Brinell method for best results.

The strength of magnesium alloys usually increases as hardness increases, but a graph of strength values determined from hardness plotted against actual strength values from the same specimens shows so much scatter that hardness cannot be used as an index for strength. Furthermore, hardness cannot indicate how an alloy will machine or how easily it can be fabricated by other methods; it is primarily a measure of the temper of the material, and normally it suffices for this purpose.

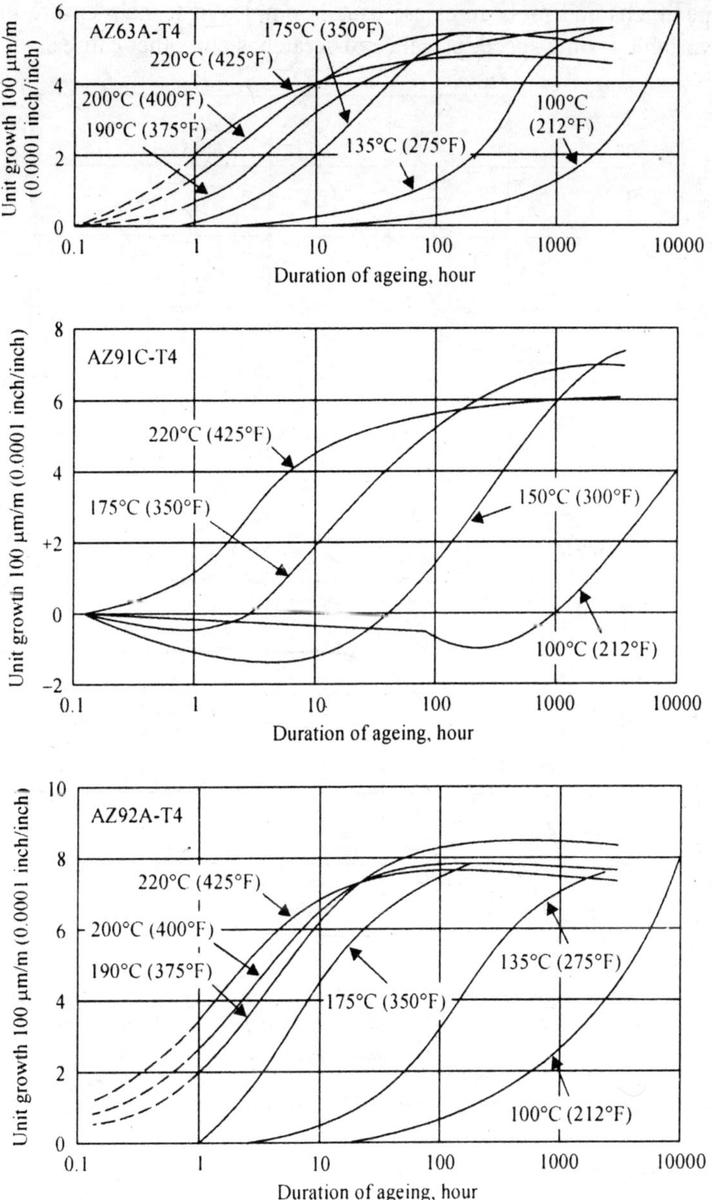

Fig. 22.8. Variation of growth with ageing time and temperature for solution-treated AZ63A, AZ92A and AZ91C.

Tensile tests

Tensile tests more accurately indicate the temper of magnesium alloys but require special test specimens. For casting alloys, these can be separately cast unmachined specimens, although specimens machined from castings are more representative of actual casting properties. Test bars must be machined from extrusions, forgings and sheet. Test bars from extrusions and forgings are machined round for convenience.

Standard ASTM specimens and procedures are usually employed to help ensure uniform results and avoid errors from variable testing speed, presence of scratches and other causes.

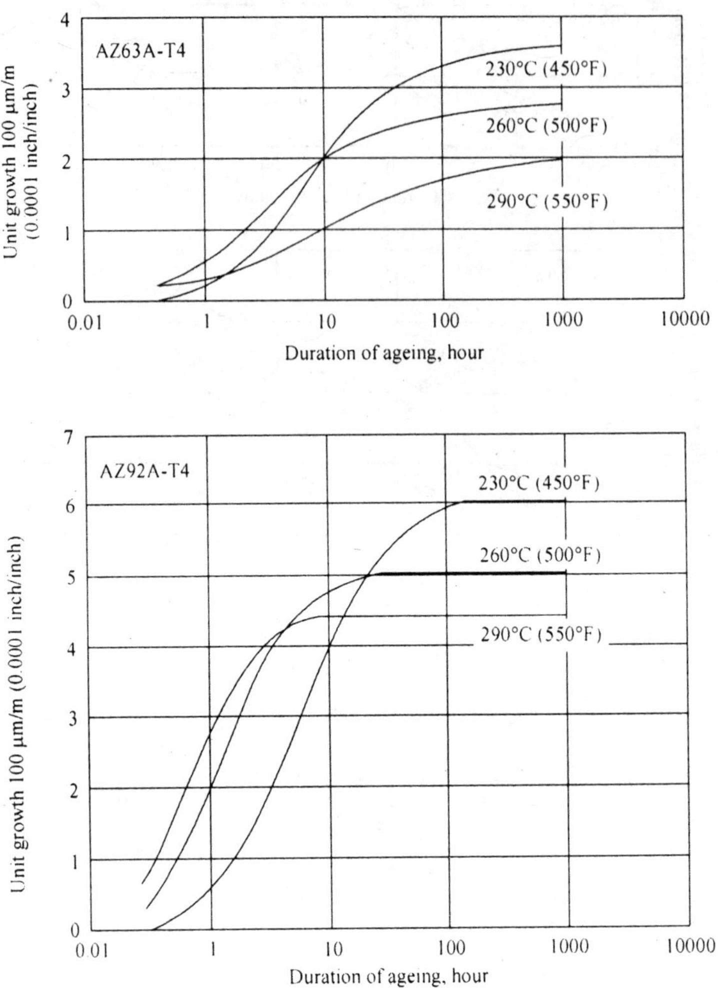

Fig. 22.9. Variation of growth with ageing time and temperature for solution-treated AZ63A and AZ92A.

Microscopic examination

A method has been devised for microscopic examination of heat treated magnesium alloys whereby even inexperienced personnel can polish specimens and rate microstructure as a series of numbers. In this method, standard microstructures are prepared for purposes of comparison. Specimens cut from magnesium alloy parts are examined and by comparing them with the standard micrographs, it is possible to rate the alloy for the following conditions:

1. Massive compound in cast alloys.
2. Percentage of pearlitic type of precipitate in cast alloys.
3. Porosity and 'burning' voids in improperly solution heat treated cast alloys.

4. Grain size in cast and wrought alloys.
5. Massive compound in extruded, forged or rolled alloys.

Determination of temper

For alloys AZ91C and AZ92A, the following solution can be applied to heat treated parts to identify the T6 conditions: 60 parts ethylene glycol; 20 parts glacial acetic acid; 19 parts distilled water; 1 part nitric acid, 42°Be. For ZE41A, the following solution can be applied to heat treated parts to identify the T5 condition: 5 parts glacial acetic acid; 95 parts distilled water. For either of these tests, the test surface should be prepared by sanding first with 180-grit paper and then with 220 to 400-grit paper to produce a smooth area about 25 mm (1 inch) square. After the test section has been wiped clean, one drop of the solution is applied with an eye-dropper. After 30 seconds, the test section is rinsed with water and blotted dry with a piece of soft cloth.

For AZ91C and AZ92A, the surface of a part in a T6 condition will appear darker where etched. For ZE41A, the surface will appear lighter where etched if the part is in the T5 condition. This alloy will show a bluish tint in the as-cast condition, whereas alloy ZE41A in the T5 temper—either 2 hours at 330°C (625°F) or 2 hours at 330°C plus 16 hours at 175°C (350°F)—tends to show a brown-to-beige tint.

Thc inspected area should be wiped clean, and a fresh type I chrome pickle should be applied to prevent corrosion.

Prevention and Control of Magnesium Fires

Improper heat treating of magnesium alloy castings not only will result in damaged castings but also may cause fire. Only clean, dry castings should be introduced into the furnace. Castings contaminated with filings, chips, oil or other foreign matter should first be cleaned. The furnace itself should be protected from these same contaminants and should be completely dry. The charge should consist of castings of only one alloy and the recommended heat treating cycle for that alloy should be closely followed.

Occasionally, because of equipment malfunction, carelessness or operator error, fires do occur. They are usually detectable either by an increase in furnace temperature without an increase in input of heat or by seepage of light-coloured smoke from the furnace.

Note: Under no circumstance should water be used to extinguish a magnesium fire.

All power, fuel and protective-atmosphere lines should be shut off immediately. This in itself may extinguish small fires, which will die from lack of oxygen in a tight furnace. Should the fire continue to burn, several methods of extinguishing it are possible, depending on the nature of the fire. If the fire is small and the burning castings are readily accessible and can be safely removed from the furnace, they should be removed to a steel container and covered with commercially available fire-extinguishing powder intended for use with magnesium. If the burning castings are not both safely and readily accessible, a pump can be used to throw a stream of powder onto the burning castings in the furnace.

When the fire is such that neither of the above methods can safely be used for extinguishing it, another method is use of boron trifluoride (BF_3) or boron trichloride (BCl_3), both of which are available as bottled gases. As with the other methods, all power, fuel and protective-atmosphere lines should be shut-off.

Boron trifluoride gas is introduced into the furnace through a pipe in the furnace door or wall so that a minimum concentration of 0.04 per cent is produced. The gas flow is continued until the fire is extinguished and the temperature of the furnace drops to 370°C (700°F). Teflon hose is suitable for transferring the pressurised boron trifluoride gas from bottle to furnace.

Boron trichloride gas also is introduced into the furnace through a pipe in the furnace door or wall, in a concentration of about 0.4 per cent. To ensure a sufficient volume of gas, a bank of infrared lamps or some other suitable device is required for heating the bottled gas. Boron trichloride gas reacts with hot magnesium to form a protective film over the castings. A supply of gas is maintained until the fire is extinguished and the temperature of the furnace drops to 370°C (700°F). In a totally enclosed furnace, it is feasible to employ the furnace fan to circulate the boron trifluoride or boron trichloride extinguishing gas around the castings. (Teflon hose can be used for transferring the boron trichloride gas from the bottle to the furnace).

Boron trichloride predates boron trifluoride as a fire extinguisher for magnesium. Boron trifluoride is preferred, because it is effective in lower concentration, it does not require a heat source to ensure an adequate supply of gas, and its reaction products are less hazardous than those of boron trichloride. The fumes of boron trichloride are irritating and are ranked with hydrochloric acid fumes as a health hazard.

If the heat treating furnace fire includes more than several hundred pounds of magnesium parts, is will advanced prior to discovery, involves a large pool of metal on the floor of the furnace or is in a furnace with excessive air leaks, these gases cannot be expected to extinguish the fire completely. However, both boron trifluoride and boron trichloride are effective in slowing or suppressing the fire until it can be extinguished with other solid materials.

Dry cast iron chips, graphite powder combined with heavy hydrocarbons and (occasionally) foundry melting flux have been used for extinguishing large magnesium fires; the effectiveness of these materials is based on smothering the fire from the oxygen of the atmosphere in the furnace.

Besides the normal safety equipment used by personnel in fire fighting, coloured glasses should be used when fighting a magnesium fire, for protection of the eyes against the characteristic intense white light.

Chapter 23

Heat Treating of Nickel and Nickel Alloys

INTRODUCTION

Nickel and nickel alloys may be subjected to one or more of five principal types of heat treatment, depending on chemical composition, fabrication requirements and intended service. These methods include:

Annealing: A heat treatment designed to produce a re-crystallised grain structure and softening in work-hardened alloys. Annealing usually requires temperatures between 705 and 1205°C (1300 and 2200°F), depending on alloy composition and degree of work hardening.

Stress relieving: A heat treatment used to remove or reduce stresses in work-hardened non-age-hardenable alloys without producing a re-crystallised grain structure. Stress-relieving temperatures for nickel and nickel alloys range from 425 to 870°C (800 to 1600°F), depending on alloy composition and degree of work hardening.

Stress equalising: A low-temperature heat treatment used to balance stresses in cold worked material without an appreciable decrease in the mechanical strength produced by cold working.

Solution treating: A high-temperature heat treatment designed to put age-hardening constituents and carbides into solid solution. Normally applied to age-hardenable materials before the ageing treatment.

Age hardening (precipitation hardening): A treatment performed at intermediate temperatures (425 to 870°C; 800 to 1600°F) on certain alloys in order to develop maximum strength by precipitation of a dispersed phase throughout the matrix.

Fig. 23.1 illustrates the effect of holding for 3 hours at various temperatures on room-temperature mechanical properties of cold drawn Monel 400 rod. Heating at 700°C (1300°F) or above produces the soft annealed condition; heating in the range 400 to 595°C (750 to 1100°F) results in stress equalisation—producing an increase in the proportional limit, a slight increase in tensile strength and no significant change in elongation.

ANNEALING

As applied to nickel and nickel alloys, annealing consists of heating the metal at a pre-determined temperature for a definite time and then slowly or rapidly cooling it, to produce a change in mechanical properties—usually, a complete softening as a result of re-crystallisation. Nickel and nickel alloys that have been hardened by cold working operations, such as rolling, deep drawing, spinning or severe bending, require softening before cold working can be continued. The thermal treatment that will produce this conditions is known as annealing or soft annealing.

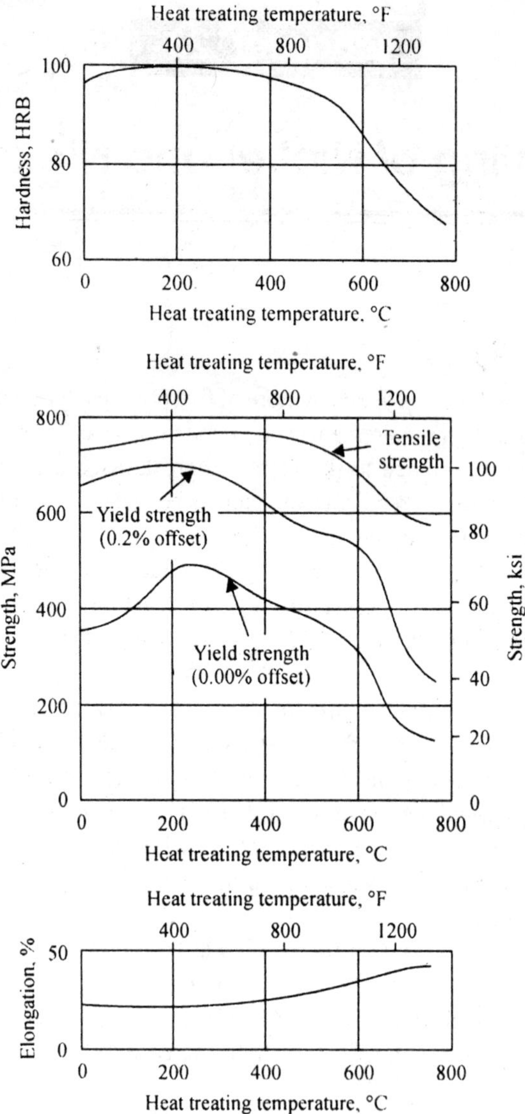

Fig. 23.1. Effects of holding 3 hours at various temperatures on room-temperature properties of cold drawn Monel 400 rod.

Annealing Practice and Methods

The differences in chemical composition among nickel and nickel alloys (Table 23.1) necessitate modifications in annealing temperatures (Table 23.2) as well as in furnace atmospheres. The precipitation-hardening alloys must be cooled rapidly after annealing if maximum softness is desired.

Three soft-annealing methods in general commercial use—open, closed and salt bath annealing—are discussed below.

Table 23.1. Nominal compositions of nickels and nickel alloys.

	Composition, wt%											
Material	*Ni*	*C*	*Mn*	*Fe*	*Si*	*Cu*	*Cr*	*Ti*	*Al*	*Mo*	*Co*	*Other*
Nickel 200	99.5	0.06	0.25	0.15	0.05	0.05	–	–	–	–	–	–
Nickel 201	99.5	0.01	0.20	0.15	0.05	0.05	–	–	–	–	–	–
Monel 400	66.0	0.12	0.90	1.35	0.15	31.5	–	–	–	–	–	–
Monel R-405	66.0	0.18	0.90	1.35	0.15	31.5	–	–	–	–	–	–
Monel K-500	65.0	0.15	0.60	1.00	0.15	29.5	–	0.50	2.80	–	–	–
Inconel 600	76.0	0.04	0.20	7.20	0.20	0.10	15.8	–	–	–	–	–
Inconel 601	60.5	0.05	0.50	14.1	0.25	–	23.0	–	1.35	–	–	–
Inconel 617	54.0	0.07	–	–	–	–	22.0	–	1.00	9.0	12.5	–
Inconel 625	61.0	0.05	0.25	2.5	–	–	21.5	–	–	9.0	–	3.65 Nb
Inconel 718	52.5	0.04	0.20	18.0	0.20	0.10	19.0	0.80	0.60	3.00	–	5.20 Nb
Inconel X-750	73.0	0.04	0.70	6.75	0.30	0.05	15.0	2.50	0.80	–	–	0.85 Nb
Hastelloy B	64.0	0.03	–	5.0	–	–	–	–	–	28.0	2.5 max	–
Hastelloy C	56.0	0.05	–	5.5	–	–	15.5	–	–	16.0	2.5 max	4.0 W
Hastelloy X	48.0	0.10	–	18.5	–	–	22.0	–	–	9.0	1.5	0.6 W

Open annealing

Open annealing is used most often. The material to bc annealed is heated at the selected temperature and protected from oxidation by the products of combustion in a fuel-heated furnace or by a reducing gas introduced into an electric furnace. Temperature control is critical, because the annealing period is short (Table 23.2). During preliminary heating of the furnace, the burners may be adjusted for optimum combustion; but, before charging, the air supply should be reduced to provide the excess of reducing gases required. Also, the top vents or dampers should be closed or partly closed to provide a positive gas pressure over the hearth and thereby prevent leakage of air.

Closed (box) annealing

Closed (box) annealing requires more time than open annealing because of the lower temperatures used (Table 23.2). Temperature control is less critical than in open annealing. In most instances, the weight of the container exceeds that of the work; consequently, the amount of fuel required, heating time and cost are greater than in open annealing. Small pressed parts, rivets, coiled wire and similar work are annealed in boxes should be sealed with fireclay, sand or other suitable refractory material. A small quantity of charcoal, pre-baked for 3 hours at 650 to 815°C (1200 to 1500°F) to remove occluded gases, should be packed in the box at points where air could enter. Reducing gas is admitted to the box immediately upon withdrawal of the box from the furnace or when the burners are shut off. Following annealing, the flow of gas should be regulated to maintain a positive pressure within the box; if the gas is combustible, it will burn gently at all outlets.

Table 23.2. Annealing, stress-relieving and stress-equalising schedules for nickels and nickel alloys.

	Soft annealing							
	Open annealing				*Closed annealing*			
	Temperature		*Time,*	*Cooling*	*Temperature*		*Time*	*Cooling*
Material	°C	°F	*min.*	*method(a)*	°C	°F	*hr.*	*method*
Nickel 200	815 to 925	1500 to 1700	½ to 5	AC or WQ	705 to 760	1300 to 1400	2 to 6	AC
Nickel 201	760 to 870	1400 to 1600	½ to 5	AC or WQ	705 to 760	1300 to 1400	2 to 6	AC
Monel 400	870 to 980	1600 to 1800	2 to 10	AC or WQ	760 to 815	1400 to 1500	1 to 3	AC
Monel R-405	870 to 980	1600 to 1800	2 to 10	AC or WQ	760 to 815	1400 to 1500	1 to 3	AC
Monel K-500	870 to 1040	1600 to 1900	5 to 15	WQ	—	Not applicable	—	
Inconel 600	925 to 1040	1700 to 1900	15 to 30	AC or WQ	925 to 980	1700 to 1800	1 to 3	AC
Inconel 601	1095 to 1175	2000 to 2150	15 to 30	AC or WQ	1095 to 1175	2000 to 2150	1 to 3	AC
Inconel 617	1120 to 1175	2050 to 2150	15 to 30	AC or WQ	1120 to 1175	2050 to 2150	1 to 3	AC
Inconel 625	980 to 1150	1800 to 2100	15 to 30	AC or WQ	980 to 1150	1800 to 2100	1 to 3	AC
Inconel 718	955 to 980	1750 to 1800	15 to 30	AC	—	Not applicable	—	
Inconel X-750	1095 to 1150	2000 to 2100	15 to 30	AC	—	Not applicable	—	
Hastelloy B	1095 to 1185	2000 to 2165	5	AC or WQ	–	–	–	–
Hastelloy C	1215	2220	5	WQ	–	–	–	–
Hastelloy X	1175	2150	5 to 15	AC or WQ	1175	2150	1	AC or WQ

(a) AC, air cool; WQ, water quench.

	Stress relieving				*Stress equalising*			
	Temperature		*Time,*	*Cooling*	*Temperature*		*Time,*	*Cooling*
Material	°C	°F	*min.*	*method*	°C	°F	*hr.*	*method*
Nickel 200	480 to 705	900 to 1300	½ to 3	AC	260 to 480	500 to 900	1 to 2	AC
Nickel 201	480 to 705	900 to 1300	½ to 3	AC	260 to 480	500 to 900	1 to 2	AC
Monel 400	540 to 565	1000 to 1050	1 to 2	AC	230 to 315	450 to 600	1 to 3	AC
Monel R-405	–	–	–	–	–	–	–	–
Monel K-500	–	–	–	–	–	–	–	–
Inconel 600	760 to 870	1400 to 1600	1 to 2	AC	–	–	–	–
Inconel 601	–	–	–	–	–	–	–	–
Inconel 617	–	–	–	–	–	–	–	–
Inconel 625	–	–	–	–	–	–	–	–
Inconel 718	–	–	–	–	–	–	–	–
Inconel X-750	–	–	–	–	–	–	–	–
Hastelloy B	1095 to 1185	2000 to 2165	1/12	AC or WQ		Not applicable		
Hastelloy C	1215	2220	1/12	WQ	—	Not applicable	—	
Hastelloy X	–	–	–	–	—	Not applicable	—	

(a) AC, air cool; WQ, water quench.

Salt bath annealing

Salt bath annealing is used for special work with small parts. Inorganic salts, such as chlorides and carbonates of sodium, potassium and barium, which are relatively stable at temperatures considerably above their respective melting points, are fused in large metallic or refractory containers at temperatures up to about 700°C (1300°F); at higher temperatures, heat-resisting Fe-Ni-Cr alloy pots or refractory containers should be used. Excessive fuming of the bath is an indication of its maximum usable temperature.

Particular care must be exercised to remove all traces of sulphur from the fused salts in order to prevent embrittlement of the work. This may be accomplished in 2 to 3 hours by adding to the fused chlorides and carbonates a small amount (1 lb or more) of a mixture consisting (by volume) of 3 parts powdered borax and 1 part powdered charcoal. If test pieces of nickel strip or wire do not embrittle after 3 or 4 hr. in the purified salt bath, the desulphurising treatment was sufficient.

The material to be annealed is placed in the molten salts and absorbs heat rapidly. After being annealed, the work metal is quenched in water to free it from particles of the salt mixture. The annealed material will not be bright and may be flash pickled to achieve a bright surface.

Bright Annealing

The temperatures required for soft annealing of nickel and nickel alloys are sufficiently high to cause slight surface oxidation unless the materials are heated in vacuum or in a furnace provided with a reducing atmosphere. Nickel 200, Monel 400 and similar alloys will remain bright and free from discolouration when heated and cooled in a reducing atmosphere. However, nickel alloys containing chromium, titanium and aluminium will form a thin oxide film. Even if oxidation is not important, the furnace atmosphere must be suitably sulphur-free and not strongly oxidising.

The protective atmosphere most commonly used in heating nickel and nickel alloys is that provided by controlling the ratio between the fuel and air supplied to burners firing directly into the furnace. A desirable reducing condition may be obtained by using a slight excess of fuel so that the products of combustion contain at least 2 per cent carbon monoxide plus hydrogen (preferably, 4 per cent) with no more than 0.05 per cent uncombined oxygen.

Another method of maintaining desired conditions of furnace atmosphere is to introduce a prepared atmosphere into the heating and cooling chambers. This can be added to the products of combustion in a direct-fired furnace; however, introduction of prepared atmosphere is more commonly practiced with indirectly heated equipment. Prepared atmospheres suitable for use with nickel and nickel alloys include: dried hydrogen, dried nitrogen, dissociated ammonia and cracked or partially reacted natural gas. Properties of various protective atmospheres are shown in Table 23.3. Nickels, modified nickels and nickel-copper alloys can be bright annealed in all of these atmospheres except the first one listed (completely burned fuel, lean atmosphere). Inconel 600 and other alloys containing chromium or molybdenum, or both, require completely dissociated ammonia or dried hydrogen for bright annealing.

Table 23.3. Prepared atmospheres suitable for annealing of nickels and nickel alloys.

Atmospheres 2 through 7 can be used for bright annealing of nickel, modified nickels and nickel-copper alloys; atmosphere 4 or atmosphere 7 must be used for bright annealing of nickel alloys that contain chromium or molybdenum or both.

Atmosphere	*Air-to-gas ratio (a)*	*Composition, vol %* H_2	CO	CO_2	CH_4	O_2	N_2	*Dew point (approx.)* °C	°F
1 Completely burned fuel, lean atmosphere	10 to 1	0.5	0.5	10.0	0.0	0.0	89.0	Saturated(b)	
2 Partially burned fuel, medium-rich atmosphere	6 to 1	15.0	10.0	5.0	1.0	0.0	69.0	Saturated(b)	

(Contd...)

Atmosphere	Air-to-gas ratio (a)	Composition, vol % H_2	CO	CO_2	CH_4	O_2	N_2	Dew point (approx.) °C	°F
3 Reacted fuel, rich atmosphere	3 to 1	38.0	19.0	1.0	2.0	0.0	40.0	+ 20	+ 70
4 Dissociated ammonia (complete dissociation)	No air	75.0	0.0	0.0	0.0	0.0	25.0	– 55 to – 75	– 70 to – 100
5 Dissociated ammonia, partially burned	1.25 to 1(c)	15.0	0.0	0.0	0.0	0.0	85.0	Saturated(b)	
6 Dissociated ammonia, completely burned	1.8 to 1(c)	1.0	0.0	0.0	0.0	0.0	99.0	Saturated(b)	
7 Electrolytic hydrogen, dried(d)	No air	100.0	0.0	0.0	0.0	0.0	0.0	– 55 to – 75	–70 to – 100

(a) Based on use of natural gas containing nearly 100 per cent methane and rated at 37 MJ/m^3 (1000 Btu/ft^3). For high-hydrogen manufactured gas (20 MJ/m^3 or 550 Btu/ft^3), ratios are about 50 per cent of values listed. For manufactured gas with lower hydrogen and high carbon monoxide contents (17 MJ/m^3 or 450 Btu/ft^3), ratios are about 40 per cent of values listed. For propane, ratios are about twice those listed. For butane, multiply listed values by three; (b) When atmosphere is cooled by tap-water heat exchangers, dew point will be about 6 to 8°C (10 to 15°F) above the temperature of the tap water. Dew point may be reduced to about 5°C (40°F) by refrigeration equipment and to –55°C (–70°F) or lower by activated-absorption equipment; (c) Ratio of air to dissociated ammonia; and (d) Dried to a dew point of –55 to –75°C (–70 to –100°F) by alumina plus molecular sieve.

The conditions for bright annealing should be maintained during heating, regardless of the method used for cooling. If facilities for cooling under protective atmosphere are not available, nickel and nickel alloys may be quenched in a 2 per cent (by volume) solution of denatured alcohol. This will reduce the 'oxide flash' formed by the oxygen of the air during transfer from the furnace to the quenching tank.

Dead-Soft Annealing

When the nickel alloys are annealed at higher temperatures and for longer periods, a condition commonly described as 'dead-soft' is obtained and hardness numbers will result that are 10 to 20 per cent lower than those of the 'soft' condition. This cannot be accomplished without increasing the grain size of the metal. Therefore, this treatment should be used only for those few applications in which grain size is of little importance.

Torch Annealing

Some large equipment is hardened locally by fabricating operations. If the available annealing furnace is too small to hold the work piece, the hardened sections can be annealed with the flames of oil or acetylene torches adjusted so as to be highly reducing. The work should be warmed gently at first, with sweeping motions of the torch and should not be brought to the annealing temperature until sufficient pre-heating has been done to prevent cracking as a result of sudden release of stress. (Note: Torch annealing is a poor method for general use, because it provides irregular and insufficient annealing and produces heavily oxidised surfaces).

PROCESS-CONTROL FACTORS IN ANNEALING

Among the more important process-control factors in annealing nickel and nickel alloys are selection of suitably sulphur-free fuels for heating, control of furnace temperature, effects of prior cold work and of cooling rates, control of grain size, control of protective atmospheres and protection from contaminations by foreign material.

Fuels for Heating

Nickel and nickel alloys are subject to intergranular attack when heated in the presence of sulphur or sulphur compounds. Fuels for heating must be low in sulphur content.

Gas is the best fuel for heating nickel alloys and should be used if available. Good heating is achieved readily with gas because of the ease with which gas can be mixed with air and its supply controlled. Gaseous fuels require little combustion space and automatic control of temperature and furnace atmosphere is easily accomplished.

Natural gas, consisting chiefly of methane (CH_4) and smaller amounts of ethane (C_2H_6), propane (C_3H_8) and butane (C_4H_{10}), and essentially free of sulphur compounds, is available in many areas.

Manufactured gases are produced from coal or oil, which may contain substantial amounts of sulphur. These gases should not be used unless the sulphur compounds are effectively removed during gas manufacture. Sulphur occurs in these manufactured gases as hydrogen sulphide (H_2S) and as organic sulphides.

Suppliers of manufactured gas endeavour to keep the total sulphur content of their product to a value below the maximum set by state regulatory agencies. Thus, manufactured gas supplied to consumers usually contains less than 6.9 gram of total sulphur per 10 m^3 of gas (less than 30 gram per 100 ft^3). This sulphur content may vary considerably from day to day, but, where adequate maintenance of sulphur removal is observed by the gas supplier, total sulphur content will average 2.3 to 3.4 gram per 10 m^3 (10 to 15 gram per 100 ft^3) or lower. This sulphur content is acceptable for heating nickel alloys; however, the generally accepted statutory limit (6.9 gram per 10 m^3) is marginal.

Two other satisfactory gaseous fuels are butane and propane, which are components of natural gas that liquefy and separate out when the gas is compressed. But fuels are stored and shipped in tank cars that range in capacity from 30 to 57 m^3 (8000 to 15000 gal) and may be distributed throughout a plant in pipelines as liquids under their own vapour pressures. Butane can be considered in a sense as an oil fuel of high volatility and proper means must be provided for gasifying it by heating, before it is mixed with air for combustion. Propane is more volatile and does not require the application of heat to convert it from liquid to gas. It is also obtainable in cylinders equipped with pressure regulators to control the flow of gas. These cylinders are useful for occasional work in heating small objects.

Furnace-Temperature Control

The importance of accurate control of annealing temperature cannot be overemphasised. Satisfactory indicating, controlling, recording and controlling-recording pyrometers are available. Iron-Constantan and Chromel-Alumel thermocouples should be changed every three months (or more often, if required); noble-metal thermocouples may be used as long as their accuracy is not impaired. All thermocouples should be checked daily for accuracy.

Effect of Prior Cold Work

The greater the amount of cold work to which the material has been subjected before annealing, the lower is the temperature required to produce the same degree of softness without increasing grain size, and the shorter is the time required at any one temperature.

The amount of any type of previous cold work also has critical influence on the ductility of nickel and nickel alloys after annealing. If only a small amount of cold work is done (for example, approximately

10 per cent reduction), full ductility for deep drawing and spinning cannot be restored by annealing, even though the hardness is reduced to that of soft material. A minimum of approximately 20 per cent cold working is required between anneals to ensure maximum ductility and softness following annealing.

Effect of Cooling Rate

Neither slow cooling, whether in or out of the furnace, nor rapid cooling by quenching, has any specific effect on the softness of the annealed, solid-solution nickel materials. Therefore, rapid cooling is preferable (except for heavy sections, in which it may set up excessive thermal stresses), both as a time-saver and to minimise the amount of oxidation. Some of the age hardenable alloys, such as Monel K-500, must be cooled rapidly by quenching from the annealing temperature.

Control of Grain Size

Coarse-grain material is unsuitable for most cold working operations. A coarse grain in the high-nickel materials cannot be refined by thermal treatment. It can be removed only by cold working sufficiently to effect re-crystallisation to a smaller grain size during a subsequent annealing treatment. Maximum workability is obtained with material that has been annealed without allowing appreciable grain growth to occur. The average grain diameter should not exceed 0.064 mm (0 0025 inch). This gives the best combinations of ductility to permit extensive deformation, strength to withstand the action of tools and surface quality to facilitate polishing.

Effect of Fluctuating Atmospheres

If nickel and nickel alloys are annealed in atmospheres that fluctuate between oxidising (excess of air) and reducing (excess of carbon monoxide or hydrogen), severe intercrystalline attack will occur with resulting embrittlement, even though the atmosphere is sulphur-free. This type of embrittlement can be prevented by maintaining a constant and sufficient excess of a reducing atmosphere during heating and cooling. Alloys containing chromium or molybdenum are affected less than nickel and nickel-copper alloys.

Protection from Contamination by Foreign Material

Many of the lubricants used for deep drawing and spinning contain sulphur or lead. Unless these elements are removed before annealing, they will cause embrittlement. They can be removed by washing with hot trisodium phosphate solution or with a volatile organic grease-removing solvent such as trichloroethylene. Lubricants of any type should be entirely removed from the material before annealing. Paints and other adherent substances that may contain sulphur, lead or similarly harmful ingredients, should be removed by appropriate methods before annealing.

STRESS RELIEVING

In stress relieving, careful regulation of time and temperature is required. These variables usually are determined experimentally for each application; some typical ranges are given in Table 23.2. Fig. 23.1 illustrates the effect of stress relief at temperatures from about 400 to about 600°C (750 to 1100°F) on the room-temperature properties of Monel 400.

Stress Equalising

Stress equalising is a low temperature heat treatment (Table 23.2) that effects what is known as 'partial recovery'. This recovery, which precedes any detectable microscopic structural changes consists of a considerable increase in the proportional limit, slight increases in hardness and tensile strength, no significant change in elongation or reduction of area, balancing of stresses and return of electrical conductivity toward its characteristic value for the alloy in the annealed condition.

The temperature required for stress equalising depends on the composition of the alloy. Fig. 23.1 shows an optimum temperature range of about 230 to 315°C (450 to 600°F) for cold drawn Monel 400 rod. A temperature of about 275°C (525°F) is recommended for commercial use. Long treatment time at this temperature has no detrimental effect.

Stress equalising is usually applied to coil springs, wire forms and flat spring stampings. If coil springs are to be given a cold 'set' or cold pressing after coiling, the stress equalisation should be carried out before the setting operation, which involves stressing the material beyond the elastic limit. Any cold working stresses set up by this operation are in such a direction that they will benefit, rather than harm, the material. If stress is equalised after cold pressing, part of the beneficial cold working stress will be removed.

AGE HARDENING

Addition of magnesium, aluminium, silicon, titanium and certain other alloying elements to nickel and nickel alloys, separately or in combinations, produces an appreciable response to age hardening. The effect is dependent on both chemical composition and ageing temperature; it is caused by precipitation of sub-microscopic particles throughout the grains, which results in a marked increase in hardness and strength.

Prior Solution Treating

Unlike precipitation hardening stainless steels and aluminium-base alloys, the nickel alloys normally do not require solution treating in the upper annealing temperature range prior to age hardening. However, solution treating may be employed to enhance special properties (Table 23.4). For example, Inconel X-750 may be solution treated for 2 to 4 hours at 1150°C (2100°F) and air cooled prior to a double (high and low temperature) ageing cycle to develop maximum creep, relaxation and rupture strength at temperatures above about 600°C (1100°F). This combination of heat treatments is considered essential for high temperature springs and turbine blades made of Inconel X-750.

Table 23.4. Solution-treating and age-hardening schedules for nickel alloys.

	Solution treating				
	Temperature		*Time,*	*Cooling*	
Alloy	*°C*	*°F*	*hr.*	*method(a)*	*Age hardening*
Monel K-500	980	1800	½ to 1	WQ	Heat to 595°C (1100°F), hold 16 hours; furnace cool to 540°C (1000°F), hold 6 hours; furnace cool to 480°C (900°F), hold 8 hour; air cool
Inconel 718	980	1800	1	AC	Heat to 720°C (1325°F), hold 8 hours; furnace cool to 620°C (1150°F), hold until furnace time for entire age-hardening cycle equals 18 hours; air cool
Inconel X-750	1150	2100	2 to 4	AC	Heat to 845°C (1550°F), hold 24 hours; air cool; re-heat to 705°C (1300°F), hold 20 hours; air cool
	980	1800	1	AC	Heat to 730°C (1350°F), hold 8 hours; furnace cool to 620°C (1150°F), hold until furnace time for entire age-hardening cycle equals 18 hours; air cool
Hastelloy X	1175	2150	1	AC	Heat to 760°C (1400°F), hold 3 hours; air cool; re-heat to 595°C (1100°F), hold 3 hours; air cool

(a) WQ, water quench; AC, air cool

Age-hardening Practices

Age-hardening practices for several nickel alloys are summarised in Table 23.4. In general, nickel alloys are soft when quenched from temperatures ranging from 790 to 1220°C (1450 to 2225°F); however, they may be hardened by holding at 480 to 870°C (900 to 1600°F) or above and then furnace or air cooling. Quenching is not a prerequisite to ageing; the alloys can be hardened from the hot worked and cold worked conditions, as well as from the soft condition.

Hardening Techniques

Nickel alloys usually are hardened in sealed boxes placed inside a furnace, although small horizontal or vertical furnaces without boxes may be used also. The box or furnace should hold the parts loosely packed, yet afford a minimum of excess space. Electric furnaces provide the optimum temperature uniformity of ± 6°C (± 10°F) and the freedom from contamination required for this work. Gas-heated furnaces, particularly those of the radiant-tube type, can be made to give satisfactory results. It is difficult to obtain good results from oil heating, even with muffle furnaces. All lubricants should be removed from the work before hardening.

Because of the long time of ageing and the difficulty of excluding air from the box or furnace, truly bright hardening cannot be accomplished commercially. For semi-bright hardening, dry hydrogen (see No. 7 in Table 23.3) or cracked and dried ammonia (see No. 4 in Table 23.3) should be used. When bright or semi-bright hardening is not required, other atmospheres may be used, such as nitrogen, cracked natural gas free of sulphur, cracked city gas, cracked hydrocarbons or a generated gas. The use of sulphur-free gases is necessary to avoid embrittlement.

Salt baths are used occasionally for small parts. The hardened material is never bright and must be flash pickled to restore the natural colour. Inorganic salts are used, such as chlorides and carbonates of sodium and potassium, which are relatively stable at temperatures considerably above their respective melting points. It is extremely important that the salts be free of all traces of sulphur, so that the work does not become embrittled.

Chapter 24

Heat Treating of Heat-Resisting Alloys

INTRODUCTION

Procedures and process control for heat treating the principal types of heat-resisting alloys and refractory metals are discussed in this chapter. Tables 24.1, 24.2, 24.3 and 24.4 list representative wrought and cast materials and include primarily those alloys most frequently used in the aerospace industry.

The first part of this chapter reviews the heat treating processes that are applied to heat-resisting alloys. Later sections describe the applicability of these processes to specific alloys.

STRESS RELIEVING

Stress relieving of heat-resisting alloys and refractory metals frequently entails a compromise; the desirability of maximum relief of residual stress must be weighed against possible effects deleterious to high-temperature properties and corrosion resistance.

True stress relieving of wrought material usually is confined to alloys that are not age-hardenable. Thus, the time and temperature cycles may vary considerably, depending on the metallurgical characteristics of the alloy and on the type and magnitude of residual stresses developed by previous fabricating processes.

Stress-relieving temperatures are usually below the annealing or re-crystallisation temperatures. Typical cycles for wrought alloys are listed in Table 24.5; temperatures at least 25°C (50°F) higher or lower than those listed are usually satisfactory.

Some heat-resisting alloy castings are placed in service in the as-cast condition. However, castings may be stress relieved: (i) when they are of an extremely complex shape that might crack during the initial heating-up period in service; (ii) when their dimensional tolerances are stringent; and (iii) after they have been welded. It is not possible to tabulate the stress relief cycles for cast alloys because they are particularly dependent on geometry and prior processing. Stress relief cycles can be developed by empirical studies of stress decay with time and temperature, as determined by non-destructive means such as X-ray diffraction for many alloys. This is not an effective technique for superalloys, where extensive material testing of critical properties and subsequent data analysis must be performed to determine the efficacy of a given cycle. Particular care must be given to such time-dependent effects as LCF hold time, crack growth and creep rate.

Table 24.1 Nominal compositions of wrought heat-resisting alloys.

Alloy	UNS number	Composition, %										
		Cr	Ni	Co	Mo	W	Nb	Ti	Al	Fe	C	Other
Iron-base and iron-nickel-chromium solid-solution alloys												
16–25–6	–	16.0	25.0	–	6.00	–	–	–	–	50.7	0.06	1.35 Mn; 0.70 Si; 0.15 N
17–14CuMo	–	16.0	14.0	–	2.50	–	0.4	0.3	–	62.4	0.12	0.75 Mn; 0.50 Si; 3.0 Cu
19–9DL	K63198	19.0	9.0	–	1.25	1.25	0.4	0.3	–	66.8	0.30	1.10 Mn; 0.60 Si
Carpenter 20Cb-3	N08020	20.0	34.0	–	2.50	–	1.0 max	–	–	42.4	0.07 max	3.5 Cu
Incoloy 800, 800H	N08800	21.0	32.5	–	–	–	–	0.38	0.38	45.7	0.05	–
Incoloy 801	N08801	20.5	32.0	–	–	–	–	1.13	–	46.3	0.05	–
Incoloy 802	–	21.0	32.5	–	–	–	–	0.75	0.58	44.8	0.35	–
Incoloy 825	–	21.5	–	–	3.0	–	–	0.9	0.1	30.0	0.03	0.5 Mn; 2.25 Cu
N-155	R30155	21.0	20.0	20.0	3.00	2.5	1.0	–	–	32.2	0.15	0.15 N; 0.02 La; 0.02 Zr
RA330	N08330	19.0	36.0	–	–	–	–	–	–	45.1	0.05	–
Cobalt-base solid-solution alloys												
Haynes 25 (L-605)	R30605	20.0	10.0	50.0	–	15.0	–	–	–	3.0	0.10	1.5 Mn
Haynes 188	R30188	22.0	22.0	37.0	–	14.5	–	–	–	3.0 max	0.10	0.90 La
S-816	R30816	20.0	20.0	42.0	4.0	4.0	4.0	–	–	4.0	0.38	–
Stellite 6B	–	30.0	1.0	61.5	–	4.5	–	–	–	1.0	1.0	–
UMCo-50	–	28.0	–	49.0	–	–	–	–	–	21.0	0.12 max	–
Nickel-base solid-solution alloys												
Hastelloy B	N10001	1.0 max	63.0	2.5 max	28.0	–	–	–	–	5.0	0.05 max	0.03 V
Hastelloy B-2	N10665	1.0 max	69.0	1.0 max	28.0	–	–	–	–	2.0 max	0.02 max	–
Hastelloy C-4	N06455	16.0	63.0	2.0 max	15.5	–	–	0.7 max	–	3.0 max	0.015 max	–
Hastelloy C-276	N10276	15.5	59.0	–	16.0	3.7	–	–	–	5.0	0.02 max	–
Hastelloy N	N10003	7.0	72.0	–	16.0	–	–	0.5 max	–	5.0 max	0.06	–
Hastelloy S	–	15.5	67.0	–	15.5	–	–	–	0.2	1.0	0.02 max	0.02 La
Hastelloy W	N10004	5.0	61.0	2.5 max	24.5	–	–	–	–	5.5	0.12 max	0.6 V
Hastelloy X	N06002	22.0	49.0	1.5 max	9.0	0.6	–	–	2.0	15.8	0.15	–
Inconel 600	N06600	15.5	76.0	–	–	–	–	–	–	8.0	0.08	0.25 max Cu

(Contd...)

Alloy	UNS number	Composition, % Cr	Ni	Co	Mo	W	Nb	Ti	Al	Fe	C	Other
Inconel 601	N06601	23.0	60.5	–	–	–	–	–	1.35	14.1	0.05	0.5 max Cu
Inconel 617	–	22.0	55.0	12.5	9.0	–	–	–	1.0	–	0.07	–
Inconel 625	N06625	21.5	61.0	–	9.0	–	3.6	0.2	0.2	2.5	0.05	–
Inconel 690	–	30.0	60.0	–	–	–	–	–	–	9.5	0.03	–
NA-224	–	27.0	48.0	–	–	6.0	–	–	–	18.5	0.50	–
Nimonic 75	–	19.5	75.0	–	–	–	–	0.4	0.15	2.5	0.12	0.25 max Cu
RA-333	N06333	25.0	45.0	3.0	3.0	3.0	–	–	–	18.0	0.05	–
Iron-base precipitation-hardening alloys												
A-286	K66286	15.0	26.0	–	1.25	–	–	2.0	0.2	55.2	0.04	0.005 B; 0.3 V
Discaloy	K66220	14.0	26.0	–	3.0	–	–	1.7	0.25	55.0	0.06	–
Haynes 556	–	22.0	21.0	20.0	3.0	2.5	0.1	–	0.3	29.0	0.10	0.50 Ta; 0.02 La; 0.002 Zr
Incoloy 903	–	0.1 max	38.0	15.0	0.1	–	3.0	1.4	0.7	41.0	0.04	–
Pyromet CTX-1	–	0.1 max	37.7	16.0	0.1	–	3.0	1.7	1.0	39.0	0.03	–
V-57	–	14.8	27.0	–	1.25	–	–	3.0	0.25	48.6	0.08 max	0.01 B; 0.5 max V
W-545	K66545	13.5	26.0	–	1.5	–	–	2.85	0.2	55.8	0.08	0.05 B
Cobalt-base precipitation-hardening alloys												
AR-213	–	19.0	0.5 max	65.0	–	4.5	–	–	3.5	0.5 max	0.17	6.5 Ta; 0.15 Zr; 0.1 Y
MP-35N	R30035	20.0	35.0	35.0	10.0	–	–	–	–	–	–	–
MP-159	–	19.0	25.0	36.0	7.0	–	0.6	3.0	0.2	9.0	–	–
Nickel-base precipitation-hardening alloys												
Astroloy	–	15.0	56.5	15.0	5.25	–	–	3.5	4.4	<0.3	0.06	0.03 B; 0.06 Zr
D-979	N09979; K66979(a)	15.0	45.0	–	4.0	4.0	–	3.0	1.0	27.0	0.05	0.01 B
IN 100	N13100	10.0	60.0	15.0	3.0	–	–	4.7	5.5	<0.6	0.15	1.0 V; 0.06 Zr; 0.015 B
IN 102	N06102	15.0	67.0	–	2.9	3.0	2.9	0.5	0.5	7.0	0.06	0.005 B; 0.02 Mg; 0.03 Zr
Incoloy 901	N09901	12.5	42.5	–	6.0	–	–	2.7	–	36.2	0.10 max	–
Inconel 702	–	15.5	79.5	–	–	–	–	0.6	3.2	1.0	0.05	0.5 Mn; 0.2 Cu; 0.4 Si
Inconel 706	N09706	16.0	41.5	–	–	–	–	1.75	0.2	37.5	0.03	2.9(Nb + Ta); 0.15 max Cu
Inconel 718	N07718	19.0	52.5	–	3.0	–	5.1	0.9	0.5	18.5	0.08 max	0.15 max Cu

(Contd...)

Alloy	UNS number	Cr	Ni	Co	Mo	W	Nb	Ti	Al	Fe	C	Other
		Composition, %										
Inconel 721	–	16.0	71.0	–	–	–	–	3.0	–	6.5	0.04	2.2 Mn; 0.1Cu
Inconel 722	–	15.5	75.0	–	–	–	–	2.4	0.7	7.0	0.04	0.5 Mn; 0.2 Cu; 0.4 Si
Inconel 751	–	15.5	72.5	–	–	–	1.0	2.3	1.2	7.0	0.05	0.25 max Cu
Inconel X-750	N07750	15.5	73.0	–	–	–	1.0	2.5	0.7	7.0	0.04	0.25 max Cu
M252	N07252	19.0	56.5	10.0	10.0	–	–	2.6	1.0	<0.75	0.15	0.005 B
Nimonic 80A	N07080	19.5	73.0	1.0	–	–	–	2.25	1.4	1.5	0.05	0.10 max Cu
Nimonic 90	N07090	19.5	55.5	18.0	–	–	–	2.4	1.4	1.5	0.06	–
Nimonic 95	–	19.5	53.5	18.0	–	–	–	2.9	2.0	5.0 max	0.15 max	+ B; + Zr
Nimonic 100	–	11.0	56.0	20.0	5.0	–	–	1.5	5.0	2.0 max	0.30 max	+ B; + Zr
Nimonic 105	–	15.0	54.0	20.0	5.0	–	–	1.2	4.7	–	0.08	0.005 B
Nimonic 115	–	15.0	55.0	15.0	4.0	–	–	4.0	5.0	1.0	0.20	0.04 Zr
Nimonic 263 (b)	–	20.0	51.0	20.0	5.9	–	–	2.1	0.45	0.7 max	0.06	–
Pyromet 860	–	13.0	44.0	4.0	6.0	–	–	3.0	1.0	28.9	0.05	0.01 B
Refractory 26	–	18.0	38.0	20.0	3.2	–	–	2.6	0.2	16.0	0.03	0.015 B
René 41	N07041	19.0	55.0	11.0	10.0	–	–	3.1	1.5	<0.3	0.09	0.01 B
Rene 95	–	14.0	61.0	8.0	3.5	3.5	3.5	2.5	3.5	<0.3	0.16	0.01 B; 0.05 Zr
Rene 100	–	9.5	61.0	15.0	3.0	–	–	4.2	5.5	1.0 max	0.16	0.015 B; 0.06 Zr; 1.0 V
Udimet 500	N07500	19.0	48.0	19.0	4.0	–	–	3.0	3.0	4.0 max	0.08	0.005 B
Udimet 520	–	19.0	57.0	12.0	6.0	1.0	–	3.0	2.0	–	0.08	0.005 B
Udimet 630	–	17.0	50.0	–	3.0	3.0	6.5	1.0	0.7	18.0	0.04	0.004 B
Udimet 700	–	15.0	53.0	18.5	5.0	–	–	3.4	4.3	<1.0	0.07	0.03 B
Udimet 710	–	18.0	55.0	14.8	3.0	1.5	–	5.0	2.5	–	0.07	0.01 B
Udimet AF2-1DA	–	12.0	59.0	10.0	3.0	6.0	–	3.0	4.6	<0.5	0.35	1.5 Ta; 0.015 B; 0.1 Zr
Waspaloy	N07001	19.5	57.0	13.5	4.3	–	–	3.0	1.4	2.0 max	0.07	0.006 B; 0.09 Zr

(a) No longer active, but shown here for reference. (b) Also known as Rolls Royce C-263.

ANNEALING

When applied to heat-resisting alloys, annealing implies full annealing—that is, complete re-crystallisation and the attainment of maximum softness. The practice is usually applied to wrought alloys of the non-hardening type. For a majority of the hardenable alloys, annealing cycles are the same as those used for solution treating. However, the two treatments serve different purposes.

Annealing is used mainly to increase ductility (and reduce hardness) to facilitate forming or machining, prepare for welding, relieve stresses after welding, produce specific microstructures, or soften age-hardened structures by resolution of second phases. Solution treating is intended to dissolve second phases to produce maximum corrosion resistance or to prepare for ageing. Additionally, it will homogenise microstructure prior to ageing.

Table 24.2 Compositions of nickel-base heat-resistant casting alloys.

Alloy designation	*Nominal composition, %*										
	C	*Ni*	*Cr*	*Co*	*Mo*	*Fe*	*Al*	*B*	*Ti*	*W*	*Zr*	*Others*
B-1900	0.1	64	8	10	6	–	6	0.015	1	–	0.10	4 Ta (a)
Hastelloy X	0.1	50	21	1	9	18	–	–	–	1	–	–
IN-100	0.18	60.5	10	15	3	–	5.5	0.01	5	–	0.06	1 V
IN-713C	0.12	74	12.5	–	4.2	–	6	0.012	0.8	–	0.1	1.75 Ta, 0.9 Nb
IN-713LC	0.05	75	12	–	4.5	–	6	0.01	0.6	–	0.1	4 TA
IN-738	0.17	61.5	16	8.5	1.75	–	3.4	0.01	3.4	2.6	0.1	2 Nb
IN-792	0.2	60	13	9	2.0	–	3.2	0.02	4.2	4	0.1	2 Nb
Inconel 718	0.04	53	19	–	3	18	0.5	–	0.9	–	–	0.1 Cu, 5 Nb
Inconel X-750	0.04	73	15	–	–	7	0.7	–	2.5	–	–	0.25 Cu, 0.9 Nb
M-252	0.15	56	20	10	10	–	1	0.005	2.6	–	–	–
MAR-M 200	0.15	59	9	10	–	1	5	0.015	2	12.5	0.05	1 Nb(b)
MAR-M 246	0.15	60	9	10	2.5	–	5.5	0.015	1.5	10	0.05	1.5 Ta
MAR-M 247	0.15	59	8.25	10	0.7	0.5	5.5	0.015	1	10	0.05	1.5 Hf, 3 Ta
NX 188 (DS)	0.04	74	–	–	18	–	8	–	–	–	–	–
Rene 77	0.07	58	15	15	4.2	–	4.3	0.015	3.3	–	0.04	–
Rene 80	0.17	60	14	9.5	4	–	3	0.015	5	4	0.03	–
Rene 100	0.18	61	9.5	15	3	–	5.5	0.015	4.2	–	0.06	1 V
TRW-NASA VIA	0.13	61	6	7.5	2	–	5.5	0.02	1	6	0.13	0.4 Hf, 0.5 Nb, 0.5 Re, 9 Ta
Udimet 500	0.1	53	18	17	4	2	3	–	3	–	–	–
Udimet 700	0.1	53.5	15	18.5	5.25	–	4.25	0.03	3.5	–	–	–
Udimet 710	0.13	55	18	15	3	–	2.5	–	5	1.5	0.08	–
Waspaloy	0.07	57.5	19.5	13.5	4.2	1	1.2	0.005	3	–	0.09	–
WAZ-20 (DS)	0.20	72	–	–	–	–	6.5	–	–	20	1.5	–

(a) B-1900 + Hf also contains 1.5 per cent Hf. (b) MAR-M 200 + Hf also contains 1.5 per cent Hf.

Table 24.3. Compositions of cobalt-base heat-resistant casting alloys.

Alloy designation	*Nominal composition, %*										
	C	*Co*	*Cr*	*Ni*	*Al*	*B*	*Fe*	*Ta*	*W*	*Zr*	*Others*
AiResist 13	0.45	62	21	–	3.4	–	–	2	11	–	0.1 Y
AiResist 213	0.20	64	20	0.5	3.5	–	0.5	6.5	4.5	0.1	0.1 Y
AiResist 215	0.35	63	19	0.5	4.3	–	0.5	7.5	4.5	0.1	0.1 Y
FSX-414	0.25	52.5	29	10	–	0.010	1	–	7.5	–	–
Haynes 21	0.25	64	27	3	–	–	1	–	–	–	5 Mo
Haynes 25; L-605	0.1	54	20	10	–	–	1	–	15	–	–
J-1650	0.20	36	19	27	–	0.02	–	2	12	–	3.8 Ti
MAR-M 302	0.85	58	21.5	–	–	0.005	0.5	9	10	0.2	–
MAR-M 322	1.0	60.5	21.5	–	–	–	0.5	4.5	9	2	0.75 Ti
MAR-M 509	0.6	54.5	23.5	10	–	–	–	3.5	7	0.5	0.2 Ti
MAR-M 918	0.05	52	20	20	–	–	–	7.5	–	0.1	–
NASA Co-W-Re	0.40	67.5	3	–	–	–	–	–	25	1	2 Re, 1 Ti
S-816	0.4	42	20	20	–	–	4	–	4	–	4 Mo, 4 Nb, 1.2 Mn, 0.4 Si
V-36	0.27	42	25	20	–	–	3	–	2	–	4 Mo, 2 Nb, 1 Mn, 0.4 Si
WI-52	0.45	63.5	21	–	–	–	2	–	11	–	2 Nb + Ta
X-40	0.50	57.5	22	10	–	–	1.5	–	7.5	–	0.5 Mn, 0.5 Si

Table 24.4. Compositions of commercially important refractory alloys.

Designation	*Nominal composition, %*
Molybdenum alloys	
Mo-0.5Ti	Mo-0.5Ti-0.02W
TZM	Mo-0.5Ti-0.1Zr-0.02W
Mo-30W	Mo-30W
Niobium alloys	
Nb-1Zr	Nb-1Zr
FS-80	Nb-0.75Zr
FS-82	Nb-33Ta-0.75Zr
FS-85	Nb-27.5Ta-11W-1Zr
SCb-291	Nb-10Ta-10W
Cb-752	Nb-10W-2.5Zr
B-66	Nb-5Mo-5V-1Zr
C-103	Nb-10Hf-1Ti
C-129Y	Nb-10W-10Hf-0.15Y
Tantalum alloys	
'63' Metal	Ta-2.5W-0.15 Nb
Ta-10W	Ta-10W

(Contd...)

Designation	*Nominal composition, %*
T-111	Ta-8W-2Hf
T-222	Ta-10W-2.5Hf-0.01C
Tungsten alloys	
W-ThO_2	W-1ThO_2; W-2ThO_2
W-Mo alloys	Various Mo contents; W-2Mo and W-15Mo are most common
W-Re alloys	Various Re contents up to 26 %; W-1.5Re, W-3Re and W-25Re are most common
Doped W	50 ppm Si, 90 ppm K, 15 ppm Al, 35 ppm O

Annealing practices vary considerably among different plants. Representative annealing temperatures, holding times and cooling procedures are given in Table 24.5. Experience with specific parts for known requirements often indicates advantageous modifications of temperature, time or cooling method.

Table 24.5. Typical stress relieving and annealing cycles for wrought heat-resisting alloys.

	Stress relieving			*Annealing(a)*		
	Temperature		*Holding time per inch of*	*Temperature*		*Holding time per inch of*
Alloy	*°C*	*°F*	*section, hr.*	*°C*	*°F*	*section, hr.*
Iron-base and iron-nickel-chromium alloys						
RA-330	900	1650	1(b)	1110(c)	2025(c)	¼(d)
19-9 DL	675(c)	1250(e)	4	980	1800	1
A-286	(f)	(f)	–	980	1800	1
Discaloy	(f)	(f)	–	1035	1900	1
Nickel-base alloys						
Astroloy	(f)	(f)	–	1135	2075	4
Hastelloy B	(f)	(f)	–	1175	2150	1
Hastelloy C	(f)	(f)	–	1215	2225	1
Hastelloy W	(f)	(f)	–	1175	2150	1
Hastelloy X	(f)	(f)	–	1175	2150	1
Incoloy 800	870	1600	1½	980	1800	¼
Incoloy 800H	–	–	–	1175	2150	–
Incoloy 825	–	–	–	980	1800	–
Incoloy 901	(f)	(f)	–	1095	2000	2
Inconel 600	900	1650	1	1010	1850	¼(d)
Inconel 601	–	–	–	980	1800	–
Inconel 625	870	1600	1	980	1800	1
Inconel 690	–	–	–	1040	1900	½
Inconel 718	(f)	(f)	–	955	1750	1

(Contd...)

	Stress relieving			Annealing(a)		
	Temperature		Holding time per inch of	Temperature		Holding time per inch of
Alloy	°C	°F	section, hr.	°C	°F	section, hr.
Inconel X-750	880(g)	1625(g)	–	1035	1900	½
Nimonic 80A	(f)	(f)	–	1080	1975	2
Nimonic 90	(f)	(f)	–	1080	1975	2
René 41	(f)	(f)	–	1080	1975	2
Udimet 500	(f)	(f)	–	1080	1975	4
Udimet 700	(f)	(f)	–	1135	2075	4
Waspaloy	(f)	(f)	–	1010	1850	4
Cobalt-chromium-nickel-base alloys						
L-605 (HS-25)	(h)	(h)	–	1230	2250	1
N-155 (HS-95)	(h)	(h)	–	1175	2150	–
S-816	(h)	(h)	–	1205	2200	1
Refractory metals(j)						
Ta-10W	1205(k)	2200(k)	1	1425(k)	2600(k)	1
FS-80	1095(k)	2000(k)	1	1315(k)	2400(k)	1
FS-82	1095(k)	2000(k)	1	1315(k)	2400(k)	1
Mo-0.5Ti	1095(m)	2000(m)	½	1315(m)(n)	2400(m)(n)	1
TZM	1205(m)	2200(m)	1	1425(m)(n)(p)	2600(m)(n)(p)	1

(a) Minimum hardness is achieved by cooling rapidly from the annealing temperature, to prevent precipitation of hardening phases. Water quenching is preferred and is usually necessary for heavy sections; air cooling is preferred for heavy sections of Waspaloy, Udimet 500, Udimet 700 and Inconel X-750, because water quenching causes cracking. However, for complex shapes subject to excessive distortion, oil quenching is often adequate and more practical. Rapid air cooling usually is adequate for parts formed from strip or sheet. Rapid cooling from the annealing or solution treating temperature does not suppress the ageing reaction of some alloys, such as Astroloy; these alloys become harder and stronger. (b) Time given is minimum; some plants use as long as 3 hour per inch. (c) Nominal temperature; 1035 to 1175°C (1900 to 2150°F) is commonly used. (d) Short time is required for prevention of grain coarsening. (e) Nominal temperature; 650 to 705°C (1200 to 1300°F) is permissible. (f) Full annealing is recommended, because intermediate temperatures cause ageing. (g) Used only for stress equalising of warm worked grades. (h) Full annealing is recommended if further fabrication is performed; otherwise, material can be stress relieved at approximately 55°C (100°F) below annealing temperature. (j) Annealing temperatures depend on prior plastic deformation, degree of cold work, alloy content and interstitial purity. Annealing temperatures given are those most frequently used for cold worked sheet or plate; in many instances, more precise determination of the re-crystallisation temperature is necessary for a specific application. (k) Heat and cool in vacuum or inert-gas atmosphere. (m) Heat and cool in hydrogen or vacuum. (n) Seldom used as finished product in annealed condition, because recrystallisation raises the ductile-brittle transition temperature, resulting in brittleness at low temperatures. (p) For vacuum arc-cast material with a minimum of 50 per cent work.

Most wrought heat-resisting alloys can be cold formed but are more difficult to form than austenitic stainless steels. Severe cold forming may require several intermediate annealing operations. Full annealing must be followed by fast cooling.

Annealing of weldments should immediately follow welding of the age-hardenable alloys where highly restrained joints are involved. If the configuration of the weldment does not permit annealing, ageing can be used for stress relieving the joints. Contact with copper chill blocks should be avoided during welding. Nickel plating is desirable.

Re-heating for hot working is an annealing practice whose aim is to promote adequate formability of the metal being deformed. Temperatures vary widely depending on alloy and working practice. Control of temperature can be critical to resultant properties, as varying degrees of re-crystallisation may be desired. In most standard operations, heating or re-heating for hot working is a full annealing step with re-crystallisation and dissolution of all or most secondary phases. Occasionally, reheating for hot working is restricted to temperatures that do not dissolve all secondary phases so that the remaining phases can be used to limit grain growth.

HEAT TREATMENT FOR HIGH STRENGTH

The strengthening of heat-resisting alloys usually requires solution treating and ageing; typical cycles for wrought alloys are given in Table 24.6, and for casting alloys, in 24.7. Note that the cooling rate from the solution temperature is critical for some alloys. Alternate heat treatments may be used to improve specific properties.

Table 24.6. Typical solution treating and ageing cycles for wrought heat-resisting alloys.

	Solution treating				*Ageing*			
	Temperature		*Time,*	*Cooling*	*Temperature*		*Time,*	*Cooling*
Alloy	°C	°F	*hr.*	*procedure*	°C	°F	*hr.*	*procedure*
Iron-base alloys								
A-286	980	1800	1	Oil quench	720	1325	16	Air cool
Discaloy	1010	1850	2	Oil quench	730	1350	20	Air cool
					650	1200	20	Air cool
N-155	1175	2150	1	Water quench	815	1500	4	Air cool
Nickel-base alloys								
Astroloy	1175	2150	4	Air cool	845	1550	24	Air cool
	1080	1975	4	Air cool	760	1400	16	Air cool
Hastelloy B	1175	2150	½	(a)	(b)	(b)	–	–
Hastelloy B-2	1065	1950	½	Rapid quench	–	–	–	–
Hastelloy C-4	1065	1950	½	Rapid quench	–	–	–	–
Hastelloy C-276	1120	2050	½	Rapid quench	–	–	–	–
Hastelloy N	1175	2150	½	Rapid quench	–	–	–	–
Hastelloy S	1065	1950	½	Rapid quench	–	–	–	–
Hastelloy C	1220	2225	1	(a)	(b)	(b)	–	–
Hastelloy W	1175	2150	1	(a)	(b)	(b)	–	–
Hastelloy X	1175	2150	1	(a)	–	–	–	–
Inconel 901	1095	2000	2	Water quench	790	1450	2	Air cool
					720	1325	24	Air cool
Inconel 600	1120	2050	2	Air cool	–	–	–	–
Inconel 601	1150	2100	1	Air cool	–	–	–	–
Inconel 617	1175	2150	2	(a)	–	–	–	–
Inconel 625	1150	2100	2	(a)	–	–	–	–

(Contd...)

	Solution treating				Ageing			
	Temperature(a)		Time,	Cooling	Temperature(b)		Time,	Cooling
Alloy	°C	°F	hr.	procedure	°C	°F	hr.	procedure
Inconel 706	925–1010	1700–1850	–	–	845	1550	3	Air cool
					720	1325	8	Furnace cool
					620	1150	8	Air cool
	925–1010	1700–1850	–	–	730	1350	8	Furnace cool
					620	1150	8	Air cool
Inconel 718	980	1800	1	Air cool	720	1325	8	Furnace col
					620	1150	8	Air cool
Inconel X-750 (AMS 5667)	855	1625	24	Air cool	705	1300	20	Air cool
Inconel X-750 (AMS 5668)	1150	2100	2	Air cool	845	1550	24	Air cool
					705	1300	20	Air cool
Nimonic 80A	1080	1975	8	Air cool	705	1300	16	Air cool
Nimonic 90	1080	1975	8	Air cool	705	1300	16	Air cool
René 41	1065	1950	½	Air cool	760	1400	16	Air cool
Udimet 500	1080	1975	4	Air cool	845	1550	24	Air cool
					760	1400	16	Air cool
Udimet 700	1175	2150	4	Air cool	845	1550	24	Air cool
	1080	1975	4	Air cool	760	1400	16	Air cool
Waspaloy	1080	1975	4	Air cool	845	1550	24	Air cool
					760	1400	16	Air cool
Cobalt-base alloys								
Haynes 25; L-605	1230	2250	1	Rapid air cool	(b)	(b)	–	–
Haynes 1881175			2150	½	Rapid air cool	–		–
–	–							
Haynes 556	1175	2150	½	Rapid air cool	–	–	–	–
S-816	1175	2150	1	(a)	760	1400	12	Air cool
Stellite 6B	1230	2250	1	Air cool	–	–	–	–

Note: Alternate treatments may be used to improve specific properties. (a) To provide an adequate quench after solution treating, it is necessary to cool below about 540°C (1000°F) rapidly enough to prevent precipitation in the intermediate temperature range. For sheet metal parts of most alloys, rapid air cooling will suffice. Oil or water quenching is frequently required for heavier sections that are not subject to cracking; and (b) Ageing occurs in service at elevated temperatures.

Table 24.7. Typical solution treating and ageing cycles for heat-resisting casting alloy.

	Solution treating				Ageing			
	Temperature(a)		Time,	Cooling	Temperature(b)		Time,	Cooling
Alloy	°C	°F	hr.	procedure	°C	°F	hr.	procedure
A-286	1095	2000	2	Rapid cool	720	1325	16	Air cool
B-1900	As-cast		–	–	–	–	–	–
FSX-414	1150	2100	4	Rapid cool	980	1800	4	Air cool

(Contd...)

Alloy	Solution treating				Ageing			
	Temperature(a)		Time,	Cooling	Temperature(b)		Time,	Cooling
	°C	°F	hr.	procedure	°C	°F	hr.	procedure
Hastelloy B	1175	2150	2	Rapid cool	(c)	(c)	–	–
Hastelloy C	1220	2225	1	Rapid cool	(c)	(c)	–	–
HS-31 (X-40)	As-cast		–	–	–	–	–	–
IN-100	As-cast		–	–	–	–	–	–
IN-713C	As-cast		–	–	–	–	–	–
IN-738	1120	2050	2	Air cool	845	1550	24	Air cool
IN-792	1120	2050	2	Air cool	845	1550	24	Air cool
IN-939	1160	2120	4	Air cool	850	1560	16	Air cool
Inconel 718	1095	2000	1	Air cool	620	1150	10	Air cool
MAR-M 200	–	–	–	–	870	1600	50	Air cool
MAR-M 200 DS	1230	2250	4	Air cool	870	1600	32	Air cool
MAR-M 246	–	–	–	–	845	1550	50	Air cool
MAR-M 247	–	–	–	–	870	1600	16	Air cool
MAR-M 302	As cast		–	–	–	–	–	–
MAR-M 509	As cast		–	–	–	–	–	–
René 41	1095	2000	½	Rapid cool	900	1650	4	Air cool
Rene 80	1220	2225	2	Air cool	1095	2000	4	Air cool
					1055	1925	4	Air cool
					845	1550	16	Air cool
Udimet 700	1150	2100	2	Air cool	760	1400	16	Air cool

(a) Furnace temperature tolerance of ± 15°C (± 25°F) is satisfactory; (b) Furnace temperature tolerance of ± 10°C (± 15°F) is recommended; and (c) Ageing occurs in service at elevated temperature. Use a vacuum or protective atmosphere for heat treating at temperatures above 1040°C (1900°F) and subsequent cooling.

Solution Treating

In some instances, the solution treating temperature employed will depend on the properties desired. This is indicated in Table 24.6 for alloys A-286, Inconel 718, René 41, Udimet 700 and Waspaloy. A higher temperature is specified for optimum creep and creep-rupture properties; a lower temperature, for optimum short-time tensile properties at elevated temperature. The higher solution treating temperature will result in some grain growth and more extensive solution of carbides in wrought alloys. The principal objective is to put gamma prime type phases into solution and dissolve some carbides.

After ageing, the resulting microstructure of these wrought alloys consists of large grains that contain the principal ageing phases and of a heavy concentration of carbides in the grain boundaries. The lower solution treating temperature dissolves the principal ageing phases without grain growth or significant high-temperature carbide solution.

For some wrought alloys, for example, Nimonic 80A and Nimonic 90 in Table 24.6, an intermediate solution treating temperature is selected to produce a compromise of the properties. For other alloys, such as Udimet 500 and Udimet 700, the intermediate temperature ageing treatment is used to tailor the grain boundaries for improved creep-rupture properties.

Quenching

The purpose of quenching heat-resisting alloys is to maintain, at room temperature, the supersaturated solid solution obtained during solution treating. Quenching permits a finer gamma prime particle size to be achieved on ageing. Cooling methods commonly used are indicated in Table 24.6 for wrought alloys and in Table 24.7 for casting alloys.

Ageing

Ageing treatments strengthen age-hardenable alloys by causing the precipitation of one or more phases from the supersaturated matrix that is developed by solution treating and retained by rapid cooling from the solution treating temperature.

Factors that influence the selection of number of ageing steps and ageing temperature include: (i) type and number of precipitating phases available; (ii) anticipated service temperature; (iii) precipitate size; and (iv) the combination of strength and ductility desired and heat treatment of similar alloys.

Principal ageing phases in the heat-resisting alloys usually include one or more of the following: gamma prime Ni_3Al or Ni_3 (Al, Ti), eta (Ni_3Ti), or gamma double prime, Ni_3Nb. Secondary phases that may be present include: carbides ($M_{23}C_6$, M_7C_3, M_6C and MC), nitrides (MN), carbonitrides (MCN), and borides (M_3B_2), as well as Laves phase (M_2Ti) and delta phase (Ni_3Nb). The above phases occur principally in nickel-base alloys. The primary phases in cobalt-base alloys are $M_{23}C_6$, M_7C_3, M_6C and MC. The primary phases in iron-base alloys will be similar to those in nickel alloys although eta is more apt to be found because of the generally higher Ti to Al ratio in iron alloys than in nickel alloys.

When more than one phase is capable of precipitating from the alloys matrix, judicious selection of a single ageing temperature may result in obtaining optimum amounts of multiple precipitating phases. Alternatively, a double ageing treatment that produces different sizes and types of precipitate at different temperatures may be employed. The ageing temperature also determines not only the type but also the size distribution of precipitate. Exposure to temperatures higher than the optimum ageing temperature results in a decrease in strength through the process of overageing; at still higher temperatures, resolution may occur. High ageing temperatures will produce coarser gamma prime particles than lower temperatures and result in higher creep-rupture properties. For optimum short-time elevated temperature properties, small, finely dispersed particles of gamma prime precipitate are desired. Therefore, lower final ageing temperatures are employed than those used to obtain high creep-rupture properties. For all gamma prime dispersions, care must be taken to ensure the correct carbide distribution. A principal reason for two-step ageing sequences, in addition to gamma prime or gamma double prime control, is the need to precipitate or control grain boundary carbide morphology.

SURFACE CONTAMINATION

Age-hardenable heat-resisting alloys may have good oxidation resistance within their normal range of service temperature, at or above their ageing temperatures—in a range of 760 to 980°C (1400 to 1800°F) depending on the alloy. Others may require coatings, particularly blade alloys for turbojet engines and compressors. However, at temperatures used for solution treating, these alloys are susceptible to intergranular oxidation, a defect that adversely affects thermal fatigue.

Intergranular oxidation

Intergranular oxidation is measured optically as depth of intergranular penetration. Fig. 24.1a shows the depth of intergranular oxidation that occurred in René 41 heated in air. Oxidation data as determined by weight change for several age-hardenable alloys and type 310 stainless steel are plotted in Fig. 24.1b. Oxidation resistance is enhanced by chromium and aluminium additions as well as by certain other elements. Chromium allows the formation of a protective chromium oxide and is most beneficial at intermediate temperatures and below ≲870°C (≲1600°F), whereas aluminium forms aluminium oxide which is more protective at the higher temperature of ≲870°C (≲1600°F). The principal mode of intergranular attack involves the preferential oxidation of chromium, aluminium, titanium, zirconium and boron. Molybdenum increases susceptibility to intergranular attack in age-hardenable alloys.

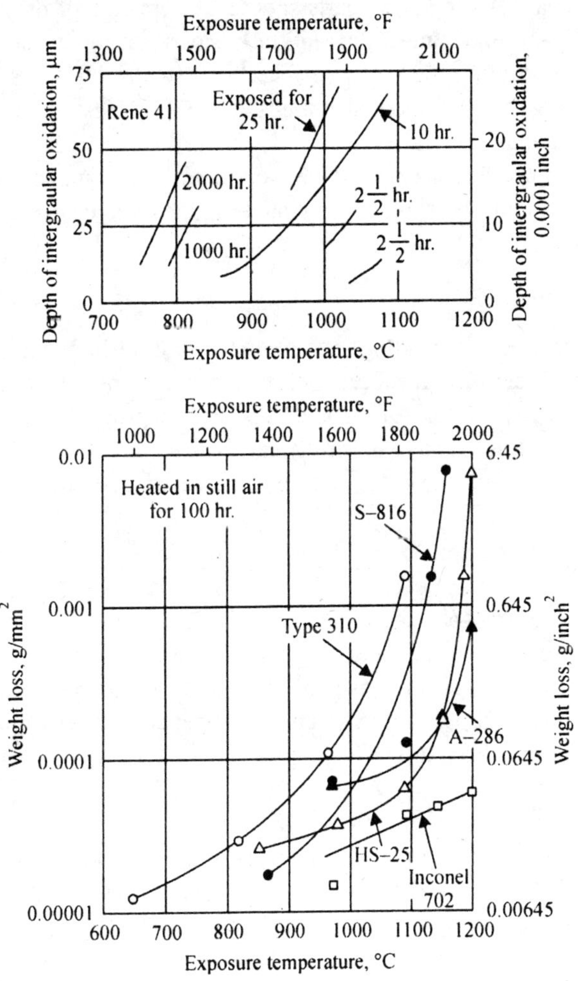

(a) Rene 41. (b) Type 310 stainless steel and the typical heat resistant alloys.

Fig. 24.1. Effect of time and temperature on oxidation.

In relation to intergranular oxidation, aluminium is preferable to titanium as a hardening element for two reasons: (i) increased aluminium reduces formation of the eta phase from gamma prime; and (ii) aluminium oxide provides a denser and less permeable barrier to the diffusion of oxygen.

Carbon Pick-up

Carbon pick-up can occur if the solution treating atmosphere has a carburising potential. For instance, the carbon content of the surface of A-286 alloy has been observed to increase from 0.05 to 0.30 per cent. The added carbon forms a stable carbide (TiC), thus removing titanium from solid solution and preventing normal precipitation hardening in the surface layers. TiN can be formed in the same manner as a result of nitrogen contamination.

Miscellaneous Contaminants

All exposed surfaces of heat-resisting alloy parts should be kept free of dirt, fingerprints, oil, grease, forming compounds, lubricants and scale. Lubricants or fuel oils that contain sulphur bearing compounds are particularly active in corroding the metal surface by first forming Cr_2S_3 and then, as the attack progresses, also forming a Ni-Ni_3S_2 eutectic that melts at 645°C (1190°F), particularly at low pressures of less than 10^{-4} torr in vacuum.

Scale and slag from furnace hearths are another source of contamination. Contact with steel scale, slag and furnace spallings should be avoided; low melting constituents can form on the metal surface and promote corrosion.

PROTECTIVE ATMOSPHERES

Protective atmospheres are used in annealing or solution treating if heavy oxidation cannot be tolerated. If oxidation can be tolerated, because of subsequent stock removal, heat-resisting alloys can be solution treated in air or in the normal mixture of air and combustion products found in gas-fired furnaces. However, refractory metals must always be heat treated in a vacuum or in an inert-gas atmosphere (argon, helium, or an ArHe mixture) or hydrogen. In some cases, ceramic coatings are used to prevent surface attack.

Exothermic Atmosphere

A lean and dilute exothermic atmosphere is relatively safe and economical. The surface scale formed in such an atmosphere can be removed by pickling or by salt bath descaling and pickling. Such an atmosphere, formed by burning fuel gas with air, contains about 85 per cent nitrogen, 10 per cent carbon dioxide, 1.5 per cent carbon monoxide, 1.5 per cent hydrogen and 2 per cent water vapour. This atmosphere will produce a scale rich in chromium oxides.

Endothermic Atmospheres

Endothermic atmospheres prepared by reacting fuel gas with air in the presence of a catalyst are not recommended, because of their carburising potential. Similarly, the endothermic mixture of nitrogen and hydrogen formed by dissociating ammonia is not used, because of the probability of nitriding.

Dry Hydrogen

Dry hydrogen (dew point, –50°C (–60°F) or lower) is used in preference to dissociated ammonia for bright annealing of heat-resisting alloys. If the hydrogen is prepared by catalytic gas reactions instead of by electrolysis, residual hydrocarbons, such as methane, should be limited to about 50 ppm, to prevent carburising. Hydrogen is not recommended for bright annealing of alloys containing significant amounts of elements (such as aluminium or titanium) that form stable oxides not reducible at normal heat treating temperatures and dew points. Hydrogen is not recommended for annealing or solution treating alloys that contain boron, because of the danger of deboronisation through formation of boron hydrides. Nor can hydrogen be used for heat treating niobium and tantalum, because of its embrittling effects.

Dry Argon

Dry argon (dew point: –50°C or –60°F or lower) should be used if no oxidation can be tolerated. It is mandatory that this type of atmosphere be used in a sealed retort or sealed furnace chamber. A purge of at least ten times the volume of the retort is recommended before the retort is placed in the furnace. The argon must be kept flowing continually during and after the treatment until the workpieces have cooled nearly to room temperature, to prevent the formation of an oxide film.

Heat-resisting alloys containing stable-oxide formers such as aluminium and titanium, with or without boron, must be bright annealed in a vacuum or in a chemically inert gas such as argon. If used, argon must be pure and dry—dew point, –50°C (–60°F) or lower. If the argon has a slightly higher dew point–but not over –40°C (–40°F)—oxidation will be limited to a thin surface film that can usually be tolerated.

Vacuum Atmosphere

Vacuum atmosphere, generally below 20 μm (2 x 10^{-3} torr) is commonly used for heat-resisting alloys above 815°C (1500°F). It is particularly desirable when parts are at or close to final dimensions.

Atmospheres for Ageing

Air is the most common ageing atmosphere. The smooth, tight oxide layer that is formed is usually unobjectionable on the finished product. However, if this oxide layer must be minimised, a lean exothermic gas (air-gas ratio, about 10 to 1) can be employed. It will not entirely prevent oxidation, but the oxide layer will be very light. The use of gases containing hydrogen and carbon monoxide for ageing cycles is dangerous because of the explosion hazard at temperatures below 760°C (1400°F).

FURNACE EQUIPMENT

Basic equipment considerations seldom differ from those influencing the selection of furnaces for heat treating steel. In general, the temperature-control limits are ±14°C (± 25°F) and temperature may range up to about 1290°C (2350°F). Belt conveyor furnaces, although widely used for production annealing, are less gastight than roller-hearth furnaces. Consequently, atmosphere costs for a belt conveyor furnace are likely to be higher than for a roller-hearth furnace of the same volume. Batch heating for annealing or solution treating is usually done in box furnaces. These may have provisions for purging, pre-heating and quenching, if the high-temperature compartment is supplemented by other chambers.

Ageing

Ageing of heat-resisting alloys, commonly in the range of 650 to 900°C (1200 to 1650°F), is usually done in box furnaces, with or without protective atmospheres. The usual operating-temperature tolerance is ±14°C (± 25°F) for wrought alloys and ± 8°C (± 15°F) for casting alloys. Continuous furnaces are seldom used, because of the long ageing cycles. Salt baths are not recommended, because reaction could occur between chloride in the bath and the alloy surface during the long-time immersion that would be required for ageing.

Vacuum Furnaces

Vacuum furnaces are used for heat treating niobium and tantalum and other heat-resisting alloys. Heating may be accomplished by resistance elements or by induction. Furnace design dictates a batch operation. Cooling can be accomplished in a vacuum retort pressurised with an inert gas that provides conductive cooling after heating is discontinued.

FIXTURING

Fixtures for holding finished parts or assemblies during heat treatment may be of either the support type or the restraint type. For alloys that must be cooled rapidly from the solution treating temperature, the best practice is to employ minimum fixturing during solution treating and quenching and to control dimensional relations by the use of restraining fixtures during ageing.

Support Fixtures

Support fixtures are used when restraint is not required or when the part itself provides sufficient self-restraint. A support fixture also functions as an aid in handling parts and helps the part to support its own weight. Long narrow pieces, such as tubes or bolts, are most easily fixtured by hanging vertically. Parts that have a large flat surface can be placed on a flat furnace tray or plate. Examples of such parts are rings, cylinders and beams. For parts of slightly asymmetrical shape, special supports can be built up from a flat tray. If these supports are fabricated by welding, they must be stress relieved prior to use.

Asymmetrical parts can be supported in several ways. One method is to lay the part on a tray of sand, making certain that most of the bottom area is well supported. Alluvial garnet sand is most commonly used as the supporting medium. Another method of support is the use of a ceramic casting formed to the shape of the part. However, this method is costly and subject to size limitations. Turbine blades and asymmetrical ducting are examples of parts that can be supported either in a sand tray or by ceramic castings.

Restraint Fixtures

Restraint fixtures are generally more complicated than support fixtures and may require machined grooves, lugs or clamps to hold parts to a given shape.

To maintain symmetry and roundness in an A-286 frame assembly during ageing, the assembly was processed on flat plate into which grooves had been machined. these grooves accepted the rims on the outer and inner shrouds and held them in restraint during heat treatment. To prevent the centre hub from rising or dropping in relation to the other shroud, both the hub and the shroud were clamped to the grooved-plate fixture.

It is possible to perform some straightening of parts in ageing fixtures of the type described above. A slightly distorted part can be forced into the fixture and clamped. Some stress relieving will occur along with ageing. However, fixtures for hot sizing are not always successful for heat-resisting alloys, because of the high creep strength of these alloys at the ageing temperature.

The use of threaded fasteners for clamping is not recommended, because they are difficult to remove after heat treatment. A slotted bar held in place by wedges is preferred.

Usually, the coefficient of expansion of both the fixture and part should be nearly the same. However, in some applications, the fixture is purposely made from a material having different expansion characteristics, in order to apply pressure to the parts as the temperature increases.

For example, for securing a René 41 turbine blade in place for brazing and solution treatment, sand or castceramic techniques could not be used because of the rapid heating and cooling rates required. Instead, a guillotine-type fixture was constructed from Hastelloy X, with 11 contour stations conforming to the correct shape for the blade. Inconel tubes standing along the edges of the blade were secured to the fixture with molybdenum wire. The wire passed through a hole in the blade and fixture and through the tube. The wire in each tube was looped and tied around a short tube placed horizontally atop the standing tube. Thus, during heating, the tubes, which have a greater expansion coefficient than the wire, produced a clamping effect on the edges of the blade.

RELATION OF FABRICATION TO HEAT TREATING

A uniformly fine-grained microstructure can be produced in the age-hardenable heat-resisting alloys if final deformation is carried out in the lower part of the hot working range. This type of structure can be solution treated to a uniform grain size. If the final working temperature is too low, or if the metal is not reduced enough during final working, a duplex structure is produced during solution treating. Alloys deformed in the upper hot working range have a coarser structure that cannot be refined by solution treating.

Cold Working

Cold working is usually performed on alloys in the solution treated condition because of the markedly lower strength and increased ductility of the material before ageing (Table 24.8). Cold working itself affects mechanical properties, through its influence on: (i) grain growth during subsequent solution treatment, and (ii) the reaction kinetics of ageing.

Table 24.8. Typical effects of ageing on room-temperature mechanical properties of solution treated heat-resisting alloys.

	Yield strength(a)				*Elongation in 50 mm (2 inch), %*	
	Not aged		*Aged*			
Alloy	*MPa*	*ksi*	*MPa*	*ksi*	*Not aged*	*Aged*
A-286	240	35	760	110	52	33
René 41	620	90	1100	160	45	15
Inconel X-750	410	60	650	92	45	24
HS-25	480	69	480	70	55	45

(a) At 0.2 per cent offset.

Effect of Grain Size

The age-hardenable heat-resisting alloys are susceptible to critical grain growth if they are solution treated after small amounts of cold or hot work. Larger amounts of cold work refine the grain size. Excessive grain growth may have deleterious effects on creep and fatigue properties in all heat-resisting alloys. Therefore, on parts subjected to cold or hot work prior to solution treating, the critical amount of work (about 1 to 6 per cent cold work, depending on the alloy, or 10 per cent hot work) must be exceeded in all areas, to avoid the growth of abnormally large grains. This rule applies to items such as cold headed bolts, spun or stretch-formed sheets and parts formed by simple bending.

Effect on ageing

Cold working accelerates the ageing reaction, causing the early appearance of precipitates at normal ageing temperature and the appearance of a precipitate at temperatures below the normal ageing temperature if the alloy is cold worked sufficiently and held at temperature long enough. In other words, cold working renders the material more readily overageable at the normal ageing temperature. Properly charted information correlating cold work with ageing response, however, can be used to shorten the required ageing period or to lower the normal ageing temperature.

Rates of Work Hardening

In cold forming operations such as sheet rolling, tube drawing, pressing, brake and stretch forming and deep drawing, the governing material property is the rate of hardening in relation to the amount of cold deformation. Materials such as A-286, Nimonic 75 and Hastelloy X can be reduced as much as 90 per cent before a softening treatment is required. For the more complex alloys such as René 41 and Udimet 500, incremental reductions of 35 to 40 per cent can be achieved without intermediate annealing.

Fig. 24.2 shows the relative work hardening rates of four heat-resisting alloys compared to type 302 stainless steel. The iron-base A-286 alloy work hardens at about the same rate as type 302. The highest work hardening rates are those of the cobalt-base alloys, such as S-816.

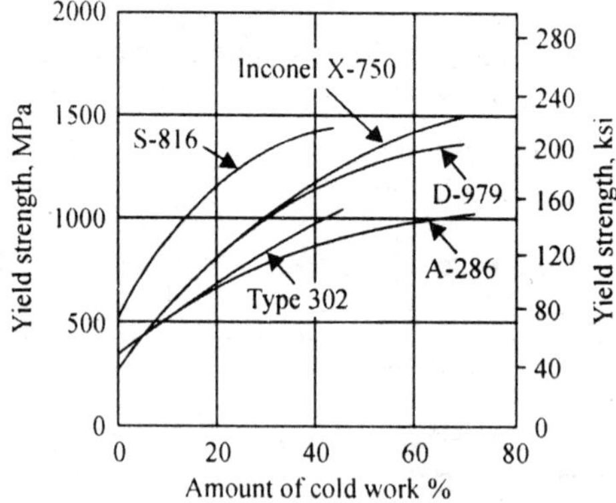

Fig. 24.2. Effect of cold work on room-temperature yield strength of some heat resistant alloys.

In general, the greater the work hardening rate of a given alloy, the less the amount of cold deformation possible before annealing is required. When a large amount of total reduction is required, as in the draw swaging of tubes, the metal is formed in small incremental reductions with interstage anneals between successive reductions.

Welding

Welding of age-hardenable heat-resisting alloys requires precise control of pre-welding and post-welding heat treatments. A typical welding sequence for wrought alloys is as follows:

1. Material received in mill-annealed condition (the specific condition is critical for certain alloys).
2. Forming by various methods (in-process anneals used if required).
3. Welding.
4. Solution treating of weldment.
5. Ageing of weldment.

Mill annealing cycles must dissolve precipitated phases without causing excessive grain growth and must not dissolve M_6C in alloys such as René 41. Large grain size and dissolved carbides may produce a continuous grain-boundary carbide precipitate that promotes cracking during welding. A maximum grain size of ASTM 3 is usually recommended for welding.

Fusion welds in age-hardenable heat-resisting alloys should be solution treated after welding to eliminate the uncontrolled ageing that occurs in the heat-affected zone of the base metal during welding and to relieve residual stresses. Special procedures, such as double ageing, have been used in large and complex components containing restrained welds, for which solution treatment is not practical.

Cast non-age-hardenable cobalt-base superalloy turbine vanes are repair welded and heat treatment procedures, such as pre-heat and post-heat, have been developed to minimise cracking and surface contamination in these alloys. Age-hardenable cast nickel-base alloys are not normally repair welded superalloy Inconel 718. Cracking and dimensional control are problems encountered with the welding of normal cast nickel-base age-hardenable alloys.

Service requirements of weldments must be considered in the selection of heat treatment. If service temperatures do not exceed the ageing temperature, no appreciable change in properties is likely in service. However, if the service temperature is above the normal ageing temperature, strength decreases and ductility increases in service.

The heat treatments that may be applied to weldments in non-hardenable heat-resisting alloys are limited to stress relieving and in some instances, stabilising. Stress relieving must be done at relatively high temperatures (815 to 980°C, or 1500 to 1800°F) to be effective. Because the stress-relieving temperature range coincides with the carbide-precipitation range, the composition of the alloys should be controlled to avoid precipitating carbides in harmful amounts.

HEAT TREATING OF IRON-BASE ALLOYS

The iron-base heat-resisting alloys (Table 24.1) include some that are hardenable by solution treating and ageing and some that are not hardenable by heat treatment. Heat treating temperatures for selected alloys are given in Tables 24.5, 24.6 and 24.7.

Stress Relieving and Annealing

Alloys (such as 19–9 DL) that depend entirely on hot and cold work to develop high strength frequently require stress relieving (Table 24.5). However, these alloys cannot be stress relieved when maximum corrosion resistance is required in service, because the recommended stress-relieving temperatures are within the sensitising range of the alloys. If these alloys are to be exposed to acid or steam in service, intergranular cracking can be minimised or eliminated by annealing instead of stress relieving.

Iron-base heat-resisting alloys, such as A-286 and Incoloy 901, which rely on gamma prime type hardening, although not susceptible to sensitisation at stress-relieving temperatures, still cannot be stress relieved, because the intermediate temperatures of stress relief result in ageing. Consequently, the restoration of ductility and reduction of stresses in cold formed parts and weldments made of these alloys is achieved by heating rapidly to the annealing temperature.

This procedure will result in partial solutioning of the gamma prime phase, reduction in residual stress and subsequent re-precipitation of gamma prime on cooling. In forging these alloys, the finishing temperature is usually above 925°C (1700°F) and stress relieving is not required for as-forged parts. These alloys usually are solution treated and aged after forging.

Castings

Iron-chromium-nickel alloys, identified by Alloy Castings Institute designations HA through HX, are usually exposed to service temperatures of 650 to 1250°C (1200 to 2200°F) in process equipment such as heat treating furnaces, oil-refinery equipment and cement kilns. Normally, they are placed in service in the as-cast condition, but when experience indicates the advisability of stress relieving, the operation is performed by placing the castings in a cold furnace and heating slowly to 980 to 1040°C (1800 to 1900°F), holding at temperature for 1 hour per inch of section thickness and then furnace cooling.

Castings usually are not annealed, because the need for restoring ductility seldom exists and slow cooling through the temperature range of 705 to 900°C (1300 to 1650°F) may cause formation of sigma phase in some grades, such as HD. Therefore, heat-resisting alloy castings are annealed only for special reasons. Some of these reasons are discussed in the following paragraphs. However, in some cases, stress relieving operations are performed to relieve residual stresses.

Iron-chromium casting alloys containing 9 to 30 per cent chromium and little or no nickel vary widely in response to heat treatment, depending on composition. Most castings of alloys in this group are used in the as-cast condition. Type HA (9 per cent Cr, 1 per cent Mo) is an exception; it transforms from austenite to ferrite on air cooling from temperatures above 815°C (1500°F). For maximum softness, castings of this alloy are annealed by furnace cooling from 885°C (1625°F) minimum (at about 30°C/hour[50°F/hour]) to below 705°C (1300°F).

The iron-chromium-nickel and iron-nickel-chromium casting alloys containing 20 to 30 per cent chromium and 10 to 20 per cent nickel (such as HE, HF and HL), as well as those containing about 12 to 20 per cent chromium and 25 to 67 per cent nickel (such as HT, HU and HX), are usually used in the as-cast condition. However, for applications that require maximum thermal fatigue properties, annealing is sometimes beneficial. Castings of alloy HT (35 per cent Ni, 15 per cent Cr), for example, may give improved performance by being heated to 1040°C (1900°F), held at temperature for about 12 hour and furnace cooled, prior to being placed in service.

Solution Treating and Ageing

The principal hardenable iron-base alloys are A-286 and Incoloy 901. Discussion and examples here will be confined to these two alloys, which contain titanium and aluminium and which are hardened mainly by precipitation of gamma prime. The gamma prime is metastable and transforms to eta phase if the aluminium to titanium ratio is too low.

Iron-base alloys such as A-286 are relatively insensitive to the rate of cooling from the solution treating temperature (Table 24.9). Thus, section thickness is not of major concern in relation to cooling practice. The effect of varying amounts of cold work on grain growth in A-286 during solution treating is illustrated in Fig. 24.3. The initial material was in the solution treated condition with a maximum grain size of ASTM 5. Cold working 14 per cent and resolution treating reproduced this same grain size. However, cold working in the range of 1 to 5 per cent caused excessive grain growth during subsequent solution treating. Cold working affects the properties after ageing, as illustrated in Fig. 24.4 for A-286. The hardness after cold working and before ageing, as well as the peak hardness after ageing, increases greatly with increasing amounts of cold rolling. The temperature at which peak hardness is attained decreases with increasing amounts of reduction. The temperature at which softening occurs also decreases with increasing amounts of deformation. (Note: In Fig. 24.4, that the material cold worked 81 per cent and then aged at 760°C or 1400°F for 16 hours is softer than material that was not cold worked before being aged).

Table 24.9. Effect of cooling rate from 980 to 535°C (1800 to 1000°F) on short-time properties of A-286(a).

Cooling rate		*Yield strength(b)*		*Tensile strength*		*Elongation,*	*Reduction*
°C/min.	*°F/min.*	*MPa*	*ksi*	*MPa*	*ksi*	*%*	*in area, %*
Tested at 20°C (70°F)							
Oil quench		680	99	1030	149	24	39
15	27	690	100	1030	150	22	42
7	15	680	99	1020	148	24	39
5	9	700	102	990	143	22	36
1	2	660	96	1030	149	23	33
Tested at 650°C (1200°F)							
Oil quench		650	94	780	113	8	10
5	27	650	94	770	112	6	7
7	15	630	92	770	112	9	16
5	9	590	86	740	108	7	17
1	2	220	90	740	108	11	18

(a) After cooling from the solution temperature, all samples were aged at 720°C (1325°F) for 16 hour and air cooled; (b) At 0.2 per cent offset.

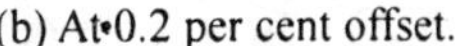

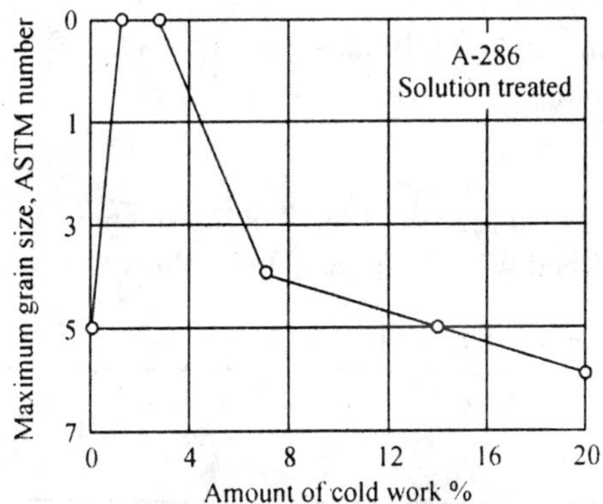

Influence of cold work on grain size of A-286 alloy solution treated at 900°C (1650°F) for 1 hour and oil quenched

Fig. 24.3. Effect of cold work.

To obtain uniformity of properties in parts that have been cold worked non-uniformly, higher-than-normal ageing temperatures are sometimes used, followed by a second ageing cycle at lower-than-normal ageing temperature. For example, in A-286 parts with varying amounts of cold work, double ageing at 760 and 705°C (1400 and 1300°F), respectively, provides more uniform hardness, short-time tensile properties and creep-rupture properties than the normal 720°C (1325°F) ageing treatment. The higher ageing temperature also improves the structural stability of the part in service.

The uniformity of properties after double ageing is illustrated by the following example.

Turbojet turbine disks made of Incoloy 901, each weighing 170 kg (375 lb), were heat treated in a gas-fired, box-type furnace by the following procedure.

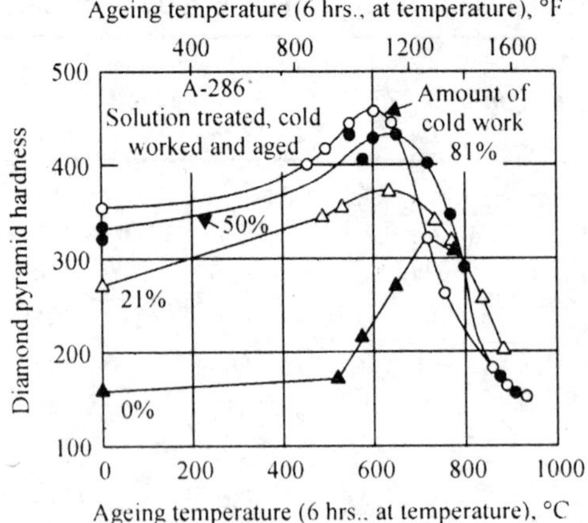

Fig. 24.4. Effect of cold work and ageing on diamond pyramid hardness on A-286.

Solution treating

Load 10 disks on tray in single layer (total weight of load, 1700 kg or 3750 lb). Heat to 1080°C (1975°F), hold 2 hours at temperature ± 10°C (± 20°F), quench rapidly in water.

First ageing treatment (stabilising)

Load 20 disks on tray, stacking in two layers (total weight of load, 3400 kg or 7500 lb). Heat to 775°C (1425°F), hold 2 hours at temperature ± 5°C (± 10°F), cool in air.

Second ageing treatment

Load as in first ageing. Heat to 720°C (1325°F), hold 2 hours at temperature ± 5°C (± 10°F), cool in air.

Uniformity of properties obtained with this procedure is shown in Table 24.10.

Table 24.10. Typical uniformity of properties obtained in double ageing of Incoloy 901 turbine disks.

	Ultimate tensile strength		*Yield strength(a)*		*Elongation in 50 mm*	*Reduction in*
Location	*MPa*	*ksi*	*MPa*	*ksi*	*(2 inch), %*	*area, %*
ID, tangential	1170	170	870	127	16	19
Midradius, radial	1180	171	890	129	16	19
OD, tangential	1180	171	880	128	17	20
Special (min)	1140	165	830	120	12	15

(a) At 0.2 per cent offset.

Practical Heat Treatment

Acceptable mechanical properties are not always accomplished as a result of the solution treating and ageing procedure first tried for a given alloy or application. To develop specified properties, changes of the following kinds are often required:

1. Adjust the solution temperature or time.
2. Adjust the ageing temperature.

3. Add an intermediate (stabilising) ageing treatment at a temperature higher than that of the final ageing.
4. Add a second (final) ageing treatment at a lower temperature.
5. Adjust one or both ageing temperatures in a double ageing cycle.
6. Add a third ageing treatment.

These modifications of practice are illustrated for A-286 and Incoloy 901 in the following eight examples. Note that for forging alloys, protective atmosphere is usually not required as the surface is subsequently machined off. However, the surface of precision investment casting must be protected during heat treatment with protective atmosphere as argon or vacuum.

Example 24.1 (A-286)

The treatment first specified—900°C (1650°F) for 2 hours oil quench; 705°C (1300°F) for 16 hours air cool—resulted in borderline values of yield strength (610 to 630 MPa, or 89 to 91 ksi for two heats). Adding a second ageing treatment (650°C, or 1200°F, for 16 hours) increased the yield strength to 630 and 700 MPa (92 and 101 ksi) for the same two heats of material.

Example 24.2 (Incoloy 901)

In one application, the alloy failed to meet a specification requiring 830 MPa (120 ksi) minimum yield strength after double ageing—1095°C (2000°F) for 2 hours water quench; 790°C (1450°F) for 2 hours air cool; 720°C (1325°F) for 2 hours air cool. The specification was met by adding a third ageing cycle at 595°C (1100°F) for 24 hours.

Example 24.3 (Incoloy 901)

Creep-rupture ductility was improved, although at some sacrifice of yield strength and creep-rupture life, by the addition of intermediate ageing to the heat treating procedure. Table 24.11 compares the effects of the two treatments on mechanical properties.

Table 24.11. Effects of intermediate ageing on typical properties of Incoloy 901.

	Ultimate tensile strength		*Yield strength(a)*		*Elongation in 50 mm (2 inch), %*	*Reduction in area, %*	*Creep-rupture life, hour*
	MPa	*ksi*	*MPa*	*ksi*			
Tested at 20°C (70°F)							
No intermediate ageing(b):							
Heat A	1050	152	790	115	12	13	–
Heat B	1080	157	790	114	17	16	–
With intermediate ageing(c):							
Heat A	1040	151	730	106	12	15	–
Heat B	1040	151	710	103	12	13	–
Tested at 650°C (1200°F)							
No intermediate ageing(b):							
Heat A	–	–	–	–	1.0	–	76
Heat B	–	–	–	–	1.5	–	118

(Contd...)

	Ultimate tensile strength		Yield strength(a)		Elongation in 50 mm (2 inch), %	Reduction in area, %	Creep-rupture life, hour
	MPa	ksi	MPa	ksi			
With intermediate ageing(c):							
Heat A	–	–	–	–	11	–	45
Heat B	–	–	–	–	7	–	54

(a) At 0.2 per cent offset; (b) Heat treatment: 1120°C (2050°F) for 2 hours water quench; 745°C (1375°F) for 24 hours air cool; and (c) Heat treatment: 1120°C(2050°F) for 24 hours water quench; 815°C (1500°F), 4 hours; air cool; 745°C (1375°F), 24 hours; air cool.

Example 24.4 (Incoloy 901)

Single ageing failed to provide sufficient yield strength to meet specifications. An additional ageing treatment at 650°C (1200°F) for 12 hours provided required properties. Table 24.12 lists specification requirements and compares mechanical properties obtained by the two treatments.

Table 24.12. Effect of single and double ageing on room-temperature properties of Incoloy 901.

	Ultimate tensile strength		Yield strength		Elongation in 50 mm (2 inch), %	Reduction in area, %
	MPa	ksi	MPa	ksi		
Specification	1140	165	827	120	12	15
Single ageing(a)	1150 to 1160	167 to 169	800 to 810	116 to 118	20 to 23	24 to 29
Double ageing(b)	1190 to 1210	173 to 175	830 to 890	121 to 129	18 to 22	24 to 29

Single and double ageing data reflect the results of four tests. (a) Solution treated at 1085°C (1985°F) for 2 hours water quenched; aged at 770°C (1450°F) for 2 hours air cooled; aged at 720°C (1325°F) for 24 hours air cooled. (b) Re-aged at 650°C (1200°F) for 12 hours and air cooled.

Example 24.5 (Incoloy 901)

A problem of low yield strength and low creep-rupture ductility was solved by increasing the temperature of the first ageing (stabilising) treatment and adding a third (final) ageing treatment. Table 24.13 lists the mechanical properties obtained from the original and revised heat treatments.

Table 24.13. Effects of adding a third ageing treatment on properties of Incoloy 901.

Ultimate tensile strength		Yield strength		Elongation in 50 mm (2 inch), %	Reduction in area, %	Creep-rupture elongation(a)
MPa	ksi	MPa	ksi			
Specification						
1140	165	830	120	12	15	4% in 23 hrs.(b)
Double ageing(c)						
1160 to 1210(d)	169 to 175(d)	810 to 900(d)	118 to 131(d)	22 to 23(d)	25 to 30(d)	4.9% in 31 hrs. to 2.8% in 85 hrs.(e)
Triple annealing (f)						
1200 to 1240(d)	174 to 180(d)	850 to 930(d)	123 to 135(d)	18 to 20(d)	23 to 29(d)	7% in 64 hrs. to 6.3% in 74 hrs.(g)

(a) At 650°C (1200°F) and 620 MPa (90 ksi). (b) Minimum values. (c) Solution treated at 1085°C (1985°F) for 2 hours, water quenched; aged at 775°F (1425°F) for 2 hours air cooled; aged at 720°C (1325°F) for 24 hours air cooled. (d) Seven tests. (e) Three tests. (f) Solution treated at 1085°C (1985°F) for 2 hours water quenched; aged at 790°C (1450 °F) for 2 hours air cooled; aged at 720°C (1325°F) for 24 hours air cooled; aged at 650°C (1200°F) for 12 hours air cooled. (g) Two tests.

Example 24.6 (Incoloy 901)

Increasing the temperature of the first ageing treatment in a double ageing cycle provided specified minimum creep-rupture properties. Results of seven tests made on specimens heat treated by the two procedures are shown in Table 24.14, with details of heat treating procedures.

Example 24.7 (A-286)

Table 24.15 shows the effects of alternative approaches to solving the problem of insufficient notch ductility in heat treated A-286 alloy—namely, the use of an additional ageing treatment and the substitution of a higher ageing temperature for single ageing. Although both approaches provided required creep-rupture properties, the use of the higher ageing temperature (with no increase in the solution treating temperature) is recommended over the additional ageing treatment.

Table 24.14. Effect of revision in ageing treatment on creep-rupture properties of Incoloy 901.

	Creep rupture(a)			
	Original treatment(b)		*Revised treatment(c)*	
Test No.	*Life, hr.*	*Elongation, %*	*Life, hr.*	*Elongation in 50 mm (2 inch), %*
1	72	4	74	13
2	126	4	115	12
3	161	4	160	13
4	111	4	110	9
5	127	4	84	9
6	76	4	84	8
7	127	4	98	9

(a) At 650°C (1200°F) and 552 MPa (80 ksi), specified minimums; life, 23 hours elongation, 5 per cent. (b) Solution treated at 1085°C (1985°F) for 2 hours cooled; aged at 715°C (1325°F) for 24 hours air cooled. (c) Same conditions as in (b), except that temperature of first ageing was 810°C (1490°F).

Table 24.15. Effects of alternative ageing treatments on rupture properties of A-286.

	Creep-rupture(a)		
Treatment	*Life, hr.*	*Elongation in 50 mm (2 inch), %*	*Location of failure*
Original heat treatment(b)			
900°C (1650°F), 2 hrs., oil quench; 720°C (1325°F), 16 hrs., air cool	7 to 69	–	Notch
Revised heat treatments(c)			
Original plus additional age at 650°C (1200°F), 12 hrs., air cool	74 to 142	5.6 to 7.7	Smooth bar
900°C (1650°F), 2 hrs., oil quench; 730°C (1350°F), 16 hrs., air cool	24 to 82	4.9 to 7.7	Smooth bar

(a) Specification requirements for creep-rupture at 650°C (1200°F) and 450 MPa (65 ksi); life, 23 hours minimum; no failure in notch permitted. (b) Ten specimens tested. (c) Five specimens for each treatment.

Example 24.8 (A-286 weldment)

For control of dimensions during heat treatment, a turbine combustion case weldment (Fig. 24.5) was loaded on radial spokes of a spider made from an HT alloy casting. Sizing was done mechanically between solution heat treating and ageing and allowed for the linear shrinkage of about 0.1 per cent during ageing. A roller-hearth furnace, utilising a rich unpurified exothermic atmosphere, was used for solution treating at 980°C (1800°F) for 1 hour. Cooling was controlled by a water-jacketed cooler that maintained a rate comparable to that of cooling in still air. Parts were aged at 720, 730, 745 or 760°C (1325, 1350, 1375 or 1400°F) in a pit furnace without protective atmosphere. Exact ageing temperature depended on test results from specimens of each heat. The following examples illustrate the use of interstage annealing or softening with welding and forming operations.

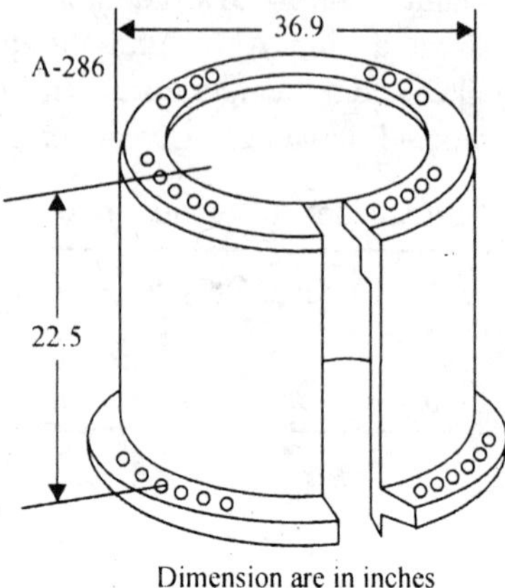

Fig. 24.5. Tubine combustion case weldment (Example 8).

Example 24.9

A part for a turbine shaft (Fig. 24.6) was machined from an A-286 alloy forging, than annealed at 900°C (1650°F) prior to forming the cone section, which was shear spun in one pass from the initial thickness of 6.4 mm (0.250 inch) to final thickness of 2 mm (0.080 inch). After forming, the part was solution treated at 900°C (1650°F) and then aged at 720°C (1325°F).

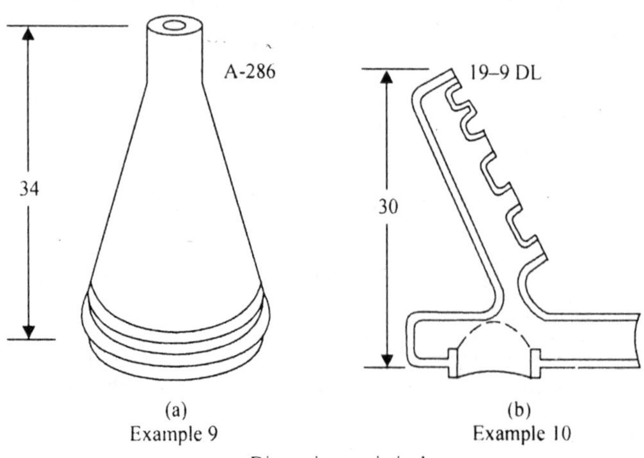

Fig. 24.6. Parts requiring interstage annealing with forming or welding operations.

Example 24. 10

An exhaust header-half stamping was made from 2.1 mm (0.082 inch) thick 19-9 DL sheet (Fig. 24.6). The part required interstage annealing at 980°C (1800°F) between forming operations and was annealed again after being welded into an assembly.

HEAT TREATING OF NICKEL-BASE ALLOYS

As indicated in Tables 24.1 and 24.2, most of the nickel-base heat-resisting alloys contain titanium and aluminium, which make the material hardenable by ageing. Heat treating temperatures for selected alloys are given in Tables 24.5, 24.6 and 24.7.

Stress Relieving and Annealing

These processes are generally applied to wrought nickel alloys of the solid solution strengthened type. Nickel-base alloy castings are seldom stress relieved or annealed. Such parts are used in the as-cast, solution treated, or solution treated and aged condition. The age hardenable nickel-base alloys (such as René 41) are more crack-sensitive than the iron-base alloys and must be annealed during fabrication to relieve forming and welding stresses. As with other wrought alloys, grain size of nickel alloys after annealing depends on the amount of prior cold or hot work and the annealing temperature. This effect is illustrated for Nimonic 90 in Fig. 24.7.

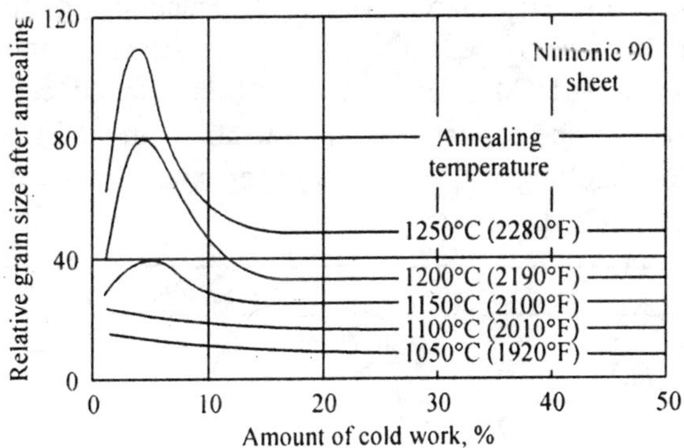

Nimonic 90 sheet was cold rolled in steps from 1.8 to 0.9 mm (0.072 to 0.036 inch) thick and annealed at five temperatures.

Fig. 24.7. Effect of cold work and annealing on grain size.

Holding time at annealing temperature ranges from 1 to 2 hours for most alloys. The more highly alloyed wrought materials such as Astroloy, Udimet 500 and Udimet 700 require longer holding times for dissolving micro-constituents. For other alloys, such as Inconel 600, time at temperature is critical because of grain growth and a holding time of about 15 minutes per inch of section thickness is commonly employed. Room-temperature yield strength is sensitive to grain size, as illustrated below for René 41 forged bar:

ASTM grain size	Average 0.2% yieldstrength	
	Mpa	ksi
3	840	122
5	900	130
7	1010	147
8	1060	154

For some wrought alloys, annealing practice can have a marked effect on subsequent response to solution treating and ageing. This is shown for René 41, one of the more sensitive alloys, in the following example. Parts formed from René 41 sheet showed strain age cracking after being solution treated at 1080°C (1975°F) for ½ hour air cooled and then aged at 760°C (1400°F) for 16 hours. Cracking was attributed to a carbide network in the grain boundaries. Cause of the carbide network was traced to an in-process anneal at 1175°C (2150°F). It was found that at 1175°C (2150°F) the M_6C carbide was dissolved. Subsequent exposure to temperatures between 760 and 870°C (1400 and 1600°F) produced $M_{23}C_6$ carbide network in the grain boundaries. This network reduced ductility to an unacceptable level.

If the annealing temperature was kept below 1095°C (2000°F), M_6C was not dissolved and ductility was improved. A similar effect can occur in weldments of nickel-base alloys if they are annealed at temperatures above 1095°C (2000°F).

A problem similar to that described in the preceding example occurred in René 41 bar stock. Grain-boundary carbide network reduced ductility and caused difficulty (sometimes cracking) during forming and welding. Investigation of the cause of the grain-boundary network indicated that the bar stock was produced with a final rolling temperature of 1175°C (2150°F). Light reductions were taken during final rolling to ensure proper size for the finished bar stock and to eliminate the possibility of surface tearing. This high rolling temperature, coupled with relatively light reductions (in the range of 2 to 3 per cent), produced boundary network because (i) the M_6C carbides were dissolved at the rolling temperature and (ii) slow cooling through the range of 870 to 760°C (1600 to 1400°F) produced $M_{23}C_6$ in an unfavourable morphology (grain boundary carbide network). Rolling temperatures of 1150°C (2100°F) maximum, coupled with a final reduction in rolling of at least 10 to 15 per cent, eliminated the network and produced bar that could be welded and formed.

	Dimension(a)							
	Original		*Draw swaged*			*Expanded*		
	mm	*inch*	*mm*	*inch*		*mm*	*inch*	
Tube diam	31.8	1.25	19.1	0.75	(– 40%)	57	2.25	(+ 80%)
Wall	1.0	0.040	1.5	0.059	(+ 48%)	0.50	0.020	(– 50%)
Length	889	35.0	895.4	35.25	(+ 0.71%)	895.4	35.25	(+ 0.71%)
Swaged transition area	0°			10°			10°	

(a) Numbers in parentheses indicate percentage change from original dimensions.

Multiple shaping and forming operations on nickel-base alloys may require numerous in-process anneals. Fabrication of the 90 cm (35 inch) long René 41 tubular truss member shown in Fig. 24.8 exemplifies fabrication by draw swaging to obtain the end reduction and hydraulic forming to obtain the expanded centre section. Both operations were necessary because the structural design of the tube exceeded the limits of the tube-reduction process alone. Shaping involved: (i) roll forming and welding to produce the tube blank; (ii) three draw swaging steps with four anneals; (iii) rotary swaging; and (iv) four hydraulic expanding steps with three anneals. Each of the seven anneals consisted of holding at 1080°C (1975°F) in an argon atmosphere for 30 minutes and then water quenching.

This method has also been applied successfully to the fabrication of HS-25 and Hastelloy X tubular parts.

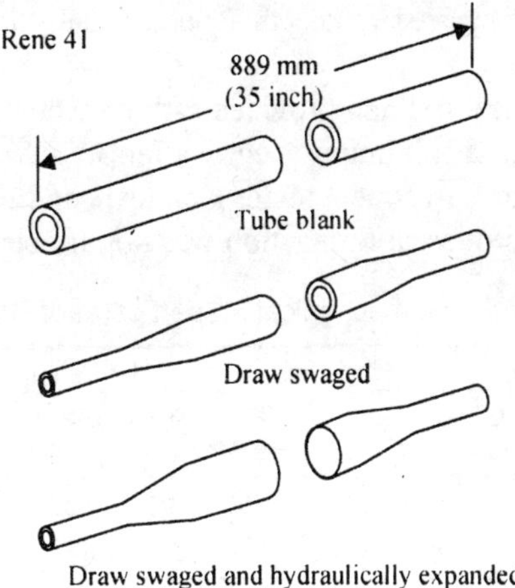

Fig. 24.8. Interstage annealing.

Atmospheres

Nickel-base alloys are susceptible to attack by sulphur-bearing atmosphere. Therefore, the gas used in a semi-muffle furnace or for manufacturing exothermic atmosphere must have a very low sulphur content. Thin sheet is especially likely to be damaged by sulphur-bearing atmospheres or by oxidation in air. Argon inert environment atmospheres frequently are used to preclude excessive oxidation. Vacuum furnaces may be used provided the necessary cooling rates can be obtained

Example 24.11

Low creep-rupture life was experienced with 0.38 mm (0.015 inch) thick Waspaloy sheet solution treated in air at 995°C (1825°F). Rupture life of 260 MPa (37.5 ksi) at 815°C (1500°F) was increased from 15 hours (heat treated in air) to 28 hours merely by using argon as a protective atmosphere. However, for thicker sheet 1.25 mm (0.050 inch) of the same alloy, there was no significant difference between specimens solution treated in air and in argon.

Solution Treating

Solution treating temperatures of nickel-base alloys are determined primarily by the properties desired after subsequent ageing. In wrought alloys, for optimum creep-rupture properties, higher solution treating temperatures are used than when optimum short-time tensile yield strength and elongation at elevated temperature are desired. For instance, René 41 and Udimet 500 are solution treated at 1175°C (2150°F) for optimum creep-rupture properties and at 1080°C (1975°F) for optimum short-time tensile properties. For cast alloys, solution treatment temperatures primarily are a function of incipient melting temperatures as well as furnace capability. Protective atmosphere or vacuum is necessary.

Because wrought nickel-base alloys are often solution-annealed below the gamma prime solubility limit and cast nickel-base alloys frequently are solutioned very close to the incipient melting temperature, control of furnace temperature is important when these materials are being heat treated. Variations in furnace temperature can result in failure to achieve specified properties. Should this occur because of an inadequate solution heat treatment, a re-heat treatment sometimes is effective for producing desired results. When incipient melting has occurred, re-heat treatments usually are not able to re-establish adequate properties in the affected part.

Table 24.16 indicates how re-heat treating provided turbine wheels of wrought René 41 alloy with sufficient yield strength to exceed a minimum specification of 896 MPa (130 ksi), which was not consistently met by a single two-stage treatment under identical conditions. Re-treatment also increased both tensile strength and reduction of area. Elongation was not affected.

Table 24.16. Effect of re-heat treatment on mechanical properties(a) of René 41 turbine wheels.

Treatment	*Ultimate tensile strength* *MPa*	*ksi*	*Yield strength(b)* *MPa*	*ksi*	*Elongation in 50 mm (2 inch), %*	*Reduction in area, %*
Heat treated(c)	1310–1380	190–200	850–920	123–133(d)	16–22	17–21
Re-heat treated(e)	1340–1420	195–206	920–1030	134–150(f)	16–22	18–24

(a) Data represent eight tests for each treatment; properties at 20°C (68°F); (b) Specified minimum; (c) Solution treated at 1065°C (1950°F), air cooled; aged at 760°C (1400°F) for 16 hours air cooled; (d) Average, 880 MPa (128 ksi); (e) Re-heat treated repeating procedure in (c); and (f) Average, 980 MPa (142 ksi).

Higher solution heat-treat temperatures and a faster quench yield more favourable properties for forming René 41 (Fig. 24.9). However, exposure to temperatures at which the carbides dissolve promotes subsequent formation of brittle films at the grain boundaries.

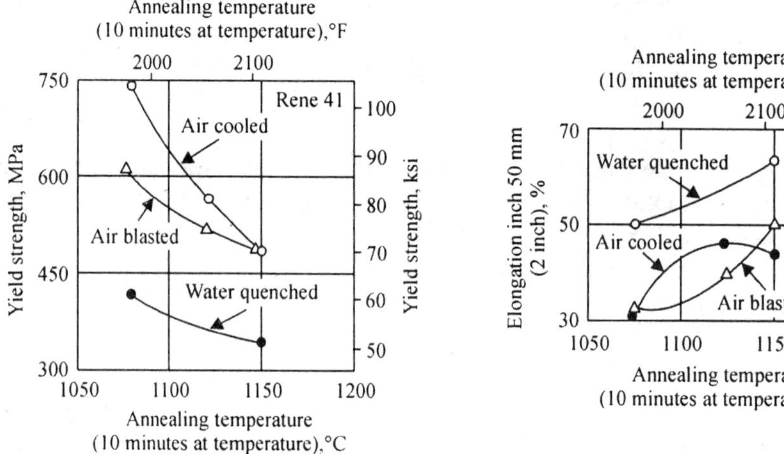

Fig. 24.9. Effects of annealing temperature and quenching rate on typical room-temperature ductility and yield strength of René 41.

In welded tubular components of René 41 solution annealed at 1150°C (2100°F), cracking is often a serious problem, even though the high-temperature annealing is followed by a solution heat treatment at 1080°C (1975°F) prior to welding. Solution treating at 1080°C (1975°F) after a 1150°C (2100°F) anneal precipitates M_6C along grain boundaries, resulting in cracking when the alloy is welded and subsequently aged. The M_6C carbides begin to dissolve between 1105 and 1150°C (2025 and 2100°F). Consequently, a solution annealing temperature has been established at 1105 ± 15°C (2025 ± 25°F) instead of 1150°C (2100°F). In addition, a modified ageing cycle has been developed to prevent cracking in welded components without adversely affecting mechanical properties. Welding and ageing tests were performed on material solution treated at 1080°C (1975°F). Several microcracks were observed in welded specimens that were aged between 760 and 845°C (1400 and 1550°F). A modified heat treatment was developed in which welded components are first aged for 1 hour at 900°C (1650°F), then for 10 hours at 760°C (1400°F). Parts subjected to mild forming, and which are not welded, are aged for 16 hours at 760°C (1400°F).

Shrinkage during Ageing

For design purposes, the contraction of components made of René 41 must be taken into consideration. An average contraction of 0.033 mm/mm (0.0013 inch/inch) can be expected after ageing for 16 hours at 760°C (1400°F) or for 1 hours at 900°C (1650°F) followed by 10 hr. at 760°C (1400°F). This average contraction value reflects both longitudinal and transverse measurements on three heats of René 41 and the following treatments prior to ageing:

1. Solution treated at 1080°C (1975°F).
2. Solution treated at 1080°C (1975°F), annealed at 1105°C (2025°F), re-solutioned at 1080°C (1975°F).

Dilatometric measurements indicate that contraction is essentially complete within 6 hours after the ageing temperature is reached.

Sheet Structures

Sheet structures experiencing moderate cold working are generally spot welded and double aged in air furnaces without a prior re-solution treatment. Although desirable, re-solutioning introduces distortion effects that are detrimental to the structure and that outweigh any benefits obtained by the treatment. Air atmosphere is used during ageing to provide a tight, adherent, dark scale useful for emissivity purposes in actual flight conditions.

Tubular Components

In fabricating tubular components that are reduced at the ends by swaging, multiple annealing is frequently required to remove cold work. Tubes are annealed in an argon atmosphere having an oxygen content of not more than 400 ppm and a dew point not higher than –20°C (–5°F). The annealing temperature is held just below the carbide solubility temperature (1105 ± 15°C or 2025 ± 25°F) for 5 minutes. The tubes are then quenched in water in less than 3 seconds. The total intergranular attack on the material after three interstage anneals is generally less than 0.3 mil. Tubular components are aged in an air atmosphere.

Effect of Rate of Cooling from the Solution Treating Temperature

Nickel-base alloys (such as René 41, Waspaloy and Astroloy) that contain large amounts of hardening elements cannot be uniformly softened throughout heavy sections even by water quenching from the solution temperature. Variations in hardness from 180 to 240 HB, caused by precipitation during cooling, have been noted from surface to centre in water quenched 968 mm^2 (1½ $inch^2$) bars of René 41. In air-cooled Astroloy forgings, higher strength and ductility values are produced in 100 to 125 mm (1 to 2 inch) sections than in 102 to 127 mm (4 to 5 inch) sections because of the uncontrolled overageing that occurs in the heavier sections.

In order to obtain more uniform hardness after solution treating and more uniform response to ageing, heavy sections of these alloys are often air cooled from the solution temperature, despite the partial ageing that occurs as a result of slower cooling.

Occasionally, a time delay may be used before quenching, particularly when parts may be susceptible to quench cracking. Component hardness will vary with time delays to some extent. The effect of time delay before quenching on the hardness of René 41 sheet is shown in Fig. 24.10. Hardening is caused by precipitation of gamma prime and carbides.

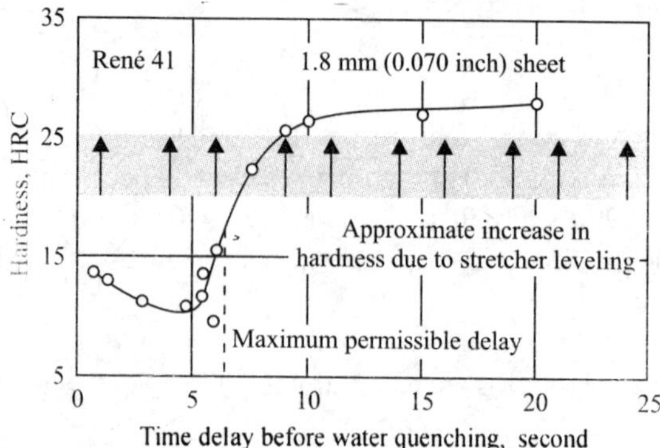

Effect of delay before water quenching on the hardness of René 41 sheet solution heat treated at 1080°C (1975°F). Shaded band shows the increase in hardness due to stretcher leveling (about 2 per cent strain) after quenching.

Fig. 24.10. Effect of time delay in quenching on hardness of René 41.

Example 24.12

High hardness (35 HRC) of semi-fabricated parts made of Waspaloy and René 41 was experienced after the parts were cooled from 1095°C (200°F) by an air blast. Water quenching at first produced no improvement, because, owing to the location of the quench tank, a time delay of 10 seconds occurred in the quenching operation. The problem was solved by moving the quench tank closer to the furnace, which allowed quenching to black heat in less than 4 seconds. The parts were then about 20 HRC as quenched and could be easily formed.

Occasionally, experimentation must be conducted to determine quenching mediums and procedures that yield an acceptable combination of cooling rate, distortion control and cost. In some applications, combined quenches, such as water immersion and water spray, are used simultaneously on a part having a heavy, thick base and a thin web.

Fig. 24.11 shows a semifinished forged turbine blade of René 41 and schematically illustrates the method by which it was simultaneously spray and immersion quenched with water after being solution treated for in-process annealing prior to cold forming. The blade was required to be as soft as possible after annealing (maximum acceptable hardness of 18 HRC), and to have a minimum of distortion and oxidation.

Parts to be heated were hand-loaded into an electrically heated full-muffle box furnace, in which an argon atmosphere was used. Processing consisted of the following:

1. Degrease, alkaline clean and dry. Handle parts with gloves to avoid contamination by fingerprints.
2. Place in handling fixture.
3. Place in furnace operating at 1080°C (1975°F) with argon flowing at 0.03 m^3/minute (1 ft^3/minute); muffle volume, 0.03 m^3 (1 ft^3).
4. Reduce argon flow to 0.01 m^3/minute (1/3 ft^3/minute) when furnace load reaches 1080°C (1975°F).
5. Hold parts at 1080°C (1975°F) for 15 minutes.
6. Place parts in quench (Fig. 24.10) within 2 seconds after withdrawal from furnace.
7. When parts are cool, remove from quench.

The annealed blades had an average hardness of 95 HRB (18 HRC). As the result of the special quenching technique, distortion was held to an acceptable minimum. Oxidation consisted only of a dark-blue surface discolouration.

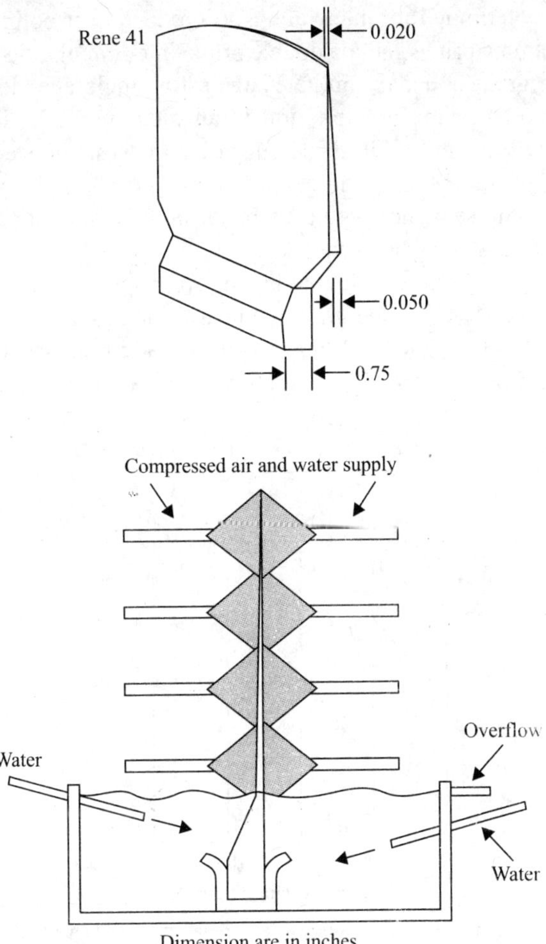

Fig. 24.11. René 41 forged turbine blade and immersion-spray arrangement for water quenching.

Air cooling rates for nickel-base blade and vane alloys generally are sufficient to produce a supersaturated gamma matrix for subsequent development of optimum hardness by ageing. Occasionally, as noted for thick section parts such as turbine wheels, air cooling may be insufficient to produce the desired structure.

Even for blade and vane components which are relatively small, cooling rates may be inhibited by other considerations, as for example in the pack aluminising of airfoil surfaces or the hot isostatic pressure closing of internal casting porosity. In such cases, either forced argon cooling may be used or the components may be slow cooled from the pack or HIP temperature and re-heated, followed by rapid cooling from a casting diffusion treatment or solution treatment temperature.

Ageing

When more than one phase is capable of precipitating from the alloy matrix, judicious selection of a single ageing temperature may result in obtaining optimum properties in the component. Although single age temperatures can be determined for many alloys to produce near optimum properties, the need to produce an appropriate gamma prime/gamma double prime intergranular dispersion with an appropriate carbide dispersion in the grain boundary militates against a single age step. Fig. 24.12 indicates the range of carbide and gamma prime precipitation in an older wrought alloy, René 41. Laboratory experiments coupled with the results of Fig. 24.12 suggested that a single-step ageing temperature which would produce the best carbide ($M_{23}C_6$) and gamma prime morphologies was 870°C (1600°F). In all likelihood, a two-step ageing sequence would be better and, in fact, almost all wrought nickel-base alloys are given a two to four-step ageing sequence.

One alternative to single ageing is a double ageing treatment. In Nimonic 80A, initial ageing at 900°C (1650°F) precipitates $M_{23}C_6$ in the grain boundaries and gamma prime within the grains. In a second ageing treatment at 700 to 760°C (1290 to 1400°F), additional precipitation of a finer gamma prime occurs within the grains, to provide an optimum combination of strength and ductility.

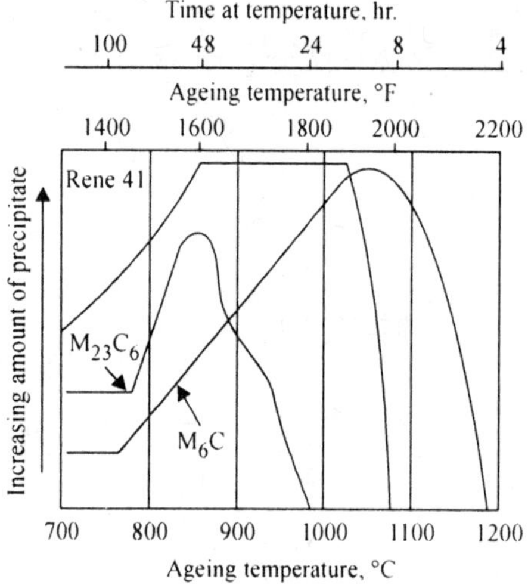

Relative amount of precipitates resulting from ageing Reneé 41 at various times and temperatures after solution treating at 1205°C (2200°F) and water quenching

Fig. 24.12. Precipitation in ageing.

Udimet 500 is typical of nickel-base-age hardenable alloys that are strengthened by gamma prime precipitation and also contain TiC and $M_{23}C_6$. The alloy is normally double aged. Solution treating at 1065 to 1175°C (1950 to 2150°F) is followed by air cooling. All phases except TiC are dissolved as a result of the solution treatment. Appreciable hardening takes place during cooling because of rapid precipitation of gamma prime. Intermediate ageing at 845°C (1550°F) precipitates $M_{23}C_6$ in a discontinuous form, mainly at the grain boundaries. Final ageing at 760°C (1400°F) determines the ultimate hardness by controlling the size of the gamma prime precipitate. The presence of $M_{23}C_6$ in grain boundaries increases the creep-rupture life; however, a continuous carbide network should be avoided because it markedly decreases ductility.

Higher ductility at room temperature can be obtained by omitting the 845°C (1550°F) ageing treatment, but undesirable cellular precipitation of $M_{23}C_6$ in the grain boundaries will cause low ductility in creep-rupture tests at elevated temperatures. Elimination of the intermediate ageing treatment proved beneficial for increasing the room-temperature strength and ductility of Udimet 500 forgings that would not be subjected to service temperatures above 730°C (1350°F). Unacceptable room-temperature mechanical properties were obtained by solution treating at 1080°C (1975°F) for 4 hours and air cooling, ageing at 845°C (1550°F) for 24 hours and air cooling and ageing at 760°C (1400°F) for 16 hours and air cooling. Acceptable properties were produced by eliminating the 845°C (1550°F) treatment only. Table 24.17 compares specified minimum properties with those obtained by the two treatments.

Weldments

Weldments are heat treated by a variety of special methods.

Example 24.13

To meet final requirements for an Inconel X-750 welded pressure vessel, 285 cm long by 46 cm (9 ft 4 inch long by 18 inch) in diameter, for a missile application, a procedure was established that incorporated the following special techniques:

1. Cradle vessel in lightweight, nonrestrictive rack (Fig. 24.13).
2. Purge interior of vessel with argon.
3. Place cradled vessel in furnace (car-bottom, elevator type) and heat at 165 to 220°C/hr. (300 to 400°F/hr.) to 955°C (1750°F), to relieve forming and welding stresses. Hold at 955°C (1750°F) for 15 minutes; maintain argon at 21 kPa (3 psi) and –50°C (–60°F) dew point within the interior of vessel, to prevent sagging and oxidation; maintain furnace atmosphere of lean exothermic gas at +20°C (+ 65°F) dew point and approximate composition of 10.5 per cent CO_2, 1.5 per cent CO, 1 to 2 per cent H_2, rem N_2.
4. Remove cradled vessel from furnace and air cool.
5. Re-heat cradled vessel to 705°C (1300°F), and hold at temperature for 6 hours under the same atmospheric conditions as described in step 3, to age that portion of the weld that was effectively solution treated in cooling from the welding temperature.
6. Air or furnace cool in cradle.

Table 24.17. Effect of eliminating intermediate ageing on typical room-temperature mechanical properties of Udimet 500.

	Ultimate tensile strength		*Yield strength(0.2%)*		*Elongation in 50 mm (2 inch), %*	*Reduction in area, %*
	MPa	*ksi*	*MPa*	*ksi*		
Specified minimum	1030	150	690	100	10	15
Obtained with intermediate ageing(a)						
Test 1	1030	149	830	120	7	11
Test 2	970	141	810	118	4	5.
Obtained without intermediate ageing(b)						
Test 1	1170	170	800	116	14.5	17
Test 2	1230	179	850	123	14	16

(a) Heat treatment: 4 hours at 1080°C (1975°F), air cool; 24 hours at 845°C (1550°F), air cool (intermediate ageing); 16 hours at 760°C (1400°F), air cool; and (b) Same as (a), but without intermediate ageing.

René 41 is solution treated at 1080°C (1975°F) before fusion welding. A higher solution treating temperature, such as 1175°C (2150°F), permits grain-boundary precipitation of $M_{23}C_6$ to occur in the heat-affected zone during welding and this precipitate causes catastrophic age cracking after welding or during subsequent heat treatment of highly restrained weldments.

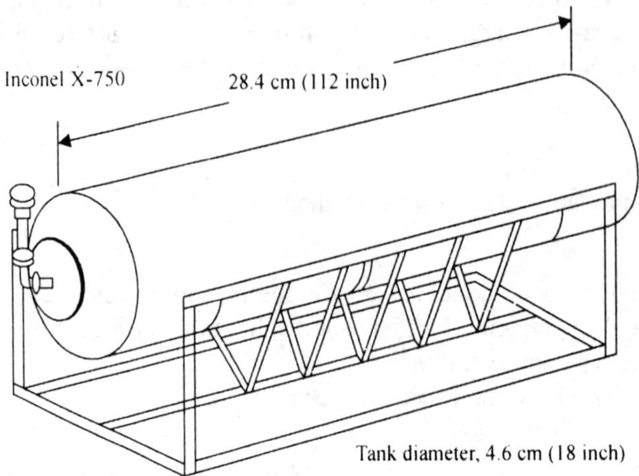

Fig. 24.13. Light-gauge (0.81mm, or 0.032 inch, wall thickness) welded pressure vessel of Inconel X-750 alloy in a non-restrictive rack for heat treating.

When it is possible to re-solution treat, at 1065 to 1080°C (1950 to 1975°F) after welding, the standard ageing treatments are used—that is, for maximum tensile strength, 760°C (1400°F) for 16 hours or, for improved creep-rupture properties, 900°C (1650°F) for 4 hours.

The re-solution treating of large and complex structural components containing restrained fusion welds is seldom feasible, on account of distortion. Single ageing of René 41 fusion welds, particularly those under high restraint, without re-solution treating, has resulted in a high incidence of strain age cracking in the heat-affected zone. One fabricator overcame this difficulty by using a duplex ageing treatment of 900°C (1650°F) for 1 hour plus 760°C (1400°F) for 10 hours.

Weld repairs of weldments may be employed to remedy defects or to repair weld casting defects in metals or to repair weld base metal cracks originating from service application. In the case of weld repair of weldments of wrought age-hardening alloys, the repair may originate either before or after ageing. For the case of René 41, for example, if weld repair recedes ageing it can be followed by the double ageing treatment cited above. If weld repair follows ageing, the problem is more severe. Assemblies that require weld repair after ageing may be classified as follows:

1. Simple assemblies that can be water quenched.
2. Complex assemblies that cannot be water quenched because of low-restraint welds.
3. Complex assemblies that cannot be water quenched because of high-restraint welds.

The first type of assembly is simply re-solution treated at 1080°C (1975°F) and water quenched prior to repair welding; the repair is then followed by double ageing. For the second type of assembly, post-weld heat treatments at 980°C (1800°F) for 1 hour and at 760°C (1400°F) for 6 hours are used. Re-solution treating and air cooling before repair welding is the correct procedures for the third type of assembly; the repair is then followed by double ageing. In all repair procedures, parts subjected to ageing after weld repair must be cooled slowly from the ageing temperature, either in the furnace or in an insulating material. Resistance welding of René 41 requires stringent process controls. For applications in which the service temperature exceeds the ageing temperature, or for weldments that do not require maximum ductility, the resistance welded assembly can be aged after welding. For all other conditions, however, René 41 should be welded in the solution treated and aged condition. Ageing after spot welding decreases the ductility of the spot weld zone. Because spot welds can develop high stress concentrations in service, good ductility is essential.

HEAT TREATING OF COBALTBASE ALLOYS

Cobalt-chromium-nickel-base alloys can be full annealed or stress relief annealed but are rarely solution treated because the gamma prime type precipitates generally do not participate in strengthening of cobalt alloys. Ageing heat treatments may be given but the purpose of these would be to modify the carbide distribution. Heat treating temperatures for selected alloys are given in Tables 24.5, 24.6 and 24.7.

Full annealing, rather than stress relieving, is recommended for most of the wrought cobalt-chromium-nickel-base alloys whenever high residual stresses are developed during fabrication. When fabrication is completed, however, these alloys can be safely stress relieved at about 55°C (100°F) below the annealing temperature (Table 24.5) provided that excessive distortion does not occur.

Annealing of the cast cobalt-base alloys is often avoided, because of the undesirable precipitation of coarse carbides that occurs during slow cooling after annealing. Investment-cast gas-turbine buckets and guide vanes that are made to close dimensional tolerances and must exhibit excellent resistance to stress cracking are occasionally stress relieved at 870 to 900°C (1600 to 1650°F) for 2 to 4 hours. There is no apparent effect of cooling rate on the properties of annealed or stress relieved cobalt-base alloys since gamma prime type precipitates are not involved in the hardening process.

Ageing reactions of hardenable cobalt-case alloys such as HS-31 (X-40) are based on the random precipitation of various complex carbides and intermetallics. HS-31, a cast alloy containing 0.4 per cent C, may contain $Cr_{23}C_6$, Cr_3C_2, MC or M_6C (or both), as well as mu (M_7W_6) phase. The ratio of chromium to carbon in the alloy determines which of the chromium carbides will appear during ageing; a high ratio favours the precipitation of $Cr_{23}C_6$ (cobalt, iron, tungsten and nickel can replace part of the chromium in $Cr_{23}C_6$. The $Cr_{23}C_6$ can be precipitated in different forms, represented by discrete, acicular or lamellar shapes. Discrete particles are obtained by ageing at temperatures near 760°C (1400°F). Ageing at higher temperatures causes the appearance of the acicular precipitate. The third form, lamellar, resembles pearlite and is nucleated by cooling from a solution treating temperature between 1205°C (2200°F) and the temperature range where $Cr_{23}C_6$ precipitates.

During elevated-temperature testing of turbine blades made from vacuum melted X-40 alloy, it was observed that the blades were elongating as a result of tensile overstress (Fig. 24.14). To increase the tensile properties, the material was subjected to ageing treatments applicable to air-melted X-40 alloy, but these were not sufficiently beneficial. Special ageing treatments were then developed and these provided satisfactory tensile properties. Average values resulting from these treatments are given in the table accompanying Fig. 24.14. The best of these treatments (50 hours at 720°C, 1325°F, air cool) resulted in increases of well over 50 per cent in the yield strength up to 595°C (1100°F), compared to the tensile properties of as-cast material and in a corresponding decrease in ductility (Fig. 24.13). Creep-rupture tests of material aged by the best treatment showed an average increase of 290 per cent in the creep-rupture life of the alloy, compared to as-cast material (Fig. 24.14). Data from tests conducted at 480°C (900°F) showed no significant change in the fatigue curve.

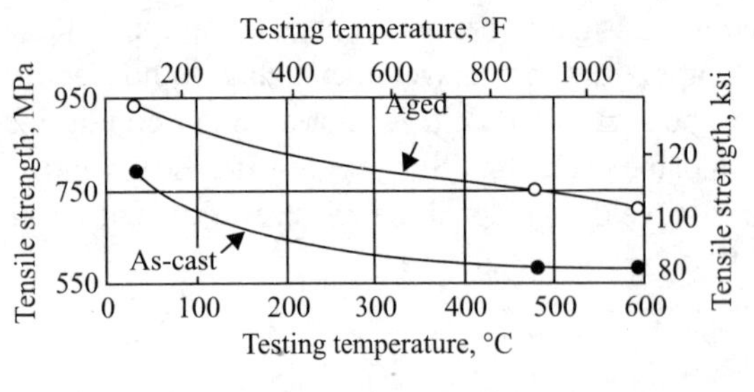

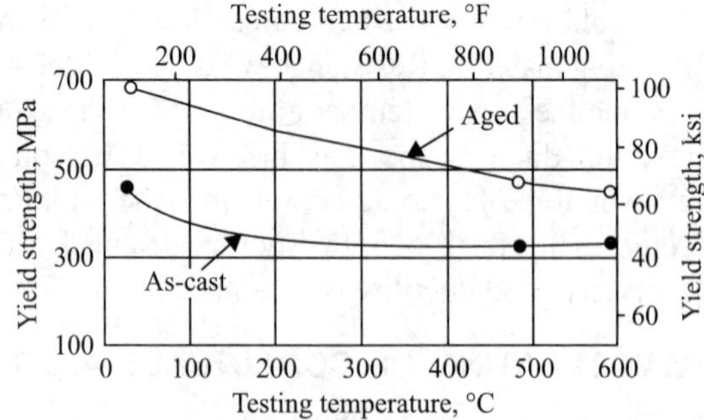

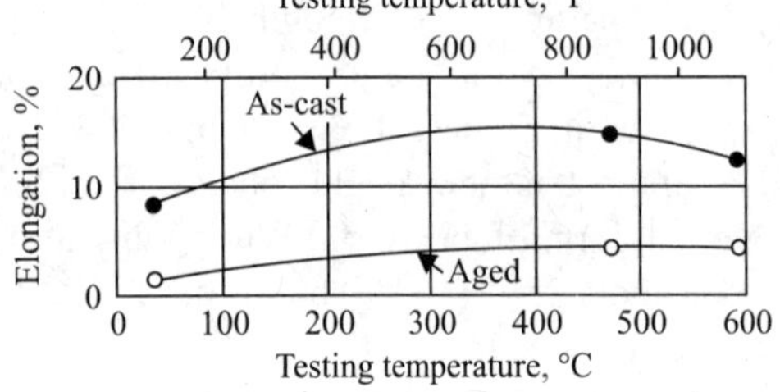

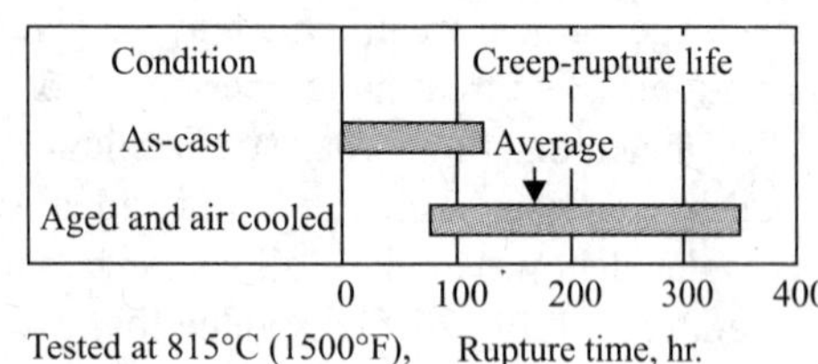

Ageing treatment(a)	*Ultimate tensile strength*		*Yield strength(b)*		*Elongation in 50 mm (2 inch), %*
	MPa	*ksi*	*MPa*	*ksi*	
Tested at room temperature					
As-cast	770	111	430	63	8
720°C (1325°F), 16 hrs.; AC	790	114	500	72	6
720°C (1325°F), 50 hrs.; AC	940	136	690	100	3
815°C (1500°F), 2 hrs.; AC	850	123	610	88	3
815°C (1500°F), 20 hrs.; AC	830	120	600	87	2
Tested at 480°C (900°F)					
As-cast	580	84	270	39	15
720°C (1325°F), 16 hrs.; AC	640	93	320	47	12
720°C (1325°F), 50 hrs.; AC	750	109	480	69	5
815°C (1500°F), 2 hrs.; AC	640	93	370	48	6
815°C (1500°F), 20 hrs.; AC	770	111	370	53	5
Tested at 595°C (1100°F)					
As-cast	590	85	270	39	13
720°C (1325°F), 16 hrs.; AC	590	86	280	40	13
720°C (1325°F), 50 hrs.; AC	720	104	440	64	4
815°C (1500°F), 2 hrs.; AC	660	96	330	48	7
815°C (1500°F), 20 hrs.; AC	720	104	330	48	5

(a) AC, air cooled. (b) At 0.2 % offset.

Fig. 24.14. Effect of ageing treatments on typical tensile properties of vacuum melted X-40 alloy.

Cobalt-base alloys generally are not susceptible to attack by furnace atmosphere and air environments can be used for heat treatment. At higher temperatures or in the case of vacuum melted alloys which contain reactive elements such as zirconium (for example, MAR-M 509), a vacuum, inert or almost neutral environment may be required. Cobalt-base alloys are welded, either as fabrications or to repair weld cracks in base alloys after service operation. Stress relief heat treatments are applied after welding but there is no uniform pre-weld heat treatment practice.

HEAT TREATING OF REFRACTORY METALS

Presently available refractory metals and alloys are not heat treated to achieve high strength, but are used in the stress-relieved or annealed (re-crystallised) condition.

Stress relieving

Stress relieving is recommended for molybdenum, molybdenum alloys and unalloyed tungsten when cold work in processing exceeds about 50 per cent. Stress relieving is performed below the re-crystallisation temperature to retain the toughness acquired during cold working. Lowering the level of residual stress permits further fabrication with less danger of cracking and delamination. Processing history markedly affects the optimum stress-relieving temperature.

Unalloyed tantalum, Ta-10W, FS-80 and FS-82 are stress relieved to increase ductility and to reduce residual stress resulting from welding and cold work. These materials may be placed in service in either the stress-relieved or the re-crystallised (fully annealed) condition, because they do not lose ductility during re-crystallisation as do molybdenum alloys and unalloyed molybdenum and tungsten.

Recrystallisation Annealing

Re-crystallisation annealing of unalloyed molybdenum, Mo-0.5Ti, TZM and unalloyed tungsten raises the ductile-brittle transition temperature of the materials and results in a nearly complete loss of ductility at room temperature. For this reason, mill products of these materials are supplied in the stress-relieved condition. Unalloyed tantalum, Ta-10W and FS-80 and FS-82 niobium alloys retain low ductile-brittle transition temperatures after re-crystallisation. Mill products of these materials are supplied in either the stress-relieved or annealed condition, depending on end use or fabrication requirements.

Annealing temperatures for refractory metals depend on prior processing history, chemical composition, amount of cold work and sequence of in-process anneals. Therefore, annealing temperatures are not standardised, although those given in Table 24.5 may be considered typical.

Unalloyed niobium and tantalum and many of their alloys may be cold formed in either the stress-relieved or the annealed (re-crystallised) condition. They are seldom formed in the as-rolled condition because of their susceptibility to cracking. A stress relief treatment at about 55°C (100°F) below the re-crystallisation temperature reduces residual stress and improves ductility enough to permit some forming. However, annealing should precede more severe operations, such as spinning, flow turning and deep drawing.

Re-crystallisation greatly improves the ductility of niobium and tantalum alloys but lowers their strength by removing the effects of strain hardening. However, when the material is used at a temperature above the re-crystallisation temperature, the effects of strain hardening are not utilised in service. Therefore, it is usually preferable to do all forming in the annealed condition.

Heat treatment after forming may consist of either stress relieving or annealing. When the formed parts are given an oxidation-resistant coating prior to use, the thermal treatment involved in the coating process usually provides adequate stress relief.

Unalloyed molybdenum, molybdenum alloys and unalloyed tungsten are nearly always formed in the stress-relieved condition. Re-crystallisation results in a marked decrease in the ductility at or near room temperature.

Surface Contamination

Tungsten, molybdenum and the alloys of molybdenum are less susceptible to surface contamination during heat treatment than are niobium and tantalum alloys. However, atmospheric contamination of the Mo-0.5Ti alloy is known to occur in sheet, with the probable formation of TiO_2 and TiN.

Absorption of oxygen, nitrogen and hydrogen by niobium and tantalum increases as temperatures increase above about 650°C (1200°F). The hard, brittle surface layers formed are detrimental to further forming and machining operations and to performance characteristics. Stress relieving or annealing of material with a contaminated surface, even in an inert atmosphere, will result in diffusion and usually in a stronger, less ductile alloy having a higher transition temperature from ductile to brittle failure.

Atmospheres

Refractory metals require a protective atmosphere during heat treatment, to avoid degradation of properties from surface contamination. Molybdenum and tungsten have an extremely low solubility for hydrogen; hence, hydrogen atmospheres are almost always used for treating these metals.

Because niobium and tantalum are severely embrittled by hydrogen, oxygen and nitrogen, these metals must be heat treated in vacuum or an inert atmosphere, such as argon with an extremely low dew point (–55°C, or –70°F or lower). Argon is preferred over helium because argon is cheaper and its greater density provides more positive gas coverage of parts. Table 24.18 lists important factors in the prevention of contamination from environment during heat treatment of refractory metals.

Table 24.18. Causes, effects, prevention and correction of contamination of refractory metals and alloys during heat treatment.

Contaminants Type and quantity	*General effect*	*Preventive action*	*Corrective action*	*Remarks*
Unalloyed molybdenum or tungsten: TZM alloy				
Oxygen and nitrogen in all quantities, including minute concentrations	Formation of surface oxides and nitrides. Diffusion into surface cause embrittlement	Use of hydrogen atmosphere, inert gas, or vacuum	Pickle in molten caustic or caustic solution to always remove surface contamination. Acid pickle to remove sub-surface contamination zone	Dry hydrogen is virtually the atmosphere selected for heat treating molybdenum and tungsten
FS-80; FS-82; Ta-10 W; unalloyed tantalum				
Oxygen, nitrogen and hydrogen in all quantities, including minute concentrations in inert gases or moderate vacuum.	Formation of surface and sub-surface oxides, nitrides and hydrides. Rapid diffusion at elevated temperature causes embrittlement	Use of high vacuum (10 mm Hg, minimum) or high-purity inert gas (preferably argon)	Remove hydrogen contamination by high-vacuum thermal treatment. Acid pickle to remove surface layers contaminated by oxygen and occurs nitrogen. No corrective action is possible when excessive diffusion of oxygen or nitrogen occurs	Diffusion of oxygen, nitrogen and hydrogen is rapid Observable surface effects are negligible, but extreme loss of ductility when impure inert atmosphere or poor vacuum is used

SECTION VI

Inspection and Quality Control In Heat Treatment

Chapter 25

Measurement and Inspection

INTRODUCTION

Measurement, that is the act of measuring or being measured, is the fundamental activity of inspection. The intent of inspection is to ensure that what is being manufactured will conform to the specifications of the product. Most products are manufactured to standard sizes and shapes.

Large-scale manufacturing based on the principles of standardisation of sizes and interchangeable parts became common practice in the early 1950s. Size control must be built into the machine tools and the workholding devices through the precision manufacture of these machines and their tooling. The output of the machines must then be carefully checked: (i) to determine the capability of specific machines; and (ii) for the control and maintenance of the quality of the product. A designer who specifies the dimensions and tolerances of a part quite often does so to enhance the function of the product, but the designer is also determining the machines and processes needed to make the part. Quite often, it is necessary for the design engineer to alter the design or the specifications to make the product easier or less costly to manufacture, assemble, or inspect, or all of these. Designers should always be prepared to do this provided they are not sacrificing functionality, product reliability, or performance.

The examination of the product either during or after manufacture, either manually or automatically, falls in the province of inspection. Basically, inspection of items or products can be done in two ways:

1. By attributes, with the use of gauges to determine if the product is good or bad resulting in a yes or no, go or no-go decision.
2. By variables, with the use of calibrated instruments to determine the actual dimensions of the product for comparison with the desired size.

In an automobile, a speedometer and an oil pressure gauge are variable types of measuring instruments and an oil pressure light is an attributes type of gauge.

As is typical of an attributes type of gauge, the driver does not know what the pressure actually is if the light goes on, only that it is not good.

Measurement is the generally accepted industrial term for inspection by variables. Gauging is the term for determining whether the dimension or characteristic is larger or smaller than the established standard or range of acceptability. Variable types of inspection generally take more time and are more expensive than attributes inspection, but they give more information because the magnitude of the characteristic is known in some standard unit of measurement.

STANDARDS OF MEASUREMENT

The four fundamental measures on which all others depend are length, time, mass and temperature. Three of these basic measures are defined in terms of material constants. These four measures, along with the ampere and the candela, provide the basis for all other units of measurement.

Most mechanical measurements involve combinations of units of mass, length and time. Thus, the Newton, a unit of force, is derived from Newton's second law of motion (f = ma) and is defined as the force that gives an acceleration of 1 metre per second to a mass of 1 kilogram.

Linear Measurements

When man first sought a unit of length, he adopted parts of the human body, mainly his hands, arms or feet. Such tools were not very satisfactory because they were not universally standard in size. Satisfactory measurement and gauging must be based on a reliable and preferably universal, standard or standards. These have not always existed. For example, although the musket parts made in Eli Whitney's shop were interchangeable, they were not interchangeable with parts made by another contemporary gun maker from the same drawings, because the two gunsmiths had different foot rulers.

Today, the entire industrialised world has adopted the international meter as the standard of linear measurement. The inch, used by both the United States and Great Britain, has been officially defined as 2.54 centimetres. Thus, the United States standard inch is 41,929.399 wavelengths of the orange-red light from krypton-86.

Although officially the United States is committed to conversion to the metric (SI) system of measurement, which uses millimetres for virtually all linear measurements in manufacturing, the English system of feet and inches is still being used by many manufacturing plants and its use will probably continue for some time.

Length Standards in Industry

Gauge blocks provide industry with linear standards of high accuracy that are necessary for everyday use in manufacturing plants. These are small, rectangular, square, or round in cross-section and are blocks made from steel or carbide with two very flat and parallel surfaces that are certain specified distances apart. Today, gauge block sets of excellent quality are produced by a number of companies all over the world.

Gauge blocks, made of alloy steel, are hardened to $65R_c$ and carefully heat-treated (seasoned) to relieve internal stresses and to minimise subsequent dimensional change. Carbide gauge blocks provide extra wear resistance. The measuring surfaces of each block are surface-ground to approximately the required dimension and are then lapped and mirror-polished to bring the block to the final dimension and to produce a very flat and smooth surface. The surface finish is 0.4 millionths of an inch (0.01 microns).

The dimensions of individual blocks are established by light beam interferometry, with which it is possible to calibrate these blocks routinely with an uncertainty as low as 1 part per million.

Gauge blocks are available in different shapes (squares, angles, rounds and pins), so that standards of high accuracy can be obtained to fill almost any need. In addition, various auxiliary clamping, scribing and base block attachments are available that make it possible to form very accurate gauging devices.

Standard Measuring Temperature

Because all the commonly used metals are affected dimensionally by temperature, a standard measuring temperature of 68°F (20°C) has been adopted for precision-measuring work. All gauge blocks, gauges, and other precision-measuring instruments are calibrated at this temperature. Consequently, when measurements are to be made to accuracies greater than 0.0001 inch (0.0025 mm), the work should be done in a room in which the temperature is controlled at standard. Although it is true that to some extent both the workpiece and the measuring or gauging device may be affected to about the same extent by temperature variations, one should not rely on this. Measurements to even 0.0001 inch (0.0025 mm) should not be relied on if the temperature is very far from 68°F (20°C).

Allowance and Tolerance

Two factors, allowance and tolerance, must be specified if one is to obtain the desired fit between mating parts. Allowance is the intentional, desired difference between the dimensions of two mating parts. It is the difference between the dimension of the largest interior-fitting part (shaft) and that of the smallest exterior-fitting part (hole). Fig. 25.1 shows pin A designed to fit into the hole in block B. This difference thus determines the condition of tightest fit between mating parts. Allowance may be specified so that either clearance or interference exists between the mating parts. With clearance fits, the largest shaft is smaller than the smallest hole, whereas with interference fits, the hole is smaller than the shaft.

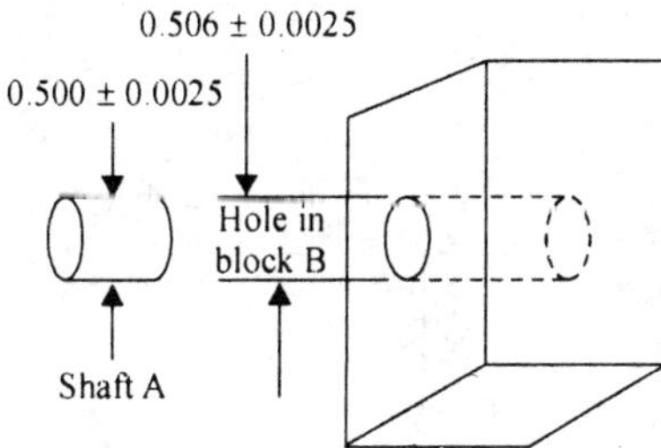

Fig. 25.1. When mating parts are designed, each shaft must be smaller than each hole for a clearance fit.

Tolerance is an undesirable, but permissible, deviation from a desired dimension in recongnition that no part can be made exactly to a specified dimension, except by chance and that such exactness is neither necessary not economical. Consequently, it is necessary to permit the actual dimension to deviate from the desired theoretical dimension (called the nominal) and to control the degree of deviation so that satisfactory functioning of the mating parts will still be ensured.

Inspection, by means of measurement techniques, provides feedback information on the actual size of the parts with reference to the size specified by the designer on the part drawing. The conventional drawing shows the nominal diameter of part A plus a positive tolerance, which is to be less than the smallest hole in part B. In other words, the minimum diameter of the hole should always be greater than the maximum diameter of the shaft if all parts are to fit or mate together.

The processes that make the shaft are different from those that make the hole, but both the hole and the shaft are subject to deviations in size because of variability in the processes and the materials. Thus, while the designer wishes ideally that all the shafts would be exactly 0.500 and all the holes 0.506, the reality of processing is that there will be deviations in size around these nominal or ideal sizes.

Most manufacturing processes result in products whose measurements of the geometrical features and sizes are distributed normally (Fig. 25.2). That is, most of the measurements are clustered around the average dimension, $\overline{X}.\overline{X}$ will be equal to the nominal dimension only if the process is perfectly centred. Parts will be distributed on either side of the average. For normal distributions, 99.73 per cent of the measurements will fall within plus or minus 3 standard deviations. ($\pm 3\sigma$), 95.46 per cent in Fig. 25.2 will be within $\pm 2\sigma$ and 68.26 per cent within $\pm 1\sigma$,

where
$$\sigma = \sqrt{\frac{\sum_{i=1}^{n} (X_i - \overline{X})^2}{n}} \quad \text{... (25.1)}$$

and
$$\overline{X} = \frac{\sum_{i=1}^{n} X_i}{n} \text{ for n items} \quad \text{... (25.2)}$$

In general, the designer applies nominal values to the mating parts according to the desired fit between the parts. Tolerances are added to those nominal values in recognition of the fact that all processes have some natural amount of variability.

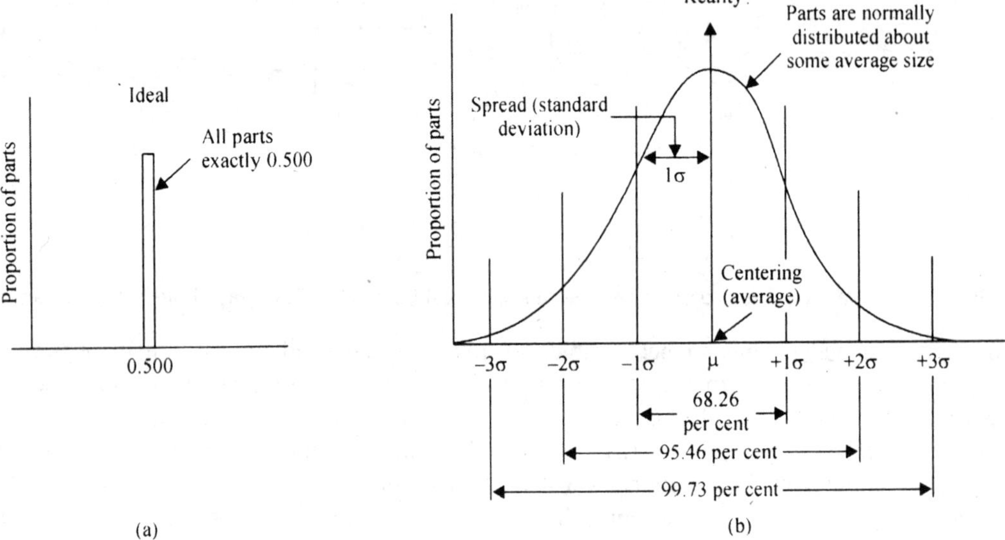

Fig. 25.2. In the ideal situation, the process would make all parts exactly of the same size. In the real world of manufacturing, parts have variability in size. The size distribution is often normal.

Assume the distributions for both the hole and the shaft are normally distributed. As shown in Fig. 25.3 the average fit of two mating parts is equal to the difference between the mean of the shaft distribution and the mean of the hole distribution. The 'range of fit' will be the difference between the minimum diameter shaft and the maximum diameter hole. The minimum 'clearance' would be the difference between the smallest hole and the largest shaft. The tigher the distribution (i.e., the more precise the process), the better the fit between the parts.

If tool wear is considered, the diametre of the shaft will tend to get larger as the tool wears. On the other hand, the diametre of the hole in part B decreases in size with tool wear. If no corrective action is taken, the distribution means will shift toward each other, and the fit will become increasingly tight (the clearance will be decreased). If the hole and the shaft distributions overlap by 2 standard deviations, as shown in Fig. 25.4, 6 parts out of every 10,000 could not be assembled. As the shaft and the hole distributions move closer together, more interference between parts will occur and the fit will become tighter, eventually becoming an interference fit.

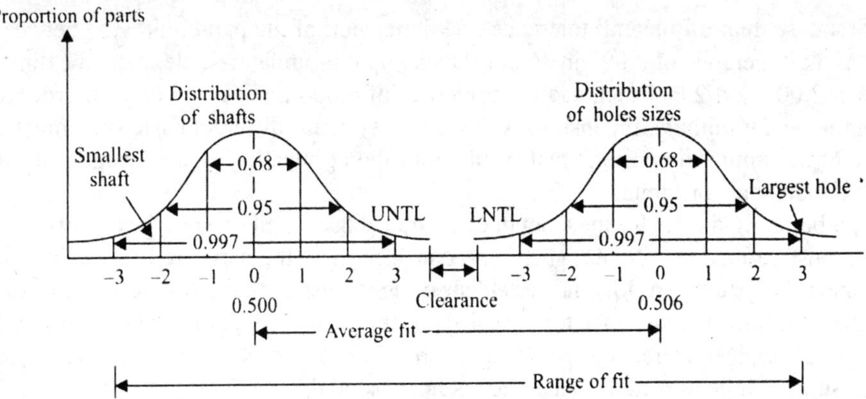

Fig. 25.3. The manner in which the distributions of the two mating parts interact determines the fit.

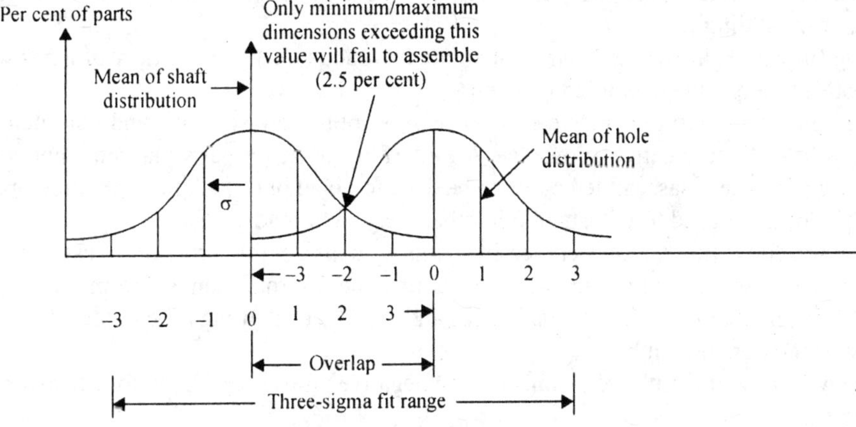

Fig. 25.4. Shifting the means of the distributions toward each other results in some Interference fits.

The designer must specify the tolerances and allowances for the mating parts according to the function. Suppose the mating parts are the cap and the body of an ink pen. The cap must fit snugly but must be easily removed by hand. A snug fit would be too tight for a dead bolt in a door lock. A snug fit is also too tight for a high-speed bearing for rotational parts, but not tight enough for permanently mounting a wheel on an axle. The next section introduces the manner in which tolerances and allowances are specified, but design engineers are expected to have a deeper understanding of this topic.

Specifying Tolerances and Allowances

Tolerance can be specified in three ways: bilateral, unilateral and limits. Bilateral tolerance is specified as a plus or minus deviation from the nominal size, such as 2.000 ± 0.002 inch.

More modern practices uses the unilateral system, where the deviation is in one direction from the basic size, such as:

	+ 0.004 inch			+ 0.1 mm	
2.000 inch		or	50.8 mm		in metric.
	– 0.000 inch			– 0.0 mm	

In the first case, that of bilateral tolerance, the dimension of the part could vary between 1.998 and 2.002 inch a total tolerance of 0.004 inch. For the example of unilateral tolerance, the dimension could vary between 2.000 and 2.004 inch again a tolerance of 0.004 inch. Obviously, in order to obtain the same maximum and minimum dimensions with the two systems, different basic sizes must be used. The maximum and minimum dimensions that result from the application of the designated tolerances are called limit dimensions, or limits.

There can be no rigid rules for the amount of clearance that should be provided between mating parts; the decision must be made by the designer, who considers how the parts are to function. The American National Standards Institute (ANSI), has established eight classes of fits that serve as a useful guide in specifying the allowance and tolerance for typical applications, and that permit the amount of allowance and tolerance to be determined merely by specifying a particular class of fit. These classes are as follows:

Class 1: Loose fit—large allowance. Accuracy is not essential.

Class 2 : Free fit—liberal allowance. For running fits where speeds are over 600 rpm and pressures are 600 psi (4.1 MPa) or over.

Class 3 : Medium fit—medium allowance. For running fits under 600 rpm and pressures less than 600 psi (4.1 MPa) and for sliding fits.

Class 4: Snug fit—zero allowance. No movement under load is intended and no shaking is wanted. This is the tightest fit that can be assembled by hand.

Class 5 : Wringing fit—zero to negative allowance. Assemblies are selective and not interchangeable.

Class 6: Tight fit—slight negative allowance. An interference fit for parts that must not come apart in service and are not to be disassembled or are to be disassembled only seldom. Light pressure is required for assembly. Not to be used to withstand other than very light loads.

Class 7: Medium force fit—an interference fit requiring considerable pressure to assemble; ordinarily assembled by heating the external member or cooling the internal member to provide expansion or shrinkage. Used for fastening wheels, crank disks and the like to shafting. The tightest fit that should be used on cast iron external members.

Class 8: Heavy force and shrink fits—considerable negative allowances. Used for permanent shrink fits on steel members.

The allowances and tolerances that are associated with ANSI classes of fits are determined according to the theoretical relationship shown in Table 25.1. The actual resulting dimensional values for a wide range of basic sizes can be found in tabulations in drafting and machine-design books.

In the ANSI system, the hole size is always considered basic, because the majority of holes are produced through the use of standard-size drills and reamers. The internal member, the shaft, can be made to any one dimension as readily as to another. The allowance and tolerances are applied to the basic hole size to determine the limit dimensions of the mating parts. For example, for a basic hole size of 2 inches and a Class 3 fit, the dimensions would be:

Allowances	0.0014 inch
Tolerances	0.0010 inch
Hole	
Maximum	2.0010 inch

Minimum	2.0000 inch
Shaft	
Maximum	1.9986 inch
Minimum	1.9976 inch

Table 25.1. ANSI recommended allowances and tolerances.

Class of fit	*Allowance*	*Average interference*	*Hole tolerance*	*Shaft tolerance*
1	$0.0025 \sqrt[3]{d^2}$		$+0.0025 \sqrt[3]{d}$	$-0.0025 \sqrt[3]{d}$
2	$0.0014 \sqrt[3]{d^2}$		$+0.0013 \sqrt[3]{d}$	$-0.0013 \sqrt[3]{d}$
3	$0.0009 \sqrt[3]{d^2}$		$+0.0008 \sqrt[3]{d}$	$-0.0008 \sqrt[3]{d}$
4	0		$+0.0006 \sqrt[3]{d}$	$-0.0004 \sqrt[3]{d}$
5		0	$+0.0006 \sqrt[3]{d}$	$+0.0004 \sqrt[3]{d}$
6		0.00025d	$+0.0006 \sqrt[3]{d}$	$+0.0006 \sqrt[3]{d}$
7		0.0005d	$+0.0006 \sqrt[3]{d}$	$+0.0006 \sqrt[3]{d}$
8		0.001d	$+0.0006 \sqrt[3]{d}$	$+0.0006 \sqrt[3]{d}$

It should be noted that for both clearance and interference fits, the permissible tolerances tend to result in a looser fit. The ISO system of limits and fits, shown in Fig. 25.5, is widely used in a number of leading metric countries. This system is considerably more complex than the ANSI system. In this system, each part has a basic size. Each limit of size of a part, high and low, is defined by its deviation from the basic size, the magnitude and sign being obtained by subtracting the basic size from the limit in question. The difference between the two limits of size of a part is called tolerance and absolute amount without sign. There are three classes of fits: (1) clearance fit; (2) transition fits (the assembly may have either clearance or interference); and (3) interference fits.

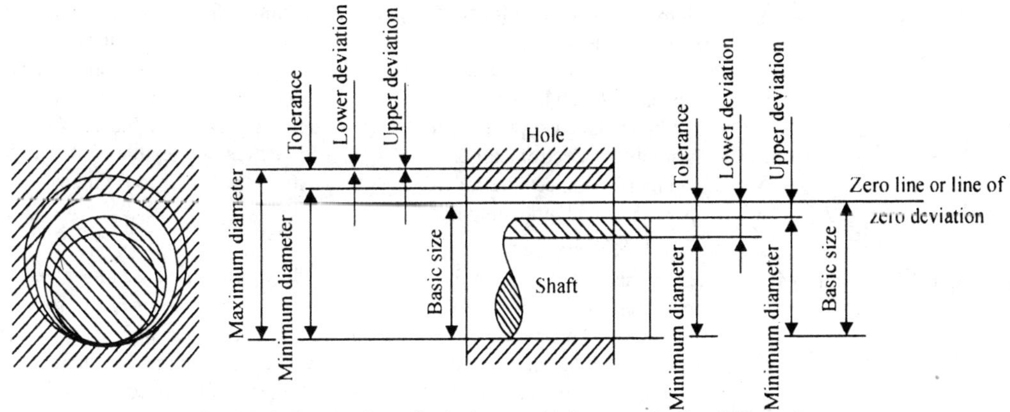

Fig. 25.5. Basic size, deviation and tolerance in the ISO system.

Accuracy Versus Precision in Processes

It is vitally important that the difference between accuracy and precision be understood. Accuracy refers to the ability to hit what is aimed at (the bull's-eye), and precision refers to the repeatability of the process.

INSPECTION METHODS OF MEASUREMENT

Table 25.2 provides a summary of inspection methods, listing four basic kinds of devices: air, optical (light and electron), electronic, and mechanical. The variety seems to be endless, but digital electronic readouts connected to any of the measuring devices are becoming the preferred method.

Table 25.2. Four basic kinds of inspeciton method.

Method	*Typical Accuracy*	*Major Applications*	*Comments*
Air	0.5–10 microinches or 2 % to 3% of scale range	Gauging holes and shafts using a calibrated difference in air pressure or airflow, with magnifications of 20,000–40,000 to 1; also used for machine control, sorting and classifying	High precision and flexibility; can measure out-of-round, taper, concentricity, camber, squareness, parallelism and clearance between mating parts; non-contact principle good for delicate parts
Optical light energy	0.5–2 microinches or better with laser interferometry 0.5–1 second of arc in autocollimation optical comparators	Interferometry; checking flatness and size of gauge blocks; finding surface flaws; measuring spherical shapes, flatness of surface plates, accuracy of rotary index tables; includes all light microscopes and devices common on plant floor.	Largest variety of measuring equipment; autocollimators are used for making precision angular measurements; lasers are used to make precision in-process measurements; laser scanning
Electron energy	100 Angstroms (Å)	Precision measurement in scanning electron microscopes of microelectronic circuits and other small precision parts	Part size restricted by vacuum chamber size; electron beam can be used for processing and part testing of electronic circuits
Electronic	0.5–10 microinches	Widely used for machine control, on-line inspection, sorting and classification; OD's, ID's, height, surface, and geometrical relationships, profile tracing for roundness, surface roughness, contours, etc.; most devices are comparators with movement of stylus or spindle producing an electronic signal that is amplified electronically; commonly connected to microprocessors and minicomputers for process adjustment	Electronic gauges come in many forms but usually have a sensory head or detector combined with an amplifier; capable of high magnification with resolution limited by size or geometry of sensory head; readouts commonly have multiple magnification steps; solid-state, digital electronics make these devices small, portable, stable and extremely flexible, with extremely fast response time
Mechanical	1–10 microinches	Large variety of external and internal measurements using dial indicators, micrometers, calipers and the like; commonly used for bench comparators for gauge calibration work	Moderate cost and ease of use make many of these, devices the workhorses of the shop floor; highly dependent on workers' skill and often subject to problems of linkages

Factors in Selecting Inspection Equipment

Many inspection devices with electronic output to communicate directly to microprocessors are being built into the processes themselves and are often computer-aided. In-process inspection generates feedback sensory data from the process or its output to the computer control of the machine, which is the first step in making the processes responsive to changes (adaptive control). In addition to in-process inspection, many other quality checks and measurements of parts and assemblies are needed. In general, six factors should be considered when selecting equipment for an inspection job by measurement techniques.

1. The rule of 10. The measurement device (or working gauge) should be 10 times more precise than the tolerance to be measured. The rule actually applies to all stages in the inspection sequence. The master gauge should be precise to 1/10 of that of the inspection device. The reference standard used to check the master gauge should be 10 times more precise than the master gauge. The application of the rule greatly reduces the probability of rejecting good parts or accepting bad components and performing additional work on them.
2. Linearity. This factor refers to the calibration accuracy of the device over its full working range. Is it linear? What is its degree on non-linearity? Where does it become non-linear, and what is, therefore, its real linear working region?
3. Repeat accuracy. How repeatable is the device in taking the same reading over and over on a given standard?
4. Stability. How well does this device retain its calibration over a period of time? Stability is also called drift. As devices become more accurate, they often lose stability and become more sensitive to small changes in temperature and humidity.
5. Magnification. The amplification of the output portion of the device over the actual input dimension. The more accurate the device, the greater must be its magnification factor, so that the required measurement can be read out (or observed) and compared with the desired standard. Magnification is often confused with resolution, but they are not the same thing.
6. Resolution. This is sometimes called sensitivity and refers to the smallest unit of scale or dimensional input that the device can detect or distinguish. The greater the resolution of the device, the smaller will be the things it can resolve and the greater will be the magnification required to expand these measurements up to the point where they can be observed by the naked eye.

Some other factors of importance in selecting inspection devices include the type of measurement information desired; the range or the span of sizes the device can handle versus the size and geometry of the workpieces; the environment; the cost of the device; and cost of installing and using the device. The last factor depends on the speed of measurement, the degree to which it can be automated and the functional life of the device in service.

MEASURING INSTRUMENTS

Because of the great importance of measuring in manufacturing, a great variety of instruments are available that permit measurements to be made routinely, ranging in accuracy from 1/64 to 0.00001 inch and from 0.5 to 0.0003 mm. Machine-mounted measuring devices (probes and lasers) for automatically inspecting the workpiece during manufacturing are beginning to compete with post-process gauging and inspection, in which the part is inspected, automatically or manually, after it has come off the machine.

In-process inspection for automatic size control has been used for some years in grinding to compensate for the relatively rapid wear of the grinding wheel. Today, touch trigger probes, with built-in automatic measuring systems, are being used on machine tools to determine cutting tool offsets and tool wear.

Linear Measuring Instruments

Linear measuring instruments are of two types: direct-reading and indirect-reading. Direct-reading instruments contain a line-graduated scale so that the size of the object being measured can be read directly on this scale. Indirect-reading instruments do not contain line graduations and are used to transfer the size of the dimension being measured to a direct-reading scale, thus obtaining the desired size information indirectly.

The simplest and most common, direct-reading linear measuring instrument are: Vernier caliper, micrometer caliper, micrometer depth gauge, super micrometers, optical projector or compactor.

Lasers provide for accurate machine tool alignment. The most widely used laser technique for inspection in process gauging is known as laser scanning with the recent emphasis on quality and 100 per cent inspection, application for inspection by machine vision have increased maskedly. Vision systems can check hundred parts per hour for multiple dimensions. Their use in an back inspection and quality control, sorting and machining tool monitoring will continue to expand.

A variety of instruments are available for measuring surface roughness and surface profiles. The majority of these devices use a diamond stylus that is moved at a constant rate across that surface, perpendicular to lay pattern.

NON-DESTRUCTIVE INSPECTION AND TESTING

A major emphasis in any manufacturing operation should be the prevention of defects that may lead to material or product failure. Care in product design, material selection, fabrication of the desired shape, heat treatment and surface treatment, as well as consideration of all possible service conditions, can do much to ensure the manufacture of a quality product. How do we determine whether these efforts have been successful? How can we be sure that the product is free from any harmful defects or flaws?

Various types of testing can be used to help evaluate the quality of products and ensure the absence of any flaws that might prevent an acceptable service performance. Destructive testing subjects selected components to test conditions that intentionally induce failure. By determining the specific conditions under which these components fail, insight can be gained into the general characteristics of a larger production lot. Statistical methods can be used to determine the probability that the entire production run will be good, provided that a certain number of test specimens prove to be satisfactory. Since each of the products tested is destroyed during the evaluation, the cost of such methods must be borne by the remaining products. In addition, there is still some degree of uncertainty about the quality of each of the remaining products, because they have never been individually evaluated.

To provide an increased measure of assurance, proof testing may be used. Here, a product is subjected to loads of a determined magnitude, generally equal to or greater than the designed capacity. If the part remains intact, then there is reason to believe that it will perform adequately in the absence of abuse or loads in excess of its rated level. Proof tests can be performed either under laboratory conditions or at the site of installation, as on large manufactured assemblies such as pressure vessels.

In some situations, hardness tests can be used to evaluate the quality of a product. Proper material and heat treatment can be reasonably ensured by requiring that all test results fall within a desired range. The tests can be performed quickly and the surface markings are often small enough so that they can be concealed or easily removed from a product. The results, however, relate only to the surface strength of the product and cannot detect serious defects, such as cracks or voids.

Non-destructive testing is the examination of a product in a manner that will not render it useless for future service. Testing can be performed directly on production items or even on parts in service and the only scrap losses are those defective parts that are detected. The entire production lot can be inspected, or representative samples can be taken. Different tests can be applied to the same item, either simultaneously or sequentially and the same test can be repeated on the same specimen for additional verification. Little or no specimen preparation is required and the equipment is often portable, permitting on-site testing in most locations.

Various objectives can be sought in non-destructive tests, including the detection of internal or surface flaws, the measurement of dimensions, a determination of the material's structure or chemistry, or the evaluation of a material, physical or mechanical properties. In general, non-destructive tests involve each of the following aspects: (i) a probing medium applied to the product; (ii) a possible modification of this probing medium by a flaw, a defect, a material property, or a specimen feature; (iii) a sensor to detect the response signal; (iv) a device to indicate or record the detector signals; and (v) a means of interpreting the response and of evaluating quality.

How you look at a material or product generally depends on what you are looking at, what you wish to see, and how finely you wish to see it. Each of the various inspection processes has its characteristic advantages and limitations. Some can be performed on only certain types of materials (such as conductors or ferromagnetic materials). Each is limited in the type, size and orientation of flaws that it can detect. Various degrees of accessibility may be required, and there may be geometric restrictions on part size or complexity. The available or required equipment, the cost of its operation, the need for a skilled operator or technician, and the possibility of producing a permanent test record are all considerations when selecting a test procedure.

Regardless of the specific method, non-destructive testing is a vital element in good manufacturing practice and its potential value is becoming more widely recognised as productivity demands increase, consumers demand higher-quality products and product liability continues to be a concern. Rather than being an added manufacturing cost, this practice can actually expand profit by ensuring product reliability and customer satisfaction. While essentially a quality-control operation, non-destructive testing can also be used to aid in product design, to provide on-line control of a manufacturing process and to reduce overall manufacturing costs.

Visual Inspection

Probably the simplest and most widely used non-destructive testing method is simple visual inspection. The human eye is a very discerning instrument and with training, the brain can readily interpret the signals. Optical aids such as mirrors, magnifying glasses and microscopes can enhance the capabilities of the system. Digital image analysers can be used to automate the testing and to make a number of quantitative geometrical evaluations. Borescopes and similar tools can provide accessibility to otherwise inaccessible locations. In the area of limitations, only the surfaces of the product can be examined.

Liquid Penetrant Inspection

Liquid penetrant testing is a simple method of detecting surface defects in metals and other non-porous material surfaces. The piece to be tested is first subjected to a thorough cleaning, often by means of solvent-type materials, and is dried before the test. Then, a penetrant, a liquid material capable of wetting the entire surface and being drawn into fine openings, is applied to the surface of the prepared workpiece by dipping, spraying, or brushing. After a period of time that permits capillary action to draw the penetrant into the surface discontinuities, the excess penetrant liquid is removed. The surface is then coated with a thin film of 'developer', an absorbent material capable of drawing traces of penetrant from the defects back onto the surface. Brightly coloured dyes or fluorescent materials that radiate in ultraviolet light are generally added to the penetrant to make these traces more visible and the developer is often selected so as to provide a contrasting background. Radioactive tracers can be added and used in conjunction with photographic paper to produce a permanent image of the defects. Cracks, laps, seams, lack of bonding, pinholes, gouges and tool marks can all be detected. After inspection, the developer and the residual penetrant are removed by a second cleaning operation. Fig. 25.6 shows a schematic representation of the liquid penetrant procedure. If previous processing involved techniques that could have induced a flow of the surface layers, such as shot peening, honing, burnishing, or various forms of cold working, a chemical etching may be required to first remove any material covering the surface defects. An alternative procedure is to penetrant-test before the final surface-finishing operation, when the defects are still open and available for detection. Penetrant inspection systems can range from a set of aerosol spray cans of cleaner, penetrant and developer (for portable applications), to automated, mass-production equipment using laser scanners and computer control.

Fig. 25.6. Liquid penetrant testing: (a) application of the penetrant to a clean surface; (b) excess penetrant is removed; and (c) developer is applied and the product inspected.

Magnetic Particle Inspection

Magnetic particle inspection is based on the principle that ferromagnetic materials (such as iron, steel, nickel and cobalt alloys), when magnetised, will have distorted magnetic fields in the vicinity of material defects. Surface and sub-surface flaws, such as cracks and inclusions, can produce magnetic anomalies that can be mapped with the aid of magnetic particles on the specimen surface.

As in the previous methods, the specimen must be cleaned before inspection. A suitable magnetic field is then established in the part, produced so as to reveal selected defects. Orientation considerations are quite important since a flaw must produce a significant disturbance of the magnetic field at or near the surface if the flaw is to be detected. For example, placing a bar of steel within an energised coil will produce a magnetic field whose lines of flux travel along the axis of the bar. Any defect perpendicular to this axis will significantly alter the field. If the perturbation is sufficiently large and close enough to the surface, it can be detected by the inspection methods. However, if the flaw is in the form of a crack running down the specimen axis, the defect is oriented in such a way as to produce little perturbation of the lines of flux and is likely to go undetected. If the same sample is magnetised by passing a current through it to create a circumferential magnetic field, the axial defect now becomes highly detectable, and the defect perpendicular to the axis may go unnoticed. Thus, a series of inspections using different forms of magnetisation may be required to fully inspect a product. Passing a current through the product between various points of contact is a popular means of inducing the desired fields. Electromagnetic coils of various shapes and sizes are also used. Alternating current methods are most sensitive to surface flaws. Direct current methods are more capable of detecting sub-surface defects, such as non-metallic inclusions.

Once the specimen is magnetised, magnetic particles are applied to the surface, as either dry powder or a suspension in a liquid carrier. The particles can be treated with a fluorescent material for observation under ultraviolet light. In addition, they are often made in and elongated form to better reveal the orientation and can be coated with a lubricant to prevent oxidation an enhance their mobility. The resultant distribution of particles is then examined and any anomalies are interpreted.

Some residual magnetisation will be retained by all parts subjected to magnetic particle inspection. Therefore, it is usually necessary to demagnetise them before further processing or before placing them in use. One common means of demagnetisation is to place the parts inside a coil powered by alternating current and then gradually to reduce the current to zero. A final cleaning operation completes the process.

In addition to in-process and final inspection of parts during manufacture, magnetic particle inspection is extensively used during the maintenance and overhaul of equipment and machinery. The testing equipment ranges from small, lightweight, portable units to heavy, complex, automated systems.

Ultrasonic Inspection

Sound has long been used to provide an indication of product quality. A cracked bell will not ring true, but a fine crystal goblet will have a clear ring when tapped lightly. Striking an object and listening to the characteristic 'ring' is an ancient art, but it is limited to the detection of large defects because the wavelength of audible sound is rather large compared to the size of most defects. By reducing the wavelength of the signal to the ultrasonic range, beyond the range of human hearing, ultrasonic inspection can detect rather small defects and flaws.

Ultrasonic inspection, therefore, involves sending high-frequency vibrations through a material and observing what happens when this ultrasonic beam hits a defect, a surface, or a change of density. At the interface, part of the ultrasonic wave is reflected and part is transmitted. If the incident beam is at an angle to the interface and the material changes across the surface, the transmitted portion of the beam is bent to a new angle by the phenomenon of refraction. By receiving and interpreting the altered signal, ultrasonic inspection can be used to detect flaws within the material, to measure thickness from only one side and to characterise metallurgical structure.

An ultrasonic inspection system begins with a pulsed oscillator and a transducer, which serves to transform electrical energy into mechanical vibrations. The pulsed oscillator generates a burst of alternating voltage—the principal frequency, duration, profile, and repetition rate being either fixed or variable. This burst is then applied to a sending transducer, which uses a piezoelectric crystal to convert the electrical oscillations into mechanical vibrations. Because air is a poor transmitter of ultrasonic waves, an acoustic coupling medium—generally a liquid such as oil or water-is required to link the transducer to the piece to be inspected and to transmit the vibrations into the part. A receiving transducer is then used to convert the received ultrasonic vibrations back into electrical signals. The receiving transducer can be identical to the sending unit and the same transducer can even be used for both functions. A receiving unit then amplifies, filters and processes the signal for display on an oscilloscope, for recording on some form of electromechanical recorder and for final interpretation. An electronic clock is generally integrated into the system to time the responses and to provide reference signals for comparison purposes.

Depending on the test objectives and the part geometry, several different inspection techniques can be used:

1. In the pulse-echo technique, a pulsed ultrasonic beam is introduced into the piece to be inspected, and the echoes from the opposing surfaces and any intervening flaws are monitored by the receiver. The time interval between the initial emitted pulse and the various echoes is displayed

on the horizontal axis of an oscilloscope screen. Defects are identified by the position and amplitude of the various echoes. Fig. 25.6b shows a schematic of this technique using a single transducer and the companion signal as it would be recorded on an oscilloscope. Fig. 25.6a shows the dual-transducer pulse-echo technique.

2. The through-transmission technique requires separate sending and receiving transducers. A pulsed beam is sent through the part by the sending transducer, and the transmitted signal is picked up by the receiver, as shown in Fig. 25.6b. Flaws in the material decrease the amplitude of the transmitted beam because of back-reflection and scattering.
3. Resonance testing can be used to determine the thickness of a plate or sheet from one side of the material. Input pulses of varying frequency are fed into the material. When resonance is detected by an increase of the energy at the transducer, the thickness can be calculated from the speed of sound in the material and the time of traverse. Some instruments can be calibrated to provide a direct digital readout of the thickness of the material.

Reference standards, consisting of a series of specimens with known thicknesses or various types and sizes of machined 'flaws', are generally used to ensure consistent results and to aid in interpreting any indications of internal discontinuities.

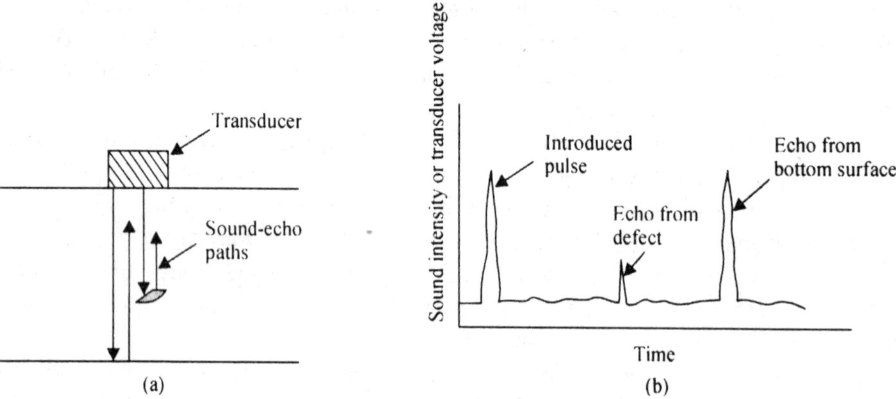

Fig. 25.7. (a) Ultrasonic inspection of flat plate with a single transducer, (b) A plot of sound intensity or transducer voltage versus time depicting the base signal and a secondary peak indicative of an intervening defect.

Radiography

Radiographic inspection uses the same principles and techniques as medical X-rays. In essence, a shadow pattern is created when certain types of radiation (X-rays, gamma rays, or neutron beams) penetrate an object and are differentially absorbed because of variations in thickness, density, or chemistry, or because of the presence of defects in the specimen. The transmitted radiation is registered on a photographic film that provides a permanent record and a means of analysing the component. Fluorescent screens can provide a direct conversion of radiation into visible light and can enable fast and inexpensive viewing without the need for film processing. The image, however, has relatively poor sensitivity compared to that possible by the photographic methods Fig. 25.8).

In order to have a reference for the image densities on a radiograph, it is standard practice to include a standard test piece, or penetrameter, in the exposure of the part being inspected. The penetrameters are made of the same material as the specimen or of similar material and contain structural features of known dimensions. The image of the penetrameter then permits direct comparison with the features present in the image of the product being inspected.

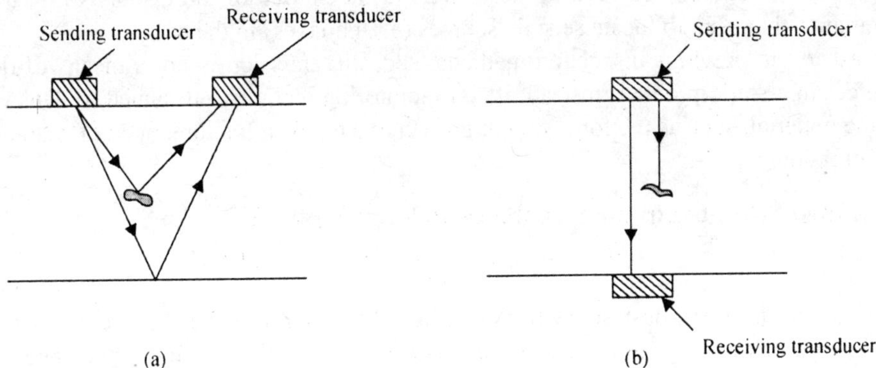

Fig. 25.8. (a) Dual transducer ultrasonic inspection in the pulse-echo mode; and (b) dual transducers in through-transmission inspection.

Density-sensitive aids have been developed to assist in detecting subtle, but important, details in the image. Television cameras connected to electronic processors with image-enhancing software can do much to accentuate density variations.

Because of the expense of testing, many users recommend extensive use of radiography during the development of a product and a process, followed by spot checks and statistical methods for subsequent production.

Eddy Current Testing

When an electrically conductive material is exposed to an alternating magnetic field, such as that generated by a coil of wire carrying an alternating current, small electric currents are generated on or near the surface of the material. These eddy currents, in turn, generate their own magnetic field, which then interacts with the magnetic field of the exciting coil to change its electrical impedance. By measuring the impedance of the exciting coil, or a separate indicating coil, eddy current testing can be used to detect any condition that would affect the current-carrying conditions (or conductivity) of the test specimen.

Eddy current testing can be used to detect surface and near-surface flaws, such as cracks, voids, inclusions, seams and even stress concentrations. Differences in metal chemistry or heat treatment will affect the magnetic permeability and conductivity of a metal and, hence, the eddy current characteristics.

Material mix-ups can be detected. Specimens can be sorted by hardness, case depth, residual stresses, or any other structure-related property. Thicknesses or variations in thickness of platings, coatings, or even corrosion can be detected or measured.

Acoustic Emission Monitoring

Materials undergoing stressing, deformation, or fracture emit sound waves in frequencies as high as 1 MHz. While these sounds are inaudible to the human ear, they are detectable through the use of sophisticated electronics. Transducers, amplifiers, filters, counters and microcomputers can be used to isolate and analyse the sonic emissions of cracking or deforming material. Much like the warning sound of ice cracking underneath skates, the acoustic emissions of materials can be used to provide a warning of impending danger. They can detect deformations as small as 10^{-12} inch/inch that occur in short intervals of time, as well as the delamination of layered materials and fibre failure in composites. In addition, multiple sensors can be coupled to accurately pinpoint the source of these sounds in a triangulation method similar to that used to locate seismic sources (earthquakes) in the earth.

In contrast to the previous inspection methods, acoustic emission is not a means of detecting an existing defect in a static product. Instead, it is a monitoring technique designed to detect a dynamic change in the material, such as the formation or growth of a crack, a fracture, or a defect, or the onset of plastic deformation.

Other Methods of Non-destructive Testing and Inspection

Leak testing

Leak testing is a form of non-destructive testing designed to determine the absence or existence of leak sites and the rate of material loss through the leaks. Various testing methods have been developed, ranging from the rather crude bubble-emission test (pressurise, immerse and look for bubbles) to advanced techniques involving tracers, detectors and sophisticated apparatus. Each has its characteristic advantages, limitations and sensitivity. Selection should be made on the basis of cost, sensitivity, and reliability.

Thermal methods

Temperature-based techniques can also be used to evaluate the soundness of engineering materials and components. The parts are heated, and various means are used to detect abnormal temperature distributions indicative of faults or flaws. Temperature-sensing tools include, thermometers, thermocouples, pyrometers, temperature-sensitive paints and coatings, liquid crystals, infrared scanners and infrared film.

Strain sensing

While primarily used in laboratories as a part of product development, strain-sensing techniques can be used to provide valuable insight into the stresses and the stress distribution within a part. Brittle coatings, photoelastic coating, or electrical-resistance strain gauges are applied to the external surfaces of the part, which is then subjected to an applied stress.

Resistivity methods

The electrical resistivity of a conductive material is a function of its chemistry, its processing history, and its structural soundness. Changes in resistivity from one sample to another can therefore be used for alloy identification, flaw detection and the assurance of proper processing (such as evaluating heat treatment, the amount of cold work, the integrity of welds, or the depth of case hardening). The development of sensitive micro-ohm-meter (microhmeter) has greatly expanded the possibilities in this area.

Chapter 26

Process Capability and Quality Control

INTRODUCTION

All manufacturing processes display some level of inherent capability or some inherent uniformity or nature. For example, suppose that we view 'shooting at a metal target' as a 'process' for putting holes in piece of metal. I hand you the gun and tell yot to take five shots at the bull's-eye. Thus, you are the operator of the process. In order to measure process capability (PC)—that is, your ability to consistently hit what you are aiming at, the bull's-eye—the target is inspected after you have finished shooting.The capability of manufacturing processes is determined by measuring the output of the process. In quality control (QC), the product is examined to determine whether or not the processing accomplished was what was specified in the design. In QC, the objective is to inspect to prevent defects from occurring, not to find the defects (deviations from the ideal or target that will make the products defective) after they have occurred (Fig. 26.1).

Process capability (PC) studies use statistical and analytical tools, as does QC, except that the results are directed at the machines used in the processing rather than the output or products from the processes. Going back to our example, a PC study would be directed towards quantifying the inherent accuracy and precision in the shooting process. The QC programme would be geared to root out the problems that can cause defective products during production.

DETERMINING PROCESS CAPABILITY

The nature of the process refers to both the variability (or inherent uniformity) and the aim of the process. Thus, in the target-shooting example, a perfect process would be capable of placing five shots right in the middle of the bull's-eye, one right on top of the other. The process would display no variability. Such performance would be very unusual in a real industrial process. The variability may have assignable causes and may be correctable if the cause can be found and eliminated. That variability to which no cause can be assigned and which cannot be eliminated is inherent in the process and is therefore its 'nature.'

Some examples of assignable causes of variation in processes include multiple machines for the same components, operator blunders, defective materials or progressive wear in the tools during machining. Sources of inherent variability in the process include variation in material properties, operator variability, vibrations and chatter and the wear of the sliding components in the machine, perhaps resulting in poorer operation of the machine. These kinds of variations, which occur naturally in processes, usually display a random nature and often cannot be eliminated. In QC terms, these are referred to as chance causes. Sometimes, the causes of assignable variation cannot be eliminated because of cost. Almost every process has multiple causes of variability occurring simultaneously, so it is extremely difficult to separate the effects of the different sources of variability during the analysis.

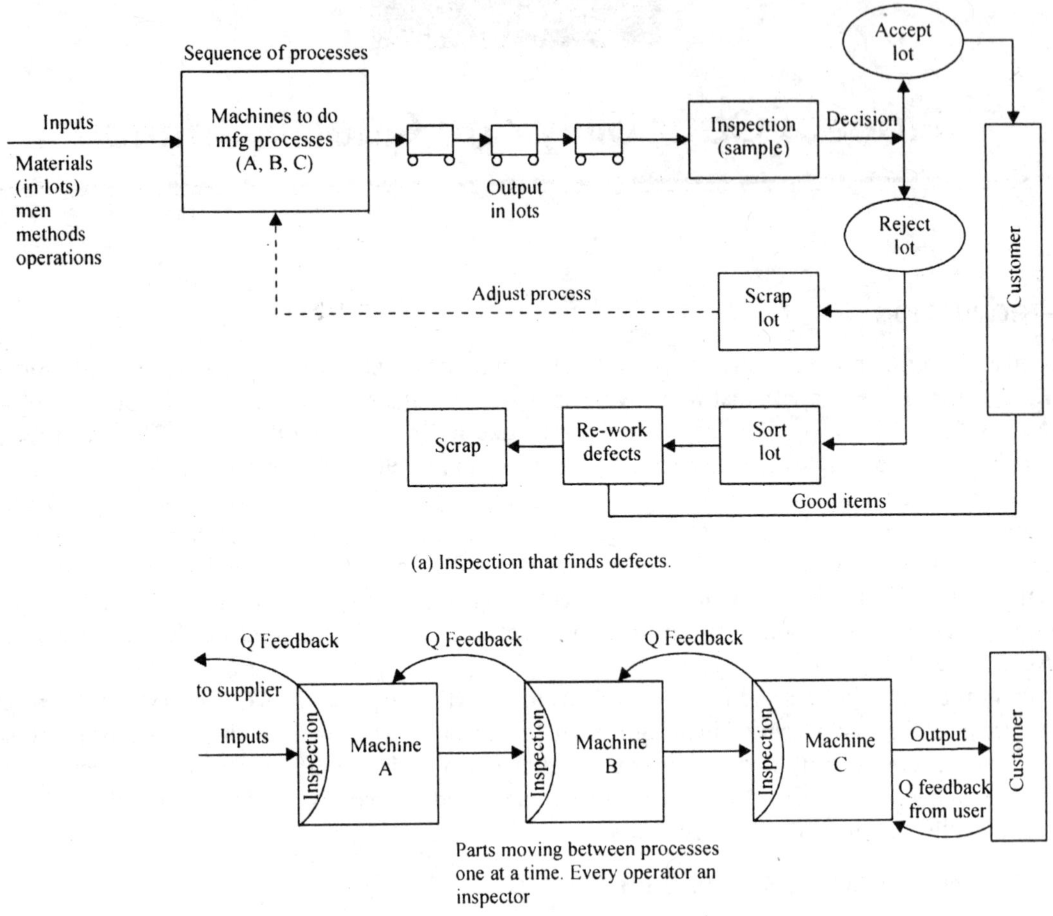

(a) Inspection that finds defects.

(b) Inspection to prevent defects.

Fig. 26.1. Don't inspect to find defects. Inspect to prevent the defect from happening.

Making Process Capability (PC) Studies

The object of the PC study is to determine the inherent nature of the process as compared to the desired specifications. The output of the process must be examined under normal conditions, or what is typically called hands-off conditions. The inputs (e.g., materials, set-ups, cycle times, temperature, pressure and operator) are fixed or standardised. The process is allowed to run without tinkering or adjusting while the output (i.e., the product or units or components) is carefully documented with respect to: (i) time; (ii) source; and (iii) order of production. A sufficient number of data have to be taken to ensure confidence in the statistical analysis of the data. The precision of the measurement system should exceed the process capability by at least one order of magnitude.

Before any data collection, these steps must be taken:

1. Design the PC experiment.
   A. Use 'normal' or 'hands-off' process conditions; specify machine settings for speed, feed, volume, pressure, material, temperature, operator and so on. (This is the standard PC study approach.)

   Alternatively A. Use specified combinations of all of the input parameters, at various levels, that are believed to influence the quality characteristics being measured. These combinations should be run with the objective of selecting the best. For example, speed levels may be high, normal and low or operators may be fast or slow. (This is the Taguchi or factorial approach.)
2. Define the inspection method and the inspection means (the procedure and the instrumentation).
3. Decide how many items will be needed to perform the statistical analysis. (For a Taguchi approach, this means deciding how many replications at each level of combination are desired).
4. For the standard PC study, use homogeneous input material, and try to contrast it with normal (more variable) input material. (For a Taguchi approach, material is an input variable specified at different levels: normal, homogeneous and highly variable. If material is not controllable, it is considered a noise factor).
5. Data sheets must be designed to record date, time, source, order of production and all the process parameters being used (or measured) while the data are being gathered. (For a Taguchi method, this is usually a simplified experimental design, called an orthogonal array).
6. Assuming that the standard PC study approach is being used, the process is run and the parts are made and measured.

What PC Studies Tell About the Process

PC studies provide the answers to these questions:

1. Does the process have the ability to meet specification?

   To answer this question, a process capability index, C_p, is often computed, where

   $$C_p = \frac{\text{Tolerance spread}}{6\sigma'} = \frac{\text{USL} - \text{LSL}}{6\sigma'} \qquad ...(26.1)$$

   A value of $C_p \geq 1.33$ is considered good.
2. Is the process well centered with respect to the desired nominal specification?

   To answer this question, the degree of bias (β) is computed, where

   $$\beta = \frac{\text{Estimated process mean} - \text{Nominal}}{1/2(\text{Tolerance spread})} = \frac{\bar{X}' - \text{Nominal}}{1/2(\text{USL} - \text{LSL}} \qquad ...(26.2)$$

The condition for not producting rejects is

$$\bar{X}' - \text{Nominal} + 3\sigma' < 1/2\,(\text{USL} - \text{LSL}) \qquad ...(26.3a)$$

or

$$C_p\,(1 - \beta) > 1 \qquad ...(26.3b)$$

Another capability index that combines both factors is

$$C_{pk} = \frac{\text{MIN}\left\{\left|\text{USL} - \overline{X}'\right|, \left|\text{LSL} - \overline{X}'\right|\right\}}{3\sigma'} \qquad \text{...(26.4)}$$

In response to the first question, the width of the histogram is compared with the specifications (Fig. 26.2). In addition, the natural specification limits are computed and are then compared with the upper and lower tolerance limits. Three situations exist:

1. $6\sigma' < \text{USL} - \text{LSL}$ or $C_p > 1$, or process variability is less than tolerance spread (Fig. 26.2a).
2. $6\sigma' = \text{USL} - \text{LSL}$ or $C_p = 1$, or process variability is equal to tolerance spread (Fig. 26.2c).
3. $6\sigma' > \text{USL} - \text{LSL}$ or $C_p < 1$, or process variability is greater than tolerance spread Fig. (26.2d)

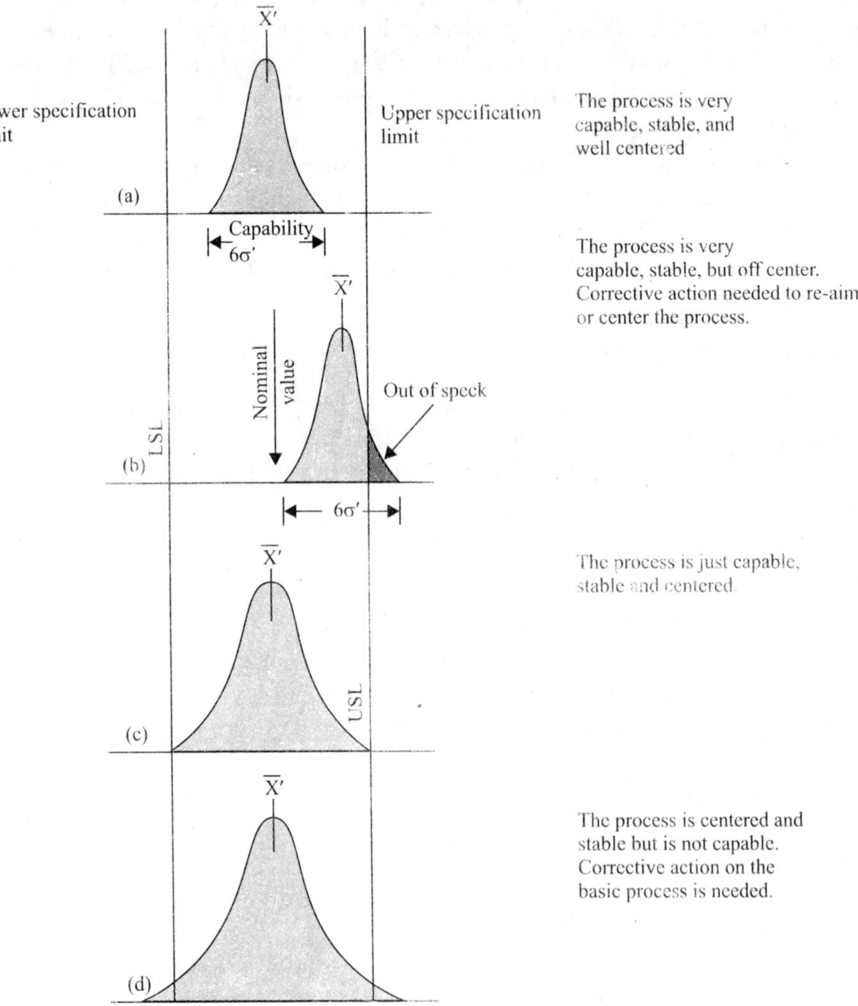

Fig. 26.2. What PC studies tell about the process.

In situation 1, the machine is capable of meeting the tolerances applied by the designer. Generally speaking, if process capability is on the order of two-thirds to three quarters of the design tolerance, there is a high probability that the process will produce all good parts over a long period of time. If the PC is on the order of one-half or less of the design tolerance, it may be that the selected process is 'too good'. That is, the company is making ball bearings when what is called for are marbles. In this case, it may be possible to trade off some precision in this process for looser specifications elsewhere, resulting in an overall economic gain. Quality in well-behaved processes can be maintained by checking the first and last part of a lot or production run. If these parts are good, then the lot is certain to be good. This is called N = 2. Naturally if the lot size is 2 or less, this is 100 per cent inspection. Sampling and control charts are also used under these conditions to maintain the process aim and variability.

In situation 3, the process is not capable of meeting the design specifications. There are a variety of alternative here, including:

1. Shifting this job to another machine with greater process capability.
2. Getting a review of the specifications to see if they may be loosened.
3. Sorting the product to separate the good from the bad. This entails 100 per cent inspection of the product, which may not be a viable economic alternative unless it can be done automatically. Automatic inspection of the product on a 100 per cent basis can ensure near-perfect quality of all the accepted parts.

Automated sorting does not determine what caused the defects, so this example is an automated defect-finder. How would one change this inspection system to make it 'inspect to prevent' the defect from occurring?

4. Determining whether the precision of the process can be improved by:
   (a) Switching cutting tools, workholding devices or materials.
   (b) Overhauling the existing process and/or developing a preventive maintenance programme.
   (c) Finding and eliminating the causes of variability.
   (d) Combination of (a), (b), and (c).
   (d) Using a factorial (Taguchi) procedure, that is, designing experiments to reduce the variability of the process.

In Fig. 26.2(c), the process capability is almost exactly equal to the assigned tolerance spread, so if the process is not perfectly centered, defective products will always result. Thus, situation 2 should be treated like situation 3 unless the process can be perfectly centered and maintained. Tool wear, which causes the distribution to shift, must be negligible. Then the situation can be treated as that in 1, particularly if a small percentage of parts just outside tolerance is acceptable.

The second question deals with the ability of the process to maintain centering so that the average of the distribution comes as close as possible to the desired nominal value (Fig. 26.2b). This process needs to be reaimed or centered so that the mean of the process distribution is at or near the nominal (print) value. Most processes can be re-aimed. Poor accuracy is often due to assignable causes, which can be eliminated. In addition to direct information about the accuracy and the precision of the process, PC studies can also tell the manufacturing engineer how pilot processes compare with production processes and vice versa. If the source and time of each product are carefully recorded, information about the instantaneous reproducibility can be found and compared with the repeatability of the process with respect to time (time-to-time variability). More important, since almost all processes are duplicated, PC studies generate information about machine-to machine variability. Suppose, going back to our target-shooting example, that five different guns were used, all of the same make and type. The results would have been different, just as having five marksmen use the same gun would have resulted in yet another outcome. Thus, PC studies generate information about the homogeneity and the differences in multiple machines and operators. It is quite often the case in such studies that one variable dominates the process. Target shooting viewed as a process is 'operator'-dominated in that the outcome is highly dependent on the skill of the 'worker.' Processes that are not well engineered nor highly automated, or in which the worker is viewed as 'highly skilled,' are usually operator-dominated. Processes that change or shift uniformly with time but that have good repeatability in the short run are often 'machine'-dominated. For example, the mean of a process ($\overline{X}'$) will usually shift after a tool change, but the variability may decrease or remain unchanged. Machines tend to become more precise (to have less variability within a sample) after they have been 'broken in' (i.e., the rough contact surfaces have smoothed out because of wear) but will later become less precise (will have less repeatability) because of poor fits between moving elements (called backlash) or the machine under varying loads. Other variables that can dominate processes are set-up, input components and even information.

In many machining processes in use today, the task of tool setting has been replaced by an automatic tool positioning capability, which means that one source of variability in the process has been eliminated and the process becomes more repeatable. In the same light, it will be very important in the future for manufacturing engineers to know the process capability of the robots they want to use in the workplace.

The discussion to this point has assumed that the parent population is normally distributed, that is, has the classic bell-shaped distribution in which the percentages shown in Fig. 26.3 are dictated by the number of standard deviations from the central value or mean. The shape of the histogram may reveal the nature of the process to be skewed to the left or the right (unsymmetrical), often indicating some natural limit in the process. Drilled holes exhibit such a trend as the drill tends to make the hole oversize. Another possibility is a bimodal distribution (two distinct peaks), often caused by two processes' being mixed together. The possibilities are endless and require a careful recording of all the source of the data to track down the factors that result in loss of percision and accuracy in the process. Rapid feedback on quality is perhaps the most important factor.

Taguchi Method

The second approach to making a PC study may involve the Taguchi method, which requires a more sophisticated analysis. The Taguchi approach uses a truncated experimental design (called an orthogonal array) to determine which process inputs have the greatest effect on process variability (i.e., precision) and which have the least. Those inputs that have the greatest influence are set at levels that minimise their effect on process variability. Those having the least influence are used to adjust or re-center the process aim. In other words, Taguchi methods seek to minimise or dampen the effect of the causes of variability and thus to reduce the total process variability (Fig. 26.3).

In machine tools, accuracy and precision in processing are affected by machine alignment, the set-up of the workholder, the design and rigidity (accuracy) of the workholder, the accuracy of the cutting tools, the design of the product, the temperature and the operating parameters. Taguchi methods provide a means of determining which of the input parameters are most influential in product quality.

INSPECTION AND QUALITY CONTROL

In virtually all manufacturing, it is extremely important that the dimensions and quality of individual parts be known and maintained. This is of particular importance where large quantities of parts, often made in widely separated plants, must be capable of interchangeable assembly. Otherwise, difficulty may be experienced in subsequent assembly or in service and costly delays and failures may result. In recent years, defective products resulting in death or injury to the user have resulted in expensive litigation and damage awards against manufacturers. Inspection is that function which controls the quality (e.g., the dimensions, the performance and the colour)—manually, by using operators or inspectors, or automatically, by using machines, as discussed previously.

The economics-based question 'how much should be inspected?' has three possible answers.

1. Inspect everyitem being made: 100 per cent checking with prompt execution of feedback and immediate corrective action can ensure perfect quality.
2. Sample: Inspect some of the product by sampling and make decisions about the quality of the process based on the sample.
3. None: Assume that everything made is acceptable or that the product is inspected by the consumer, who will exchange it if it is defective. (This is not a recommended procedure).

The reasons for not inspecting all of the product (i.e., for sampling) include:

1. The test is destructive.
2. There is too much product for all of it to be inspected.
3. The testing takes too much time or is too complex or too expensive.
4. It is not economically feasible to inspect everything even though the test is simple, cheap and quick.

Some characteristics are non-dissectible; that is, they cannot be measured during the manufacturing process because they do not exist until after a whole series of operations have taken place. The final edge geometry of a razor blade is a good example, as is the yield strength of a rolled bar of steel.

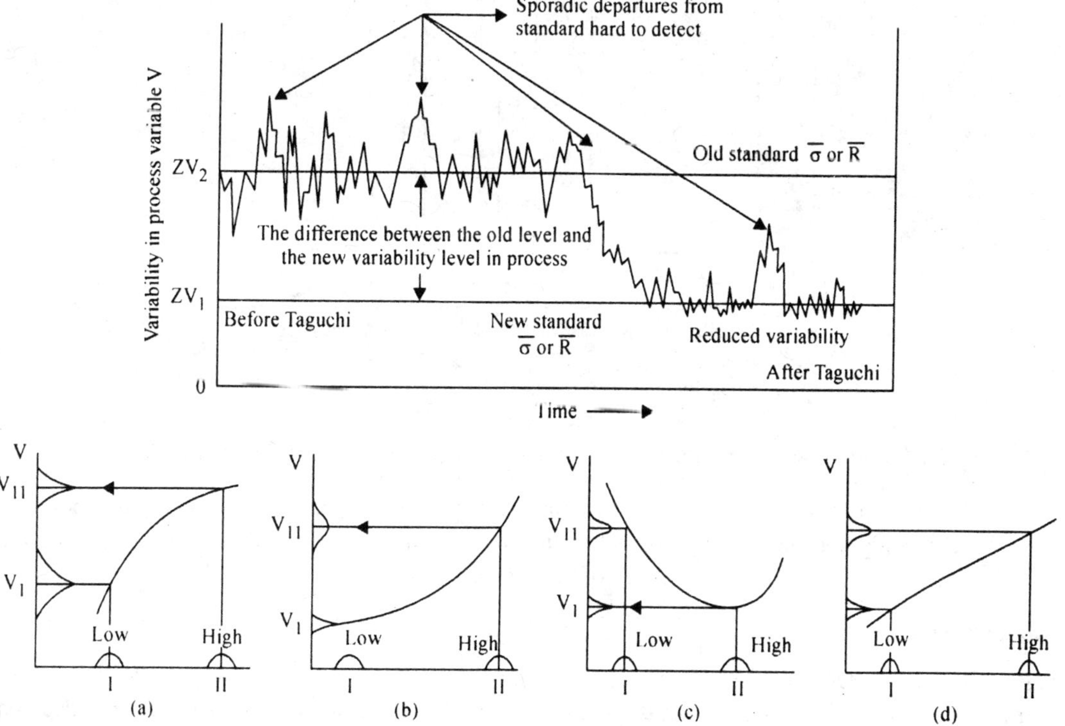

Fig. 26.3. The Taguchi method can reduce the process variability, as measured by range values (R) or standard deviations (σ).

Sampling or looking at some percentage of the whole requires the use of statistical techniques that permit decisions about the acceptability of the whole based on the quality found in the sample. This is known as statistical process control (SPC).

Statistical Process Control

Looking at some (sampling) and deciding about the behaviour of the whole (the parent population) is common in industrial inspection operations. The basic SPC techniques are the histogram and control charts. The histogram (Fig. 26.4) could be used to classify the range of shaft diameters after turning. The X axis represents the diameter measurement, and the Y axis represents the frequency with which these measurements were found. The natural process capability limits of the distribution can be compared with the engineering specifications to determine if the process is centered at the print nominal value and to estimate if diameters are being produced within tolerance values (or product specifications).

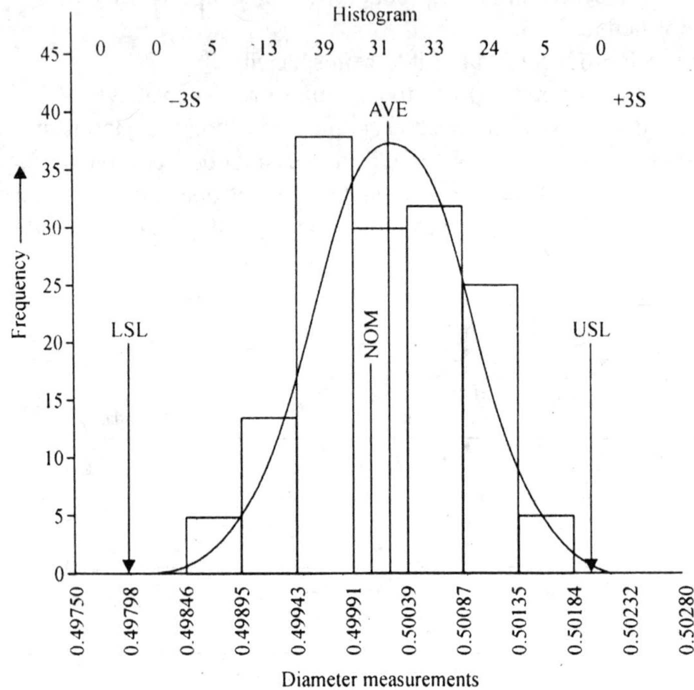

Fig. 26.4. Histogram of a process to compare measured sizes with designer specifications.

Control charts for variables are used to monitor the output of a process by sampling (looking at some), by measuring selected quality characteristics, by plotting the sample data on the chart and then by making decisions about the performance of the process.

Sampling Errors

It is important to understand that in sampling, two kinds of decision errors are always possible. Suppose that the process is running perfectly but the sample data indicate that something is wrong. You decide to stop the process to make adjustments. This is a Type (1) error. Suppose that the process was not running perfectly and was making defective products. However, the sample data did not indicate that anything was wrong. You did not stop the process and set it right. This is a Type (2) error. Both types of errors are possible in sampling. For a given sample size, reducing the chance of one type of error will increase the chance of the other. Increasing the sample size or the frequency of sampling reduces the probability of errors but increases the cost of inspection. Many companies determine the size of the errors they are willing to accept according to the overall cost of making the errors plus the cost of inspection. If, for example, a Type (2) error is very expensive in terms of product recalls or legal suits, the company may be willing to make more Type (1) errors, to sample more, or even to go to 100 per cent inspection on very critical items to ensure that the company is not accepting defective materials as good and passing them on to the customer. The inspection should take place immediately after the processing.

As mentioned earlier, in any continuing manufacturing process, variations from established standards are of two types: (i) assignable cause variations, such as those due to malfunctioning equipment or personnel, or to defective material, or due to a worn or broken tool; and (ii) normal chance variations, resulting from the inherent non-uniformities that exist in materials and in machine motions and operations. Deviations due to assignable causes may vary greatly. Their magnitude and occurrence are unpredictable, and one thus wishes to prevent their occurrence. On the other hand, if the assignable causes of variation are removed from a given operation, the magnitude and frequency of the chance variations can be predicted with great accuracy. Thus, if one can be assured that only chance variations will occur, the quality of the product will be better known and manufacturing can proceed with assurance about the results. By using statistical process control procedures, one may detect the presence of an assignable-cause variation and remove the cause, often before it causes quality to become unacceptable. Also, the astute application of statistical experimental design methods (Taguchi experiments) can help to identify some assignable causes.

To sum up, use PC analysis and process improvements to get the process in the conditions described in Fig. 26.2 (a). Use control charts to keep the process centered ($\overline{X}$ chart) with no change in variability ($\bar{R}$ chart).

Quality Control Charts

Quality control charts are widely used as aids in maintaining quality and in achieving the objective of detecting trends in quality variation before defective parts are actually produced. These charts are based on the previously discussed concept that if only chance causes of variation are present, the deviation from the specified dimension or attribute will fall within pre-determined limits.

When sampling inspection is used, the typical sample sizes are from 3 to about 12 units.

With regard to control charts in general, it should be kept in mind that the charts are capable of indicating only that something has (or has not) happened, and that a certain amount of detective work will be necessary to find out what has occurred to cause a break from the random, normal pattern of sample points on the charts. Keeping careful track of when and where the sample was taken is very helpful in such investigations, but the best procedure is to have the operator do the inspection and run the chart. In this way, quality feedback is very rapid and the causes of defects readily found.

DETERMINING CAUSES FOR PROBLEMS IN QUALITY

The best way to isolate quality problems quickly is to make everyone an inspector. This means that every worker, foreman, supervisor, engineer, manager, and so on is responsible for making it right the first time and every time. One very helpful tool in this effort is the fishbone chart (Fig. 26.5). The problem (bad welds) can have many causes. Generally, the cause lies in the process, the operators, the materials, or the method, the four main branches on the chart. Every time a quality problem is caused by one of these events, it is noted by the observer and corrective action is taken. As before, experimental design procedures can help to identify the causes that affect performance.

In summary, the data gathered to develop the PC study can be used to prepare the initial control charts from the process after the removal of all assignable causes for variability and the proper setting of the process average. After the charts have been in place for some time and a large number of data have been obtained from the process output, the PC study can be re-done to give better estimates of the natural spread of the process during actual production. On-line or in-process methods like SPC, and off-line (Taguchi) methods are key elements in a total quality control programme.

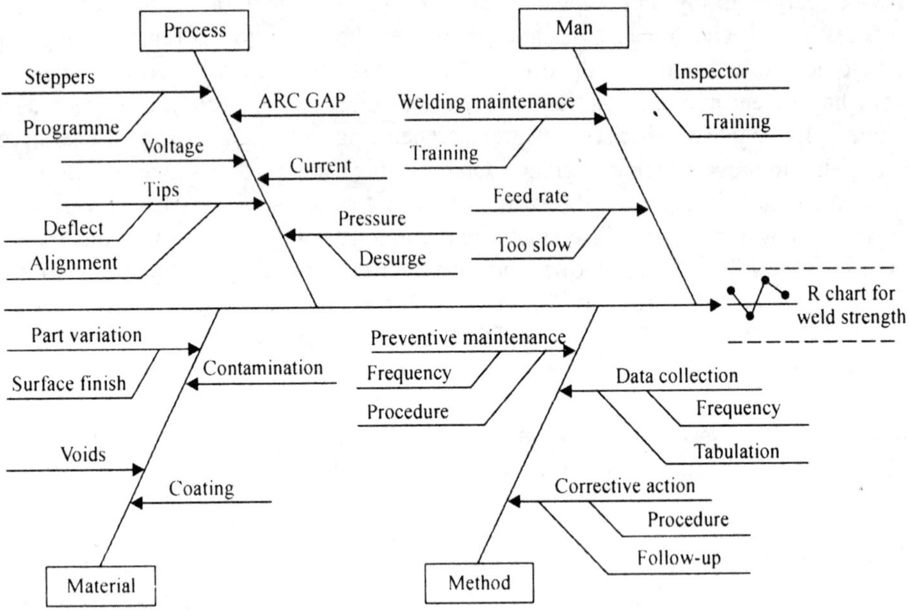

Fig. 26.5. Fishbone diagram: used to determine the causes of quality problems. The problem is bad welds.

Total Quality Control (TQC)

Quality begins with optimising the process design. The emphasis is on making it right the first time. Few companies have embarked on implementing Taguchi methods in order to reduce the inherent variability in their processes. A change in company attitude is needed in order to give the responsibility for quality to the worker, along with the authority to stop the process when something goes wrong. An attitude of defect prevention and a habit of constant improvement of quality are fundamental to the Integrated Manufacturing Production System (IMPS). Good companies have accomplished TQC by extensive education of the workers, giving them the analysis tools they need (control charts with cause-and-effect diagrams) to find and expose the problems. Workers are encouraged to correct their own errors and 100 per cent inspection (often done automatically) is the rule. Passing defective products on to the next process is not allowed. The goal is perfection. Quality circles, now popular in the United States, are just one of the methods used by Japanese industries to achieve perfection.

The designer of the product must have quality in mind during the design phase, seeking the least costly means to ensure the quality of the desired functional characteristics. Major factors that can be handled during the early stages of the product design cycle include temperature, humidity, power variations, and the deterioration of materials and tools. Compensation for these factors is difficult or even impossible to implement after the product is in production. The distinction between superior quality and poor quality products can be seen in their variability in the face of internal and external causes. This is where Taguchi parameter design methods are critical.

The secret to successful process control is putting the control of quality in the hands of the workers. Many companies in this country are currently engaged in SQC (statistical quality control), but they are still inspecting to find defects. The number of defects will not be reduced if the inspection stage is merely made better or faster or is automated. Then, one is simply more efficient in discovering defects. The trick is to inspect to prevent defects. How can this be done? *S. Shingo* suggested these basic ideas: Use source inspection techniques that control quality at the stage where defects originate. Use 100 per cent inspection with immediate feedback, rather than sampling. Make every worker an inspector. Minimise the time it takes to carry out corrective action. Remember that people are human and not infallible, so devise methods and devices that prevent them from making errors. Concentrate on making processes efficient, not on simply making the operators and operations more proficient. The latter approach seems to have been the American stumbling block. Industrial engineers can do operations improvement work, called operations research, but not process improvement work. We devise fancy, complex computerised solutions to solve complex manufacturing system problems. Why not simplify the manufacturing system so that the need for complex solutions disappears?

SECTION VII

Special Topics

Chapter 27

Energy-Efficient Operations in Heat Treatment

INTRODUCTION

Savings can be achieved in a variety of ways, including the introduction of more advanced technological approaches whereby expenditure of the necessary capital can yield corresponding energy savings.

THEORETICAL EVALUATION OF ENERGY REQUIREMENTS

Perhaps the most fundamental definition of energy relates various types of energy to the amount of work being done or to the amount of potential work that energy in a particular form is capable of doing. Heat is a form of energy that, when applied to a body, increases the energy of that body. Because molecules are in constant motion, they possess kinetic energy. Because neither increases in potential energy nor increases in kinetic energy can be measured for that body as a whole, it is concluded that heat energy must have been imparted to the molecules of the body. Moreover, almost all solids and liquids expand when heated. This means that work (energy) must have been exerted on the molecules to separate them in opposition to the forces of cohesion. Heat is, therefore, a form of energy that exists as kinetic energy, especially in gases. In solids and liquids, heat also includes potential energy due to expansion.

Kinetic energy can be demonstrated by heating one end of a steel rod. The entire rod warms as heat is conducted along the rod. According to the kinetic theory of heat, the molecules of the metal in the heated end of the rod are set into rapid motion. These molecules strike neighbouring molecules and impart kinetic energy to them and this process continues throughout the length of the rod. The amount of kinetic energy is greatest at the heated end and least at the opposite end.

The potential theory of heat is demonstrated by the expansion and contraction of solids with temperature changes. In many instances, this property of metals is put to work in products and in production processes. The application of heat energy to a body can produce other dramatic effects, although these effects may not be as visible as expansion or contraction. In some instances, the temperature of an object does not increase when the object is heated, although the over-all internal energy of the body is increased. Basic examples of this phenomenon are the formation of steam when water is heated continuously at 100°C (212°F) or the formation of liquid when heat is applied to ice at 0°C (32°F) . Thus, when heat is applied to an object at a specific temperature, a change of state can occur without an increase in temperature—a change from solid to liquid or from liquid to vapour, for example.

Heat treatment of metals deals with a material in its solid state. Thus, energy requirements to produce changes of state, although substantial in quantity, are of little consequence in a normal heat treating process. Further, except in the broadest concepts of heat treatment, the energy requirements related to expansion or contraction of the solid are not major concerns. In heat treating, the desired changes in properties or conditions result from changes in structure. Energy is used to change residual stress within a structure. More frequently, the changes desired in heat treating involve changes in the nature, form, size or distribution of the structural constituents. Changes in the nature of the constituents result from the effect of temperature on phase equilibrium or from changes in the composition of material, as in carburising or nitriding. The forms of constituents are affected by the conditions under which they separate from the solid solution and by their tendencies to assume certain shapes at temperatures that permit diffusion and rearragement, as in spheroidising. Changes in the arrangement or distribution of constituents can be brought about by solution and re-precipitation.

The energy requirements of any particular heat treatment are finite, predictable and usually unalterable. The requirements are proportional to the temperature level at which that treatment is carried out. Further, the quantity of heat that a substance can contain or hold for each degree of temperature rise is known as its heat capacity. The ratio of this heat capacity to that of water is known as the specific heat of a substance. By use of this value, the level of energy (temperature) and the quantity of energy (heat content or thermal capacity) in a process can be measured. By use of these energy requirements in conjunction with the time to achieve the energy level (temperature) desired, the time the material is held at that level awaiting change of structure, and the prescribed time to achieve cooling of the substance, the total energy requirements for a particular heat treating process can be calculated.

The potential energy due to any given phase change or other structural change is calculable. Although such a change is calculable. Although such a change may require extended time to be accomplished, the kinetic energy due to the specific heat capacity of the material and the temperature level of the process is calculable and is also not time-related. The function of time relates to the over-all energy requirement because of thermal inefficiencies in heat treating processes. The inherent inefficiencies of a process continue throughout the time required to complete changes during the heat treatment. Generally, the inefficiency of a particular heat treatment, and the amount of energy consumed, is proportional to the treatment temperature.

An equipment designer or heat treater cannot alter the energy requirement of a specific heat treatment, but a great deal of latitude exists in selection of heat treatment processes. In this choice lies the opportunity for energy conservation and cost control.

ACCOUNTING PRACTICE

Because it is an intensive user of energy, the heat treating industry remains a virtual bonanza of energy savings awaiting energy-conscious heat treatment operators. Part of the reason for inaction often relates to accounting systems. In many shops, cost systems do not account for the costs of process energy as a separate entity. These costs often are obscured within burden rates or machine centre costs. If listed as a utility cost, the cost of process energy frequently is not separated from the costs of building heat and electricity. The energy cost for a particular heat treating operation in the past typically has been a rather insignificant part of the total product cost. In some situations, nearer the end-use product form, even the total cost of heat treatment became insignificant when viewed against the total cost of the product. Even with high energy prices, energy savings are measured in terms of cost effectiveness relative to other means of reducing costs rather than being seen as ends in themselves. Many corporations and large, privately held companies have reorganised their approaches to energy utilisation, however. Through the energy manager approach, they have isolated and analysed energy costs in great detail. Energy requirements are analysed for each operation and for each phase of an operation.

ENERGY-SAVING STRATEGIES

Although many methods of reducing energy may seem elementary, experience has shown that simple or obvious solutions are often overlooked—particularly where quality and productivity are paramount goals, as in the heat treating industry. The following areas should be examined for possible improvements or changes:

1. Product, process or specification restraits.
2. Modification of process.
3. Modification of equipment.
4. Alternate equipment.

Product or Process

When faced with the task of improving any operation, an obvious question should be asked: 'why perform the operation at all'? Examples of the results of this type of analysis can be found in almost every sector of the heat treating industry. At one time, for example, all manufacturers of large gears specified that all forging be double austenitised—that is, normalised, re-austenitised, quenched and tempered. When the mechanical properties of normalised, re-austenitised, quenched and tempered large rolled gear blanks or gear tyres were compared with those of normalised, quenched, and tempered gear tyres, however, absolutely no difference in properties was found. Now, large marine gear tyres are normalised, quenched and tempered. The net savings in energy resulting from elimination of a heat treating operation have been sizable. In many such instances, product and process specifications were written when energy was comparatively inexpensive, and it was common practice to specify that certain operations be done twice to ensure that they were done once correctly.

Design Criteria

Design criteria sometimes can be altered to reduce the energy required for heat treatment. For example, a case depth of about 0.75 mm (0.030 inch) can be attained in a 0.1 to 0.15 per cent carbon steel in 3 hours in a gas-carburising atmosphere, whereas a case depth of 0.25 mm (0.010 inch) can be attained in the same steel in less than 1 hour of exposure in the same atmosphere. Costly energy can be conserved through use of shallower case depths, even considering that the designer may have to specify a base material with a slightly higher carbon content and consequently higher cost. Further savings could accrue if designers would standardise carburised case depths: deep, medium and shallow, for example. Commercial heat treaters often process batch loads with less than optimum loading simply because other parts awaiting carburisation require slightly different case depths. Any sort of over-specification usually results in increased use of energy and increased cost. Carburising is just one example. Marked savings also can be achieved by changing combinations of both material and process. For instance, ball-bearing screws have been changed from carburised 4615 steel to 1045 steel with induction hardened threads.

Cleaning Specifications

Sometimes, energy can be saved by making changes in cleaning specifications. Considerable development work has been done recently on room-temperature cleaners, which usually are aqueous solutions for removing soil from the product without the aid of heat. Not all types of work can be cleaned acceptably with room-temperature cleaners, but all such cleaners achieve significant savings in energy.

Conservation Checklist

All areas of a heat-treatment-related specification should be examined for energy-saving opportunities. The following can be used as a checklist to help locate these opportunities:

1. Are the service requirements of the part clearly known?
2. In design of the part, has every effort been made to either simplify or eliminate one or more heat-processing operations?
3. Is selection of material based on actual property requirements rather than on traditional selection practices?
4. Have the possibilities of using alternative materials been investigated, for reasons of lowering cost or increasing availability?
5. Has selective heat treating been considered?
6. Has the possibility of using pretreated material been considered?
7. Has the required case depth been reviewed?
8. Is unnecessary normalising being done?
9. Are multiple tempering operations of questionable value being performed?
10. Can tempering be eliminated without sacrificing quality?
11. Is the heat treating equipment being upgraded in line with other equipment?
12. Is a sound maintenance programme being enforced in the heat-processing shop?
13. Has quenching practice been thoroughly reviewed, and has consideration been given to use of alternative quenching media?
14. If double quenching is being used, has the necessity for it been thoroughly reviewed?
15. Is direct quenching used when possible?

MODIFICATION OF HEAT TREATING PRACTICES

A major equipment manufacturer modified his link manufacturing operation to allow the link steel to transform fully in air at 540°C (1000°F). This modification permitted additional continuous operations to be performed at that temperature, thereby utilising the heat in the link. The project was subsequently abandoned for mechanical reasons, but the change was metallurgically sound, and the energy savings were significant. Use of residual heat from a previous operation should be explored as a conservation strategy. Material can be fully transformed after a forging or casting operation and still retain substantial amounts of energy. It could be scheduled for immediate entry into a heat treating furnace as soon as the transformation has been fully completed. New process lines often are built with these features in mind, but practices in existing plants can also be changed to achieve the same energy savings. Large savings in energy can be accomplished by optimising loading of batch furnaces. Too often, batch furnaces are cycled with only a portion of the full design load. In most batch-type furnaces without recuperative devices, 65 per cent or more of the maximum energy input for the cycle is required for the furnace, regardless of the size of the load. Sometimes loads can be combined even with large differences in material thicknesses. The total furnace time is set for the largest cross-section, but the reduction in energy consumption can be considerable, when compared with the energy required for processing two separate loads. A careful review of heat treating records may reveal that combining loads or improving load density could save three or more additional cycles per month.

Lowering of Cycle Temperatures

A heat treating operation that is carried out at the lowest temperature normally will require the least energy, and the temperature of a cycle should be reduced whenever possible. A traditional practice in metallurgy has been to normalise at a temperature 110°C (200°F) above the upper critical temperature and to austenitise at 55°C (100°F) above the upper critical, but this may not be necessary. Instead of normalising at 110°C above the upper critical, a temperature 55°C above it may be sufficient for a particular job. For hardening, instead of 55°C above the upper critical, a temperature 28°C (50°F) above it may be appropriate. The furnace operator must carry out instructions as to which temperature to use, but reduced furnace temperatures can be specified.

Energy can also be saved by minimising heating times or holding times (soak times). Many specifications require holding times for austenitising of one hour or more per inch of cross-section. This is a conservative requirement, compared with other hardening processes (such as flame hardening and laser hardening, for example). From a metallurgical viewpoint, all that is required is that the metal be heated to a temperature above the hardening temperature. Unfortunately, design specifications are written to include excessive holding times. The metallurgist wants to achieve a certain microstructure or a certain combination of properties in a material. In repetitive situations, it would be worthwhile to discover the minimum time required to produce that microstructure or those properties.

Newer instruments for control of furnace temperature permit pre-setting of the desired heat-up rate. Not all material can or should be heated to temperature at the same rate or within the same time span. When the heat-up rate is specified, maximum energy savings can be realised by bringing the work-pieces to temperature as quickly as possible within the constraints of the capability of the furnace and the necessity of avoiding deleterious metallurgical effects.

Trays and Fixtures

In any heat treating operation, the furnace charge often is considered to be limited to the parts being heat treated. The furnace charge, however, consists of all material that must be brought to temperature, including trays and fixtures. In job-shop operations, it often would be worthwhile to have several sets of trays and fixtures. Lighter parts could be treated in lighter-weight trays or baskets, and heavier parts could be handled in more substantial containers. This problem has been observed in batch treating of titanium parts where the net weight of the high-alloy furnace trays exceeded the net weight of the parts being treated. In another shop, shaft-type parts were being treated on a tray in a batch furnace with a grooved hearth when the workpieces could have been treated adequately without a tray. The most energy-efficient load is the one whose net-charge weight most nearly approaches the gross-charge weight.

Furnace Utilisation

Most industrial furnaces are designed to be used at a specific production rate. At that rate, they will achieve the expected design thermal efficiency, but any departure from that specific design rate will result in a decrease in thermal efficiency. Once a furnace reaches a steady-state operating condition, the furnace losses will remain constant provided that the operating conditions are unchanged, regardless of whether the furnace is being operated empty or with a capacity load. This constant loss reduces the relative efficiency of the furnace when it is operated at less than its rated capacity. The curve in Fig. 27.1 is based on data obtained from metering of the gas used in a pusher-type heat treating furnace operating at 950°C (1750°F) . This curve shows that furnace efficiency dropped when the furnace was operated below its design capacity.

In certain instances, heat treaters conserve energy by stacking, nesting or otherwise consolidating workpieces so as to use a furnace at its design capacity without any deleterious effect on the product. In other instances, energy can be saved by selecting the proper furnace for the job at hand. For instance, a batch furnace should never be used when a fully loaded continuous furnace can be used. Savings result not only from full utilisation of a furnace but also from heating in an inherently more efficient furnace. Batch furnaces are always less efficient than continuous furnaces, in which some of the exhaust-gas heat is imparted to incoming cold workpieces.

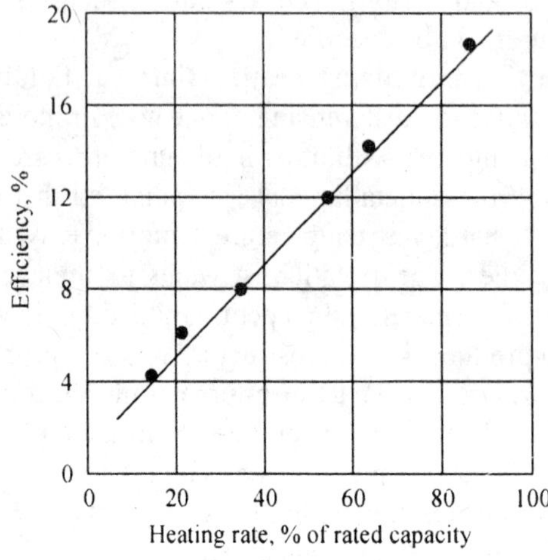

Fig. 27.1. Relationship between furnace efficiency and heating rate in percentage of rated capacity.

Idling of Furnaces

Another practice that often can be modified for energy savings is idling of furnaces when not in use. For example, a batch-type furnace idling at 815°C (1500°F) requires 9.3 m^3/hr. (327 ft^3/hr.) of natural gas. Reducing the temperature to 760, 705 or 650°C (1400, 1300 or 1200°F) reduces gas consumption to 6.9, 6.1 or 5.2 m^3/hr. (245, 215 or 184 ft^3/hr.), respectively. Thus, a 165°C (300°F) reduction in idling temperature reduces gas consumption and can produce a savings of 43.9 per cent each weekend, or an annual savings of 8.6 per cent. In this example, the furnace could produce 515 load cycles in 55 day weeks of operation, requiring 220 m^3 (7776 ft^3) of natural gas per load. If the same furnace were operated seven days a week, the same 515 loads could be processed in 30 weeks. Furthermore, the gas required per load would be reduced to 198 m^3 (6995 ft^3) an 11.2 per cent savings. In addition, there would be 20 weeks of free time for repairs or for other work.

In another example, in an industrial plant with a captive heat treating facility, operators stored all flanges forged each month in front of the heat treating furnace. In the first week of the next month, the previous month's production was heat treated on a 24-hr.-a-day, 7-days-a-week basis and then sent to the machine shop. This energy-efficient operation was possible because this manager could stage his in-process inventory at whatever point in the process that provided the lowest over-all manufacturing cost. Although other factors, such as premium pay and urgent delivery schedules, may preclude 24-hr.-a-day, 7-days-a-week operation, plant managers often have introduced such a practice and have reaped significant energy savings.

If intermittent operation is necessary, the relative costs of idling a furnace and of re-heating it after shutdown can be determined with the following empirical formulas:

$$Q/Q_1 = B$$

and $$B/K = t$$

Where Q is the total amount of energy consumed in bringing the furnace from ambient temperature to operating temperature; Q_1 is an average of the combined fuel consumed per hour while idling at a given temperature and the fuel used to return from idling temperature to operating temperature; B is break-even time; K is a constant (1.7 is suggested for tube-fired furnaces, and 1.5 for direct-fired furnaces); and t is the maximum idling time in hours, that will still result in energy savings.

Thus, if it took 1223 m^3/hr. (43200 ft^3/hr.) of natural gas to heat a direct-fired furnace from ambient temperature to operating temperature, and 25.2 m^3/hr. (890 ft^3/hr.) to idle it and to return it from idling temperature to operating temperature, the break-even point would be 1223/25.2 (43200/890), or 48.5 hours Because the furnace is direct-fired, the optimum idling time would be 48.5/1.5,or 32.33 hours. The break-even point in terms of energy savings is 48.5 hours; but, considering damage to the furnace, alloy life and all other factors, the optimum idling time is 32.33 hours.

EQUIPMENT CONSIDERATIONS

Improved Use of Equipment

In industry, if high quality and good production are being maintained, there is a natural reluctance to tamper with the equipment. But this reluctance can postpone correction of unnecessary losses of energy. Production equipment should be examined in a search for opportunities to conserve energy. With fuel-fired furnaces, the importance of proper fuel-to-air ratios cannot be over-emphasised. In products of combustion, the presence of 1 per cent CO or H_2 corresponds to a 5 per cent loss in heat release.

The problem of energy remaining in the unburned fuel can be corrected through simple adjustments of the ratio-control equipment. The effects of improper ratios (both excess fuel and excess air) are shown in Fig. 27.2. If an insufficient amount of air is supplied, not all of the energy will be extracted from the fuel, as shown in Fig. 27.2. If too much air is supplied, additional fuel will be required to heat the excess air. If either condition exists, a substantial loss of energy results.

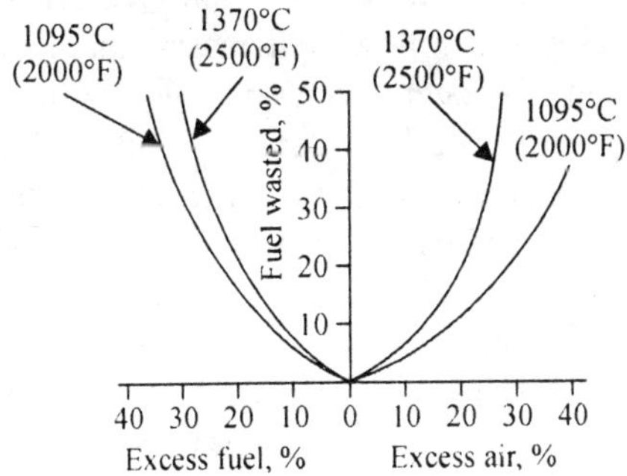

Fig. 27.2. Effects of incorrect air-to-fuel ratio on fuel economy.

Too often, furnace doors are opened too wide or left open too long, or both, when furnace charges are moved in and out. A high rate of heat loss, primarily by radiation, occurs when a furnace door is opened, and this rate of heat loss can be calculated; the following formula is often used:

$$q = 0.173\ Ae\left[\left(\frac{T_o}{100}\right)^4 - \left(\frac{T_a}{100}\right)^4\right]$$

where q is rate of heat loss, A is effective area of door opening, e is emissivity, T_0 is absolute temperature and T_a is absolute temperature of air.

Examination of this formula reveals that the amount of heat lost is directly proportional to the effective area of the door opening. The advantages of limiting door openings to that area actually required for proper access to the furnace are obvious.

Convective heat losses also are proportional to the effective area of the door opening. This loss can be accentuated when both doors are opened at the same time on a continuous furnace, because a tunnel effect is created. Convective losses also result around closed doors and through other openings in fuel-fired furnaces. To conserve energy, all unintentional openings should be sealed. The number of intentional openings should be reduced to the absolute minimum, and these openings should be sealed when not in use. Even when not permitting convective losses, openings in furnaces can permit infiltration of cold air, which must be heated to exit-gas temperature. Proper control of furnace pressure will help prevent these convective losses through necessary openings.

Proper application of refractories in re-worked furnaces often can reduce or eliminate constant heat losses that may occur in water-cooled members. Water-cooled skid pipes can be replaced with alloy load supports if maximum temperature of operation permits, or with skid blocks made of high-strength, high-temperature refractories. Walking beam rails sometimes can be topped with refractory shoes rather than with non-insulating alloy shoes.

Marked energy savings sometimes can be achieved by improving heat treating quality controls. Production scrap results in a two-fold energy loss; not only the energy required to treat the replacement part but also the energy required to produce and process the part to the point of heat treatment. Re-treatments due to poor quenching, missed draw temperatures and any other irregularity are very expensive in terms of both time and energy.

Equipment Replacement or Modification

As the cost of energy increases, the number of economically feasible opportunities to replace or alter heat treating equipment also increases. The following items are not intended to be all-inclusive, but they will point the reader in the direction of greater savings and more efficient operation.

Heat treating equipment that is sturdy and substantially built will undoubtedly provide longer life with less downtime. Such equipment may be slightly higher in cost than equipment of less substantial design, but the difference in cost often can be justified through avoidance of energy costs that continue when equipment is out of service for minor, repetitive repairs. The most efficient furnace is one requiring the least fuel per operating hour, not necessarily the one with the best hourly design efficiency. Fuel cost per unit weight of product produced should be the critical measure instead of cost per furnace-cycle hour. For instance, a cold-wall vacuum furnace uses energy only during the actual heat treating cycle, whereas a conventional box furnace uses energy from the time it is lighted until it is shut down at the end of the week. When the opportunity arises, much energy can be saved by altering the gases used in atmosphere heat treating. In some cases, an inert carrier gas can be added to the normal working gas. Perhaps a 100 per cent substitution with an inert gas is possible during periods when working gas is not required, but the furnace must be kept 'bright'.

Other energy-saving opportunities may be realised by improving the insulation systems in heat treating furnaces. The energy lost through a furnace wall is a function of area, operating temperature and composition of the insulation. The first two factors are fixed, but heat flow through and heat storage in the insulating system can be reduced by addition of insulation. Heat storage, which is the amount of heat contained in the wall, can be greatly reduced by using newly developed insulating materials, principally ceramic fibres and mineral wools. At the same time, the thickness of a wall for a given heat flow (loss) also can be changed significantly. One major heat treating firm not only reduced energy requirements by using ceramic-fibre insulating materials, but also was able to heat treat larger rolls because of the decrease in required wall thickness, which in turn provided greater furnace work-zone width.

Addition of alloy fans to existing furnaces sometimes can change a stagnant atmosphere into a high-velocity stream. This increase in ambient velocity breaks up boundary layers of furnace gases that surround the work-pieces and shortens the heating time. This reduced heating time is a result of the change from heating only by radiation heat transfer to a combined radiation and convection transfer. The energy savings accrue through reduced furnace time required per cycle.

Thermodynamic Thermal Efficiency

Thermodynamic thermal efficiency can also be improved by use of waste combustion gases. A particular batch furnace, heated to 980°C (1800°F) with natural gas, with the flue gas exiting at or near that temperature, operates at only about 35 per cent efficiency. Through use of a recuperator, the incoming combustion air can be heated by the waste products of combustion. If the combustion air is heated to 425°C (800°F), a 23 per cent savings in fuel can be achieved. This method is perhaps the least efficient way of re-using heat from the products of combustion; sometimes, however, it is the only practical way.

Because the coefficient of heat transfer between combustion products and a solid is twice as high as between combustion products and incoming combustion air, it is better to use waste gases to pre-heat the workpieces, as is done in properly designed continuous direct-fired furnaces. Other commonly used recuperative methods also are employed successfully on continuous furnaces, but some of these methods employ a lower heat-transfer coefficient. Methods may vary, but increases in the cost of energy are likely to move the heat treating industry more and more toward use of waste gases.

Chapter 28

Testing Processes to Evaluate Metallic Conditions

INTRODUCTION

There are many processes that can be helpful in making an evaluation of a metal. Evaluation processes can be established.

1. To determine specific mechanical properties.
2. To reveal discontinuities and flaws in metals.
3. To determine the behaviour of a metal with respect to force, temperature, corrosive environment, and other external influences.
4. To study changes in microstructure to review exposure conditions, analyse failure, inspect quality of fabrication, and develop new alloys.
5. To qualify specific processes and techniques.

WHAT YOU CAN LEARN

1. Application of mechanical tests to qualify a material or process.
2. Application of non-destructive tests to determine quality level.

Testing processes to evaluate metallic conditions will be divided into the following two major groups:

1. Destructive testing—a process that causes an alteration of the surface or of the microstructure.
2. Non-destructive testing—a test that can be conducted without altering the usefulness of the material.

Processes in the destructive group include:

1. Tensile test.
2. Compression test.
3. Hardness test.
4. Axial fatigue test.
5. Rotating beam fatigue test.
6. Creep test.
7. Stress rupture test.
8. Impact test.
9. Bend test.
10. Thermal analysis test.

11. Burst test.
12. Flare test.
13. Stress-corrosion test.
14. Linear thermal expansion test.
15. Grain size test.

Processes in the non-destructive group include:

1. Visual test.
2. Penetrant test.
3. Magnetic particle test.
4. Eddy current test.
5. Radiographic test.
6. Ultrasonic test.
7. Acoustic-emission test.

DESTRUCTIVE MECHANICAL TESTING

Tensile Test

The tensile test is one of the most widely accepted means of obtaining valid data about the mechanical properties of a metal. Standard tensile coupons are round or rectangular in cross-section with a carefully prepared centre section. A selected distance in the centre is then marked for the gauge length (Fig. 28.1).

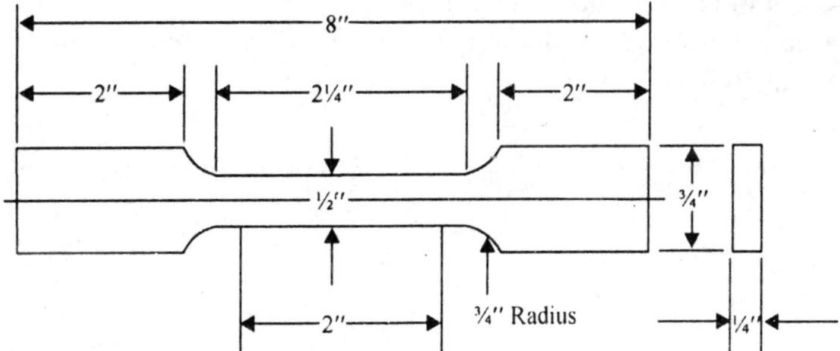

Fig. 28.1. Typical dimensions of tensile testing coupons. Round coupons are also used that have a 0.25 inch dia. which gives an area of approximately 0.05 $inch^2$.

The specimen profiles are machined and given a required surface finish. The next step is to compute the cross-sectional area by measuring the diameter or the thickness and width. The specimen is given a gauge length mark and mounted in the tensile grips. An extensometer, if used, is mounted, and the tensile machine is adjusted for the test. Adjustments include selection of load range, rate of loading, and magnification of extension. Usually a recorder plots the behaviour of the test piece by indicating stress in pounds and strain in inches. Stress will be directly proportional to strain while the material is exhibiting elastic behaviour. If the load were to be released while in the elastic range, the material would regain its original length.

As more load is applied, the metal will start to elongate and show little increase in load capacity. This is called the zone of plastic deformation. During plastic deformation the metal will tend to work harden. Some metals elongate rapidly near maximum load. This elongation reduces the cross-sectional area.

During the process of elongation on a permanent basis, the metal is becoming smaller but stronger. The load indicator may even indicate a lower load while this is happening.

Later, the load indicator reaches a maximum value and drops slowly to a point where failure occurs. The fracture load is usually much lower than the maximum load when testing ductile metals. After fracture has caused a complete separation, the specimen is taken from the machine. One of the parts is measured across the face of the fracture to obtain final dimensions. From the final dimensions the final area is computed. At this time the two parts are fitted back together and the final gauge length is measured.

A typical report on the tensile test includes the yield strength, tensile strength, ductility, and modulus of elasticity.

Yield strength

The yield strength is found by dividing the yield load by the original area. The formula is written as:

$$YS = \frac{L_y}{A_o}$$

YS = Yield strength (psi)

L_y = Load at yield point
A_o = Area original ($inch^2$)

To better understand the formula, work the following problem:

Given: Yield load of 8000 lb.
Thickness of cross-section is 0.25 inch.
Width of cross-section is 0.75 inch.
Area is thickness × width or $0.25 \times 0.75 = 0.19$ $inch^2$

Find: Yield strength

$$YS = \frac{L_y}{A_o}$$

$$= \frac{8000}{0.19}$$

$$= 42105 \text{ psi}$$

There are different ways to arrive at the value for yield load. The most simple is called drop of the beam which is especially applicable on alloys like structural steel. The observer watches the load indicator (while loading) and records the load value when the indicator either changes to a reduced pace or actually drops a small percentage of the load. This method requires experience and knowledge of the metal under test. A more scientific and valid procedure is called the offset method. With the offset method a plot is made of the stress-strain curve. On this diagram a second line is constructed parallel to the original line but to the right a specified amount. The specified amount is called the offset and is usually equal to 0.2 per cent of the gauge length, as shown in Fig. 28.2.

Tensile strength

Tensile strength is found by dividing the maximum load by the original area. The formula is written as:

$$TS = \frac{L_m}{A_o}$$

TS – Tensile strength (psi)

L_m – Load at maximum value
A_o – Area original (inch^2)

To better understand the formula, work the following problem:

Given: Load at maximum value is 10,400 lb.
Area = 0.19 inch^2

Find: Tensile strength

$$TS = \frac{L_m}{A_o}$$

$$= \frac{10400}{0.19}$$

$$= 54737 \text{ psi}$$

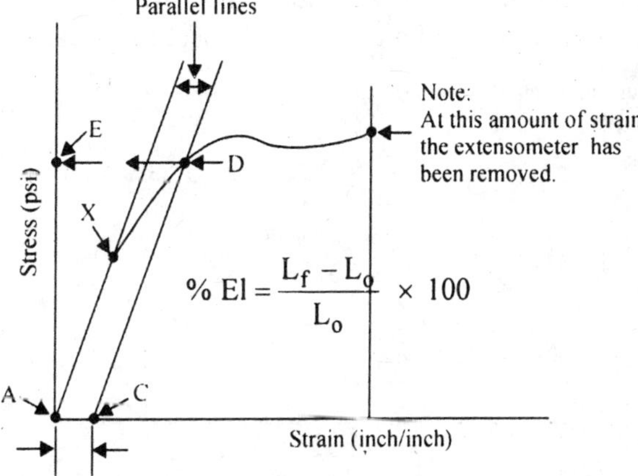

Fig. 28.2. The offset method to determine yield load. Offset (A to C) is determined by a percentage of the gage length, usually 0.2 per cent of the gage length. Line A-X has been made by machine recorder. The operator computers offset and locates point C. Line C-D is constructed. Point D (intersection) is projected to the left to point E which is the yield load in psi.

Ductility

Ductility is found by two methods. One is based on the change in length, and the other is based on the change in area. To find per cent elongation use the formula:

$$\% \text{ El} = \frac{\text{Change in length}}{\text{Original length}} \times 100$$

Determine the change in length by subtracting the original length from the final length. Then the actual formula becomes

$$\% \text{ El} = \frac{L_f - L_o}{L_o} \times 100$$

% El = Per cent elongation

L_f = Length final
L_o = Length original

To better understand the formula, work the following problem:

Given: Original gauge length was 2.0 inch.
After fracture gauge marks had spread to 2.25 inch.

Find: Per cent elongation

$$\%El = \frac{L_f - L_o}{L_o} \times 100$$

$$= \frac{2.25 - 2.00}{2.0} \times 100$$

$$= 12.5 \text{ per cent}$$

A very small amount of strain (elongation in inches per inch) will take place during the elastic behaviour compared to the plastic behaviour. An example of this could be observed on a test with a structural steel specimen. It will usually elongate about 0.001 inch/inch of gauge length in the elastic mode. In the plastic behaviour range, it will increase by almost 250 times, as shown in Fig. 28.3.

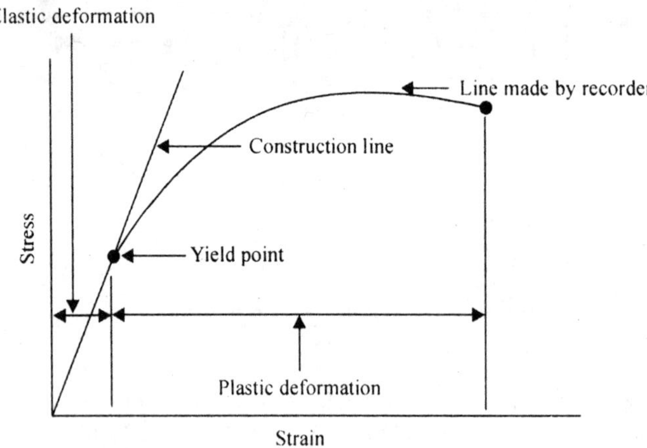

Fig. 28.3. A comparison of the elongation as measured during elastic and plastic deformation of a structural steel specimen with a 2.0 inch gauge length. The steel was cold worked before testing.

Modulus of elasticity

The modulus of elasticity is determined by dividing stress (psi) by strain (inch/inch). Remember that when the item calls for strain (inch/inch), this means inches of elongation divided by the gauge length. Thus, modulus of elasticity is merely a ratio of stress to strain. Some designers consider the numerical value of modulus of elasticity as a factor of how rigid a material will be when subjected to a load. The modulus of elasticity value depends on what the material is made of more than how strong or hard it is. In fact, the modulus of elasticity of a steel is not changed by annealing or quench hardening. Typical stiffness ratios of selected metals are listed in Fig. 28.4.

Compression Test

The compression test is similar to the tensile test except that most metals exhibit a tendency to hold increased loads as stress is applied. This is simple to understand because as metals are compressed the area holding the load becomes larger. Of course this is not true for those brittle metals that shatter rather than deform plastically.

To conduct a test of this type, a specimen is machined and placed in a fixture as seen in Fig. 28.5.

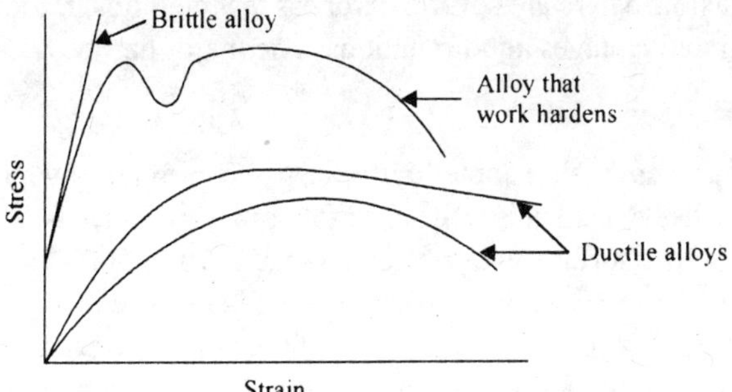

Fig. 28.4. A comparison of different metals as they elongate under load in tension.

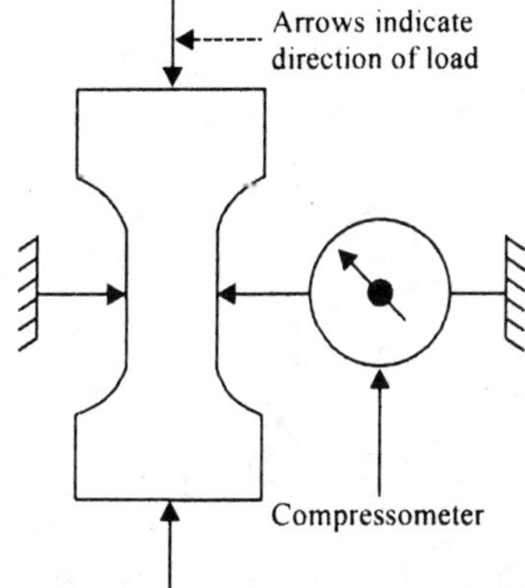

Fig. 28.5. Typical set-up for compression testing of materials. Compressometer measures the diameter increase as load is applied.

Several report procedures are suggested to obtain data for this test. The method used may be selected by the customer. Yield strength and ultimate compression strength values may be obtained by specifying compressometer readings for each. For example, when the compressometer reading indicates 0.00075 inch yield, the load is recorded. For the ultimate compressive strength, the compressometer would indicate 0.001 inch. Specimen sizes may vary according to type of material or customer preference. Some profiles of compression test specimens are given in Table 28.1.

Table 28.1. Typical solid cylindrical specimen for compression testing.

Specimen description	*Diameter (inch)*	*Tolerance in diameter (inch)*	*Length (inch)*	*Tolerance in length (inch)*
Short	1.125	± 0.01	1.0	± 0.05
Medium	0.50	± 0.01	1.5	± 0.05
Long	0.80	+ 0.01	6.375	± 0.125

Hardness Test

The hardness value of a metal in this discussion is considered to be its ability to resist indentation. A hardness test is significant of changes in certain mechanical properties. These changes may include strength, ductility, impact, and abrasion. There are several hardness test methods in prominent use. Each of these hardness tests have certain advantages and limitations. The major hardness tests are discussed below.

Brinell hardness test

This is one of the first hardness tests adopted by industry. It provides average values of hardness by using a large indenter and a heavy load. It is well suited for non-ferrous metals. The test is quite simple. A 10 mm steel ball is forced into the surface with a selected load (500 to 3000 kg.). The load is left constant for a specified time such as 15 to 30 seconds. Then the load is removed from the indenter and the crater is measured with a portable microscope. Two readings of the impression (crater) are made at right angles to each other. These two readings are averaged, and that value is converted into a Brinell hardness number. Additional information is given in Fig 28.6

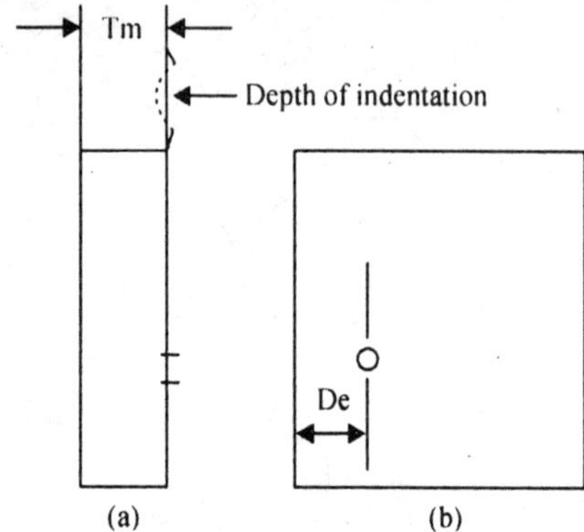

Fig. 28.6. Details of the procedure for Brinell hardness testing. (a) Side view. Tm (thickness of material) must be 10 times the depth of indentation. (b) Top view. De (distance from centre of impression to edge of material) must be two and one-half times the diameter of impression or more.

The Brinell hardness number (BHN) is computed by the formula

$$\text{BHN} = \frac{\text{Load}}{(\pi D/2)(D - \sqrt{D^2 - d^2})}$$

where, D = diametєr ball indenter

d = diameter of impression

The problem is easy with the aid of a hand calculator. Try this practice problem.

Given: 3,000-kg. load

10-mm. ball indenter

4.90-mm. diameter of impression

$$\text{BHN} = \frac{\text{Load}}{(\pi D/2)(D - \sqrt{D^2 - d^2})} = \frac{3000}{(3.14 \times 10/2)(10 - \sqrt{10^2 - 4.9^2})}$$

$$\text{BHN} = 149$$

When reading the Brinell microscope, the technician should attempt to obtain a value to the nearest 0.05 mm. If the test is conducted in a professional manner, the operator should be able to get repeatable accuracy to within ± 3 per cent of actual hardness. Additional information is given on reading the Brinell microscope in Fig. 28.7.

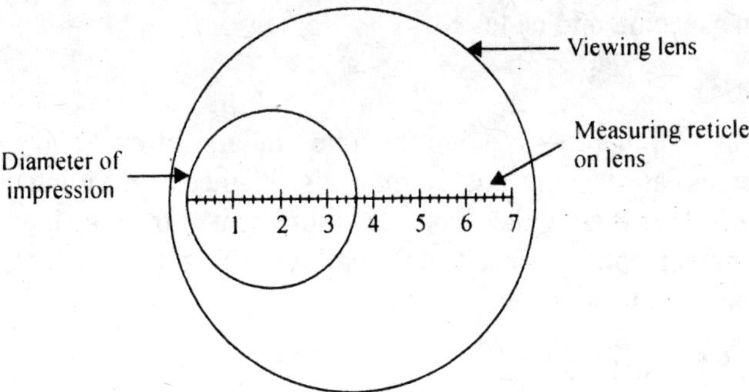

Fig. 28.7. A view of what the operator sees when placing the Brinell microscope over an impression.

Rockwell hardness test

The Rockwell hardness test is somewhat similar to the Brinell hardness test, but there is a significant difference in the equipment. The Rockwell system uses indenters ($\frac{1}{8}$ inch ball, $\frac{1}{16}$ inch ball and a diamond cone with 120° face angle) and load values of a combination obtainable with 40, 50 and 60 kg weights. Additional information on the procedure is seen in Fig. 28.8.

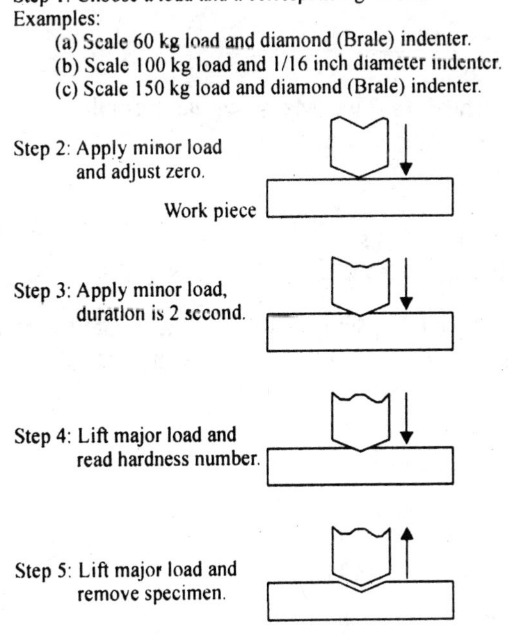

Fig. 28.8. A diagram of the Rockwell hardness test procedure.

By observing the procedure steps, it should be apparent that the Rockwell system measures hardness by determining the difference in penetration from the minor load position to the major load position. Each number on the Rockwell scale represents a distance of 0.00008 inches (8×10^{-5}). A standard micrometer measures accurately 10^{-3}. This means that the Rockwell system is 100 times more sensitive than the standard micrometer. Because of the sensitivity the Rockwell test requires careful set-up and proper preparation of the specimen to be tested.

Microhardness testing

The two major systems in microhardness testing are quite similar. Both use a diamond pyramid indenter. The Knoop system indenter has one long axis. Hardness is obtained by dividing the load in kilograms by the area of the indentation. Area is computed from a measurement of the long indentation. Measurements are conducted by high-power optics. For a sample problem use a 1 gram load and an indentation of 6.2 microns (μ) in length. The formula:

$$HK = \frac{c \times P}{d^2} = \frac{14229 \times 1}{6.2^2}$$

$$= 370.2$$

where, HK = Hardness Knoop system
c = Constant developed for this test
d = Distance across long diagonal
P = Force in grams

The other microhardness system uses a square-based diamond indenter and is called Vickers hardness test. Hardness values are determined by dividing the load (kg.) by the surface area of the indentation. The original formula was:

$$HV = \frac{2 \times P \times \text{Sin}(a/2)}{d^2}$$

This seems complex and difficult to solve. Many original formulas are like this until the surplus steps are eliminated. A simplified version now becomes

$$HV = 1.8544 \times P/d^2$$

where, HV = Hardness Vickers system
P = Load in kilograms
d^2 = Mean diagonal squared

The mean diagonal is found by adding both diagonals of the square-based indentation and dividing by 2.

To better understand the formula, work the following problem:

Given: Load of 1 gram
First diagonal is 2.41 μ
Second diagonal is 2.39 μ

Find: HV

$$HV = 1.8544 \times P/d^2$$

$$= \frac{1.8544 \times (1 \times 1000)}{\dfrac{(\text{diagonal } 1 + \text{diagonal } 2)^2}{2}}$$

$$= \frac{1.8544 \times (1 \times 1000)}{\frac{(2.41 + 2.39)^2}{2}}$$

$$= \frac{1854.4}{2.4^2}$$

$$HV = 321.9$$

Equotip hardness testing

The Equotip process of hardness testing relies upon a new principle. To conduct the test a small probe unit carrying a spring loaded test tip formed of tungsten carbide is caused to strike the test surface. The impact and rebound velocities are measured by an electronic battery-powered digital readout device. Hardness numbers are in L-values. L-value means Leeb-value, according to the name of the inventor. Conversion tables are available to compare L-values with standard hardness measurements in Brinell, Rockwell, and Vickers.

The technique is simple in that the probe is either in the vertical or horizontal position when placed against the workpiece. The test tip is released and L-value of hardness is indicated instantly. Testing time is less than 3 seconds usually. Little experience is required. Accuracy has been found to deviate as little as ± 0.8 per cent (related to L = 800). Suggestions to increase accuracy are:

1. Surface should be ground with fine abrasive grit.
2. Three test should be conducted over a ½ inch square area.
3. Work piece should weigh 12 lb minimum and be of compact shape.
4. A coupling paste and anvil are provided for the materials.
5. Minimum radius of work piece should be 1.25 inch (unless a compensating factor is used).

The measuring range of the Equotip hardness test covers a greater span than either Rockwell or Brinell hardness tests, as shown in Fig. 28.9.

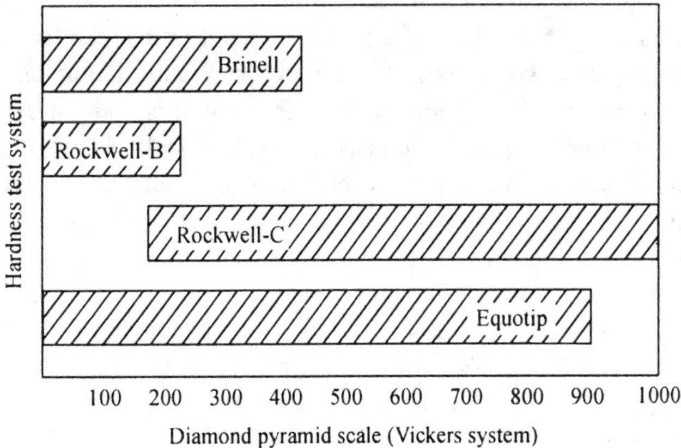

Fig. 28.9. Comparing the diamond pyramid (Vickers scale) hardness test to Brinell, Rockwell and Equotip hardness tests.

Axial Fatigue Test

This test is used on metals to obtain data about how the material will behave when subjected to many repeated loads in tension. Usually there will be several specimen to be tested. It is customary that the first will have a high stress level and stay in service only a few thousand cycles. Gradually the stress level should be reduced as future specimens fail. At some lower load or stress value, the test piece may be in service after 10^8 cycles. This is sometimes considered the endurance limit. In other words, at that stress level and a service life of 10^8 cycles the metal specimen would last indefinitely.

The term axial means that the specimen is to be loaded equally about its axis. Three standard systems are in use in axial fatigue testing. These systems are:

1. Tension-tension (usually maximum stress at full load reduced to 1/10 load).
2. Tension-compression (maximum stress in tension to maximum compression).
3. Compression-compression (maximum compression to 1/10 maximum compression).

Standard specimens for this test are 0.25 inch diameter and 1.0 inch gauge length with threads on each end to screw into the grips. Load cell adjustment is computed to apply a specified stress level. A timing mechanism is zeroed and the test started. There will be some relaxation or creep of the material in long-term testing. The load cell can be set to maintain the stress level throughout the test. See typical data from an axial fatigue test in Fig. 28.10.

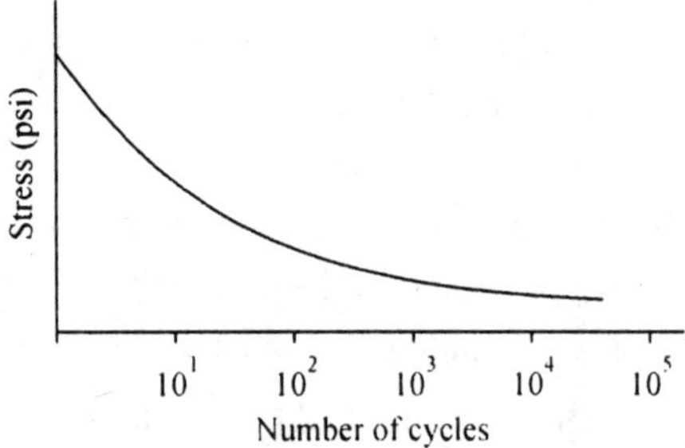

Fig. 28.10. Fatigue test diagram showing stress level versus number of cycles. This is called an S-N diagram.

Fatigue tests provide information about materials ability to resist low loads under many repeated cycles of stress. To understand the behaviour of a metal, it might be helpful to visualise its infinitely small structure—down in the atomic world. A metal is not just a solid blob of material as it might appear on the surface. Although it appears as a continuous solid structure, it is really more space than mass.

Most solid metal alloys at room temperature consist of crystals packed against each other, and each crystal has an axis slightly inclined from adjacent neighbouring crystals. As stress is applied, atoms within the crystals are maneuvered to position so that one axis is parallel to the stress and the other axis is perpendicular. As stress becomes great enough to cause plastic behaviour, groups of atoms shift along a 45° plane. This permits the metal to become longer in the direction of stress but narrower in width. If the stress continues beyond the elastic limit, the slip planes may continue in the original direction or new slip planes may proceed at 90° to the original planes. If the stress is less than the yield load, the atoms start to move from their original plane to the next. Stress less than yield loads may be repeatedly applied, and when removed the metal will regain its original dimension.

In fatigue testing the stress level is selected at some percentage value of the yield stress. This will permit the metal to be loaded and unloaded through many thousands or millions of cycles.

You will recall the solidification defect entitled dislocation. The dislocation occurred between two sections of perfect atomic alignment. When fatigue stress is in the area of the edge of a dislocation, atomic movement tends to displace the dislocation. As the dislocation is moved, more ideal atomic alignment exists. As more ideal atomic alignment is arranged, two properties change. The metal becomes stronger, and ductility is reduced. With continued fatigue cycles the metal will fail in a brittle manner as seen in Fig. 28.11.

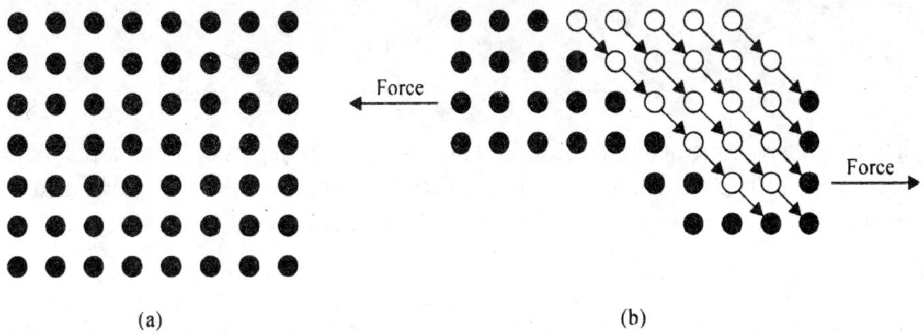

Fig. 28.11. Schematic diagram of applied stress upon a crystal. (a) Atoms in a crystallographic plane with no stress. (b) Typical path of slippage as oriented with the direction of stress.

Initial or beginning slip is not harmful to the metal. Continued slip forced by some particular stress level and continued long enough causes the formation of sub-microscopic cracks. These cracks begin at the atomic level and progress to complete failure with repeated stress. This type of failure is not at all like the normal material failure. It may occur with little warning, with very little elongation or reduction of area. A typical axial fatigue machine will have the components shown in Fig. 28.12.

Rotating Beam Fatigue Test

This type of fatigue test causes stress to appear at the outer section of round shafts. The test specimen is machined and polished to remove any areas that might act as stress raisers. The section is measured with a micrometer to determine the diameter. It is now mounted into a collet on each end. One collet is attached to the drive motor, and the other collet is connected to a load capable of applying the specified stress. The motor is started and adjusted to the required RPM. With the motor running at the RPM specified, the load is applied. The RPM counter or timer is set, and the test is started (Fig. 28.13).

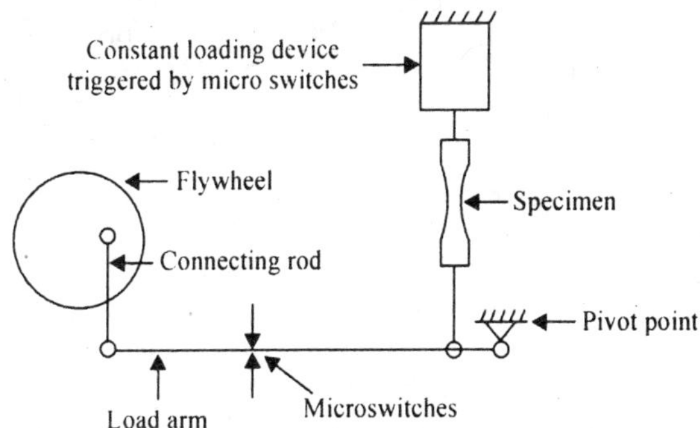

Fig. 28.12. Schematic diagram of an axial fatigue machine.

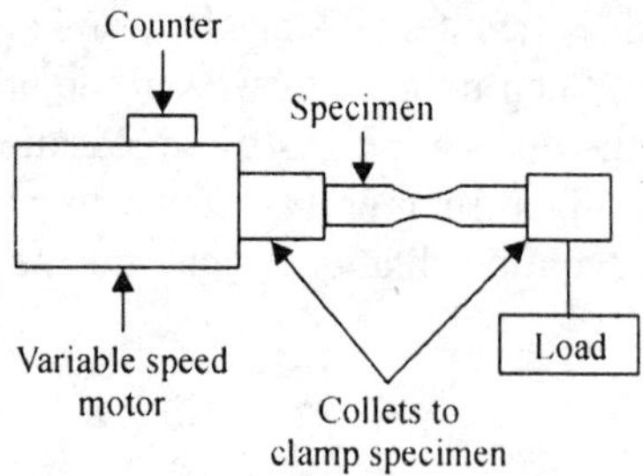

Fig. 28.13. A schematic diagram of a rotating beam machine.

The objective of the rotating beam test is to find a stress level for a particular material that will permit the required service life. The test can be helpful in making a selection of material and also in designing adequate size cross-sections to permit specified service life at a designated stress level.

Creep Test

The purpose of this test is to determine the amount of deformation caused by a specific stress level as a function of time. More simply stated, how much elongation in inches per inch of gauge length can be detected with a particular stress in a given period of time. The test is usually conducted at some specified temperature. A standard creep test machine is shown in Fig. 28.14.

A typical performance for an alloy during a creep test is seen in Fig. 28.15.

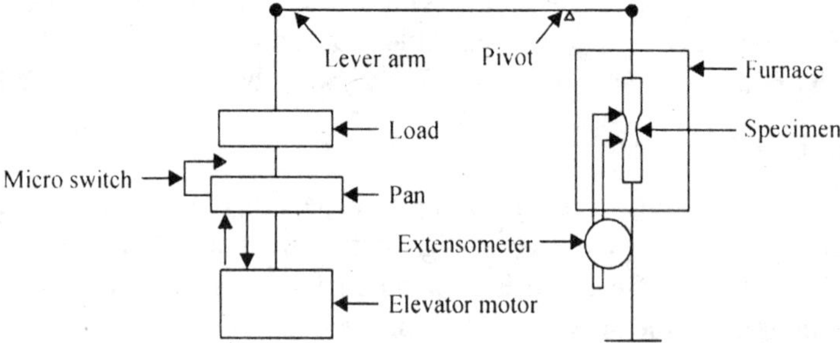

Fig. 28.14. Schematic diagram of a creep testing machine.

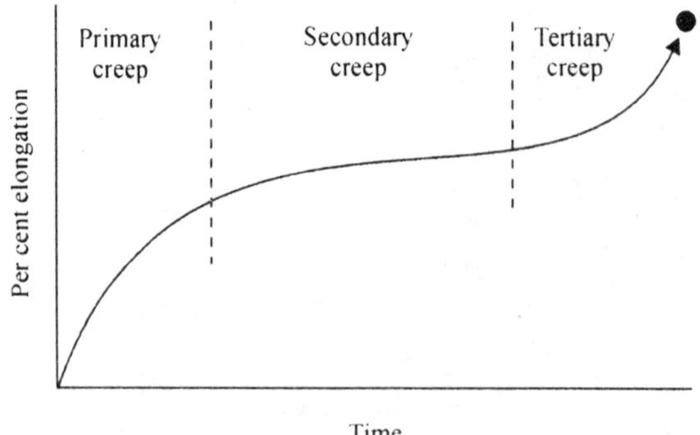

Fig. 28.15. A diagram showing the behaviour of an alloy under constant stress versus time.

Stress Rupture Test

In recent times aerospace industries have required some extreme service conditions of stress and temperature. In service, because of the unusual stress, the part may fail after a few hundred hours. The short time can be acceptable, but the part must not fail in service. In other words replace it before it fails. Designers have asked technicians to quantify the life of certain parts in hours. To do this the stress rupture test is required rather than long-term creep test. Deformation values may not be of interest in the stress rupture test. Test specimens are machined to small-diameter threaded end parts similar to a tensile test coupon. The diameter of the specimen is read to the closest 0.0001 inch and the area is computed. Then the specimen is mounted in the grips, and two thermocouples are installed on the specimen (top and bottom). The furnace is installed and brought up to the required temperature. The load is then applied, and the test is started. Several specimens may be used to generate a curve characteristic of the material, as shown in Fig. 28.16. The machine for stress rupture and creep are the same. The difference is that on stress rupture an extensometer is not usually used.

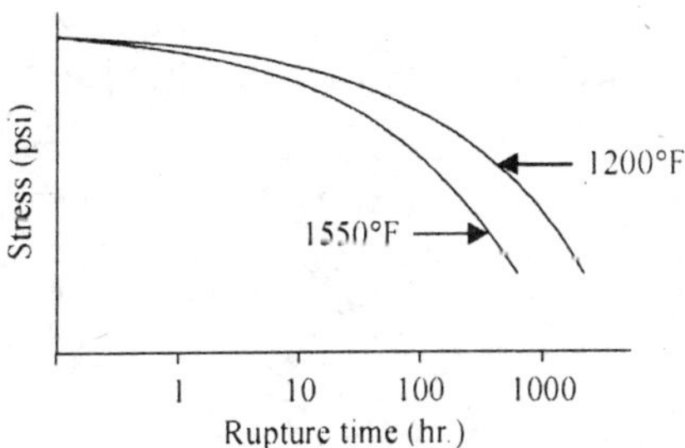

Fig. 28.16. Data from a stress rupture test showing stress level and hours when rupture occurred. Note the effect of higher temperature.

Impact Test

There are several types of impact tests in prominent use with the main objective of assessing the behaviour of metals as affected by temperature. Some impact tests are conducted at room temperature and are designed to provide data about the type of failure. This section will cover the drop-weight test and the notched-bar test.

The drop-weight test was developed at the Naval Research Laboratory. This test was to investigate conditions required for initiation of brittle fractures in structural steels. Since the Naval test was developed for a special purpose, it was given the name nil-ductility transition temperature test. The American Society for Testing Materials (ASTM) has developed a standard for conducting this test (ASTM E208-69). The test employs a simple beam-type specimen and a falling weight. Failure may occur on the bottom side where a rapid tensile stress is imposed. The object is to determine the minimum temperature at which a specimen cracks. Specimens are bent only a small amount, as shown in Fig. 28.17.

The usual sequence is to estimate temperature near the nil-ductility transition temperature and proceed to establish limits with 10°F (5°C) for crack and no-crack performance. It is recommended that a second specimen be run to confirm the no-crack performance. The crack or break results are visually examined. The inspector will look for a small cleavage crack to propagate from the crack-starter weld to the base metal. The specimen is considered cracked if a fracture on one or both tension surfaces appears. Complete separation of the specimen is not required. Typical results are shown in Fig. 28.18.

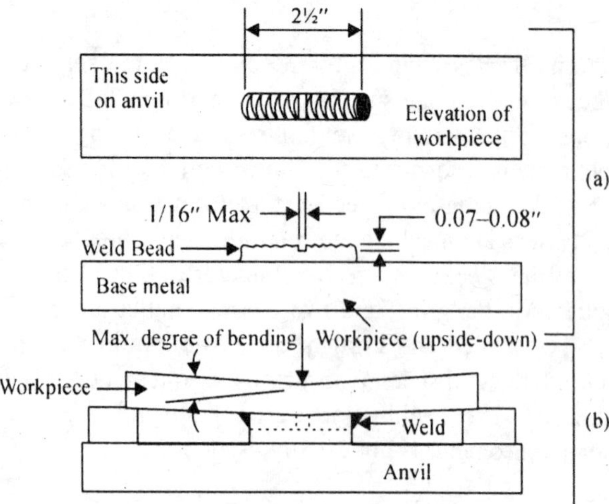

Fig. 28.17. Drop weight test to determine the nil-ductility transition temperature. (a) Coupon and weld bead; and (b) Anvil assembly.

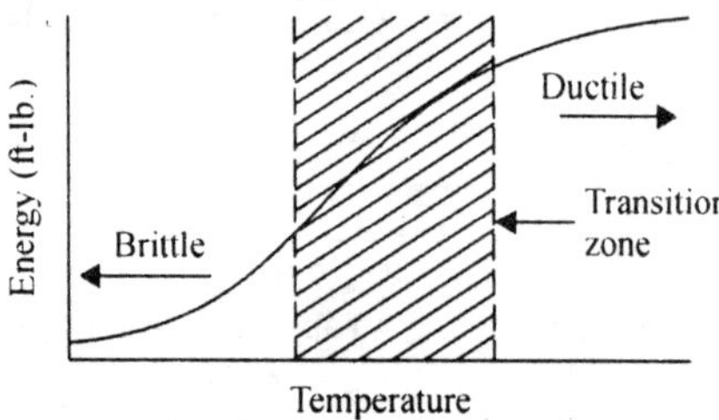

Fig. 28.18. Typical results of how temperature affects structural steel. Transition zone is the temperature range where ductile metals start to become brittle.

The notched-bar test is simple to conduct if the procedures are followed. Two types of specimens may be machined for the test. The machine uses a swinging pendulum and a rigid vise to hold the specimens. Once calibrated, the machine will provide valid results for many specimens. The specimens are held in a liquid bath at the required temperature. When ready, the specimens are placed in the fixture and the pendulum (hammer) is released. The hammer will strike the specimen and lose part of its energy. This energy is recorded in foot pounds (ft lb). Specimen profiles are shown in Fig. 28.19.

When the hammer strikes the specimen in the centre and it is supported at each end, the test is referred to as a charpy (simple-beam) impact test. When the specimen is clamped in a vise from one end and the hammer strikes the other end, it is called the izod (cantilever-beam) impact test. Typical results of a test are shown in Fig. 28.20.

Bend Test

The bend test has been devised for the purpose of determining the ductility of a welded metal ¼ inch thick and over. It takes high-quality welds to pass the bend test. This is because the metal on the outside surface of the bend is severely elongated. Weld metal and base metal that have porosity, stress cracks, slag inclusions, and unfused zones will not usually pass the bend test. It is not uncommon to find a shipment of steel direct from the mill (no welds) that will not pass the bend test. There are some X-ray-quality weld samples that have failed during bend testing.

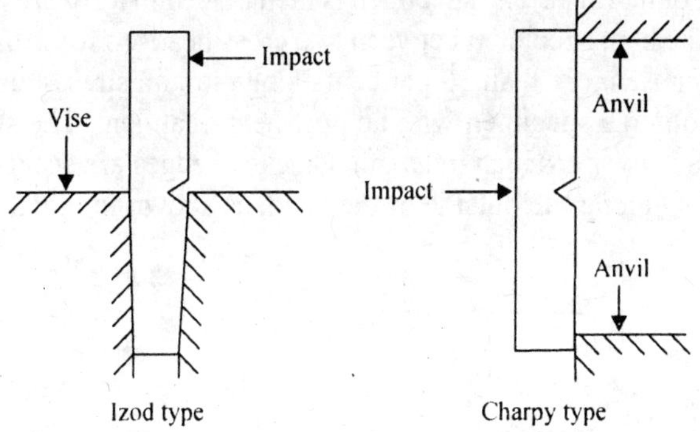

Fig. 28.19. Profiles of standard notched-bar specimens. All specimens are approximately 2.165 inch (55 mm) in length.

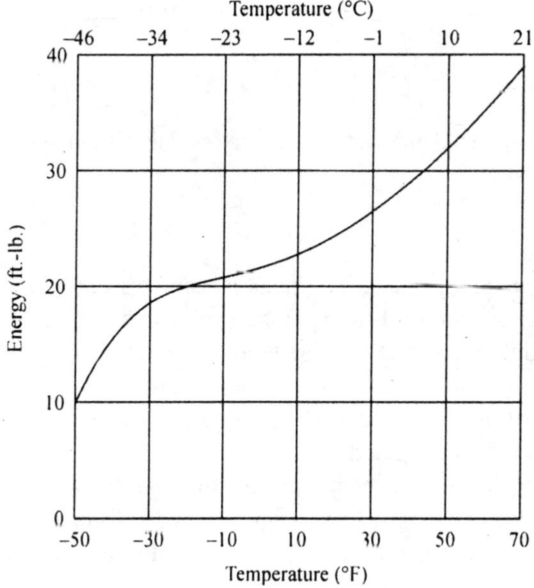

Fig. 28.20. Graphical results of a charpy impact test showing how low temperature affects impact energy values on structural steel.

The American Society for Testing Materials (ASTM) and the American National Standards Institute (ANSI) have approved three types of bend tests for welded structures. The names are descriptive of the testing conditions. These bend tests are:

1. Free bend test.
2. Guided bend test.
3. Semi-guided bend test.

Free bend test

On the free bend test the specimen shall have the weld build-up removed to the height of the base metal—in other words, ground flush. The specimen is then placed in a fixture and bent about 30° to 45°. The next step is to place the bent specimen between two cross heads to force the ends together. The final bending radius shall not cause more than 30 per cent elongation on stress-relieved welds and not more than 25 per cent elongation on a specimen with no post-heat treatment. The structure does not pass the test if a crack appears that is over 1/8 inch in length. Cracks at edges are acceptable. Per cent elongation is determined by measuring along the outside of the bend, as shown in Fig. 28.21.

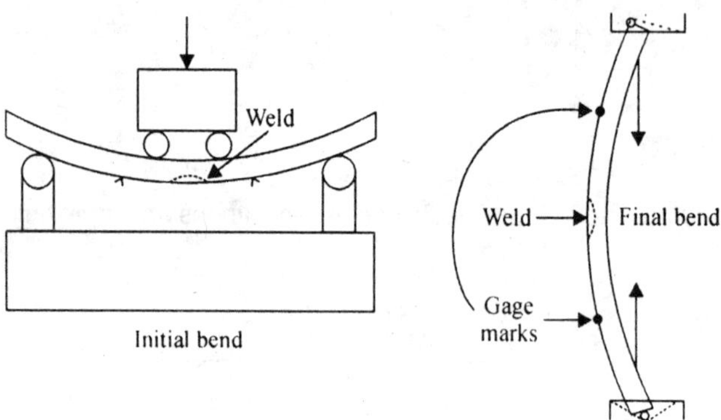

Fig. 28.21. Fixtures for initial and final bending for the free-bend test. Elongation is found by a measurement of the gauge marks after bending.

Thermal Analysis Test

This test is used to investigate phase changes in metallic elements and alloys. One of the most common phase changes is the transformation on heating from a solid to a liquid. Also the transformation from liquid to solid. A person might anticipate that the temperature at which an element would melt during heating would be the same as that at which an element would solidify upon cooling. This is not so because it is possible to heat or cool the element beyond a particular temperature due to the time lag required for transformation to be completed.

A metal at almost melting has far less internal energy than nearly the same temperature after melting. This was attributed to two kinds of energy. The energy required to force atoms far enough apart to satisfy the laws of a liquid is called potential energy. The second kind of energy is called kinetic energy and is related to the speed and possibly the vibration of the particles of an atom.

Any metal that exhibits a phase change during heating or cooling can be given a thermal analysis test. The difficulty may be to obtain sophisticated instruments to record the event.

Alloys with lower melting temperatures are less complex to work with but will demonstrate the principle on an equal basis. On an experimental basis an alloy that melts in hot water can be used. A beaker of hot tap water then could be used for the heating cycle and cold tap water for the cooling cycle. Notice that the transformation on heating begins at a temperature that is higher by about 20 per cent than that at the beginning of liquid to solid transformation on cooling in Fig. 28.22. Also note that if cooling is rapid enough, the first areas to solidify will be re-melted as the atoms release the internal energy required for the original melting to occur. This transfer of energy is called latent heat of fusion.

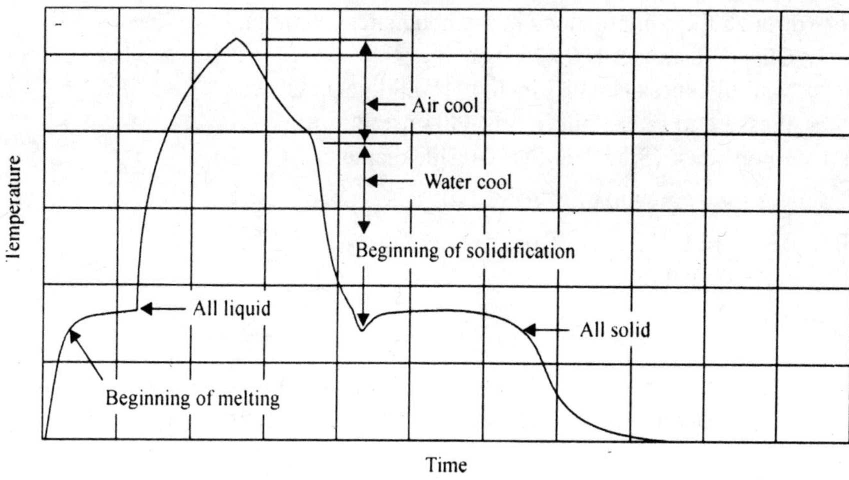

Fig. 28.22. Showing the alloy's reaction to rapid heating and cooling.

Burst Test

This test was intended to evaluate maximum working stress in pipe, tubing, and some types of cylindrical tanks. The test is started by filling the article to be tested with a liquid (oil or water). When all the air inside has been replaced with the non-compressible liquid, pressure is applied. The pressure is carefully monitored until failure occurs. This data is then put into a formula to determine the stress value of the container (Fig. 28.23).

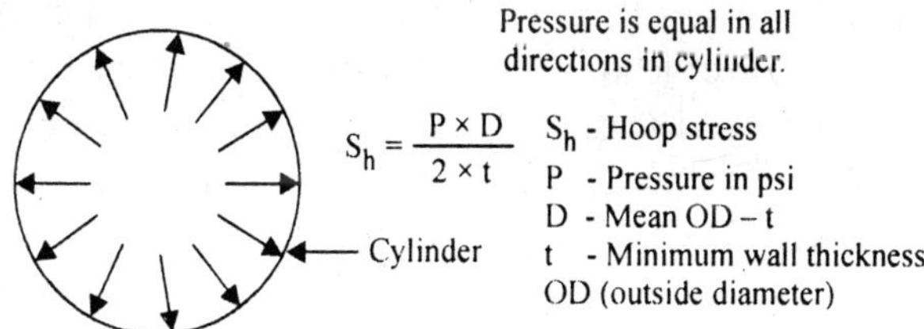

Fig. 28.23. A schematic diagram of a burst test in operation.

An example of an actual test using copper tubing with an 0.625 inch outside diameter and a wall thickness of 0.037 inch is shown in Fig. 28.24.

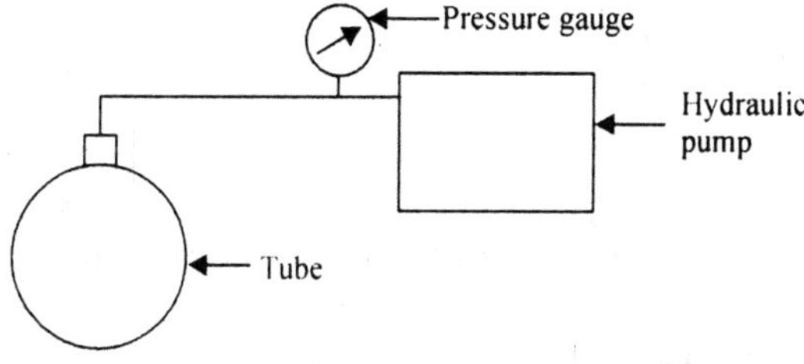

Fig. 28.24. An application of the hoop stress formula in burst testing.

Please work through the computations in the hoop stress formula:

Given: Outside diameter is 0.625 inch

Wall thickness is 0.037 inch, t = Wall thickness

Pressure to cause failure was 4404 psi, P = psi

Find: Hoop stress (S_h) when D = Outside diameter – t

Formula: $$S_h = \frac{P \times D}{2 \times t} = \frac{4404 \times (0.625 - 0.037)}{2 \times 0.037}$$

$$S_h = 35{,}000 \text{ psi}$$

Flare Test

This test provides an indication of the ductility of a specific material to permit an end enlargement for a fitting. For example, copper tubing may be attached with the use of proper flare nut and accessories, as seen in Fig. 28.25.

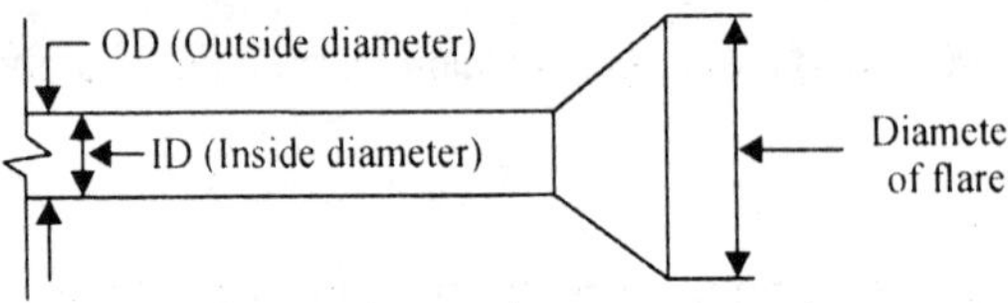

Fig. 28.25. A cross-section of a tube with one end flared to be joined to an accessory fitting.

The theory is that as the tube is flared, the end of the flare has to elongate to accommodate the larger diameter. The flare should have a certain minimum diameter to make a satisfactory leak-tight serviceable joint. The tendency is to make too large a flare. The larger the flare the greater the ductility requirement. The basic reason for the test is to correlate the overhang of the tubing on the flaring vise as minimum and maximum dimensions—in other words determine the overhang for a specific diameter and type of material. As an application of this test, select a tubing that is capable of an 18 per cent elongation factor. The technician is attempting to make a flare with a $\frac{3}{4}$ inch outside diameter. Can this be achieved? Apply the formula:

$$\%E_1 = \frac{(\pi \times F_d) - (\pi \times T_d)}{(\pi \times T_d)} \times 100$$

$$= \frac{(3.14 \times 0.75) - (3.14 \times 0.625)}{(3.14 \times 0.625)} \times 100 = 20 \text{ per cent}$$

where, E_1 = Elongation = 18 per cent

π = Pi = 3.14

F_d = Flare diameter = $\frac{3}{4}$ inch

T_d = Tube diameter = $\frac{5}{8}$ inch

Since the answer indicates that the material does not have the required ductility, this flare would probably cause failure. As a suggestion the material could be heat treated (annealed), and it would easily possess the ductility to make the flare.

Stress-Corrosion Test

This test was designed to study the problem of why normally ductile metals that had been in service would fail in a brittle manner. The metals had been subjected to a stress of less than yield strength value and in a mildly corrosive environment. Many names have been given to the failures such as stress-corrosion cracking, caustic embrittlement, and season cracking. It seems that with the metal under stress there was a chemical attack at the surface grain boundary. The original surface crack appeared to follow the grain boundary (intergranular). On a few cases surface cracks appeared through the surface grains (transgranular) but never as a mixture of the two on the same metal. The test is also used to determine the compatibility of welding dissimilar metals. For example, malleable cast iron was welded with a nickel alloy filler rod—a recommended practice in weld joining. After the weld was made, a 25 per cent dilution factor (base metal to filler) was estimated. The working stress was about 32,500 psi and wet with a liquid of 0.4 per cent HCl and water. The service life was two months. With just the stress alone, the structure would last for an indefinitely long time.

The stress-corrosion test is not effective unless a stress magnitude of at least 2000 psi is used. Typical fixtures for this test are shown in Fig. 28.26.

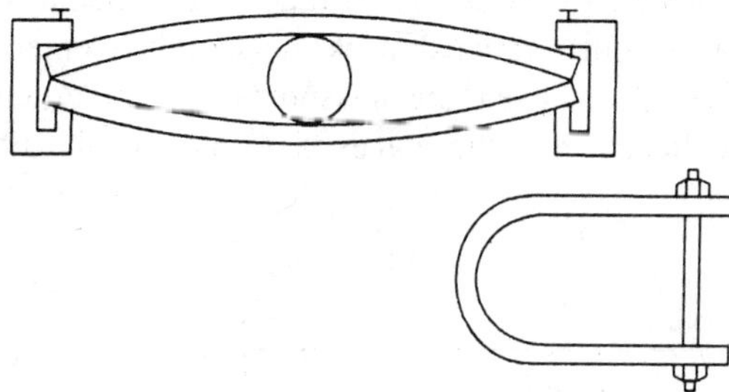

Fig. 28.26. Fixtures to maintain a pre-determined stress on test specimen during stress-corrosion testing.

After a designated exposure time the test specimen is inspected for damage. The inspection may be destructive or non-destructive (Fig. 28.27). Two widely used inspection methods to determine extent of damage to test specimen are tensile test and metallographic evaluation.

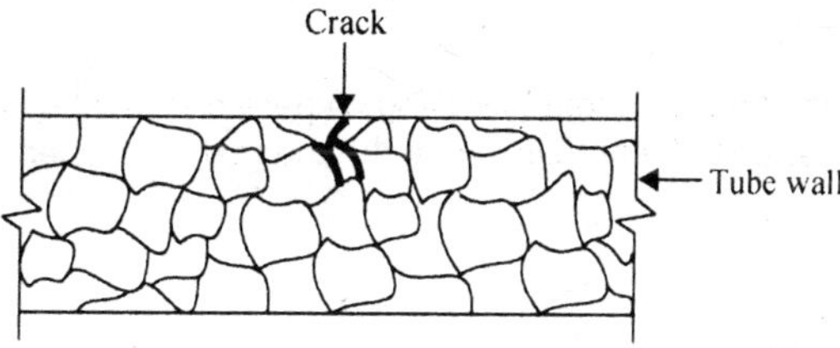

Fig. 28.27. An intergranular crack caused by stress-corrosion in the wall of a copper alloy tube. Crack follows the grain boundaries.

Linear Thermal Expansion Test

As metals are heated they undergo a change in density. In some temperature regions this change is positive and in others it is negative. Over a wide temperature range most metals when heated become less dense and the outward result is expansion. From the reaction of expansion with temperature change comes the term co-efficient of thermal expansion. It is a physical property of all elements and alloys, as shown in Table 28.2.

Table 28.2. Physical properties of selected metals.

Element	*Microinches of expansion (°F)*	*Density (g/cm^3)*	*Melting Temperature (°F)*
Aluminium	13.3	2.71	1215
Carbon	0.3–2.4	2.22	6700
Copper	9.2	8.96	1981
Iron	6.5	7.87	2800
Lead	16.3	11.34	618
Tin	13.0	7.30	449
Tungsten	2.4	19.30	6170

As shown in Table 28.2, elements have a known ratio of expansion per unit of temperature rise. Alloys can be made to meet a particular expansion. While specific elements have a known amount of expansion with temperature rise, alloys have to be measured if it is important to know the co-efficient of expansion. The instrument to collect data on thermal expansion is called a dilatometer. Details are shown in Fig 28.28.

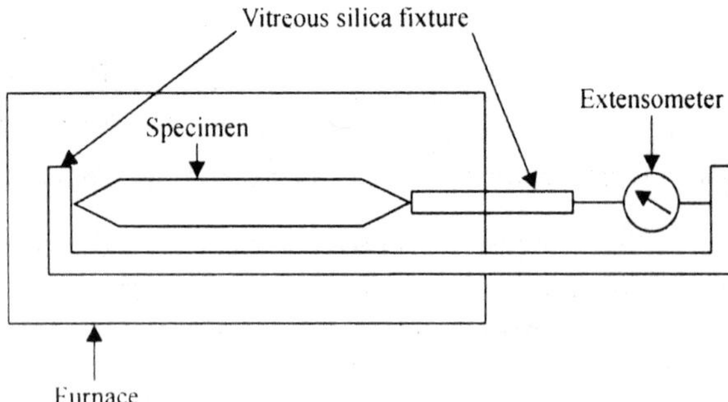

Fig. 28.28. A schematic diagram of a dilatometer. Heating a specific amount causes the specimen to change length. Expansion per degree per unit of length is called the thermal co-efficient of expansion.

Grain Size Test

Metals have a wide range of grain sizes. These grains are caused to be a certain size because of such factors as thermal exposure, rate of temperature change, and alloys present. We are concerned about grain size because of the correlation of grain size to specific properties. In some metals the grain sizes are relatively easy to change. Some grain changes are affected by heat treatment, and in other alloys they may have to be worked and then heat treated. The process of cold working an alloy and then heating to a temperature high enough to alter the grain structure is called re-crystallisation. Also, the cooling rate during solidification of a liquid will regulate grain size. Fast cooling during solidification will result in small grain size. The slower the cooling rate as a metal liquid transforms to a solid the larger the grains will be. Tests have been devised to estimate average grain sizes.

Probably one of the most simple grain-size tests especially for annealed mild steel is to make a sharp bend and compare surface roughness. The large grain sizes will have coarse 'orange peel' type surface. The smoother and more polished surfaces are easier to estimate grain size. Reasonably good results can be obtained from this test if reference standards are available for comparison. More valid grain size tests are:

1. Intercept method.
2. Planametric method.
3. Comparison method.

Intercept method

This method consists of drawing a line across the face of a 100X photomicrograph. Then a count is made of the number of grains touching the line. When the end of the line extends into a grain, it counts as one-half. The length of the line in millimeters divided by the average number of grains touched gives the average grain diameter (Fig. 28.29).

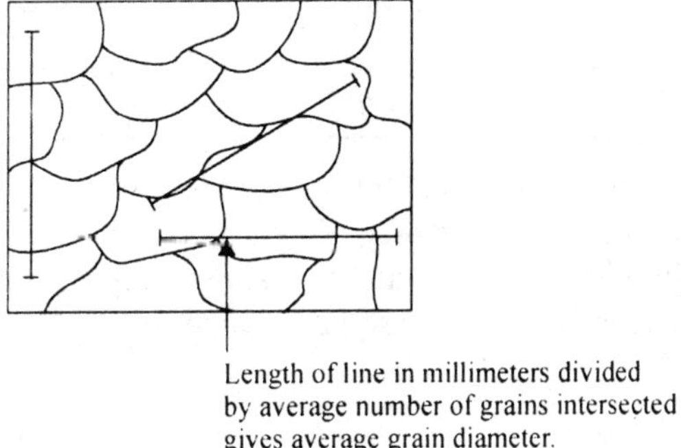

Fig. 28.29. Grain size by the intercept method. Grains touched by the end of the line count as half grains.

Planametric method

The planametric method requires a circle of known area to be drawn on a photomicrograph. The magnification of the photomicrograph should provide about 50 grains on the average in the areas to be counted. Count all grains within the circle and one-half for each grain touched by the circle. The answer is in number of grains per square millimeter.

Comparison method

This method is the most simple and requires no computations. The specimen is prepared for viewing with a microscope at 100X magnification. Then a measuring reticle is attached to the eyepiece of the microscope. The reticle has a grid system and corresponding grain-size number attached. The technician merely selects the grid pattern that most closely matches the actual grain boundaries, and the grain size is read directly. (Fig. 28.30). ASTM grain sizes are based on grains per square inch of a specimen at 100X magnification Table 28.3.

The ASTM grain-size number may be computed by the formula

$$N = 2^{n-1}$$

N = Mean grains per square inch (at 100X)

n = Grain-size number

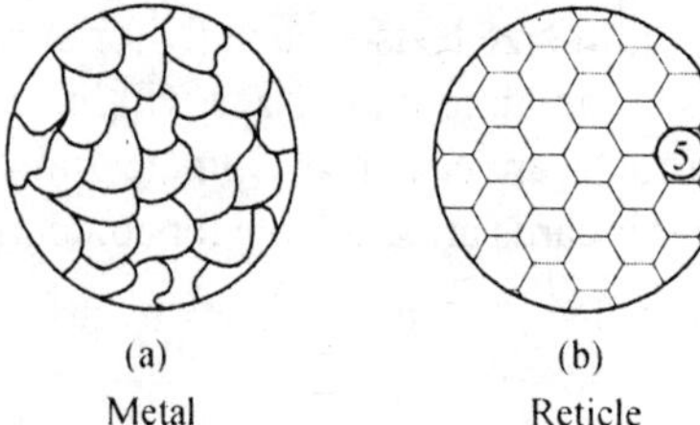

Fig. 28.30. Grain size can be determined by using a reticle eyepiece. (a) Shows a view of the specimen at 100X magnification; and (b) Shows a typical grain pattern on the reticle and corresponding grain-size number.

Table 28.3. ASTM grain sizes as related to grains/square inch.

ASTM grain Size	*Approximate actual Size*	*Range Grains/inch²@100X*
1	0.0105	1
4	0.0035	6–12
8	0.0009	96–192

To better understand this formula, work the following problem:

Given: A photomicrograph at 100X
A number 8 reticle matches grain boundaries

Find: Number of grains per square inch @ 100X

Formula: $N = 2^{(n-1)}$

$= 2^{(8-1)}$

$= 2^7$ 2 to the seventh power is 2 times itself seven times

$= 2 \times 2 \times 2 \times 2 \times 2 \times 2 \times 2 = 128$ grains/inch^2 (at 100X)

Non-destructive testing

Quality control is now becoming an expected process in manufacturing. Consumer liability, coupled with the development of lower cost NDT processes, is bringing the day closer when all manufactured materials will be completely evaluated. Designs for articles of tomorrow are asking for unbelievable service conditions. Equipment failure has already proven that no one can afford it. The answer to this dilemma is greater usage of non-destructive testing.

Visual Test

Technicians making a visual inspection of a workpiece may get some of the most important feedback that will help with any test to follow. One of the first principles of non-destructive testing is that quality is only relative. Materials are only better or worse than a specified reference standard. Without a reference standard there are no limits for acceptability. For some types of testing the reference standard can be a written description.

For example a welded section may have surface profile limitations such as undercutting or reinforcement of weld height or weld width. Many of the non-destructive testing processes rely on a visual summary of discontinuities. It is for this reason alone that technicians should develop a practice of visually evaluating fine details.

Penetrant Test

This type of testing is in great demand by many industries because

1. It is one of the most acceptable processes for locating cracks and similar types of discontinuities on any non-porous material.
2. The equipment requirement for basic testing is inexpensive.
3. The process is simple, and it is easy for the technician to develop a high level of skill in conducting the test.

Penetrant tests are not used by some industries because the total inspection costs are greater than other NDT processes. Another limitation of the penetrant test is that a record of the results is difficult to obtain.

Penetrant tests consist of the following five major steps:

Step 1: Pre-cleaning. During pre-cleaning the method selected should be determined by the surface condition of the workpiece. Methods available include abrasive grit blasting, wire wheel brushing, chemical bath immersion, and ultrasonic cleaning. When the pre-cleaning step is complete and the surface is dry, the part is ready for the next sequence.

Step 2: Penetrant application. Several methods are available to apply the penetrant, such as brush, roller, spray and dip. Two major types of penetrant are available. They are the visible dye and the fluorescent. The visible dye is generally the least sensitive but the most simple. The fluorescent system requires the use of a black light for cleaning and inspection. The penetrant must be on the surface long enough to be pulled into surface cracks by capillary attraction. The time that the penetrant is left on the surface may range from 2 minutes to 30 minutes. This is called dwell time. Time factors depend upon material, type of discontinuity, and sensitivity level. As an aid in making the workpiece more sensitive, it may be heated to a maximum of 100°F before application of the penetrant (Fig. 28.31).

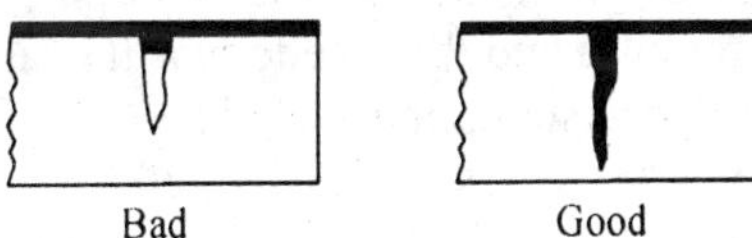

Fig. 28.31. Penetrant dwell time permits the internal surface of the discontinuity to fill by capillary attraction. (a) Dwell time too short; and (b) Dwell time adequate.

Step 3: Penetrant removal. The method selected is related to the type of penetrant used. Some surface penetrants may be removed by first wiping with a clean towel and then dampening the towel with solvent and wiping until all traces of surface penetrant have disappeared. This is for solvent base penetrants. Another method of penetrant removal may consist of applying an emulsifier to the surface penetrant. The emulsifier is left on from 15 seconds to 2 minutes.

The emulsifier will mix with the solvent base penetrant and render it soluble in water. Therefore the time the emulsifier is on the surface is critical (Fig. 28.32).

The third penetrant removal method is merely a water wash. This works because the penetrant is water soluble. In all the penetrant processes, time and temperature factors should be established to develop the required sensitivity. Average rinse time should vary from 15 seconds to 2 minutes. The water pressure should be about 50 psi. Adjust the spray nozzle for a coarse pattern with the temperature at about 105°F. After washing, the surfaces must be dried completely. This may be done in still air, hot forced air, or in an oven. Oven temperatures may vary from 175°F to 220°F. It is recommended that the part temperature not exceed 175°F.

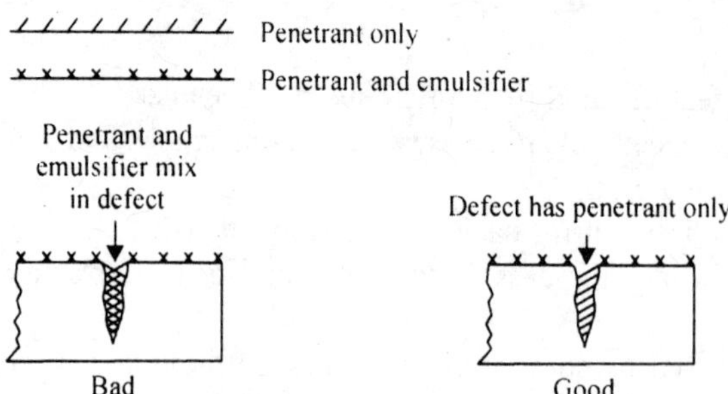

Fig. 28.32. Emulsifier must mix with surface penetrant only. If emulsifier mixes with penetrant in surface opening, it will wash out with water.

Step 4: Developer application. The developer may be a liquid or a powder. Rough surfaces such as sand castings are better for use with the powder method. The liquid may have a water base or a solvent base. The developer is sprayed on at just the right coverage—thick enough to blot the undersurface penetrant but not enough to mask fine detail (28.33).

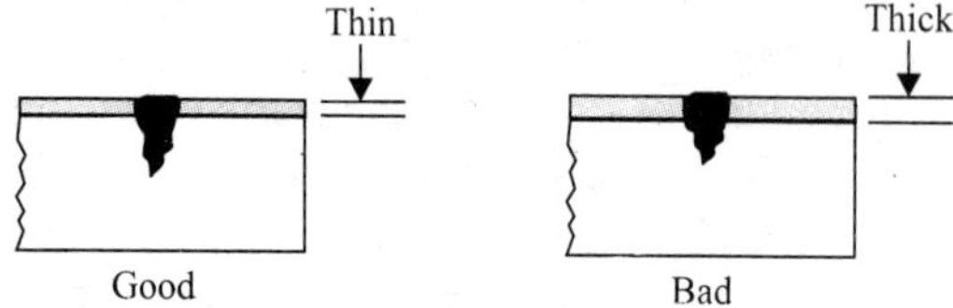

Fig. 28.33. Developer sprayed on the surface can be a critical factor in masking small defects.

The time that the developer remains on the surface is important. Generally, 10 minutes is the recommended minimum. Over 30 minutes and the bleedout will exaggerate the details.

Step 5: Inspection. For visible dye systems the technician should be provided with about 50 ft-candles of illumination. If the fluorescent penetrant is used, 800 microwatt/cm^2 lamps should be used. The black light lamp should warm up for 10 minutes before being used. It is recommended that the technician remain in the dark room for 3 minutes before making critical inspections.

Magnetic Particle Test

This is an inspection process used to locate surface and sub-surface discontinuities in ferromagnetic alloys. Magnetic particle tests are ideally suited for locating very small surface cracks. The major limitation of this testing process is that it is good only for iron, nickel and cobalt alloys. Precautions in using the process are needed because some electrical connections may cause arcing and localised heating of the workpiece. Also, the workpiece for some types of service will have to be demagnetised. A disadvantage of the process is that it requires a certain level of skill to interpret the particle pattern. The basic theory of magnetic particle testing is that as a surface or near-surface discontinuity appears in a metal it causes a leakage field to appear. This surface leakage field has a small North and South Pole to collect the particles. Magnetic fields may be created to travel in a circular pattern around the outside of a workpiece or in a linear direction along the surface parallel to the axis of the workpiece, as shown in Fig. 28.34.

Magnetic particle inspection can be simple and fast. It consists of the following five steps:

Step 1: Pre-cleaning. The surface of the workpiece needs to be free of contaminants such as corrosion, oil, and grease. Even paint will need to be removed if the test is to be completely reliable. After cleaning, the workpiece is to be completely dry.

Step 2: Selecting equipment and magnetising. The choice of equipment in magnetising may be important to the success of locating the discontinuity. Several types of equipment are available for creating the proper magnetic field in a workpiece (Fig. 28.35).

As part of the magnetising step the technician needs to understand the function of different types of current. Additional information on this is in Table 28.4.

Step 3: Application of particles. So far the discussion has provided information on selecting a method, direction of magnetic field, and type of current. The next item of importance is the selection of the type of particles. The technician needs to understand what the advantages will be with different types of particles (Table 28.5).

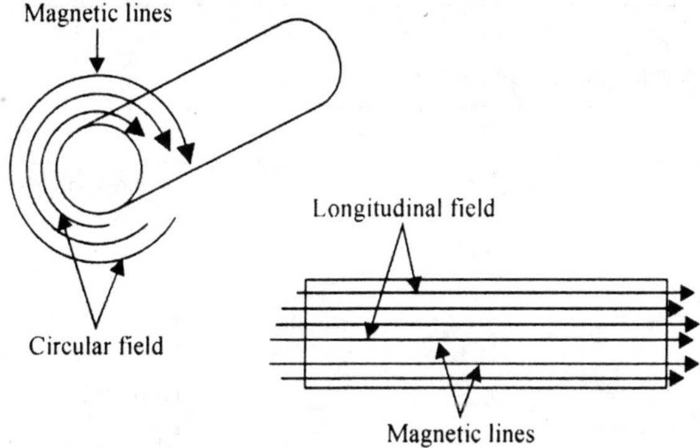

Fig. 28.34. Direction of magnetic fields. The two principal directions are circular (around the part) and longitudinal (along the axis of the part).

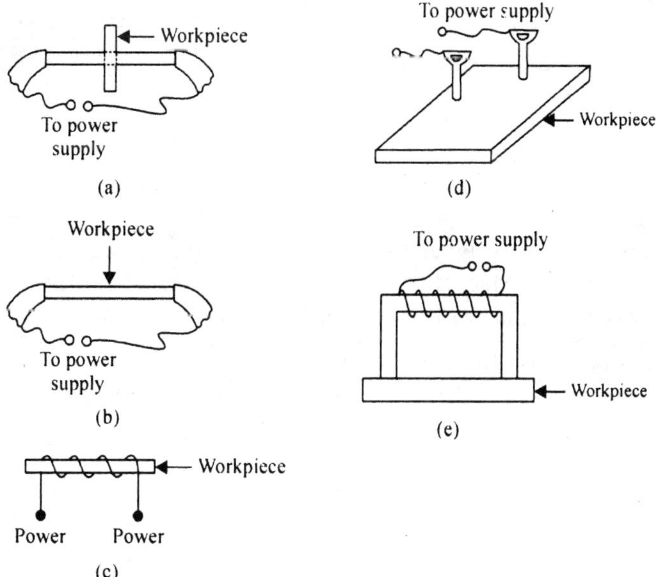

Fig. 28.35. Equipment for establishing a magnetic field in a workpiece. (a) Head shot with central conductor (circular field); (b) Head shot with workpiece (circular field); (c) Coil with workpiece (longitudinal field); and (d) Prods on workpiece (circular field around prods). (e) Yoke on workpiece (longitudinal field).

Table 28.4. Listing types of current and related advantages.

Type of current	Advantages
Alternating current (AC)	Maximum field strength at surface
	Good mobility of particles
	Highest sensitivity for surface cracks
Direct current (DC)	Good depth on wet particles
	Better depth on dry particles
Half-wave rectified alternating current (HWDC)	Best depth of any with dry particles

Table 28.5. Listing particles and related advantages.

Type of particle	*Current*	*Advantage*
Dry fluorescent	HWDC	Greatest depth and medium sensitivity
Dry visible	DC	Good depth
Wet fluorescent	AC	Greatest sensitivity on fine surface cracks
		Most adaptable to mass production
Dry fluorescent	AC	Fast clean up for medium size cracks
Wet visible	AC	Good for mass production.
		Recommended for medium or larger cracks
Dry visible	AC	Good for medium sensitivity and field work
Wet fluorescent	–	Greatest sensitivity.
		Ultimate in contrast visibility
Dry fluorescent	–	More positive evaluation than dry visible
Dry visible	–	Less contamination than wet particles
Black colour	–	Best on light coloured surfaces
Gray colour	–	Best on darkened surfaces
Red colour	–	Highest visibility on silvery polished surfaces

By now all test variables have been reduced except one. The final choice is how much current to use. High currents could cause heat build-up and more possibility of arcing. Also with high current, more residual magnetism could remain. Magnetic field strength tends to show little increase as it reaches the saturation point, regardless of the amount of current potential. Additional information is given in Table 28.6.

Table 28.6. Current requirements for magnetic particle testing.

Type of Equipment	*Amperage*	
Coil	$I = \dfrac{45000/(L/D)}{N}$	I = Amperage
		L = Length of part
		D = Diameter or width of part
		N = Number of turns in coil

(Contd...)

Type of Equipment	*Amperage*
Heat shot (current thru part)	800–1000 amp./inch of diameter
Prod	100–125 amp./inch of prod spacing (measure prod spacing centre to centre)
Yoke (AC)	1540 ampere turns
	(512 turns on 115 v. and 3 amp.)
Yoke (DC)	7200 ampere turns
	(900 turns on 115 v. H.W.D.C and 8 amp.)

Step 4: Inspecting for discontinuities. This is the most important part of the test. Regardless of how well the other sequences have been completed, the validity may be destroyed by improper recognition of details. This is one phase of magnetic particle testing that has to be learned by experience.

Step 5: Demagnetising and cleaning. This is probably the simplest part of the test. There are several techniques to demagnetise the workpiece. The one discussed here uses a coil that provides more ampere turns that the largest value used in causing a magnetic field. The coil is powered by alternating current. The workpiece is placed inside the coil, the coil is energised, and then the workpiece is slowly moved out to about 18 inch. Movement time should be about 5 seconds. As shown in Fig 28.36, this causes the magnetic field to build up and collapse each time to a lower value. As the workpiece leaves the coil, a cycle of alternating current causes a reversal of the magnetic field. Each magnetic field intensity is lower than the last. Final cleaning could be an application of either a degreaser or a water wash.

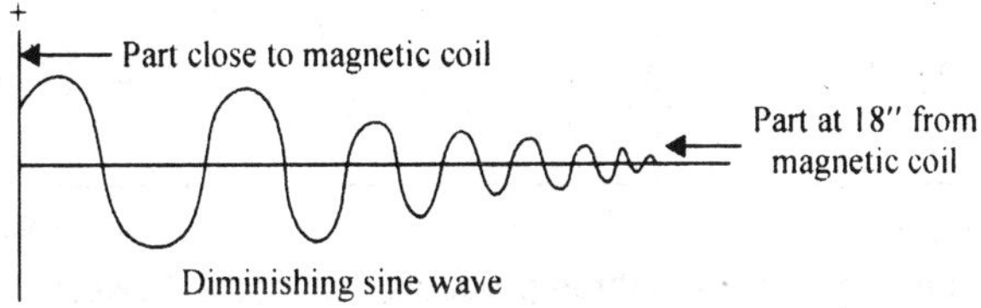

Fig. 28.36. Current flow causes magnetic flux to build up and collapse as current changes polarity. Each time the magnetic flux collapses, some magnetic domains are randomly distributed. When the total number of poles is equally divided (north and south), the part is magnetically neutral (demagnetised).

Eddy Current Test

This is a testing system that can measure conductivity of metals or probe lift-off distance. In testing for conductivity, surface cracks or openings and near-surface inclusions can be located. Also, when set up for conductivity, alloys can be separated and heat treatment can be monitored. If the system is adjusted for lift-off, coating thicknesses such as paint, varnish, and anodising can be measured.

The system is composed of three major units—the AC power source, a work coil (probe), and a monitor to read the reactions between the probe and metal to be tested. Here is how the system works (conductivity mode). Two reference standards are selected. One is a quality standard and the other should be representative of the lowest quality that will be accepted. Using one reference standard at a time, the probe is attached and the R control dial, which is the resistance control, is adjusted for minimum reading. Then with the probe still in place, the X control dial, which is the inductive reactance control, is also adjusted for minimum reading. Now a graph called an impedance plane can be constructed. This graph will show the material points related to the X and R value of each reference standard (Fig. 28.37).

So that the image quality may be verified, a device called a penetrameter is placed on another section of metal equivalent in thickness to the workpiece, as seen in Fig. 28.42.

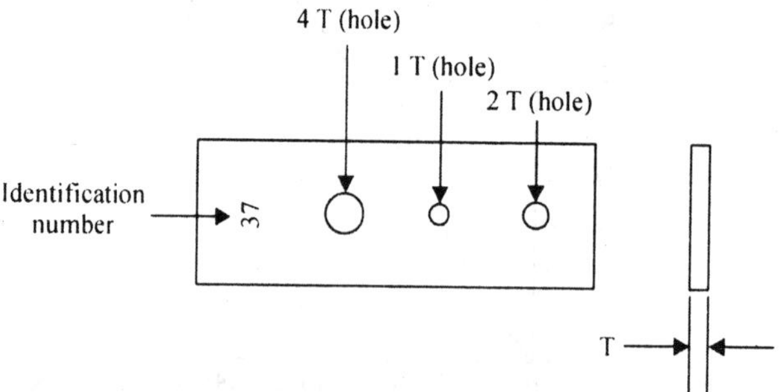

Fig. 28.42. A penetrameter as used in radiography. T is the thickness of the penetrameter, which is related to the identification number. Each hole diameter is an increment of the thickness.

The film is inspected for changes in density. Any density change that is equal to or greater than the penetrameter image is a possible discontinuity.

Ultrasonic Test

This is a process that uses high frequency sound waves in a metal to detect discontinuities. Here is how it works. A thin crystal called a transducer is selected for its special properties. The transducer will vibrate when connected to alternating current, or if the surfaces of the crystal are vibrated it will generate alternating current. This is called the piezoelectric effect. When vibrations (sounds) are made by the transducer and coupled to a metal surface, the sound will travel in a straight line. The travel will continue through to the back side if there are no discontinuities. On the back side it is reflected (assuming parallel surfaces) and will go back through the metal again hitting its point of origin. Some of the sound energy strikes the transducer as it hits the point of origin and is immediately transformed into electrical energy. The electrical energy is amplified and fed to a cathode ray tube or oscilloscope.

An image will appear on the screen of the cathode ray tube indicating a time factor and a voltage factor. With proper adjustment a vertical pip will appear on the left side of the screen indicating top surface and on the right side for bottom surface. If a discontinuity lies between the surfaces it will be located between the two pips, as shown in Fig. 28.43.

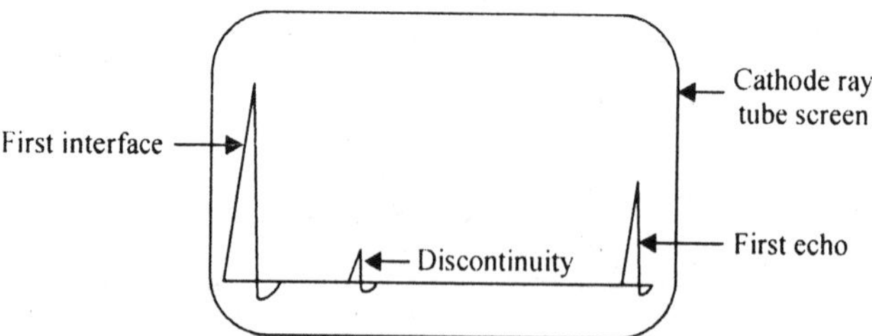

Fig. 28.43. Image pattern on the cathode ray tube when a discontinuity is located. Distance along the scan line (interface to discontinuity) is a ratio of the distance from top of piece where transducer is located to discontinuity.

Many adaptations of this principle are used. For example, in testing weldments one of the problems encountered is the irregular surface of the weld. To overcome this, the crystal is placed on a plastic wedge. Now the sound may enter at an angle and be reflected from the bottom to the weld zone. Any discontinuity or interface may cause the sound to echo back along the original path and be picked up by the transducer, as shown in Fig. 28.44.

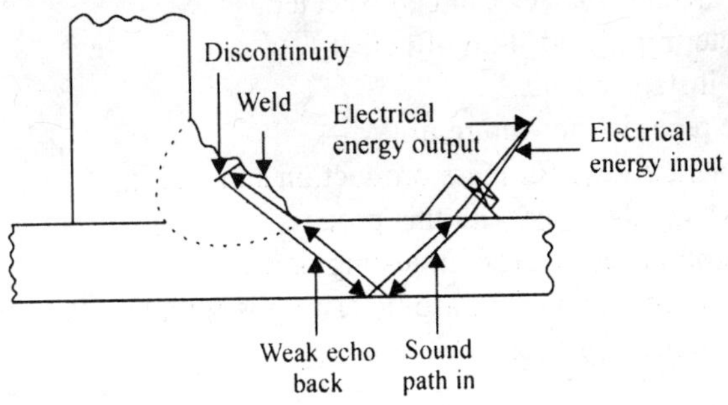

Fig. 28.44. Weld testing with an angle beam transducer.

In ultrasonic testing it is of major importance that the technician interpret the CRT (an abbreviation for cathode ray tube, which is the screen of an ultrasonic instrument) indication in a valid manner. In an attempt to standardise quality control, reference standards are used to calibrate the instrument. The reference standard must be the same alloy and microstructure as the workpiece. A flat-bottom hole is machined in from the back side at a specified distance. The ultrasonic instrument is adjusted so that the hole provides a certain percentage of the amplitude of the back reflection. The quality of the workpiece is based on the amplitude of the discontinuity (Fig.28.45).

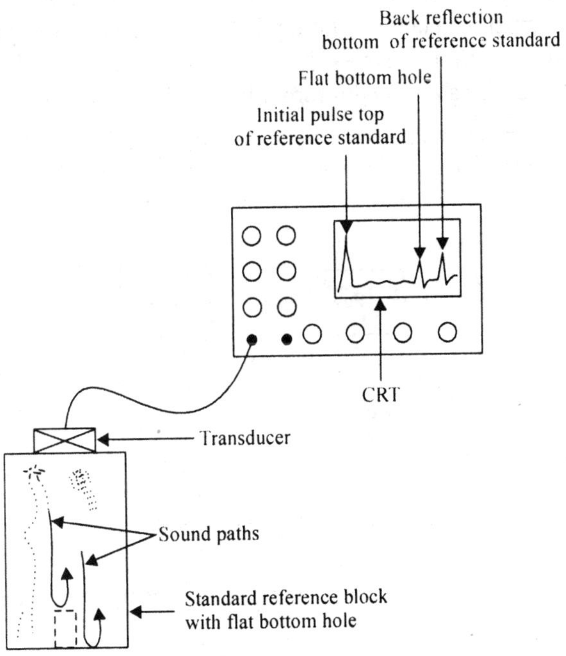

Fig. 28.45. A standard reference block as used in ultrasonic testing and the CRT (cathode ray tube) pattern.

Ultrasonic testing has been one of the fastest growing of any of the non-destructive testing processes. Advantages found with ultrasonic testing are:

1. It is highly sensitive to locating small flaws.
2. Almost unlimited metal distances can be inspected.
3. It is accurate in determining position of defect.
4. Results of test are instantaneous.
5. Test results can be recorded for future use.
6. Automated equipment allows for mass production schedules.
7. No environmental hazards occur with this process.
8. Field tests can be conducted with small testing units.

Perhaps the most serious limitation of ultrasonic testing is found in its application to materials with rough, non-parallel, and irregular surfaces.

Acoustic-Emission Test

This process relies upon the theory that, as flaws propagate in a material under stress, a sound is emitted. This sound is analysed by a monitor and interpreted as to severity of defect. Material inspection consists of attaching sensors to the workpiece, applying a specific level of stress, and determining if any flaws become active. As flaws become active, acoustic energy will be emitted. Static defects cannot be located.

Acoustic-emission tests have detected very small active flaws. The sound energy is radiated out in a spherical direction from the flaw. Multiple sensors can be used to analyse these signals and provide information on the location. Flaws moving a small dimension (on the order of 1×10^{-12} inch) in the proper time range can be recorded. Clusters of atoms with a size range of 1×10^{-6} inch can be detected as they move or relocate under stress. As the movement of crystallographic planes becomes more dynamic, sound energy level warns of impending failure. The applied stress can be of any type that causes movement in the workpiece. For example, thermal stress that appears as a result of the heat from welding may cause a flaw to become active. For additional information on this test system, refer to Fig. 28.46.

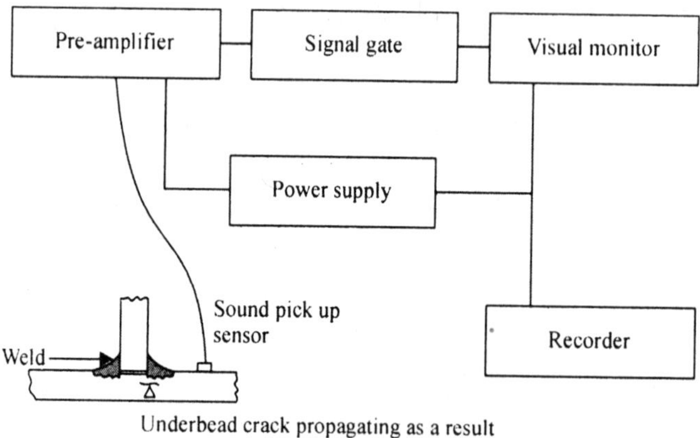

Fig. 28.46. A schematic diagram representing an acoustic-emission testing system.

References

Allen, L.K., *Heat Treatment of Metals*, Butterworths, London.
Alvarez, J.L., *Mechanical Properties of Steel*, John Wiley & Sons, New York.
Arceivala, K.J., *Case Hardening of Steel*, Marcel Dekker Inc., New York.
Barsom, K.G., *Heat Treatment of Metals*, Swedish Institute of Metal Research.
Batterman, S.A., *Diffusion in Solids*, McGraw-Hill, Tokyo.
Brown, M.H., *Alloying Elements in Steel*, Cambridge University Press, Cambridge.
Cambell, K.E. and Lemer, H.A., *Iron and Steel Alloys*, Academic Press, London.
Chang, J.C., *Fundamentals of Mettallography*, John Wiley & Sons, New York.
Coolingwood, R.W., *Temperability of Steel,* John Wiley & Sons, New York.
Cox, C.S., *Theory of Heat Treatment of Steel*, Marcel Dekker, New York.
Downe, S.A., *Hardenability concepts with Applications of Steel*, John Wiley & Sons, New York.
Friedel, J., *Tempering of Steel*, Van Nostrand, London.
Gould, G.W., *Physical Metallurgy Principles*, D. Van Nostrand, New York.
Hillert, M., *Decomposition of Austenite by Diffusional Processes, Intersciences*, New York, London
Huff, C.B., *Cast Iron and Ductile Iron*, Elsevier Scientific Publishing Co., Amsterdam.
Jackwerth, E., *Heat Treatment Furnaces and Equipments*, Academic Press, London.
James, A. and Evison L., *Annealing of Steel*, John Wiley & Sons, New York.
Jencks, W.P., *Salt and Fluidised Bed Equipments*, John Wiley & Sons, New York.
Kiessling, R., *Heat Treatment of Non-ferrous Metals*, The Metals Society, London.
Kim, C.K., *Diffusio in Solids, Liquids and gases*, Marcel Dekker, New York.
Lechevallier, M.W., *Dimensional Stability of Steel*, Academic Press, London.
Lewis, R.L., *Heat Treatment Furnaces*, Chilton Book Co., USA.
Mason, C.F., *Testing of Metals*, Longman Group Ltd., London.
McCaull, J. and Crossland, J., *Hardening of Steel*, Harcourt Brace Jovanovich, New York.
Mitchell, R., *Temperature Control,* McGraw-Hill, New York.

Nyer, E.K., *Normalising of Steel*, Van Nostrand Reinhold, New York.
Phillips, D.J.H., *Testing of Metals*, Applied Science Publishers, London.
Randak, A.M., *Carbon and Tool Steels*, Academic Press, London.
Reid, G.K., *Carburising and Nitriding*, Reinhold Publishing Corporation, New York.
Samson, R.A., *Physicalmetallurgy*, Pergamon Press, Oxford, London.
Schrowebel, J., *Induction Hardening of Steel*, Pergamon Press, New York.
Scott. H., *Influence of Tempering Temperature on Quenched Steel*, Progress Publishers, Moscow.
Snell, I.D. and Snell, C.T., *Chemical Heat Treatment*, D. Van Nostrand, New York.
Southwood, T.R.E., *Cyaniding and Carbonitriding,*, Swedish Institute of Metal Research, London.
Stephens, W.D., *Metallography*, John Wiley & Sons, New York.
Stumm, W. and Stumm-Zollineger, E., *Distortion in Tool Steel*, Wiley Interscience, New York.
Teal, J.M., *Nature of Metals and Alloys*, Pergamon Press, New York.
Thelning, K.E., *Steel and its Heat Treatments*, Butterworth, U.K.
Wyatt, G.M., *Non-metallic Inclusions in Steel*, Reston Publishing Co., Reston, Virginia.

Index

D

E

READERS NOTES

READERS NOTES

READERS NOTES